Lecture Notes in Artificial Intelligence 12320

Subseries of Lecture Notes in Computer Science

More information about this series at http://www.springer.com/series/1244

Ricardo Cerri · Ronaldo C. Prati (Eds.)

Intelligent Systems

9th Brazilian Conference, BRACIS 2020
Rio Grande, Brazil, October 20–23, 2020
Proceedings, Part II

Editors
Ricardo Cerri
Federal University of São Carlos
São Carlos, Brazil

Ronaldo C. Prati
Federal University of ABC
Santo Andre, Brazil

ISSN 0302-9743 ISSN 1611-3349 (electronic)
Lecture Notes in Artificial Intelligence
ISBN 978-3-030-61379-2 ISBN 978-3-030-61380-8 (eBook)
https://doi.org/10.1007/978-3-030-61380-8

LNCS Sublibrary: SL7 – Artificial Intelligence

This Springer imprint is published by the registered company Springer Nature Switzerland AG
The registered company address is: Gewerbestrasse 11, 6330 Cham, Switzerland

Preface

The Brazilian Conference on Intelligent Systems (BRACIS) is one of Brazil's most meaningful events for students and researchers in Artificial and Computational Intelligence. Currently, In its 9th edition, BRACIS originated from the combination of the two most important scientific events in Brazil in Artificial Intelligence (AI) and Computational Intelligence (CI): the Brazilian Symposium on Artificial Intelligence (SBIA), with 21 editions, and the Brazilian Symposium on Neural Networks (SBRN), with 12 editions. The conference aims to promote theory and applications of artificial and computational intelligence. BRACIS also aims to promote international-level research by exchanging scientific ideas among researchers, practitioners, scientists, and engineers.

BRACIS 2020 received 228 submissions. All papers were rigorously double-blind peer-reviewed by an International Program Committee (an average of three reviews per submission), followed by a discussion phase for conflicting reports. At the end of the reviewing process, 90 papers were selected for publication in two volumes of the *Lecture Notes in Artificial Intelligence* series, an acceptance rate of 40%.

We are very grateful to Program Committee members and reviewers for their volunteered contribution in the reviewing process. We would also like to express our gratitude to all the authors who submitted their articles, the general chairs, and the Local Organization Committee, to put forward the conference during the COVID-19 pandemic. We want to thank the Artificial Intelligence and Computational Intelligence commissions from the Brazilian Computer Society for the confidence in serving as program chairs for BRACIS 2020.

We are confident that these proceedings reflect the excellent work in the fields of artificial and computation intelligence communities.

October 2020

Ricardo Cerri
Ronaldo C. Prati

Organization

General Chairs

Hélida Salles Santos	Universidade Federal do Rio Grande, Brazil
Graçaliz Dimuro	Universidade Federal do Rio Grande, Brazil
Eduardo Borges	Universidade Federal do Rio Grande, Brazil
Leonardo Emmendorfer	Universidade Federal do Rio Grande, Brazil

Program Committee Chairs

Ricardo Cerri	Federal University of São Carlos, Brazil
Ronaldo C. Prati	Federal University of ABC, Brazil

Steering Committee

Leliane Barros	Universidade de São Paulo, Brazil
Heloisa Camargo	Universidade Federal de São Carlos, Brazil
Flavia Bernardini	Universidade Federal Fluminense, Brazil
Jaime Sichman	Universidade de São Paulo, Brazil
Karina Delgado	Universidade de São Paulo, Brazil
Kate Revoredo	Universidade Federal do Estado do Rio de Janeiro, Brazil
Renata O. Vieira	Pontifícia Universidade Católica do Rio Grande do Sul, Brazil
Solange Rezende	Universidade de São Paulo, Brazil
Ricardo Prudencio	Universidade Federal de Pernambuco, Brazil
Anne Canuto	Universidade Federal do Rio Grande do Norte, Brazil
Anisio Lacerda	Universidade Federal de Minas Gerais, Brazil
Gisele Pappa	Universidade Federal de Minas Gerais, Brazil
Gina Oliveira	Universidade Federal de Uberlândia, Brazil
Renato Tinós	Universidade de São Paulo, Brazil
Paulo Cavalin	IBM, Brazil

Program Committee

Adenilton da Silva	Universidade Federal de Pernambuco, Brazil
Adrião Dória Neto	Universidade Federal do Rio Grande do Norte, Brazil
Albert Bifet	LTCI, Télécom ParisTech, France
Alberto Paccanaro	Royal Holloway, University of London, UK
Alex Freitas	University of Kent, UK
Alexandre Delbem	Universidade de São Paulo, Brazil
Alexandre Ferreira	Universidade Estadual de Campinas, Brazil

Alexandre Plastino	Universidade Federal Fluminense, Brazil
Aline Paes	Universidade Federal Fluminense, Brazil
Alvaro Moreira	Universidade Federal do Rio Grande do Sul, Brazil
Ana Carolina Lorena	Instituto Tecnológico de Aeronáutica, Brazil
Ana Vendramin	Universidade Tecnológica Federal do Paraná, Brazil
Anísio Lacerda	Centro Federal de Educação Tecnológica de Minas Gerais, Brazil
Anderson Soares	Universidade Federal de Goiás, Brazil
André Coelho	Universidade de Ceará, Fortaleza, Brazil
André Carvalho	Universidade de São Paulo, Brazil
André Rossi	Universidade de São Paulo, Brazil
André Grahl Pereira	Universidade Federal do Rio Grande do Sul, Brazil
Andrés Salazar	Universidade Tecnológica Federal do Paraná, Brazil
Anne Canuto	Universidade Federal do Rio Grande do Norte, Brazil
Araken Santos	Universidade Federal Rural do Semi-árido, Brazil
Aurora Pozo	Universidade Federal do Paranáv, Brazil
Bianca Zadrozny	IBM, Brazil
Bruno Nogueira	Universidade Federal de Mato Grosso do Sul, Brazil
Bruno Travencolo	Universidade Federal de Uberlândia, Brazil
Carlos Ferrero	Instituto Federal de Santa Catarina, Brazil
Carlos Silla	Pontifical Catholic University of Paraná, Brazil
Carlos Thomaz	Centro Universitário da FEI, Brazil
Carolina Almeida	State University in the Midwest of Paraná, Brazil
Celia Ralha	Universidade de Brasília, Brazil
Celine Vens	KU Leuven, Belgium
Celso Kaestner	Universidade Tecnológica Federal do Paraná, Brazil
Cesar Tacla	Universidade Tecnológica Federal do Paraná, Brazil
Cleber Zanchettin	Universidade Federal de Pernambuco, Brazil
Daniel Araújo	Federal University of Rio Grande do Norte, Brazil
Danilo Sanches	Universidade Tecnológica Federal do Paraná, Brazil
David Martins-Jr	Universidade Federal do ABC, Brazil
Delia Farias	Universidad de Guanajuato Lomas del Bosque, Mexico
Denis Fantinato	Universidade Federal do ABC, Brazil
Denis Mauá	Universidade de São Paulo, Brazil
Diana Adamatti	Universidade Federal do Rio Grande, Brazil
Diego Furtado Silva	Universidade Federal de São Carlos, Brazil
Eder Gonçalves	Universidade Federal do Rio Grande, Brazil
Edson Gomi	Universidade de São Paulo
Edson Matsubara	Fundação Universidade Federal de Mato Grosso do Sul, Brazil
Eduardo Borges	Universidade Federal do Rio Grande, Brazil
Eduardo Costa	Corteva Agriscience, USA
Eduardo Goncalves	Escola Nacional de Ciências Estatísticas, Brazil
Eduardo Palmeira	Universidade Estadual de Santa Cruz, Brazil
Eduardo Spinosa	Universidade Federal do Paraná, Brazil
Elaine Faria	Federal University of Uberlândia, Brazil

José Antonio Sanz	Universidad Publica de Navarra, Spain
Julio Nievola	Pontifícia Universidade Católica do Paraná, Brazil
Karina Delgado	Universidade de São Paulo
Kate Revoredo	Universidade Federal do Estado do Rio de Janeiro, Brazil
Krysia Broda	Imperial College London, UK
Leandro Coelho	Pontifícia Universidade Católica do Paraná, Brazil
Leliane Barros	Universidade de São Paulo
Leonardo Emmendorfer	Universidade Federal do Rio Grande, Brazil
Leonardo Ribeiro	Technische Universität Darmstadt, Germany
Livy Real	B2W Digital Company, Brazil
Lucelene Lopes	Roberts Wesleyan College, USA
Luciano Barbosa	Universidade Federal de Pernambuco, Brazil
Luis Garcia	Universidade de Brasília, Brazil
Luiz Carvalho	Universidade Tecnológica Federal do Paraná, Brazil
Luiz Coletta	Universidade Estadual Paulista, Brazil
Luiz Merschmann	Universidade Federal de Lavras, Brazil
Luiza Mourelle	State University of Rio de Janeiro, Brazil
Marcela Ribeiro	Universidade Federal de São Carlos, Brazil
Marcella Scoczynski	Universidade Tecnológica Federal do Paraná, Brazil
Marcelo Finger	Universidade de São Paulo, Brazil
Marcilio de Souto	Université d'Orléans, France
Marcos Domingues	Universidade Estadual de Maringá, Brazil
Marcos Quiles	Federal University of São Paulo, Brazil
Marilton Aguiar	Universidade Federal de Pelotas, Brazil
Marley Vellasco	Pontifícia Universidade Católica do Rio de Janeiro, Brazil
Mauri Ferrandin	Universidade Federal de Santa Catarina, Brazil
Márcio Basgalupp	Universidade Federal de São Paulo, Brazil
Mário Benevides	Universidade Federal Fluminense, Brazil
Moacir Ponti	Universidade de São Paulo, Brazil
Murillo Carneiro	Federal University of Uberlândia, Brazil
Murilo Naldi	Universidade Federal de São Carlos, Brazil
Myriam Delgado	Federal University of Technology of Paraná, Brazil
Nádia Felix	Universidade Federal de Goiás, Brazil
Newton Spolaôr	Universidade Estadual do Oeste do Paraná, Brazil
Patricia Oliveira	Universidade de São Paulo
Paulo Cavalin	IBM Research, Brazil
Paulo Ferreira Jr.	Universidade Federal de Pelotas, Brazil
Paulo Gabriel	Universidade Federal de Uberlândia, Brazil
Paulo Quaresma	Universidade de Évora, Portugal
Paulo Pisani	Universidade Federal do ABC, Brazil
Priscila Lima	Universidade Federal do Rio de Janeiro, Brazil
Rafael Bordini	Pontifícia Universidade Católica do Rio Grande do Sul, Brazil
Rafael Mantovani	Federal Technology University of Paraná, Brazil

Rafael Parpinelli	Universidade do Estado de Santa Catarina, Brazil
Rafael Rossi	Federal University of Mato Grosso do Sul, Brazil
Reinaldo Bianchi	Centro Universitario FEI, Brazil
Renato Assuncao	Universidade Federal de Minas Gerais, Brazil
Renato Krohling	Universidade Federal do Espírito Santo, Brazil
Renato Tinos	Universidade de São Paulo, Brazil
Ricardo Cerri	Universidade Federal de São Carlos, Brazil
Ricardo Silva	Universidade Tecnológica Federal do Paraná, Brazil
Ricardo Marcacini	Universidade de São Paulo, Brazil
Ricardo Prudêncio	Universidade Federal de Pernambuco, Brazil
Ricardo Rios	Universidade Federal da Bahia, Brazil
Ricardo Tanscheit	Pontifícia Universidade Católica do Rio de Janeiro, Brazil
Ricardo Fernandes	Federal University of São Carlos, Brazil
Roberta Sinoara	Instituto Federal de Ciência, Educação e Tecnologia de São Paulo, Brazil
Roberto Santana	University of the Basque Country, Spain
Robson Cordeiro	Universidade de São Paulo, Brazil
Rodrigo Barros	Pontifícia Universidade Católica do Rio Grande do Sul, Brazil
Rodrigo Mello	Universidade de São Paulo
Rodrigo Wilkens	University of Essex, UK
Roger Granada	Pontifícia Universidade Católica do Rio Grande do Sul, Brazil

Contents – Part II

Agent and Multi-agent Systems, Planning and Reinforcement Learning

Knowledge Representation, Logic and Fuzzy Systems

Machine Learning and Data Mining

Contents – Part I

Neural Networks, Deep Learning and Computer Vision

Agent and Multi-agent Systems, Planning and Reinforcement Learning

A Multi-level Approach to the Formal Semantics of Agent Societies

Alison R. Panisson(✉), Rafael H. Bordini, and Antônio Carlos da Rocha Costa

Pontifical Catholic University of Rio Grande do Sul (PUCRS), Porto Alegre, Brazil
alison.panisson@acad.pucrs.br, rafael.bordini@pucrs.br,
ac.rocha.costa@gmail.com

Abstract. Typically, an agent society is composed of at least three main components, namely, the set of its individual agents, the set of multi-agent organisations that help coordinate the actions of those agents, and the environment that those agents share. However, other levels have also been pointed out in the agent society literature, for example, the internal normative systems that regulate the structure and operation of the organisations themselves and, more generally, the overarching cultural system of the society. Many types of formal semantics for such systems, such as for example operational semantics, are not immediately suitable for systems that have ongoing operations at several architectural levels and requiring that those levels synchronise when particular events occur. This paper addresses this problem by proposing an approach to giving formal semantics to multi-level agent societies. The approach allows for the reuse of existing semantics for the separate levels, improves readability, and reduces the number of inference rules to be written.

Keywords: Multi-agent systems · Multi-level semantics · Operational semantics

1 Introduction

In multi-agent systems (MAS) there are multiple different levels of specification, each one corresponding to a different conceptual level in the system, and playing important roles in a more general framework for programming multi-agent systems. For example, one concrete framework to develop multi-agent system is JaCaMo [1], which integrates technologies supporting agent programming, organisation programming, and environment programming (i.e., approaches that emerged from multi-agent system research). One of the most desirable characteristics of multi-agent systems is agent autonomy, which boosted the development of organisation-centred systems as an approach towards *open systems* [14]. Open systems might be regulated by an organisational specification which defines the goals, norms, and social structure of the organisation, in order to regulate agent behaviour. The organisations themselves may be regulated by superior levels of regulation, and so on. However, even with an organisational specification,

R. Cerri and R. C. Prati (Eds.): BRACIS 2020, LNAI 12320, pp. 3–17, 2020.
https://doi.org/10.1007/978-3-030-61380-8_1

agents can autonomously act in ways that may compromise the system's functioning. Therefore, it is important to guarantee that *processes*, which describe the functioning of the organisations by means of inserting a sort of ordering for actions/events, will occur as specified, in order to satisfy the system's goals and prevent undesired overall behaviour of the system.

In order to ensure the desired behaviour of multi-agent system, it has been common in the Agents literature to give formal semantics to such systems as transition systems, expressing the possible states of the system and the necessary conditions for the system to move from one such state to another. For example, in [28] the authors give operational semantics to an agent-oriented programming language called AgentSpeak [22] and some of its extensions found in [3]. Operational semantics [21] models systems as transition systems, which are established through inference rules that have conditions on the configurations (i.e., states) of the system as premises and infer when transitions to next states may occur. It provides clean and easy-to-understand formalisations of the semantics of programming languages, but in multi-agent systems, operational semantics may become rather involved. Still, the naturalness with which such formalisations can be implemented often makes the formalism very attractive for multi-agent software development.

Given the complexity of multi-agent systems abstractions, we propose a multi-level operational semantics with vertical (i.e., inter-level) integrity constraints in order to specify those systems. Such vertical integrity constraints aim to ensure that the transition systems giving separate semantics to the individual levels of abstraction are combined in a way that preserves the required interrelations of those levels. Furthermore, they allow some of the required semantic rules to be automatically generated from compact representations of such integrity constraints. Particularly, in this work, we focus on processes occurring in multi-agent organisations, modelling them as multiple, independent but interrelated, transition systems. Our approach allows us to specify all dimensions/levels evidently present in multi-agent systems abstractions, making clear the effects and relations among the components at each level. Furthermore, we define simple expressions with which to specify the vertical integrity constraints, that is, a particular way to restrict transitions between states that are caused by transitions at different levels. In particular, we express the specification in terms of *count-as* relations between levels, i.e., relations that express certain combinations of actions—i.e., which we here call *processes*—executed at a given abstraction level *count as* actions being executed at an upper level[1].

The paper is organised as follows. First, in Sect. 2, we describe approaches to developing multi-agent systems that consider different levels of abstractions

[1] We have taken the expression *count-as* from [26], but we use it in a weaker and objective way, just to mean that some X, occurring at a given level of abstraction, *operationally realises* or *brings about* Y at an upper level provided a context C of that upper level is in place, independently of there being a subjective collective agreement about that among the agents involved with the occurrences of X and Y as required by Searle.

in those systems, just to evince the existence of such multiplicity of levels. After that, in Sect. 3, we review previous work on multi-level semantics and the notion of *count-as* relation that we adopt. In Sect. 4, we introduce the notion of processes in multi-agent systems. In Sect. 5, the main section of this paper, we introduce the *multi-level semantics with vertical integrity constraints* that is being proposed here. In Sect. 6, we describe the scenario with which we exemplify this work by instantiating the general semantic rule presented previously. Finally, we discuss some related work and conclude the paper.

2 Multiple Levels in Agent Societies

There are various approaches to developing multi-agent systems that consider different levels of abstraction. In this section, we discuss some of such approaches in order to clarify the existence of such multiple levels in multi-agent systems. One well-known practical multi-agent programming framework is JaCaMo [1]. JaCaMo is built upon three existing platforms: Jason [3] for programming autonomous agents, MOISE [16] for programming agent organisations, and CArtAgO [24] for programming shared environments. JaCaMo emphasises the existence of three levels, in that work called "dimensions", in a single framework so that all levels can be programmed using the languages provided by the combined platforms mentioned above. Also, the development of multi-agent systems based on Electronic Institutions [17,29] considers the existence of multiple levels in multi-agent systems, emphasising that, while electronic institutions focus on the so-called macro-level (societal) aspects, the so-called micro-level has its focus on the agents [27]. Further, the model of agent societies adopted in [5–9]) also aims to structure the multi-agent systems in various levels of abstraction. The model focuses on the dynamics of the different organisational levels of agent societies[2]. In addition, the work in [4] introduces so-called *social sub-systems*, which account for the structuring of organisational entities into sub-systems within the multi-agent system. Other aspects that emphasises the existence of multiple levels in multi-agent systems can be found in the survey [15].

3 Multi-level Semantics

Preliminary versions of multi-level semantics appeared in [18] and in [19,20], formalising the effects of speech-acts used in argumentation-based dialogues, which affect not only the agents' individual mental states but also the sequence of claims that the agents make, in a public way, at the social level. Although concerned particularly with communication, the work presented in [19] already introduced a multi-level operational semantics, separating two main levels of abstraction of multi-agent systems (the agent and social levels).

[2] The *count-as relations* used in the present work are called *implementation relations* in [5].

In those works, the authors have focused on speech acts and argumentation, assuming that each of the actions executed by an agent, in an argumentation-based dialogue, had a direct effect on the social level. Here, we follow the same general approach, exploring multi-level semantic representations of multi-agent systems, but we extend it in several ways. First, we are not interested in the specification of any particular communication method. We here propose a more general specification of processes for the various multi-agent system levels, defining vertical integrity constraints to ensure appropriate overall system behaviour. Further, we consider multi-agent systems composed of multiple abstraction levels (rather than only two), and how these levels are connected by events occurring at different levels of the multi-agent system. Thus, some actions that are executed at a given level may be directly mapped to actions executed at higher levels (for instance, actions executed by the organisation, i.e., the action directly counts as an organisational action). However, some situations require that *a number of actions* be executed in particular orders for the whole *process* to count as a higher-level action. These cases often require a (possibly partial) ordering in the execution of the actions, which may have to be executed by several different performers (*agents*, at the lowest system-level, or *organisational units*, at some intermediate system level).

In order to specify this type of multi-level dynamics of multi-agent systems, we propose a multi-level semantic representation, inspired by [19], to make clear both the working of each level and how each level affects the others. Furthermore, the aim is to allow for "vertical modularity", that is, the modular reuse of previously existing formalisations of specific system levels, by supporting their

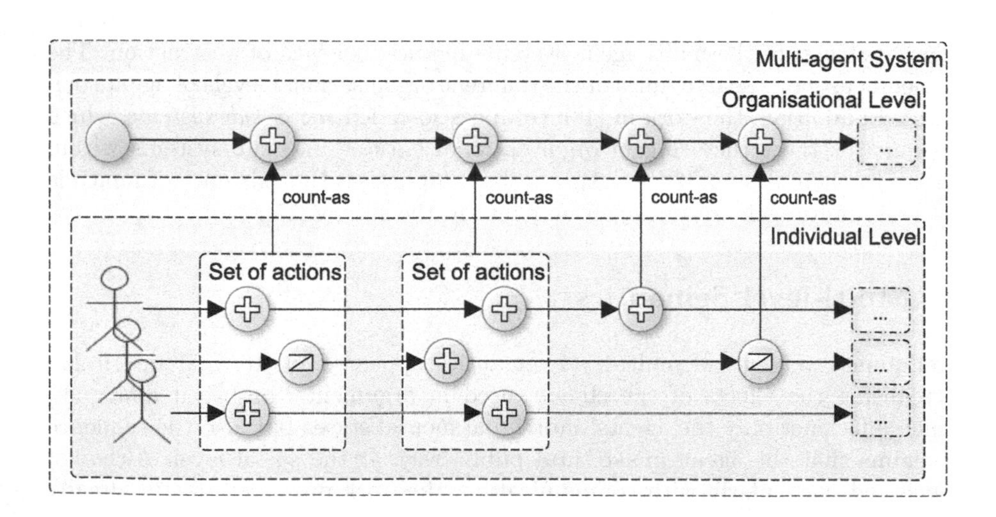

Fig. 1. Set of Actions and *count-as* Relations.

integration with each other and with new ones, in a way that is more clear than the specification of the whole system as a single transition system[3].

Figure 1 shows different possibilities of the counts as relation, where first we have two sets of ordered actions that count as specific actions of the organisation, but for each set of actions, only the execution of *all* actions (and respecting the appropriate order) at the individual level counts as a corresponding organisational action (i.e., an action or a meaningful change at the organisational level). A point to note is that "actions" here are not necessarily actions performed on objects, they can also be communicative actions or more generally represent state changes or goal achievement at any level of a system. Another point to note is that, for multi-agent systems with multiple organisational levels (for example, with the level of social sub-systems [4], which are taken to regulate the functioning and interactions among the organisations in such multi-agent systems), we can have a *count-as* relation from the organisational level to that higher-level social sub-system level (for example, a process performed by one or more organisations that counts as a single action that is subject to some norm of the *normative sub-system* that regulates the overall functioning of the multi-agent system). Finally, it is important to note that it is not always the case that relations are from an inferior level to a superior one; there are situations where events at a superior level have effects over the inferior ones.

4 Processes

For the purposes of this work, and for the sake of simplicity, we left out important concepts in multi-agent systems such as goals, beliefs, etc. We assume that the state of the system is described only by the state of actions, which are the main idea behind the count-as relations as relevant for this work.

When we consider different levels of a multi-agent system, as described before, an action of a level can be a part of a process which counts as an action at another level of the multi-agent system in a particular context. However, each individual agent action alone may not count as an organisational action in a direct way. With that in mind, we use a particular form for specifying possible orderings of interest for the execution of actions, which is used in our approach for the so called integrity constraints. This notation is inspired by the *functional specification* for missions in the MOISE organisational model [16], where missions are sets of goals and the root goal is achieved by achieving subgoals structured using sequence, parallel, or choice operators. The notation used is as follows:

- *sequence* ',': the notation $\textit{count-as}((a', a''), a, C)$ means that the sequence of actions a', a'' (in this exact order) needs be executed for it to count as the action a at another level of the system, in the context C. The context itself is also a process but one that needs to be checked against the actions that already took place at the level of a.

[3] *Vertical modularity* is the main reason for the neat separation between the populational and the organisational system levels, and between the micro-, meso-, and macro-organisational levels, in [5].

- *choice* '|': the notation $count\text{-}as((a' \mid a''), a, C)$ means that either action a' or action a'' being executed counts as the action a at another level of the system, in that particular context C.
- *parallelism* " || ": the notation $count\text{-}as((a' \parallel a''), a, C)$ means that both actions a' and a'' need to be executed (regardless of any order between them) for them to count as the action a at the other level, in that particular context C—i.e., the actions can be executed in parallel or in any order.

It is important to note that the *count-as* relation used is abstract; we explicitly intend to use a generalised form of such relation, for example, allowing one to specify: (i) actions executed by agents which *count as* organisational actions, and (ii) actions executed by organisations which *count as* social sub-system actions (or vice-versa). Note also that for simplicity we are assuming only these three levels in this work—the *individual* (agents), *organisational* and *social sub-system* levels—but of course the approach can be applied to any number of levels. Furthermore, in order to describe complex processes occurring in the multi-agent system, the operators described above can be combined.

4.1 Vertical Integrity Constraints and Count-As Relations

In this section, we introduce the idea of Vertical (i.e., inter-level) Integrity Constraints (VIC) used in our approach. As described above, we have well-defined processes, which correspond to count-as relations existing between the multiple levels of multi-agent systems. Thus, using the concepts of process introduced above, we are able to check the vertical integrity constraints between pairs of levels of the multi-agent system, which guarantees that such *count-as* relations are followed in accordance to their definitions.

In the examples used here we consider three levels in the multi-agent system, the *agents*, the *organisations*, and the *social sub-system*, so we introduce a generic vertical integrity constraint for count-as relations between those levels, corresponding to the relation from an inferior to a superior level (due to space constraints, we leave the relations from superior to inferior levels for future work). It is important to note that, although we use three levels in this example, vertical integrity constraints can be applied to any number of levels.

The generic vertical integrity constraint is called $\#VIC_i^s \Uparrow$ and constraints which actions at level i (inferior) count as actions at level s (superior). Checking vertical integrity constraints is modelled as a function that has as input the set of actions already executed at the superior level AS_s, the actions yet (i.e., expected) to be executed at the superior level A_s, the set of actions already executed at the inferior level AS_i, and the actions that have just been executed at the inferior level Act_i. The funtion returns a set S of superior-level actions that can be deemed as executed (through a completed process) by satisfying any of the existing *count-as* relations; this needs to be checked against the history of executed actions (i.e., *traces* at both levels involved) and the actions that have just been executed Act_i. Algorithm 1 shows how the vertical integrity constraints are checked.

$$\#VIC_i^s \Uparrow [AS_s, A_s, AS_i, Act_i] = S$$

```
Algorithm #VIC_i^s ⇑ [AS_s, A_s, AS_i, Act_i]
    S = {};
    AS_temp = AS_i ∪ Act_i;
    for each a ∈ A_s do
        for each p, C such that count-as(p, a, C) do
            if (satisfies(AS_s, C) ≠ −1 and satisfies(AS_temp, p) ≠ −1) then
                S = S ∪ {a};
    return S;
```

Algorithm 1: VIC

```
Function satisfies(AS, p)
    if p = a, with a an atomic action then
        if ⟨a, time⟩ ∈ AS then
            return time
        else
            return −1
    else
        p = e1 op e2, op ∈ {'|','||',','};
        t1 = satisfies(AS, e1);
        t2 = satisfies(AS, e2);
        if (op = '|') then
            if t1 = −1 and t2 = −1 then
                return −1
            if t1 ≠ −1 and t2 ≠ −1 then
                return { t1, t2 }
            if ((t1 = −1 and t2 ≠ −1) or (t1 ≠ −1 and t2 = −1)) then
                return max(t1, t2)
        if (op = '||') then
            if t1 ≠ −1 and t2 ≠ −1 then
                return max(t1, t2)
            else
                return −1
        if (op = ',') then
            if ((t1 ≠ −1 and t2 ≠ −1) and (min(t1) < max(t2))) then
                if min(t1) < min(t2) then
                    return min(t2)
                else
                    return max(t2)
            else
                return −1
```

Algorithm 2: Satisfies Function

5 Multi-level Semantics with VIC

In this section we discuss the main contribution of this work, the multi-level semantics with vertical integrity constraints. We have just one abstract semantic rule interpreting all *count-as* relations in the multi-agent system, where the corresponding actions can be instantiated to particular cases. Before we show that rule, we first define the configuration of the transition system we use in the operational semantics, which is abstracted away in order to facilitate the understanding of this work.

In our semantics, the social sub-system is a tuple $\langle ss_{id}, \mathcal{O}, \mathcal{A}_{ss} \rangle$ with ss_{id} the social sub-system identifier, $\mathcal{O}$ a set of organisation identifiers (each one representing the organisation populating the social sub-system), and $\mathcal{A}_{ss}$ the set of actions of the social sub-system. The organisations are represented by the tuple $\langle org_{id}, Ag, \mathcal{A}_{org} \rangle$ with org_{id} the organisation identifier, Ag a set of agents identifiers (which are populating that organisation), and $\mathcal{A}_{org}$ the set of organisational actions. Each agent in the multi-agent system is represented by a tuple $\langle ag_{id}, \mathcal{A} \rangle$ with ag_{id} the agent identifier and $\mathcal{A}$ the set of actions that the agent is able to execute. As mentioned before, for simplicity we left out of this work the representation of goals, beliefs and other attitudes often used in multi-agent systems. We assume that agents are committed to execute the actions in order to achieve either individual or organisational and social sub-system goals that for uniformity are here treated also as higher-level "actions". At each level of the system (i.e., *agents*, *organisations* and *social sub-systems*), we maintain a history (i.e., trace) of actions already executed at that level; $\mathcal{AS}_{ag}$ represents the agent actions, $\mathcal{AS}_{org}$ the organisational actions, and $\mathcal{AS}_{ss}$ the social sub-system actions already executed in that particular instance of the multi-agent system.

Further, we use a function called *continuation*[4] with the following parameters: $c(A, AS, a)$, where A is the set of actions expected to be executed at that particular level, AS is the set of actions already executed at that particular level, and a is the action being executed at the current time/transition of the multi-agent system[5]. This function defines when the action being executed (i.e., a) is removed or not from the set of actions A, and if other actions are included in the set of actions given the ones that were executed. Thus, we are able to treat both: achievement and maintenance tasks for multi-agent systems. This function is clearly domain-dependent so we assume it as given in this work. The abstract semantic rule[6] is shown in `AbstractSemanticRule`.

$$
\frac{\begin{array}{c} execute(\texttt{act}) \qquad \texttt{act} \in \mathcal{A} \\ \#VIC_{ag}^{org} \Uparrow [\mathcal{AS}_{org}, \mathcal{A}_{org}, \mathcal{AS}_{ag}, \{\texttt{act}\}] = \mathcal{S}_{org} \\ \#VIC_{org}^{ss} \Uparrow [\mathcal{AS}_{ss}, \mathcal{A}_{ss}, \mathcal{AS}_{org}, \mathcal{S}_{org}] = \mathcal{S}_{ss} \end{array}}{\begin{array}{llcl} (a) & \langle ss_l, \mathcal{O}, \mathcal{A}_{ss} \rangle \ldots \mathcal{AS}_{ss} & \rightarrow_{ss} & \langle ss_l, \mathcal{O}, \mathcal{A}'_{ss} \rangle \ldots \mathcal{AS}'_{ss} \\ (b) & \langle org_n, Ag, \mathcal{A}_{org} \rangle \ldots \mathcal{AS}_{org} & \rightarrow_{org} & \langle org_n, Ag, \mathcal{A}'_{org} \rangle \ldots \mathcal{AS}'_{org} \\ (c) & \langle ag_m, \mathcal{A} \rangle \ldots \mathcal{AS}_{ag} & \rightarrow_{ag} & \langle ag_m, \mathcal{A}' \rangle \ldots \mathcal{AS}'_{ag} \end{array}} \quad (\text{AbstractSemanticRule})
$$

$$
\begin{array}{lll} \textit{where:} & & \\ (a) & \mathcal{AS}'_{ss} & = \mathcal{AS}_{ss} \cup \mathcal{S}_{ss} \\ & \mathcal{A}'_{ss} & = c(\mathcal{A}_{ss}, \mathcal{AS}_{ss}, \mathcal{S}_{ss}) \\ (b) & \mathcal{AS}'_{org} & = \mathcal{AS}_{org} \cup \mathcal{S}_{org} \\ & \mathcal{A}'_{org} & = c(\mathcal{A}_{org}, \mathcal{AS}_{org}, \mathcal{S}_{org}) \\ (c) & \mathcal{AS}'_{ag} & = \mathcal{AS}_{ag} \cup \{\texttt{act}\} \\ & \mathcal{A}' & = c(\mathcal{A}, \mathcal{AS}_{ag}, \{\texttt{act}\}) \end{array}
$$

[4] We borrowed the term *continuation* from the area of computer programming where it refers to the abstract representation of a program's control state [23].

[5] Note that in case multiple actions can be executed simultaneously in the system, action a above could be simply substituted by a set of actions.

[6] We use the notation "$\langle ag_m, \mathcal{A} \rangle \ldots \mathcal{AS}_{ag}$" as a simplified representation for "$\{\langle ag_1, A_{ag_1} \rangle, \ldots, \langle ag_n, A_{ag_n} \rangle\}, \mathcal{AS}_{ag}$", being $\{ag_1, ..., ag_n\}$ all agents on the multi-agent system.

In Sect. 6, we present one scenario to illustrate the use of the multi-level semantics approach we introduce in this paper.

6 Organisational Processes

To illustrate this work, we use a scenario of a multi-agent system inspired by the relations between Human Resources (HR) departments of two companies[7](organisations). In particular, we model the activities related to the exchange of employees between such companies, i.e., when an employee leaves a company to join another. The organisations (companies) have their exchanges regulated by a social sub-system [4], describing the labour rights of employees. Therefore, the organisation must follow some particular process in order to do such exchange of employees, e.g., first a company must dismiss the employee before the other company can appoint that same employee. The social sub-system has its own goal within the multi-agent system, which is to maintain the exchanges between the companies in accordance with the labour laws, having a process to be followed that define such norm-conforming exchanges.

6.1 Organisations

In our multi-agent system scenario, we consider two companies, which we will call $\texttt{org}_1$ and $\texttt{org}_2$, respectively. At the upper level, regulating the organisations, we have the social sub-systems, the one of interest here we call `ssys`. Therefore, the social sub-system `ssys` is populated by the organisations $\texttt{org}_1$ and $\texttt{org}_2$. Further, the organisations are populated by agents, which, for the purposes of this paper, we can abstract away from their name an refer to their roles only. Both organisations have three main internal roles: the `interviewer`, `proc_administrator` and `manager`. We consider two main processes for which the organisations are responsible. The first process is formed by the sequence {`interview`, `admission_request`, `admission_processing`, `appoint`} where the agents playing the role of `interviewer`, `proc_administrator` and `manager` are responsible for the actions `interview`, `admission_request` and `admission_processing`, and `appoint`, respectively. This sequence composes the process of `appointing` which *counts-as* the organisation action of appointing an employee $\texttt{appoint}_{\texttt{org}}$. The second process is composed of the sequence {`give_notice`, `dismissal_request`, `dismissal_processing`, `dismiss`} where the agents playing the role of `manager` are responsible for the actions `give_notice` and `dismiss`, and the role of `proc_administrator` for the actions `dismissal_request` and `dismissal_processing`. This sequence composes the process of `dismissing` which *counts-as* the organisation action of dismissing an employee $\texttt{dismiss}_{\texttt{org}}$. The social sub-system `ssys` has a process in order to regulate the exchange of employees between the companies $\texttt{org}_1$ and $\texttt{org}_2$, this process is composed of the sequence {$\texttt{dismiss}_{\texttt{org}}$, $\texttt{appoint}_{\texttt{org}}$}, where the actions are the responsibility of the organisations. This sequential process *count-as* a social sub-system action of a norm-conforming exchange, i.e., `norm-conf_exchange`.

[7] We use only two companies for the sake of simplicity.

6.2 Semantics

In this section, we give semantics for the example scenario described above. All semantic rules presented in this section are instances of the abstract semantic rule `AbstractSemanticRule` described in Sect. 5; this means that when our approach is used in a system formalisation, the rules below no longer need to be written because they can be deduced from the integrity constraints written by the user of our approach together with the one general rule in the previous section and the existing operational semantics of each level separately given. The execution of simple actions by agents can be represented by the semantic rule:

$$\frac{\begin{array}{c} execute(\texttt{interview}) \qquad \texttt{interview} \in \mathcal{A} \\ \#VIC_{ag}^{org} \Uparrow [\mathcal{AS}_{org}, \mathcal{A}_{org}, \mathcal{AS}_{ag}, \{\texttt{interview}\}] = \{\ \} \\ \#VIC_{org}^{ss} \Uparrow [\mathcal{AS}_{ss}, \mathcal{A}_{ss}, \mathcal{AS}_{org}, \{\ \}] = \{\ \} \end{array}}{\begin{array}{ll} (a) & \langle \texttt{ssys}, \mathcal{O}, \mathcal{A}_{ss} \rangle \ldots \mathcal{AS}_{ss} \rightarrow_{ss} \langle \texttt{ssys}, \mathcal{O}, \mathcal{A}_{ss} \rangle \ldots \mathcal{AS}_{ss} \\ (b) & \langle \texttt{org}_2, Ag, \mathcal{A}_{org} \rangle \ldots \mathcal{AS}_{org} \rightarrow_{org} \langle \texttt{org}_2, Ag, \mathcal{A}_{org} \rangle \ldots \mathcal{AS}_{org} \\ (c) & \langle \texttt{interviewer}, \mathcal{A} \rangle \ldots \mathcal{AS}_{ag} \rightarrow_{ag} \langle \texttt{interviewer}, \mathcal{A}' \rangle \ldots \mathcal{AS}'_{ag} \end{array}} \quad (\textsc{ExInterviewAction})$$

$$\begin{array}{lll} \textit{where:} & & \\ (c) & \mathcal{AS}'_{ag} & = \mathcal{AS}_{ag} \cup \{\texttt{interview}\} \\ & \mathcal{A}' & = c(\mathcal{A}, \mathcal{AS}_{ag}, \{\texttt{interview}\}) \end{array}$$

The semantic rule above shows that actions that do not satisfy the vertical integrity constraints, do not have any effects over any other level of the system. In this case, the effects of the agent `interviewer` executing the action `interview`, which updates its internal state marking the action `interview` as executed—i.e., the *continuation* function just removes the action `interview` from the set of actions $\mathcal{A}$—and adds it to the history of executed actions at the agent own level. Single actions have a simple semantics, it is easy to observe that whenever agents execute some action their internal state will be updated. This agent update does not interfere in the configuration of the transition systems at higher levels, which helps with the modularity we want to achieve. This is also the case for the actions `admission_request`, `admission_processing`, `give_notice`, `dismissal_request` and `dismissal_processing`, because all of them are part of a sequential process, therefore the processes are completed only when the last action of each sequential process is executed, in our case `appoint` and `dismiss`, respectively. Therefore, for simplicity we will not present the semantic rules for the actions mentioned, given that all the semantic rules are similar to the `ExInterviewAction` semantic rule demonstrated above, the only difference is that the action `interview` will be substituted by the corresponding action and respective agent, the one executing the action (the same for the organisation in which the agent plays a role).

Observe that the vertical integrity constraints $\#VIC_{ag}^{org} \Uparrow$ and $\#VIC_{org}^{ss} \Uparrow$ are not satisfied, therefore the organisational actions $\mathcal{A}_{org}$ and social sub-system actions $\mathcal{A}_{ss}$ are not achieved/updated. However, when an agent executes the last action of those sequential process, an organisational action is achieved. In order to exemplify this idea, we will define the operational semantics for the $\texttt{dismiss}_{\texttt{org}_1}$ action, which is resulting from a process of `dismissing` = {`give_notice`, `dismissal_request`, `dismissal_processing`, `dismiss`}. In the semantic rule presented below, we consider that the other actions `give_notice`, dismissal_request and dismissal_processing have been already executed by the agents playing the roles that are entitled to execute them, as described above.

$$\frac{\begin{array}{c} execute(\mathtt{dismiss}) \quad \mathtt{dismiss} \in \mathcal{A} \\ \#VIC^{org}_{ag} \Uparrow [\mathcal{AS}_{org}, \mathcal{A}_{org}, \mathcal{AS}_{ag}, \{\mathtt{dismiss}\}] = \{\mathtt{dismiss}_{\mathtt{org_1}}\} \\ \#VIC^{ss}_{org} \Uparrow [\mathcal{AS}_{ss}, \mathcal{A}_{ss}, \mathcal{AS}_{org}, \{\mathtt{dismiss}_{\mathtt{org_1}}\}] = \{\ \} \end{array}}{\begin{array}{llll} (a) & \langle \mathtt{ssys}, \mathcal{O}, \mathcal{A}_{ss}\rangle \ldots \mathcal{AS}_{ss} & \rightarrow_{ss} & \langle \mathtt{ssys}, \mathcal{O}, \mathcal{A}_{ss}\rangle \ldots \mathcal{AS}_{ss} \\ (b) & \langle \mathtt{org_1}, Ag, \mathcal{A}_{org}\rangle \ldots \mathcal{AS}_{org} & \rightarrow_{org} & \langle \mathtt{org_1}, Ag, \mathcal{A}'_{org}\rangle \ldots \mathcal{AS}'_{org} \\ (c) & \langle \mathtt{manager}, \mathcal{A}\rangle \ldots \mathcal{AS}_{ag} & \rightarrow_{ag} & \langle \mathtt{manager}, \mathcal{A}'\rangle \ldots \mathcal{AS}'_{ag} \end{array}} \ (\textsc{ExDismissAction})$$

$$\begin{array}{lll} \textit{where:} & & \\ (b) & \mathcal{AS}'_{org} & = \mathcal{AS}_{org} \cup \{\mathtt{dismiss}_{\mathtt{org_1}}\} \\ & \mathcal{A}'_{org} & = c(\mathcal{A}_{org}, \mathcal{AS}_{org}, \{\mathtt{dismiss}_{\mathtt{org_1}}\}) \\ (c) & \mathcal{AS}'_{ag} & = \mathcal{AS}_{ag} \cup \{\mathtt{dismiss}\} \\ & \mathcal{A}' & = c(\mathcal{A}, \mathcal{AS}_{ag}, \{\mathtt{dismiss}\}) \end{array}$$

All mentioned actions do not satisfy the vertical integrity constraints, therefore they do not cause any effect over the organisational actions, because the organisational action of $\mathtt{dismiss}_{\mathtt{org_1}}$ is only achieved by the correct sequence of actions defined in the process described above, the last action execution being necessary to satisfy the process, i.e., counting as the $\mathtt{dismiss}_{\mathtt{org_1}}$ organisational action, as shown in the semantics rule ExDismissAction.

$$\frac{\begin{array}{c} execute(\mathtt{appoint}) \quad \mathtt{appoint} \in \mathcal{A} \\ \#VIC^{org}_{ag} \Uparrow [\mathcal{AS}_{org}, \mathcal{A}_{org}, \mathcal{AS}_{ag}, \{\mathtt{appoint}\}] = \{\mathtt{appoint}_{\mathtt{org_2}}\} \\ \#VIC^{ss}_{org} \Uparrow [\mathcal{AS}_{ss}, \mathcal{A}_{ss}, \mathcal{AS}_{org}, \{\mathtt{appoint}_{\mathtt{org_2}}\}] = \{\mathtt{norm\text{-}conf\_exchange}\} \end{array}}{\begin{array}{llll} (a) & \langle \mathtt{ssys}, \mathcal{O}, \mathcal{A}_{ss}\rangle \ldots \mathcal{AS}_{ss} & \rightarrow_{ss} & \langle \mathtt{ssys}, \mathcal{O}, \mathcal{A}'_{ss}\rangle \ldots \mathcal{AS}'_{ss} \\ (b) & \langle \mathtt{org_2}, Ag, \mathcal{A}_{org}\rangle \ldots \mathcal{AS}_{org} & \rightarrow_{org} & \langle \mathtt{org_2}, Ag, \mathcal{A}'_{org}\rangle \ldots \mathcal{AS}'_{org} \\ (c) & \langle \mathtt{manager}, \mathcal{A}\rangle \ldots \mathcal{AS}_{ag} & \rightarrow_{ag} & \langle \mathtt{manager}, \mathcal{A}'\rangle \ldots \mathcal{AS}'_{ag} \end{array}} \ (\textsc{ExAppointAction})$$

$$\begin{array}{lll} \textit{where:} & & \\ (a) & \mathcal{AS}'_{ss} & = \mathcal{AS}_{ss} \cup \{\mathtt{norm\text{-}conf\_exchange}\} \\ & \mathcal{A}'_{ss} & = c(\mathcal{A}_{ss}, \mathcal{AS}_{ss}, \{\mathtt{norm\text{-}conf\_exchange}\}) \\ (b) & \mathcal{AS}'_{org} & = \mathcal{AS}_{org} \cup \{\mathtt{appoint}_{\mathtt{org_2}}\} \\ & \mathcal{A}'_{org} & = c(\mathcal{A}_{org}, \mathcal{AS}_{org}, \{\mathtt{appoint}_{\mathtt{org_2}}\}) \\ (c) & \mathcal{AS}'_{ag} & = \mathcal{AS}_{ag} \cup \{\mathtt{appoint}\} \\ & \mathcal{A}' & = c(\mathcal{A}, \mathcal{AS}_{ag}, \{\mathtt{appoint}\}) \end{array}$$

Note that the vertical integrity constraint $\#VIC^{org}_{ag} \Uparrow$ returns the actions that are deemed executed from the set of organisational actions given the history of actions of the lower level (agents' actions), and the particular context, represented by the history of organisational actions already executed[8]. In this way, it is guaranteed that the process was followed in accordance with the rules and the organisational action of $\mathtt{dismiss}_{\mathtt{org_1}}$ was completed. Observe also that, as the organisation executes an action of $\mathtt{dismiss}_{\mathtt{org_1}}$, the history of organisational actions is updated accordingly. However, as the vertical integrity constraint $\#VIC^{ss}_{org} \Uparrow$ is not satisfied, no social sub-system action is deemed as executed. Further, the *continuation* function, in this case, removes the action of $\mathtt{dismiss}_{\mathtt{org_1}}$ from the set of action $\mathcal{A}_{org}$, but adds the action $\mathtt{appoint}_{\mathtt{org_1}}$, because to appoint and to dismiss employees is a continuous process in the organisations in our scenario.

The semantic rule for the action $\mathtt{appoint}_{\mathtt{org_2}}$, which is resulting from a process of appointing = {interview, admission_request,

[8] We assume that the context here is satisfied, and it consist of the employee that has been appointed earlier, i.e., $\mathtt{appoint}_{\mathtt{org_1}} \in \mathcal{AS}_{org}$.

admission_processing, appoint} is presented above. As before, the semantic rule considers that the other agents have already executed the actions interview, admission_request and admission_processing when playing the respective roles. All previous actions do not have effects over the organisational actions, because the organisational action of $\mathtt{appoint}_{\mathtt{org}_2}$ is only achieved with the correct sequence of actions defined by the process. Furthermore, the operational semantics above demonstrates the social sub-system action of norm-conf_exchange being achieved. Note that the vertical integrity constraints $\#VIC^{ss}_{org} \Uparrow$ between the organisational and the social sub-system levels is satisfied, i.e, the history of the organisational actions, correctly satisfies the process to achieve the social sub-system action of norm-conf_exchange. Recall that such an action is considered executed by the sequence of $\mathtt{dismiss}_{\mathtt{org}_1}$ and $\mathtt{appoint}_{\mathtt{org}_2}$, both represented by the two semantic rules above. Further, the *continuation* function does not remove the norm-conf_exchange action from $\mathcal{A}_{ss}$, because it is a repeating action available to the social sub-system.

7 Related Work

In [12], the authors propose an *Agent Infrastructure Layer* (AIL) for BDI-style programming languages. The aims are: (i) to provide a common semantic basis for a number of BDI languages, and; (ii) to support formal verification by developing a *model-checker* optimised for agent programs. The authors propose the design of AIL using an extensive operational semantics presented in [11]. The authors argue that AIL captures all major features of common BDI languages. In [10], the authors propose an operational semantics introducing the concept of modules. The authors argue that modularisation facilitates the implementation of agents, agents roles, and agents profiles, besides being an essential principle in structured programming, and so the authors assume that it must be so in agent programming too. Operational semantics is given to creating, executing, testing, updating, and realising module instances, i.e., module-related actions. In [13], an operational semantics is given to cover the details of operations that may be applied to goals (dropping, aborting, suspending, and resuming goals). The authors argue that the semantics clarifies how an agent can manage its goals, based on the decisions that it chooses to make. The semantics, according to the authors, further provides a foundation for correctness verification of agent behaviour. The authors emphasise that the work contributes to the development of a rich and detailed specification of the appropriate operational behaviour when a goal is being pursued, has succeeded or failed, or has been aborted, suspended, or resumed by the agent. In [4], the author proposes an operational semantic framework for legal systems that are (structurally and operationally) situated in *agent societies*. The work uses operational semantics for modelling the *structure* and *dynamics* of legal systems. The work presented in [4] is concerned with actions associated with legal systems (internal legal acts, external legal acts, social acts, etc.), i.e., under an *action-based dynamics* where the transitions between configurations of the system are determined by the performance of actions by the *legal organs* and *legal subjects* of the studied legal system. It is an example of work that uses operational semantics to formalise multi-level systems. In [25], the authors propose a semantic framework for MOISE+ [16] and an accompanying linear temporal logic (LTL) to express its properties. Following [25], the organisation should make the agents more effective in attaining their purpose, or prevent certain undesired behaviour from occurring. An organisational specification achieves this by imposing *organisational constraints* on the behaviour of agents playing some role in the organisation, and the agents are expected to take these constraints into account in their decision-making. The work in [25] aims to make organisational constraints precise by defining the formal semantics of MOISE+. That work is concerned with a semantics that allows the agent to respect: (i) preconditions for the execution of organisational actions, and (ii) acquaintance and communication links.

Our work differs from all those in that we present a new style for the semantics of multi-level systems in which can use vertical integrity constraints to specify relations between the multiple levels in multi-agent systems. We are not focused on a special semantics for BDI languages as in [12], in semantics for module-related actions [10], in semantics for goals [13], in semantics for legal systems [4], or semantics for the MOISE model [25] specifically. We have used "counts as" to refer to relations between the multiple levels of our operational semantics style in order to exemplify our approach to vertical integrity constraints. Further, we argue that in other domains various other inter-level relations may make more sense than the *count-as* relation.

8 Conclusion

In this work, we introduced an approach to the formalisation of multi-agent systems based on operational semantics; we call it multi-level semantics with vertical integrity constraints. The approach allows the representation of the interactions between components of different system-levels. Given the complexity and ubiquity of such multiple levels in multi-agent systems, the approach seems to allow for a clearer understanding of such complex semantics. Furthermore, we demonstrate, using count-as relations applied to *processes*, how the proposed style for multi-level operational semantics can be strengthened through the definition of vertical integrity constraints. Such multi-level semantics with vertical integrity constraints allows the independent specification of the various levels typical of the multi-agent system, so that each level is formalised through its own transition system, and the vertical integrity constraints between these transition systems help guarantee that the overall system operates coherently in all possible executions.

In our future work, we intend to explore other relations between the multiple levels of multi-agent system, including relations from events occurring at higher levels and affecting the lower ones, e.g. considering some changes in the social sub-systems that affect all organisations and the respective agents playing some role in those organisations. Also, we intend to investigate how our multi-level semantics and the vertical integrity constraints can play a part in research towards model checking of multi-agent systems [2].

Acknowledgements. This research was partially funded by CNPq and CAPES.

References

1. Boissier, O., Bordini, R.H., Hübner, J.F., Ricci, A., Santi, A.: Multi-agent oriented programming with JaCaMo. Sci. Comput. Program. **78**(6), 747–761 (2013)
2. Bordini, R.H., Fisher, M., Visser, W., Wooldridge, M.: Verifying multi-agent programs by model checking. Auton. Agent. Multi-Agent Syst. **12**(2), 239–256 (2006)
3. Bordini, R.H., Hübner, J.F., Wooldridge, M.: Programming Multi-Agent Systems in Agent Speak Using Jason (Wiley Series in Agent Technology). Wiley (2007)
4. Rocha Costa, A.C.: Situated legal systems and their operational semantics. Artif. Intell. Law **23**(1), 43–102 (2015). https://doi.org/10.1007/s10506-015-9164-z
5. da Rocha Costa, A.C.: A Variational Basis for the Regulation and Structuration Mechanisms of Agent Societies. Springer, Cham (2019). https://doi.org/10.1007/978-3-030-16335-8
6. da Rocha Costa, A.C., Dimuro, G.P.: A basis for an exchange value-based operational notion of morality for multiagent systems. In: Neves, J., Santos, M.F., Machado, J.M. (eds.) EPIA 2007. LNCS (LNAI), vol. 4874, pp. 580–592. Springer, Heidelberg (2007). https://doi.org/10.1007/978-3-540-77002-2_49
7. da Rocha Costa, A.C., Dimuro, G.P.: Semantical concepts for a formal structural dynamics of situated multiagent systems. In: Sichman, J.S., Padget, J., Ossowski, S., Noriega, P. (eds.) COIN -2007. LNCS (LNAI), vol. 4870, pp. 139–154. Springer, Heidelberg (2008). https://doi.org/10.1007/978-3-540-79003-7_11
8. Costa, A.C.R., Dimuro, G.P.: Introducing social groups and group exchanges in the poporg model. In: Proceedings of the 8th International Conference on Autonomous Agents and Multiagent Systems, vol. 2, pp. 1297–1298. International Foundation for Autonomous Agents and Multiagent Systems (2009)
9. Costa, A.C.R., Dimuro, G.P.: A minimal dynamical mas organization model. In: Handbook of Research on Multia-Agent Systems: Semantics and Dynamics of Organizational Models, pp. 419–445 (2009)
10. Dastani, M., Steunebrink, B.R.: Operational semantics for BDI modules in multi-agent programming. In: Dix, J., Fisher, M., Novák, P. (eds.) CLIMA 2009. LNCS (LNAI), vol. 6214, pp. 83–101. Springer, Heidelberg (2010). https://doi.org/10.1007/978-3-642-16867-3_5
11. Dennis, L.A.: Agent infrastructure layer (AIL): design and operational semantics v1. Technical report, Technical Report ULCS-07-001, Department of Computer Science, University of Liverpool (2007)
12. Dennis, L.A., Farwer, B., Bordini, R.H., Fisher, M., Wooldridge, M.: A common semantic basis for BDI languages. In: Dastani, M., El Fallah Seghrouchni, A., Ricci, A., Winikoff, M. (eds.) ProMAS 2007. LNCS (LNAI), vol. 4908, pp. 124–139. Springer, Heidelberg (2008). https://doi.org/10.1007/978-3-540-79043-3_8
13. Harland, J., Morley, D.N., Thangarajah, J., Yorke-Smith, N.: An operational semantics for the goal life-cycle in BDI agents. Auton. Agent. Multi-Agent Syst. **28**(4), 682–719 (2014)

14. Hewitt, C.: Offices are open systems. ACM Trans. Inf. Syst. (TOIS) **4**(3), 271–287 (1986)
15. Horling, B., Lesser, V.: A survey of multi-agent organizational paradigms. Knowl. Eng. Rev. **19**(4), 281–316 (2004)
16. Hubner, J.F., Sichman, J.S., Boissier, O.: Developing organised multiagent systems using the moise+ model: programming issues at the system and agent levels. Int. J. Agent-Oriented Softw. Eng. **1**(3–4), 370–395 (2007)
17. Noriega, P.: Agent mediated auctions: the fishmarket metaphor. Institut d'Investigació en Intelligència Artificial (1999)
18. Panisson, A.R., Bordini, R.H., Costa, A.C.R.: Towards multi-level semantics for multi-agent systems. In: 3th Workshop-School on Theoretical Computer Science (2015)
19. Panisson, A.R., Meneguzzi, F., Fagundes, M., Vieira, R., Bordini, R.H.: Formal semantics of speech acts for argumentative dialogues. In: Thirteenth International Conference on Autonomous Agents and Multiagent Systems, pp. 1437–1438 (2014)
20. Panisson, A.R., Meneguzzi, F., Vieira, R., Bordini, R.H.: Towards practical argumentation in multi-agent systems. In: 2015 Brazilian Conference on Intelligent Systems, BRACIS 2015 (2015)
21. Plotkin, G.D.: A structural approach to operational semantics. Computer Science Department, Aarhus University Denmark (1981)
22. Rao, A.S.: AgentSpeak(L): BDI agents speak out in a logical computable language. In: Van de Velde, W., Perram, J.W. (eds.) MAAMAW 1996. LNCS, vol. 1038, pp. 42–55. Springer, Heidelberg (1996). https://doi.org/10.1007/BFb0031845
23. Reynolds, J.C.: The discoveries of continuations. Lisp Symb. Comput. **6**(3–4), 233–247 (1993)
24. Ricci, A., Piunti, M., Viroli, M.: Environment programming in multi-agent systems: an artifact-based perspective. Auton. Agent. Multi-Agent Syst. **23**(2), 158–192 (2011)
25. van Riemsdijk, M.B., Hindriks, K.V., Jonker, C.M., Sierhuis, M.: Formalizing organizational constraints: a semantic approach. In: 9th International Conference on Autonomous Agents and Multiagent Systems (AAMAS 2010), Toronto, Canada, 10–14 May 2010, vol. 1–3, pp. 823–830 (2010)
26. Searle, J.R.: The Construction of Social Reality. Simon and Schuster (1995)
27. Sierra, C., Rodriguez-Aguilar, J.A., Noriega, P., Esteva, M., Arcos, J.L.: Engineering multi-agent systems as electronic institutions. Eur. J. Inform. Prof. **4**(4), 33–39 (2004)
28. Vieira, R., Moreira, A., Wooldridge, M., Bordini, R.H.: On the formal semantics of speech-act based communication in an agent-oriented programming language. J. Artif. Int. Res. **29**(1), 221–267 (2007)
29. Vivanco, M.E., García, C.S.: Electronic Institutions: from specification to development. Consell Superior d'Investigacions Científiques, Institut d'Investigació en Intelligéncia Artificial (2003)

A Reinforcement Learning Based Adaptive Mutation for Cartesian Genetic Programming Applied to the Design of Combinational Logic Circuits

Frederico José Dias Möller(✉), Heder Soares Bernardino, Luciana Brugiolo Gonçalves, and Stênio Sã Rosário Furtado Soares

Computer Science Departament, Federal University of Juiz de Fora (UFJF), Juiz de Fora, Brazil
fredericomollerper@gmail.com

Abstract. The design of digital logic circuits involves transforming logical expressions into an electronic circuit seeking to minimize certain attributes such as the number of transistors. There are deterministic algorithms for this task, but they are limited to small problems. On the other hand metaheuristics can be used for solving this type of problem and Cartesian Genetic Programming (CGP) is widely adopted in the literature. In CGP, mutation is commonly the only operator for generating new candidate solutions. Thus, the performance of this metaheuristic is dependent on the proper choice of mutation and its parameters. Normally, CGP mutates the candidate solutions according to a uniform distribution. Thus, any modification has the same chance to occur. In order to improve the performance of CGP an adaptive mutation operator using an ϵ-greedy strategy for bias the selection of gates is proposed here. The proposal is evaluated using problems of a benchmark from the literature. The results obtained indicate that the proposed adaptive mutation is promising and that its relative performance increases with the size of the problem.

Keywords: Cartesian Genetic Programming · Reinforcement learning · Combinational logic circuit

1 Introduction

The design of electronic digital circuits is not a trivial task [16], so deterministic algorithms, such as Karnaugh's map [9] and Quine - McCluskey [12] have practical application only for small circuits. As presented in [20], this is a two-level logical optimization problem whose decision version is NP-Complete.

This study was financed in part by the Coordenação de Aperfeiçoamento de Pessoal de Nível Superior - Brasil (CAPES) - Finance Code 001. Also, the authors thank the support of CNPq (312682/2018-2), FAPEMIG (APQ-00337-18), and UFJF.

R. Cerri and R. C. Prati (Eds.): BRACIS 2020, LNAI 12320, pp. 18–32, 2020.
https://doi.org/10.1007/978-3-030-61380-8_2

To deal with larger circuits, an alternative is to use heuristic approaches. Among such, ESPRESSO [1], a method that seeks to build a circuit reducing the complexity of logical instructions, stands out. There is no way to guarantee that the solution found by ESPRESSO is optimal, but it finds good solutions with a low consumption of memory and time.

Another option is the use of evolvable hardware [5], which is the application of evolutionary techniques to hardware development. Such techniques, although they usually have a higher computational cost than ESPRESSO, can be used both to generate a circuit, as well as to reduce a circuit generated by another technique [5,19] and usually design circuits with less transistors. Among these methods is the Cartesian Genetic Programming (CGP) [13]. It has already shown that CGP can design circuits with fewer transistors than ESPRESSO [19].

The main operators of CGP are traditionally elitism and mutation, and its mutation operator is usually applied with uniform probability, with no bias. However, an intelligent mutation operator can learn which changes of gates are able to improve the quality of the candidate circuits and thereby guide this the evolutionary process. Some studies on the mutation of gate types in the CGP suggest that certain changes favor the evolutionary process more than others [3,4]. A biased mutation for CGP was proposed in [17], where the probabilities of changing the logical gates were modified in order to generate better individuals. Reinforced learning techniques have been applied in other evolutionary computing works with the proposal to choose which operator will be used [11] and the result was promising. If the bias of a mutation operator is interpreted as variations of that operator, there is no reason to imagine that such a technique cannot be applied successfully in CGP. Such an intelligent operator is expected to improve CGP's performance.

In this work, CGP is used to design optimal combinational circuits. The goal is to create circuits with as few transistors as possible. For this, the CGP mutation operator acts on the connections between the circuit nodes and on the logic gates that represent each node. Our mutation operator uses the ϵ-greedy strategy to determine which mutations are bringing the greatest reward to the circuit and thereby give preference to these changes. In the tests we performed, there were signs of improvement in performance in relation to the equitable mutation operator as the budget given to the problem grows.

2 Methods

The methods used here are Cartesian Genetic Programming with the single active mutation variant of its mutation operator and the ϵ-greedy strategy. These methods are described in the following sections.

2.1 Cartesian Genetic Programming

Cartesian Genetic Programming is a method of genetic programming developed by Miller [13]. It gets its name because each individual is made up of nodes

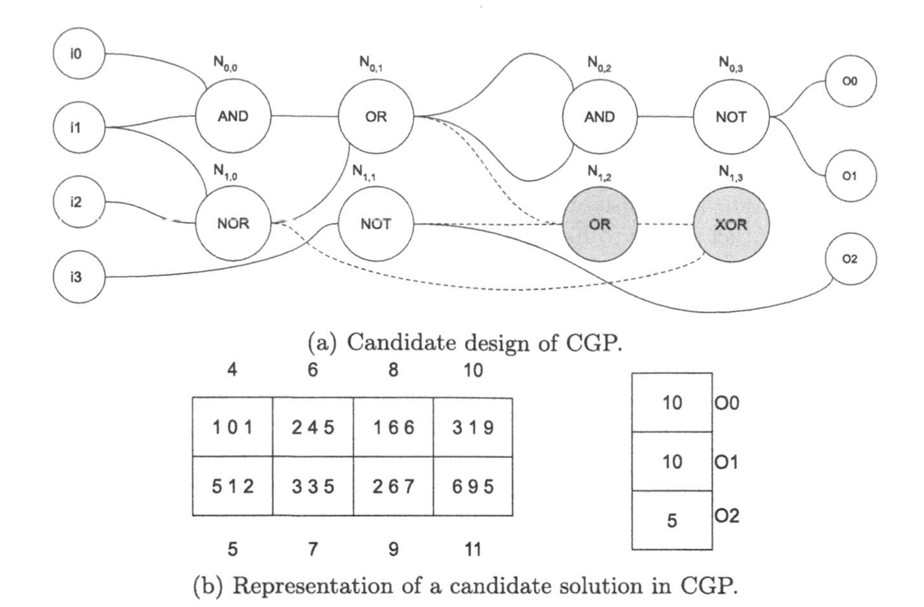

(a) Candidate design of CGP.

(b) Representation of a candidate solution in CGP.

Fig. 1. An example of an individual from the CGP, applied to logic circuits, encoded in a 2×4 matrix. A node may not be connected to any circuit output and a node may receive both inputs from the same node.

organized in n rows and m columns. Although Miller adopts the possibility of cycles [14], traditionally the nodes of a column can only receive input from previous columns, limited to a range of columns called levelsback. Thus, CGP individuals can be interpreted as acyclic digraphs.

The nodes carry information such as the address of their inputs and their function. In the case of digital circuits design application, nodes commonly have two inputs and the logic gate. When the gate has only one entry, such as NOT, the second is ignored. An example of an individual encoded as 2×4 matrix is show in Fig. 1a. Not all nodes are connected to the individual's output. These nodes are called inactive nodes and do not affect the phenotype. Active nodes have at least one path to the exit and affect the phenotype. Genetically, each node is a cell in a matrix of r rows and c columns. In the case of CGP applied to the design of logic circuits, each node carries the coding of the type of gate and the reference to the nodes that are its inputs, as can be seen in Fig. 1b.

The evolutionary process of CGP consists of generating λ new individuals from each of the μ parents. After that, the new individuals are evaluated by the fitness function and the score is compared with those of the parents. A parent is discarded every time he generates an child with a score better than or equal to his. Within an offspring, if more than one child are the best, only one is selected to be a parent in the next generation. The values normally used are $\mu = 1$ and $\lambda = 4$ [14], we use these values in this work. Algorithm 1 shows how CGP works.

Algorithm 1. Pseudo-code of CGP.

```
 1: procedure CGP(μ,λ)
 2:     for i in 0 to μ * λ do
 3:         Generate a random individual i
 4:     Select the μ best individuals and promote them as parents
 5:     while stop criteria is not met do
 6:         for i in 0 to μ do
 7:             for j in 0 to λ do
 8:                 Mutate the parent i to generate the offspring i,j
 9:         for i in 0 to μ do
10:             if An offspring in i group has a score equals or better then i then
11:                 One of the better offspring in group i will be promoted to a parent
12:             else
13:                 The parent i still in the parents group
```

Initially, CGP was proposed with Point Mutation (PM) where a percentage of the individual's chromosome is altered. According to Miller, the number of inactive nodes far exceeds that of active nodes [14]. Therefore, it is common for PM not to generate phenotypically different individuals.

To solve this, alternative methods of mutation were developed for the CGP. The Single Active Mutation (SAM) [7,8] consists of mutating nodes until an active node mutates. Thus, except on occasions when null mutation occurs, the individuals generated are phenotypically different from the parent.

2.2 K-Armed Bandits

The K-Armed Bandits is a problem where a player has a slot machine with several levers. Each lever generates a different average reward, however the player is unaware of such values [10]. In the reinforcement learning conjecture, the K-armed bandits problem serves to illustrate situations where an agent can take different actions within a finite set and he needs to maximize his gains looking for the most advantageous action [2,6,15].

The simplest strategy is the greedy, where the agent chooses a certain action as long as the average reward obtained by it is the highest. This type of strategy aims to increase the chances of a positive final reward, but does not guarantee its maximization. On the other hand, a purely exploratory approach will give the agent the knowledge of the actions with the best reward, but it will not be useful for the objective, if such knowledge is not applied in decision making.

The ϵ-greedy strategy combines exploration and exploitation characteristics. For each event, there is a ϵ% chance for the agent to adopt the exploration strategy. In other cases, the agent adopts the greedy strategy, using the knowledge that accumulated from both phases. This helps the agent to more quickly detect the action that has the highest average reward and stick to it in most events. There are some variations of the ϵ-greedy [15] strategy. For this work it is relevant the start with Optimistic Initial Values.

In this method, the vector with the average reward obtained for each action is initialized at a value equal to or above the optimum. So, as each action is taken, the value of the average reward obtained is reduced until it approaches the real average reward of the action. This forces the agent to take different actions more often during the onset of the problem, finding the best action more quickly [15]. To illustrate this, in a problem of minimizing transistors in a logic circuit, the ideal value is zero transistors. No logic circuit will have a lower value than that. Given the circuit of origin, the actor can choose k different actions. Each causes a different average rate of reduction in this circuit. Regardless of this rate, the generated circuits will always have more than zero transistors. Thus, when taking any k action, the first update of the average reward obtained for this will be a value between zero and the R_1 reward that the actor obtained when taking the k_1 action for the first time. With the average reward obtained from k_i it will be greater than zero, the other actions will be more advantageous for the actor, until he takes them and decreases his average rewards obtained. After taking each of the actions a few times, the action that generates the highest average reward will become evident.

3 Proposed Method

The CGP mutation operator can act on the inputs or the logical gate of each node. It acts unbiased and the chance of modifying any of these attributes is the same. When acting on the logic gate, it can replace the current gate with any of the gates that can be used in the problem. This exchange is also made with uniform odds. The proposal consists of applying the ϵ-greedy strategy to the mutation operator when it acts on the type of logic gate.

For that, there are two square matrices of size equal to the number of gates types used in the problem. One of the matrices records the occurrence of a certain type of gate mutation, while the other registers the average reward obtained. The rows of the matrix refer to the parent's gates and the columns to the gates used in the change for the generation of the children. A pseudo-code of the mutation is presented in Algorithm 2.

The ϵ-greedy strategy is applied when a gate mutation occurs. In $\epsilon\%$ of the cases, the mutation is random, as in traditional CGP. In other cases, the line of the average reward matrix referring to the gate that is going to be mutated is accessed and the type of gate is changed to the gate corresponding to the column that contains the first occurrence of the best average value for that line. A conditional mechanism prevents the values contained in the main diagonal from being chosen, thus preventing the algorithm from making non-trivial exchanges.

The occurrence matrix is updated when a gate mutation occurs. The cell corresponding to that mutation is updated and the average reward matrix is modified according to

$$R_{i,j,t} = R_{i,j,t-1} + \frac{Q_{i,j} - R_{i,j,t-1}}{O_{i,j,t}} \tag{1}$$

Algorithm 2. Pseudo-code of the CGP-RL's mutation operator.

```
1: procedure NODE-MUTATION-RL(individual i,ϵ)
2:     Choose randomly among modifying the gate, input 1, or input 2.
3:     if The gate type was chosen then
4:         Generate a value V between 0 and 1 randomly and evenly.
5:         if V is greater than ϵ then
6:             Go to line P of AREM for the type of the parent gate.
7:             Find the first occurrence F of the lowest value in row P (except P)
8:             Mutate gate P by the port corresponding to that of value F
9:         else
10:            Mutate the parent's logic gate P through an F gate, randomly and even
   within the possible gate types.
11:        Record that such individual has mutated from a P gate to an F gate
12:    else
13:        Perform the input type mutation normally.
14:        Record in the individual that he has not mutated the type of gate.
```

where R corresponds to the average reward obtained, Q the reward obtained in the iteration, O the number of occurrences of the referred port exchange, i the parent gate, j the child gate and t the current iteration.

As the proposal uses SAM, only the information from the last modified node is considered. Since the number of types of logic gates is much smaller than the budget of the problems, the proposed algorithm uses initial optimistic values. In this way, the average reward matrix is initialized to zero. A pseudo-code of the proposed CGP-RL is shown in Algorithm 3.

4 Computational Experiments

Computational experiments were performed in order to evaluate the proposed adaptive mutation. The problems in [19] were used, and their features are presented in Table 1, where one can found: the number of inputs and outputs, the output balance factor, the simplification rate, the number of calls to the objective function allowed for the search algorithm, the number of baseline's transistors (obtained using ESPRESSO) and functionality.

The SAM mutation operator performed better than SAM-GAM [18] and Point Mutation [14] in [19]. Thus, the mutation proposed here is used with SAM. Two strategies were defined for the initial population of CGP in [19]: a randomly generated candidate solution ('R') or a design created by ESPRESSO ('E'). SAM-R and SAM-E are labeled, respectively, as SAM-RRL and SAM-ERL when the reinforcement learning based adaptive mutation is used. The parameters adopted for the search algorithms are the same used in [19]: $\mu = 1$, $\lambda = 4$. 25 independent executions of each method applied to each problem were made. For the proposed mutation, the exploration rate ϵ was set to 10%, since it is a common exploration rate as seen in [15].

Algorithm 3. Pseudo-code of CGP-RL.

```
1: procedure CGP-RL(μ,λ,ε)
2:     Initializes occurrence matrix (OM) and average reward earned matrix (AREM)
3:     for i in 0 to μ * λ do
4:         Generate a random individual i
5:     Select the μ best individuals and promote them as parents
6:     while stop criteria is not met do
7:         for i in 0 to μ do
8:             for j in 0 to λ do
9:                 Mutate parent i to generate offspring (i,j), using ε as exploration rate
10:        for i in 0 to μ do
11:            if Individual i underwent a logic gate type mutation. then
12:                Makes p the value for the parent gate
13:                Makes f the value for the child gate.
14:                Increments OM_{p,f}
15:                Updates the average of AREM, considering the fitness obtained by i.
16:            if An offspring in i group has a score equals or better then i then
17:                One of the better offspring in group i will be promoted to a parent
18:            else
19:                The parent i still in the parents group
```

In [19], problems `alu4`, `cordic`, `t481` and `vda` were performed for 24 seeds in the SAM-E algorithm, as it suffered an interruption due to execution time. Except for `t481`, such problems were not performed for the SAM-ERL variation, due to restrictions on the processing capacity of the available hardware.

The experiments were performed on a PC with an Intel i5 3230m @ 2.60 GHz dual-core processor, 6 GB DDR3 DIMM, and Ubuntu 18.04.4 LTS operational system. The proposed approach was implemented in C programming language and is publicly available[1]. This implementation extends that provided in [19].

4.1 Analysis of the Results

Table 2 show data from the series of runs for each problem for each algorithm. The results with the least amount of transistors obtained by each algorithm (Best) were recorded, their success rate, which indicates how many executions managed to generate a feasible solution (SR), the average number of transistors obtained by the solutions (Avg) and the deviation standard (SD).

As can be seen in Table 2, SAM-RRL was able to obtain the best solutions in six of the eight small problems, five of the eight medium problems and four of the nine large problems, being the method that obtained the best solutions in most of the problems. SAM-RRL also achieved better averages in three, five and four of the small, medium and large problems, respectively, with the best performance in solving the medium and large problems. However, the this method proved to

[1] https://github.com/FMoller/cgp-rl.

Table 1. Data of the problems solved here [19].

Group	Name	In	Out	Balancing	Simplification	Evaluations	Baseline	Functionality
1	C17	5	2	0.93750	3.000E−1	1.20E+7	26	Logic
	cm42a	4	10	0.93750	1.921E−2	1.28E+7	156	Logic
	cm82a	5	3	0.50000	7.872E−2	8.00E+6	159	Logic
	cm138a	6	8	0.98437	3.758E−3	1.92E+7	148	Logic
	decod	5	16	0.96875	2.292E−1	1.20E+7	132	Decoder
	f51m	8	8	0.50000	3.510E−3	1.92E+7	638	Arithmetic
	majority	5	1	0.65625	6.207E−2	8.00E+6	24	Voter
	z4ml	7	4	0.50000	7.622E−3	1.68E+7	503	2-bit Adder
2	9symml	9	1	0.82031	7.585E−3	2.16E+7	1039	Count Ones
	alu2	10	6	0.63363	9.658E−3	1.60E+7	1790	ALU
	alu4	14	8	0.56482	5.250E−4	3.36E+7	10336	ALU
	cm85a	11	3	0.73437	9.110E−4	1.76E+7	610	Logic
	cm151a	12	2	0.75000	4.480E−4	2.88E+7	154	Logic
	cm162a	14	5	0.78125	3.900E−5	4.48E+7	200	Logic
	cu	14	11	0.92436	8.800E−5	4.48E+7	261	Logic
	x2	10	7	0.80915	5.520E−4	2.40E+7	174	Logic
3	cc	21	20	0.70703	1.137E−7	5.04E+7	256	Logic
	cmb	16	4	0.99976	1.300E−5	3.84E+7	144	Logic
	cordic	23	2	0.91595	3.423E−7	7.36E+7	27669	Logic
	frg1	28	3	0.73570	9.255E−9	6.72E+7	1605	Logic
	pm1	16	13	0.80183	4.000E−6	3.84E+7	2084	Logic
	sct	19	15	0.74722	8.180E−7	4.56E+7	466	Logic
	t481	16	1	0.64111	2.055E−3	2.56E+7	9518	Logic
	tcon	17	16	0.50000	1.564E−6	4.08E+7	49	Logic
	vda	17	39	0.78359	2.100E−5	4.08E+7	10829	Logic

have greater difficulty in generating feasible solutions, having a lower success rate than SAM-R in three problems and a better rate in one.

On the other hand, SAM-ERL showed four of the best results of all problems and achieved the best average in one. The performance differences between SAM-RRL and SAM-ERL are in line with the differences observed between SAM-R and SAM-E in [19].

The mean differences and the standard deviations measured already indicated that there could be no statistical difference between the results obtained in the proposed algorithms and their counterparts. To verify this, the Kruskal-Wallis test was performed on the data obtained for each problem and then these results were treated as a multiple testing problem, so the Bonferroni correction was applied. In comparisons between SAM-RRL and SAM-R and considering an $\alpha = 0.05$, no problem showed a significant difference between the samples. The closest problem to showing such a difference was `decod`, with an H value of 6.48 in the Kruskall-Wallis test p-value of 0.011, however, not below the value of 0.002 stipulated by the Bonferroni correction. In the comparisons between the SAM-ERL and the SAM-E, two problems showed a significant difference between the samples, namely `cm82a` and `majority`, with H values of 9.14 and 12.55 and

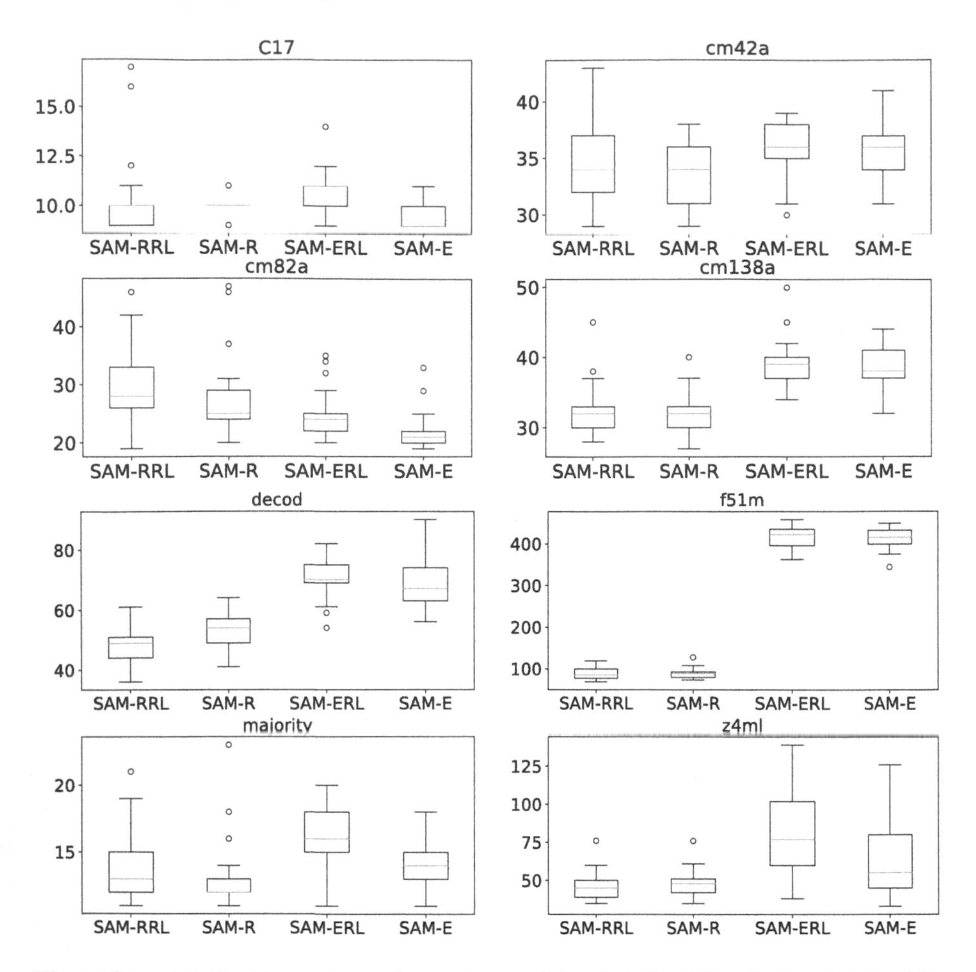

Fig. 2. Boxplots for the number of transistors of the final feasible solution obtained by each method for the problems in group 1.

p-values of 0.002 and 0.001 respectively. In the comparisons involving all the problems, `C17`, `cm42a`, `cm85a`, `x2`, `cm162a` and `tcon` did not present significant differences, but this was already expected, since significant differences between SAM-R and SAM-E had already checked at [19]. Since SAM-RRL was unable to generate feasible solutions in `alu2`, `cmb`, `cordic` and `vda`, as SAM-R in these last three problems and SAM-ERL was not performed for `alu4`, `cordic` and `vda`, comparisons between all algorithms were not performed for these problems (Figs. 2, 3 and 4).

To measure the relative performance, the ratio of the averages obtained by SAM-RRL and SAM-R in each problem was made and subtracted from 1. Thus, positive values indicate that SAM-R obtained better averages and negative results indicate where SAM-RRL obtained better averages.

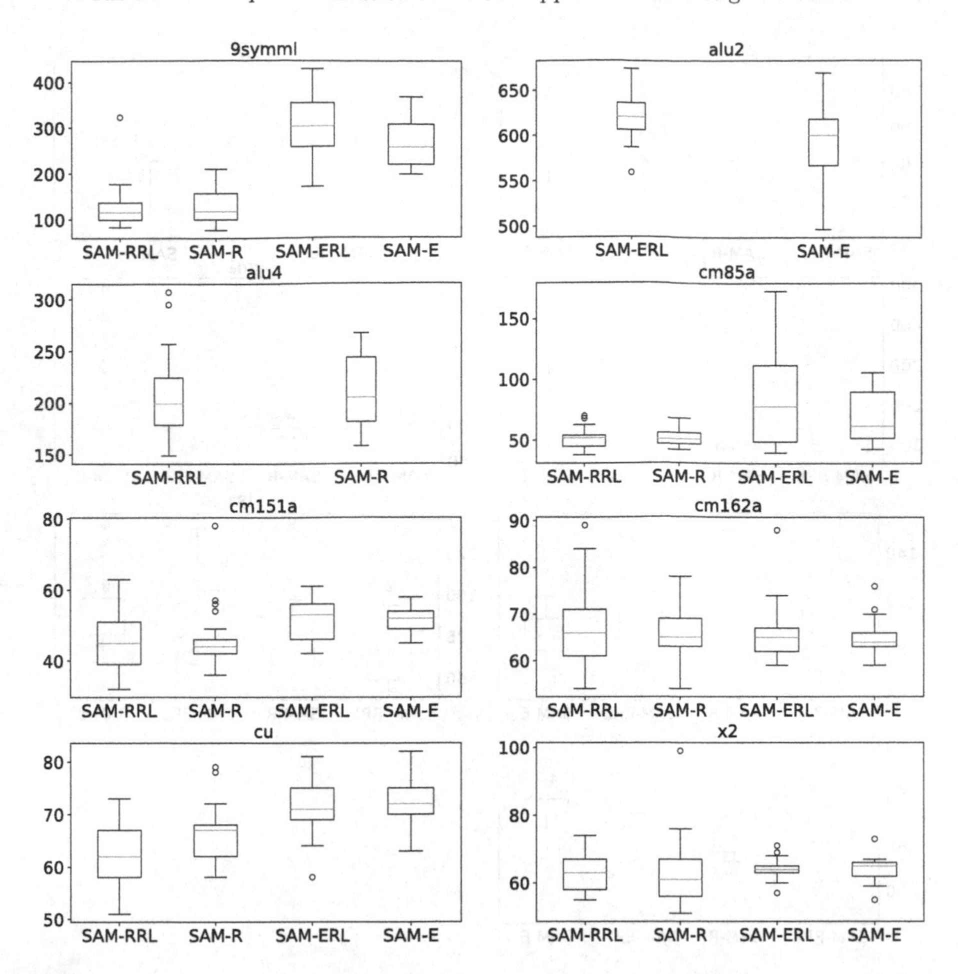

Fig. 3. Boxplots for the number of transistors of the final feasible solution obtained by each method for the problems in group 2.

Considering the data presented in Table 2, one can observe a trend of the superiority of the performance of the proposed SAM-RRL to grow with the budget provided to solve the problem and, consequently, with its difficulty. We use the relative performance of SAM-RRL in relation to SAM-R as a way of measuring the relative variation in performance between these techniques. Relative performance is defined here as

$$P_{RRL,R} = \frac{\overline{x}_{RRL}}{\overline{x}_R} - 1 \tag{2}$$

where $\overline{x}_a$ is the average of the number of transistors obtained in a given problem by technique a.

The budgets given in [19] are proportional to the size and complexity of the problems. In Fig. 5 it is possible to observe a trend of improvement in the

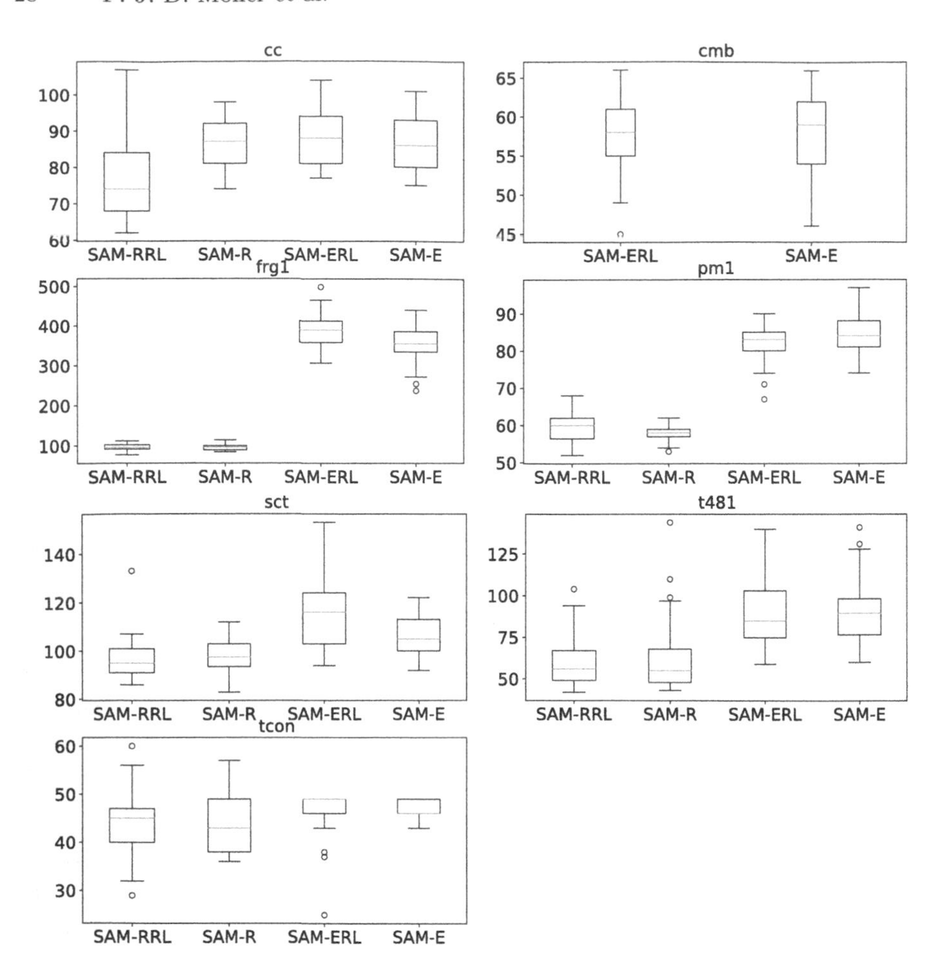

Fig. 4. Boxplots for the number of transistors of the final feasible solution obtained by each method for the problems in group 3.

performance of SAM-RRL over SAM-R as the problem budget grows. To assess this relative improvement in performance when the number of objective function calculations grows, the correlation between $P_{RRL,R}$ and the budget for each problem was calculated, with a value of -0.46. This indicates evidence that the proposed algorithm performs better as the problem's budget grows and, consequently, its difficulty. Looking at Fig. 5, the problems `decod` and `frg1` are different from the observed correlation. The problem `decod` is a relatively low budget problem and where SAM-RRL has managed significantly smaller individuals than SAM-R and `frg1` is a problem with big budget, but that the performance of SAM-RRL was slightly better than the SAM-R. Removing those from the correction calculation, it goes to the value of -0.68, which can be considered a good correlation, reinforcing the indication that the SAM-RRL algorithm will be a good choice for large and complex problems.

Table 2. Results found by SAM-RRL, SAM-R, SAM-ERL, and SAM-E when solving the problems.

Problem	SAM-RRL				SAM-R				SAM-ERL				SAM-E			
	Best	SR	Avg	SD	Best	SR	Avg	SD	Best	SR	Avg	SD	Best	SR	Avg	SD
C17	**9**	**100%**	10.32	1.95E+00	**9**	**100%**	9.80	**4.90E-01**	**9**	**100%**	10.48	1.17E+00	**9**	**100%**	**9.60**	8.00E-01
cm42a	**29**	**100%**	34.84	3.70E+00	**29**	**100%**	**33.48**	2.65E+00	30	**100%**	35.96	**2.41E+00**	31	**100%**	35.56	2.65E+00
cm82a	**19**	**100%**	30.00	6.66E+00	20	**100%**	27.36	6.67E+00	20	**100%**	24.76	3.85E+00	**19**	**100%**	**22.40**	**3.36E+00**
cm138a	28	**100%**	32.48	3.44E+00	**27**	**100%**	**32.36**	2.94E+00	34	**100%**	39.00	3.41E+00	32	**100%**	38.64	**2.91E+00**
decod	**36**	**100%**	**48.04**	6.83E+00	41	**100%**	53.12	**5.52E+00**	54	**100%**	70.44	6.84E+00	56	**100%**	67.80	8.31E+00
f51m	**69**	88%	**89.41**	1.46E+01	74	**100%**	89.60	**1.22E+00**	363	**100%**	414.92	2.78E+01	344	**100%**	412.32	2.49E+01
majority	**11**	**100%**	13.76	2.64E+00	**11**	**100%**	**12.92**	2.56E+00	**11**	**100%**	16.52	2.53E+00	**11**	**100%**	14.00	1.67E+00
z4ml	35	**100%**	**46.08**	9.12E+00	35	**100%**	48.08	**8.86E+00**	38	**100%**	83.00	2.84E+01	**33**	**100%**	65.32	2.56E+01
9symml	83	**100%**	**128.64**	4.69E+01	**77**	**100%**	128.84	**3.95E+01**	171	**100%**	309.40	6.56E+01	197	**100%**	262.92	5.03E+01
alu2	–	0%	–	–	**306**	8%	**327.50**	**2.15E+01**	508	**100%**	600.16	4.77E+01	497	**100%**	595.76	4.20E+01
alu4	**149**	80%	**207.00**	4.20E+01	158	92%	209.70	**3.40E+01**	–	–	–	–	8333	**100%**	8685.25	2.33E+02
cm85a	**38**	**100%**	**51.56**	8.01E+00	42	**100%**	51.84	**7.14E+00**	40	**100%**	87.80	3.96E+01	43	**100%**	69.12	2.07E+01
cm151a	**32**	**100%**	**45.52**	8.40E+00	36	**100%**	46.00	8.11E+00	42	**100%**	51.48	6.05E+00	45	**100%**	51.36	**3.45E+00**
cm162a	**54**	**100%**	66.60	8.48E+00	**54**	**100%**	66.00	5.97E+00	59	**100%**	65.96	5.85E+00	59	**100%**	**64.84**	**3.58E+00**
cu	**51**	**100%**	**62.16**	5.84E+00	58	**100%**	66.28	5.12E+00	58	**100%**	71.44	5.16E+00	63	**100%**	71.88	**4.48E+00**
x2	55	**100%**	63.16	5.69E+00	**51**	**100%**	**62.80**	9.51E+00	57	**100%**	64.36	**3.21E+00**	55	**100%**	63.76	3.39E+00
cc	**62**	**100%**	**77.20**	1.18E+01	74	**100%**	86.76	**6.27E+00**	77	**100%**	87.80	7.34E+00	75	**100%**	86.56	7.26E+00
cmb	–	0%	–	–	–	0%	–	–	**45**	**100%**	**57.64**	**4.69E+00**	46	**100%**	57.76	5.57E+00
cordic	–	0%	–	–	–	0%	–	–	–	–	–	–	190	**100%**	13156.16	8.44E+03
frg1	**77**	88%	**96.14**	8.90E+00	84	**100%**	96.20	**7.27E+00**	307	**100%**	387.40	4.43E+01	238	**100%**	354.32	5.34E+01
pm1	**52**	92%	60.04	4.35E+00	53	**100%**	**57.84**	**2.09E+00**	67	**100%**	82.00	5.29E+00	74	**100%**	84.52	5.22E+00
sct	86	84%	**90.90**	9.78E+00	**83**	72%	97.72	**6.93E+00**	94	**100%**	115.92	1.37E+01	92	**100%**	105.60	8.47E+00
t481	**42**	**100%**	**60.72**	**1.49E+01**	43	**100%**	63.60	2.40E+01	59	**100%**	92.16	2.44E+01	60	**100%**	91.71	2.06E+01
tcon	29	**100%**	44.04	7.17E+00	36	**100%**	**43.80**	6.31E+00	**25**	**100%**	46.04	5.37E+00	43	**100%**	46.60	**1.70E+00**
vda	–	0%	–	–	–	0%	–	–	–	–	–	–	7394	100%	8218.46	3.81E+02

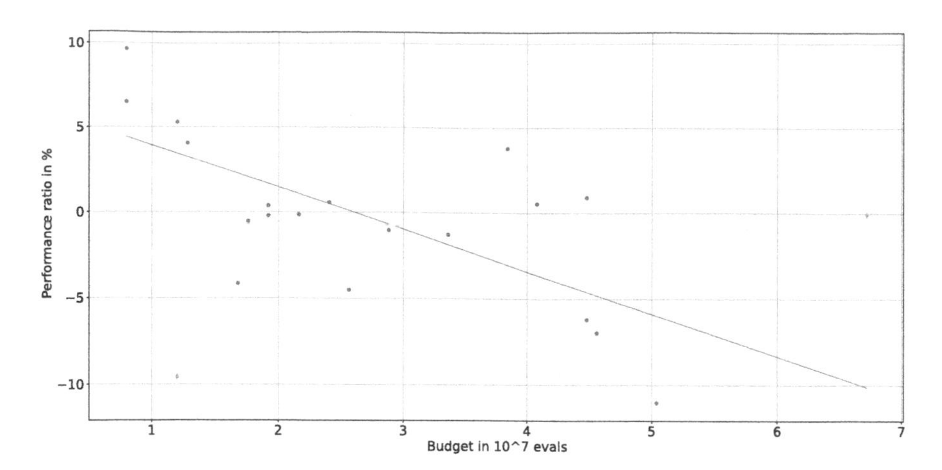

Fig. 5. Performance ratios $P_{RRL,R}$ and computational budgets of the problems solved here. The problems `decod` and `frg1` are marked as orange diamonds.

5 Conclusion

In this work, we proposed an adaptive mutation operator for CGP applied to the development of combinational logic circuits. The proposal bias the mutations of the gates taking into account the mutations that have improved the individuals in previous generations using reinforced learning.

When initialized with a randomized population (SAM-RRL), the proposed method obtained the best solutions in 15 of the 25 problems, and the best average results in 12. Also, a high correlation between its performance relative to SAM-R and the budget allowed for solving the problems, indicating that SAM-RRL may stand out in larger and more complex problems. When started with a feasible solution, the proposed technique (SAM-ERL) found better individuals than the SAM-E algorithm, but worse average. Compared with all the techniques covered in this work, SAM-ERL was, in general, the one with the worst performance. Despite the results obtained using the adaptive mutation are better than those found with the baseline mutation, the proposed algorithms showed no statistically significant difference with respect to SAM-R and SAM-E in most problems.

The results obtained here are promising. For improving the performance of the adaptive mutation, it is necessary to understand what happens during the evolutionary process. A crucial point is to understand how the structural changes in the circuit affects the learning element of the mutation operator. It is intended to incorporate this data into the adaptive process and to enable the frequency adjustment of the type of mutation. Finally, it is possible to change the reinforcement learning method used here by a specific strategy for non-stationary rewards, for instance, by increasing the weights to the most recent information.

References

1. Brayton, R., Hachtel, G., McMullen, C., Sangiovanni-Vincentelli, A.: Logic Minimization Algorithms for VLSI Synthesis, vol. 2, 1st edn. Springer, Heidelberg (1987). https://doi.org/10.1007/978-1-4613-2821-6
2. Berry, D.A., Fristedt, B.: Bandit Problems: Sequential Allocation of Experiments. Monographs on Statistics and Applied Probability, 1st edn. Springer, Heidelberg (1985). https://doi.org/10.1007/978-94-015-3711-7
3. Manfrini, F.A.L., Bernardino, H.S., Barbosa, H.J.C.: A novel efficient mutation for evolutionary design of combinational logic circuits. In: Handl, J., Hart, E., Lewis, P.R., López-Ibáñez, M., Ochoa, G., Paechter, B. (eds.) PPSN 2016. LNCS, vol. 9921, pp. 665–674. Springer, Cham (2016). https://doi.org/10.1007/978-3-319-45823-6_62
4. Manfrini, F., Bernardino, H., Barbosa, H.: On heuristics for seeding the initial population of cartesian genetic programming applied to combinational logic circuits. In: Genetic and Evolutionary Computation Conference (GECCO), pp. 105–106 (2016)
5. Greenwood, G.W., Tyrrell, A.M.: Introduction to Evolvable Hardware: A Practical Guide for Designing Self-Adaptive Systems, 1st edn. Wiley-IEEE Press (2006)
6. Gittins, J., Glazebrook, K., Weber, R.: Multi-Armed Bandit Allocation Indices, vol. 33, 2nd edn. Wiley, Hoboken (2011)
7. Goldman, B.W., Punch, W.F.: Reducing wasted evaluations in cartesian genetic programming. In: Krawiec, K., Moraglio, A., Hu, T., Etaner-Uyar, A.Ş., Hu, B. (eds.) EuroGP 2013. LNCS, vol. 7831, pp. 61–72. Springer, Heidelberg (2013). https://doi.org/10.1007/978-3-642-37207-0_6
8. Goldman, B., Punch, W.: Analysis of cartesian genetic programming's evolutionary mechanisms. IEEE Trans. Evol. Comput. **19**, 1 (2014)
9. Karnaugh, M.: The map method for synthesis of combinational logic circuits. Trans. Am. Inst. Electr. Eng. Part I: Commun. Electron. **72**(5), 593–599 (1953)
10. Katehakis, M., Veinott Jr., A.F.: The multi- armed bandit problem: decomposition and computation. Math. Oper. Res. **12**, 262–268 (1987)
11. Maturana, J., Fialho, A., Saubion, F., Schoenauer, M., Sebag, M.: Extreme compass and dynamic multi-armed bandits for adaptive operator selection. In: Congress on Evolutionary Computation (CEC), pp. 365–372 (2009)
12. McCluskey, E.: Minimization of Boolean functions. Bell Labs Tech. J. **35**(6), 1417–1444 (1956)
13. Miller, J., Thomson, P., Fogarty, T., Ntroduction, I.: Designing electronic circuits using evolutionary algorithms. arithmetic circuits: a case study. In: Genetic Algorithms and Evolution Strategies in Engineering and Computer Science (1999)
14. Miller, J.F.: Cartesian Genetic Programming. Natural Computing Series. Springer, Heidelberg (2011)
15. Sutton, R.S., Barto, A.G.: Reinforcement Learning: An Introduction. Adaptive Computation and Machine Learning Series, 2nd edn. A Bradford Book (1998)
16. da Silva, J., Bernardino, H.: Cartesian genetic programming with crossover for designing combinational logic circuits. In: Brazilian Conference on Intelligent Systems (BRACIS), pp. 145–150 (2018)
17. da Silva, J.E., Manfrini, F., Bernardino, H.S., Barbosa, H.: Biased mutation and tournament selection approaches for designing combinational logic circuits via cartesian genetic programming. In: Anais do Encontro Nacional de Inteligência Artificial e Computacional, pp. 835–846. SBC (2018)

18. da Silva, J.E.H., de Souza, L.A.M., Bernardino, H.S.: Cartesian genetic programming with guided and single active mutations for designing combinational logic circuits. In: Nicosia, G., Pardalos, P., Umeton, R., Giuffrida, G., Sciacca, V. (eds.) LOD 2019. LNCS, vol. 11943, pp. 396–408. Springer, Cham (2019). https://doi.org/10.1007/978-3-030-37599-7_33
19. de Souza, L.A.M., da Silva, J.E.H., Chaves, L.J., Bernardino, H.S.: A benchmark suite for designing combinational logic circuits via metaheuristics. Appl. Soft Comput. **91**, 106246 (2020)
20. Umans, C., Villa, T., Sangiovanni-Vincentelli, A.: Complexity of two-level logic minimization. IEEE Trans. Comput.-Aided Des. Integr. Circuits Syst. **25**(7), 1230–1246 (2006)

AgentDevLaw: A Middleware Architecture for Integrating Legal Ontologies and Multi-agent Systems

Fábio Aiub Sperotto[1,2](✉) and Marilton Sanchotene de Aguiar[2]

[1] Sul-rio-grandense Federal Institute of Education, Science and Technology, Camaquã, RS, Brazil
fabio.sperotto@camaqua.ifsul.edu.br

[2] Postgraduate Program in Computer Science, Federal University of Pelotas, Pelotas, RS, Brazil
marilton@inf.ufpel.edu.br

Abstract. In social simulation, agents play different roles with different behaviors. Each has its goals and can cooperate or not with others and consume resources. Sometimes it is necessary to regulate these agents with mechanisms like norms that explain action limitations and how the society works for the agents. To explain how the world works for the agents, in different domains, ontologies have been used to provide human knowledge. In this sense, a more special kind of ontology is built with the legal information of a society: legal ontologies. These legal ontologies operate with concepts and data collected from human society and this can be provided to the agents. With this type of information, the agents can be assisted in monitoring and evaluating actions, comparing their beliefs with existing laws. Platforms into programming multi-agent systems provide an environment to develop the social simulations, but not always provided a connection with ontologies resources. This study proposes the AgentDevLaw middleware that integrates platforms of agents' simulation with an existent legal ontology. The application of the software component is exemplified and tested with simulation scenarios in two agent's environments: JaCaMo and JADE.

Keywords: Multi-agent systems · Legal ontology · Brazilian legislation · Social simulation

1 Introduction

The social simulations have an important place in multi-agent systems (MAS), due to the fact of agents' capability in modelling the human beliefs and desire.

This study was financed in part by the Coordenação de Aperfeiçoamento de Pessoal de Nível Superior - Brasil (CAPES) - Finance Code 001.

R. Cerri and R. C. Prati (Eds.): BRACIS 2020, LNAI 12320, pp. 33–46, 2020.
https://doi.org/10.1007/978-3-030-61380-8_3

The social simulation contains different fields of research like social science, computer simulation and agent-based computing [4] offering an opportunity to simulate human interactions.

Each agent in a social simulation can be modelled and execute different internal intentions. For this, each one can have different behaviors and norms act in regulations to avoid problems when agent break rules [2]. Currently, discussions about laws and policies are made to provide these mechanisms to MAS [16] platforms in order to provided sanctions systems to the agents who violate rules or the laws of society.

To provide human knowledge about laws in a software way, legal ontologies [8, 9,15,19] has been trying to solve in different domains or countries the necessity of a computer-based legal knowledge model.

In our previous work [19], we described a legal ontology to solve the providing legal necessary knowledge to agents in a simulation. The addressed problem is in the possibilities of modeling Brazilian legal information and openness of this knowledge. The model proposed an ontology structure to query the agent's actions to return possible rules about the behaviors. In this case, since the publication, the legal ontology has been continuously improving[1] to deal with the newest legal understanding but the access from multiagent systems was an open issue. This open issue remains to provide a middleware mechanism to deal with access to different agents platforms.

The platforms of agent programming like JaCaMo, JADE[2] and many others [13] may provide or not components that agents can interact with ontologies. The aim of this study is to provide a software component that makes the interoperability of the law's knowledge from legal ontology and different MAS platforms. This study proposes to the agents' developers a middleware that can use inside the simulation and can access legal information with less knowledge in how ontologies work. This paper initially describes the related works and the necessary bases for the elaboration of the approach. Subsequently, the approach is presented and a test simulation scenario is designed and applied on two agents' platforms.

2 Related Work

The application of middleware in MAS is made in different ways, always relating the necessity of access to an external resource or obtain more knowledge about the environment. One related study provides the context recognition where the agents create a layer to solve the communications problems between human users and environment sensor devices [14].

Another perspective in building middleware for agents in their platforms is the temporal synchronization layer [17]. This layer provides a mechanism to deal with temporal aspects of simulation execution. Improve the temporal features not native in some simulation platforms. In the same interoperability vision,

[1] https://github.com/fabiosperotto/agentdevlaw/tree/master/ontologies.

[2] https://jade.tilab.com.

another type of middleware is the MISIA [7]. This component act between a platform with the behavior's agents and another environment of simulation that operate with them.

Another middleware options focus on ontology resources and simulation components. One study is the knowledge intermediate component that provides treatment of imprecise information [18]. This study provides an ontology model with a software component dealing with information and its synonyms concepts, in agent communications.

Other studies advocate the access of agents to ontologies inside the simulation's platforms without external middleware. An example of this kind of access with the JaCaMo platform (this will be more explore in Sect. 4.1) can be seen in [6]. The study uses the CArtAgO's artifacts inside the platform to implement access to ontologies through API OWL [10].

The PlaSMA system [21] provides the integration between ontologies and agents in one platform. This platform is based on JADE and focus in logistic simulations (can be applied to other domains). The PlaSMA architecture group agents in the environment and actors' types, due to the fact the first type is responsible to give ontological information to the agents. The ontology provided by PlaSMA has a top-level of concepts about logistics and other ontologies-levels with more information about physical and non-physical objects. The ontology, in this case, can be reused in other systems or applied with other domains ontologies, but in this case, a platform of simulation is provided, not a middleware to reused in different platforms (our scope).

The SeSAm platform is a visual programming system for development agents' simulations [11]. There are plugins that extend the system functionalities and one of them works to import ontologies to the platform [12]. This platform plugin converts the ontology structures in data attributes or classes to be used inside the platform. Agents and messages can be constructed based on ontology information.

The JADE (JAVA Agent DEvelopment Framework) platform is a framework implemented in Java and builds by FIPA[3] (Foundation for Intelligent Physical Agents) specifications [1]. With many internal middlewares and other components, JADE provides a peer-to-peer environment capable of any distributed agent simulation, including mobile platforms with Android systems. This platform has the internal components in support to operate and check content languages and ontologies by a content manager component[4]. The content manager can operate and check information from ontologies, in this way, native mechanisms can be achieved to deal with the conversion information, directly operating with concepts and the relations from an ontology. In this case, the ontologies classes and their predicates need to be code in Java classes or use other plugins to generate classes from ontologies files.

The way in how the agent's simulation platform is conceived is possible to provide an agent interacting directly with an ontology, but this is not the main

[3] http://www.fipa.org.
[4] https://jade.tilab.com/doc/tutorials/CLOntoSupport.pdf.

rule. Some platforms have native methods (or specific extensions) to deal with ontology, many of them not [13]. Sometimes other intermediary components are needed to solve this connection. These components can help to translate the ontology to build agents attributes, focusing on edit the agents' specification and protocols, but sometimes not focus on their knowledge base.

Other questions are about the agents' and the developer's capacity. The first one is the ontology understanding[5], how the agents can access and process the concepts relations, the individual data types, and the SPARQL[6] queries commonly used with ontologies. The second question is the level of understanding from agents developers. They need to know how to code and integrate agents and ontologies technologies to translate and work with the information flow. Sometimes it's necessary a component that can be attached with the simulation system and access an ontology knowledge with less effort (with less ontology development knowledge).

3 The Approach AgentDevLaw

Trying to solve the questions describe at the end of Sect. 2 and the necessity to provide better communication with the resource of legal knowledge, we propose a middleware AgentDevLaw. This component needs to associate with a legal ontology to work with, trough an OWL[7] file attached or a web service providing an endpoint to query ontologies.

The legal ontology is a model of knowledge about the Brazilian legislation and the details can be seen in our previous work [omitted for review]. The structure follows the Brazilian law specification but the ontology can be reused for other legislative domains. The Legislation concept has the description of laws and their individuals' concepts are used to provide restrictions actions information for the agents. The law specification is described by the Norm concept, which applies sanctions, explaining consequences to the agents' actions. There are different Norms concepts for different sanctions or rewards and they are all connected and applied with the Law and the agent role.

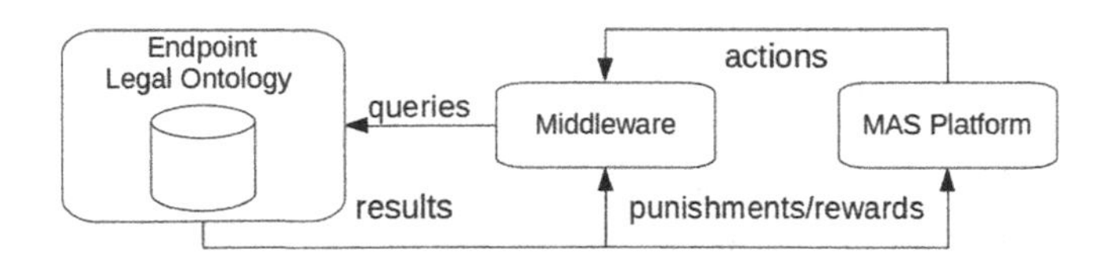

Fig. 1. The middleware proposed [omitted for review].

[5] Because we can make the ontology with different technologies not used in agent communication techniques.

[6] https://www.w3.org/TR/rdf-sparql-query.

[7] https://www.w3.org/OWL.

The Fig. 1 shows the middleware proposed that describes the main flow between the MAS system and the legal ontology. The specification of this middleware is inspired by the vertical integration with domain [22]. The vertical architecture integrates domain-specific services upwards with the agents, providing a better offer of ontology services for the agent developers. In this way, it is possible to offer a mechanism to any agent regardless of their structure. This type of architecture focuses in patterns of specification and use of the mechanism by the agents.

```
1 process string to regex filter
2 check if the agent role is received
3 prepare query with agent role and action
4 verifies against legal ontology
5 while exists results
6       build list of laws and their consequences
7 return list of laws
```

Fig. 2. The searchAction method in AgentDevLaw.

Initially the AgentDevLaw middleware receives the agent's actions and searches by-laws inside the ontology. The Fig. 2 describes the pseudo-code of the main method in middleware (accessed by agents). The method receives the agent action and the agent role. The agent role is necessary because some legislation can be addressed to a specific type of agent, but this is not obligatory and the legal ontology uses a default role for all agents (all agents are citizens).

The search follows the ontology OWL structure with SPARQL queries, checking if the agent's action combines with one or more laws. If exists any law-related, a list of objects of type Law (and internally Norms) are created in middleware to return all data to MAS. To configure the middleware is only needed the ontology location and the location type. Two options are provided: use a SPARQL web service or directly an ontology OWL file (more technical information can be obtained in the project's repository). After this, just need to ask the middleware to search for the agent's actions. In the next section, this will be more explored with the application of this component in two different agents' platforms.

4 Applications and Simulations

For application and tests of this AgentDevLaw, is necessary to describe a scenario simulation and check this in one or more agent's simulation platforms. The example scenario is derived from the same Brazilian law specification in [19]. It's about the law 7653, article 27 of water resources protection regulating the fishing activity:

> Article 27 of this law defines punishable crime with imprisonment from 2 (two) to (5) (five) years when occurring violation of the prescribed articles 2, 3, 17 and 18 of this law. Paragraph 4 - It's prohibited to fish in the period in which the spawning season, from October 1 to January 30, in rivers or in

standing water and territorial sea, in the period that takes place to spawn and/or the reproduction of fish; anyone who violates this rule is subject to the following penalty:

a. if a professional fisherman, a fine of 5 (five) to 20 (twenty) National Treasury Obligations (OTN) and suspension of professional activity for a period of 30 (thirty) to 90 (ninety) days;
b. if it is a company that exploits fishing, a fine of 100 (one hundred) to 500 (five hundred) OTN and suspension of its activities for a period of 30 (thirty) to 60 (sixty) days; and,
c. if an amateur fisherman, a fine of 20 (twenty) to 80 (eighty) OTN and loss of all instruments and equipment used in fishing.

From this law description, is possible to determine the minimum existence of two agents' types, as show in Fig. 3. The fisherman type who acts with the fishing activity. We can have other types like police, inspector or others who supervise the agent's actions. This work will maintain with one more agent type: the government. This agent will be responsible to determine the law for all simulation and can check the agent's actions. This government agent can act alone or have support from other agent or mechanism from the simulation platform.

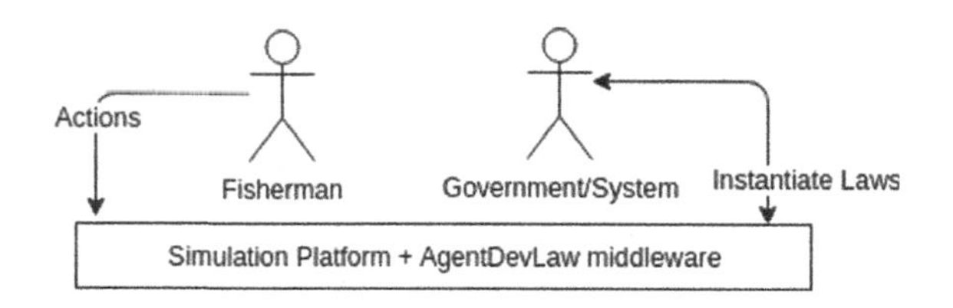

Fig. 3. The organization of agents in scenario simulation.

In the scenario, if an amateur fisherman fish in the *piracema*[8] period, he needs to pay a fine and loses all his equipment. In this work, we will not provide all the details of monetary and equipment aspects. The main focus is on how this legal information can be accessible inside the simulation platforms. The operation of verifying agent actions can be different between the simulation platforms and this will be explained in the next Sections.

4.1 JaCaMo

JaCaMo has been recognized as a widely used environment in real scenarios through a MAS oriented programming platform [3]. The platform brings together three technologies: Jason[9] to programming autonomous agents, CArtAgO[10] in environment artifacts programming and Moise[11] in agents' organization.

[8] Brazilian name given to that period of fish reproduction.
[9] http://jason.sourceforge.net.
[10] http://cartago.sourceforge.net.
[11] http://moise.sourceforge.net.

There are some possibilities in using this study inside the JaCaMo agents' simulations. The option using two of technologies provided by the platform is the application of the middleware through CArtAgO's artifacts. In Fig. 4 we show the code integrating JaCamo with AgentDevLaw.

The artifact is an example of how can be programmed Java inside the platform with classes extending the Artifact from CArtAgO package. The function in line 17 of Fig. 4 check the agent action against the ontology through the middleware. This is made when are send to the ontology the action and the agent type, or role (line 23). If some law is found, in line 36 a signal is sent with the consequence of infraction and their type.

In order to use this artifact are needed agents in Jason technology simulating the society. Initially, we can have the government agent with the responsibility in instantiate the laws in the simulation (in this platform the artifacts are provided by the *makeAartifact*[12] functionality).

```
14 public class Legislation extends Artifact {
15
16      @OPERATION
17      void checkAction(String action) {
18
19          OntologyConfigurator ontology = new OntologyConfigurator();
20          ontology.setOrigin(OntologyConfigurator.MODEL);
21          QueryProcess middleware = new QueryProcess(ontology);
22
23          List<Law> laws =  middleware.searchAction(action,
24                                getCurrentOpAgentId().getAgentName());
25
26          if(!laws.isEmpty()) {
27
28              for(int i = 0; i < laws.size(); i++) {
29                  System.out.print("Law found -> ");
30                  System.out.println(laws.get(i).getIndividual());
31                  List<Norm> norms = laws.get(i).getNorms();
32                  System.out.print("Norm needed to apply -> ");
33                  System.out.println(norms.get(0).getIndividual());
34                  for(int j = 0; j < norms.size(); j++) {
35
36                      signal("legal",
37                              norms.get(j).getConsequenceType(),
38                              norms.get(j).getConsequence() );
39                  }
40              }
41          }
42      }
43 }
```

Fig. 4. JaCaMo artifact with the middleware access.

As shown in Fig. 5 on left, we have an extra agent called system. This system agent is used as support the government agent. This agent can be something like a government's legislation system or more specific ramifications of law like inspector, policemen or others. In this case, the system stays watching if a law exists (if the legislation CArtAgO's artifact is instantiated by the government). If legislation exists (line 13 on left of Fig. 5), he maintains the focus on any information generated from the artifact. The signal sent from the artifact is

[12] http://cartago.sourceforge.net/?page_id=69.

received by this agent in function `+legal` (Fig. 5 line 22 on left) when the agent act about this legal information.

The `+legal` function process two data received from ontology (through middleware): concept and their value. These two data describe the norm applied to a specific law regulating the fish activity at the moment. The concept is the type of ontology individual that describes any of Norms concepts in the legal ontology model. The value is the individual related or the specific consequence of the law. In Fig. 6 this is more detailed.

Another agent, fisherman, have the fish action. But the problem is how we can collect the agent's actions. In this example we prefer gives to the same fisherman the responsibility in if exists some legislation belief (line 14 of Fig. 5 on right) he needs to tell always about their action[13]. In the same Fig. 5, on line 15, the action function in legislation artifact (Fig. 4) is called and the action of the agent is sent. Then, the artifact will use the middleware, search about laws and fire signals to the system about sanctions or violations.

```
7 //system agent code
8 !observe.
9 /* Plans */
10 +!observe : true <-
11     ?existsLaw(L).
12
13 +?existsLaw(LegId): true <-
14     lookupArtifact(legislation, LegId);
15     focus(LegId);
16     println("I found a legislation").
17
18 -?existsLaw(LegId): true <-
19     .wait(10);
20     ?existsLaw(LegId).
21
22 +legal(Concept, Value) : true <-
23     println("I received a communication about
24         a sanction \nof type => ", Concept,
25         " with value = ", Value
26     );
27     +sanction(Value);
28     .perceive.
```

```
6 //fisherman agent code
7 !start.
8 /* Plans */
9 +!start : true <-
10     .print("Amateur fisherman online").
11     //eval.action("fish", A);
12     //+stateSanction(A).
13
14 +legislation: true <-
15     checkAction("fish").
```

Fig. 5. AgentSpeak code of government system (on left) and fisherman agent (on right) in JaCaMo.

The Fig. 6 show the simulation results. When the system found the legislation artifact and knows about a law violation, he writes in the console about it. For example, the value is the individual `pay-a-fine` from the concept `PayAFine`. This and other samples of code can be accessed in the example project repository. After this, a lot of simulations options can be coded like: the agent believes that he doesn't have any more the equipment and need to pay a fine to State.

Another option is about the insertion of a new agent, called police, that act about the sanctions and helps the government in law maintenance. This may have a lot of different details depending on the simulation necessity. For example, it is

[13] We consider the agents don't have the capability of lie.

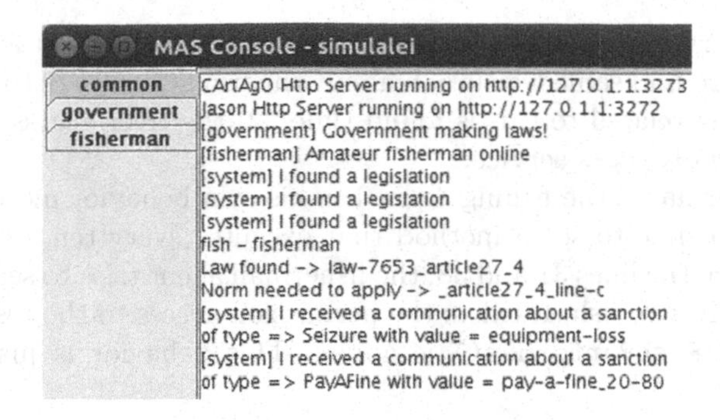

Fig. 6. Results of JaCaMo simulation.

possible to use a public politic framework [16] to maintain a list of laws updated with the legal ontology. With this, automatic action plans based on the agent's actions can be made to act about sanctions. The observable properties and events of CArtAgO can be mapped into agents' beliefs database, similar to what is done in Fig. 5 in line 27 on left.

Another example to give direct access to the agents can know about the ontology information is with internal actions. The codification of an internal action can be seen in the project repository but is similar to the artifacts. In Fig. 5 at line 11, on the right, the agent fisherman can access a method provided by a class that extends `DefaultInternalAction`, where use the middleware in the same way. After this, the agent can use in his belief base, the example at line 12: belief in a sanction received by a State (`+stateSanction`).

4.2 JADE

In this platform, we choose to implement two agents: Government and Fisherman. This platform differs from JaCaMo because it doesn't have a technology-based with artifacts and don't follow the same beliefs architecture. But the principles of how to use the proposed middleware is similar a JaCaMo. In this platform we have another kind of simulation, the river, for example, is provided as a resource in the called *yellow pages* of the platform. The *yellow pages* are services directory specifications where agents can register their services and resources while others can search to consume the same things.

In this simulation, a government agent provides the access permission to the river if it has the legal knowledge allowing the fishing in the moment. The fisherman tries to fish searching by the river resource, if not found, wait a moment, and try again, until can execute the activity. It differs from previously simulation in JaCaMo because it simulates the government administrative management, not the agents' actions evaluation.

A standard feature of this platform is each agent may have one or more behaviors. Each behavior can be executed depending on some actions or in a time interval during the simulation. The government agent has a behavior that will execute every six seconds (see Fig. 7, line 24). This method is in the agent setup and tries to find some legislation about the fish action in ontology law (line 30). If something related to this is found (line 32) the river access permission is registered in *yellow pages* service.

On the other hand, the fishing agent has his own behavior method. In Fig. 8 in line 27, is possible to see a method that executes every ten seconds. In this method, between the lines 31 and 38, the fisherman agent tries to search in *yellow pages* about a river to fish. If exists the permission access to the resource, a new behavior called `FishPerformer()` is added. This behavior is just the fishing activity itself.

```
24addBehaviour(new TickerBehaviour(this, 6000) {
25    protected void onTick() {
26
27        OntologyConfigurator ontology = new OntologyConfigurator();
28        ontology.setOrigin(OntologyConfigurator.MODEL);
29        QueryProcess middleware = new QueryProcess(ontology);
30        List<Law> laws =  middleware.searchAction("fish", "fisherman");
31
32        if(laws.isEmpty()) {
33            dfd.setName(getAID());
34            sd.setType("river");
35            sd.setName("River access permission");
36            dfd.addServices(sd);
37            try {
38                DFService.register(myAgent, dfd);
39            }
40            catch (FIPAException fe) {
41                fe.printStackTrace();
42            }
43
44        }else {
45            dfd.removeServices(sd);
46            System.out.println("Fishing is not allowed");
47        }
48    }
49});
```

Fig. 7. The behaviour of government agent in JADE.

The *yellow pages* was used in a sense of legal system resources. If the resource exists, the permission of use is released to the society by the government. So, it's not illegal to use the resource. The Fig. 9 shows the results of this simulation. After the agent is ready, the behaviors are executed. If the prohibition of fishing is active, the permission access is not provided. It's almost as the government saying "the resource is legally closed and you cannot enter" to the fisherman.

Other kinds of simulations could be offered in a sense of freedom of knowledge. As done before in JaCaMo simulation, the fisherman agent may access directly the legal information with middleware and check about permissiveness.

More details about this simulation can be found in the project repository. The JADE platform has a robust ontology API[14]. Whole components could be implemented and reflect all the legal ontology of this study[15]. But this demands more ontology knowledge for whose want work with this study and the code will work only in the JADE environment. Both items are not the aim of our approach.

[14] https://jade.tilab.com/doc/api/jade/content/onto/Ontology.html.

[15] Without requiring extra middleware.

```
25 protected void setup() {
26     System.out.println("Agent "+this.role+" is ready.");
27     addBehaviour(new TickerBehaviour(this, 10000) {
28         protected void onTick() {
29             System.out.println("Trying to fish");
30
31             DFAgentDescription template = new DFAgentDescription();
32             ServiceDescription sd = new ServiceDescription();
33             sd.setType("river");
34             template.addServices(sd);
35             try {
36                 DFAgentDescription[] result = DFService.search(myAgent, template);
37
38                 if(result.length > 0) {
39                     System.out.println("Exists river access permission!");
40                     myAgent.addBehaviour(new FishPerformer());
41
42                 }else {
43                     System.out.println("River access denied");
44                 }
45             }
46             catch (FIPAException fe) {
47                 fe.printStackTrace();
48             }
49         }
50     });
51 }
```

Fig. 8. The behaviour of fisherman agent in JADE.

```
Agent fisherman is ready.
Government Agent is ready!
Law -> law-7653_article27_4
Norm -> _article27_4_line-c
Consequence -> pay-a-fine_20-80 (PayAFine)
Consequence -> equipment-loss (Seizure)
Fishing is not allowed
Trying to fish
River access denied
```

Fig. 9. The results from the simulation in JADE.

4.3 Discussions

The problem about collect the agents' actions without them communicate what they are doing, remains the same. For years the theories of agency describe the communication acts that affect the world like the physical acts [5]. The architecture of that simulation's platforms could permit, at a low level, intercept these messages to make the necessary legal evaluation of the agent action. In this study case, the preference stays with the action's evaluation in some way by other agents, asking about their actions, or the agents declaring their self-actions. To provide a more sophisticated implementation, a "big brother" or similar component is needed to follow the agents' actions. But in that case, other problems in agent's society like surveillance or privacy [20] can arise and this is not the scope of our study.

One important detail is the capacity of middleware in returning data from the ontology. As seen in all examples in previous Sections, the AgentDevLaw always returns to the agent system a list of laws. This is important because the law search mechanism uses regular expression filter to find laws related to the agent action. Enhancements are in conduct, trying to solve the issues with the better processes of expression regular filters and text similarities.

The individual concepts of laws and norms consequences may have data details to better describe their information. For example, the ontology individual *pay-a-fine-20-80* have the 20.0 as a float value. This describes that it is necessary to pay 20 monetary units as a sanction. Another example is for individuals like *detention-2_5*, have two specification values of the integer type: minimum value of 2 and a maximum of 5 (describing years in detention). All data types inside the OWL ontology is specified by an XML schema[16] the middleware has methods that process this information, offer the option to return together with the value and the data type, in favor of any agents' developer intention. This is important because all information in the ontology is identified by URI (Uniform Resource Identifier) resources and sometimes it is necessary to know what type is each value for better computational purposes.

5 Conclusion and Further Work

The simulation platforms like JaCaMo and JADE provide a good environment for researchers to place their projects and execute many types of simulations. To provide this middleware for the same platforms was easy in the sense that all have adherents with Java technology (partially or completely). We learn with these simulations' software and more tests with other platforms can be conducted in a future. The complexity is in bring together components and provides a good "relationship" between the agents with the ontology. The study comprised a good practice of this with the used platforms.

Ontologies have been used to provide human knowledge for the agents. In this study, the legal ontology can be fully connected with the multi-agent system and provide legal knowledge for the agents. This approach is focused on Brazilian legislation, in how the legislators structure their laws. But the use of common concepts in agents systems field like *Norm*, *Consequence* and *Law* may help with a model for other legislative bodies of other countries. The agents' actions monitoring, privacy and the use of more than one legal ontology are important issues and will be addressed in future works.

The middleware provides a better sense of the communication of legal information for agents and the agents' developers. While the legal ontology is attached to agents' platform, the developers can write methods in agents to stay ready to ask for any legal information. This fact decreases the necessity of more knowledge in ontology technology and increases the information access for the agents even they need to change between different simulations platforms. The various agents in different platforms can check their behaviors against the legal rules make evaluations about probable violations.

The proposed work is a software component that needs to be attached to agents platforms to be operative. There is future planning to transform this middleware into a web service that provides the same functionalities with fewer configurations demands.

[16] http://www.w3.org/2001/XMLSchema.

Another issue is the legal categories: it is possible to categorize Brazilian law between civil and criminal legislation. Another type is the international legislation that can be necessary inside globalization legal information. For all the cases, the proposed middleware doesn't deal with this categorization of information. The middleware needs improvements in recognize behaviors and patterns, searching for different groups of legislation, understanding automatically what legal categories need to be processed.

So, the main contribution of this work is to provide a mechanism to evaluate the software component's actions against society legislation and explain for those agents what is permitted or not. Other types of software components can benefit from this information access. Shortly, this mechanism will provide ways to operate the laws inside the ontology. The agents will be capable of creating and editing laws allowing the agents not only to understand but rationalize and build their legislation.

Acknowledgments. This study was financed in part by the Coordenação de Aperfeiçoamento de Pessoal de Nível Superior - Brasil (CAPES) - Finance Code 001.

References

1. Bellifemine, F., Bergenti, F., Caire, G., Poggi, A.: Jade – a Java agent development framework. In: Bordini, R.H., Dastani, M., Dix, J., El Fallah Seghrouchni, A. (eds.) Multi-Agent Programming. MSASSO, vol. 15, pp. 125–147. Springer, Boston, MA (2005). https://doi.org/10.1007/0-387-26350-0_5
2. Boella, G., van der Torre, L., Verhagen, H.: Introduction to normative multiagent systems. Comput. Math. Organ. Theory **12**(2), 71–79 (2006)
3. Boissier, O., Bordini, R.H., Hübner, J.F., Ricci, A., Santi, A.: Multi-agent oriented programming with JaCaMo. Sci. Comput. Program. **78**(6), 747–761 (2013)
4. Davidsson, P.: Agent Based Social Simulation: a computer science view. J. Artif. Soc. Soc. Simul. **5**(1), 1–7 (2002)
5. Dignum, F., Greaves, M.: Issues in agent communication: an introduction. In: Dignum, F., Greaves, M. (eds.) Issues in Agent Communication. LNCS (LNAI), vol. 1916, pp. 1–16. Springer, Heidelberg (2000). https://doi.org/10.1007/10722777_1
6. Freitas, A., Panisson, A.R., Hilgert, L., Meneguzzi, F., Vieira, R., Bordini, R.H.: Integrating ontologies with multi-agent systems through cartago artifacts. In: 2015 IEEE/WIC/ACM International Conference on Web Intelligence and Intelligent Agent Technology (WI-IAT), vol. 2, pp. 143–150 (2015)
7. García, E., Rodríguez, S., Martín, B., Zato, C., Pérez, B.: MISIA: middleware infrastructure to simulate intelligent agents. In: Abraham, A., Corchado, J.M., González, S.R., De Paz Santana, J.F. (eds.) International Symposium on Distributed Computing and Artificial Intelligence. AISC, vol. 91, pp. 107–116. Springer, Heidelberg (2011). https://doi.org/10.1007/978-3-642-19934-9_14
8. Ghosh, M.E., Naja, H., Abdulrab, H., Khalil, M.: Towards a middle-out approach for building legal domain reference ontology. Int. J. Knowl. Eng. **2**(3), 109–114 (2016)
9. Gómez-Pérez, A., Ortiz-Rodríguez, F., Villazón-Terrazas, B.: Legal ontologies for the Spanish e-Government. In: Marín, R., Onaindía, E., Bugarín, A., Santos, J. (eds.) CAEPIA 2005. LNCS (LNAI), vol. 4177, pp. 301–310. Springer, Heidelberg (2006). https://doi.org/10.1007/11881216_32

10. Horridge, M., Bechhofer, S.: The OWL API: a Java API for OWL ontologies. Semant. Web **2**(1), 11–21 (2011)
11. Klügl, F., Herrler, R., Fehler, M.: SeSAm: implementation of agent-based simulation using visual programming. In: Proceedings of the Fifth International Joint Conference on Autonomous Agents and Multiagent Systems, pp. 1439–1440 (2006)
12. Klügl, F., Herrler, R., Oechslein, C.: From simulated to real environments: how to use SeSAm for software development. In: Schillo, M., Klusch, M., Müller, J., Tianfield, H. (eds.) MATES 2003. LNCS (LNAI), vol. 2831, pp. 13–24. Springer, Heidelberg (2003). https://doi.org/10.1007/978-3-540-39869-1_2
13. Kravari, K., Bassiliades, N.: A survey of agent platforms. J. Artif. Soc. Soc. Simul. **18**(1), 11 (2015)
14. Olaru, A., Florea, A.M., Fallah Seghrouchni, A.: A context-aware multi-agent system as a middleware for ambient intelligence. Mob. Netw. Appl. **18**(3), 429–443 (2013)
15. Palmirani, M., Martoni, M., Rossi, A., Bartolini, C., Robaldo, L.: Legal ontology for modelling GDPR concepts and norms. In: Palmirani, M. (ed.) Legal Knowledge and Information Systems - JURIX 2018: The Thirty-first Annual Conference, Groningen, The Netherlands, 12–14 December 2018. Frontiers in Artificial Intelligence and Applications, vol. 313, pp. 91–100. IOS Press (2018)
16. Santos, I.A.S., Costa, A.C.R.: Simulando a execução de políticas públicas através de jason e cartago. In: Proceedings of the 6th Workshop-Escola de Sistemas de Agentes, seus Ambientes e aplicações (WESAAC), pp. 81–91. UFSC, Florianopolis (2013)
17. Schuldt, A., Gehrke, J.D., Werner, S.: Designing a simulation middleware for FIPA multiagent systems. In: Proceedings of the IEEE/WIC/ACM International Conference on Web Intelligence and Intelligent Agent Technology, vol. 2, pp. 109–113 (2008)
18. Sperotto, F.A., Adamatti, D.F.: A proposal for interoperability to agent communication using synonyms. In: 3rd Brazilian Workshop on Social Simulation, pp. 39–43 (2012)
19. Sperotto, F.A., Belchior, M., de Aguiar, M.S.: Ontology-based legal system in multi-agents systems. In: Martínez-Villaseñor, L., Batyrshin, I., Marín-Hernández, A. (eds.) MICAI 2019. LNCS (LNAI), vol. 11835, pp. 507–521. Springer, Cham (2019). https://doi.org/10.1007/978-3-030-33749-0_41
20. Such, J.M., Espinosa, A., García-Fornes, A.: A survey of privacy in multi-agent systems. Knowl. Eng. Rev. **29**(3), 314–344 (2014)
21. Warden, T., Porzel, R., Gehrke, J.D., Herzog, O., Langer, H., Malaka, R.: Towards ontology-based multiagent simulations: the plasma approach. In: ECMS, pp. 50–56 (2010)
22. Weyns, D., Helleboogh, A., Holvoet, T., Schumacher, M.: The agent environment in multi-agent systems: a middleware perspective. Multiagent Grid Syst. **5**(1), 93–108 (2009)

An Argumentation-Based Approach for Explaining Goals Selection in Intelligent Agents

Mariela Morveli-Espinoza[(✉)], Cesar A. Tacla, and Henrique M. R. Jasinski

Program in Electrical and Computer Engineering (CPGEI), Federal University of Technology of Parana (UTFPR), Curitiba, Brazil
morveli.espinoza@gmail.com, tacla@utfpr.edu.br, henriquejasinski@alunos.utfpr.edu.br

Abstract. During the first step of practical reasoning, i.e. deliberation or goals selection, an intelligent agent generates a set of pursuable goals and then selects which of them he commits to achieve. Explainable Artificial Intelligence (XAI) systems, including intelligent agents, must be able to explain their internal decisions. In the context of goals selection, agents should be able to explain the reasoning path that leads them to select (or not) a certain goal. In this article, we use an argumentation-based approach for generating explanations about that reasoning path. Besides, we aim to enrich the explanations with information about emerging conflicts during the selection process and how such conflicts were resolved. We propose two types of explanations: the partial one and the complete one and a set of explanatory schemes to generate pseudo-natural explanations. Finally, we apply our proposal to the cleaner world scenario.

Keywords: Goals selection · Explainable agents · Formal argumentation

1 Introduction

Practical reasoning means reasoning directed towards actions, i.e. it is the process of figuring out what to do. According to Wooldridge [20], practical reasoning involves two phases: (i) deliberation, which is concerned with deciding what state of affairs an agent wants to achieve, thus, the outputs of deliberation phase are goals the agent intends to pursue, and (ii) means-ends reasoning, which is concerned with deciding how to achieve these states of affairs. The first phase is also decomposed in two parts: (i) firstly, the agent generates a set of pursuable

Partially supported by CAPES.

R. Cerri and R. C. Prati (Eds.): BRACIS 2020, LNAI 12320, pp. 47–62, 2020.
https://doi.org/10.1007/978-3-030-61380-8_4

goals[1], and (ii) secondly, the agent chooses which goals he will be committed to bring about. In this paper, we focus on the first phase, that is, goals selection.

Given that an intelligent agent may generate multiple pursuable goals, some conflicts among these goals could arise, in the sense that it is not possible to pursue them simultaneously. Thus, a rational agent selects a set of non-conflicting goals based in a criterion or a set of criteria. There are many researches about identifying and resolving such conflict in order to determine the set of pursued goals (e.g., [1,14,18,19,21]). However, to the best of our knowledge, none of these approaches gives explanations about the reasoning path to determine the final set of pursued goals. Thus, the returned outcomes can be negatively affected due to the lack of clarity and explainability about their dynamics and rationality.

In order to better understand the problem, consider the well-know "cleaner world" scenario, where a set of robots (intelligent agents) has the task of cleaning a dirty environment. The main goal of all the robots is to have the environment clean. Besides cleaning, the robots have other goals such as recharging their batteries or being fixed. Suppose that at a given moment one of the robots (let us call him `BOB`) detects dirt in slot (5, 5); hence, the goal "cleaning (5, 5)" becomes pursuable. On the other hand, `BOB` also auto-detects a technical defect; hence, the goal "be fixed" also becomes pursuable. Suppose that `BOB` cannot commit to both goals at the same time because the plans adopted for each goal lead to an inconsistency. This means that only one of the goals will become pursued. Suppose that he decides to fix its technical defect instead of cleaning the perceived dirt. During the cleaning task or after the work is finished, the robot can be asked for an explanation about his decision. It is clear that it is important to endow the agents with the ability of explaining their decisions, that is, to explain how and why a certain pursuable goal became (or not) a pursued goal.

Thus, the research questions that are addressed in this paper are: (i) how to endow intelligent agents with the ability of generating explanations about their goals selection process? and (ii) how to improve the informational quality of the explanations?

In addressing the first question, we will use arguments to generate and represent the explanations. At this point, it is important to mention that in this article, argumentation is used in two different ways. Firstly, argumentation will be used in the goals selection process. The input to this process is a set of possible conflicting pursuable goals such that each one has a preference value and a set of plans that allow the agent to achieve them, and the output is a set of pursued goals. We will base on the work of Morveli-Espinoza et al. [14] for this process. One important contribution given in [14] is the computational formalization of three forms of conflicts, namely terminal incompatibility, resource incompatibility, and superfluity, which were conceptually defined in [5]. The identification of conflicts is done by using plans, which are represented by instrumental argu-

[1] Pursuable goals are also known as desires and pursued goals as intentions. In this work, we consider that both are goals at different stages of processing, like it was suggested in [5].

ments[2]. These arguments are compared in order to determine the form of conflict that may exist between them. The set of instrumental arguments and the conflict relation between them make up an Argumentation Framework (AF). Finally, in order to resolve the conflicts, an argumentation semantics is applied. This semantics is a function that takes as input an AF and returns those non-conflicting goals the agent will commit to. Secondly, argumentation is used in the process of explanation generation. The input to this process is the AF mentioned above and the set of pursued goals and the output is a set of arguments that represent explanations. The arguments constructed in this part are not instrumental ones, that is, they do not represent plan but explanations. Regarding the second question, we will use the information in instrumental arguments for enriching explanations about the form(s) of conflict that exists between two goals.

Next section focuses on the knowledge representation and the argumentation process for goal selection. Section 3 presents the argumentation process for generating explanations. Section 4 is devoted to the application of the proposal to the cleaner world scenario. Section 5 presents the main related work. Finally, Sect. 6 is devoted to conclusions and future work.

2 Argumentation Process for Goals Selection

In this section, we will present part of the results of the article of Morveli-Espinoza et al. [14], on which we will base to construct the explanations. Since we want to enrich the explanations, we will increase the informational capacity of some of the results.

Firstly, let $\mathcal{L}$ be a first-order logical language used to represent the mental states of the agent, $\vdash$ denotes classical inference, and $\equiv$ the logical equivalence. Let $\mathcal{G}$ be the set of pursuable goals, which are represented by ground atoms of $\mathcal{L}$ and $\mathcal{B}$ be the set of beliefs of the agent, which are represented by ground literals[3] of $\mathcal{L}$. In order to construct instrumental arguments, other mental states are necessary (e.g. resources, actions, plan rules); however, they are not meaningful in this article. Therefore, we will assume that the knowledge base (denoted by $\mathcal{K}$) of an agent includes such mental states, besides his beliefs.

According to Castelfranchi and Paglieri [5], three forms of incompatibility could emerge during the goals selection: terminal, due to resources, and superfluity[4]. Morveli-Espinoza et al. [14] tackled the problems of identifying and resolving these three forms of incompatibilities. In order to identify these incompatibilities the plans that allow to achieve the goals in $\mathcal{G}$ are evaluated.

[2] An instrumental argument is structured like a tree where the nodes are planning rules whose premise is made of a set of sub-goals, resources, actions, and beliefs and its conclusion or claim is a goal, which is the goal achieved by executing the plan represented by the instrumental argument.

[3] Literals are atoms or negation of atoms (the negation of an atom a is denoted $\neg a$).

[4] Hereafter, terminal incompatibility is denoted by t, resource incompatibility by r, and superfluity by s.

Considering that in their proposal each plan is represented by means of instrumental arguments, as a result of the identification problem, they defined three AFs (one for each form of incompatibility) and a general AF that involves all of the instrumental arguments and attacks of the three forms of incompatibility.

Definition 1 ***(Argumentation frameworks).*** *Let* $\texttt{ARG}_{\texttt{ins}}$ *be the set of instrumental arguments that an agent can build from* $\mathcal{K}$[5]*. A x-AF is a pair* $\mathcal{AF}_x = \langle \texttt{ARG}_x, \mathcal{R}_x \rangle$ *(for* $x \in \{t, r, s\}$*) where* $\texttt{ARG}_x \subseteq \texttt{ARG}_{\texttt{ins}}$ *and* $\mathcal{R}_x$ *is the binary relation* $\mathcal{R}_x \subseteq \texttt{ARG}_{\texttt{ins}} \times \texttt{ARG}_{\texttt{ins}}$ *that represents the attack between two arguments of* $\texttt{ARG}_{\texttt{ins}}$*, so that* $(A, B) \in \mathcal{R}_x$ *denotes that the argument A attacks the argument B.*

Since we want to improve the informational quality of explanations, we modify the general AF proposed in [14] by adding a function that returns the form of incompatibility that exists between two instrumental arguments. Thus, an agent will not only be able to indicate that there is an incompatibility between two goals but he will be able to indicate the form of incompatibility.

Definition 2 ***(General AF).*** *Let* $\texttt{ARG}_{\texttt{ins}}$ *be a set of instrumental arguments that an agent can build from* $\mathcal{K}$*. A general AF is a tuple* $\mathcal{AF}_{gen} = \langle \texttt{ARG}_{\texttt{ins}}, \mathcal{R}_{gen}, \texttt{f\_INCOMP} \rangle$*, where* $\mathcal{R}_{gen} = \mathcal{R}_t \cup \mathcal{R}_r \cup \mathcal{R}_s$ *and* $\texttt{f\_INCOMP} : \mathcal{R}_{gen} \rightarrow 2^{\{t,r,s\}}$.

Example 1. Recall the cleaner world scenario that was presented in Introduction where agent BOB has two pursuable goals, which can be expressed as $clean(5, 5)$ and $be(fixed)$ in language $\mathcal{L}$. Consider that there are two instrumental arguments whose claim is $clean(5, 5)$, namely A that has a sub-argument E whose claim is $pickup(5, 5)$ and C that has a sub-argument D whose claim is $mop(5, 5)$. Besides, there are two instrumental arguments whose claim is $be(fixed)$, namely B that has a sub-argument H whose claim is $be(in\_workshop)$ and F that does not have any sub-argument.

Recall also that terminal incompatibility was also exemplified. In order to exemplify the other forms of incompatibility and generate the general AF for this scenario, consider the following situations:

- BOB has 90 units of battery. He needs 60 units for achieving C, he needs 70 units for achieving A, he needs 30 units for achieving B, and he does not need battery for achieving F because the mechanic can go to his position. We can notice that there is a conflict between A and B and consequently between their sub-arguments.
- As can be noticed, there are two instrumental arguments whose claim is $clean(5, 5)$ and two instrumental arguments whose plan is $be(fixed)$. It would be redundant to perform more than one plan to achieve the same goal, this means that arguments with the same claim are conflicting due to superfluity. This conflict is also extended to their sub-arguments.

[5] For further information about how instrumental arguments are built, the reader is referred to [14].

We can now generate the general AF for the cleaner world scenario: $\mathcal{AF}_{gen} = \langle \{A, B, C, D, E, F, H\}, \mathcal{R}_{gen}, \texttt{f\_INCOMP} \rangle$ where $\mathcal{R}_t = \{(A, B), (B, A), (E, B), (B, E), (E, H), (H, E), (A, H), (H, A), (C, B), (B, C), (D, B), (B, D), (D, H), (H, D), (C, H), (H, C)\}$, $\mathcal{R}_r = \{(A, B), (B, A), (E, B), (B, E), (A, H), (H, A), (E, H), (H, E)\}$, and $\mathcal{R}_s = \{(C, A), (A, C), (E, D), (D, E), (C, E), (E, C), (A, D), (D, A), (F, B), (B, F), (F, H), (H, F)\}$. Figure 1 shows the graph representation.

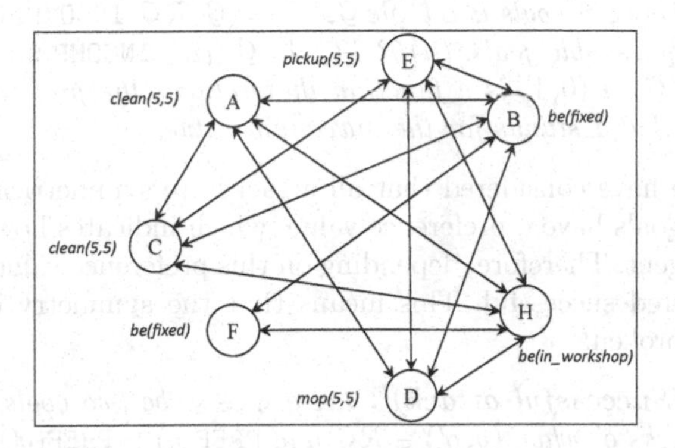

Fig. 1. (Obtained from [13]) The general AF for the cleaner world scenario. The nodes represent the arguments and the arrows represent the attacks between the arguments. The text next to each node indicates the claim of each instrumental argument.

So far, we have referred to instrumental arguments – which represent plans – however, since the selection is at goals level, it is necessary to generate an AF where arguments represent goals. In order to generate this framework, it is necessary to define when two goals attack each other. This definition is based on the general attack relation $\mathcal{R}_{gen}$, which includes the three kinds of attacks that may exist between arguments. Thus, a goal g attacks another goal g' when all the instrumental arguments for g (that is, the plans that allow to achieve g) have a general attack relation with all the instrumental arguments for g'. This attack relation between goals is captured by the binary relation $\mathcal{RG} \subseteq \mathcal{G} \times \mathcal{G}$. We denote with (g, g') the attack relation between goals g and g'. In other words, if $(g, g') \in \mathcal{RG}$ means that goal g attacks goal g'.

Definition 3 ***(Attack between goals).*** *Let* $\mathcal{AF}_{gen} = \langle \texttt{ARG}_{\texttt{ins}}, \mathcal{R}_{gen}, \texttt{f\_INCOMP} \rangle$ *be a general AF,* $g, g' \in \mathcal{G}$ *be two pursuable goals,* $\texttt{ARG\_INS}(g)$[6]$, \texttt{ARG\_INS}(g') \subseteq \texttt{ARG}_{\texttt{ins}}$ *be the set of arguments for* g *and* g'*, respectively. Goal* g *attacks goal* g' *when* $\forall A \in \texttt{ARG\_INS}(g)$ *and* $\forall A' \in \texttt{ARG\_INS}(g')$ *it holds that* $(A, A') \in \mathcal{R}_{gen}$ *or* $(A', A) \in \mathcal{R}_{gen}$.

[6] $\texttt{ARG\_INS}(g)$ denotes all the instrumental arguments that represent plans that allow to achieve g.

Once the attack relation between two goals was defined, it is also important to determine the forms of incompatibility that exist between any two conflicting goals. The function $\texttt{INCOMP\_G}(g, g')$ will return the set of forms of incompatibility between goals g and g'. Thus, if $(g, g') \in \mathcal{RG}$, then $\forall(A, A') \in \mathcal{R}_{gen}$ and $\forall(A', A) \in \mathcal{R}_{gen}$ where $A \in \texttt{ARG\_INS}(g)$ and $A' \in \texttt{ARG\_INS}(g')$, $\texttt{INCOMP\_G}(g, g') = \bigcup \texttt{f\_INCOMP}((A, A')) \cup \texttt{f\_INCOMP}((A', A))$. We can now define an AF where arguments represent goals.

Definition 4 ***(Goals AF).*** *An argumentation-like framework for dealing with incompatibility between goals is a tuple* $\mathcal{GAF} = \langle \mathcal{G}, \mathcal{RG}, \texttt{INCOMP\_G}, \texttt{PREF} \rangle$, *where: (i)* $\mathcal{G}$ *is a set of pursuable goals, (ii)* $\mathcal{RG} \subseteq \mathcal{G} \times \mathcal{G}$, *(iii)* $\texttt{INCOMP\_G} : \mathcal{RG} \to 2^{\{t,r,s\}}$, *and (iv)* $\texttt{PREF} : \mathcal{G} \to (0, 1]$ *is a function that returns the preference value of a given goal such that 1 stands for the maximum value.*

Hitherto, we have considered that all attacks are symmetrical. However, as can be noticed goals have a preference value, which indicates how valuable each goal is for the agent. Therefore, depending on this preference value, some attacks may be considered successful. This means that the symmetry of the relation attack may be broken.

Definition 5 ***(Successful attack)***[7]**.** *Let* $g, g' \in \mathcal{G}$ *be two goals, we say that* g *successfully attacks* g' *when* $(g, g') \in \mathcal{RG}$ *and* $\texttt{PREF}(g) > \texttt{PREF}(g')$.

Let us denote with $\mathcal{GAF}_{sc} = \langle \mathcal{G}, \mathcal{RG}_{sc}, \texttt{INCOMP\_G}, \texttt{PREF} \rangle$ *the AF that results after considering the successful attacks.*

The next step is to determine the set of goals that can be achieved without conflicts, which can also be called acceptable goals and in this article, they can be explicitly called pursued goals. With this aim, it has to be applied an argumentation semantics. Morveli-Espinoza et al. did an analysis about which semantics is more adequate for this problem. They reached to the conclusion that the best semantics is based on conflict-free sets, on which a function is applied. Next we present the definition given in [14] applied to the Goals AF.

Definition 6 ***(Semantics).*** *Given a* $\mathcal{GAF}_{sc} = \langle \mathcal{G}, \mathcal{RG}_{sc}, \texttt{INCOMP\_G}, \texttt{PREF} \rangle$. *Let* $\mathcal{S}_{\mathcal{CF}}$ *be a set of conflict-free sets calculated from* $\mathcal{GAF}_{sc}$. $\texttt{MAX\_UTIL} : \mathcal{S}_{\mathcal{CF}} \to 2^{\mathcal{S}_{\mathcal{CF}}}$ *determines the set acceptable goals. This function takes as input a set of conflict-free sets and returns those with the maximum utility for the agent in terms of preference value.*

Let $\mathcal{G}' \subseteq \mathcal{G}$ *be the set of goals returned by* $\texttt{MAX\_UTIL}$. *This means that* $\mathcal{G}'$ *is the set of goals the agent can commit to, which are called pursued goals or intentions.*

Regarding the function for determining acceptable goals, there may be many ways to make the calculations; for example, one way of characterizing $\texttt{MAX\_UTIL}$ is by summing up the preference value of all the goals in an extension. Another way may be by summing up the preference value of just the main goals without considering sub-goals. We will use the first characterization in our scenario.

[7] In other works (e.g., [11,12]), it is called a defeat relation.

Example 2. Consider the general AF of Example 1, the agent generates: $\mathcal{GAF}_{sc} = \langle \{clean(5,5), pickup(5,5), mop(5,5), be(in\_workshop), be(fixed)\}, \{(mop(5,5), pickup(5,5)), (clean(5,5), be(in\_workshop)), (mop(5,5), be(in\_workshop)), (pickup(5,5), be(in\_workshop))\}, \mathtt{INCOMP\_G}, \mathtt{PREF}\rangle$. Figure 2 shows this GAF, the preference values of each goal, and the form of incompatibilities that exists between pairs of goals.

From $\mathcal{GAF}_{sc}$, the number of conflict-free extensions is: $|\mathcal{S}_{\mathcal{CF}}| = 14$. After applying MAX_UTIL, the extension with the highest preference is: $\{clean(5,5), mop(5,5), be(fixed)\}$. This means that $\mathcal{G}' = \{clean(5,5), mop(5,5), be(fixed)\}$ are compatible goals that can be achieved together without conflicts.

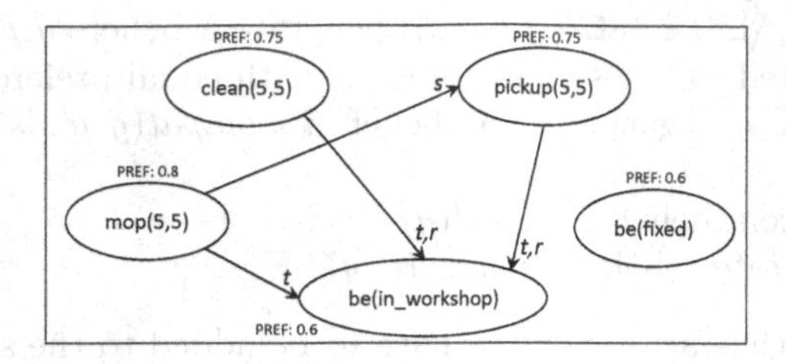

Fig. 2. GAF for the cleaner world scenario. The text next to each arrow indicates the form of incompatibility.

3 Argumentation Process for Explanations Generation

In this section, we present explanatory arguments and the process for generating explanations for a goal become pursued or not.

First of all, let us present the types of questions that can be answered:

- WHY(g): it is required an explanation to justify why a goal g became pursued[8].
- WHY_NOT(g): it is required an explanation to justify why a goal g did not become pursued.

3.1 Explanatory Arguments and Argumentation Framework

As a result of the above section, we obtain a Goals Argumentation Framework (GAF) and a set of pursued goals. Recall that in a GAF, the arguments represent goals; hence, in order to generate an explanation from a GAF, it is necessary to generate beliefs and rules – that reflect the knowledge contained in it – from which, explanatory arguments can be constructed. Before presenting the beliefs and rules, let us present some functions that will be necessary for the generation of beliefs:

[8] In order to better deal with goals, we map each goal to a constant in $\mathcal{L}$.

- $\texttt{COMPS}(\mathcal{GAF}_{sc}) = \{g \mid \nexists(g,g') \in \mathcal{RG}_{sc}$ (or $(g',g) \in \mathcal{RG}_{sc})$, where $g, g' \in \mathcal{G}\}$. This function returns the set of goals without conflicting relations.
- $\texttt{EVAL\_PREF}(\mathcal{GAF}_{sc}) = \{(g,g') \mid (g,g') \in \mathcal{RG}_{sc}$ and $(g',g) \notin \mathcal{RG}_{sc}\}$. This function returns all the pairs of goals in $\mathcal{RG}'$ that represent non-symmetrical relations between goals. When the relation is not symmetrical, it means that one of the goals is preferred to the other.

Using these functions, the set of beliefs generated from a $\mathcal{GAF}_{sc} = \langle \mathcal{G}, \mathcal{RG}_{sc}, \texttt{INCOMP\_G}, \texttt{PREF}\rangle$ are the following:

- $\forall g \in \texttt{COMPS}(\mathcal{GAF}_{sc})$ generate a belief $\neg incomp(g)$
- $\forall(g,g') \in \texttt{EVAL\_PREF}(\mathcal{GAF}_{sc})$, if $\texttt{PREF}(g) > \texttt{PREF}(g')$, then generate $pref(g,g')$ and $\neg pref(g',g)$.
- $\forall(g,g') \in (\mathcal{RG}_{sc} \backslash \texttt{EVAL\_PREF}(\mathcal{GAF}_{sc}))$ generate a belief $eq\_pref(g,g')$. These beliefs are created for those pairs of goals with equal preference.
- $\forall(g,g') \in \mathcal{RG}_{sc}$ generate a belief $incompat(g,g',ls)$ where $ls = \texttt{INCOMP\_G}(g,g')$
- $\forall g \in \mathcal{G}'$ generate a belief $max\_util(g)$
- $\forall g \in \mathcal{G} \backslash \mathcal{G}'$ generate a belief $\neg max\_util(g)$

All the beliefs that are generated have to be added to the set of beliefs $\mathcal{B}$ of the agent. These beliefs are necessary for triggering any of the following rules:

- $r1 : \neg incomp(x) \rightarrow pursued(x)$
- $r2 : incompat(x,y,ls) \wedge pref(x,y) \rightarrow pursued(x)$
- $r3 : incompat(x,y,ls) \wedge \neg pref(y,x) \rightarrow \neg pursued(y)$
- $r4 : incompat(x,y,ls) \wedge eq\_pref(x,y) \rightarrow pursued(x)$
- $r5 : max\_util(x) \rightarrow pursued(x)$
- $r6 : \neg max\_util(x) \rightarrow \neg pursued(x)$

Let $\mathcal{ER} = \{r1, r2, r3, r4, r5, r6\}$ be the set of rules necessary for constructing explanatory arguments.

Definition 7 ***(Explanatory argument).*** *Let $\mathcal{B}$, $\mathcal{ER}$, and $g \in \mathcal{G}$ be the set of beliefs, set of rules, and a goal of an agent, respectively. An explanatory argument constructed from $\mathcal{B}$ and $\mathcal{ER}$ for determining the status of g is a pair $A = \langle \mathcal{S}, h\rangle$ such that (i) $\mathcal{S} \subseteq \mathcal{B} \cup \mathcal{ER}$, (ii) $h \in \{pursued(g), \neg pursued(g)\}$, (iii) $\mathcal{S} \vdash h$, and (iv) $\mathcal{S}$ is consistent and minimal for the set inclusion[9].*

Let $\texttt{ARG}_{exp}$ be the set of explanatory arguments that can be built from $\mathcal{B}$ and $\mathcal{ER}$. We call $\mathcal{S}$ the support of an argument A (denoted by $\texttt{SUPPORT}(A)$) and h its claim (denoted by $\texttt{CLAIM}(A)$).

We can notice that rules in $\mathcal{ER}$ can generate conflicting arguments because they have inconsistent conclusions. Thus, we need to define the concept of attack. In this context, the attack that can exist between two explanatory arguments is the well-known rebuttal [3], where two explanatory arguments support contradictory claims. Formally:

[9] Minimal means that there is no $\mathcal{S}' \subset \mathcal{S}$ such that $\mathcal{S} \vdash h$ and consistent means that it is not the case that $\mathcal{S} \vdash pursued(g)$ and $\mathcal{S} \vdash \neg pursued(g)$ [9].

Definition 8 ***(Rebuttal).*** *Let $\langle \mathcal{S}, h\rangle$ and $\langle \mathcal{S}', h'\rangle$ be two explanatory arguments. $\langle \mathcal{S}, h\rangle$ rebuts $\langle \mathcal{S}', h'\rangle$ iff $h \equiv \neg h'$.*

Rebuttal attack has a symmetric nature, this means that two arguments rebut each other, that is, they mutually attack. Recall that the semantics for determining the set of pursued goals is based on conflict-free sets and on a function based on the preference value of the goals. This function is decisive in the selection of the extension that includes the goals the agent can commit to. Thus, it is natural to believe that arguments related to such function are stronger than other arguments. This difference in the strength of arguments turns out in a defeat relation between them, which breaks the previously mentioned symmetry.

Definition 9 ***(Defeat Relation - $\mathcal{D}$).*** *Let $\mathcal{ER}$ be the set of rules and $A = \langle \mathcal{S}, h\rangle$ and $B = \langle \mathcal{S}', h'\rangle$ be two explanatory arguments such that A rebuts B and vice versa. A defeats B iff $r5 \in \mathcal{S}$ (or $r6 \in \mathcal{S}$).*

We denote with (A, B) the defeat relation between A and B. In other words, if $(A, B) \in \mathcal{D}$, it means that A defeats B.

Once we have defined arguments and the defeat relation, we can generate the AF. It is important to make it clear that a different AF is generated for each goal.

Definition 10 ***(Explanatory AF).*** *Let $g \in \mathcal{G}$ be a pursuable goal. An Explanatory AF for g is a pair $\mathcal{XAF}_g = \langle \mathtt{ARG}^g_{exp}, \mathcal{D}^g\rangle$ where:*

- *$\mathtt{ARG}^g_{exp} \subseteq \mathtt{ARG}_{exp}$ such that $\forall A \in \mathtt{ARG}^g_{exp}$, $\mathtt{CLAIM}(A) = pursued(g)$ or $\mathtt{CLAIM}(A) = \neg pursued(g)$.*
- *$\mathcal{D}^g \subseteq \mathtt{ARG}^g_{exp} \times \mathtt{ARG}^g_{exp}$ is a binary relation that captures the defeat relation between arguments in $\mathtt{ARG}^g_{exp}$.*

The next step is to evaluate the arguments that make part of the AF. This evaluation is important because it determines the set of non-conflicting arguments, which in turn determines if a goal becomes pursued or not. Recall that for obtaining such set, an argumentation semantics has to be applied. Unlike the semantics for goals selection, in this case we can use any of the semantics defined in literature. Next, the main semantics introduced by Dung [6] are recalled[10].

Definition 11 ***(Semantics).*** *Let $\mathcal{XAF}_g = \langle \mathtt{ARG}^g_{exp}, \mathcal{D}^g\rangle$ be an explanatory AF and $\mathcal{E} \subseteq \mathtt{ARG}^g_{exp}$:*

- *$\mathcal{E}$ is* ***conflict-free*** *if $\forall A, B \in \mathcal{E}$, $(A, B) \notin \mathcal{D}^g$*
- *$\mathcal{E}$* ***defends*** *A iff $\forall B \in \mathtt{ARG}^g_{exp}$, if $(B, A) \in \mathcal{D}^g$, then $\exists C \in \mathcal{E}$ s.t. $(C, B) \in \mathcal{D}^g$.*
- *$\mathcal{E}$ is* ***admissible*** *iff it is conflict-free and defends all its elements.*
- *A conflict-free $\mathcal{E}$ is a* ***complete extension*** *iff we have $\mathcal{E} = \{A|\mathcal{E}$ defends $A\}$.*

[10] It is not the scope of this article to study the most adequate semantics for this context or the way to select an extension when more than one is returned by a semantics.

- $\mathcal{E}$ *is a* ***preferred extension*** *iff it is a maximal (w.r.t. the set inclusion) complete extension.*
- $\mathcal{E}$ *is a* ***grounded extension*** *iff is a minimal (w.r.t. set inclusion) complete extension.*
- $\mathcal{E}$ *is a* ***stable extension*** *iff* $\mathcal{E}$ *is conflict-free and* $\forall A \in \mathtt{ARG}^g_{exp}$, $\exists B \in \mathcal{E}$ *such that* $(B, A) \in \mathcal{D}^g$.

Finally, a goal g becomes pursued when $\exists A \in \mathcal{E}$ such that $\mathtt{CLAIM}(A) = pursued(g)$.

3.2 Explanation Generation Process

In this article, an explanation is made up of a set of explanatory arguments that justify the fact that a pursuable goal becomes (or not) pursued. Recall that there is a different explanatory AF for each pursuable goal. Thus, we can say that an explanation for a given goal g is given by the explanatory AF generated for it, that is $\mathcal{XAF}_g$. Besides, if $g \in \mathcal{G}'$, the explanation is required by using $\mathtt{WHY}(g)$; otherwise, the explanation is required by using $\mathtt{WHY\_NOT}(g)$. Finally, we can differentiate between partial and complete explanations depending on the set of explanatory arguments that are employed for the justification:

- A *complete explanation* for g is: $\mathcal{CE}_g = \mathcal{XAF}_g$
- A *partial explanation* for g is: $\mathcal{PE}_g = \mathcal{E}$, where $\mathcal{E}$ is an extension obtained by applying a semantics to $\mathcal{XAF}_g$.

We can now present the steps for generating explanations. Given a $\mathcal{GAF}_{sc} = \langle \mathcal{G}, \mathcal{RG}_{sc}, \mathtt{INCOMP\_G}, \mathtt{PREF} \rangle$ and a set of pursued goals $\mathcal{G}'$, the steps for generating an explanation for a goal $g \in \mathcal{G}$ are:

1. From $\mathcal{GAF}_{sc}$ generate the respective beliefs and add to $\mathcal{B}$
2. Trigger the rules in $\mathcal{ER}$ that can be unified with the beliefs of $\mathcal{B}$
3. Construct explanatory arguments based on the rules and beliefs of the two previous items
4. $\forall g \in \mathcal{G}$ do
   (a) Generate the respective explanatory AF (that is, $\mathcal{XAF}_g$) with the arguments whose claim is $pursued(g)$ or $\neg pursued(g)$ and the defeat relation
   (b) Calculate the extension $\mathcal{E}$ from $\mathcal{XAF}_g$

3.3 From Explanatory Arguments to Explanatory Sentences

Like it was done in [7], in this sub-section we present a pseudo-natural language for improving the understanding of the explanations when the agents are interacting with human users. Thus, we propose a set of *explanatory schemes*, one for each rule in $\mathcal{ER}$. This means that depending on which rule an argument was constructed, the explanation scheme is different. In this first version of the scheme, we will generate explanatory sentences only for partial explanations.

Recall that goals are mapped to constants of $\mathcal{L}$, in order to improve the natural language let $\texttt{NAME}(g)$ denote the original predicate of a given goal g. Besides, let $\texttt{RULE}(A)$ denote which of the rules in $\mathcal{ER}$ was employed in order to construct A.

Definition 12 ***(Explanatory Schemes).*** *Let $A = \langle \mathcal{S}, h \rangle$ be an explanatory argument. An explanatory scheme* exp_sch *for A is:*[11]

- *If* $\texttt{RULE}(A) = r1 : \neg incomp(x) \rightarrow pursued(x)$, *then* exp_sch $= \langle \underline{\texttt{NAME}(x)}$ *has no incompatibility, so it became pursued.*$\rangle$
- *If* $\texttt{RULE}(A) = r2 : incompat(x, y, ls) \wedge pref(x, y) \rightarrow pursued(x)$, *then* exp_sch $= \langle \underline{\texttt{NAME}(x)}$ *and* $\underline{\texttt{NAME}(y)}$ *have the following conflicts:* $\underline{ls}$. *Since* $\underline{\texttt{NAME}(x)}$ *is more preferable than* $\underline{\texttt{NAME}(y)}$, $\underline{\texttt{NAME}(x)}$ *became pursued.*$\rangle$
- *If* $\texttt{RULE}(A) = r3 : incompat(x, y, ls) \wedge \neg pref(y, x) \rightarrow \neg pursued(y)$, *then* exp_sch $= \langle \underline{\texttt{NAME}(x)}$ *and* $\underline{\texttt{NAME}(y)}$ *have the following conflicts:* $\underline{ls}$. *Since* $\underline{\texttt{NAME}(y)}$ *is less preferable than* $\underline{\texttt{NAME}(x)}$, $\underline{\texttt{NAME}(y)}$ *did not become pursued.*$\rangle$
- *If* $\texttt{RULE}(A) = r4 : incompat(x, y, ls) \wedge eq\_pref(x, y) \rightarrow pursued(x)$, *then* exp_sch $= \langle \underline{\texttt{NAME}(x)}$ *and* $\underline{\texttt{NAME}(y)}$ *have the following conflicts:* $\underline{ls}$. *Since* $\underline{\texttt{NAME}(x)}$ *and* $\underline{\texttt{NAME}(y)}$ *have the same preference value,* $\underline{\texttt{NAME}(x)}$ *became pursued.*$\rangle$
- *If* $\texttt{RULE}(A) = r5 : max\_util(x) \rightarrow pursued(x)$, *then* exp_sch $= \langle$*Since* $\underline{\texttt{NAME}(x)}$ *belonged to the set of goals that maximize the utility, it became pursued.*$\rangle$
- *If* $\texttt{RULE}(A) = r6 : \neg max\_util(x) \rightarrow \neg pursued(x)$, *then* exp_sch $= \langle$*Since* $\underline{\texttt{NAME}(x)}$ *did not belong to the set of goals that maximizes the utility, it did not become pursued*$\rangle$.

4 Application: Cleaner World Scenario

Let us consider the $\mathcal{GAF}_{sc} = \langle \mathcal{G}, \mathcal{RG}_{sc}, \texttt{INCOMP\_G}, \texttt{PREF} \rangle$ presented in Example 2, whose graph is depicted in Fig. 2. Recall also that $\mathcal{G}' = \{clean(5,5), mop(5,5), be(fixed)\}$.

Firstly, we map the goals in $\mathcal{G}$ into constants of $\mathcal{L}$ in the following manner: $g_1 = clean(5,5)$, $g_2 = pickup(5,5)$, $g_3 = mop(5,5)$, $g_4 = be(in\_workshop)$, and $g_5 = be(fixed)$. We will also map the beliefs and rules to constants in $\mathcal{L}$.

We can now follow the steps to generate the explanations:

1. Generate beliefs

- $b_1 : \neg incomp(g_5)$
- $b_2 : incompat(g_3, g_2, \text{`}s\text{'})$
- $b_3 : incompat(g_3, g_4, \text{`}t\text{'})$
- $b_4 : incompat(g_1, g_4, \text{`}t, r\text{'})$
- $b_5 : incompat(g_2, g_4, \text{`}t, r\text{'})$

$b_{10} : \neg max\_util(g_4)$
$b_{11} : pref(g_3, g_4)$
$b_{12} : \neg pref(g_4, g_3)$
$b_{13} : pref(g_1, g_4)$
$b_{14} : \neg pref(g_4, g_1)$

[11] Underlined characters represent the variables of the schemes, which depend on the variables of rules.

- $b_6 : max\_util(g_1)$
- $b_7 : max\_util(g_3)$
- $b_8 : max\_util(g_5)$
- $b_9 : \neg max\_util(g_2)$

$b_{15} : pref(g_2, g_4)$
$b_{16} : \neg pref(g_4, g_2)$
$b_{17} : pref(g_3, g_2)$
$b_{18} : \neg pref(g_2, g_3)$

2. Trigger rules

- $r_1 : \neg incomp(g_5) \rightarrow pursued(g_5)$
- $r_2 : incompat(g_3, g_2, \text{'s'}) \wedge pref(g_3, g_2) \rightarrow pursued(g_3)$
- $r_3 : incompat(g_3, g_2, \text{'s'}) \wedge \neg pref(g_2, g_3) \rightarrow \neg pursued(g_2)$
- $r_4 : incompat(g_3, g_4, \text{'t'}) \wedge pref(g_3, g_4) \rightarrow pursued(g_3)$
- $r_5 : incompat(g_3, g_4, \text{'t'}) \wedge \neg pref(g_4, g_3) \rightarrow \neg pursued(g_4)$
- $r_6 : incompat(g_1, g_4, \text{'t, r'}) \wedge pref(g_1, g_4) \rightarrow pursued(g_1)$
- $r_7 : incompat(g_1, g_4, \text{'t, r'}) \wedge \neg pref(g_4, g_1) \rightarrow \neg pursued(g_4)$
- $r_8 : incompat(g_2, g_4, \text{'t, r'}) \wedge pref(g_2, g_4) \rightarrow pursued(g_2)$
- $r_9 : incompat(g_2, g_4, \text{'t, r'}) \wedge \neg pref(g_4, g_2) \rightarrow \neg pursued(g_4)$
- $r_{10} : max\_util(g_1) \rightarrow pursued(g_1)$
- $r_{11} : max\_util(g_3) \rightarrow pursued(g_3)$ - $r_{12} : max\_util(g_5) \rightarrow pursued(g_5)$
- $r_{13} : \neg max\_util(g_2) \rightarrow \neg pursued(g_2)$ - $r_{14} : \neg max\_util(g_4) \rightarrow \neg pursued(g_4)$

3. Construct explanatory arguments

- $A_1 = \langle \{b_1, r_1\}, pursued(g_5)\}\rangle$ - $A_2 = \langle \{b_2, b_{17}, r_2\}, pursued(g_3)\}\rangle$
- $A_3 = \langle \{b_2, b_{18}, r_3\}, \neg pursued(g_2)\}\rangle$ - $A_4 = \langle \{b_3, b_{11}, r_4\}, pursued(g_3)\}\rangle$
- $A_5 = \langle \{b_3, b_{12}, r_5\}, \neg pursued(g_4)\}\rangle$ - $A_6 = \langle \{b_4, b_{13}, r_6\}, pursued(g_1)\}\rangle$
- $A_7 = \langle \{b_4, b_{14}, r_7\}, \neg pursued(g_4)\}\rangle$ - $A_8 = \langle \{b_5, b_{15}, r_8\}, pursued(g_2)\}\rangle$
- $A_9 = \langle \{b_5, b_{16}, r_9\}, \neg pursued(g_4)\}\rangle$ - $A_{10} = \langle \{b_6, r_{10}\}, pursued(g_1)\}\rangle$
- $A_{11} = \langle \{b_7, r_{11}\}, pursued(g_3)\}\ \rangle$ - $A_{12} = \langle \{b_8, r_{12}\}, pursued(g_5)\}\rangle$
- $A_{13} = \langle \{b_9, r_{13}\}, \neg pursued(g_2)\}\rangle$ - $A_{14} = \langle \{b_{10}, r_{14}\}, \neg pursued(g_4)\}\rangle$

4. For each goal, generate an explanatory AF and extension

- For g_1: $\mathcal{XAF}_{g_1} = \langle \{A_6, A_{10}\}, \{\}\rangle$, $\mathcal{E} = \{A_6, A_{10}\}$
- For g_2: $\mathcal{XAF}_{g_2} = \langle \{A_3, A_8, A_{13}\}, \{(A_3, A_8), (A_{13}, A_8)\}\rangle$, $\mathcal{E} = \{A_3, A_{13}\}$
- For g_3: $\mathcal{XAF}_{g_3} = \langle \{A_2, A_4, A_{11}\}, \{\}\rangle$, $\mathcal{E} = \{A_2, A_4, A_{11}\}$
- For g_4: $\mathcal{XAF}_{g_4} = \langle \{A_5, A_7, A_9, A_{14}\}, \{\}\rangle$, $\mathcal{E} = \{A_5, A_7, A_9, A_{14}\}$
- For g_5: $\mathcal{XAF}_{g_5} = \langle \{A_1, A_{12}\}, \{\}\rangle$, $\mathcal{E} = \{A_1, A_{12}\}$

Thus, the – partial or complete – explanations for justifying the status of each goal were generated. Next, we present the query, set of arguments of the partial explanation, and the explanatory sentences for the status of each goal:

- For the query `WHY`(g_1), we have $\mathcal{PE} = \{A_6, A_{10}\}$, which can be written:
  * *clean(5, 5) and be(in_workshop) have the following conflicts: 't, r'. Since clean(5, 5) is more preferable than be(in_workshop), clean(5, 5) became pursued*
  * *Since clean(5, 5) belonged to the set of goals that maximizes the utility, it became pursued*
- For the query `WHY_NOT`(g_2), we have $\mathcal{PE} = \{A_3, A_{13}\}$, which can be written:

  * *mop(5, 5) and pickup(5, 5) have the following conflicts: 's'. Since pickup(5,5) is less preferable than mop(5,5), pickup(5,5) did not become pursued*
  * *Since pickup(5,5) did not belong to the set of goals that maximizes the utility, it did not become pursued*
- For the query WHY(g_3), we have $\mathcal{PE} = \{A_2, A_4, A_{11}\}$, which can be written:
  * *mop(5,5) and pickup(5,5) have the following conflicts: 's'. Since mop(5,5) is more preferable than pickup(5,5), mop(5,5) became pursued*
  * *mop(5,5) and be(in_workshop) have the following conflicts: 't'. Since mop(5,5) is more preferable than be(in_workshop), mop(5,5) became pursued*
  * *Since mop(5,5) belonged to the set of goals that maximizes the utility,* it *became pursued*
- For the query WHY_NOT(g_4), we have $\mathcal{PE} = \{A_5, A_7, A_9, A_{14}\}$, which can be written:
  * *mop(5,5) and be(in_workshop) have the following conflicts: 't'. Since be(in_workshop) is less preferable than mop(5,5), be(in_workshop) did not become pursued*
  * *clean(5,5) and be(in_workshop) have the following conflicts: 't, r'. Since be(in_workshop) is less preferable than clean(5,5), be(in_workshop) did not become pursued*
  * *pickup(5,5) and be(in_workshop) have the following conflicts: 't, r'. Since be(in_workshop) is less preferable than pickup(5,5), be(in_workshop) did not become pursued*
  * *Since be(in_workshop) did not belong to the set of goals that maximizes the utility, it did not become pursued*
- For the query WHY(g_5), we have $\mathcal{PE} = \{A_1, A_{12}\}$, which can be written:
  * *be_fixed has no incompatibility, so it became pursued*
  * *Since be_fixed belonged to the set of goals that maximizes the utility,* it *became pursued*

For all the queries, except WHY_NOT(g_2), the complete explanation is the same. In the case of WHY_NOT(g_2), the complete explanation includes the attack relations between some of the arguments of its explanatory AF.

We are also working in a simulator – called ArgAgent[12] – for generating explanations. In it first version, just partial explanations are generated. Figure 3 shows the explanation for query WHY(g_1).

5 Related Work

Since XAI is a recently emerged domain in Artificial Intelligence, there are few reviews about the works in this area. In [2], Anjomshoae et al. make a Systematic Literature Review about goal-driven XAI, i.e., explainable agency for robots and agents. Their results show that 22% of the platforms and architectures have not

[12] Available at: https://github.com/henriquermonteiro/BBGP-Agent-Simulator/.

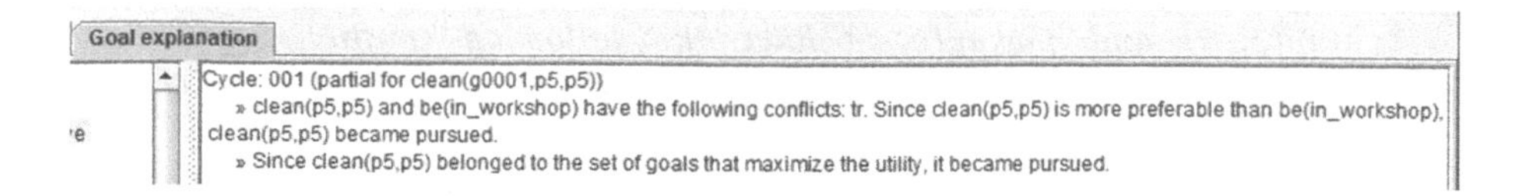

Fig. 3. Partial explanation for query WHY(g_1). Obtained by using the simulator ArgAgent.

explicitly indicate their method for generating explanations, 18% of papers relied on *ad-hoc* methods, 9% implemented their explanations in BDI architecture.

Some works relied on the BDI model are the following. In [4] and [8], Broekens et al. and Harbers et al., respectively, focus on generating explanations for humans about how their goals were achieved. Unlike our proposal, their explanations do not focus on the goals selection. Langley et al. [10] focus on settings in which an agent receives instructions, performs them, and then describes and explains its decisions and actions afterwards.

Sassoon et al. [17] propose an approach of explainable argumentation based on argumentation schemes and argumentation-based dialogues. In this approach, an agent provides explanations to patients (human users) about their treatments. In this case, argumentation is applied in a different way than in our proposal and with other focus, they generate explanations for information seeking and persuasion. Finally, Morveli-Espinoza et al. [15] propose an argumentation-based approach for generating explanations about the intention formation process, that is, since a goal is a desire until it becomes an intention; however, the generated explanations about goals selection are not detailed and they do not present a pseudo-natural language.

6 Conclusions and Future Work

In this article, we presented an argumentation-based approach for generating explanations about the goals selection process, that is, giving reasons to justify the transition of a set of goals from being pursuable (desires) to pursued (intentions). Such reasons are related to the conflicts that may exist between pursuable goals and how that conflicts were resolved. In the first part of the approach, argumentation was employed to deal with conflicts and in the second part it was employed to generate explanations. In order to improve the informational quality of explanations, we extended the results presented in [16]. Thus, explanations also include the form of incompatibility that exists between goals. Besides, we proposed a pseudo-natural language that is a first step to generate explanations for human users. Therefore, our proposal is able generate explanations for both intelligent agents and human-users.

As future work, we aim to further improve the informational quality of explanations by allowing information seeking about the exact point of conflict between two instrumental arguments (or plans) and information about the force of the arguments. The pseudo-natural language was only applied to partial explanations, we plan to extend such language in order to support complete explanations.

References

1. Amgoud, L., Devred, C., Lagasquie-Schiex, M.-C.: A constrained argumentation system for practical reasoning. In: Rahwan, I., Moraitis, P. (eds.) ArgMAS 2008. LNCS (LNAI), vol. 5384, pp. 37–56. Springer, Heidelberg (2009). https://doi.org/10.1007/978-3-642-00207-6_3
2. Anjomshoae, S., Najjar, A., Calvaresi, D., Främling, K.: Explainable agents and robots: results from a systematic literature review. In: Proceedings of the 18th International Conference on Autonomous Agents and MultiAgent Systems, pp. 1078–1088 (2019)
3. Besnard, P., Hunter, A.: Argumentation based on classical logic. In: Simari, G., Rahwan, I. (eds.) Argumentation in Artificial Intelligence, pp. 133–152. Springer, Boston (2009). https://doi.org/10.1007/978-0-387-98197-0_7
4. Broekens, J., Harbers, M., Hindriks, K., van den Bosch, K., Jonker, C., Meyer, J.-J.: Do you get it? User-evaluated explainable BDI agents. In: Dix, J., Witteveen, C. (eds.) MATES 2010. LNCS (LNAI), vol. 6251, pp. 28–39. Springer, Heidelberg (2010). https://doi.org/10.1007/978-3-642-16178-0_5
5. Castelfranchi, C., Paglieri, F.: The role of beliefs in goal dynamics: prolegomena to a constructive theory of intentions. Synthese **155**(2), 237–263 (2007). https://doi.org/10.1007/s11229-006-9156-3
6. Dung, P.M.: On the acceptability of arguments and its fundamental role in non-monotonic reasoning, logic programming and n-person games. Artif. Intell. **77**(2), 321–357 (1995)
7. Guerrero, E., Nieves, J.C., Lindgren, H.: An activity-centric argumentation framework for assistive technology aimed at improving health. Argument Comput. **7**(1), 5–33 (2016)
8. Harbers, M., van den Bosch, K., Meyer, J.J.: Design and evaluation of explainable BDI agents. In: 2010 IEEE/WIC/ACM International Conference on Web Intelligence and Intelligent Agent Technology, vol. 2, pp. 125–132. IEEE (2010)
9. Hunter, A.: Base logics in argumentation. In: COMMA, pp. 275–286 (2010)
10. Langley, P., Meadows, B., Sridharan, M., Choi, D.: Explainable agency for intelligent autonomous systems. In: Twenty-Ninth IAAI Conference, pp. 4762–4763 (2017)
11. Martínez, D.C., García, A.J., Simari, G.R.: Progressive defeat paths in abstract argumentation frameworks. In: Lamontagne, L., Marchand, M. (eds.) AI 2006. LNCS (LNAI), vol. 4013, pp. 242–253. Springer, Heidelberg (2006). https://doi.org/10.1007/11766247_21
12. Modgil, S., Prakken, H.: The ASPIC+ framework for structured argumentation: a tutorial. Argument Comput. **5**(1), 31–62 (2014)
13. Morveli-Espinoza, M., Nieves, J.C., Possebom, A., Puyol-Gruart, J., Tacla, C.A.: An argumentation-based approach for identifying and dealing with incompatibilities among procedural goals. Int. J. Approx. Reason. **105**, 1–26 (2019)
14. Morveli-Espinoza, M.M., Nieves, J.C., Possebom, A.T., Puyol-Gruart, J., Tacla, C.A.: An argumentation-based approach for identifying and dealing with incompatibilities among procedural goals. Int. J. Approx. Reason. **105**, 1–26 (2019). https://doi.org/10.1016/j.ijar.2018.10.015

15. Morveli-Espinoza, M., Possebom, A., Tacla, C.A.: Argumentation-based agents that explain their decisions. In: 2019 8th Brazilian Conference on Intelligent Systems (BRACIS), pp. 467–472. IEEE (2019)
16. Morveli-Espinoza, M., Possebom, A.T., Puyol-Gruart, J., Tacla, C.A.: Argumentation-based intention formation process. DYNA **86**(208), 82–91 (2019)
17. Sassoon, I., Kökciyan, N., Sklar, E., Parsons, S.: Explainable argumentation for wellness consultation. In: Calvaresi, D., Najjar, A., Schumacher, M., Främling, K. (eds.) EXTRAAMAS 2019. LNCS (LNAI), vol. 11763, pp. 186–202. Springer, Cham (2019). https://doi.org/10.1007/978-3-030-30391-4_11
18. Thangarajah, J., Padgham, L., Winikoff, M.: Detecting and avoiding interference between goals in intelligent agents. In: Proceedings of the 23rd International Joint Conference on Artificial Intelligence. Morgan Kaufmann Publishers (2003)
19. Tinnemeier, N.A.M., Dastani, M., Meyer, J.-J.C.: Goal selection strategies for rational agents. In: Dastani, M., El Fallah Seghrouchni, A., Leite, J., Torroni, P. (eds.) LADS 2007. LNCS (LNAI), vol. 5118, pp. 54–70. Springer, Heidelberg (2008). https://doi.org/10.1007/978-3-540-85058-8_4
20. Wooldridge, M.J.: Reasoning About Rational Agents. MIT Press, Cambridge (2000)
21. Zatelli, M.R., Hübner, J.F., Ricci, A., Bordini, R.H.: Conflicting goals in agent-oriented programming. In: Proceedings of the 6th International Workshop on Programming Based on Actors, Agents, and Decentralized Control, pp. 21–30. ACM (2016)

Application-Level Load Balancing for Reactive Wireless Sensor Networks: An Approach Based on Constraint Optimization Problems

Igor Avila Pereira[1,2(✉)], Lisane B. de Brisolara[1], and Paulo Roberto Ferreira Jr.[1]

[1] Programa de Pós Graduação em Computação (PPGC), UFPel, Pelotas, Brazil
[2] Instituto Federal do Rio Grande do Sul (IFRS), Rio Grande, Brazil
{igor.pereira,lisane,paulo}@inf.ufpel.edu.br

Abstract. As Wireless Sensor Networks - WSNs are being used more and more in applications with high demand for the processing of videos and images, the processing load of the nodes also must be considered by energy-saving strategies. Usually, these strategies focused only on communication. The overloading of nodes can result in a degraded performance regarding the network lifetime and availability of the provided services. A few algorithms, based on heuristics, have been proposed in the literature to balance the load at the application level. These algorithms drawnback are the lack of quality assurance. Thus, this work presents a novel semi-distributed solution, applying techniques to solve Constraint Optimization Problem - COP on the load balancing in a WSN. Our approach guarantee that all events are processed during the lifetime of the network and that there is a fixed number of exchanged messages per event, only reducing the lifetime of the network by an estimable time. Experimental results pointed out that the proposed approach allows the network, while alive, to sense all happened events each day, and its performance is better than the one achieved by the state-of-the-art for lower density networks.

Keywords: Wireless sensor networks · Application-level load balancing · Event-triggered · Constraint optimization problems

1 Introduction

Wireless Sensor Networks (WSNs) are composed by several sensor nodes and one or more sinks that work together to monitor and extract data from an environment [21]. A node can send and receive messages from neighbors inside its communication range and sense events in the region of its coverage range [14]. The sink acts as a gateway for the information collected and processed by the nodes and is the first external connected device of the WSNs. Usually, the

R. Cerri and R. C. Prati (Eds.): BRACIS 2020, LNAI 12320, pp. 63–76, 2020.
https://doi.org/10.1007/978-3-030-61380-8_5

nodes use batteries as a source of energy, and it can not be easily replaced or recharged due to these are deployed into a harsh environment.

Thus, the battery discharge of the nodes influences the lifetime of the whole network and the availability of the provided services by it. Furthermore, this discharge does not occur homogeneously in the WSNs. The workload of the nodes may vary depending on the location or even on the kind of application. Some nodes can run out of battery more quickly than the others, leaving lacks in the covering, or also isolating nodes, when they depend on inactive nodes to communicate with the sink.

Reactive WSNs operate by capturing data only when a given event is triggered, which can reduce the energy consumption of the sensors and the traffic on the network. In these networks, the use of balancing techniques that consider the processing load can improve the efficiency of the network. This type of approach that works at the application level has gained attention from the WSN community with the growth of more complex monitoring applications with high demand for video and image processing.

Performance optimizations in WSNs are considered a significant challenge, specifically in terms of managing the energy consumption of the nodes [18]. The limited resources of the devices composing the nodes make it impossible to use computationally complex algorithms [15] and motivate the use of heuristics and approximate solutions. The use of Artificial Intelligence techniques for load balancing in reactive WSNs has been widely explored, manly through bio-inspired solutions [4,8,12,20].

The most recent bio-inspired solutions presented in the literature, Pheromone Signaling (PS) [5] and Ant-based [12], works in a decentralized manner, enabling nodes to independently decide who will be the one responsible for processing a detected event. Despite presenting improvements in the WSNs overall performance, as these techniques are heuristic-based ones, they do not guarantee any kind of optimally. Both approaches are hardly dependent on their configuration parameters, and usually, they drive the network to lose events.

The most prominent multi-agent coordination and distributed resource allocation approach are based on the Distributed Constraint Optimization Problems - DCOP, which are a distributed version of the Constraint Optimization Problems [17]. In the DCOPs, agents must associate values to its variables in a distributed fashion, according to constraints, so that a cost function is optimized.

One can map the load balancing problem in WSNs we focus here as a multi-agent coordination problem [9]. However, DCOPs algorithms tend to be prohibitively expensive for large scale scenarios due to the cost associated with the communication [11]. The polynomial growth or even exponentially increase in the number of messages exchanged among agents is a known hard limitation to its real-world application.

To avoid this problem, here we propose a semi-distributed approach based on the original Constraint Optimization Problems. This approach allows the definition, among the nodes that detected an event, the most suitable node for processing it, adopting an algorithm to deal with the Constraint Optimization

Problem. The deliberation takes place in a single node through the Branch-and-Bound (BnB) algorithm [13]. This node is elected to lead the process using a limited number of exchanged messages.

The remain of the paper is organized as follows: Sect. 2 discusses related works; Sect. 3 details the proposed semi-centralized load balancing approach; Sect. 4 presents and discusses the experimental results, comparing our approach with the state-of-the-area; and Sect. 5 presents the conclusions and future works.

2 Related Works

The problem of load balancing for the application level in reactive WSNs has recently addressed by the use of bio-inspired heuristic techniques in [5] and [12]. Both methods decide in a decentralized way which node will process an event considering only the nodes that detected it.

The Pheromone Signaling (PS) [5] technique is based on the bees' hormonal system and allows the nodes to be periodically classified into two roles. Nodes called queen are capable of processing the detected events, while the worker nodes, remain at rest, performing only message forwarding. The strategy of this algorithm is based on the periodic transmission of the pheromone by the queen nodes. The pheromone level of the nodes declines over time and with distance from the source. If at any given moment, a node's pheromone level is below a threshold, it will be converted to a queen node. Changes in pheromone levels are used to define the role of the node and thus balance the load [5].

In the Ant-based technique proposed by [12] the nodes decide, probabilistically, which one will process an event. In this approach, events produce stimuli for nodes that, based on their internal thresholds, determine different probabilities for processing these events. When the nodes detect an event, each node will decide whether or not to process this event considering its calculated internal probability. The approach works by modifying the probability of a node processing an event according to the number of nodes that detected this event at the same time and the number of times that the same node was previously devoted to processing previous events [12].

Compared to PS, Ant-based technique achieved results varying from better to almost the same, according to the density of the network. The authors also compared these two approaches with a greedy (non-practical) approach and a random technique, in which the node that will handle the event is drawn. This greedy approach serves as an upper-bound for other approaches and is also used in our comparisons here.

Recently in [3], the parameters of these bio-inspired techniques were tuned using a Genetic Algorithm, exploring optimal settings for each density to improve the efficiency of these techniques. From here, one can see the dependency of the parameters of these.

In comparison with the previous solutions, our proposed approach works similarly to the Ant-based technique in two points. Firstly, because the entire load balancing process occurs when a event-triggered is detected by nodes in a region of the WSN, unlike the PS in which the definition of roles occurs periodically. Secondly, both use the number of events previously processed for the deliberation of the node responsible for processing the event.

The employment of DCOPs in the WSN domain has been investigated in [10] and in [6]. Firstly, authors use a graph coloring problem to discuss the optimization of energy consumption in devices and secondly, adopted the DCOP algorithm MAXSUM for coordination of the movement of mobile sensors. More recently, in [9], an algorithm called MAXSUM was adopted to coordinate the sensing and hibernation cycles of the nodes and thus balance the load on the WSN. Their goal is guarantee the covering of the area of interesting keeping at least one node activated in a given region, while saving energy and increase network lifetime. In their experiments, authors compared the MAXSUM-based approach to a random solution, a DSA-based solution and a Simulating Annealing based solution. However, in the same way as the PS and Ant-based techniques, solutions based on MAXSUM do not guarantee the quality of the solution.

An alternative to the lack of quality guarantee would be the use of an optimal DCOP algorithm to solve the problem such as, for instance, the one called ADOPT [17]. However, these algorithms, as we previously discussed, are known to be very expensive regarding communication [11]. In the proposed approach, the decision is centralized in a single node which minimizes the number of messages exchanged among nodes in the network.

3 Proposed Approach

Formally a COP is defined as a tuple $\{V, D, F\}$, where V is the set of variables $\{v_1, v_2, ..., v_m\}$, D the domain set $\{d_1, d_2, ..., d_n\}$ and C the set of constraints $\{c_1, c_2, ..., c_u\}$. Each constraint $\{c_u\} \in C$ involves a pair of variables v_x and v_y $\in V$ and defines a cost function $f(x, y)$ associated to each value of the domain D these variables may assume. The objective of the problem is to choose values to the variables in order to minimize the sum of the costs given by all constraints [16].

In the proposed approach each node that perceives an event is considered as a variable of the set V. All variables in V have the same domain $D = \{0, 1\}$. In D, 1 represents that the node will process the event and 0 that the node will stay in hibernation. The cost function $f()$ is given by the combination of all possible values of the domain assumed by the variables considering the number of previous processed events by the nodes they represent. In the load balancing problem, we are focused, the cost functions must represent the situation regarding energy availability by nodes. That why we adopt the number of processed events. As the number of processed events increases, the amount of energy available on the battery decreases. So, the chosen node to process the event will be the one with the highest battery level.

For instance, lets suppose there are two nodes 1 and 2 that perceives an event. They have processed 10 and 15 events in the past, respectively. Each node will be represented by the variables v_1 and v_2. Thus, constraint c_{1-2} has a cost function

$f(0,0) = \infty, f(0,1) = 15, f(1,0) = 10$, and $f(1,1) = 25$. If there are only these two nodes, the solution to the COP that minimizes the constraint cost is $v_1 = 1$ and $v_2 = 0$.

The algorithm that details our approach is presented in Algorithm 1. When an event occurs on the network area of interest, one or more nodes detect it. A node is randomly defined as the leader of load balancing, which, internally, will be responsible for the deliberation process (line 1). Neighboring nodes must exchange messages with the leader node so that they can effectively notify him if they have also detected an event too (lines 4 to 11). With information from neighbors, the leader must define the best possible values of the variables of each participant in the process solving the COP running a Branch-and-Bound algorithm (lines 12 to 14). After defining the values of the variables to each node, the leader notifies the other neighbors through broadcasting (line 14). Individually, each neighbor changes the value of its variable and, consequently, its state of processing or hibernation (lines 15 to 24). Only the chosen node will process the event (lines 19–20).

Algorithm 1. Proposed Algorithm

```
1: node ← firstNodeEventDetected();
2: tempEvent ← node.eventSensed();
3: node.sendMessageToNeighbours(tempEvent);
4: for all n from neighboursOfNode do
5:    tempEvent ← n.receiveMessage();
6:    temp ← n.eventSensed();
7:    if temp == tempEvent then
8:       numEventSensed ← n.numEventSensed();
9:       n.sendMessageToNeighbours(numEventSensed);
10:   end if
11: end for
12: node.receiveAllMessage();
13: loadBalancing ← node.deliberate();
14: node.sendMessageToNeighbours(loadBalancing);
15: for all n from neighboursOfNode do
16:   loadBalancing ← n.getReceiveMessage();
17:   chosenNode ← loadBalancing.getResponsible();
18:   if chosenNode.getName() == n.getName() then
19:      n.setStatus(on);
20:      n.sensorNodeAction();
21:   else
22:      n.setStatus(off);
23:   end if
24: end for
```

Figure 1 illustrates the operation of the proposed algorithm in a WSN with 9 nodes. The overlapping maximum coverage is of 4 nodes, which means that an event is detected by 4 nodes almost simultaneously. In the example, node 1 was chosen. Any neighbor node that also detected the same event (node 2, 4, or 5) could also lead the process. The set of four nodes will be part of the load balancing process (Figs. 1b–1d). Each neighbor that also detected this same event sends a confirmation message to the leader containing the number of events that this node was previously engaged in processing (Figure 1d). In this way, the proposed approach has a semi-distributed operation. It merges a centralized deliberation stage with two distributed stages: the first one for defining the participants through messages; and the last one, responsible for forwarding the deliberation scheme to the other participants in the process of load balancing (Fig. 1e). These other nodes should follow the deliberation, enabling or disabling its processing capability (Fig. 1f). In the example, node 1 running the deliberation process defined node 5 as responsible for the processing of the event. Thus, node 1 (leader), as well as nodes 2 and 4, must evade this responsibility by momentarily deactivating their processing capacity and preserving their battery charge. In contrast, node 5 should work on the processing task.

4 Experiments and Results

The framework for modeling and simulating WSNs named EBORACUM[1] [1–3] is used in our experiments. EBORACUM was chosen as a simulation environment because it is possible to compare the proposed approach against the results obtained by the PS and Ant-based heuristic approaches, previously incorporated into this simulator. In addition to the Ant-based and PS approaches, the analysis compared our semi-centralized approach against to a centralized (non-practical) Greedy technique.

In the simulations, we used WSNs deployed in mesh with 49, 64, 81 and 100 nodes implanted on a square 810 km^2 area (900 m × 900 m) and one sink located on the side as illustrated in Figs. 2a, 2b, 2c and 2d. The nodes are connected to one of its four neighbors (the closest to the sink) or directly to the sink. The deployment scheme presents redundant sensing coverage equal to 4 nodes, in maximum, in the 49 nodes configuration and the redundancy increases in more dense WSNs. A energy cost of cpu = 50 for task, battery with initial charge of 2700000, radius of sensing of 120 and communication radius of, respectively, 160, 140, 120 and 120 (varying to keep the same 4 neighborhood nodes for all densities).

Events appear in the space following a Uniform distribution and the events frequency is given by a Poisson distribution with a function interval between events from [1, 120] s. Each sensed event generates a processing workload equivalent to 14 tasks and the sending of messages with 3 Bytes. This application model generates the WSN workload that combined to energy costs is used to simulate the battery discharge. These configurations are detailed in Table 1.

[1] http://sourceforge.net/projects/eboracum/.

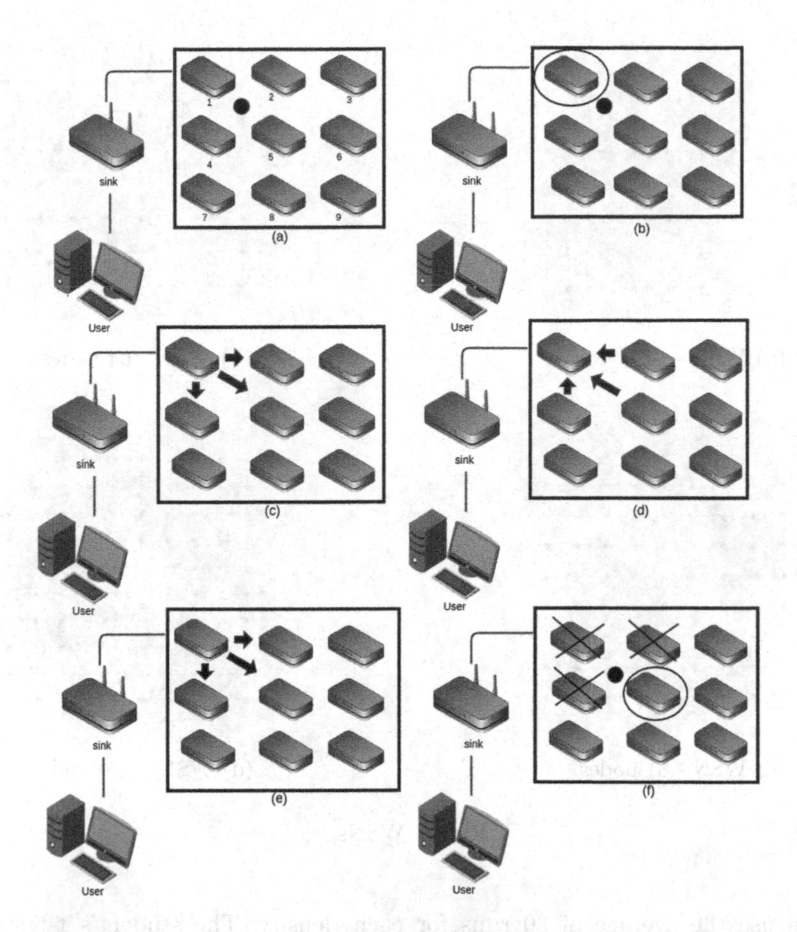

Fig. 1. COP approach: (a) triggering the event; (b) leader of the load balancing scheme; (c) and (d) neighbors exchanging messages with the leader to confirm or not the sensing of the same event; (e) the leader forwarding the defined balancing scheme to neighbors and, (f) only the node chosen by the balancing scheme will remain with its active processing capacity.

Table 1. Energy costs.

Energy-related parameter	Value
Battery capacity	5400000 mAs
Idle discharge rate	0.3 mAs
Task computation discharge rate	3.57 mAs
Discharge rate per message (3 bytes) at 30 kbps	0.0018 mAs

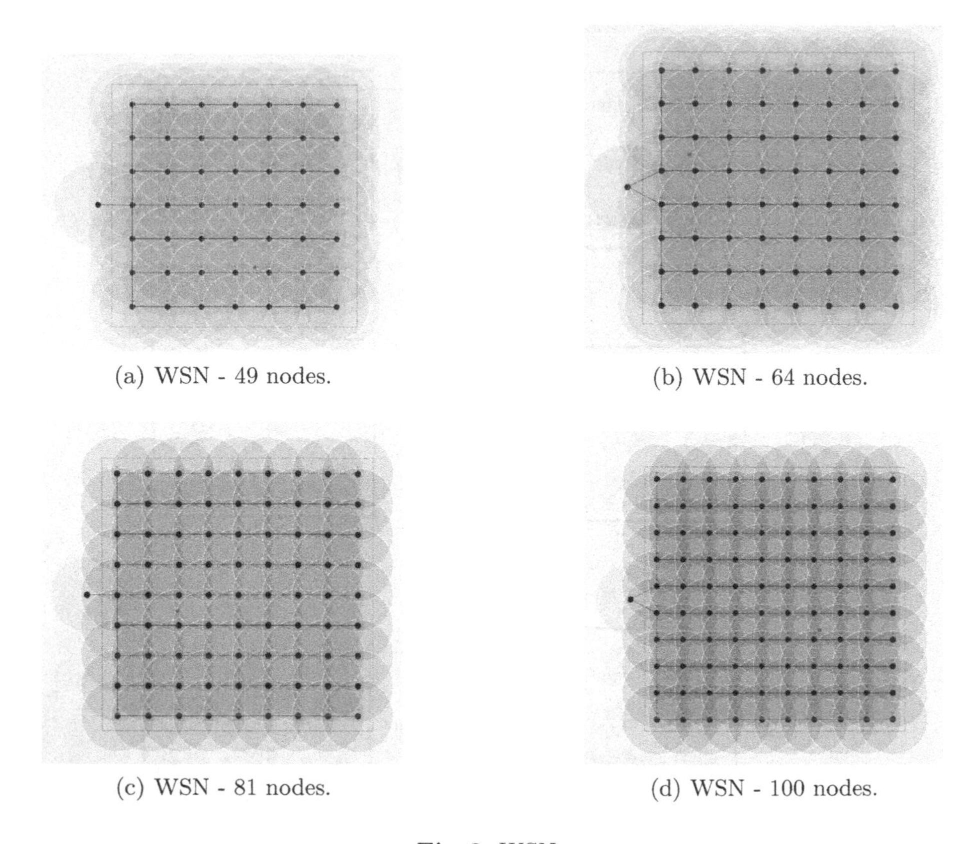

(a) WSN - 49 nodes.

(b) WSN - 64 nodes.

(c) WSN - 81 nodes.

(d) WSN - 100 nodes.

Fig. 2. WSNs.

We use the average of 30 runs for each density. The student's t-test was adopted to analyze the statistical significance between the difference of the average result obtained by each approach. Our proposed approach is called COP here for the sake of simplification.

Firstly, we present the results depicting the average number of events processed per day for each approach. The charts of Figs. 3, 4, 5 and 6 depict the results for the different experimented densities, with 49, 64, 81 and 100 nodes, respectively.

These show that the proposed approach (COP) was able to achieve the same number of events sensed by day that the Greedy, our upper bound. In other words, all events are sensed successfully by the network while it is alive, the best possible result over this perspective. This behavior is a remarkable improvement compared with the best previously presented one, the Ant-based. Ant-based in the lowest density network loses, by each day, around 7% of the happening events.

One also can see in Figs. 2a, 2b, 2c and 2d that our approach drives the network to has a shorter lifetime in all scenarios. COP demands more commu-

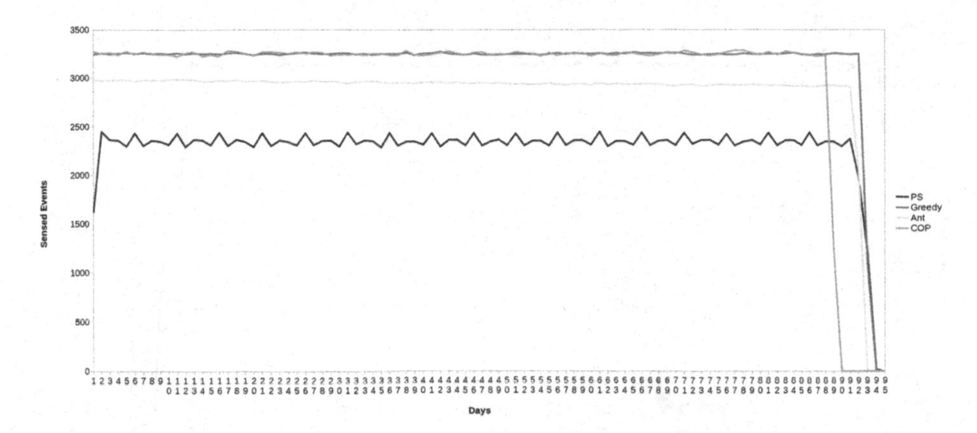

Fig. 3. Average of processed events by day - WSN of 49 nodes.

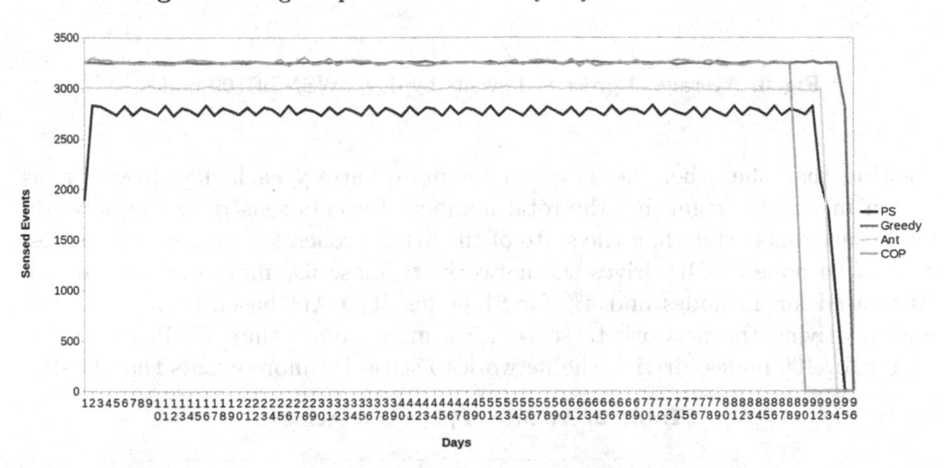

Fig. 4. Average of processed events by day - WSN of 64 nodes.

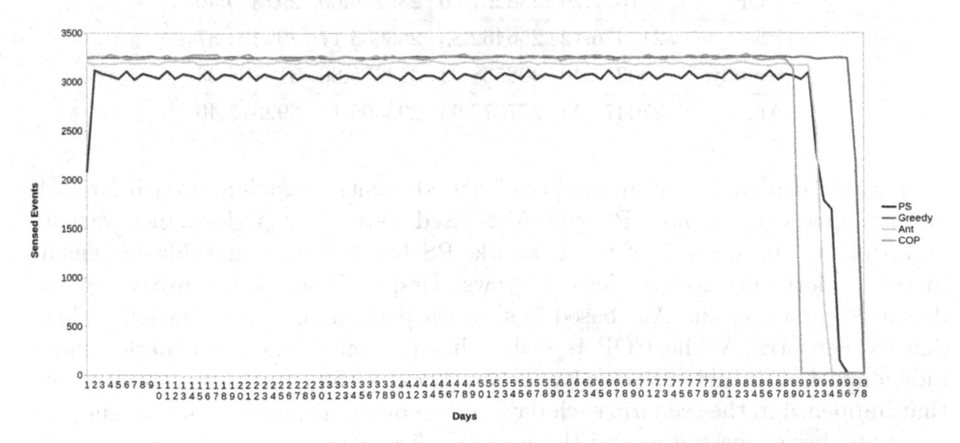

Fig. 5. Average of processed events by day - WSN of 81 nodes.

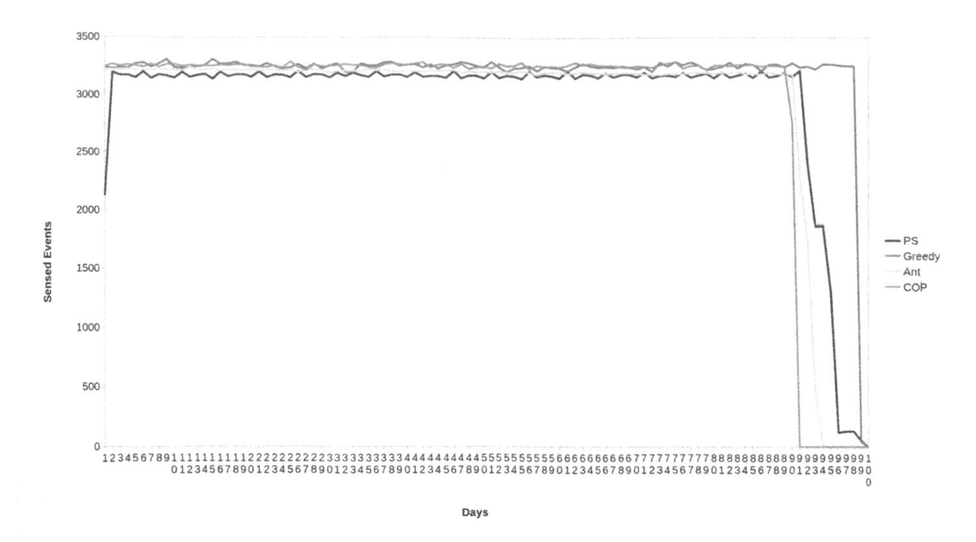

Fig. 6. Average of processed events by day - WSN of 100 nodes.

nication than the others, so it consumes more battery each day. However, as shown in Table 2, regarding the total number of events sensed by the network, COP performs better than the state-of-the-art approach for the lower densities, 49 and 64 nodes. COP drives the networks to sense 6% more events than the Ant-based for 49 nodes and 4% for 61 nodes. The Ant-based is better for 81 nodes, driving the network to sense 1.5% more events than COP, and PS is better for 100 nodes, driving the network to sense 1% more events than COP.

Table 2. Average of processed events.

Approach	49 nodes	64 nodes	81 nodes	100 nodes
COP	287577.20	288291.00	289236.00	292389.40
PS	216476.12	256462.83	283683.17	295401.87
Greedy	300317.40	308538.33	314375.86	318704.33
Ant	270477.44	277377.93	293565.13	292815.40

Figure 7 depicts the total number of sensed events by each approach for each density. One can see how PS and Ant-based have their performance varying according to the density of the network. PS has the most notable increment in the performance as the density grows. Despite being less sensitive to the densities variation, the Ant-based has also a performance degradation in low-density networks. While COP is stable, having almost the same performance independent of the network density. That is because the network senses all events that happened in the scenario each day, nodes communicate always changing the same number of messages, and the network dies after a reasonably predictable amount of time.

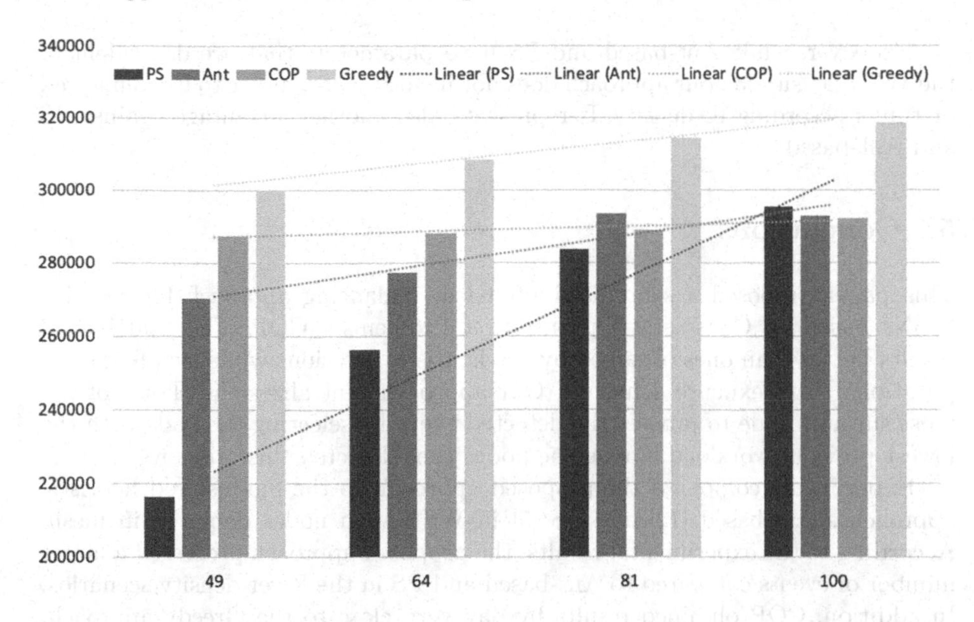

Fig. 7. Comparing the performance of the approaches for all experimented scenarios.

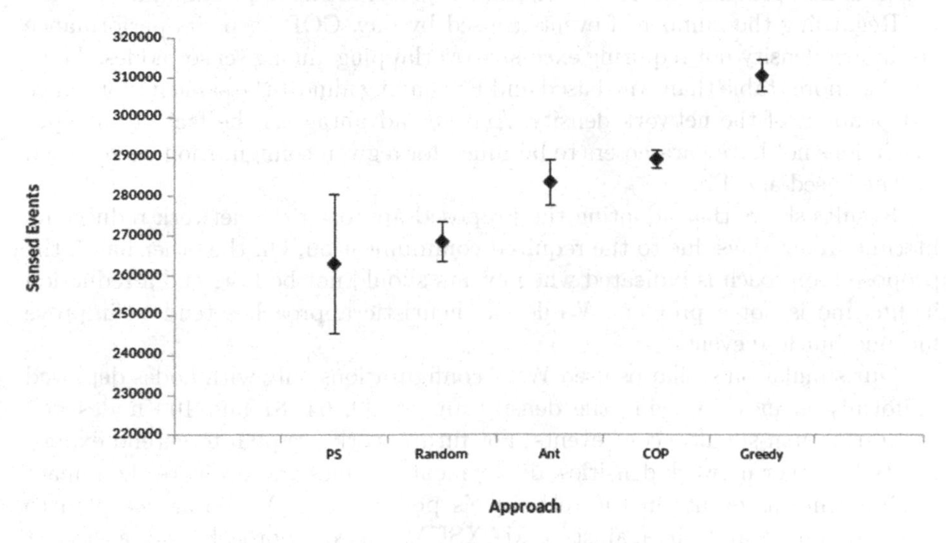

Fig. 8. Average events sensed by the approaches in the four density scenarios.

To better analyze the difference in the quality of overall results achieved by the evaluated approaches, Fig. 8 illustrates the average processed events considering the four densities. COP and Ant-based achieve averages with no statistically significant difference. Despite it, if the density of the network is previously unknown, through our approach lower density networks will perform better than the ones adopting Ant-based, as we discuss above.

Moreover, while Ant-based and PS have parameters that are dependent of the WSN density [3], our approach does not have any parameter to be configured or tuned according to density. It represents also another advantage against PS and Ant-based.

5 Conclusions

This paper proposed a semi-distributed load balancing approach for reactive WSNs based on Constraint Optimization Problems - COP. The goal is find results better than ones obtained by the heuristic solutions while avoiding computational high expense solutions. Our approach centralizes the choice of the most suitable node to process the detected event by selecting the node with the lowest previous workload among the nodes that detected this same event.

Experiments compared the proposed approach to the bio-inspired heuristic approaches Ant-based [12] and PS [5] in WSN with nodes deployed in mesh. According to the experimental results, the proposed approach processed a larger number of events compared to Ant-based and PS in the lower density scenarios. In addition, COP obtained results by day very close to the Greedy approach, which is non-practical for real WSNs but represents our upper bound.

Regarding the number of events sensed by day, COP keeps its performance yet in low density not requiring excessive overlapping among sensor nodes. Thus, COP is more stable than Ant-based and PS, having almost the same performance independent of the network density. Another advantage is the fact of our approach does not have parameters to be tuned for a given configuration, as required in Ant-based and PS.

Results shows that adopting the proposed approach the network reduces its lifetime in few days due to the required communication. On the other hand, the proposed approach is indicated when events should not be lost, and a reduction in lifetime is not a problem. While the heuristic approaches tend to improve lifetime but lose events.

Our simulation scenarios used WSN configurations only with nodes deployed uniformly in mesh, varying the density among 49, 64, 81 and 100 nodes and generating non-simultaneous events. For future work, we plan to extend experiments for other network densities, deployment schemes and evaluate the impact of simultaneous events in the technique's performance. Moreover, we plan to compare our approach against a MAXSUM based approach and a newest DCOP algorithm like, for example, Particle Swarm Based F-DCOP (PFD) [7] or Weighted-DSA [19] to explore wide the modeling of the problem.

Acknowledgments. This study was financed in part by the Coordenação de Aperfeiçoamento de Pessoal de Nível Superior - Brasil (CAPES).

The author thanks the Federal Institute of Education, Science and Technology of Rio Grande do Sul (IFRS) for the support granted by the institution.

References

1. Brisolara, L., Ferreira, P.R., Indrusiak, L.S.: Impact of temporal and spatial application modeling on event-triggered wireless sensor network evaluation. In: Brazilian Symposium on Computing Systems Engineering (SBESC). IEEE (2015). https://doi.org/10.1109/sbesc.2015.13
2. Brisolara, L., Ferreira, P.R., Indrusiak, L.S., Ferreira Jr., P.R., Indrusiak, L.S.: Application modeling for performance evaluation on event-triggered wireless sensor networks. Des. Autom. Embedded Syst. **20**(4), 269–287 (2016). https://doi.org/10.1007/s10617-016-9177-1
3. Brisolara, L., Lorenzatto Braga, M., Braga, A., Roberto Ferreira, P.: Parameter tuning in load balancing techniques for wireless sensor networks through genetic algorithms. In: 8th Brazilian Conference on Intelligent Systems (BRACIS), pp. 42–47. IEEE, October 2019. https://doi.org/10.1109/BRACIS.2019.00017, https://ieeexplore.ieee.org/document/8923923/
4. Caliskanelli, I., Broecker, B., Tuyls, K.: Multi-robot coverage: a bee pheromone signalling approach. In: Headleand, C.J., Teahan, W.J., Ap Cenydd, L. (eds.) ALIA 2014. CCIS, vol. 519, pp. 124–140. Springer, Cham (2015). https://doi.org/10.1007/978-3-319-18084-7_10
5. Caliskanelli, I., Harbin, J., Indrusiak, L.S., Mitchell, P., Polack, F., Chesmore, D.: Bioinspired load balancing in large-scale WSNs using pheromone signalling. Int. J. Distrib. Sens. Netw. **9**(7), 172012 (2013). https://doi.org/10.1155/2013/172012
6. Chachra, S., Marefat, M.: Distributed algorithms for sleep scheduling in wireless sensor networks. In: IEEE International Conference on Robotics and Automation. ICRA 2006, pp. 3101–3107 (2006). https://doi.org/10.1109/ROBOT.2006.1642173
7. Choudhury, M., Mahmud, S., Khan, M.M.: A Particle Swarm Based Algorithm for Functional Distributed Constraint Optimization Problems. Advancement of Artificial Intelligence, pp. 1–8 (2019). http://arxiv.org/abs/1909.06168
8. Derakhshan, F., Yousefi, S.: A review on the applications of multiagent systems in wireless sensor networks. Int. J. Distrib. Sens. Netw. **15**(5) (2019). https://doi.org/10.1177/1550147719850767
9. Farinelli, A., Rogers, A., Jennings, N.R.: Agent-based decentralised coordination for sensor networks using the max-sum algorithm. Autonomous Agents Multi-Agent Syst. **28**(3), 337–380 (2013). https://doi.org/10.1007/s10458-013-9225-1
10. Farinelli, A., Rogers, A., Petcu, A., Jennings, N.R.: Decentralised coordination of low-power embedded devices using the max-sum algorithm. In: 7th International Joint Conference on Autonomous Agents and Multiagent Systems-Volume 2, vol. 2, pp. 630–637. International Foundation for Autonomous Agents and Multiagent Systems (2008)
11. Ferreira, P.R., Boffo, F.S., Bazzan, A.L.C.: Using swarm-GAP for distributed task allocation in complex scenarios. In: Jamali, N., Scerri, P., Sugawara, T. (eds.) AAMAS 2007. LNCS (LNAI), vol. 5043, pp. 107–121. Springer, Heidelberg (2008). https://doi.org/10.1007/978-3-540-85449-4_8
12. Ferreira, P.R., Brisolara, L., Indrusiak, L.S.: Decentralised load balancing in event-triggered WSNs based on ant colony work division. In: 41st Euromicro Conference on Software Engineering and Advanced Applications. IEEE (2015). https://doi.org/10.1109/seaa.2015.52
13. Fukunaga, K., Narendra, P.M.: A Branch and bound algorithm for computing k-nearest neighbors. IEEE Trans. Comput. **C-24**(7), 750–753 (1975). https://doi.org/10.1109/T-C.1975.224297

14. Houssaini, D.E., Khriji, S., Besbes, K., Kanoun, O.: Wireless sensor networks in agricultural applications. Energy Harvesting for Wireless Sensor Networks: Technology, Components and System Design, pp. 323–342 (2018). https://doi.org/10.1515/9783110445053-019
15. Caliskanelli, I.: A Bio-inspired load balancing technique for wireless sensor networks. Ph.D. thesis, The University of York (2014). http://etheses.whiterose.ac.uk/id/eprint/7030
16. Meisels, A.: Constraints optimization problems - COPs. Distributed Search by Constrained Agents. AIKP, pp. 19–26. Springer, London (2008). https://doi.org/10.1007/978-1-84800-040-7_3
17. Modi, P.J.: Distributed constraint optimization for multiagent systems. Ph.D. thesis, University of Southern California (2003)
18. Rajpoot, P., Dwivedi, P.: Optimized and load balanced clustering for wireless sensor networks to increase the lifetime of WSN using MADM approaches. Wirel. Netw. **26** (2020)
19. Van Der Lee, T., Exarchakos, G., De Groot, S.H.: Distributed wireless network optimization with stochastic local search. In: 2020 IEEE 17th Annual Consumer Communications and Networking Conference, CCNC 2020 (2020). https://doi.org/10.1109/CCNC46108.2020.9045189
20. Wohwe Sambo, D., Yenke, B.O., Förster, A., Dayang, P.: Optimized clustering algorithms for large wireless sensor networks: a review. Sensors (Switzerland) **19**(2) (2019). https://doi.org/10.3390/s19020322, https://www.mdpi.com/1424-8220/19/2/322
21. Yick, J., Mukherjee, B., Ghosal, D.: Wireless sensor network survey. Comput. Netw. **52**(12), 2292–2330 (2008). https://doi.org/10.1016/j.comnet.2008.04.002

Cooperative Observation of Smart Target Agents

Matheus S. Araújo, Thayanne F. da Silva, Vinícius A. Sampaio, Gabriel F. L. Melo, Raimundo J. Ferro Junior, Leonardo F. da Costa(✉), Joao P. B. Andrade, and Gustavo A. L. de Campos

Graduate Program in Computer Science, State University of Ceara, Fortaleza, Brazil
{math.araujo,thayanne.silva,vinicius.sampaio,gabriel.lins, junior.ferro,leonardo.costa,joao.bernardino}@aluno.uece.br, gustavo.campos@uece.br

Abstract. The Cooperative Multi-Robot Observation of Multiple Moving Targets (CMOMMT) considers two types of robots, observers and targets, in a partially observable 2D environment, in which the task of the observers is to monitor the target robots under a limited radial range of the sensor, minimizing the total time the targets escape observation. The Cooperative Target Observation (CTO), a variant of the CMOMMT problem, considers the environment to be fully observable, where targets cooperate with observers by reporting their locations. These problems are at the center of many problems that occur in real-world wildlife research, crowd social movement, and surveillance situations. Regarding research related to the field of surveillance, this work proposes an approach to the extended CTO problem, what we call the COSTA problem: Cooperative Observation of Smart Target Agents. The approach employs genetic algorithm and recurrent deep neural networks to improve the performance of target robots and, therefore, to improve the research of new decision-making strategies in the context of observer robots. The first results were auspicious, as the average number of target evasion increased, considering the other recent approaches to the problem.

Keywords: Cooperative Target Observation · Genetic algorithm · Deep neural networks

1 Introduction

The Cooperative Multi-Robot Observation of Multiple Moving Targets (CMOMMT) considers two kinds of robots, the observers and the targets, in a 2D field environment partially observably, in which the observers' task is to search and observe the target robots under a radial limited sensor range, minimizing the total time in which targets escape from observation of the team of observers [16]. The Cooperative Target Observation (CTO) considers that the environment is fully observable and that the targets cooperate with the observers informing

R. Cerri and R. C. Prati (Eds.): BRACIS 2020, LNAI 12320, pp. 77–92, 2020.
https://doi.org/10.1007/978-3-030-61380-8_6

their locations. The first approach to CTO compared the Hill-Climbing search and the K-means clustering strategy, to locate the observers such that they can observe the maximum number of moving targets [13]. Recently, we improved the approach based on K-means and introduced the notion of an organization in the team of observers, to model its functional, structural, and behavioral dimensions, which must be present in a rational team of observers [2].

The CMOMMT/CTO problems are the core of many problems in real-world situations. In wildlife research employing visual monitoring of endangered animals carrying GPS locators in ecological reserves, the task is to assign some drones to observe as much as possible animals [19]. As a result of the crowd social movement research, drones are being used for surveillance and monitoring of significant events, protests, and other crowd situations [6]. One or more moving entities need to be continuously maintained under observation or surveillance by other moving entities [8].

In many of the approaches to CMOMMT/CTO problems, the focus was on the rationality of the observer agents' movement. The targets move randomly by the environment or only with basic movements, like the straight-line movement and a controlled randomization movement [3]. In a recent approach, we tried to improve the rationality of the target agents in two manners. We compared two organizational paradigms in the group of target agents, and a quite simple neural network incorporated in each independent target, to predict and avoid the observer movements [17]. We use sophisticated neural network techniques because they considered the behavior of observers and can map it very efficiently. Behavior is not considered in the previous decentralized approaches for targets in literature what is an unrealistic representation. We evaluated four strategies for the target team, i.e., Multilayer Perceptron and Radial Basis Function, Elman, and Jordan networks [7]. These were the first approaches to the problem of Cooperative Observation of Smart Target Agents (COSTA).

To evaluate the proposed approaches in agent-based simulations, we measured the average number of target evasion (ANTE) subtracting the total of targets from the average number of observed targets [7]. The results showed that modeling targets as intelligent agents harm the performance of observers more than previous decentralized approaches in the literature, especially when targets employ strategies based on recurrent networks. However, although these nets obtained the best error performance, in some scenarios, the ANTE metric values obtained by the four basic models were extremely close, and in other scenarios, the model with the best training and testing error values was not the model with the best ANTE value. Recently, we evaluated the recurrent deep neural nets (RDNN) Long-Short Term Memory (LSTM) and Gated Recurrent Unit (GRU) in similar scenarios, and we noticed that the disagreements between the error metric and the ANTE metric persisted. This paper tries to show how to deal with this disagreement problem.

Deep learning is one of the most popular techniques in artificial intelligence. Recurrent deep neural networks like LSTM and GRU, and the Convolutional Neural Networks (CNN) have gained a remarkable success on many real-world problems in recent years [4,9,12]. The performance of these deep neural networks is a consequence of their architecture. In the construct a deep learning model, various components must be set up, including activation functions and optimization methods, i.e., to the selection of some hyperparameters to design and train the deep models. These hyperparameters play a crucial role in the forward and backward processes of model learning, but they are set up in a heuristic way and based only on the network output. For some state-of-the-art deep neural networks, their architectures are manually designed with human expertise and the investigated problems. Then, it is difficult for the researchers who have no extended expertise in the neural network, to explore it to solve their problems. Researchers are working hard to select optimal hyperparameters to solve their problems with deep learning, to attain better performances.

The rationality to improve the target team performance in old approaches to the COSTA problem divided the decision-making system of each target agent in two subsystems. The first one employs a neural network, trained with data collected from the agent-based simulations of a lot of CTO scenarios, to predict the next position of its nearest neighbor observer, based on the current position of the observer and the current position of the observer's five nearest targets. In the second subsystem, to escape, the target agent under observation could move in the opposite direction of its nearest observer. In summary, the COSTA problem we decided to improve the first subsystem of each target. More specifically, during the RDNN design process, employing a simple genetic algorithm oriented in real time by the ANTE metric [7,17] of target team in the environment, we tried to select the hyperparameters for the RDNNs that attained the best performance, considering a CTO scenario simulated to the testing phase of target agents.

The paper is organized in more four sections. Section 2 presents some related works to improve the observer and target agents and on the automatic design of deep neural networks to solve different CTO problems. Section 3 presents the old approach and a new approach to Cooperative Observation of Smart Target Agents. Section 4 presents the experiments elaborated to validate the approaches and their results considering the error metric and ANTE metric. Section 5 presents the main conclusions and the next steps in the context of the COSTA problem.

2 Related Works

The CMOMMT problem considers a team of m robots with 360° of view observation sensors, that are noisy and limited range, in a two-dimensional, bounded, enclosed spatial region, without entrances/exits, where there is a team of n targets [16]. The robot team's objective is to maximize the collective time that each target is being observed by at least one robot during the duration of the mission. The A-CMOMMT algorithm was the first approach to the problem. It combines low-level with high-level multi-robot control. The low-level control is described in terms of force fields emanating from targets. Mechanisms provide the higher-level control for cooperative control and adjustments in the low-level actions. This higher-level control is presented in a formalism called ALLIANCE [15].

In the CTO problems, observers collectively attempt to stay within an "observation range" of as many targets as possible, and the targets wander randomly and are slower than the observers [13]. In the original problem, an observer does not have a global view of all available targets to observe. In some reformulation, the environment is fully observable. The first approach to the problem [13] evaluated three algorithms for controlling the observers: K-means clustering [14], hill-climbing search, and a combination of K-means clustering followed by hill-climbing. The three algorithms were tunable decentralized by adjusting a parameter that dictates how many subsets the observers are divided. All observers within a given subset collectively participate in a separate, concurrent decision-making process. Thus, one unified set yields a centralized algorithm, whereas many small subsets are decentralized.

Recently, the surveillance domain was modeled as a decentralized CTO problem [3], such that each observer takes its decision independently with the help of its local knowledge. The work modified the assumption that the targets' movement is randomized and presented two target strategies. The observers' strategy based on the K-means algorithm was modified. As the observers may themselves be a subject of observation, the work considers randomizing the observer's actions to help to make their target observation strategy less predictable. Although the strategies of straight-line strategy and controlled randomization movement improved the behavior of the targets, they are out of the reality of competitive multi-agent environments.

In a recent approach, we assumed that observers and targets' behavior could be modeled in the context of a formal organization [2,7,17]. The first approach introduced the idea of organizations in the team of observers [2]. We performed a comparison of the organizational approach with the first approach to the prediction of observers' movement. Four strategies for the target team were compared, three involving clustering algorithms and two organizational paradigms, and one involving MLP neural networks to prediction [17]. The results showed that the target team performance increased when employing these new behaviors. Then we intensified our investigations employing other basic neural network models [7].

Recent works propose an automatic architecture design method for Convolutional Neural Networks (CNN) by using genetic algorithms [18]. The method could discover a promising architecture of a CNN on handling image classification tasks. It was validated on widely used benchmark datasets by comparing to the state-of-the-art peer competitors covering eight manually designed CNNs, four semi-automatically designed CNNs, and four automatically designed CNNs. The method achieved the best classification accuracy among manually and automatically designed CNNs.

Researchers are working to find optimal hyperparameters for deep learning networks. The work described in [10] proposes a method based on GA to find the optimal activation function and optimization techniques. The fitness is computed by the performance of the network with the activation function and optimization technique represented in an individual. The model accuracies were 82.59% and 53.04% for the CIFAR-10 and CIFAR-100 datasets, which outperforms the conventional methods.

3 Approaches to Improve the Performance of Target Teams

As in the CMOMMT, the environment in the CTO problem is a continuous 2D non-toroidal rectangular field, obstacle-free, containing N observer agents and M target agents, such that $N < M$. The velocity of target agents is slower than the velocity of observing agent. The observer agents try collectively to move and remain within a "range of observation" of as many target agents as possible. In the task of Cooperative Observation of Smart Target Agents, the next position of each observing agent is computed by a coordinator agent that computes and sends the new destination position to each observer agent every γ time steps, based on the k-means clustering algorithm. Each observer agent moves toward the computed destination and continuously waits until a new destination point is sent. If one observer reaches its destination point in less than γ time steps, then it waits until a new destination is received.

The objective of the observer agents is to maximize the average number of target agents observed (ANOT) during the observation time interval. The COSTA problem is a reformulation of other approaches to the CTO, i.e., those that allow targets to act as rational agents throughout the task, which we are calling the Cooperative Observation of Smart Target Agent (COSTA) problem, i.e., in cases where the target agents have sensors like the observer agents and can move in the same way they do, but trying to escape from their observation ranges. Thus, in the COSTA problem, the objective of the target agents is to maximize the average number of evasions of the target agents (ANTE), that is, $ANTE = N - ANOT$. We seek to improve the performance of the target team by improving the rationality incorporated in the decision-making system of each target agent of the team. The following two subsections show the two approaches we have developed to improve the performance of the target team.

3.1 The Old Approaches

In the old approaches to the COSTA problem [7,17], we divided the decision-making system of each target agent in two principal subsystems, mentioned in the final of the Section I. In These first approaches to the problem, the first subsystem of each target in the team incorporated an artificial neural network trained with data collected from agent-based simulations of CTO scenarios, to predict the next position of its nearest neighbor observer, based on the current position of the observer and the current positions of the observer's five nearest neighbors targets. The second information processing subsystem, based on the prediction of the first subsystem, tries to escape deciding to move in the opposite direction of its nearest observer. Figure 1 illustrates partially the first information processing subsystem incorporated in the decision-making system of all target agents of the old approaches to the COSTA problem.

Figure 1 (a) focuses on the training and testing processes of a manual designed neural network utilizing an error measure in the output layer of that network.

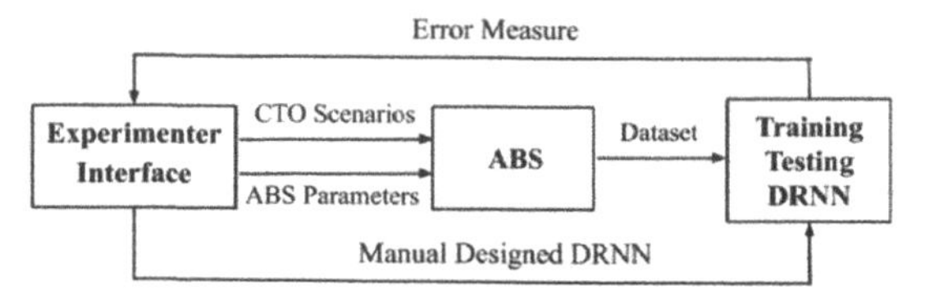

(a) Learning to predict the next position of observers

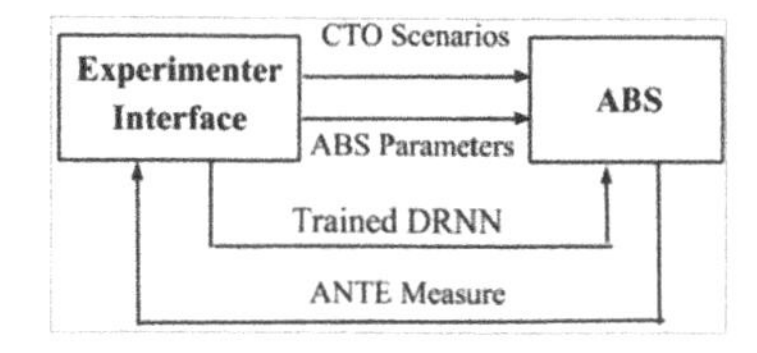

(b) Validating predicted next position of observers

Fig. 1. Designed Information processing subsystems in each target's decision-making system in old approaches.

It considers that there is an interface to elaborate experiments employing techniques of agent-based simulation (ABS), such that it is possible collect relevant datasets containing K observations of the behavior of target and observer agents, $\Psi = \{(x^k, y^k)\}_{k=1...k}$, where x^k and the current positions of the five closest neighboring targets to the observer, and where y^k is a numerical vector defined in R^m employed to represent the next position of the observer. The dataset collected must be normalized and divided into the training and testing sets of the neural network. Figure 2 illustrates the format in which the dataset is continually collected as the ABS is being carried out.

The blue agents are observers, and the red ones are targets. For instance, we see that the observer agent 1 has five target agents pointed out by the arrows as well as the observer agent 2. We first annotate the current coordinate position X_o and Y_o of the observer in question and its orientation, ie, the direction the agent is pointing. We collected the coordinate position of the five closest targets as well as their current speed. The first 13 (thirteen) columns constitute the x^k input numerical vector, whose values must be reported to the neural network input layer. The last two columns constitute the y^k output numerical vector, the next position X and Y of the observer, whose values must be reported to the network's output layer. The table in Fig. 2 shows the input-output information considering two observer agents.

Figure 1(b) focuses on the process of neural network validation, predicting the next positions of online observer agents, new scenarios configured at the experimenter's interface, and using as performance measure the ANTE metric. Thus, in this work, the higher the number of evasions of the target agents, the more efficient will be the strategy adopted by the targets and the higher the team's ANTE values. For each time in the simulation, we fill a number of rows equal to the number of observers, as the data is collected for every observer

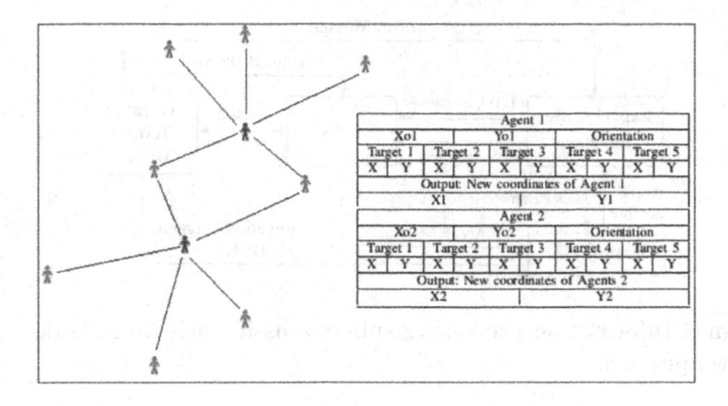

Fig. 2. Data set collection scheme in CTO. (Color figure online)

agent. The table in Fig. 2 shows the required input and its output information formatting the collected data, for instance, when time is 3.000, and the number of observers is 12, we collected approximately 36.000 (thirty-six thousand) rows. Before being read by the networks, the dataset is normalized with mean zero and variance one. Then, we divided using simple cross-validation 80% of the dataset for training and 20% for the test to performance of the models and the overfitting verification.

3.2 The New Approach

Like the old approaches, in this new approach to the COSTA problem, we divide the decision-making system of each target agent into two main information processing subsystems. Considering the first subsystem mentioned in the last subsection, similar to the old approaches, each target in the team incorporates an artificial neural network to predict the next position of its closest observer, to decide later, through its second information processing subsystem, its next position. Figure 3 partially illustrates the information processing subsystem incorporated in all target agents. This new figure integrates the learning process to predict the next positions of the observers with the process of validation of these predicted positions, which we separate respectively in Fig. 1(a) and Fig. 1(b) in the last subsection.

The role of a simple genetic algorithm (GA) is central to the new approach. This GA was specially designed to solve the problem of hyperparameter selection for a DRNN, formulated as: given the collected data set $\Psi = \{(x^k, y^k)\}_{k=1...k}$, the task is to find optimal values for the learning parameters ($w*$), as a consequence of a previous selection of the related hyperparameters that maximizes the ANTE value, the metric of the target team, during the observation time interval, instead of minimizing the error measure between y^k and y_c^k in the DRNN output layer. Each candidate DRNN solution is composed of a finite number of neurons and their connections, generated from the selection of global (structural)

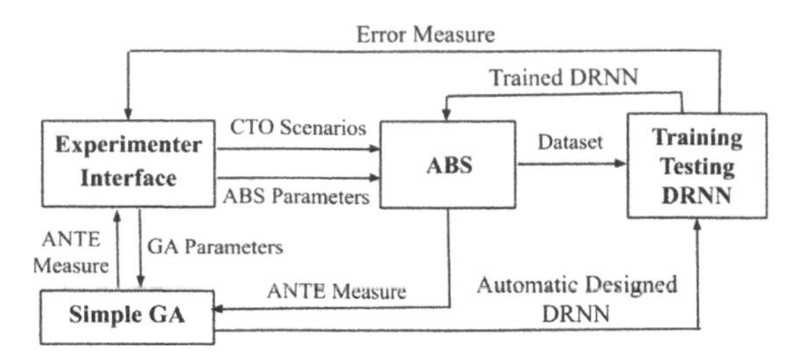

Fig. 3. Designed Information processing subsystems in each target's decision-making system in new approach.

hyperparameters and neuron hyperparameters (learning). Table 1 presents the main hyperparameters considered in the LSTM and GRU recurrent neural networks.

Table 1. Hyperparameters in LSTM-GRU neural networks.

Node Hyperparameters

Hyperparameter	Range
Learning Rate	[0.01, 0.1]
Recurrent Dropout	[0.2, 0.7]
Weight Initialization	[Random Uniform, Glorot Normal]
Use Bias	[True, False]

Global Hyperparameters.

Hyperparameter	Range
Number of Layers	[1, 2, 3, 4 ,5]
Layer Type	[LSTM, GRU]
Layer Size	[16, 32, 64, 128]
Layers Activation	[ReLU, Linear, hard sigmoid]
Layers Dropout	[0, 0.7]
Optimizer	[adam, rmsprop, sgd, nadam]

At the beginning of the search process, the experimenter should encode the values of the hyperparameters as genes on each chromosome in an initial population set. The values of hyperparameters on one chromosome allow the design of a corresponding DRNN. Thus, each DRNN in the current population should be trained and tested with the current data set, collected during the evolution in agent-based simulations of CTO scenarios. Then, each DRNN must be evaluated considering the target team's ANTE metric as a fitness function of the GA, i.e., incorporating the corresponding DRNN into each target agent and performing ABS, considering CTO scenarios and simulation parameters similar and different from those used previously. We use for the evolution environment the highest values scenario of targets velocity and sensor range of observers configured by [13] and [7]. Table 2 presents the parameters used in performing this simulated scenario.

Considering the initial population set and the ANTE value associated to each chromosome in the initial population, to find the best hyperparameters,

Table 2. ABS and CTO parameters.

Parameters	Value
Environment dimension	150 × 150 units
Number of time steps	400
Observer velocity	1 step/time interval
Target velocity	0.9 step/time interval
Sensor-range	25 units
Number of targets	24
Number of observers	12
k-means parameter γ	15

the GA implements an adaptation of the programming technique of generation and testing. More specifically, the simple GA performs a local search in the state space of the hyperparameters selection problem, by executing a cycle composed of two main actions:

1. Generation of a new population: by modifying the current chromosome population, i.e., selecting chromosome pairs oriented by the ANTE value of chromosomes, crossing the selected pairs, and mutating selected chromosomes.
2. Testing the new population generated: checking whether the GA can stop the local search process and return the best DRNN solution found so far, using a Boolean function to check the satisfaction of a stop condition, relating current generation counter, the maximum number of generations and an ideal ANTE value for a DRNN can be considered a solution to the problem.

In case the stop condition is not satisfied, as is common in the case of the initial population, first each DRNN of the new current population must be trained and tested with a new set of data, collected by the various ABSs performed until the current generation in which the GA is found. Subsequently, each DRNN trained in the new population must be evaluated considering the performance of the target team, measuring its ANTE value through the execution of new simulations of scenarios and parameters similar and different from those previously used in the training and testing phases of other neural networks. Considering the current population and the ANTE values associated with the chromosomes, the GA continues the local search in the state space of the hyperparameter selection problem, rerunning the cycle, until stop condition has been satisfied. Table 3 shows some parameters to configure the GA.

As was the case for the ABS-CTO parameter ranges in Table 2, the GA parameter values in the second column of Table 3 are illustrative. The retention length parameter defines an elite subset of surviving individuals equal in size to 15% of the population size parameter value. In addition, there is a random selection set of individuals equal in size to 10% of the chromosomal population. Finally, all chromosomes in the population have a 30% probability of mutation.

Table 3. Parameters in the simple GA.

Parameters	Value
Number of generations	10
Population size	20
Retain length	0.15
Random selection	0.1
Mutate chance	0.3

4 Experiments and Results

The approaches to calculate the next position of a target were evaluated through computational experiments on the NetLogo platform [21], that is, using agent-based simulation to measure the performance of the target team in a virtual environment that can be programmed to represent different complexities of tasks in CTO scenarios. The NetLogo platform was chosen because it allows parallel experiments and easy integration with Python [20] for implementation of GA, and more specifically, the open-source libraries Keras [5] and Tensorflow [1] for the implementation of deep neural networks. The LSTM-GRU hyperparameters and the ABS-CTO-GA parameters were configured according to the values in Tables 1, 2, and 3.

In the old and new approaches, the next position of the observing agents is computed employing the k-means clustering algorithm, where, every time steps, a coordinator observer agent computes and sends the new destination position to each observer agent that moves toward the destination waiting for receiving a new destination. Considering the behavior of the observer agents, the target agents employ their DRNN to predict the observer's next position and thus allow the target to move in the opposite direction. More specifically, this section compares the targets' decision-making systems designed according to the old and new approaches, mainly when incorporating the LSTM and GRU networks.

4.1 Old Approach to Improving the Target Agents Team Performance

This subsection presents the results obtained when the target agents incorporate in their decision-making system the LSTM and GRU neural networks manually designed. Beyond these two networks, we programmed the other four basic models of neural networks to compare the results with that two deep neural networks, i.e., the feed forward networks Multilayer Perceptron (MLP) and Radial Basis Function (RBF), and the recurrent networks Elman and Jordan. The parameters related to each basic model were suitably chosen during the experiments. After successive tests, the best hyperparameters related to the LSTM and GRU networks were chosen: a topology with a single hidden layer with 64 neurons in the form of hard sigmoid activation functions, trained with the Adaptive Moment Estimation (Adam) algorithm [11].

We compared the RMSE (Root Mean Square Error) of the predictions of future observers' positions of the neural networks to better measure forecasts accuracy. This value is computed by the square root of the mean difference between predicted values and the actual position (Euclidean distance) of the observer's X and Y next positions. Table 4 shows the mean RMSE values of the predictions computed in the output layer of all six neural networks. The values vary from 3 to 11. The best mean RMSE was attained by the GRU network, and the RBF network attained the worst. Figure 4 shows the values of the ANTE metric attained by the best neural networks in 5 different CTO scenarios, where the speed of target agents is 0.9. The values of the ANTE metric of the approaches with neural networks were better than the values attained by the adjusted-randomization and straight-line strategies, employed to implement the behaviors of targets in other approaches to the CTO problem. In contrast, the LSTM and GRU networks attained the best performance in almost all scenarios considered.

Table 4. Values of mean RMSE for all neural networks.

GRU	LSTM	Jordan	Elman	MLP	RBF
3.983	4.522	6.768	7.701	9.790	10.291

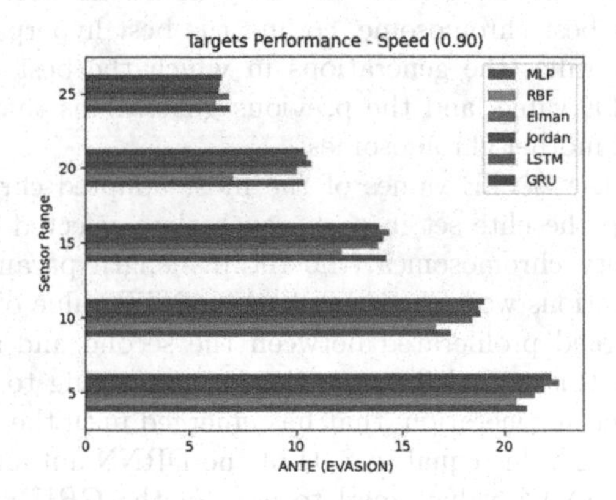

Fig. 4. Values of ANTE metric for all networks in five different CTO scenarios.

In total there are 25 scenarios described in [7], the LSTM network performance was better in seven scenarios, and the GRU network performance, which attained the best mean RMSE value, was better in nine scenarios. Figure 4 presents only the five scenarios with the highest speed (0.9). In the first scenario (sensor range 25) in the figure the manually designed LSTM attained an ANTE value equal to 6.42, and the manually GRU equal to 6.51. However, two basic neural network models were better than the DRNN models in this scenario, i.e., the Elman network attained a value equal to 6.87 and the MLP equal to 6.90. At the lowest speeds in the remaining scenarios, the performances of the two DRNNs prevail and were better than the performance of the basic models, reflecting their lower RMSE values in Table 4 again.

In addition, the performance of the Jordan network was better than the performance of the MLP, RBF, and Elman networks, reflecting their terrible RMSE values in Table 4. The performance of the Elman network oscillated, but in most simulations, it was better than the performance of MLP and RBF networks. Curiously, at lower speeds, the value of ANTE of the RBF network outperformed the value of the MLP network, despite the value of RMSE of MLP being smaller than the value of RBF. However, the MLP network was better in only one scenario and the RBF network in two scenarios.

4.2 New Approach to Improving the Target Agents Team Performance

This subsection presents the results obtained when the target agents incorporate the LSTM and GRU networks selected by a simple GA, which carries out a process of searching the state space of the values of the hyperparameters associated with these two DRNs, oriented by a fitness function defined by the ANTE metric. The best hyperparameters of DRNNs were selected after ten executions (generations) of the implicit generate-and-test cycle in the GA. Figure 4 shows the evolution of the ANTE measure as the generations were performed until the selection of the best chromosome, coding the best hyperparameters for the DRNN. Table 5 identifies the generations in which the best chromosome was generated, its ANTE value, and the previous generations that were generated from its father and mother chromosomes.

Figure 5 shows the ANTE values of the most adapted chromosomes, those three that make up the elite set in each generation, selected from the current population of twenty chromosomes (the retains-length parameter is equal to 0.15). As the generations were carried out, the ANTE value of the best DRNN automatically designed proliferated between the second and sixth generation, in such a way that this growth became slower, converging to an optimal local solution after the tenth generation, that has emerged until the ninth generation. In Table 5, the ANTE value equal to 8.04 of the DRNN automatically designed is better than the ANTE value equal to 6.51 of the GRU network manually designed, described in the last subsection. Table 6 shows the hyperparameter codified in the chromosome selected to generate the best DRNN.

Table 5. ANTE value and main generations related to the chromosome solution birth.

Father Birth	Mother Birth	Birth	ANTE
7	8	9	8.0425

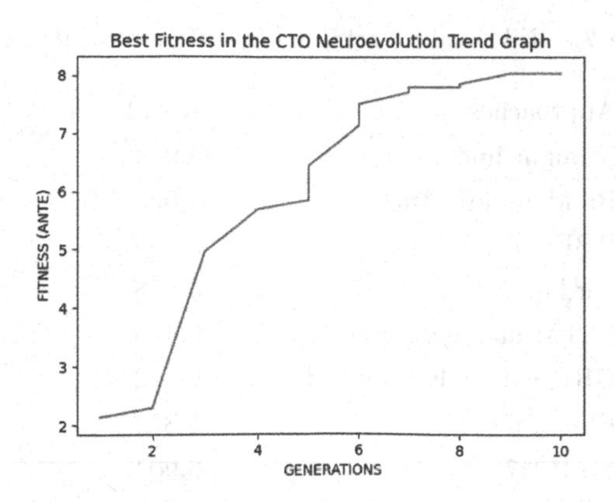

Fig. 5. Values of ANTE measures through 10 generations.

Table 6. Hyperparameters to generate the best DRNN.

Node Hyperparameters

Hyperparameter	**Range**
Learning Rate	0.02
Recurrent Dropout	0.6
Weight Initialization	Glorot Normal
Use Bias	True

Global Hyperparameters.

Hyperparameter	**Range**
Number of Layers	5
Layer Type	LSTM
Layer Size	[64, 32, 128, 16, 16, 64]
Layers Activation	hard sigmoid
Layers Dropout	0.1
Optimizer	Rmsprop

The hyperparameters selected gave rise to LSTM network with five layers of neurons, each one as a hard sigmoid as activation function, trained with the rmsprop optimizer algorithm. These hyperparameters were selected still in the first generations. The other hyperparameters, like dropout adjustments and the number of neurons in each layer, emerged almost in the final of the search process performed by the GA. Table 7 summarizes the ANTE value attained by the different approaches in the specific scenario of CTO with target speed 0.9 and observer sensor range of 25. The approaches that do not use neural networks were the worst ones. The RBF network was the worst neural networks. The Jordan network attained a low value compared to the LSTM and GRU networks manually designed. The MLP and Elman networks achieved the best performance in this particular scenario, although the DRNNs were generally best in the most scenarios. The automatically projected DRNN surpassed all other approaches, showing the adaptation of the target agents to the scenario.

Table 7. ANTE values attained by different approaches.

Approaches	ANTE values
Straight line [3]	3.983
Random adjustment [3]	4.522
RBF [7]	6.250
Jordan [7]	6.358
LSTM manually designed	6.416
GRU manually designed	6.512
Elman [7]	6.872
MLP [17]	6.904
DRNN automatically designed	8.043

5 Conclusions

This work examined an evolutionary approach to the extended CTO problem, what we call the COSTA problem, i.e., an initiative to improve the performance measurement of the target agent team at the CTO and to improve the research of new decision-making strategies in the context of observer agents. The results of the evolutionary approach have proved to be better than other simple approaches and other approaches that employ hand-drawn neural networks.

Although there are similar approaches in the literature to the problem of the automatic design of deep neural networks, there is no approach of the same kind applied to a domain in the context of CTO, much less in the context of the extended problem we named COSTA. The evolutive approach, specially designed to the problem, always converges for the most suitable solution, and the search process does not run the risk of getting stuck on a local minimum. The diversity caused by chromosome crossovers and mutations in populations culminates in the selection of ideal hyperparameters for an RDNN to be incorporated into the target agents since this RDNN obtained the best ANTE values in CTO scenario.

Although the evolutionary approach has performed well, the values of the ANTE metric have not increased significantly due to the intrinsic limitations of approaches that adopt a decentralized decision-making scheme, in which targets are independent and do not share their information, since each agent decides on a new position, that theoretically allows him to escape from his closest observer, without taking into account the future decisions (or their causes) of other target agents in the team.

Another limitation was the long training time that the DRNNs take, making the simulations quite costly. In the future we intend to work with scenarios with obstacles, to carry out the evolution of the weights of the DRNNs as well and to make comparisons with other approaches of the real time learning such as Multi-agent Reinforcement Learning.

References

1. Abadi, M., et al.: Tensor-flow: Large-scale machine learning on heterogeneous systems (2016)
2. Andrade, J.P., Oliveira, R., da Silva, T.F., Maia, J.E.B., de Campos, G.A.: Organization/fuzzy approach to the CTO problem. In: 2018 7th Brazilian Conference on Intelligent Systems (BRACIS), pp. 444–449. IEEE (2018)
3. Aswani, R., Munnangi, S.K., Paruchuri, P.: Improving surveillance using cooperative target observation. In: Thirty-First AAAI Conference on Artificial Intelligence (2017)
4. Cho, K., et al.: Learning phrase representations using RNN encoder-decoder for statistical machine translation. arXiv preprint arXiv:1406.1078 (2014)
5. Chollet, F., et al.: Keras (2015). https://keras.io
6. Cognetti, M., De Simone, D., Lanari, L., Oriolo, G.: Real-time planning and execution of evasive motions for a humanoid robot. In: 2016 IEEE International Conference on Robotics and Automation (ICRA), pp. 4200–4206. IEEE (2016)
7. Costa, L., Araújo, M., Silva, T., Junior, R., Andrade, J., Campos, G.: Comparative study of neural networks techniques in the context of cooperative observations. In: Anais do XVI Encontro Nacional de Inteligência Artificial e Computacional, pp. 563–574. SBC (2019)
8. Freund, E., Schluse, M., Rossmann, J.: Dynamic collision avoidance for redundant multi-robot systems. In: Proceedings 2001 IEEE/RSJ International Conference on Intelligent Robots and Systems. Expanding the Societal Role of Robotics in the Next Millennium (Cat. No. 01CH37180), vol. 3, pp. 1201–1206. IEEE (2001)
9. Hochreiter, S., Schmidhuber, J.: Long short-term memory. Neural Comput. **9**(8), 1735–1780 (1997)
10. Kim, J.Y., Cho, S.B.: Evolutionary optimization of hyperparameters in deep learning models. In: 2019 IEEE Congress on Evolutionary Computation (CEC), pp. 831–837. IEEE (2019)
11. Kingma, D., Ba, J.: Adam: a method for stochastic optimization. In: 3rd International Conference on Learning Representations, San Diego, USA, arXiv preprint arXiv:1412.6980 (2015)
12. LeCun, Y., Bottou, L., Bengio, Y., Haffner, P.: Gradient-based learning applied to document recognition. Proc. IEEE **86**(11), 2278–2324 (1998)
13. Luke, S., Sullivan, K., Panait, L., Balan, G.: Tunably decentralized algorithms for cooperative target observation. In: Proceedings of the Fourth International Joint Conference on Autonomous Agents and Multiagent Systems, pp. 911–917 (2005)
14. McQueen, J.B., Hall, D.: Some methods of classification and analysis in multivariate observations. In: Proceedings of Fifth Barkeley Symposium on Mathematical and Probability, pp. 281–297 (1967)
15. Parker, L.E.: Alliance: an architecture for fault tolerant, cooperative control of heterogeneous mobile robots. In: Proceedings of IEEE/RSJ International Conference on Intelligent Robots and Systems (IROS 1994), vol. 2, pp. 776–783. IEEE (1994)
16. Parker, L.E.: Cooperative robotics for multi-target observation. Intell. Autom. Soft Comput. **5**(1), 5–19 (1999)
17. França da Silva, T., et al.: Smart targets to avoid observation in CTO problem. In: Proceedings of the 18th International Conference on Autonomous Agents and MultiAgent Systems, pp. 1958–1960 (2019)
18. Sun, Y., Xue, B., Zhang, M., Yen, G.G., Lv, J.: Automatically designing CNN architectures using the genetic algorithm for image classification. IEEE Trans. Cybern. (2020)

19. Tseng, H., Asgari, J., Hrovat, D., Van Der Jagt, P., Cherry, A., Neads, S.: Evasive manoeuvres with a steering robot. Veh. Syst. Dyn. **43**(3), 199–216 (2005)
20. Van Rossum, G., Drake, F.: Python 3 reference manual (2009)
21. Wilensky, U., Stroup, W.: Netlogo (1999)

Finding Feasible Policies for Extreme Risk-Averse Agents in Probabilistic Planning

Milton Condori Fernandez[1(✉)], Leliane N. de Barros[1(✉)], Denis Mauá[1(✉)], Karina V. Delgado[2(✉)], and Valdinei Freire[2(✉)]

[1] IME – Universidade de São Paulo, São Paulo, Brazil
{miltoncf,leliane}@ime.usp.br, denis.maua@gmail.com
[2] EACH – Universidade de São Paulo, São Paulo, Brazil
{kvd,valdinei.freire}@usp.br

Abstract. An important and often neglected aspect in probabilistic planning is how to account for different attitudes towards risk in the process. In goal-driven problems, modeled as Shortest Stochastic Path (SSP) problems, risk arises from the uncertainties on future events and how they can lead to goal states. An SSP agent that minimizes the expected accumulated cost is considered a risk-neutral agent, while with a different optimization criterion it could choose between two extreme attitudes: risk-aversion or risk-prone. In this work we consider a Risk Sensitive SSP (called RS-SSP) that uses an expected exponential utility parameterized by the risk factor λ that is used to define the agent's risk attitude. Moreover, a λ-value is feasible if it admits a policy with finite expected cost. There are several algorithms capable of determining an optimal policy for RS-SSPs when we fix a feasible value for λ. However, so far, there has been only one approach to find an extreme λ feasible i.e., an extreme risk-averse policy. In this work we propose and compare new approaches to finding the extreme feasible λ value for a given RS-SSP, and to return the corresponding extreme risk-averse policy. Experiments on three benchmark domains show that our proposals outperform previous approach, allowing the solution of larger problems.

Keywords: Probabilistic planning · Risk sensitive MDP · Dual linear programming

1 Introduction

Automated planning is the branch of artificial intelligence concerned with sequential decision making problem i.e. how to find a policy (a mapping from states to actions) that takes the agent to its best behaviour. Probabilistic planning deals with sequential decision making in stochastic environments and is usually modeled as a Markovian Decision Process (MDP), representing the interaction between an agent and its environment: at each time step, the agent is at

R. Cerri and R. C. Prati (Eds.): BRACIS 2020, LNAI 12320, pp. 93–107, 2020.
https://doi.org/10.1007/978-3-030-61380-8_7

state s, executes an action a, with a cost, that takes the agent (with a known probability) to a next state s'. The agent objective is to find an optimal policy that satisfies a given optimization criteria (e.g. minimizes the expected accumulated cost). An MDP agent that must achieve a goal state can be modeled as a Stochastic Shortest Path (SSP), which makes the assumption of no dead-end states, i.e. the agent of an SSP will eventually achieve its goal, despite the number of interactions.

An important aspect in probabilistic planning is how to consider the risk in the process. In SSPs risk arises from the uncertainties on future events and how they can lead to goal states. An agent that minimizes the expected accumulated cost is considered a *risk-neutral agent*, while with a different optimization criterion an agent could choose between two attitudes: risk-aversion or risk-prone [1,2]. There are different approaches to quantify risk in MDPs and SSPs, e.g.: (i) the use of an expected exponential utility with a given risk factor [1,3–7]; (ii) the use of a piece-wise linear transformation function with a discount factor [8]; (iii) weighted sum between expectation and variance [9,10]; and (iv) the estimation of performance in a confidence interval [11–14]. However, due to the complexity of the mentioned risk approaches, finding optimal policies for them are computationally more costly than solving risk-neutral SSPs [15].

The exponential utility approach to model risk for SSP problems with no dead-ends has been used in many works [7,16,17]. An RS-SSP (risk sensitive SSP) problem uses the expected exponential of accumulated cost, weighted with a risk factor λ. Within this approach, we can define agents with risk-prone or risk-averse attitudes, i.e.: $\lambda < 0$ implies in a risk-prone agent; $\lambda > 0$ implies in a risk-averse agent and $\lambda \rightarrow 0$ corresponds to a risk-neutral attitude.

There are several algorithms capable of finding optimal policies for RS-SSPs considering a fix value for the risk factor λ. However, when working with averse risk attitude ($\lambda > 0$), there is not always a feasible optimal policy for every λ and therefore there is an interest to define the extreme risk-averse λ value, for which there is an optimal feasible policy. We call this value as *extreme λ-feasible* for risk-averse attitude (from now on, the term *extreme λ-feasible* refers to extreme risk-averse λ-feasible). A sequential search algorithm using Policy Iteration (PI) [18] that finds this extreme λ-feasible value was proposed in [19] but it is not scalable. In this work, we propose three new algorithms to find the extreme-averse λ value: two algorithms based on adaptive sequential search, and one algorithm based on binary search. The algorithm we called SEQSEARCH+LRTDP, were able to find the extreme λ-feasible with up 2 orders of magnitude faster than the previous state-of-the-art solution in two domains, River and Navigation, and up to 1 order of magnitude in the Triangle Tire World Domain.

2 Risk-Prone and Risk-Averse Policy Illustrative Examples

Consider the River domain [16] where there is a grid with N_x lines and N_y columns with columns 1 and N_y being the riversides; line 1 being a bridge and line

N_x is a waterfall (Fig. 1). The initial state (yellow cell) is on the opposite riverside of the goal state (red cell). The agent may cross the river by: (i) swimming from any point of the riverside, or (ii) walking along the riverside and take the bridge. However, the river flows to a waterfall. We encode the agent actions as: north(↑), south (↓), west (←) and east (→).

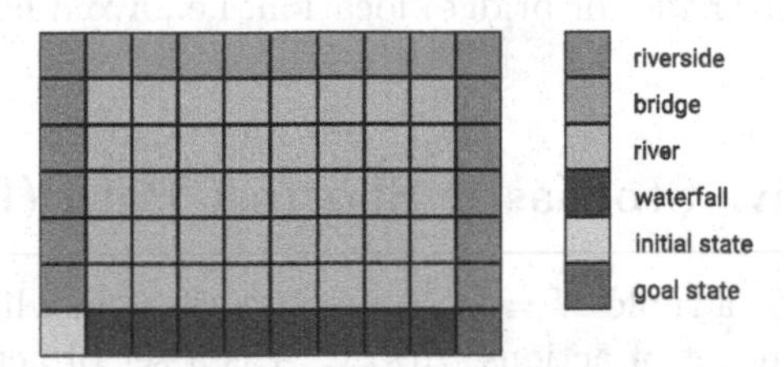

Fig. 1. Instance 7 × 10 of the River domain represented as a grid with riversides (first and last columns), brigde (first row), an initial state (yellow cell) and a goal state (red cell). (Color figure online)

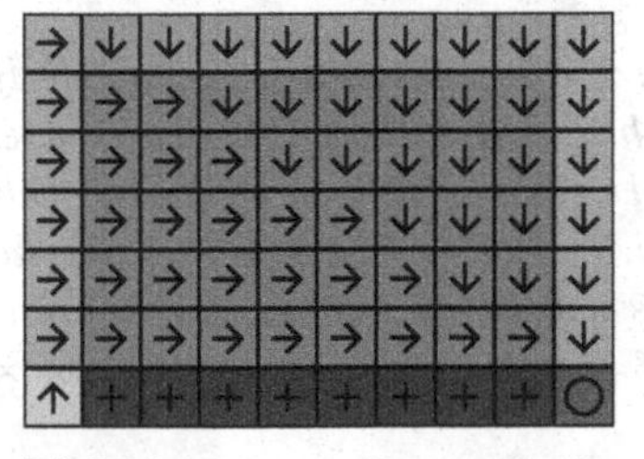

(a) Risk prone policy attitude.

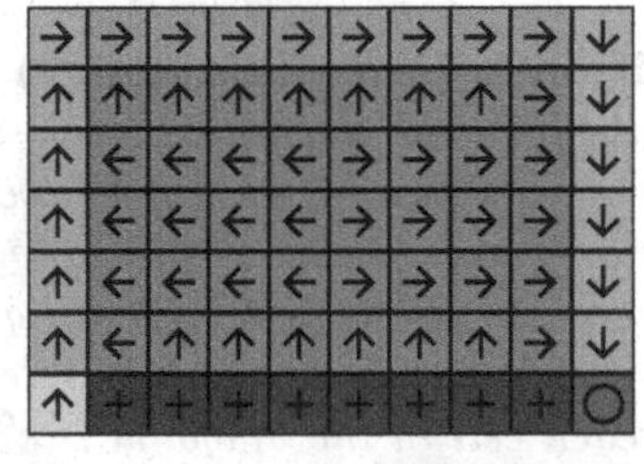

(b) Risk averse policy attitude.

Fig. 2. Illustrative examples of two extreme risk policies for instance 7 × 10 of the River Domain where (a) is the extreme risk-prone policy ($\lambda < 0$), and (b) is the extreme risk-averse policy ($\lambda > 0$).

Actions executed on riversides or the bridge have 99% chance of success and 1% chance of failing (i.e., the agent remains in the same location). Actions executed inside the river have 80% chance of being dragged by the flow (towards the south) and 20% chance of success. We make the assumption that the goal state is an absorbing state and if the agent falls into the waterfall, it returns to the initial state (so we do not consider dead-end states). Figure 2 depicts two policies for the River domain with clear predicted behaviour in terms of risk attitude: **Risk-prone agent** in the River domain always goes towards the goal location as fast as possible no matter the risk. If the agent is in a west riverside cell he would jump into the river towards the goal; and if he is in the east riverside he goes towards the goal along the riverside. When the agent is inside the river he would either tries to go toward east or south (take the advantage of the flow) i.e., he seeks for the goal even under the risk of falling into the waterfall (see Fig. 2a). **Risk-averse agent** in the River domain goes towards the goal walking along the riversides or the bridge. If the agent is in the west (east) riverside, he tries to go towards north (south) always along the riverside; if he is in the bridge he goes to east. If the agent falls inside the river he would always try to achieve the nearest riverside (or bridge) location, i.e., avoiding the waterfall (see Fig. 2b).

3 Risk Sensitive Stochastic Shortest Path (RS-SSP)

Formally, an RS-SSP is a tuple $M = \langle S, A, P, C, s_0, G, \lambda \rangle$ in which S is a finite set of states; A is a finite set of actions, $A(s) \subseteq A$ is a set of actions applicable in state s; $P(s'|s,a)$ is the probability of taking action $a \in A(s)$ in state s ending in state s'; $C(s,a)$ is the cost function that gives the immediate cost value for taking an action $a \in A(s)$ in state s; $s_0 \in S$ is the initial state; G is the set of goal states and λ is the risk-attitude factor. In sum, an RS-SSP is an SSP with an optimization criterion that takes into account a risk factor λ indicating if the agent is averse, neutral or prone to risk. In this work, we consider a solution for an RS-SSP based on an exponential utility function.

Definition 1 (λ-feasible policy). *A policy π of an* RS-SSP *is λ-feasible if the probability of not reaching a goal state vanishes faster than the expected exponential accumulated cost. Let $\mathbf{P}^\pi$ be the matrix representation of the probability transition function P constrained to policy π, i.e., $P(s'|s,\pi(s))$. $\mathbf{P}^\pi_G$ is the matrix $\mathbf{P}^\pi$ where the columns representing the states s belonging to G are assigned to* 0 *(i.e. the goals are absorbing states). Let $\mathbf{D}^\pi$ be the diagonal matrix $|S| \times |S|$ where each cell in the diagonal is $e^{C(s,\pi(s))}$.*

$$if\, \rho((\mathbf{D}^\pi)^\lambda \mathbf{P}^\pi_G) < 1 \; then\, policy\, \pi \; is \; \lambda - feasible,$$

where ρ represents the spectral radius function [7].

We say that a λ is feasible if there exists a λ-feasible policy. If λ is unfeasible, then the cost of any policy is unbounded (it grows to infinity). If $\lambda < 0$ (risk prone attitude) and π is a proper policy (i.e., $\forall s \in S$ the policy reaches the goal with probability 1), then π is also λ-feasible [16]. If π is λ-feasible, then the value function of a policy π can be computed by solving the following system of equations $\forall s \in S$ [16]:

$$V^\pi(s) = \begin{cases} sgn(\lambda) & \text{if } \; s \in G, \\ e^{\lambda C(s,\pi(s))} \sum\limits_{s' \in S} P(s'|s,\pi(s)) V^\pi(s') & \text{otherwise.} \end{cases} \tag{1}$$

Moreover, if λ is feasible then $V^*(s) = \min_{\pi \in \Pi} V^\pi(s)$ is the optimal solution of the following Bellman equation:

$$V^*(s) = \begin{cases} sgn(\lambda) & \text{if } s \in G, \\ \min_{a \in A} \left[e^{\lambda C(s,a)} \sum_{s' \in S} P(s'|s,a) V^*(s') \right] & \text{otherwise.} \end{cases} \quad (2)$$

The optimal policy can then be obtained from the greedy policy:

$$\pi^*(s) = \underset{a \in A}{\operatorname{argmin}} \left[e^{\lambda C(s,a)} \sum_{s' \in S} P(s'|s,a) V^*(s') \right]. \quad (3)$$

Note that "*when λ is unfeasible, the Bellman equation is unsolvable*" [16]. There are many algorithms to solve RS-SSPs for a given fixed feasible λ; in the rest of this section we describe two of the most popular and efficient: Policy Iteration and LRTDP.

3.1 Policy Iteration for RS-SSP

The Policy Iteration Algorithm for RS-SSPs is called RS-PI [7] and consists of two main steps: *policy evaluation* and *policy improvement*. Given a fixed λ value, RS-PI starts with a λ-feasible policy π_0. At each iteration i the algorithm computes the value of that policy solving the system of equations in Eq. 1 (the policy evaluation step) and then it computes a new improved policy based on previous value function (the policy improvement step). This is repeated until reaching a fixed-point, i.e. $\pi_i = \pi_{i-1}$. For the purposes of this paper, we describe the main steps of the Policy Iteration (PI) algorithm for RS-SSPs as a procedure, called POLICY-EVAL-IMPROVE, that includes the policy evaluation of policy π_{i-1} and the policy improvement at iteration i.

---

procedure POLICY-EVAL-IMPROVE(π, λ)
  Policy evaluation: compute V^π using Equation 1 until convergence.
  Policy improvement: find a better policy π' by:

$$\pi'(s) \leftarrow \underset{a \in A}{argmin} \left[\exp(\lambda C(s,a)) \sum_{s' \in S} P(s'|s,a) V^\pi(s') \right]$$

  return π'
end procedure

Since an RS-SSP problem is not well defined for every λ, i.e. we must guarantee there is a λ-feasible policy. For that, we must first find an extreme-averse value that is feasible and choose a λ value less or equal to this extreme.

3.2 LRTDP for RS-SSP

The asynchronous dynamic programming (DP) algorithm *Labeled Real Time Dynamic Programming* (LRTDP) [20] is a state-of-the-art solution for probabilistic planning. It applies asynchronous DP with a heuristic to prioritize the update of relevant states. This approach is also known as a special class of heuristic search, called Find&Revise algorithms [21]. The adaptation of this algorithm for RS-SSPs, called RS-LRTDP (Algorithm 1), simply modifies the original version with an exponential updating procedure called REVISE(s, λ) (lines 12–16) [22].

This algorithm performs trials that simulate the execution of the greedy policy starting from the initial state (lines 5–9). Given a state s, the agent revise the value function $V(s)$ (calling REVISE(s, λ)) and selects a greedy action a (line 6) and computes the successor state s' (line 7) by sampling according to $P(s'|s, a)$. This operation is repeated until a *solved* state is found (i.e. a state for which $V(s)$ has converged) (line 9).

Finally, the algorithm converges when the value function for the initial state s_0 converges (line 11), which guarantees the convergence for all states reachable from s_0. In the experiments performed in this work, we use the h^{rs}_{local} heuristic proposed by [17]. One can use any other heuristic that takes the risk into account, e.g., the heuristic based on occupation measure variables proposed in [22].

When considering the expected exponential utility, a main limitation is the high numerical value resulting from the exponential value function computation. Depending on the parameters of the problem, the exponential values can become so large that they can not be processed using a 64-bit floating-point variable. Even when the computation of intermediate values is possible, the variation of exponents can cause errors of precision. To solve this type of numerical precision problems, in this work, we use the RS-LRTDP with the *LogSumExp* technique [17,23] (by modifying line 13) which transforms the exponential growth of a function into an arithmetic growth through a logarithmic function.

Algorithm 1. RS-LRTDP

Input: RS-SSP M, feasible λ, admissible heuristic function h
Output: optimal value function V^*
1: initialize V with the heuristic function
2: **repeat**
3: $s \leftarrow s_0$
4: label all goal states as solved
5: **repeat**
6: $a_{greedy} \leftarrow$ REVISE(s, λ)
7: $s' \leftarrow$ FIND: sample s' not solved from $P(s'|s, a_{greedy})$
8: $s \leftarrow s'$
9: **until** s is solved
10: **for all** states s in the trial try to label s as solved
11: **until** s_0 is solved
12: **procedure** REVISE(s, λ)
13: $V(s) \leftarrow \underset{a \in A}{min} [\exp(\lambda C(s, a)) \sum_{s' \in S} P(s'|s, a)V(s')]$
14: $a_{greedy} \leftarrow \underset{a \in A}{argmin} [\exp(\lambda C(s, a)) \sum_{s' \in S} P(s'|s, a)V(s')]$
15: **return** a_{greedy}
16: **end procedure**

4 Finding the Extreme Feasible λ

Given an RS-SSP, let us call λ-extreme the extreme risk-averse λ value, for which there is a feasible policy. We can use the algorithms described in the previous section to compute an optimal policy for any $\lambda \leq \lambda$-extreme. However, if such λ is unknown, the algorithms discussed in the next sections are capable of finding this value and also return a extreme λ-feasible policy.

4.1 Sequential Search with Policy Evaluation and Improvement

Freire et al. [16] suggested exploiting the definition of λ-feasible policies (Definition 1) to find the extreme feasible λ by an adaptive sequential search. Algorithm 2, that we call Sequential *Search with Policy Evaluation and Improvement* (SEQSEARCH+PEI), starts with an arbitrary policy π_0 and $\lambda < 0$ (which is always feasible). The algorithm alternates between finding the extreme feasible λ for a fixed policy (line 6), and finding an improved policy (i.e., one with smaller cost) with a fixed λ (line 7). The ADAPTIVE-STEP procedure performs the adaptive sequential search for the extreme λ, that uses the criteria in Definition 1 as stopping rule. The β value in line 12 is an input parameter that specifies the maximum approximation error. The algorithm terminates when the policy improvement step is optimal for the given λ, which is then a β-approximation to the extreme feasible λ, i.e. the λ value that corresponds to the extreme risk-averse attitude. Note that if we remove line 6, Algorithm 2 reduces to the RS-PI algorithm (Sect. 2), for a given fixed λ.

Algorithm 2. SEQSEARCH+PEI : Compute the extreme λ-feasible value

Input: RS-SSP: M, $\epsilon > 0$, $\beta > \epsilon$
Output: optimal policy π^* and extreme λ-feasible

1: Choose an initial policy π_0 arbitrarily and V^{π_0}
2: $i \leftarrow 1$
3: $\lambda \leftarrow -1$
4: $\pi_i(s) \leftarrow \underset{a \in A}{argmin} \, [\exp(\lambda C(s,a)) \sum_{s' \in S} P(s'|s,a) V^{\pi_0}(s')], \forall s \in S$
5: **while** $\pi_i \neq \pi_{i-1}$ **do**
6: $\quad \lambda \leftarrow$ ADAPTIVE-STEP(M, $\pi_i, \epsilon, \lambda, \beta$)
7: $\quad \pi_{i+1} \leftarrow$ POLICY-EVAL-IMPROVE(π_i, λ)
8: $\quad i \leftarrow i + 1$
9: **end while**
10: **return** π_i, λ
11: **procedure** ADAPTIVE-STEP(M, $\pi, \epsilon, \lambda, \beta$)
12: $\quad$ **while** $\rho((\mathbf{D}^\pi)^\lambda \mathbf{P}_G^\pi) \leq (1-\beta)$ **do**
13: $\quad\quad \lambda \leftarrow \lambda + \frac{ln(1-\epsilon) - ln(\rho((\mathbf{D}^\pi)^\lambda \mathbf{P}_G^\pi))}{max_{s \in S, a \in A} \, C(s,a)}$
14: $\quad$ **end while**
15: $\quad$ **return** λ
16: **end procedure**

This algorithm has two large drawbacks. First, the ADAPTIVE-STEP procedure is called for every small improvement of the policy, which requires computing a cost quantity $\rho((\mathbf{D}^{\pi})^{\lambda}\mathbf{P}_{G}^{\pi}))$; this search has also slow convergence as the step sizes become increasingly smaller, which can be very time consuming. Second, POLICY-EVAL-IMPROVE involves a fixed-point algorithm to evaluate the policy in the entire space, which scales poorly for large state spaces.

4.2 Sequential Search with Optimal Policy

As discussed, the SEQSEARCH+PEI alternates between finding the extreme λ value (up to an approximation error), and performing one step of the optimization of the incumbent policy π_i through POLICY-EVAL-IMPROVE. A simple extension to the algorithm is to allow also for several steps of the policy optimization step. In particular, the policy improvement step can be ran until convergence, thus finding an optimal policy for the current feasible λ. This can be done efficiently for example using RS-LRTDP. Algorithm 3 formalizes this idea, where line 7 (POLICY-EVAL-IMPROVE) is replaced with an optimal policy solver (for the current feasible λ). As our experiments show, taking larger steps in the policy optimization leads to less calls of ADAPTIVE-SEARCH, and an overall reduced runtime.

Algorithm 3. SEQSEARCH+OP : Compute the extreme λ-feasible value

Input: RS-SSP M, ϵ, β
Output: optimal policy π^* and extreme λ-feasible

1: Choose an initial policy π_0 arbitrarily
2: $i \leftarrow 1$
3: $\lambda \leftarrow -1$
4: $\pi_i(s) \leftarrow \underset{a \in A}{argmin} \, [\exp(\lambda C(s,a)) \sum_{s' \in S} P(s'|s,a) V^{\pi_0}(s')]$
5: **while** $\pi_i \neq \pi_{i-1}$ **do**
6: $\quad \lambda \leftarrow$ ADAPTIVE-STEP(M, $\pi_i, \epsilon, \lambda, \beta$)
7: $\quad \pi_{i+1} \leftarrow$ FIND-OPTIMAL-POLICY(λ)
8: $\quad i \leftarrow i+1$
9: **end while**
10: **return** π_i, λ

4.3 Checking λ-feasibility by Linear Programming

Another improvement to SEQSEARCH+PEI proposed in this paper is to replace the sequential search for the extreme λ for a feasible policy by a more efficient search scheme, for example, by binary search. Doing so however requires us to establish a criteria for efficiently determining if a given λ is feasible. Note that the criteria in Definition (1) only allows one to decide if a given policy is λ-feasible. Verifying feasibility of λ in this way thus amounts to exhaustive search in the policy space.

We show here the λ-feasibility can be cast as linear programming feasibility, which can then be solved by high performing commercial solvers. Algorithm 4 performs binary search for the extreme λ, where the decision to increase/decrease the current value uses linear programming to verify the feasibility of a candidate λ value (line 6). If the RS-SSP M has a feasible policy we update the bottom search value, otherwise we update the top search value (lines 7–10). The search continues until the interval is sufficiently small (line 2), and contains a feasible λ at the left endpoint (line 3).

Algorithm 4. BINSEARCH+DLP : Compute the extreme λ-feasible value

Input: RS-SSP DLP, error ϵ, values $b < t$ such that the extreme feasible $\lambda \in [b, t]$
Output: extreme λ-feasible

```
1: function BINSEARCH (DLP, b, t)
2:     if (t − b ≤ ε) then
3:         return λ ← b
4:     end if
5:     λ ← (b + t)/2
6:     RESP ← LP-SOLVER(DLP, λ)
7:     if RESP is feasible then
8:         return BINSEARCH (DLP, λ, t)
9:     else
10:        return BINSEARCH (DLP, b, λ)
11:    end if
12: end function
```

In the rest of this section we show how to verify λ-feasibility by recasting the optimization problem as a (primal and dual) linear program of the RS-SSP.

A well-known alternative approach to solving an SSP is to encode the Bellman Eq. 2 as a linear program. Thus, we can adapt the same approach to solve RS-SSPs, producing the linear program of Definition 2.

Definition 2 (Primal Linear Program of RS-SSP). *The primal linear program of an* RS-SSP *M is defined by the following linear optimization problem:*

$$\begin{aligned} \underset{v}{\textit{maximize}} \quad & \sum_{s \in S} v_s \\ \textit{subject to} \quad & v_s = sgn(\lambda) \qquad \forall s \in G \qquad \textit{(LP1)} \\ & v_s \leq e^{\lambda C(s,a)} \sum_{s' \in S} P(s'|s,a) v_{s'}, \quad \forall s \in S \setminus G, a \in A. \end{aligned}$$

The variables v_s in the linear program *LP1* represent the value function $V^*(s)$ in the Bellman Equation (2). At the optimum, the constraints are satisfied with equality, thus computing the same value as the Bellman Equation. If λ is unfeasible, then the above program is unbounded (i.e., any v_s is a solution). According to the weak duality property [24], a feasible linear program is unbounded if and only if its dual linear program is unfeasible (i.e., it has no solutions). Following the works [25–27], note that if $\lambda > 0$, *LP1* can be rewritten as the following linear program:

$$\begin{aligned} \underset{v}{\text{maximize}} \quad & \sum_{s \in S} v_s \\ \text{subject to} \quad & v_s \leq e^{\lambda C(s,a)} P(s_g|s,a) + \sum_{s' \in S \setminus G} e^{\lambda C(s,a)} P(s'|s,a) v_{s'}, \forall s \in S \setminus G, a \in A. \end{aligned}$$

As we discuss next, this allows us to interpret the respective dual program as a maximum flow problem (Definition 3). To the best of our knowledge, this is the first dual formulation of SSPs with an exponential utility function.

Definition 3 (Dual Linear Program of RS-SSP). *The dual linear program of an* RS-SSP *M is the following linear optimization problem:*

$$\begin{aligned} \underset{x}{\textit{minimize}} \quad & sgn(\lambda) \sum_{s \in S, a \in A(s)} x_{s,a} \; e^{\lambda C(s,a)} P(s_g|s,a) && (LP2) \\ \textit{subject to} \quad & x_{s,a} \geq 0 \qquad \forall s \in S, a \in A(s) && (F1) \\ & out(s) = \sum_{a \in A(s)} x_{s,a} \qquad \forall s \in S && (F2) \\ & in(s) = \sum_{\substack{a \in A(s') \\ s' \in S \setminus G}} x_{s',a} \, e^{\lambda C(s',a)} P(s'|s,a), \forall s \in S, a \in A(s) && (F3) \\ & out(s) - in(s) = 0 \qquad \forall s \in S \setminus \{G \cup s_0\} && (F4) \\ & out(s_0) - in(s_0) = 1 && (F5) \end{aligned}$$

The dual linear program (*LP2*) can be interpreted as a maximum flow [28,29] as follows. The variables $x_{s,a}$ are known as *occupation variables*, and measures the proportion of time that action a is applied in state s. If the optimal policy is unique (hence deterministic), these variables take on 0/1 values. Constraint F2 and F3 define expected outgoing and incoming flows entering a state $x_{s,a}$. Constraint F3 modifies the way of the incoming flow: instead of a directed cost, the incoming flow is pondered to the exponential cost, allowing model risk. Constraint F4 establishes a flow conservation principle: all flows reaching s must leave s (for all states $s \in S$ that are neither the initial state s_0, called *source*, nor a goal state, called *sink*). Constraints F5 defines the equation for the source state. Finally, the objective function gives the total cost to reaching the goal from the initial state. An optimal policy can be extracted from the optimal solution to the program by $\pi^* = a$, where a is the only action such that $x^*_{s,a} \neq 0$. This formulation also allows for efficient heuristics that takes into account the probabilities [28,29].

5 Empirical Analysis

In this section we compare the SEQSEARCH+PEI algorithm [16] with the three algorithms proposed in this paper to find the extreme λ-feasible when solving instances of three benchmark planning domains.

We run 10 times the algorithms for each instance of the analysed domains and compute the average time, with a maximum of 60 min and 6 GB of memory in an intel i7 processor at 2.7 Hz. To solve the LPs we used the Gurobi7.5 solver. We apply the *LogSumExp* transformation [17,23] to avoid errors of numerical precision, only for the algorithms based on value function. For the LP based algorithms, there is not yet a way to apply a similar transformation, and we leave that as a future work.

Following, we specify the three benchmark domains analysed in this work and show how they were modified to become RS-SSPs with no dead-ends.

Triangle Tire World Domain. In this domain, a car can move to a different location through routes. The objective is to go from an initial state s_0 to a goal location s_g. However, in each movement, there is a probability of puncturing a tire. Since the domain has no dead-ends, states with flat-tire have always a spare-tire. The probability of puncturing a tire is much higher when the agent moves along the shortest path to the goal.

River Domain. This is the domain described in the Sect. 2. Actions can be taken in any of the cardinal directions: N, S, E and W. If actions are taken on the river bank or in the bridge then transitions are deterministic to the cardinal directions; if actions are taken in the river then transitions are probabilistic and follows the chosen cardinal directions with probability p or follows down the river with probability $1 - p$.

Navigation Domain. In this domain a robot moves through a grid world by executing four actions: N, S, E and W. When the robot reaches some rows in the grid (called *disappearing rows*), there is a probability to return to the initial location (instead of disappearing, as in the original version, which would configure a dead-end state). Each instance of this domain is parameterized by the total number of columns ($ncol$) and rows ($nrow$). The robot starts at location $(1, ncol)$ and the goal location is $(nrow, ncol)$. In the disappearing rows the probability to return to the initial state is min_p in the first column and max_p in the last column. The probability of the robot returning to the initial state in any column $j \in [1, ncol]$ of a disappearing row, is given by: $probability\_return = (min_p) + (j-1)\frac{max_p - min_p}{ncol-1}$.

The average computational time are shown in Tables 1, 2 and 3. Note that "-" indicates instances that were not solved due to numerical precision error. The results show that the algorithms based on the dual linear formulation (BINSEARCH+DLP and SEQSEARCH+DLP) are faster than SEQSEARCH+PEI and SEQSEARCH+LRTDP in small and medium instances. However they can not scale up efficiently due to the problem of numerical precision error.

On the other hand, SEQSEARCH+PEI and SEQSEARCH+LRTDP can use the *LogSumExp* strategy [17,23] which allows them to solve all instances of the

Table 1. Average computational Time (sec) to find the extreme λ-feasible for 40 instances in the **Triangle Tire World Domain**.

		BINSEARCH	SEQSEARCH		
Instances	#states	DLP	PEI	DLP	LRTDP
2	15	0.12	0.07	0.01	0.01
4	66	0.33	0.24	0.07	0.16
6	153	0.79	0.96	0.12	0.54
8	276	1.62	1.94	0.25	0.61
10	435	2.83	4.05	0.32	0.79
12	630	5.10	7.74	0.98	1.26
14	861	7.95	13.62	1.75	2.61
16	1128	12.45	21.64	2.48	3.98
18	1431	17.42	35.42	3.37	6.40
20	1770	25.90	57.64	5.21	7.35
22	2145	31.25	67.13	6.12	8.91
24	2556	-	85.03	-	10.45
26	3003	-	115.94	-	12.95
28	3486	-	134.54	-	15.90
30	4005	-	180.56	-	18.42
32	4560	-	237.28	-	22.03
34	5151	-	289.07	-	28.61
36	5778	-	324.55	-	31.46
38	6441	-	375.86	-	33.45
40	7140	-	434.64	-	40.26

Table 2. Average computational Time (sec) to find the extreme λ-feasible for 18 instances in the **River Domain**.

		BINSEARCH	SEQSEARCH		
Instances	#states	DLP	PEI	DLP	LRTDP
5×6	30	0.33	113.56	0.06	0.09
5×8	40	0.48	159.16	0.16	1.65
5×10	50	0.69	237.60	0.38	2.15
5×12	60	0.82	360.48	0.72	3.75
5×14	70	1.05	488.07	1.35	4.91
5×16	80	1.13	591.05	1.76	6.04
5×18	90	1.29	643.85	2.09	7.49
5×20	100	1.45	780.96	2.98	8.83
5×22	110	2.36	862.35	4.09	9.71
5×24	120	3.78	978.06	5.26	11.48
5×26	130	5.90	1062.45	6.71	15.08
5×28	140	7.94	1196.45	8.49	19.67
5×30	150	-	1325.76	-	28.19
5×32	160	-	1483.65	-	34.94
5×34	170	-	1605.45	-	41.08
5×36	180	-	1763.65	-	47.98
5×38	190	-	1912.65	-	55.19
5×40	200	-	2506.45	-	61.98

Table 3. Average computational Time (sec) to find the extreme λ-feasible for 20 instances in the **Navigation Domain**.

		BINSEARCH	SEQSEARCH		
Instances	#states	DLP	PEI	DLP	LRTDP
6×5	30	0.27	38.45	0.05	0.13
8×5	40	0.34	56.83	0.09	0.34
10×5	50	0.54	76.14	1.13	0.79
12×5	60	0.78	121.36	1.26	1.38
14×5	70	1.17	172.60	1.47	1.90
16×5	80	1.28	214.64	1.65	2.76
18×5	90	1.37	268.48	2.46	3.07
20×5	100	1.59	315.47	3.01	3.58
22×5	110	2.56	375.14	3.45	4.87
24×5	120	3.19	425.86	3.95	5.98
26×5	130	4.89	493.45	4.78	6.04
28×5	140	5.64	519.65	5.21	7.56
30×5	150	6.98	594.82	6.09	8.90
32×5	160	7.95	648.15	6.94	11.73
34×5	170	11.65	705.65	7.14	13.95
36×5	180	-	779.08	-	16.48
38×5	190	-	816.54	-	19.24
40×5	200	-	892.61	-	21.08
42×5	210	-	1022.76	-	23.89
44×5	220	-	1134.02	-	27.46

three analysed domains. The SEQSEARCH+LRTDP were able to find the extreme λ-feasible with up 2 orders of magnitude faster than SEQSEARCH+PEI in two domains, River and Navigation, and up to 1 order of magnitude in the Triangle Tire World Domain, where instances have the largest number of states. Note that, although BINSEARCH+DLP performed in general worst than the sequential search algorithms, it has the lower time performance for all solved instances in the two of the three domains. Our intuition is that it could not solve larger instances due to problems with numerical precision that was mitigated with the *LogSumExp* technique by the other algorithms.

6 Conclusions

In this paper we propose three new algorithms to find the extreme risk-averse λ factor and feasible policies for RS-SSPs, and compare them with an existing solution, called in this paper SEQSEARCH+PEI [16]. This algorithm makes an adaptive sequential search and performs a policy evaluation and improvement at each iteration. Although SEQSEARCH+PEI guarantees to find the extreme λ-feasible for RS-SSPs, it has two main drawbacks. First, the search has slow convergence as the adaptive step sizes become increasingly smaller when approaching the extreme. Second, the algorithm requires running policy evaluation in the entire state space for each λ value, which scales poorly for large search spaces. Thus, to overcome these shortcomings, we propose two extensions. The first extension consists of replacing the policy improvement step by a full policy optimization, either using the RS-LRTDP algorithm or linear programming. The second extension performs binary search to find the extreme λ. This is not trivial as Dynamic Programming cannot be used with an infeasible λ value (consequently, we cannot use it to check the direction to move to in binary search). Thus, we take advantage of a new dual linear programming reformulation of the problem to efficiently verify the feasibility of any given λ. To the best of our knowledge, this is the first dual formulation of stochastic shortest path problems with an exponential utility function.

We compare the approaches in three benchmark planning domains with varying instances size. The results show the algorithms that use the dual linear programming are faster than SEQSEARCH+PEI in small and medium instances. They however face numerical issues as the instances size increase. For these instances, the extension of SEQSEARCH+PEI that uses RS-LRTDP to (optimally) improve policies finding the extreme feasible λ with up 2 orders of magnitude faster. As a future work we intend to implement the *LogSumExp* technique in the dual linear program formulation to increase its scalability.

Acknowledgments. We thank the CNPq (*Conselho Nacional de Desenvolvimento Científico e Tecnológico*) and CAPES (*Coordenação de Aperfeiçoamento de Pessoal de Nível Superior*) for the financial support.

References

1. Marcus, S.: Risk sensitive Markov decision processes. Systems and Control in the 21st Century (1997)
2. Shen, Y., Tobia, M.J., Sommer, T., Obermayer, K.: Risk-sensitive reinforcement learning. Neural Comput. **26**(7), 1298–1328 (2014)
3. Howard, R.A., Matheson, J.E.: Risk-sensitive Markov decision processes. Manag. Sci. **18**(7), 356–369 (1972)
4. Jaquette, S.C.: A utility criterion for Markov decision processes. Manag. Sci. **23**(1), 43–49 (1976)
5. Denardo, E.V., Rothblum, U.G.: Optimal stopping, exponential utility, and linear programming. Math. Program. **16**(1), 228–244 (1979)
6. Rothblum, U.G.: Multiplicative Markov decision chains. Math. Oper. Res. **9**(1), 6–24 (1984)
7. Patek, S.D.: On terminating Markov decision processes with a risk-averse objective function. Automatica **37**(9), 1379–1386 (2001)
8. Mihatsch, O., Neuneier, R.: Risk-sensitive reinforcement learning. Mach. Learn. **49**(2), 267–290 (2002)
9. Sobel, M.J.: The variance of discounted Markov decision processes. J. Appl. Probab. **19**(4), 794–802 (1982)
10. Filar, J.A., Kallenberg, L.C.M., Lee, H.-M.: Variance-penalized Markov decision processes. Math. Oper. Res. **14**(1), 147–161 (1989)

11. Filar, J.A., Krass, D., Ross, K.W., Ross, K.W.: Percentile performance criteria for limiting average Markov decision processes. IEEE Trans. Autom. Control **40**(1), 2–10 (1995)
12. Yu, S.X., Lin, Y., Yan, P.: Optimization models for the first arrival target distribution function in discrete time. J. Math. Anal. Appl. **225**(1), 193–223 (1998)
13. Hou, P., Yeoh, W., Varakantham, P.: Revisiting risk-sensitive MDPs: new algorithms and results. In: International Conference on Automated Planning and Scheduling (ICAPS) (2014)
14. Hou, P., Yeoh, W., Varakantham, P.: Solving risk-sensitive POMDPs with and without cost observations. In: Proceedings of the Thirtieth AAAI Conference on Artificial Intelligence, pp. 3138–3144 (2016)
15. García, J., Fernández, F.: A comprehensive survey on safe reinforcement learning. J. Mach. Learn. Res. **16**, 1437–1480 (2015)
16. Freire , V., Delgado, K.V.: Extreme risk averse policy for goal-directed risk-sensitive Markov decision process. In: Brazilian Conference on Intelligent System (BRACIS), pp. 79–84 (2016)
17. de Freitas, E., Delgado, K., Freire, V.: Risk sensitive probabilistic planning with ILAO* and exponential utility function. In: Anais do XV Encontro Nacional de Inteligência Artificial e Computacional, (Porto Alegre, RS, Brasil), pp. 401–412. SBC (2018)
18. Puterman, M.L.: Markov Decision Processes: Discrete Stochastic Dynamic Programming, 1st edn. Wiley, New York (1994)
19. Freire, V.: The role of discount factor in Risk Sensitive Markov Decision Processes. In: 2016 5th Brazilian Conference on Intelligent Systems (BRACIS), pp. 480–485, October 2016
20. Bonet, B., Geffner, H.: Faster heuristic search algorithms for planning with uncertainty and full feedback. In: International Joint Conference on Artificial Intelligence (IJCAI), pp. 1233–1238 (2003)
21. Bonet, B., Geffner, H.: Planning as heuristic search. Artif. Intell. **129**(1–2), 5–33 (2001)
22. Fernandez, M.C.: Heuristics based on projection occupation measures for probabilistic planning with dead-ends and risk. Master's thesis, USP (2019)
23. William, N., Ross, D., Lu, S.: Non-linear optimization system and method for wire length and delay optimization for an automatic electric circuit placer, no. US **6301693**, B1 (2001)
24. Boyd, S., Vandenberghe, L.: Convex Optimization. Cambridge Press, Cambridge (2004)
25. Altman, E.: Constrained Markov Decision Processes. Chapman & Hall/CRC (1999)
26. d'Epenoux, F.: A probabilistic production and inventory problem. Manag. Sci. **10**(1), 98–108 (1963)
27. Trevizan, F., Teichteil-Königsbuch, F., Thiébaux, S.: Efficient Solutions for Stochastic Shortest Path Problems with Dead Ends, pp. 1–10 (2017)
28. Trevizan, F., Thiébaux, S., Haslum, P.: Occupation measure heuristics for probabilistic planning background: SSPs. In: International Conference on Automated Planning and Scheduling (ICAPS), pp. 306–315 (2017)
29. Fernandez, M.C., de Barros, L.N., Delgado, K.V.: Occupation measure heuristics to solve stochastic shortest path with dead ends. In: 7th Brazilian Conference on Intelligent Systems, BRACIS 2018, São Paulo, Brazil, pp. 522–527 (2018)

On the Performance of Planning Through Backpropagation

Renato Scaroni(✉), Thiago P. Bueno, Leliane N. de Barros, and Denis Mauá

Instituto de Matemática e Estatística, Universidade de São Paulo, São Paulo, Brazil
renato.scaroni@usp.br

Abstract. Planning problems with continuous state and action spaces are difficult to solve with existing planning techniques, specially when the state transition is defined by a high-dimension non-linear dynamics. Recently, a technique called *Planning through Backpropagation* (PtB) was introduced as an efficient and scalable alternative to traditional optimization-based methods for continuous planning problems. PtB leverages modern gradient descent algorithms and highly optimized automatic differentiation libraries to obtain approximate solutions. However, to date there have been no empirical evaluations comparing PtB with Linear-Quadratic (LQ) control problems. In this work, we compare PtB with an optimal algorithm from control theory called LQR, and its iterative version iLQR, when solving linear and non-linear continuous deterministic planning problems. The empirical results suggest that PtB can be an efficient alternative to optimizing non-linear continuous deterministic planning, being much easier to be implemented and stabilized than classical model-predictive control methods.

Keywords: Gradient based optimization · Deep learning · Continuous deterministic planning

1 Introduction

Planning through Backpropagation (PtB) [20,21] has been recently introduced as a scalable alternative to traditional optimization-based methods for deterministic continuous planning problems. Automated planning is a subarea of AI focused on solving a sequential decision making process where an agent has to select the best action to be taken at each decision step in order to achieve a desired goal in a minimum number of steps [9].

PtB exploits a key feature of many planning problems: if all functional dependences in the transition and cost functions are almost-everywhere differentiable then the planning problem can be cast as a gradient-based optimization task and efficiently solved by deep-learning techniques. Although the approach has

Supported by CNPq and FAPESP.

R. Cerri and R. C. Prati (Eds.): BRACIS 2020, LNAI 12320, pp. 108–122, 2020.
https://doi.org/10.1007/978-3-030-61380-8_8

showed impressive results on a suite of planning and control problems, we notice that it has never been properly compared with state-of-the-art optimal control algorithms.

Following the current trend in reinforcement learning of evaluating approaches on benchmarks from the optimal control literature such as Linear-Quadratic Regulation (LQR) problems [8,12], in this work we seek to better characterize PtB in terms of its empirical performance in LQR-like problems. Our goal is to demonstrate the ability of PtB in leveraging the modern gradient-descent optimization tool-sets to finding good approximate solutions, despite the well-known limitations of those descent algorithms. With that objective in mind, we turn ourselves to LQR problems, precisely because their optimal solutions can be easily obtained via dynamic programming.

Moreover, for non-linear dynamical systems, a comparison between a gradient-based optimization solution such as PtB to Differential Dynamic Programming [18] can highlight the advantages of an approach that approximately solve the original non-linear planning problem, relative of optimally solving locally-linearized approximations [15] of the original problem, which is the standard approach of state-of-the-art control algorithms such the iterative Linear-Quadratic Regulator (iLQR) [17].

In this work, we formulate the planning and control problems as Deterministic Continuous Markov Decision Processes (DC-MDPs) that are amenable to be optimized by backpropagating gradients of the total cost with respect to the actions. Our empirical results suggest that PtB is competitive with iLQR, while being much easier to implement and stabilize than classical model-predictive control methods.

The paper is organized as follows. First, we introduce the notation and mathematical background of DC-MDPs and the theory of Linear-Quadratic problems and present the optimal control algorithms of LQR and iLQR. In Sect. 3 we develop the main concepts of the PtB approach. Finally, we discuss the empirical analysis of results.

2 Mathematical Foundation

We start by fixing the notation and reviewing the basics of deterministic planning problems with continuous state and action spaces.

2.1 Deterministic Continuous Markov Decision Process

We consider single-agent planning problems formulated as continuous state and action Markov Decision Process with a finite horizon $H \in \mathbb{N}$. Such processes describe a finite number of iterations of the agent with the environment such that at each decision time step the agent observes the current state s_t, and performs an action a_t. That iteration incurs a cost c_t and takes the agent to a future state s_{t+1}, and the process repeats. We assume actions have deterministic effects, thus s_{t+1} is a deterministic function of s_t and a_t. More formally, we have that:

Definition 1 (DC-MDP). *A Deterministic Continuous Markovian Decision Process (DC-MDP) problem is defined by the tuple $\langle \mathcal{S}, s_0, \mathcal{A}, \mathcal{R}, \mathcal{T} \rangle$, where:*

- $\mathcal{S} \subseteq \mathbb{R}^m$ *is a set of states;*
- $s_0 \in \mathcal{S}$ *is the initial state;*
- $\mathcal{A} \subseteq \mathbb{R}^n$ *is a set of actions;*
- $\mathcal{C} : \mathcal{S} \times \mathcal{A} \rightarrow \mathbb{R}$ *is a cost function; and*
- $\mathcal{T} : \mathcal{S} \times \mathcal{A} \rightarrow \mathcal{S}$ *is a deterministic state transition function.*

We assume that $\mathcal{T}$ and $\mathcal{C}$ are almost-everywhere differentiable.

Definition 2 (DC-MDP plan). *A plan is an ordered sequence of actions, denoted by the sequence $a_{0:H} = (a_0, a_1, ..., a_{H-1}) \in \mathcal{A}^H$, where $\mathcal{A}^H$ is the set of plans of size H to be executed over the time-steps $t \in \{0, 1, ..., H-1\}$.*

To evaluate a plan $a_{0:H}$ we define the value function that assigns a real value to each state while following this plan.

Definition 3 (Value function). *The value function $V_K^{a_{K:H}} : \mathcal{S} \rightarrow \mathbb{R}$, where K is the remaining number of steps, returns the total cost obtained by executing the plan $a_{K:H}$ starting at state $s \in \mathcal{S}$, i.e.:*

$$V_K^{a_{K:H}}(s) = \mathcal{C}(s_K, a_K) + V_{K-1}^{a_{K-1:H}}(s), \qquad V_0(s) = 0. \tag{1}$$

Definition 4 (DC-MDP optimal plan). The optimal plan for a DC-MDP, is the plan $a^*_{0:H} = (a_0, a_1, ..., a_{H-1}) \in \mathcal{A}^H$ that has the minimum accumulated cost through horizon H, i.e.:

$$a^*_{0:H} = \underset{a_{0:H} \in \mathcal{A}^H}{\arg\min} V_H^{a_{0:H}}(s) = \underset{a_{0:H}}{\arg\min} \sum_{t=0}^{H-1} \mathcal{C}(s_t, a_t), \tag{2}$$

where $s_{t+1} = \mathcal{T}(s_t, a_t)$ for $t = 1, \ldots, H-1$.

2.2 Linear Quadratic Regulator (LQR)

The field of optimal control also involves acting in a dynamic system at minimum cost. A simple Linear Quadratic (LQ) control problem assumes deterministic state transition function of the form:

$$s_{t+1} = \mathbf{A}s_t + \mathbf{B}a_t, \tag{3}$$

where $\mathbf{A}$ and $\mathbf{B}$ are real matrices, and the cost function is of the form:

$$\mathcal{C}(s_t, a_t) = s_t^\top \mathbf{Q}\, s_t + a_t^\top \mathbf{R}\, a_t, \qquad \text{if } t < H, \tag{4}$$

$$\mathcal{C}(s_t, a_t) = s_t^\top \mathbf{Q}_H s_t, \qquad \text{if } t = H, \tag{5}$$

where $\mathbf{Q}$ and $\mathbf{R}$ are a semidefinite positive and definite positive real matrices, respectively. The optimal solution for this class of LQ control problems is given by Eq. 2 subject to Eqs. 3, 4 and 5. This LQ problem can be solved by a *Linear-Quadratic Regulator* (LQR) approach [3], an optimal feedback controller based on dynamic programming.

Algorithm 1 shows a pseudo-code of the LQR controller that takes as input the matrices $\mathbf{A}, \mathbf{B}, \mathbf{Q}$ and $\mathbf{R}$ and the initial state s_0. The algorithm works in two phases: **backward pass** and **forward pass**. The **backward pass** starts at the last time-step $H-1$ and solves a set of Riccati[1] equations [4] that obtains the optimal total cost of executing an action a_t and then following the optimal plan in the future. The **forward pass** computes the optimal plan $a^*_{0:H}$: it starts by computing the optimal action a^*_0 in s_0 and generates the next state s_1 applying action a^*_0; the process repeats until it reaches the last time-step $H-1$.

Algorithm 1: Linear-Quadratic Regulator (LQR)

Input: An LQR specification $(\mathbf{A}, \mathbf{B}, \mathbf{Q}, \mathbf{R}, s_0)$
Output: An optimal plan $\mathbf{a}^*_{0:H}$

1 **Backward pass:**
2 $\quad \mathbf{V}_H \leftarrow \mathbf{Q}_H$
3 $\quad$ **for** $t = H-1, \ldots, 0$ **do**
$\qquad$ // Solve discrete algebraic Riccati equations
$\qquad$ // Minimize $s_t^\top \mathbf{Q} s_t + a_t^\top \mathbf{R} a_t + s_{t+1}^\top \mathbf{V}_{t+1} s_{t+1}$
$\qquad$ // subject to $s_{t+1} = \mathbf{A} s_t + \mathbf{B} a_t$
4 $\qquad \mathbf{K}_t \leftarrow (\mathbf{R} + \mathbf{B}^\top \mathbf{V}_{t+1} \mathbf{B})^{-1} \mathbf{B}^\top \mathbf{V}_{t+1} \mathbf{A}$
5 $\qquad \mathbf{V}_t \leftarrow \mathbf{Q} + \mathbf{A}^\top \mathbf{V}_{t+1} \mathbf{A} - (\mathbf{A}^\top \mathbf{V}_{t+1} \mathbf{B})(\mathbf{R} + \mathbf{B}^\top \mathbf{V}_{t+1} \mathbf{B})^{-1} (\mathbf{B}^\top \mathbf{V}_{t+1} \mathbf{A})$
6 **Forward pass:**
7 $\quad$ **for** $t = 0, 1, \ldots, H-1$ **do**
8 $\qquad a^*_t \leftarrow -\mathbf{K}_t s_t$
9 $\qquad s_{t+1} \leftarrow \mathbf{A} s_t + \mathbf{B} a^*_t$
10 **return** $\mathbf{a}^*_{0:H}$

In this work, we use the LQR optimal solution as a benchmark to evaluate linear-quadratic DC-MDPs, and we use the iterative version of LQR (iLQR) to evaluate non-linear DC-MDPs, as described next.

2.3 Iterative Linear-Quadratic Regulator (iLQR)

When we use the LQR controller, we assume the model dynamics is linear. The iLQR (iterative LQR) [11,17,18] algorithm can handle non-linear models by computing a linear approximation of the dynamics and a quadratic approximation of the cost around the current trajectory and then applying LQR (Algorithm 1) to find the optimal solution for this approximation. Then it proceeds

[1] The Riccati equations can be seen as the analytical counterpart of the value iteration for LQRs.

Algorithm 2: Iterative Linear Quadratic Regulator (iLQR)

Input: A DC-MDP specification $(\mathcal{T}, \mathcal{C}, \mathcal{A}, s_0)$
Output: A sub-optimal plan $\hat{\mathbf{a}}_{0:H}$

1. Uniformly sample a valid initial plan $\hat{\mathbf{a}}_{0:H} \sim \mathcal{U}(\mathcal{A}^H)$
2. Generate an initial state trajectory $\hat{s}_{t+1} = \mathcal{T}(\hat{s}_t, \hat{a}_t)$
3. **while** not converged **do**
4.   **Model Approximation:**
5.   Compute linear transition $(\tilde{\mathbf{A}}_t, \tilde{\mathbf{B}}_t) \leftarrow \nabla_{s_t, a_t} \mathcal{T}(s_t, a_t)|_{s_t = \hat{s}_t, a_t = \hat{a}_t}$
6.   Compute quadratic costs $(\tilde{\mathbf{Q}}_t, \tilde{\mathbf{R}}_t) \leftarrow \nabla^2_{s_t, a_t} \mathcal{C}(s_t, a_t)|_{s_t = \hat{s}_t, a_t = \hat{a}_t}$
7.   **Backward pass:**
8.   Find LQR policy $\pi_{\mathbf{K}, \mathbf{k}} \leftarrow$ LQR.BACKWARD$(\tilde{\mathbf{A}}, \tilde{\mathbf{B}}, \tilde{\mathbf{Q}}, \tilde{\mathbf{R}})$
9.   **Forward pass:**
10. $\hat{s}_0 \leftarrow s_0$
11.   **for** $t = 0, 1, \ldots, H-1$ **do**
12.   Select action $a_t \leftarrow \hat{a}_t + \mathbf{k}_t + \mathbf{K}_t(s_t - \hat{s}_t)$
13.   Apply action $s_{t+1} \leftarrow \mathcal{T}(s_t, a_t)$
14.   Update trajectory $(\hat{a}_t, \hat{s}_{t+1}) \leftarrow (a_t, s_{t+1})$
15. **return** $\hat{\mathbf{a}}_{0:H}$

iteratively refining the current best solution until no improvement can be made. Algorithm 2 take as input a DC-MDP and returns an approximate (valid) plan $\hat{\mathbf{a}}_{0:H} \sim \mathcal{U}(\mathcal{A}^H)$. It works in three phases: **model approximation**, **backward pass** and **forward pass**. First, during the model approximation step, the state and action Jacobian matrix of the transition function $\nabla_{s_t, a_t} \mathcal{T}(s_t, a_t)|_{s_t = \hat{s}_t, a_t = \hat{a}_t}$ and the Hessian matrix $\nabla^2_{s_t, a_t} \mathcal{C}(s_t, a_t)|_{s_t = \hat{s}_t, a_t = \hat{a}_t}$ are obtained from the symbolic model through automatic differentiation. Then, the backward pass solves the Riccati equations to obtain the controller gain $\mathbf{K}_t$ and the forward pass apply the controller to the transition function to generate an approximately optimal plan. iLQR is a well-known method for locally-optimal feedback control of non-linear dynamical systems. For further technical details the reader is invited to check the thorough survey in [15].

3 Planning Through Backpropagation (PtB)

Formulating the planning problem as an optimization task can widen the range of techniques amenable to be applied to solve it, e.g., mixed-integer linear programming [16], gradient-based optimization [5,20] or cross-entropy methods [2]. The key observation is that any planning problem can be naturally defined in terms of a state-action constraints given by the state-transition function and initial state, and an objective function given by the total cost.

3.1 Planning as Gradient-Based Optimization of Recurrent Models

The advancements of gradient based optimization techniques, fairly popular on machine learning applications, can lead to more scalable solutions due to the efficiency of the backpropagation algorithm. Leveraging from gradient descent algorithms and automatic differentiation libraries, Scott et al. (2017) [20] propose a framework called *Planning through Backpropagation* (PtB), for solving non-linear continuous planning through backpropagation.

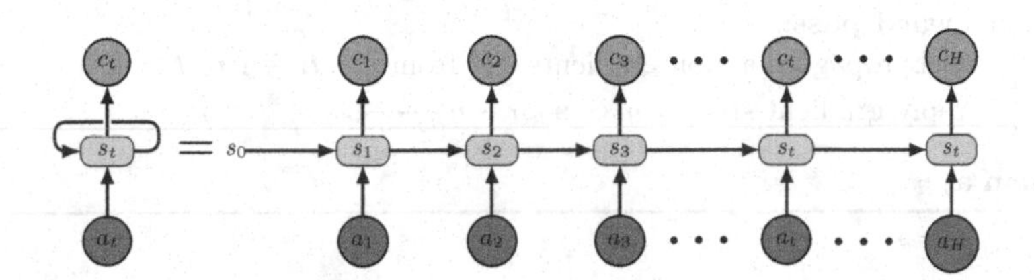

Fig. 1. Planning through backpropagation

A central piece of this framework consists of representing a DC-MDP as a recurrent model in which the inputs are the actions for each time-step, the hidden states are the Markovian states and the outputs are the costs on taking each action. Let a_t be the action taken at time-step, $s_{t+1} = \mathcal{T}(s_t, a_t)$ be the hidden state of the recurrent model and $c_t = \mathcal{C}(s_t, a_t)$ the output of each cell of the recurrent model. Figure 1 shows a representation of this model.[2] In other words, the PtB approach transforms the original *planning as optimization* problem into action-constrained optimization problem over the composition of transition and cost functions given by Definition 5.[3]

Definition 5 (Planning through Backpropagation).

$$\min_{\mathbf{a}_{0:H}} \; \mathcal{C}(s_0, a_0) + \mathcal{C}(\mathcal{T}(s_0, a_0), a_1) + \cdots + \mathcal{C}(\mathcal{T}(\ldots \mathcal{T}(s_0, a_0), a_1), a_2), \ldots a_{H-1})$$

$$\text{s.t. } \forall t: \quad a_t \in \mathcal{A}, \quad s_0 = \bar{s}.$$

The optimization of the plan consists of applying gradient descent on each action according to an objective function. Wu et al. [20] adopt as objective function the surrogate loss $\mathcal{L} = \sum_{t=1}^{H}(0 - c_t)^2 = \sum_{t=1}^{H} c_t^2$. Once the gradient of the loss function with respect to the actions is computed, an optimization step is applied on the current plan $\mathbf{a} = (a_0, a_1, ..., a_H) \in \mathcal{A}^H$ according to the learning rate η. This update rule is given by:

$$\mathbf{a}' = \mathbf{a} - \eta \frac{\partial \mathcal{L}}{\partial \mathbf{a}}, \tag{6}$$

where $\frac{\partial \mathcal{L}}{\partial \mathbf{a}}$ are called the *action gradients.*

[2] Notice that, since the functions represented in the model cells are not parameterized, this model is not exactly a recurrent neural network (RNN), despite its resemblance.

[3] This is akin to a shooting formulation in optimal control, but solved through gradient-based optimization instead of dynamic programming or other methods.

Algorithm 3: Planning through Backpropagation (PtB)

Input: A DC-MDP specification $(\mathcal{T}, \mathcal{C}, \mathcal{A}, s_0)$
Output: A valid plan $\mathbf{a}_{0:H}$

1. Uniformly sample valid initial plan $\mathbf{a}_{0:H} \sim \mathcal{U}(\mathcal{A}^H)$
2. **while** not converged **do**
3.   **Forward pass:**
4.   Symbolically unroll recurrent model $s_{t+1} = \mathcal{T}(s_t, a_t)$, $c_t = \mathcal{C}(s_t, a_t)$
5.   Evaluate surrogate loss $\mathcal{L} = \sum_{t=0}^{H} c_t^2$
6.   **Backward pass:**
7.   Backpropagate action gradients $\frac{\partial \mathcal{L}}{\partial a_t}$ from $t = H - 1$ to $t = 0$
8.   Apply gradient step $\mathbf{a}_{0:H} \leftarrow \mathbf{a}_{0:H} - \eta \frac{\partial}{\partial \mathbf{a}_{0:H}} \mathcal{L}$
9. **return** $\mathbf{a}_{0:H}$

Therefore, applying PtB amounts to: (1) sample an initial plan, and (2) run multiple optimization steps until the plan has converged, i.e. when the action optimization step is no longer relevant due to a local optimum or the maximum optimization epochs are reached. As described in Algorithm 3, PtB optimizes actions by updating them in the direction that minimizes the surrogate loss function $\mathcal{L}$.

3.2 Backpropagation over Long Horizons

Running backpropagation on recurrent models has some particularities once an output for a given time-step depends on the hidden state computed on the previous time-step. This implies that computing the gradient of the objective function is actually computing the gradient of a number of composed functions. This process of computing gradients in recurrent models is called *backpropagation through time*. For better understanding the process of backpropagation through time we are going to look deeper into the gradient $\frac{\partial \mathcal{L}}{\partial \mathbf{a}} = \sum_{t=1}^{H} \frac{\partial c_t^2}{\partial \mathbf{a}}$. We are particularly concerned with the computation of $\partial L_t / \partial \mathbf{a_t}$ where $L_t = \sum_{\tau=t}^{H-1} c_\tau^2$ is the cost-to-go from time-step t:

$$\frac{\partial L_t}{\partial \mathbf{a_t}} = \frac{\partial}{\partial \mathbf{a_t}} \sum_{\tau=t}^{H-1} c_\tau^2 = \frac{\partial}{\partial \mathbf{a_t}} \left(c_t^2 + \sum_{\tau=t+1}^{H-1} c_\tau^2 \right) \tag{7}$$

$$= \frac{\partial c_t^2}{\partial \mathbf{a_t}} + \sum_{\tau=t+1}^{H-1} \frac{\partial c_\tau^2}{\partial \mathbf{a}_t} = \frac{\partial c_t^2}{\partial \mathbf{a_t}} + \sum_{\tau=t+1}^{H-1} \frac{\partial \mathbf{s}_{t+1}}{\partial \mathbf{a}_t} \frac{\partial c_\tau^2}{\partial \mathbf{s}_{t+1}} \tag{8}$$

$$= \frac{\partial c_t^2}{\partial \mathbf{a_t}} + \frac{\partial \mathbf{s}_{t+1}}{\partial \mathbf{a}_t} \sum_{\tau=t+1}^{H-1} \frac{\partial c_\tau^2}{\partial \mathbf{s}_{t+1}} \tag{9}$$

$$= \frac{\partial c_t^2}{\partial \mathbf{a_t}} + \frac{\partial \mathbf{s}_{t+1}}{\partial \mathbf{a}_t} \sum_{\tau=t+1}^{H-1} \left(\prod_{\kappa=t+1}^{\tau-1} \frac{\partial \mathbf{s}_{\kappa+1}}{\partial \mathbf{s}_\kappa} \right) \frac{\partial c_\tau^2}{\partial \mathbf{s}_\tau}. \tag{10}$$

We see from Eq. (10) that longer horizon plans leads to longer products of the transition Jacobians $\partial \mathbf{s}_{\kappa+1}/\partial \mathbf{s}_{\kappa}$ and then are more likely to suffer from the *vanishing gradients* problem [13] if the largest eigenvalue is less than one. Conversely, if the smallest eigenvalue of this Jacobian is greater than one then the opposite problem of *exploding gradients* can happen.

However, an important observation is that in a large class of planning problems (including all the problems used in this work) the transition function has the form $s_{t+1} = \mathcal{T}(s_t, a_t) = s_t + \phi(s_t, a_t)$. This implies that $\frac{\partial \mathbf{s}_{t+1}}{\partial \mathbf{s}_t} = I + \frac{\partial \phi(s_t, a_t)}{\partial s_t}$, where I denotes the identity matrix. Therefore, if $\partial \phi(s_t, a_t)/\partial s_t$ is bounded and not ill-conditioned then the computation of the gradients should not exhibit problems for moderate horizons (i.e., hundreds of time-steps).

3.3 Overcoming Ill-Conditioned Optimization Problems

Solving the planning task as proposed by Definition 5 involves a global optimization of a plan, i.e., all actions in the plan are jointly updated. This can cause the optimization to be ill-conditioned as the norm of the action gradients may be widely distributed. This is intuitive if we consider the relative influence on the total cost of early actions in comparison with later actions.

These problems may be mitigated by replacing the simple gradient descent update with more sophisticated optimizers such as Adam [10] or RMSprop [19]. These optimizers leverage the momentum of the gradients, i.e., they maintain an exponential moving average of gradients in order to approximate 2nd order information of the local optimization problem and thus under certain conditions can accelerate the training. We note that in a PtB approach the use of these modern optimizers from the deep learning literature is essential for the convergence of the planning algorithm. Previous experiments have showed that amongst the optimizers routinely used in practice, RMSProp [19] performs the best.

3.4 Mitigating Local Optima

Figure 2 portraits the objective function $\mathcal{L}$ as the green curve. Assuming that the optimization does not overshoot, it is expected that the initial plan given by point p_2 will converge to point l which is a local optimum. However if the initial plan was p_1, it would ideally converge to point g, which is the objective function global optimum.

This example shows that as any gradient-based optimization technique, PtB may yield to sub-optimal results. To mitigate the sub-optimality problem the recurrent model is implemented to accept a batch of random initial plans and optimize them in parallel, returning the one that yields best results.

4 Experiment Domains

In this section we describe the two domains used in our experiments to compare PtB with LQR and iLQR, in their linear and non-linear versions.

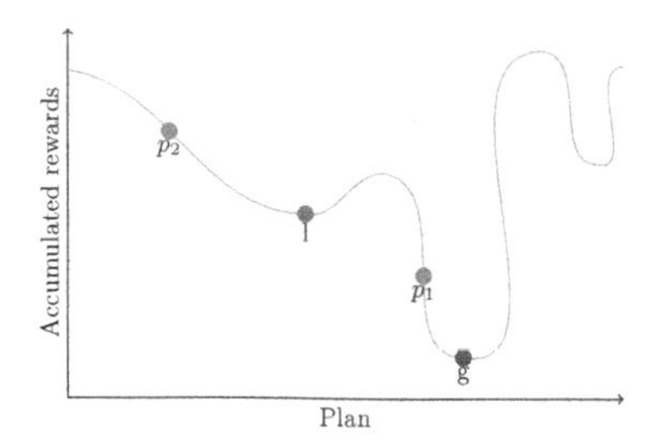

Fig. 2. Mitigating global and local optima.

4.1 Navigation

In the Navigation domain [7] an agent moves at each time-step in a 2-dimensional continuous space trying to reach a goal location as fast as possible, while avoiding deceleration zones. Each deceleration zone j is characterized by its center position z_j. At each time-step t the agent location s_t is changed in the direction of the agent's moving action a_t.

We consider two variations of the Navigation instances: a linear and a non-linear one. The simple linear transition function version has no deceleration zones and is given by:

$$s_{t+1} = s_t + a_t.$$

A non-linear transition function variation considers multiple deceleration zones and the transition function in the time-step t is given by:

$$s_{t+1} = s_t + \lambda_t(s_t)\ a_t,$$

where the *joint deceleration factor* λ_t is computed as:

$$\lambda_t = \prod_j \frac{2}{1 + \exp(-\alpha_j||s_t - z_j||_2)} - 1.$$

For both variations, the cost function is the Euclidean distance from the current agent location to the goal:

$$\mathbf{C}(s_t, a_t) = ||s_t - \mathbf{g}||_2.$$

4.2 Heating, Ventilation and Air-Conditioning

The Heating, Ventilation and Air Conditioning (HVAC) domain involves the problem of temperature regulation of an installation with the objective of minimizing energy consumption [1,6]. Since thermal conductance between walls and convection properties of rooms are nearly impossible to derive from architectural layouts, this is considered a difficult problem, close to real applications, and therefore a challenge for planning and gradient-based optimization solutions. In this work, we adopt the following formulation of the HVAC problem. The state vector represent the temperatures of each room and the action variables correspond to heated air sent to each room via vent actuation. We denote by $s^{(r)}$ and $a^{(r)}$ the r-th element of the state and action vectors, respectively. The objective is to maintain the temperature in a nominal interval defined by lower and upper bounds, $\mathbf{l}$ and $\mathbf{u}$. The transition function is given by:

$$s_{t+1}^{(r)} = s_t^{(r)} + \frac{1}{C^{(r)}} \left(a_t^{(r)} + \sum_{r'} \frac{(s_t^{(r')} - s_t^{(r)})^2}{R^{(r,r')}} \right), \tag{11}$$

where $C^{(r)}$ is the heat capacity of room r, $R^{(r,r')}$ represent the thermal conductance between rooms r and r', and the sum is defined over adjacent rooms.

The cost function penalizes deviation from nominal values while encouraging the controller to keep the temperature as close as possible to the mean point of the nominal temperature interval as well as to reduce the use of heated air in each room:

$$\mathcal{C}(s_t, a_t) = \sum_r [10.0(\max(0.0, l^{(r)} - s_t^{(r)}) + \max(0.0, s_t^{(r)} - a^{(r)}))] \tag{12}$$

$$+0.1 \left| \frac{(\mathbf{l} + \mathbf{u})}{2} - s_t^{(r)} \right| + K a_t^{(r)}. \tag{13}$$

5 Experimental Results

To implement the PtB we used Pytorch [14] which allowed symbolic representation of a sequential plan to be directly optimized via gradient descent leveraging its automatic differentiation feature. The following results were all generated using the RMSProp [20] optimizer. Table 1 shows all hyperparameters used on PtB for each experiment. For the iLQR we show results of 10 runs.

Table 1. PtB hyperparameters.

Domain	Learning-rate	Training epochs	Batch-size
Navigation (linear)	$1E-2$	300	1
Navigation (non-linear)	$1E-2$	300	100
HVAC (non-linear)	$1E-1$	500	20

5.1 Comparing LQR with PtB on a Linear DC-MDP

As mentioned in Sect. 2.2, LQR has been considered an ideal benchmark to analyze the quality of plans generated by the PtB approach for linear-quadratic

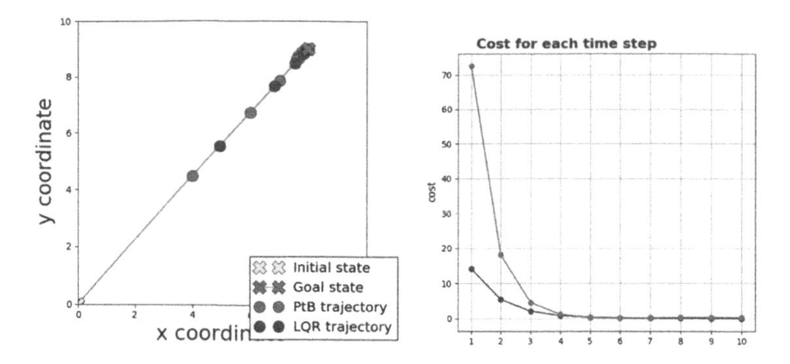

Fig. 3. Results for a linear-quadratic instance (10×10) of the Navigation domain with horizon 10. Trajectory (figure on the left) and cost per time-step (on the right) computed through PtB (green) and LQR (red). (Color figure online)

problems, since it provides a globally optimal solution [8, 12]. For this experiment, we modeled a 10×10 instance of the Navigation domain, with a linear transition function to be solved using LQR and PtB. In each test we consider the initial state (0,0), finishing at goal state (8,9) and the horizon of 10 time-steps.

Figure 3 (left) shows the optimal trajectory computed by LQR in red dots as well as the trajectory computed by PtB in green dots. This result shows that PtB is able to approximate results very close to the optimal.

Figure 3 (right) shows the cumulative cost of each time-step for the optimal plan computed by LQR (red) and for the approximated plan computed by our PtB implementation (green). Note that after the fourth time-step, the costs were very close which corroborates with the intuition given by the trajectories that PtB can return plans with low costs and very close to the optimal for linear-quadratic DC-MDP problems.

5.2 Comparing iLQR with PtB on a Non-linear DC-MDP

As described in Sects. 2.3 and 3, the iLQR and PtB algorithms, respectively, attempts the solution of non-linear continuous problems by different approaches. While iLQR solves the problem by computing the exact solution for several linear-quadratic approximations, PtB aims to solve approximately the exact problem.

This section focus on comparing solutions computed with both iLQR and PtB for two instances of the non-linear versions of HVAC and Navigation domains, described in Sect. 4. Each instance was solved with both PtB and iLQR methods, using the hyperparameters described in Table 1. A summary of those results in terms of total average costs is presented at Table 2, showing they are similar, as expected. Notice that the goal of our evaluation is to analyze the quality of the solutions and not the scalability of the methods in terms of state and action space sizes, as done in [20, 21].

Table 2. Total average costs for non-linear domains.

Domain/instance	PtB	iLQR
Navigation (2 zones, H = 15)	85.81 ± 1.73	83.80 ± 1.31
HVAC (6 rooms, H = 40)	2.332E+06 ± 797.1616	2.356E+06 ± 5.825E+04

Figure 4 portraits the results for Navigation domain defined with two deceleration zones, initial state (0,0), goal state (9,8) and horizon 15. The figure on the left shows the trajectories for PtB in green dots and LQR in red dots. The figure on the right shows the cumulative costs for each time-step, for both solutions. Those results give us the evidence that the performance for both iLQR and PtB are fairly similar, yielding to very similar trajectories and cumulative costs as expected. Since iLQR has been proved to have the convergence of quasi-Newton methods [11], while PtB has no guarantees.

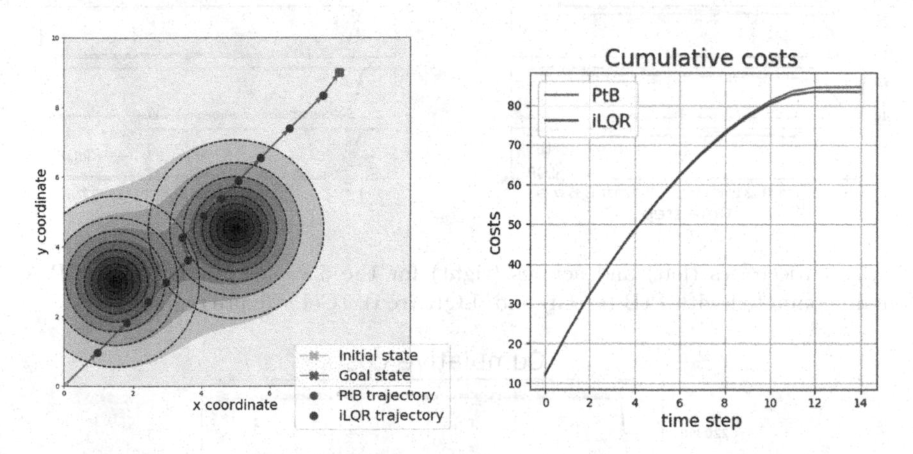

Fig. 4. Results for the non-linear version of the Navigation domain; (left) Visualization of the trajectories generated by PtB (green plot) and iLQR (red plot); (right) Cumulative costs for each time-step PtB (green plot) and iLQR (red plot). (Color figure online)

Figure 5 shows the trajectories and actions for an HVAC instance with 6 rooms, target temperature interval of $[20, 23.5]$ degrees and actions with heat volume up to 10 (volume units). In both figures we have results for PtB in green and for iLQR in red. The figure on the left portraits the temperature of each room over 40 time-steps, and the one on the right shows the actions chosen at each time-step. Note that, after 7 time-steps the temperature of each room seems to be constant, however since there is a small loss of temperature due to the thermal conductance among rooms, the returned plan maintains a small vent volume in all rooms. Notice also that room temperatures of the iLQR plan is closer to the target temperature resulting in less energy consumption.

These two graphics give us an intuition that both solutions are fairly close and that the performance of both approaches are similar, despite the differences between the two them.

Finally, Fig. 6 shows two curves of the cumulative costs of the plans: the one in green contains the results for PtB and the one in red the results for iLQR. This result confirms the similarity of both approaches even for more intricate domains like HVAC.

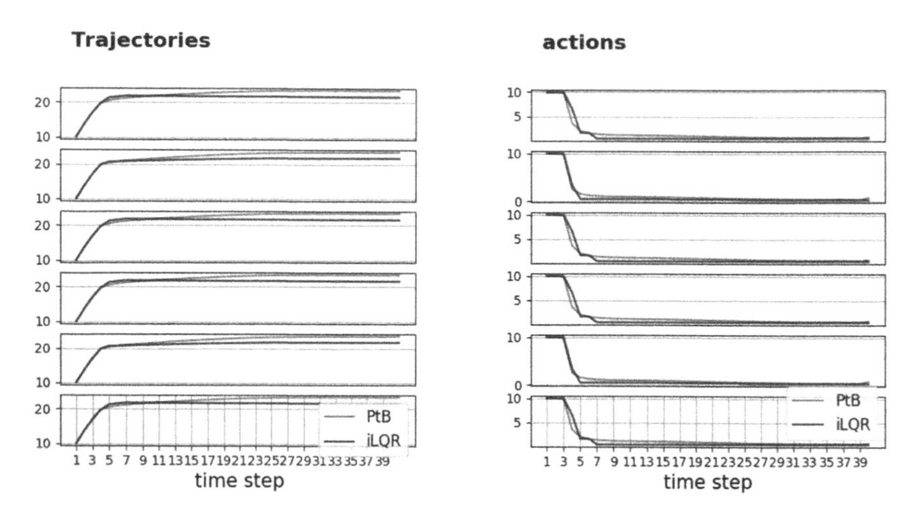

Fig. 5. Trajectories (left) and actions (right) for the 6 rooms for non-linear HVAC domain computed with PtB (green) and iLQR (red). (Color figure online)

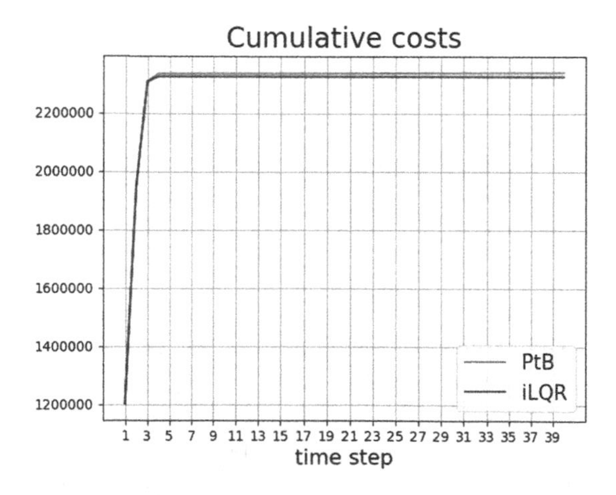

Fig. 6. Cumulative cost at each time-step of plans generated by PtB (green plot) and iLQR (red plot) for the HVAC (non-linear) domain. (Color figure online)

6 Conclusion

In this paper we considered planning problems modeled as Markov Decision Processes with continuous state and action spaces and non-linear deterministic transitions. To date, there has not been any comparison of specialized solvers such as iLQR, commonly used in the Control Theory community, and Gradient Descent Optimizers such as PtB, recently proposed by the AI planning community. To fill this gap partly, in this work we empirically evaluated them in two benchmark domains with non-linear transitions, Navigation and HVAC.

We first compared the two approaches in a linear version of the Navigation domain, where the optimal plan can be computed through LQR. Our results confirm the good performance of PtB, despite its lack of guarantees. We then compared the PtB and iLQR methods in non-linear DC-MDPs. Our empirical results show that both techniques obtain very similar results for both domains. One can understand these results as suggesting that optimally solving a local approximated version of the problem (which is done by iLQR) achieves similar results as to approximately solving the exact version (done by PtB), despite the dissimilarities in the approaches. An advantage of PtB is that, being a gradient-based optimization technique, it can leverage modern automatic differentiation packages such as Pytorch and Tensorflow, which are much easier to be implemented and stabilized than classical model-predictive control methods.

References

1. Agarwal, Y., Balaji, B., Gupta, R., Lyles, J., Wei, M., Weng, T.: Occupancy-driven energy management for smart building automation. In: Proceedings of the 2nd ACM Workshop on Embedded Sensing Systems for Energy-Efficiency in Building, pp. 1–6 (2010)
2. Amos, B., Yarats, D.: The differentiable cross-entropy method. CoRR abs/1909.12830 (2019)
3. Anderson, B.D.O., Moore, J.B.: Optimal Control: Linear Quadratic Methods. Prentice-Hall Inc., Upper Saddle River (1990)
4. Bertsekas, D.P.: Dynamic programming and optimal control, 3rd edn. Athena Scientific (2005)
5. Bueno, T.P., de Barros, L.N., Mauá, D.D., Sanner, S.: Deep reactive policies for planning in stochastic nonlinear domains. In: Proceedings of the Thirty-Third AAAI Conference Artificial Intelligence, pp. 7530–7537 (2019)
6. Erickson, V.L., et al.: Energy efficient building environment control strategies using real-time occupancy measurements. In: Proceedings of the First ACM Workshop on Embedded Sensing Systems for Energy-Efficiency in Buildings, pp. 19–24 (2009)
7. Faulwasser, T., Findeisen, R.: Nonlinear model predictive path following control. Nonlinear Model Predictive Control **384**, 335–343 (2009)
8. Fazel, M., Ge, R., Kakade, S.M., Mesbahi, M.: Global convergence of policy gradient methods for the linear quadratic regulator. In: Proceedings of the 35th International Conference on Machine Learning, pp. 1466–1475 (2018)
9. Ghallab, M., Nau, D., Traverso, P.: pp. i–iv. Cambridge University Press, Cambridge (2016)

10. Kingma, D.P., Ba, J.: Adam: A method for stochastic optimization. arXiv preprint arXiv:1412.6980 (2014)
11. Li, W., Todorov, E.: Iterative linear quadratic regulator design for nonlinear biological movement systems. In: Proceedings of the First International Conference on Informatics in Control, Automation and Robotics, pp. 222–229 (2004)
12. Mania, H., Tu, S., Recht, B.: Certainty equivalence is efficient for linear quadratic control. In: Proceedings of the 33rd Conference on Neural Information Processing Systems, pp. 10154–10164 (2019)
13. Pascanu, R., Mikolov, T., Bengio, Y.: On the difficulty of training recurrent neural networks. In: Proceedings of the 30th International Conference on Machine Learning, pp. 1310–1318 (2013)
14. Paszke, A., et al.: Pytorch: an imperative style, high-performance deep learning library. In: Advances in Neural Information Processing Systems, pp. 8024–8035 (2019)
15. Roulet, V., Drusvyatskiy, D., Srinivasa, S.S., Harchaoui, Z.: Iterative linearized control: stable algorithms and complexity guarantees. In: Proceedings of the 36th International Conference on Machine Learning, pp. 5518–5527 (2019)
16. Say, B., Wu, G., Zhou, Y.Q., Sanner, S.: Nonlinear hybrid planning with deep net learned transition models and mixed-integer linear programming. In: Proceedings of the Twenty-Sixth International Joint Conference on Artificial Intelligence, IJCAI-17, pp. 750–756 (2017)
17. Tassa, Y., Erez, T., Todorov, E.: Synthesis and stabilization of complex behaviors through online trajectory optimization. In: IEEE/RSJ International Conference on Intelligent Robots and Systems, pp. 4906–4913 (2012)
18. Tassa, Y., Mansard, N., Todorov, E.: Control-limited differential dynamic programming. In: IEEE International Conference on Robotics and Automation, pp. 1168–1175 (2014)
19. Tieleman, T., Hinton, G.: Lecture 6.5-RmsProp: Divide the gradient by a running average of its recent magnitude. COURSERA: Neural Netw. Mach. Learn. **4**(2), 26–31 (2012)
20. Wu, G., Say, B., Sanner, S.: Scalable planning with tensorflow for hybrid nonlinear domains. In: Advances in Neural Information Processing Systems 30, pp. 6273–6283 (2017)
21. Wu, G., Say, B., Sanner, S.: Scalable nonlinear planning with deep neural network learned transition models. CoRR abs/1904.02873 (2019)

Risk Sensitive Stochastic Shortest Path and LogSumExp: From Theory to Practice

Elthon Manhas de Freitas[1], Valdinei Freire[2], and Karina Valdivia Delgado[2(✉)]

[1] FIAP, São Paulo, Brazil
profelthon.freitas@fiap.com.br
[2] University of São Paulo, São Paulo, Brazil
{valdinei.freire,kvd}@usp.br

Abstract. Stochastic Shortest Path (SSP) is the most popular framework to model sequential decision-making problems under stochasticity. However, decisions for real problems should consider risk sensitivity to provide robust decisions taking into account bad scenarios. SSPs that deal with risk are called Risk-Sensitive SSPs (RSSSPs), and an interesting framework from a theoretical perspective considers Expected Utility Theory under an exponential utility function. However, from a practical perspective, exponential utility function causes overflow or underflow in computer implementation even in small state spaces. In this paper, we make use of LogSumExp technique to solve RSSSPs under exponential utility in practice within Value Iteration, Policy Iteration, and Linear Programming algorithms. Experiments were performed on a toy problem to show scalability of the proposed algorithms.

1 Introduction

Stochastic Shortest Path (SSP) is a mathematical model for sequential decision making [1,30]. At each time step, an agent observes the state, executes an action, pays a cost, and transits to a next state. The objective is to find a policy that minimizes some criterion. The most common criterion is to minimize the expected cumulative cost. Minimizing expected cumulative cost is especially natural when many repetitions (executions) are required and the results of such repetitions are considered together [6].

However, there are real-life problems that can only be executed once. For example, a vehicle with autonomous navigation has to consider that each route is unique and will not be repeated, so the process cannot simply restart in the case of a failure. Other problems have such a long time span that they cannot be executed multiple times, such as making a trip to Mars or investing in life retirement [26]. These are some examples where mitigating, avoiding and even eliminating environmental risks are much more important than minimizing the expected cumulative cost.

Except for extreme cases, daily life is ruled by risk aversion or prone attitudes. A race driver is willing to push his car a little further on the last lap to improve

R. Cerri and R. C. Prati (Eds.): BRACIS 2020, LNAI 12320, pp. 123–139, 2020.
https://doi.org/10.1007/978-3-030-61380-8_9

his position, just as people are willing to visit a new place in search of new experiences. We must also consider that different people have different levels of accepting risk, even different context can affect and make the person more prone or more averse to risk. In the following, we describe a problem to introduce the intuition about risk.

Problem 1: The River Problem – Adapted from [12]**.** *In the river problem, an agent is on one side of a river and wants to get to the other side. This problem is modeled as a $n \times m$ grid world, where n is the number of rows and m, the number of columns. The agent has two options: to swim across the river, or to walk to the northernmost row where she can find a bridge. When swimming across the river, there exists a probability of 0.8 of not going to the planned direction and flowing down the river, where in the southernmost row the agent falls in a waterfall and returns to the initial state. When walking north, the agent has a probability 0.99 of being successful and a probability 0.01 of staying in the same place due to some unforeseen problem encountered along the way. Note that in this problem the more a person walks to the north and then swims, the lower the risk of falling into the waterfall.*

If the distance to reach the bridge is much greater than the width of the river, it is quite likely that a neutral agent chooses to cross the river to get the lowest average cost. If the river is too wide, this same agent may choose to go to the bridge if the expected cost is lower. An extremely risk-prone agent would choose to cross the river regardless of its width, since it is not concerned with safety. On the other hand, an extremely risk-averse agent would choose to use the bridge because of the safety this route provides.

To deal with this type of problems, there are works in the literature that assesses sensitivity and tolerance to risk and somehow consider these parameters in their models, the so-called Risk Sensitive Markov Decision Processes (RSMDPs). There are several criteria that can be used to deal with risk, some of which are the criterion that uses the expected exponential utility [7,10,15,16,29,33], the weighted sum between mean and variance [9,35] and the performance estimation in a confidence interval [4,8].

Expected exponential utility criterion has some advantages over other RSMDP criteria: (*i*) it is based on a normative decision theory; (*ii*) it is an integral evaluation; (*iii*) optimal solution is stationary; and (*iv*) it is sensitive to symmetrical distributions. However, it presents two major drawbacks; first, the risk averse parameter is bounded and depends on the problem itself; second, because of the use of exponential function, intermediate calculations may overflow or underflow.

The β_{ext}-PI algorithm [10] gives a solution for the first drawback. This algorithm is a policy iteration algorithm for RSMDPs that can be used to compute the extreme value of the risk averse parameter, such that a solution for the problem exists. On the other hand, the second drawback prevents the expected exponential utility criterion from being useful in practice [13] and to the best of our knowledge there are no studies that address this problem.

Here we propose the use of the LogSumExp [27] to avoid overflows and underflows so that RSMDPs with Exponential Utility can be used for large problems, in this paper we focus on Risk Sensitive Stochastic Path problems with Exponential Utility (RSSSP). We show how LogSumExp can be used to practically implementing adapted versions of the traditional MDPs algorithms for RSSSPs such as: Value Iteration, Policy Iteration and Linear Programming.

2 Background

In this section we introduce Shortest Stochastic Path (SSP). We also define the attitudes regarding risk and Risk Sensitive Shortest Stochastic Path with Exponential Utility (RSSSP).

2.1 Shortest Stochastic Path

A Shortest Stochastic Path (SSP) problem is defined as a tuple $\mathcal{SSP} = \langle \mathcal{S}, \mathcal{A}, T, c, \mathcal{G} \rangle$, where $\mathcal{S}$ is a finite state set; $\mathcal{A}$ is a finite action set; $T : \mathcal{S} \times \mathcal{A} \times \mathcal{S} \rightarrow [0, 1]$ is a transition function that represents the probability of visiting a state $s' \in \mathcal{S}$ after the agent executes an action $a \in \mathcal{A}$ in a state $s \in \mathcal{S}$, i.e., $\Pr(s_{t+1} = s' | s_t = s, a_t = a) = T(s, a, s')$; $c : \mathcal{S} \times \mathcal{A} \rightarrow \mathbb{R}^+$ is a non-negative cost function that represents the cost when action $a \in \mathcal{A}$ is applied in a state $s \in \mathcal{S}$; and $\mathcal{G} \subseteq \mathcal{S}$ is a set of absorbing goal states, i.e., $T(s, a, s) = 1$ and $c(s, a) = 0$ for each $a \in \mathcal{A}, s \in \mathcal{G}$.

The SSP problem defines a discrete dynamic process. At any time t, the agent observes a state s_t, executes an action a_t, transits to a state s_{t+1} and pays a cost c_t. The process ends after reaching any goal state in $\mathcal{G}$. The objective is to reach the goal with the minimum expected cumulative cost, which is considered a risk-neutral criterion.

A solution to an SSP is a stationary policy defined by $\pi : S \rightarrow A$. The set of stationary policies is represented by Π. A policy maps the action to be executed in each state at any time t. The execution of a policy and the dynamics of a process define a random variable C^π which stands for total cost for policy π and is defined by:

$$C^\pi = \lim_{M \to \infty} \sum_{t=0}^{M} c_t = \lim_{M \to \infty} \sum_{t=0}^{M} c(s_t, \pi(s_t)). \tag{1}$$

Following the Expected Utility Theory, to define optimal policies, a utility function $u : \mathbb{R}^+ \rightarrow \mathbb{R}$ must be chosen [17]. Then, optimal policies minimize the value $V^\pi(s)$ of a policy π at state s defined by the negative expected utility[1], i.e.:

$$V^\pi(s) = -\mathrm{E}[u(C^\pi) | \pi, s_0 = s]. \tag{2}$$

[1] Usually, Expected Utility Theory considers a value function to be maximized, the positive expected utility. However, because we are considering cost functions, the SSP literature minimizes expected cost, then we chose to follow the SSP literature to avoid any misunderstanding.

A policy π^* is optimal if and only if $V^{\pi^*}(s) \leq V^{\pi}(s)$ for every policy $\pi \in \Pi$ and every state $s \in \mathcal{S}$. An SSP evaluates a policy π by considering the negative identity utility function $u(x) = -x$ and by defining the value function by:

$$V^{\pi}(s) = \mathrm{E}\left[C^{\pi} \middle| \pi, s_0 = s\right] = \lim_{M \to \infty} \mathrm{E}\left[\sum_{t=0}^{M} c(s_t, \pi(s_t)) \middle| \pi, s_0 = s\right]. \tag{3}$$

Definition 1 (Proper policy) [1]. *A policy π is proper if $\lim_{t \to \infty} \Pr(s_t \in \mathcal{G}|\pi) = 1$, i.e., by following π, an absorbing state in $\mathcal{G}$ is reached with probability 1.*

The value function of a policy π is well-defined if there exists at least one proper policy and every improper policy has infinite cost. In this case, the value function can be found by solving the following system of equations:

$$V^{\pi}(s) = \begin{cases} 0 & \text{, if } s \in \mathcal{G} \\ c(s, \pi(s)) + \sum_{s' \in \mathcal{S}} T(s, \pi(s), s') V^{\pi}(s') & \text{, otherwise.} \end{cases} \tag{4}$$

Then the optimal value $V^*(s) = \min_{\pi \in \Pi} V^{\pi}(s)$ is the solution of the Bellman equation:

$$V^*(s) = \begin{cases} 0 & \text{, if } s \in \mathcal{G} \\ \min_{a \in \mathcal{A}} \left[c(s, a) + \sum_{s' \in \mathcal{S}} T(s, a, s') V^*(s') \right] & \text{, otherwise.} \end{cases} \tag{5}$$

An optimal policy can be obtained from the optimal value function by:

$$\pi^*(s) \in \arg\min_{a \in \mathcal{A}} \left[c(s, a) + \sum_{s' \in \mathcal{S}} T(s, a, s') V^*(s') \right]. \tag{6}$$

2.2 Attitudes Regarding Risk

Since C^{π} is a random variable, we may consider three general attitudes regarding risk [17]: neutral, prone and averse. First, we need to define the certainty equivalent of a policy π.

Intuitively certainty equivalent is the amount of cost paying for sure to not play a lottery (making decisions on a SSP), i.e., when the agent has a chance of paying a smaller cost, but he also has a chance of paying a bigger cost. If $V^{\pi}(s) < \infty$ and there exists the inverse function $u^{-1} : \mathbb{R} \to \mathbb{R}^+$, the certainty equivalent $\overline{C}^{\pi}(s)$ of a policy π is defined by:

$$\overline{C}^{\pi}(s) = u^{-1}(-V^{\pi}(s)), \tag{7}$$

and the expected cost $\widetilde{C}^{\pi}(s)$ of a policy π is defined by:

$$\widetilde{C}^{\pi}(s) = \mathrm{E}[C^{\pi}|\pi, s_0 = s].$$

An agent is risk prone if $\overline{C}^{\pi}(s) < \widetilde{C}^{\pi}(s)$, risk averse if $\overline{C}^{\pi}(s) > \widetilde{C}^{\pi}(s)$, and risk neutral if $\overline{C}^{\pi}(s) = \widetilde{C}^{\pi}(s)$ for every state $s \in \mathcal{S}$ and policy $\pi \in \Pi$. Intuitively, if you are risk-prone (optimistic), you would pay less than the expected cost-to-go to not play the lottery because you are focused on better results (low cost) of the lottery (i.e. certainty equivalent is less than the expected cost-to-go). If you are risk-averse (pessimistic), you would pay more than the expected cost-to-go to not play the lottery because you are focused on worse results (high cost) of the lottery (i.e. certainty equivalent is bigger than the expected cost-to-go). Considering these definitions, the SSP that uses the negative identity function as a utility function characterizes a neutral attitude.

2.3 SSP and Risk Sensitive SSP

Formally, a Risk Sensitive SSP [29] is defined by the tuple $\mathcal{RSSSP} = \langle \mathcal{SSP}, \beta \rangle$ where $\mathcal{SSP}$ is a SSP and β is the risk-attitude factor. RSMDPs consider the utility function:

$$u(x) = -\mathrm{sgn}(\beta)\exp(\beta x), \tag{8}$$

and model arbitrary risk attitude by considering a risk-attitude factor β. If $\beta < 0$ the agent considers a risk-prone attitude, if $\beta > 0$ the agent considers a risk-averse attitude and, in the limit, if $\beta \to 0$ the agent considers a risk-neutral attitude.

In RSSSPs, the value function of a policy π is defined by:

$$V^{\pi}(s) = \lim_{M\to\infty} \mathrm{E}\left[\mathrm{sgn}(\beta)\exp\left(\beta \sum_{t=0}^{M} c(s_t, \pi(s_t))\right) \middle| \pi, s_0 = s\right]. \tag{9}$$

Definition 2 (β-feasible policy) [29]. *A policy π is β-feasible if the probability of not being in an absorbing state vanishes faster than the exponential accumulated cost, i.e., the value function $V^{\pi}(s)$ is bounded.*

When $\beta < 0$ (risk prone), any policy π is β-feasible. On the other hand, if a policy π is β-feasible and $\beta > 0$ (risk averse), then the policy π is also proper. If π is β-feasible, then the value function of a policy π can be calculated by solving the following system of equations:

$$V^{\pi}(s) = \begin{cases} \mathrm{sgn}(\beta) & \text{, if } s \in \mathcal{G} \\ \exp\Big(\beta c(s, \pi(s))\Big) \sum_{s' \in \mathcal{S}} T(s, \pi(s), s') V^{\pi}(s') & \text{, otherwise.} \end{cases} \tag{10}$$

If a β-feasible policy exists, then the optimal value function $V^*(s) = \min_{\pi\in\Pi} V^{\pi}(s)$ is the solution of the following equation:

$$V^*(s) = \begin{cases} \mathrm{sgn}(\beta) & \text{, if } s \in \mathcal{G} \\ \min_{a \in \mathcal{A}} \left[\exp(\beta c(s,a)) \sum_{s' \in \mathcal{S}} T(s,a,s') V^*(s') \right] & \text{, otherwise.} \end{cases} \tag{11}$$

An optimal policy can be obtained from the optimal value function by:

$$\pi^*(s) \in \arg\min_{a \in \mathcal{A}} \left[\exp(\beta c(s,a)) \sum_{s' \in \mathcal{S}} T(s,a,s') V^*(s') \right]. \tag{12}$$

3 Algorithms to RSSSPs

Theoretically, algorithms designed for SSPs can be easily adapted for RSSSPs, if we consider the difference between proper policy (that is necessary for SSP) and β-feasible policy (that is necessary for RSSSPs).

3.1 Risk Sensitive Policy Iteration Algorithm

Similar to the Policy Iteration algorithm [30] to solve SSPs, at each iteration i, the Risk Sensitive Policy Iteration algorithm [29] executes two steps: *policy evaluation* and *policy improvement*. The policy evaluation step uses Eq. 10 to compute the value of $V^{\pi_i}(\cdot)$ and the policy improvement step improves π_i obtaining π_{i+1}. If π_0 is a β-feasible policy, then there exits an optimal policy and the Risk Sensitive Policy Iteration algorithm finds an optimal policy π^* [29].

Since the risk averse parameter is bounded and we do not known this extreme value a priori, the β_{ext}-PI algorithm [10] was proposed to solve this problem. β_{ext}-PI is a policy iteration algorithm for RSSSP that begins with an arbitrary risk factor $\beta_0 < 0$ and increases the risk factor β_i iteratively. Given an arbitrary risk factor $\beta > 0$, the β_{ext}-PI algorithm can be used to decide if there is a β-feasible policy, then finding the optimal policy, otherwise calculating an ϵ-extreme β_{ext}, such that there is a β_{ext}-feasible policy.

3.2 The Risk Sensitive Value Iteration Algorithm

In the Risk Sensitive Value Iteration algorithm [29], at each iteration i, the value $V^i(s)$ can be computed based on the value $V^{i-1}(s)$ for each state $s \in \mathcal{S}$, i.e.:

$$V^i(s) = \min_{a \in \mathcal{A}} \left[Q^i(s,a) \right], \tag{13}$$

where:

$$Q^i(s,a) = \exp(\beta c(s,a)) \sum_{s' \in \mathcal{S}} T(s,a,s') V^{i-1}(s'). \tag{14}$$

One possible stopping criterion is to consider the residual $\max_{s \in \mathcal{S}} |V_i(s) - V_{i-1}(s)|$, iterating while the residual is greater than a minimum error ϵ. If there is at least one β-feasible policy, this algorithm finds the solution.

3.3 Risk Sensitive Linear Programming

Another important method to solve SSPs is Linear Programming [1,36,37]. A Linear Program is constructed based on the Bellman equation, and, since RSSSPs also present an equivalent Bellman equation (Eq. 11), it is straightforward to create a Linear Program to RSSSPs. We have the following formulation:

$$\begin{aligned} \underset{V(s)}{\text{maximize}} & \sum_{s\in\mathcal{S}} V(s) \\ \text{s.t. } & V(s) \leq \exp(\beta c(s,a)) \sum_{s'\in\mathcal{S}} T(s,a,s')V(s'), \forall s \subset \mathcal{S}\setminus\mathcal{G}, a \subset A \\ & V(s) = \text{sgn}(\beta), \qquad \forall s \in \mathcal{G}. \end{aligned} \tag{15}$$

4 LogSumExp Strategy for RSSSP

Although risk sensitivity is desired in real-world problems, solutions that implement the exponential utility function in MDPs are not widely used in the literature when compared to other types of utility functions.

One of the main drawbacks is the large numerical value of the value function V^π, which may cause a numeric overflow. The main components responsible for the final value of V^π are: (i) the number of states; (ii) the cost function $c(s,a)$; and (iii) the risk factor β. Remember that the value of a state is the exponential of the sum of cost times the risk factor β (Eq. 9). If $\beta > 0$, the value of a state grows exponentially with the distance to the goal, and if $\beta < 0$ the value of a state decreases exponentially with the distance to the goal. When we implement any of the algorithms (discussed in Sect. 3) in a computer, $\beta > 0$ may cause overflow, while $\beta < 0$ may cause underflow. In this section we propose a solution for this problem.

This type of problem is becoming more common in machine learning algorithms [32]. There are techniques that have been used to deal with this type of problem such as *LogSumExp* [27], *Gordian-L* [34] and *Lp-Norm* [19]. In particular, the *LogSumExp* technique, initially proposed by Naylor, Donelly, and Sha (2001), has been used successfully in several recent works [3,28]. This type of problem also happens in Hidden Markov Models and some techniques used are a scaling technique [31] and the *LogSumExp* technique [24]. We chose the *LogSumExp* strategy because this strategy uses a logarithmic function on an exponential function.

Given two numbers A and B, the *LogSumExp* technique considers the following:

$$\log\Big(\exp(A)+\exp(B)\Big) = \log\left(\exp(A)\left(1+\frac{\exp(B)}{\exp(A)}\right)\right) = A + \log\Big(1+\exp(B-A)\Big).$$

If $A > B$, then exponential function is only applied to negative numbers, avoiding overflow.

4.1 Certainty Equivalent of RSSSPs

Although the value of policies in RSSSPs grows exponentially, its certainty equivalent grows linearly Note that from Eq. 8 we have: $u^{-1}(-y) = \frac{1}{\beta}\log(\text{sgn}(\beta)y)$. Following Eq. 7, let the certainty-equivalent function $L^\pi(s)$ be defined by:

$$L^\pi(s) = \frac{1}{\beta}\log[\text{sgn}(\beta)V^\pi(s)] \tag{16}$$

$$= \begin{cases} 0 & \text{, if } s \in \mathcal{G} \\ \frac{1}{\beta}\log\left[\text{sgn}(\beta)\exp(\beta c(s,\pi(s)))\sum_{s'\in\mathcal{S}} T(s,\pi(s),s')V^\pi(s')\right] & \text{, otherwise.} \end{cases} \tag{17}$$

The system of equations 17 can be solved for all $s \notin \mathcal{G}$ by following:

$$\begin{aligned} L^\pi(s) &= \frac{1}{\beta}\log\left[\exp(\beta c(s,\pi(s)))\sum_{s'\in\mathcal{S}} T(s,\pi(s),s')\text{sgn}(\beta)V^\pi(s')\right] \\ &= c(s,\pi(s)) + \frac{1}{\beta}\log\left[\sum_{s'\in\mathcal{S}} T(s,\pi(s),s')\text{sgn}(\beta)V^\pi(s')\right] \\ &= c(s,\pi(s)) + \frac{1}{\beta}\log\left[\sum_{s'\in\mathcal{S}} \exp(\log[T(s,\pi(s),s')])\exp(\log[\text{sgn}(\beta)V^\pi(s')])\right] \\ &= c(s,\pi(s)) + \frac{1}{\beta}\log\left[\sum_{s'\in\mathcal{S}} \exp\left(\log[T(s,\pi(s),s')] + \beta L^\pi(s')\right)\right]. \end{aligned} \tag{18}$$

Because $u^{-1}(\cdot)$ is a monotonically increasing function, minimizing $L^\pi(s)$ is the same as minimizing $V^\pi(s)$. Therefore, when using the function $L^\pi(s)$ in the Risk Sensitive Value Iteration, Risk Sensitive Policy Iteration and Risk Sensitive Linear Programming algorithms, the policies obtained must be the same as using $V^\pi(s)$ in these algorithms, i.e., if $L^{\pi'}(s) > L^{\pi''}(s)$ then $V^{\pi'}(s) > V^{\pi''}(s)$.

4.2 Risk Sensitive Value Iteration with LogSumExp

The Risk Sensitive Value Iteration with LogSumExp iterates over the function $Q^{i+1}_{\log}(\cdot)$ defined by:

$$Q^{i+1}_{\log}(s,a) = u^{-1}(Q^{i+1}(s,a)) \tag{19}$$

We have that:

$$
\begin{aligned}
Q_{\log}^{i+1}(s,a) &= u^{-1}(Q^{i+1}(s,a)) = \frac{1}{\beta}\log\Big(\operatorname{sgn}(\beta)Q^{i+1}(s,a)\Big)\\
&= \frac{1}{\beta}\log\left(\operatorname{sgn}(\beta)\exp(\beta c(s,a))\sum_{s'\in\mathcal{S}}T(s,a,s')V^i(s')\right)\\
&= c(s,a) + \frac{1}{\beta}\log\left(\operatorname{sgn}(\beta)\sum_{s'\in\mathcal{S}}T(s,a,s')V^i(s')\right)\\
&= c(s,a) + \frac{1}{\beta}\log\left(\sum_{s'\in\mathcal{S}}\exp\left(\log\left(T(s,a,s')\operatorname{sgn}(\beta)V^i(s')\right)\right)\right)\\
&= c(s,a) + \frac{1}{\beta}\log\left(\sum_{s'\in\mathcal{S}}\exp\left(\log\left(T(s,a,s')\right)+\log\left(\operatorname{sgn}(\beta)V^i(s')\right)\right)\right)\\
&= c(s,a) + \frac{1}{\beta}\log\left[\sum_{s'\in\mathcal{S}}\exp\left(\log\left(T\left(s,a,s'\right)\right)+\beta L^i(s')\right)\right].
\end{aligned}
\tag{20}
$$

The *LogSumExp* strategy consists in identifying the largest term of an exponential sum. In this paper, we consider the two auxiliary functions $k_{s,s'}^{a,i}$ and $K_s^{a,i}$, defined by an arbitrary certainty equivalent function $L^i(s)$:

$$
k_{s,s'}^{a,i} = \log\big(T\left(s,a,s'\right)\big) + \beta L^i(s') \tag{21}
$$

$$
K_s^{a,i} = \max_{s'\in\mathcal{S}}\left(k_{s,s'}^{a,i}\right). \tag{22}
$$

Introducing $k_{s,s'}^{a,i}$ and $K_s^{a,i}$ in Eq. 20, we have:

$$
\begin{aligned}
Q_{\log}^{i+1}(s,a) &= c(s,a) + \frac{1}{\beta}\log\left[\sum_{s'\in\mathcal{S}}\exp\left(k_{s,s'}^{a,i}\right)\right]\\
&= c(s,a) + \frac{1}{\beta}\log\left[\sum_{s'\in\mathcal{S}}\exp(k_{s,s'}^{a,i}-K_s^{a,i})\exp(K_s^{a,i})\right]\\
&= c(s,a) + \frac{1}{\beta}\log\left[\exp(K_s^{a,i})\sum_{s'\in\mathcal{S}}\exp(k_{s,s'}^{a,i}-K_s^{a,i})\right]\\
&= c(s,a) + \frac{1}{\beta}K_s^{a,i} + \frac{1}{\beta}\log\left[\sum_{s'\in\mathcal{S}}\exp(k_{s,s'}^{a,i}-K_s^{a,i})\right].
\end{aligned}
\tag{23}
$$

The use of the *LogSumExp* strategy prevents overflow by avoiding the application of the exponential function over large numbers, in fact, exponential function is only applied over negative numbers.

4.3 Risk Sensitive Policy Iteration and Linear Programming with LogSumExp

In the heart of the Policy Iteration Algorithm is the solution of Eq. 9 which can be solved, for example, by using Gauss elimination [5] or an approximate iterative policy evaluation algorithm. On the other hand, Linear Program in Eq. 15 can be solved, for example, by using the Simplex Algorithm [5]. The core of both algorithms, Gauss elimination and Simplex, consists in keeping a system of equation:

$$\mathbf{A}\mathbf{x} = \mathbf{b}, \tag{24}$$

where $\mathbf{A}$ is a $m \times n$ matrix, $\mathbf{x}$ is a column-vector of size n, and $\mathbf{b}$ is a column vector of size m. At each iteration a cell (i, j) of $\mathbf{A}$ is chosen and a pivoting operation on $\mathbf{A}$ and $\mathbf{b}$ is done around cell (i, j). The pivoting operation repeats algebraic operations such as sum, subtraction, multiplication and division [5].

In Eq. 24, column vector $\mathbf{x}$ represents states values $V(s)$ and these values grow exponentially. As we have done previously, to avoid overflow or underflow, it is better working with certainty equivalent values, in this case, we should work with equation:

$$\log(\mathbf{A}\mathbf{x}) = \log(\mathbf{b}). \tag{25}$$

Our solution consists in the following adaptation of Gauss elimination and Simplex algorithms by:

1. keeping $\mathbf{A}^{\text{sgn}}$, $\mathbf{b}^{\text{sgn}}$, $\mathbf{A}^{\log}$, and $\mathbf{b}^{\log}$, where

$$\begin{aligned} \mathbf{A}^{\text{sgn}} &= \text{sgn}(\mathbf{A}) \;\; \mathbf{A}^{\log} = \log\big(\text{sgn}(\mathbf{A}) \odot \mathbf{A}\big), \\ \mathbf{b}^{\text{sgn}} &= \text{sgn}(\mathbf{b}) \;\;\; \mathbf{b}^{\log} = \log\big(\text{sgn}(\mathbf{b}) \odot \mathbf{b}\big), \end{aligned} \tag{26}$$

   the operator $\odot$ performs element-by-element multiplication. Note that because $\mathbf{A}$ and $\mathbf{b}$ may have non-positive cells, the sign of each cell must be kept separated and $0^{\text{sgn}} = 0$ and $0^{\log} = -\infty$;
2. because the log of a multiplication (division) is a sum (subtraction) of logs, multiplication and division of numbers p and q (cells of matrices $\mathbf{A}$ or $\mathbf{b}$) are done by:

$$\begin{aligned} (pq)^{\log} &= p^{\log} + q^{\log} \\ (p/q)^{\log} &= p^{\log} - q^{\log} \end{aligned}$$

   and the sign of the product (division) is computed by:

$$\begin{aligned} (pq)^{\text{sgn}} &= p^{\text{sgn}} q^{\text{sgn}} \\ (p/q)^{\text{sgn}} &= p^{\text{sgn}} q^{\text{sgn}} \end{aligned}$$

3. the log of a summation (subtraction) is done by making use of *LogSumExp* technique (Eq. 16):

$$(p+q)^{\log} = \begin{cases} -\infty & \text{, if } p^{\log} = q^{\log} = -\infty \\ p^{\log} + \log(1 + p^{\text{sgn}} q^{\text{sgn}} \exp(p^{\log} - q^{\log})), & \text{if } p^{\log} > q^{\log} \\ q^{\log} + \log(1 + p^{\text{sgn}} q^{\text{sgn}} \exp(q^{\log} - p^{\log})), & \text{otherwise} \end{cases}$$

and the sign of the sum (difference) is computed by:

$$(p+q)^{\text{sgn}} = \begin{cases} p^{\text{sgn}} & \text{, if } p^{\text{sgn}} q^{\text{sgn}} = 1 \\ p^{\text{sgn}} & \text{, if } p^{\log} > q^{\log} \\ q^{\text{sgn}} & \text{, if } q^{\log} > p^{\log} \\ 0 & \text{, otherwise} \end{cases}$$

4. pivoting are done over $\mathbf{A}^{\text{sgn}}$, $\mathbf{b}^{\text{sgn}}$, $\mathbf{b}^{\log}$, and $\mathbf{A}^{\log}$ following the algebraic operations previously defined.

Again, the LogSumExp strategy avoids overflow by only calculating exponential of negative numbers.

5 Analysis of the Main Advantages and Drawbacks of the Exponential Utility Function

In this section, we make a simple analysis of why the exponential utility function is desirable and we also present its main drawbacks. Exponential utility functions are not the only way of modeling risk, there are many other alternatives in the literature. However, the exponential utility criterion presents some characteristics that are not shared by alternative criteria, such as:

- **Symmetrical Distribution Sensibility:** The expected total cost criterion is risk insensitive because it does not differentiate symmetrical distribution around the same mean. Although Mean-Variance criterion [9,35] differentiates most symmetrical distributions, it is easy to construct different symmetrical distributions which present the same Mean-Variance value. It is not the case with the exponential utility function.
- **Stationary Optimal Policy:** A desired mathematical characteristic for SSPs is that stationary solutions (that are optimal) exist, which is not presented in every risk sensitive criterion. For example, CVaR criterion presents non-Markovian policies as optimal solutions [4]. To guarantee the existence of stationary optimal policies, in [25], Mihatsch and Neuneier make use of a discount factor, but a discount factor is known to be related to risk-prone attitudes [11].
- **Normative Decision Theory:** Expected Utility Theory is an axiomatic decision theory and is supposed to design rational decision under stochasticity. CVaR and VaR criteria are also designed over coherent risk theory [4], but no rational theory is behind Mean-Variance criterion or the criterion of [25].
- **Integral Evaluation:** Finally, because the exponential utility criterion is based on expectation, every potential history followed by a policy contributes to the value of such policy. This is not the case with CVaR and VaR criteria, which may produce unreasonable decisions, because they make use of hard constrains[2].

[2] For example, a policy that pays for sure M arbitrarily bigger, may be chosen over a policy that pays $M + \varepsilon$ with probability α and ε with probability $1 - \alpha$.

Despite the advantages presented, the exponential utility function presents three main drawbacks:

- **Bounded Risk Averse Parameter:** As discussed in Sect. 1, given an RSSSP, the risk parameter β is bounded above and depends on the RSSSP problem. The use of the β_{ext}-PI algorithm [10] gives a solution for such a problem, however β_{ext}-PI requires solving a RSSSP several times.
- **Non-Deterministic Cost:** The formulation for RSSSPs here presented and also used in [29] considers a cost function $c : \mathcal{S} \times \mathcal{A} \rightarrow \mathbb{R}$, which may not model properly every problem. Two possible alternatives are using: (i) $c : \mathcal{S} \times \mathcal{A} \times \mathcal{S} \rightarrow \mathbb{R}$ which depends on the next state; and (ii) $c : \mathcal{S} \times \mathcal{A} \rightarrow \mathcal{X}$, where $\mathcal{X}$ is the space of random variables and the cost received by the agent is probabilistic. One solution for such a problem is considering an augmented state space $\mathcal{S}^+ = \mathcal{S} \times \mathcal{S}$ [23], in this case the number of states is the square of the original space in the worst case.
- **Intermediate Calculation Overflow:** While both previous drawbacks are mainly a problem of modelling, in general, the use of exponential utility function prevents scalability because the underflow and overflow problems. In this paper, we propose a solution for this problem.

6 Related Work

Some works in the literature report that the exponential utility framework suffers from computational deficiencies that prevent it from being useful in practice for example in reinforcement learning [13,25] or in planning [21,22].

A simple numerical example with significant variability in rewards and extreme risk averse parameter value (where the overflow problem happens) is shown in [13]. However they do not solve directly the problem, instead they propose a new criterion, the variance-adjusted (VA) approach that has two objectives: maximize expected returns and minimize variance. On the other hand, instead of transforming the return of the process or solve the overflow/underflow problem in the exponential utility framework, Mihatsch and Neuneier (2002) transform the temporal differences during learning using a linear transformation. Different from them, in this paper, we propose a solution for the overflow/underflow problem.

Risk-sensitive problems with one-switch utility function and the Functional Value Iteration (FVI) algorithm to solve it were proposed in [21,22]. FVI is designed to deal with the issue of unbounded wealth values and the use of wealth values to augment the state; unbounded wealth values is not an issue in exponential utility function since optimal policies are stationary and it is not necessary to augment the state space. Theorem 4 in [21] is related to β-feasibility in the exponential utility function. Therefore, FVI presents the same problem of calculating large exponential values. We stress that large exponential values appear mainly when the risk factor is close to extreme risk-averse value. Specifically, in the case of FVI, the piecewise representation of the utility function does not avoid calculating the exponential of the certainty equivalent. Thus, the *LogSumExp* strategy can also be used in the FVI algorithm to avoid overflow.

7 Experiments

We performed experiments on MATLAB R2016a with a processor Intel Core i3 at 2.7 GHz and 6 GB RAM. We make use of the River domain [12] to compare algorithms with and without *LogSumExp* strategy.

7.1 River Domain

The River Domain [12] was described in the introduction. In this domain it is possible to move to the four cardinal directions (North, West, East, South). If the waterfall is reached, the agent goes back to the starting point.

We experiment within different grid sizes: 10×3, 20×4, 30×5, 40×6, 50×7, 60×8, 70×9, 80×10, 90×11, 100×12, 150×17, and 200×22. Note that the number of states in those grids varies between 30 and 4400 states. Because in the safest path there is a chance of 0.99 going forward, the extreme risk factor can be calculate by: $\beta_{ext} = \log\left(\frac{1}{1-0.99}\right) = 4.6052$. We conducted experiment with two risk-factor values: $\beta = 0.9\beta_{ext}$ and $\beta = 0.99\beta_{ext}$.

7.2 Algorithms Implementation

We compare the Value Iteration and Linear Programming algorithms with and without the *LogSumExp* strategy. We do not report test on the Policy Iteration algorithm since it requires an initial proper policy [29].

Linear Programming without the *LogSumExp* strategy was implemented simply by using the MATLAB's solver. However, it was not able to solve any problem. The solver reports 'unbounded', even for the smallest problem (10×3).

Linear Programming with the *LogSumExp* (LSE-LP) strategy was implemented in MATLAB by adapting the Simplex algorithm following rules on Sect. 4.3. Because numbers must be kept structured into sign and log, no matrix operation was used.

Value Iteration (VI) and Value Iteration with the *LogSumExp* strategy (LSE-VI) were implemented in MATLAB making use of matrix operation in every iteration. We use $\epsilon = 10^{-4}$ for the stopping criterion of the Value Iteration algorithm.

7.3 Evaluation of the LogSumExp Strategy

Figure 1 shows the convergence time of the LSE-VI, VI and LSE-LP algorithms for the River problem. We set a time limit of 10 000 s. The LSE-VI algorithm was able to solve every 12 instances of the River Domain. The LSE-LP algorithm timed out for instance 90×11 in both scenarios ($\beta = 0.9\beta_{ext}$ and $\beta = 0.99\beta_{ext}$). The VI algorithm overflowed from instance 60×8 (scenario $\beta = 0.9\beta_{ext}$) and from instance 50×7 (scenario $\beta = 0.99\beta_{ext}$).

The LSE-VI and LSE-LP algorithms seem to show a computational cost below exponential. We also see that LSE-LP was not influenced by the choice

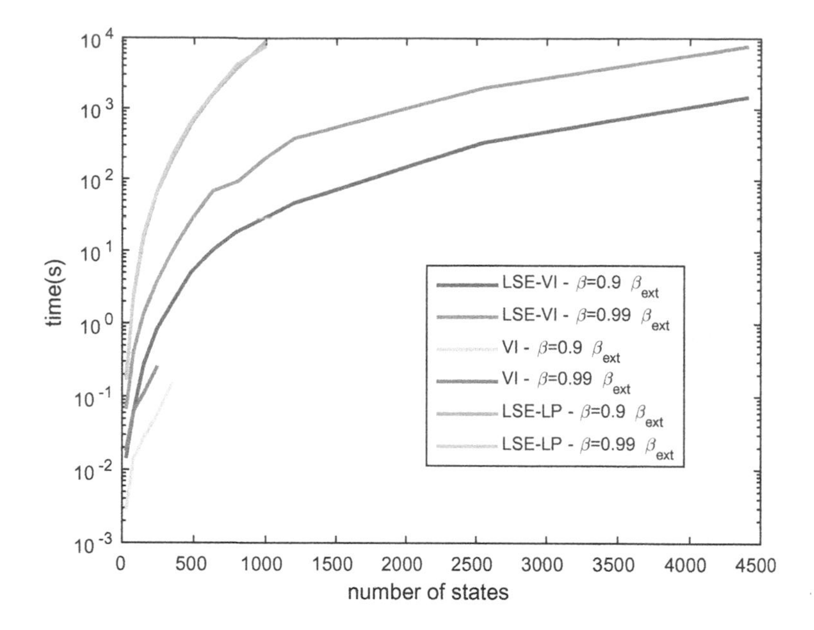

Fig. 1. Convergence time of the LSE-VI, VI and LSE-LP algorithms for the River problem with different grid sizes.

of the risk factor β, presenting similar convergence times in both scenarios. The LSE-VI and VI algorithms presented an increase of almost an order of magnitude when change from scenario $\beta = 0.9\beta_{ext}$ to scenario $\beta = 0.99\beta_{ext}$. This was expected, since the bigger the risk factor, the bigger the value function. The value iteration algorithm iterates over values, while Linear Programming works by searching solutions in the vertices of the convex polytopes.

When comparing both versions of Value Iteration, note that the LSE-VI algorithm is one order of magnitude slower than the VI algorithm. Besides using the log operation, the LSE-VI algorithm has the overhead of computing $k_{s,s'}^{a,i}$ and $K_s^{a,i}$. LSE-LP proved to be slower than LSE-VI in our implementation, we remember that this can be explained because of the use of matrix operations in LSE-VI, but not in LSE-LP.

8 Conclusion

To the best of our knowledge there are no studies that address numbers with high exponents in RSSSPs, which limits its practical use in current computational environments due to the precision errors. Thus, in this work we identify, evaluate and implement the *LogSumExp* strategy in RSSSPs which allows working with numbers with high exponents.

The experiments show that the use of the *LogSumExp* strategy in algorithms that solve RSSSPs allows obtaining policies for instances that previously diverged due to limitations of floating-point representations in the computational environment.

According with [20], the use of non-traditional objective functions (for example exponential utility) is critical for the success of AI planning and reinforcement learning and warrant much more research focus than they have received so far. Thus with this work, we contribute with these areas, we see a field of opportunities open to the exploration of the exponential utility function as risk modeler, thus far little explored in the AI area.

We conjecture that techniques developed here can be applied with small effort to many state-of-the-art algorithms such as: PROST [18], ILAO* [14], LRTDP [2] and i-dual [37].

References

1. Bertsekas, D.P., Tsitsiklis, J.N.: An analysis of stochastic shortest path problems. Math. Oper. Res. **16**(3), 580–595 (1991)
2. Bonet, B., Geffner, H.: Labeled RTDP: improving the convergence of real-time dynamic programming. In: ICAPS 2003, pp. 12–21. AAAI Press (2003)
3. Chen, Y., Gao, D.Y.: Global solutions to nonconvex optimization of 4th-order polynomial and log-sum-exp functions. J. Glob. Optim. **64**(3), 417–431 (2014). https://doi.org/10.1007/s10898-014-0244-5
4. Chow, Y., Tamar, A., Mannor, S., Pavone, M.: Risk-sensitive and robust decision-making: a CVaR optimization approach. In: Advances in Neural Information Systems (2015)
5. Chvatal, V.: Linear Programming. Freeman Press, New York (1983)
6. Delage, E., Mannor, S.: Percentile optimization for Markov decision processes with parameter uncertainty. Oper. Res. **58**(1), 203–213 (2010)
7. Denardo, E.V., Rothblum, U.G.: Optimal stopping, exponential utility, and linear programming. Math. Program. **16**(1), 228–244 (1979). https://doi.org/10.1007/BF01582110
8. Filar, J.A., Krass, D., Ross, K.W., Ross, K.W.: Percentile performance criteria for limiting average Markov decision processes. IEEE Trans. Autom. Control **40**(1), 2–10 (1995)
9. Filar, J.A., Kallenberg, L.C.M., Lee, H.M.: Variance-penalized Markov decision processes. Math. Oper. Res. **14**(1), 147–161 (1989)
10. Freire, V., Delgado, K.V.: Extreme risk averse policy for goal-directed risk-sensitive Markov decision process. In: 5th Brazilian Conference on Intelligent Systems, pp. 79–84 (2016)
11. Freire, V.: The role of discount factor in risk sensitive Markov decision processes. In: 5th Brazilian Conference on Intelligent Systems, pp. 480–485 (2016)
12. Freire, V., Delgado, K.V.: GUBS: a utility-based semantic for goal-directed Markov decision processes. In: Proceedings of the 16th Conference on Autonomous Agents and MultiAgent Systems, pp. 741–749 (2017)
13. Gosavi, A., Das, S.K., Murray, S.L.: Beyond exponential utility functions: a variance-adjusted approach for risk-averse reinforcement learning. In: 2014 IEEE Symposium on Adaptive Dynamic Programming and Reinforcement Learning (ADPRL), pp. 1–8 (2014)

14. Hansen, E.A., Zilberstein, S.: LAO*: A heuristic search algorithm that finds solutions with loops. Artif. Intell. **129**, 35–62 (2001)
15. Howard, R.A., Matheson, J.E.: Risk-sensitive Markov decision processes. Manag. Sci. **18**(7), 356–369 (1972)
16. Jaquette, S.C.: A utility criterion for Markov decision processes. Manag. Sci. **23**(1), 43–49 (1976)
17. Keeney, R.L., Raiffa, H.: Decisions with Multiple Objectives: Preferences and Value Tradeoffs. Wiley, New York (1976)
18. Keller, T., Eyerich, P.: Prost: probabilistic planning based on UCT. In: Twenty-Second International Conference on Automated Planning and Scheduling, pp. 1–9 (2012)
19. Kennings, A.A., Markov, I.L.: Analytical minimization of half-perimeter wire-length. In: Proceedings of the 2000 Asia and South Pacific Design Automation Conference, pp. 179–184. ACM (2000)
20. Koenig, S., Muise, C., Sanner, S.: Non-traditional objective functions for MDPs. In: IJCAI-18 Workshop on Goal Reasoning, pp. 1–8 (2014)
21. Liu, Y., Koenig, S.: Risk-sensitive planning with one-switch utility functions: value iteration. In: Proceedings of the 20th National Conference on Artificial Intelligence, pp. 993–999. AAAI Press (2005)
22. Liu, Y., Koenig, S.: Functional value iteration for decision-theoretic planning with general utility functions. In: Proceedings of the 21st National Conference on Artificial Intelligence, pp. 1186–1186. AAAI Press (2006)
23. Ma, S., Yu, J.Y.: State-augmentation transformations for risk-sensitive reinforcement learning. In: The Thirty-Third AAAI Conference on Artificial Intelligence. The Thirty-First Innovative Applications of Artificial Intelligence Conference. The Ninth AAAI Symposium on Educational Advances in Artificial Intelligence, pp. 4512–4519 (2019)
24. Mann, T.P.: Numerically stable hidden Markov model implementation. In: An HMM Scaling Tutorial, pp. 1–8 (2006)
25. Mihatsch, O., Neuneier, R.: Risk-sensitive reinforcement learning. Mach. Learn. **49**(2), 267–290 (2002). https://doi.org/10.1023/A:1017940631555
26. Moldovan, T.M., Abbeel, P.: Risk aversion in Markov decision processes via near optimal Chernoff bounds. In: Advances in Neural Information Processing Systems, NIPS 2012, pp. 3131–3139 (2012)
27. Naylor, W.C., Donelly, R., Sha, L.: Non-linear optimization system and method for wire length and delay optimization for an automatic electric circuit placer (2001). US Patent 6,301,693
28. Nielsen, F., Sun, K.: Guaranteed bounds on information-theoretic measures of univariate mixtures using piecewise log-sum-exp inequalities. Entropy **18**(12), 442 (2016)
29. Patek, S.D.: On terminating Markov decision processes with a risk-averse objective function. Automatica **37**(9), 1379–1386 (2001)
30. Puterman, M.L.: Markov Decision Processes: Discrete Stochastic Dynamic Programming. Wiley, Hoboken (2014)
31. Rabiner, L.R.: A tutorial on hidden Markov models and selected applications in speech recognition. Proc. IEEE **77**(2), 257–286 (1989)
32. Robert, C.: Machine Learning, A Probabilistic Perspective. Taylor & Francis, Milton Park (2014)
33. Rothblum, U.G.: Multiplicative Markov decision chains. Math. Oper. Res. **9**(1), 6–24 (1984)

34. Sigl, G., Doll, K., Johannes, F.M.: Analytical placement: a linear or a quadratic objective function. In: 28th ACM/IEEE Design Automation Conference, pp. 427–432 (1991)
35. Sobel, M.J.: The variance of discounted Markov decision processes. J. Appl. Probab. **19**(4), 794–802 (1982)
36. Trevizan, F., Thiébaux, S., Santana, P., Williams, B.: I-dual: solving constrained SSPs via heuristic search in the dual space. In: Proceedings of the 26th International Joint Conference on AI (IJCAI) (2017)
37. Trevizan, F., Thiébaux, S., Santana, P., Williams, B.: Heuristic search in dual space for constrained stochastic shortest path problems. In: Twenty-Sixth International Conference on Automated Planning and Scheduling (2016)

Solving Multi-Agent Pickup and Delivery Problems Using a Genetic Algorithm

Ana Carolina L. C. Queiroz[1(✉)], Heder S. Bernardino[1(✉)], Alex B. Vieira[1(✉)], and Helio J. C. Barbosa[1,2(✉)]

[1] Universidade Federal de Juiz de Fora, Juiz de Fora, Brazil
{anacarolina,heder,alex.borges}@ice.ufjf.br, hcbm@lncc.br
[2] Laboratório Nacional de Computação Científica, Petrópolis, Brazil

Abstract. In the Multi-Agent Pickup and Delivery (MAPD) problem, agents must process a sequence of tasks that may appear in the system in different time-steps. Commonly, this problem has two parts: (i) task allocation, where the agent receives the appropriate task, and (ii) path planning, where the best path for the agent to perform its task, without colliding with other agents, is defined. In this work, we propose an integer-encoded genetic algorithm for solving the task allocation part of the MAPD problem combined with two-path planning algorithms already known in the literature: the Prioritized Planning and the Improved Conflict-Based Search (ICBS). Computational experiments were carried out with different numbers of agents and the frequency of tasks. The results show that the proposed approach achieves better results for large instances when compared to another technique from the literature.

Keywords: Multi-Agent Pickup and Delivery · Genetic algorithm · Real-world application

1 Introduction

Stock levels and the number of orders are rising worldwide recently. One of the reasons is the trend of increasing the number of purchases made online, either wholesale or retail. Warehouses and distribution centers are responsible for storing items and the flow of goods. These depots receive a large number of orders and need to attend them quickly, optimizing resources. In the traditional system, this problem is solved manually: the orders are allocated to the appropriate carriers. This costs time, resources, and generates errors, such as exchanges or loss of objects. One of the solutions is the automation of warehouses using autonomous mobile robots, called here "agents". These agents perform the task of moving through the warehouse to pick up products and prepare them for delivery [21].

The authors thank the financial support provided by CAPES, CNPq (grants 312337/2017-5, 312682/2018-2, 311206/2018-2, and 451203/2019-4), FAPEMIG, FAPESP, and UFJF.

R. Cerri and R. C. Prati (Eds.): BRACIS 2020, LNAI 12320, pp. 140–153, 2020.
https://doi.org/10.1007/978-3-030-61380-8_10

There are several other applications besides distribution centers that agents need to operate efficiently in a known environment, such as autonomous aircraft towing vehicles [13], rescue robots, office robots [19], and video game characters [11]. Currently, hundreds of robots already navigate autonomously in Amazon's service centers facilitating the movement of inventory [2]. The coexistence of multiples agents and the high demand for this service turns necessary one to find high-quality and collision-free paths in real-time.

In this context, Multi-Agent Path Finding (MAPF) searches for paths for different agents guaranteeing that they all reach their destination without conflicts. MAPF works as a "one-shot" system, in which all tasks are known a priori, and the problem is closed when the last agent reaches its target position [17].

In this work, we consider an "*online*" version of MAPF, known in the literature as the Multi-Agent Pickup and Delivery (MAPD) [3,16,17]. In this problem, tasks are added to the system during the process, making it necessary for agents to adapt to the task flow and execute them in the previously known environment. Tasks are characterized by a pickup and a delivery location. Its execution consists of moving the robot from a start position to the point of pickup and, in the sequence, to the point of delivery. After the item is delivered, the agent is released to receive another task. The agent must choose the best task to be performed considering the shortest path to the goal position without colliding with other agents. This version is more faithful to the real characteristics of a distribution center, where new tasks appear continuously.

The online MAPF problem can be divided into two sub-problems: (i) the task allocation and (ii) path planning. We propose here the combination of a genetic algorithm (GA) with the Improved Conflict Based Search (ICBS) and Prioritized Planning to solve this problem. In this proposal, the tasks are allocated by the GA, while ICBS and Prioritized Planning are used to define the paths. GAs are efficient for solving optimization problems and widely used for solving problems with single or multiple objectives in production and operations management [3]. ICBS is a modified version of Conflict-Based Search (CBS) designed for the path planning of several agents.

Also, we proposed here two integer encoding structures: (i) tasks, agents and allocations are represented, and (ii) tasks are represented and a heuristic is used for scheduling the agents to the tasks. In the latter case, the delivery time of the agent and the distance from the agent to the task are considered. To conclude our contributions, a heuristic approach is used to initialize the GA with a better individual (when compared to a random initialization). Computational experiments are performed and the proposed hybrid approach using a GA with heuristics for generating the initial population performed better than the other methods in most of the problems tested.

2 Related Work

The classical algorithms for MAPF problems assume that each action (the moving from one cell to the next one) takes one timestep. Moreover, it is assumed that one cell (or vertex) cannot have more than one agent occupying it (and each agent occupies one cell at a given instant). These assumptions often fail to capture the characteristics of the real world and, as a consequence, there are several variations of the MAPF problem to address real-world issues. For example, [20] presents a version of the problem where agents' actions may take more than one time step to complete the movement from one vertex to another. This version of the problem can be called "MAPF with non-unit edge cost".

In real situations, the agents move in Euclidean space and the agents may have a specific geometric shape. Thus, the agents may conflict as their geometric shapes overlap. In [6], this problem is referred to as "MAPF with large agents" when agents can occupy more than a single cell in large space. The tasks are the focus of the study presented in [9]. The authors assume that the system knows all the tasks, but a release time is added to the problem formulation.

Here, we are interested in an online version of MAPF, called Multi-Agent Pickup and Delivery (MAPD). Three algorithms to solve this problem, namely, TP, TPTS, and Central are proposed in [12]. The best results of that study were obtained by Central and, thus, we used its results to comparatively evaluate the techniques proposed here. Multi-Label A* and the h-value-based centralized heuristic were proposed in [4] to solve the same problem, and they achieve better results compared to TP.

Other works can be found in the literature dealing with the online MAPF problem, where agents are constantly engaged with new goal locations. For example, the online MAPF problem is modeled in [10] adding kinematic constraints to the robot. Also, an online version of MAPF is considered in [18], where new agents appear during the execution of the problem. The focus of that paper is the intersection problem, as each agent already has an associated task. On the other hand, we are interested here in the allocation of tasks.

A framework is proposed in [7] to solve online MAPF, decomposing the problem into a sequence of Windowed MAPF instances and re-planing paths once every h timesteps. The framework has demonstrated success in terms of throughput on fulfillment warehouse maps and sorting center maps.

The use of genetic algorithms for task allocation was successfully applied to allocate tasks to agents for inspection of industrial plants [5,8]. In [5] the solution in the GA is represented by an array of bits (binary coded) and integer coded. The part of the string integer-coded is for the task sequence and the other with binary encoding shows the number of tasks assigned to each robot.

3 Problem Definition

In this work, a 2D environment is mapped into an undirected graph $G = (V, E)$, where the vertices $v \in V$ correspond to the locations and the edges $e \in E$ correspond to the connections between these locations (vertices). Time is discretized, and in every timestep each agent can move to an adjacent vertex (up, down, left and right) or wait at the current vertex. Timestep is an artificial unit of time that represents the time required for the agents to perform a movement.

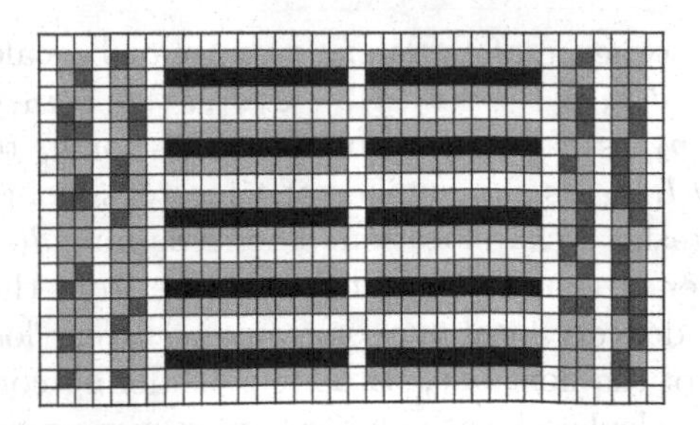

Fig. 1. Example of an environment with 50 agents (adapted from [12]). Black cells are blocked positions, green cells are task endpoints (pickup and delivery locations), and red cells are non-task endpoints (start and parking locations of agents). (Color figure online)

The location of agent a_i at the timestep t is denoted by $l_i(t) \in V$ and the pair $(l_i(t), l_i(t+1)) \in E$ represents a movement (along an edge), where i is an agent index.

In this formulation of the problem, the agents cannot collide with each other. Thus, two or more agents cannot occupy the same vertex or edge at the same timestep. Formally, all agents a_i and a_j, i and j being the indexes of the agents and $i \neq j$, must comply with the following conditions at timestep t:

1. To avoid a vertex collision: two agents cannot be in the same location at the same timestep t: $l_i(t) \neq l_j(t)$.
2. To avoid an edge collision: two agents cannot move along the same edge in opposite directions at the same time step, $l_i(t) \neq l_j(t+1)$ and $l_i(t+1) \neq l_j(t)$.

A set τ contains all tasks where each task τ_j, $j \in \{1, \ldots, |\tau|\}$, is defined by a pickup location s_j and a delivery location g_j. Moreover, tasks can be added to the system at any time. Task endpoints (s_j, g_j) are all possible pickup and delivery locations. We also define here non-task endpoints, which are the agent's parking locations. An example of a warehouse is shown in Fig. 1.

Each agent must complete a subset of tasks from τ and can only deal with one single task at each timestep. The agent moves from its current location to the task s_j when it is assigned to a task. When the agent arrives at the location s_j, it starts to execute the task τ_j. We define this agent as a "busy agent" and this status is changed to "free agent" when it arrives at the task delivery point. Only free agents can be assigned to tasks. Finally, we define the agent's path as the sequence of positions/vertices visited when solving a given task.

We define the number of timesteps an agent takes from their current to their target location as path cost. An agent moves from a vertex (position) to its neighbor with a timestep. Thus, the time for an agent to travel a given path is the distance value, which is calculated using the Manhattan distance.

The objective function here is the makespan, which is the instant of time the last task is concluded. Each agent has its own sequence of tasks and, consequently, its makespan. For instance, the makespan Ma_i of a given agent a_i

which performs the sequence of tasks $[\tau_6, \tau_1, \tau_3]$ can be calculated as $Ma_i = d(l_i, s_6) + d(s_6, g_6) + d(g_6, s_1) + d(s_1, g_1) + d(g_1, s_3) + d(s_3, g_3)$, where l_i is its start position, $d(p_1, p_2)$ is the number of timesteps for a_i to move from position p_1 to p_2, and $d(l_i, s_6)$ is the number of timesteps from the agent's initial location to the first task's pickup location. The remaining $d(\cdot, \cdot)$ calculations are the minimum number of timesteps needed to move from the pickup of a task to its corresponding delivery locations, or from a delivery location of a task to the pickup location of the next one, without considering conflicts. Finally, the objective function is calculated as the largest makespan among all agents.

4 Path Planning

Two path planning techniques are used here, namely, Prioritized Planning and Improved Conflict-Based Search (ICBS). Descriptions of these methods follow.

4.1 Prioritized Planning

A* is a path search algorithm widely used to search for the lowest cost path between two points on a given map. A modification of A*, called Cooperative A*, focuses on solving the multi-agent problem, where each agent has full knowledge of the other agents and their planned routes. The algorithm solves the search problem by optimizing the path of every single agent. After planning, it fills a "reservation table" with information on which vertex is occupied in a certain time so that another agent in the list can plan its path efficiently, checking such table and avoiding future collisions. The reservation table is treated as a three-dimensional grid, being two spatial dimensions and a third dimension representing the time [15].

In this work, we use the Prioritized Planning algorithm. This algorithm schedules the agents, giving priority to those with larger estimated timesteps to execute the task. The first agent to choose its path in Cooperative A* has higher priority as the table is empty and, thus, there are fewer obstacles to be avoided (fewer collisions). This way, agents with longer estimated time to conclude its task have fewer restrictions, which may result in the smallest makespan.

4.2 Improved Conflict-Based Search (ICBS)

The ICBS [1] is an improved version of the Conflict Based Search (CBS). The CBS is an algorithm that guarantees an optimal solution for the MAPF problem [14]. It splits the problem into a set of constraints and finds paths that fit those constraints to solve the problem as single-agent pathfinding. The problem is transformed into a sequence of constrained single-agent pathfinding problems.

The CBS is a two-level algorithm. At the higher level, the algorithm processes the node of a constraint tree (CT) that provides a set of constraints imposed on the lower level search. A CT is a tree where each node has a constraint and a set of paths. A constraint is a (a_i, v, t) tuple where the agent a_i is prohibited from occupying vertex v in time t. The set of paths contains each path for each agent. At the lower level, the algorithm searches for the smallest path for each agent without violating the constraints imposed by the other level. A Cooperative A* algorithm is used at the lower level.

At the higher level, CBS chooses to expand the CT node with the smallest cost at each iteration. The algorithm goes from one CT node to the other by checking the conflicts, generating the constraints, and calling the search procedure of the lower level to re-plan the path of the agents.

The algorithm works as follows: first, the root of CT is generated with the set of paths for all agents which were found in the low-level search without any restrictions. The solution is valid if no conflict occurs in the root. Thus, there is no collision between the paths of all the agents, and the search ends. On the other hand, if there is a conflict in the paths of agents a_i and a_j, in location v and at time t, two new CT nodes are generated: n_i and n_j. These nodes are added as children of the root node. Respectively, the constraints (i, v, t) and (j, v, t) are added to nodes n_i and n_j. The cost of the CT node is the sum of the costs of the set of paths.

At a higher level, the CT node with the smallest cost is expanded. Expanding a node means to call a search at the lower level and to resolve the conflict, i.e., the agent's path must undergo a re-planning and, if there are new conflicts, two new nodes are generated considering the new restrictions found. The search ends when no constraint violations are found in the CT node. That corresponds to a solution with no conflict in the path of all agents. CBS places the nodes in a priority queue according to the sum, over all agents, of the number of timesteps they took to reach their destination positions.

CBS has several extensions and improvements, and the Improved Conflict Based Search (ICBS) [1] is used here. ICBS tries to reduce the size of CT by prioritizing and solving the conflicts. The conflicts are classified into cardinal, semi-cardinal and non-cardinal. A cardinal conflict occurs when two agents use the same vertex or edge in the shortest path. The cardinal conflicts are solved first, as they increase the cost of the solution.

A semi-cardinal conflict occurs when the shortest path for an agent includes the same edge or vertex as the shortest path for another agent, but solving this conflict for that 2nd agent doesn't increase the cost. A non-cardinal conflict occurs when the solution to a conflict does not imply in a change in the cost.

4.3 Deadlock Avoidance

During the execution of some methods, the agent is kept at the delivery position after concluding a task and until a new task is allocated. Thus, other agents cannot use this position. Instead, agents executing tasks that require to move via this position should wait for a new task to be assigned to the first agent. A new task can be assigned to the agent and a delay occurs. However, the tasks cannot be concluded when this agent does not receive new tasks. In this last case, a deadlock occurs.

Algorithm 1. Pseudo-code of the proposed procedures using the GA.

```
1: timestep = 0
2: while the stop criterion is not met do
3:    if a task has entered the system then
4:       add task to the taskset
5:       call the task assignment algorithm: GA-TA or GA-E
6:    end if
7:    if an agent has finished executing a task then
8:       set agent as free
9:       set final position of an agent as the pickup location of its next task
10:   end if
11:   if an agent has arrived at the pickup location of a task then
12:      set agent as busy
13:      set final position of the agent as the delivery location of the task
14:      remove task from the task set
15:   end if
16:   if the final position of an agent has changed then
17:      call the path planning algorithm: Prioritized Planning or ICBS
18:   end if
19:   timestep = timestep + 1
20: end while
```

As a way to avoid this type of deadlock, the agents return to their initial position, the parking location, as soon as they have concluded a given task. One can notice that the agent will not return to its initial location when the number of new tasks is large. In this case, the agent always has tasks to execute.

5 Proposed Approach

The proposed combination of a genetic algorithm (GA) with the ICBS and Prioritized Planning for solving MAPD problems is described here. In the proposal, the GA starts whenever new tasks are added (online problem). The GA defines which tasks the agents will execute. Two GA variants are proposed here: GA-TA, where the candidate solutions are represented by a vector of task indexes and a vector of agent indexes; and GA-E, where only the tasks are represented. For both, GA-TA and GA-E, the chromosome length is the number of tasks currently in the problem.

A pseudo-code of the proposed methods is presented in Algorithm 1. The algorithm starts with $timestep = 0$ and all agents are free. As long as the stop criterion is not satisfied, the timestep is increased each time the loop runs. The stop criterion is met when there is no more task to be added.

When a task is added to the task set, the GA is used to assign the tasks to the agents. The final position of a busy agent is the delivery location of the task being executed. The GA considers for the busy agent the time and location that the agent is free again, choosing the best task for the agent to execute after it has finished its task. Also, the GA can modify the assignments of the non-concluded

[Candidate solution of the GA-TA.]

Tasks	7 1 3 9 5 4 8 10 2 6
Agents	3 2 0 4 3 2 4 1 4 2

[Order one crossover (OX).]

Parent 1	0 1 2 3 ~~4 5 6 7~~ 8 9
Parent 2	9 8 7 6 ~~5 4 3 2~~ 1 0
Child 1	9 8 2 3 4 5 7 6 1 0
Child 2	0 1 7 6 5 4 2 3 8 9

Fig. 2. Illustration of a candidate solution (a) and OX recombining tasks (b).

tasks and the ones that are not being executed. The GA returns a sequence of tasks to be performed for each agent.

The path planning occurs every time the final position of an agent is modified (when they are assigned to the pickup location and when they finish the task and are assigned to the delivery location). The goal location of a free agent is the pickup position of the task defined by the GA. For a busy agent, the goal location is the delivery location of its respective current task. ICBS or Prioritized Planning path planning are only performed when a goal location is set or changed.

At each iteration, the algorithm checks if the agent has already completed a task or if it has already reached the pickup position of a task. The agents can be assigned to any task when they arrive at the delivery position. Otherwise, the agent is busy and its current task cannot be modified.

The evolutionary process considered here is composed of the parental selection, crossover, mutation, and replacement strategy. Makespan is the objective function (as defined in Sect. 3), the solutions are selected to reproduce by a tournament selection, and the replacement procedure keeps the current best individuals in the population (elitism).

5.1 The GA-TA

Here, the candidate solutions are represented by a vector of task indexes and a vector of agent indexes. Agents and tasks are associated with their positions in the vectors. Figure 2 illustrates the representation of a candidate solution of GA-TA when solving a problem with 10 tasks and 5 agents. According to the vectors of tasks and agents, agent 3 executes tasks 7 and 5, agent 2 executes task 1, 4 and 6, and so on. Also, each agent performs the tasks in the order they appear in the chromosome.

The GA-TA recombines task and agent vectors separately. The one-point crossover is used for the agent vector, as the same agent is allowed to perform more than one task and agents can be unused. On the other hand, the task vectors are recombined using the order one crossover (OX), a popular procedure for permutations. In OX, two cutoff points are selected and the segment between the two cuts of Parent 1 is copied to Child 1. To complete Child 1, the tasks not presented in its chromosome are copied from Parent 2 in the same order. In the illustrative example presented in Fig. 2, Child 1 is created by (i) copying tasks

$[2,3,4,5]$ from Parent 1 and placing them in the same positions in Child 1, and (ii) copying the tasks $[7,6,1,0,9,8]$ from Parent 2. A second child is similarly generated.

The type of mutation is randomly selected among task or agent mutation. For the task mutation, the pairwise interchange neighborhood mutation (IM) is used, where two different tasks are selected and their respective positions are exchanged on the chromosome. For agent mutation, a position of the chromosome is randomly chosen and its value is replaced by an agent index randomly generated within a variation of 30% of the number of agents. For instance, considering 10 agents, an agent index 5 in the mutated position is replaced by another index between 2 and 8.

5.2 The GA-E

In the GA-E, only the tasks (the first part of the GA-TA individuals) are represented. A heuristic is used here to schedule agents to perform tasks. For each task in the chromosome, this scheduling function performs a greedy search by the agents. The scheduler chooses the agent according to two criteria: (i) the agent's final time and (ii) the distance from the agent to the pickup location of the task. The agent with the lowest value of the summation of (i) and (ii) is selected to perform the task. The current location and final time of the agent are updated after completing the task. The position is replaced by the delivery location of the task and the number of timesteps to conclude the task is added to the final time. One can notice that this heuristic is called before the candidate solutions are evaluated (makespan is calculated).

OX is also used in the GA-E to recombine the parents (Fig. 2). In addition, the mutation of the tasks is IM.

5.3 Initialization of the Population

Normally, the populations in the GA techniques are randomly initialized and this simple strategy is considered here. However, improvements can be observed in the search when using a heuristic approach to generate the initial population. Thus, we also propose a heuristic to assist the initialization of the population of both GA-TA and GA-E when solving MAPD problems.

The proposed greedy heuristic uses a distance matrix between agents and tasks. The distances are calculated from the current position of the agent to the pickup location of the task. At each iteration, the shortest distance is used to select an agent and the task this agent will perform. As the task is already associated with an agent, this task is removed from the distance matrix. Also, the current location of the agent and its final time are updated. As a consequence, the distance values involving this agent are recalculated. This procedure creates one individual and the remaining candidate solutions of the initial population are randomly generated.

6 Computational Experiments

We have performed computational experiments (i) to solve MAPD problems simulating a real-world situation, (ii) to evaluate the two proposed representations, (iii) to analyze the performance of the GA when a heuristic is used to create its initial population, and (iv) to comparatively evaluate the proposals with Central [12]. The environment is a 21 × 35 grid, as shown in Fig. 1. The agents can move to the four adjacent cells or stay in their current location. For each problem, 500 tasks must be concluded. These tasks are randomly created with their task-endpoints (pickup and delivery locations). As the problem is online, the tasks are deployed during the execution at a given rate, and here 6 different values are considered: 0.2, 0.5, 1, 2, 5, and 10. A task rate equal to 10 means that 10 new tasks are added per timestep. Also, we consider 5 numbers of agents: 10, 20, 30, 40 and 50.

The proposed approaches combine (i) two GA variants (GA-TA or GA-E), (ii) random (labeled with 'b') initialization of the population or using a heuristic (labeled with 'a'), and (iii) Prioritizing Planning or ICBS as the path planning method. As a result, 8 proposed methods are analyzed here. The parameters used for the GAs were 1% for selection by tournament and for the creation of the new population, selection by elitism is used in which 4% of the best individuals in the population are copied without changes to the new population. An 80% chance of crossover in both algorithms was used, in GA-TA the crossover occurs in the task string and in the agents' string, and in GA-E it occurs only in the tasks. And for the mutation, GA-TA has a 50% chance of being a task mutation or an agent mutation, and for GA-E, an agent mutation occurs or no mutation occurs. A maximum number of 50 generations is used as the stopping criterion, population size is equal to 20 for GA-TA and for GA-E, and 30 independent runs were performed. The parameters were defined empirically. The source-codes (in C++), the instances used in the experiments, and the results obtained are available[1].

Table 1 presents the makespan obtained by the proposed approaches and Central [12]. Similarly to [12], a time limit of five minutes is defined here for each call of the ICBS path planning. A timeout occurs when this time limit is violated and the number of timeouts (among the independent runs) is presented in parentheses.

ICBS presented the best results in relation to path planning, but it only works very well for a small number of agents, as in the case of 10 agents and frequencies of 10, and 1 in ICBS+GA-TA+a had no execution that occurred timeout. As the number of agents increases, the number of timeouts also increases. 'Prioritized Planning + a' achieves better results than Central when the frequency is equal to 0.5, and 30, 40, and 50 agents are used, and is equal to 1 and 40 and 50 agents are used.

Regarding task assignment, the GA-TA algorithm obtained better results for high task rates when compared with GA-E and Central, while GA-E obtained

[1] https://github.com/carolladeira/MAPD.

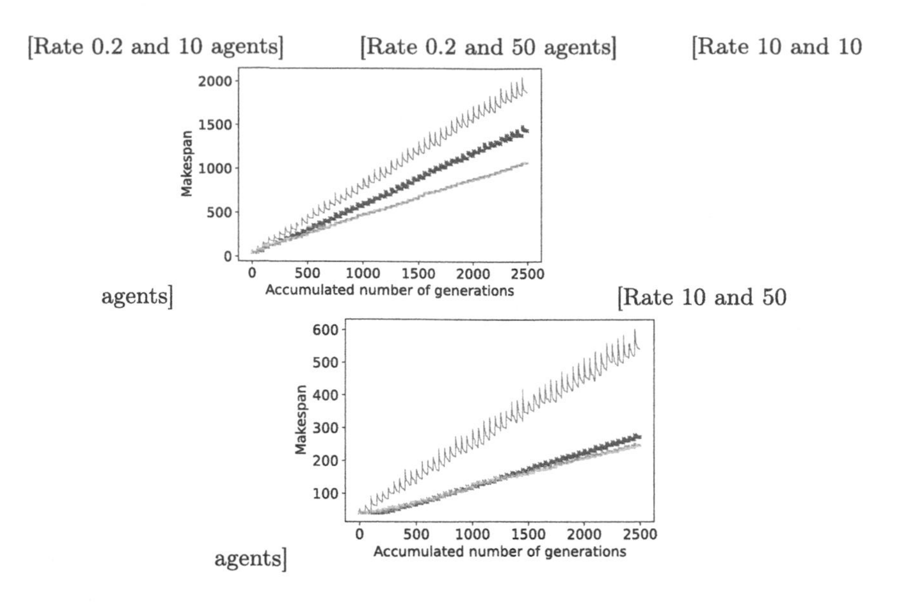

Fig. 3. Makespan values of the proposed methods without the occurrence of conflicts by the accumulated number of the GA generations. In these plots, the number of generations is increased for any GA iteration.

better results for low task rates. Central obtained the best results when rate is 0.2, rate is 0.5 and 10 agents are used, rate 1 and 30 agents are used, and rate is 2 and 50 agents are used.

The Wilcoxon signed-rank test is a non-parametric test that was used to check the difference between the ICBS and Prioritized Planning algorithms against Central. Due to a requirement of the implementation[2], the tests were only performed when the sample size is greater than 20. The stars (*) in Table 1 indicate that the results are statistically significantly better than those achieved by Central (null hypothesis is rejected).

Figure 3 presents the makespan of "Without conflicts" algorithms. We show results for task rates equal to 0.2 and 10, and the number of agents equal to 10 and 50. One can see that the makespan decreases when the GA is executed and increases when a new set of tasks is added. Using a heuristic initialization reduces the makespan of GA-TA, but it has a small effect on GA-E. Also, the superior performance of GA-TA+a is reinforced.

[2] https://docs.scipy.org/doc/scipy/reference/generated/scipy.stats.wilcoxon.html.

Table 1. Average of the makespan values obtained by the search techniques. "F" is the task rate, "Ag" is the number of agents, and "Central" is the technique proposed in [12]. The best results for each instance (rate and number of agents) are highlighted in **boldface**. Labels 'a' and 'b' refer to the use of the heuristics for initialization of the population and random population seeding, respectively.

F	Ag	Central	Prioritized Planning				ICBS				Without conflicts			
			GA-TA		GA-E		GA-TA		GA-E		GA-TA		GA-E	
			a	b	a	b	a	b	a	b	a	b	a	b
0.2	10	**2513**	2519.7	3476.9	2519.8	2520.3	2517.3 (0)	3210.4 (1)	2516.9 (0)	2516.2 (0)	2517.7	3627.2	2514.7	2514.0
	20	**2513**	2518.7	2621.7	2520.7	2520.2	2517.2 (0)	2516.8 (0)	2521.0 (0)	2521.0 (0)	2514.5	2514.8	2513.0	2513.0
	30	**2513**	2518.4	2575.7	2521.1	2521.0	2516.8 (0)	2517.0 (0)	2520.0 (0)	2520.0 (0)	2513.4	2513.4	2511.2	2511.0
	40	**2511**	2516.4	2558.6	2515.0	2515.0	2516.6 (0)	2516.1 (0)	2515.0 (0)	2515.0 (0)	2512.8	2513.5	2511.0	2511.0
	50	**2511**	2517.1	2559.6	2515.0	2515.0	2515.3 (0)	2516.3 (0)	2515.0 (0)	2515.0 (0)	2512.9	2513.0	2511.0	2511.0
0.5	10	**1242**	1255.0	2777.0	1334.1	1913.6	1250.6 (2)	2743.6 (7)	1326.4 (1)	1903.7 (1)	1234.2	2734.0	1310.5	1873.6
	20	1031	1049.8	1940.2	1025.2*	1025.4*	1051.4 (15)	Timeout	1025.1* (1)	**1025.0*** (0)	1045.5	1895.2	1021.6	1022.0
	30	1034	1028.5*	1651.3	1021.9*	1021.9*	1026.8 (18)	Timeout	**1021.1*** (1)	1021.4* (0)	1023.9	1616.7	1020.0	1020.0
	40	1034	1025.4*	1512.1	1021.2*	1021.0*	1022.0* (9)	Timeout	1020.1* (1)	**1020.0*** (0)	1021.1	1479.8	1020.0	1020.0
	50	1031	1025.4*	1421.9	1020.5*	1020.4*	1022.3* (9)	Timeout	1019.9* (0)	**1019.8*** (0)	1020.5	1400.9	1020.0	1020.0
1	10	1143	1144.0	2373.2	1174.8	1790.8	**1121.7*** (0)	2341.2 (13)	1155.4 (2)	1757.8 (2)	1111.5	2317.4	1144.6	1746.4
	20	673	669.4*	1522.3	741.6	965.1	**656.0*** (9)	Timeout	731.3 (7)	939.1 (7)	641.8	1471.3	712.0	920.0
	30	**557**	571.8	1241.4	595.0	623.6	567.1 (21)	Timeout	574.5 (26)	584.0 (26)	558.8	1186.5	542.8	546.2
	40	556	546.2*	1101.0	527.8	527.8	549.7 (23)	Timeout	527.0 (12)	**526.7** (15)	537.7	1049.2	526.0	526.1
	50	552	539.0*	1010.6	527.9	527.8	538.0 (23)	Timeout	**526.5** (13)	**526.5*** (9)	530.8	960.0	526.0	526.0
2	10	1121	1120.3	2136.5	1132.2	1626.1	**1093.9*** (1)	2103.5 (5)	1109.7* (0)	1597.6 (0)	1083.2	2075.7	1098.2	1583.6
	20	628	611.3*	1288.4	634.7	844.5	**590.9** (10)	1239.5 (28)	615.4* (7)	818.9 (9)	573.9	1226.0	599.4	803.4
	30	466	445.8*	1003.2	496.4	587.3	**433.1** (21)	Timeout	474.4 (23)	564.1 (19)	411.7	942.7	460.3	540.9
	40	385	372.3*	864.0	430.3	453.8	**361.2** (24)	Timeout	419.6 (27)	431.7 (26)	346.1	797.7	392.0	411.0
	50	**320**	332.4	773.7	361.8	366.1	329.8 (26)	Timeout	362.0 (28)	Timeout	313.9	712.0	316.8	319.7
5	10	1105	1103.7	1994.1	1106.3	1527.8	**1081.5*** (1)	1968.5 (4)	1084.7 * (0)	1499.3 (0)	1068.0	1943.6	1074.0	1482.9
	20	594	594.9	1151.7	599.5	770.4	**570.6** (15)	1110.1 (14)	575.5* (7)	742.5 (7)	553.9	1097.4	563.0	722.0
	30	426	430.3	866.0	473.4	529.2	**411.3** (26)	828.6 (25)	413.8 (22)	503.2 (19)	388.0	805.0	400.4	477.3
	40	334	347.0	721.5	363.2	410.5	**318.0** (29)	698.0 (29)	343.0 (28)	378.0 (29)	304.2	661.6	321.4	359.3
	50	295	294.4	637.7	321.7	339.2	**275.0** (29)	624.0 (29)	299.0 (29)	Timeout	251.7	575.2	274.6	290.4
10	10	1090	1103.7	1943.6	1103.7	1490.8	**1077.4*** (0)	1915.8 (1)	1073.5* (2)	1468.6 (1)	1063.0	1895.3	1068.9	1449.5
	20	607	588.0*	1099.0	590.8	742.9	**564.4** (14)	1062.5 (6)	560.9 (12)	719.7 (13)	548.8	1040.0	552.3	693.1
	30	414	423.6	815.3	423.3	509.2	**399.3** (26)	790.5 (14)	397.3 (18)	482.0 (21)	380.9	759.1	384.9	456.5
	40	341	344.8	678.4	342.3	394.4	Timeout	636.8 (21)	**315.3** (27)	360.5 (28)	297.2	613.7	303.1	340.1
	50	277	286.4	591.0	301.9	326.3	**273.0** (29)	568.0 (26)	Timeout	Timeout	244.2	532.4	252.1	272.1

7 Conclusions

We proposed here the combination of a genetic algorithm (GA) with two path planning methods already known in the literature for solving Multi-Agent Pickup and Delivery (MAPD) problems. The GA is applied to the task allocation part of this problem, and the Prioritized Planning and the Improved Conflict-Based Search (ICBS) are used for path planning. Also, we considered two representations for the GA and a heuristic for the initialization of the population.

Computational experiments were performed and the proposed approaches were compared with Central, a well-known technique from the literature for MAPD problems. ICBS presented the best path-planning performance, but it only works very well for a small number of agents. Prioritized Planning scaled well to a larger number of agents and achieved better results than Central. Moreover, GA-TA presents the best task assignment performance, in particular for high task rates, with the two path planning algorithms and with the heuristic for the initialization. This heuristic assists the search process of GA-TA, reducing the makespan values of the obtained solutions.

For future work, we intend to extend the application of the proposed techniques to situations with a larger number of agents, different (and even more realistic) environments and considering additional constraints in the tasks.

References

1. Boyarski, E., et al.: ICBS: improved conflict-based search algorithm for multi-agent pathfinding. In: Proceedings of the 24th International Joint Conference on Artificial Intelligence (IJCAI), Buenos Aires, Argentina, pp. 740–746, July 2015
2. D'Andrea, R.: Guest editorial: a revolution in the warehouse: a retrospective on kiva systems and the grand challenges ahead. IEEE Trans. Autom. Sci. Eng. **9**(4), 638–639 (2012)
3. Felner, A., et al.: Search-based optimal solvers for the multi-agent pathfinding problem: summary and challenges. In: Proceedings of the 10th International Symposium on Combinatorial Search, SOCS 2017, USA, 16–17 June 2017, pp. 29–37. AAAI Press, Pittsburgh, June 2017
4. Grenouilleau, F., van Hoeve, W., Hooker, J.N.: A multi-label A* algorithm for multi-agent pathfinding. In: Benton, J., Lipovetzky, N., Onaindia, E., Smith, D.E., Srivastava, S. (eds.) Proceedings of the 29th International Conference on Automated Planning and Scheduling, ICAPS, pp. 181–185 (2019)
5. Jose, K., Pratihar, D.K.: Task allocation and collision-free path planning of centralized multi-robots system for industrial plant inspection using heuristic methods. Robot. Auton. Syst. **80**, 34–42 (2016)
6. Li, J., Surynek, P., Felner, A., Ma, H., Kumar, T.K.S., Koenig, S.: Multi-agent path finding for large agents. In: The 33rd AAAI Conference on Artificial Intelligence, AAAI 2019, 31st Innovative Applications of Artificial Intelligence Conference, IAAI 2019, The Ninth AAAI Symposium on Educational Advances in Artificial Intelligence, EAAI 2019, 27 January–1 February 2019, pp. 7627–7634. AAAI Press, Honolulu (2019)

7. Li, J., Tinka, A., Kiesel, S., Durham, J.W., Kumar, T.K.S., Koenig, S.: Lifelong multi-agent path finding in large-scale warehouses. CoRR abs/2005.07371 (2020). https://arxiv.org/abs/2005.07371
8. Liu, C., Kroll, A.: A centralized multi-robot task allocation for industrial plant inspection by using A* and genetic algorithms. In: Rutkowski, L., Korytkowski, M., Scherer, R., Tadeusiewicz, R., Zadeh, L.A., Zurada, J.M. (eds.) ICAISC 2012. LNCS (LNAI), vol. 7268, pp. 466–474. Springer, Heidelberg (2012). https://doi.org/10.1007/978-3-642-29350-4_56
9. Liu, M., Ma, H., Li, J., Koenig, S.: Task and path planning for multi-agent pickup and delivery. In: Proceedings of the 18th International Conference on Autonomous Agents and MultiAgent Systems, AAMAS 2019, 13–17 May 2019, pp. 1152–1160. International Foundation for Autonomous Agents and Multiagent Systems, Montreal (2019)
10. Ma, H., Hönig, W., Kumar, T.K.S., Ayanian, N., Koenig, S.: Lifelong path planning with kinematic constraints for multi-agent pickup and delivery. In: The 33rd AAAI Conference on Artificial Intelligence, AAAI 2019, The 31st Innovative Applications of Artificial Intelligence Conference, IAAI 2019, The Ninth AAAI Symposium on Educational Advances in Artificial Intelligence, EAAI, Honolulu, Hawaii, USA, pp. 7651–7658 (2019)
11. Ma, H., et al.: Overview: generalizations of multi-agent path finding to real-world scenarios. CoRR abs/1702.05515, pp. 1–4 (2017)
12. Ma, H., Li, J., Kumar, T.S., Koenig, S.: Lifelong multi-agent path finding for online pickup and delivery tasks. In: Proceedings of the 16th Conference on Autonomous Agents and MultiAgent Systems, AAMAS 2017, pp. 837–845. International Foundation for Autonomous Agents and Multiagent Systems, Richland (2017)
13. Morris, R., et al.: Planning, scheduling and monitoring for airport surface operations. In: Planning for Hybrid Systems, Papers from the 2016 AAAI Workshop, 13 February 2016, vol. WS-16-12, pp. 608–614. AAAI Press, Phoenix (2016)
14. Sharon, G., Stern, R., Felner, A., Sturtevant, N.R.: Conflict-based search for optimal multi-agent pathfinding. Artif. Intell. **219**, 40–66 (2015)
15. Silver, D.: Cooperative pathfinding. In: AIIDE, vol. 1, pp. 117–122 (2005)
16. Standley, T.S.: Finding optimal solutions to cooperative pathfinding problems. In: Proceedings of the 24th AAAI Conference on Artificial Intelligence, AAAI 2010, 11–15 July 2010, pp. 173–178. AAAI Press, Atlanta (2010)
17. Stern, R., et al.: Multi-agent pathfinding: definitions, variants, and benchmarks. In: Proceedings of the 12th International Symposium on Combinatorial Search, SOCS, Napa, California, pp. 151–159 (2019)
18. Svancara, J., Vlk, M., Stern, R., Atzmon, D., Barták, R.: Online multi-agent pathfinding. In: The 33rd AAAI Conference on Artificial Intelligence, AAAI, The 31st Innovative Applications of Artificial Intelligence Conference, IAAI 2019, The 9th AAAI Symposium on Educational Advances in Artificial Intelligence, EAAI, Honolulu, Hawaii, USA, pp. 7732–7739 (2019)
19. Veloso, M.M., Biswas, J., Coltin, B., Rosenthal, S.: CoBots: robust symbiotic autonomous mobile service robots. In: Proceedings of the 24th International Joint Conference on Artificial Intelligence, IJCAI 2015, 25–31 July 2015, p. 4423. AAAI Press, Buenos Aires (2015)
20. Walker, T.T., Sturtevant, N.R., Felner, A.: Extended increasing cost tree search for non-unit cost domains. In: Proceedings of the 27th International Joint Conference on Artificial Intelligence, IJCAI, Stockholm, Sweden, pp. 534–540 (2018)
21. Wurman, P.R., D'Andrea, R., Mountz, M.: Coordinating hundreds of cooperative, autonomous vehicles in warehouses. AI Mag. **29**(1), 9–20 (2008)

Testing Multiagent Systems Under Organizational Model $\mathcal{M}$oise Using a Test Adequacy Criterion Based on State Transition Path

Ricardo Arend Machado(✉) and Eder Mateus Gonçalves

Center for Computational Sciences, Universidade Federal do Rio Grande - FURG, Rio Grande, RS, Brazil
ricardoarend@gmail.com, edergoncalves@furg.br

Abstract. The test phase is a crucial step to seek the correction of the entire software system to provide guarantees of operation and safety for users. However, applications based on multiagent systems have greater difficulty of being tested due to their properties such as autonomy, reactivity, pro-activity, and social skills. An organizational model imposes some restrictions on the behavior of agents since they are a set of behavioral constraints. This paper presents a method for systematic testing of multiagent systems specified under the $\mathcal{M}$oise$^+$ organizational model. First, a mapping is performed on a colored Petri net to measure the number of test cases needed to verify the system using adequacy criteria called the state transition path. Then each test case is specified from the Petri Net model in detail to be used as a guide in a test environment. The results indicate that the methodology can measure the testability and generate use cases for a $\mathcal{M}$oise$^+$ specification, guaranteeing a correction degree for a social level of a multiagent system.

Keywords: Testability · $\mathcal{M}$oise · Organization · Petri nets

1 Introduction

Multiagent Systems (MAS) are gaining increasing importance in the most different areas due to the unique characteristics of the agents, such as reactivity, proactivity, autonomy, and social capacity [18]. However, specific applications require a minimum of reliability so that the system does not present any risk to the user, and testing is essential. In summary, software testing consists of dynamic verification of a program's behavior in a set of suitably selected test cases [1]. The problem is that testing a MAS is not a simple task as these systems are often programmed to be autonomous and deliberative, and operate in an open world, which makes them sensitive to the context [10].

In general, testing in MAS can be classified in five different levels [17]: (i) **unity:** testing codes and modules unity that compose agents such as plans, goals,

R. Cerri and R. C. Prati (Eds.): BRACIS 2020, LNAI 12320, pp. 154–168, 2020.
https://doi.org/10.1007/978-3-030-61380-8_11

sensors, reasoning mechanisms and others; (ii) **agent:** testing the different modules of an agent and testing the agent resources to reach its goals and to detect the environment; (iii) **integration:** testing the agent's interaction, communication protocols and semantic, agents interacting with the environment and also with the shared resources; (iv) **system:** testing the whole MAS in an operational environment of destiny; testing for quality properties such as adaptation, openness, failure tolerance, and performance; (v) **acceptance:** testing of MAS in its execution environment and verifying if it reaches the system goals.

In [3] is presented a specific strategy for testing an embedded MAS based on three primary levels: agent test, collective resources test, and acceptance test. For each level, it is necessary to define the goals for a well-succeeded test. In [20], it is proposed a method to evaluate the testability of BDI agents using an adequacy criterion called all edges, which is satisfied if the set of tests covers all edges in a control-flow graph. The testability degree indicates the number of test cases necessary to validate a BDI program. Another contribution in the agent level was proposed in [16], where an architecture based on signatures simplifies the MAS project, including also an integration level of the test. On the other hand, in [15] proposes an approach covering the agent, integration, and system levels with functional tests using a model based on Petri Nets. In this case, the test cases are automatically generated using a behavioral system model as input.

One way to restrict the behavior of MAS is to represent them as a group through organizational models. Such models coordinate agents in groups and hierarchies in addition to establishing specific behavioral rules. An organizational structure can reduce the scope of interactions between agents, reduce or explicitly increase redundancy of a system, or formalize high-level system goals, of which a single agent may be not aware [19]. From another perspective, an organization can be seen as a set of behavioral constraints that a group of agents adopts to control the agent's autonomy and easily achieve their goals purposes [4].

One of the most referenced organizational models is $\mathcal{M}$oise$^+$, a model that proposes an organizational modeling language that explicitly decomposes a specification into structural, functional, and deontic dimensions. The structural dimension describes organizations using concepts such as *roles*, *groups* and *links*. The functional dimension describes a system by global collective goals that must be achieved. The deontic dimension realize the binding between the structural and functional dimensions, where it defines *permissions* for each role and *obligations* for missions [11].

To achieve good test coverage it is necessary to choose an appropriate test adequacy criteria for the system to be tested. Formally, an adequacy criterion can be defined as a predicate which asserts whether a set of test cases is adequate for testing a program against a specification [21]. According to the test criteria selected, each technique can reach a certain coverage in a search space of correctness for a system [2]. Dealing with Petri Nets as a mapping tool for testing procedures, [22] indicates four different methods, to know: (i) transition-oriented testing; (ii) state-oriented testing; (iii) flow-oriented testing and; (iv) specification-oriented testing. In [6], it was proposed the first approach of a testing methodology based on flow-oriented technique specific for organization models under $\mathcal{M}$oise$^+$ specification.

This paper aims to propose an increase in the coverage of testing procedures, including a state-oriented technique based on Coloured Petri Nets (CPN). Since the distributed and asynchronous nature of an organizational model, the method is also based on CPN [13] that can integrate the three dimensions of a $\mathcal{M}$oise$^+$ description. Employing a test adequacy criterion called *state transition path*, the proposed method indicates the number of test paths (or test cases) necessary to verify a $\mathcal{M}$oise$^+$ specification. In the end, test cases are generated individually for each test path found. We apply the method in a classic example of $\mathcal{M}$oise$^+$ in the literature, demonstrating that it can manage the most common scenarios for this kind of description.

2 Testing $\mathcal{M}$oise$^+$ Using State Transition Path Method

In a global view of the development of a MAS, we propose a test approach based on a perspective in which a MAS is seen from three dimensions [7]: (i) an *organization*, which corresponds to the description of the social level, where the elements that enable interaction and coordination between agents are formally or informally structured; (ii) a *communication* which deals with aspects related to protocols and information exchanges between agents and; (iii) the *agents* which correspond to the specification of the individual level where the demands of the organization are instantiated based on the individual capacities and skills of each agent. In this context, this article deals with tests at the level of the organization of a multiagent system.

In a MAS, an organization is a roles collection, relationships, and authority structures that regulate behaviors. The organization exists, even implicitly and informally, or by formal organizational models. These regulations influence authority relations, data flow, resources allocation, coordination standards, or any other system feature [9].

One of the most referenced organizational models is $\mathcal{M}$oise$^+$. It conceives a MAS as an organization, i.e., a normative set of rules that constrains the agent's behaviors. Three main concepts represent these organizational constraints: roles, plans, and norms [8]. Its structure has three levels (individual, collective, and social) and three types of specification (structural, functional, and deontic). The individual level defines constraints about the possible actions of each agent. The collective level constrains the set of agents that can cooperate. The social level imposes constraints about the kind of interactions that the agents can perform.

The choice of Petri Nets for mapping a MAS is due to the following reasons [5]: (i) they provide a graphically and mathematically founded modeling formalism which is a strong requirement for system development processes that need graphical as well as algorithmic tools; (ii) they are able to integrate the different levels and types of the $\mathcal{M}$oise$^+$ specification as a result of its mechanisms for abstraction and hierarchical refinement; (iii) there exists a huge variety of algorithms for the design and analysis of Petri Nets and powerful tools have been developed to aid in these process; (iv) there are different variants of Petri Net models that are all related by the basic formalism which they build upon. Besides the basic model, the Ordinary Petri Net, there are extensions such as timed, stochastic, high-level, and others, meeting the specific needs for almost every application area.

However, the ordinary Petri Net model is not sufficient to represent a complete $\mathcal{M}$oise$^+$ model, once it integrates different dimensions of representation, expressed by a Structural Specification (SS), a Functional Specification (FS), and a Deontic Specification (DS) [8]. Since it demands a more robust representation resource, our approach has a mapping procedure using a model based on Coloured Petri Nets [13].

Therefore we propose the Coloured Petri Net for $\mathcal{M}$oise$^+$ (*CPN4M*) description as a nine-tuple, $CPN4M = (P, T, A, \Sigma, V, C, G, E, I)$ where P is a finite set of places with dimension n which represent goals in a $\mathcal{M}$oise$^+$ organization; T is a finite set of transitions with dimension m which define plans to achieve the goals in a organization; A is a set of arcs and $A \subseteq P \times T \cup T \times P$, Σ is a finite nonempty color set which represents roles in a $\mathcal{M}$oise$^+$ organization; V is a set of typed variables such that $Type[v] \in \Sigma$ for all variables $v \in V$, which define the deontic specification in a $\mathcal{M}$oise$^+$ organization; $C : P \to \Sigma$ is a set of colors functions, which assigns color sets to places, defining which roles are responsible by which goals in a $\mathcal{M}$oise$^+$ organization; $G : T \to EXPR_V$ is a guard function which assigns a guard to each transition t such that $Type[G(t)] = Bool$; and $E : A \to EXPR_V$ is a arc expression function that assigns an arc expression to each arc a such that $Type[E(a)] = C(p)_{MS}$, where p is the place connected to the arc a; and $I : P \to EXPR_\emptyset$ is an initialization function which assigns an initialization expression to each place p such that $Type[I(p)] = C(p)_{MS}$, defining an order of execution for goals in a $\mathcal{M}$oise$^+$ organization.

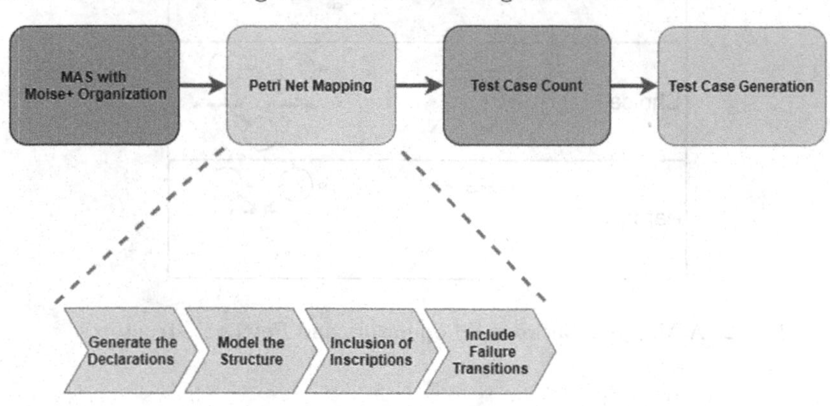

Fig. 1. Sequence of steps for the proposed approach

Figure 1 shows an overview of the steps contained in the testing methodology model proposed in this work. The first step is to analyze the $\mathcal{M}$oise$^+$ specification mapping it in a *CPN4M* specification. The net mapping consists of the following steps: (i) Generate Declarations, (ii) Model the Structure, (iii) Inclusion of

Inscriptions, (iv) Include Failure Transitions. With the *CPN4M* generated, the second step consists of counting the test cases of the network, thus ending the testability part of $\mathcal{M}$oise$^+$ specification. The third and last step is the generation of test cases related to the paths found, using them to perform tests on the system.

2.1 The Mapping

The first step in our method is mapping a $\mathcal{M}$oise$^+$ specification in a *CPN4M*. Figure 1 describes the mapping steps, consisting of the following: (i) generate declarations, (ii) model the structure, (iii) include inscriptions, (iv) include failure transitions. Each of these steps is described in detail below.

Generate Declarations. The declarations are based on the Structural Specification. For each role of the SS, a specific color must be created. For example, if in an SS we have a role called *writer*, we create a color also called *writer* and then assign that color to a set of colors. In addition to the colors, it is also necessary to declare the variables that are used to portray those same colors in the arcs.

Model the Structure. The modeling of the *CPN4M* structure should be based on the functional specification of the $\mathcal{M}$oise$^+$ model. Using the list of operators described in Fig. 2, it is possible to transform each plan operator (sequence, choice, and parallelism) of an FE into a corresponding *CPN4M* structure.

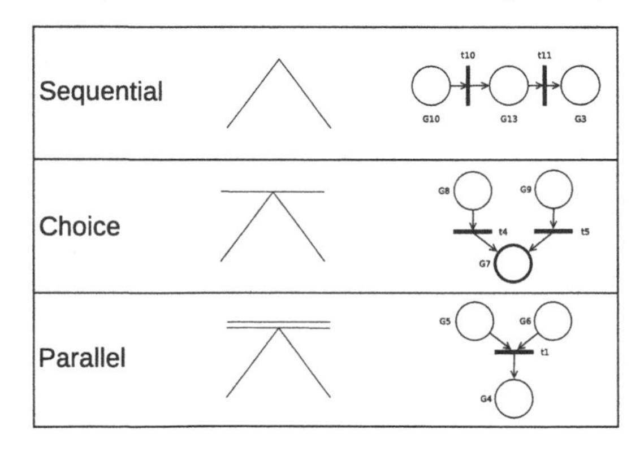

Fig. 2. A Mapping between FS operators and Petri net structures.

Including Inscriptions. The *CPN4M* inscriptions must obey the assignment of roles to each mission, which defines which colors should be assigned to each place in the net, according to the Deontic Specification. The same logic must define arc transitions and inscriptions. At this point, it is already possible to perform some simulations, and the complete relationship between SS and FS is already defined.

Goals that do not have a linked mission and therefore do not need a role that satisfies them, depending only on the completion of other goals to fulfill them, receive the color type *BOOL* which represents the Boolean type that accepts two types of tokens: *true* and *false*.

Failure Transitions. The last part of the network mapping is the inclusion of transitions activated in case of failure that can occur during the execution of a goal. A goal can fail because the actions needed to complete the plan, which the agent needs to perform to meet a given goal, may not go as planned. Once the net specifies the organization, if two or more agents have the same role, the failure occurs if none of them manage to reach the defined goal. As mentioned in [20], on a MAS, there is a possibility of failure executing a plan due to environmental interference. Summarizing, only places that are not *BOOL* type can receive a failure path.

2.2 Test Case Counting

After finishing the mapping process, the next step consists of assessing the testability of the model. Then, test cases are enumerated using test adequacy criteria. These criteria define how the test cases are extracted from the testing system mapped in a CPN4M.

In order do cover a different search space of correction than those proposed on [6], this work proposes an alternative approach using the adequacy criterion called *state transition path* [22]. The different markings of a CPN are transformed into a directed graph called state space. In this graph, the nodes represent the net markings, and the arcs represent the connecting elements between these markings. The *state transition path* criterion takes into account all possible execution paths for a state space graph in a CPN, so we must identify each of these paths.

Therefore, to count the test cases, it generates a state space graph corresponding to our *CPN4M*. After, it must perform the path count from the initial node to all possible final nodes of this graph. Then, the number of paths identified is the number of test cases, since each test path of the state space corresponds to a test case.

2.3 Test Case Generation

The system test consists of a test set where each different test path is a sequence of goals execution. In the test environment, simulations indicate where an agent with the role x being committed to the goal y must perform actions that satisfy the individual plan to fulfill the goal y following the sequence of the path of the goal until the end or until an error occurs.

The accounted test cases must be described individually to be used as a guide when performing system tests. Each test path properly identified by the method already presented corresponds to a test case. The test case description is made through a table containing the respective node number, goal, role, and a brief description of that goal.

3 Application Scenario

To demonstrate how the proposed method in the previous section works, we use a scenario that represents a soccer team whose primary goal is to score a goal, adapted from [12]. In Fig. 3, we can see the structural specification where we can identify the roles goalkeeper, back, middle, and attacker that are sub-roles of a player.

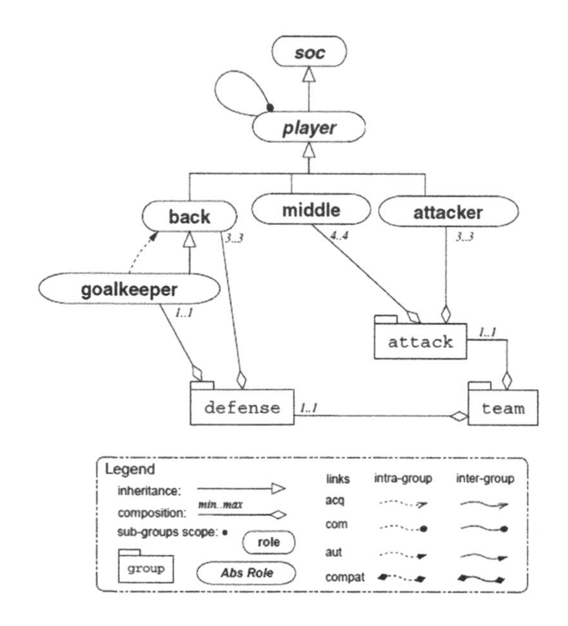

Fig. 3. Structural specification of the *Soccer Team*

The functional specification defines the coordination between the groups of agents to achieve their goals, as shown in Fig. 4. Here, a goal decomposition tree is presented and the reading is done as follows: the goal $g6$ is done; the choice $g7$ or $g8$; the choice $g16$ or $g17$; $g18$ and $g19$ in parallel; the goal g14; the choice $g20$ or $g21$. Table 1 describes all goals for this example.

Table 2 represents the deontic specification for this scenario and defines the permissions and obligations of the roles that assume the missions. The missions are defined by $m1$ referring to transfer the defense ball to the midfield, $m2$ referring to transfer the middle ball to the attack, $m3$ referring to position the attackers, and kick the ball on goal.

Now that we present the scenario, it is the moment to implement the method.

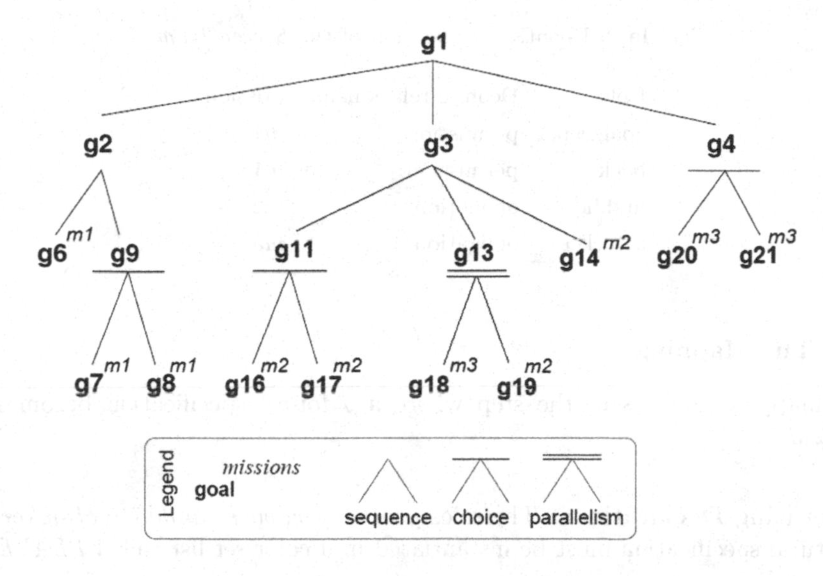

Fig. 4. Functional specification of the *Soccer Team*

Table 1. Goals descriptions *Soccer Team*

Goal	Description
g1	a goal was scored
g2	the ball is in the midfield
g3	the ball is in the attack field
g4	the ball was kicked into the goal
g6	a defensive player received the ball
g7	the ball was passed to the left midfielder
g8	the ball was passed to the right midfielder
g9	the ball was passed to the midfield
g11	a middle is with the ball
g13	an attacker is in a good position
g14	a middle passes the ball to an attacker
g16	a middle receives the ball on the left side
g17	a middle receives the ball on the right side
g18	an attacker positions himself to receive the ball
g19	a midfielder prepares to pass the ball
g20	an attacker kicks the ball on the left side of the goal
g21	an attacker kicks the ball on the right side of the goal

Table 2. Deontic specification of the *Soccer Team*

Role	Deontic relationship	Mission
goalkeeper	permission	$m1$
back	permission	$m1$
middle	obligation	$m2$
attacker	obligation	$m3$

3.1 The Mapping

The mapping consists of the step where a $\mathcal{M}$oise$^+$ specification becomes a *CPN4M*.

Generating Declarations. The roles *goalkeeper*, *back*, *middle* e *attacker* of structural specification must be instantiated in a color set list called *PLAYER*.

```
colset players = with goalkeeper | back | attacker | middle ;
colset PLAYER = list players;
var p, p1,p2: PLAYER;
```

Modeling the Structure. In this step, it is generated the net structure. The goal decomposition tree in Fig. 4 is used as a reference for the primary *CPN4M* structure. The mapping in Fig. 2 is a guide to generate the corresponding net. In the end, the net structure shown in Fig. 5, is generated.

Including Inscriptions. This step constitutes the inclusion of inscriptions in the net, as shown in Fig. 5, where places represent the goals and the tokens (colors) identify the roles. The inscriptions include arc expressions, variables, place names, color sets, transition names, and the initial marking.

Failure Transitions. For the inclusion of failure paths, all places in the network that are not of the *BOOL* type are considered, that is, leaving only the goals that have a linked mission and, therefore, a role to carry out actions that satisfy that goal. Figure 5 presents the complete version of the *Soccer Team* example mapping in the *CPN4M* format, with the place of failure and their respective paths included.

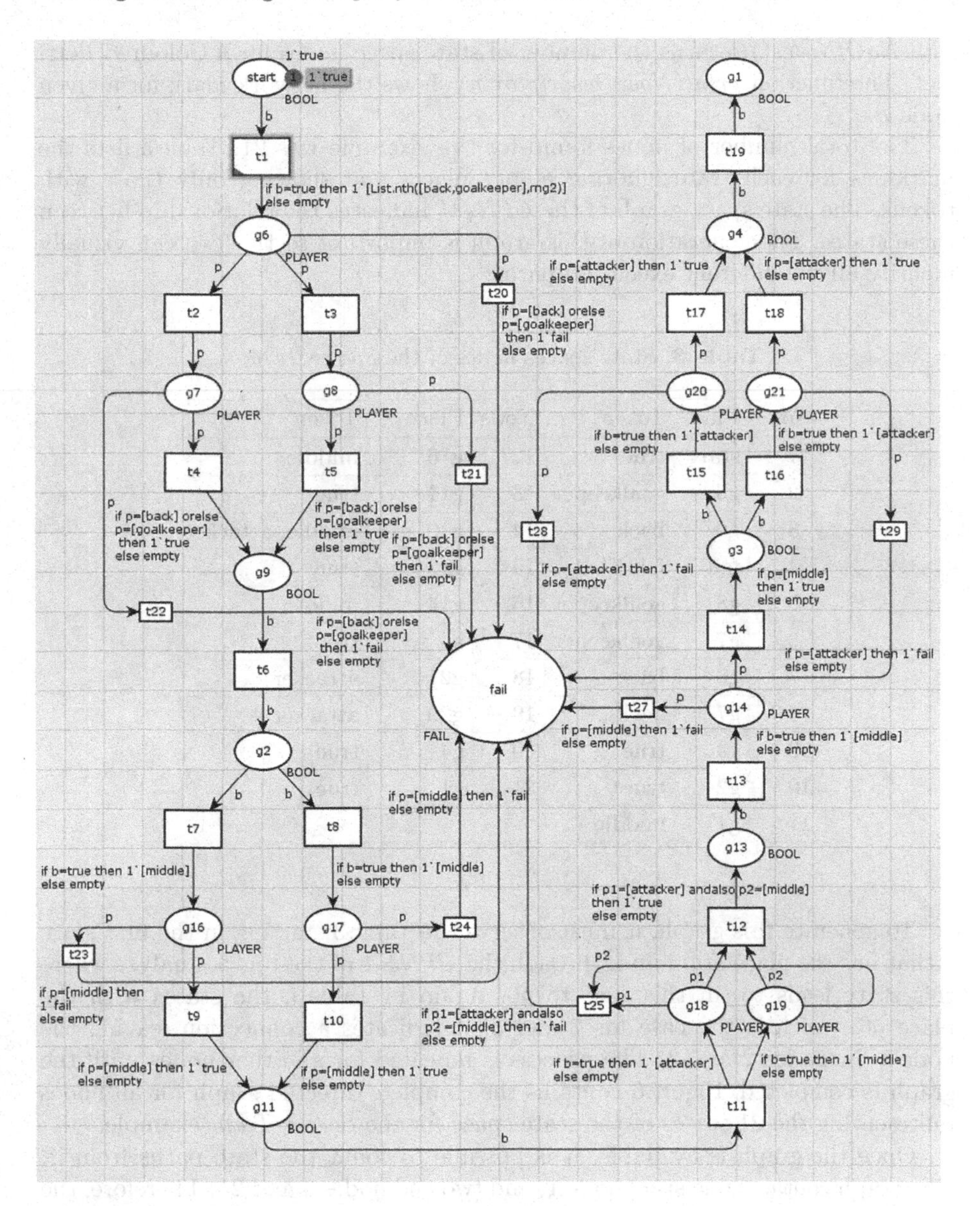

Fig. 5. *Soccer Team* in *CPN4M* format completed

3.2 Test Case Counting

Now that the net mapping is complete, the test paths are counted using the adequacy criterion *state transition path*. The first step is the identification of all state nodes that make up the state spaces. The CPN Tools, a Petri Net modeling tool [14], has some functions for helping in this process. The function *NoOfNodes*() returns the number of state space nodes for a Coloured Petri Net. The function *print*(*NodeDescriptor n*) shows the network mark for a given node *n*.

The total number of nodes found for this example was 21. Table 3 lists the markings for each state ignoring empty places and showing only those with tokens. The state space graph of the *CPN4M* indicates the relationship between these states. The generation of this graph is important so that we can visually identify all the different execution paths.

Table 3. State spaces nodes of the *Soccer Team*

Node	Place	Token	Node	Place	Token
1	start	true	12	g16	middle
2	g6	goalkeeper	13	g11	true
3	g6	back	14	g19, g18	middle, attacker
4	fail	fail	15	g13	true
5	g8	goalkeeper	16	g14	middle
6	g7	goalkeeper	17	g3	true
7	g8	back	18	g21	attacker
8	g7	back	19	g20	attacker
9	g9	true	20	g4	true
10	g2	true	21	g1	true
11	g17	middle			

To generate this graph, it must observe the Table 3 starting in the first state 1 that has the place *start* and, through the *CPN4M* of the Fig. 5, analyze where this state leads to, in this case to *g*6. Again in Table 3, the states with the place *g*6, which is the case for 2 and 3, it indicates a connection leaving the node 1 for nodes 2 and 3. The process is repeated for all other nodes until the graph is completed. Figure 6 contains the complete directed graph for all nodes representing the 21 states of the state space for the *Soccer Team* example.

Once the graph is available, it is possible to count the state paths from it. The graph contains one start node 1, and two end nodes 4 and 21. Therefore, the number of paths corresponding to all possible systems executions must include the sum of all paths from node 1 to 4 as well from node 1 to node 21. A backtracking algorithm[1] can be used to perform this count. For our scenario, there are 62 different paths, 46 correspondings to those that end at node 4 and 16 at node 21.

[1] https://www.geeksforgeeks.org/count-possible-paths-two-vertices/.

3.3 Test Case Generation

After identifying and counting the test cases, the last step is the detailed description of each one. Since the number of paths found is large, it is challenging to present them all, so we show two examples of these test cases, one for when the execution is successful and another when it fails. First, we identify what these paths arc in our graph of Fig. 6:

1. $1 \rightarrow 3 \rightarrow 8 \rightarrow 9 \rightarrow 10 \rightarrow 12 \rightarrow 13 \rightarrow 14 \rightarrow 15 \rightarrow 16 \rightarrow 17 \rightarrow 18 \rightarrow 20 \rightarrow 21$.
2. $1 \rightarrow 3 \rightarrow 8 \rightarrow 4$.

Using the Table 3 to identify which place and token each node represent, identifying the goals and roles of our test case, along with Table 1 which describes these goals, the Table 4 is generated with a description of the test case for the chosen successful path.

Table 4. A test case example for *Soccer Team* without fail

Node	Goal	Role	Description
1	start	bool	start the test
3	g6	*back*	a *back* gets the ball in the defense field
8	g7	*back*	a *back* passes the ball to the left midfielder
9	g9	bool	the ball was passed to the midfield
10	g2	bool	the ball is in the midfield
12	g16	*middle*	a *middle* receives the ball on the left side
13	g11	bool	a *middle* is with the ball
14	g19, g18	*middle*, *attacker*	a *middle* prepares to pass the ball and an *attacker* positions himself to receive the ball
15	g13	bool	an *attacker* is in a good position
16	g14	*middle*	a *middle* passes the ball to an attacker
17	g3	bool	the ball is in the attack field
18	g21	*attacker*	an *attacker* kicks the ball on the left side of the goal
20	g4	bool	the ball was kicked into the goal
21	g1	bool	a goal was scored

Repeating the procedure for path 2, where a failure in the network's execution occurs, the intention is not to simulate that failure. However, for predicting this possibility, in case it happens during the execution of the tests. Table 5 contains the description for this test case.

Table 5. A test case example for *Soccer Team* with fail

Node	Goal	Role	Description
1	*start*	bool	start the test
3	g6	*back*	a *back* gets the ball in the defense field
8	g7	*back*	a *back* passes the ball to the left midfielder
4	fail	bool	a failure occurs in the execution of goal g7

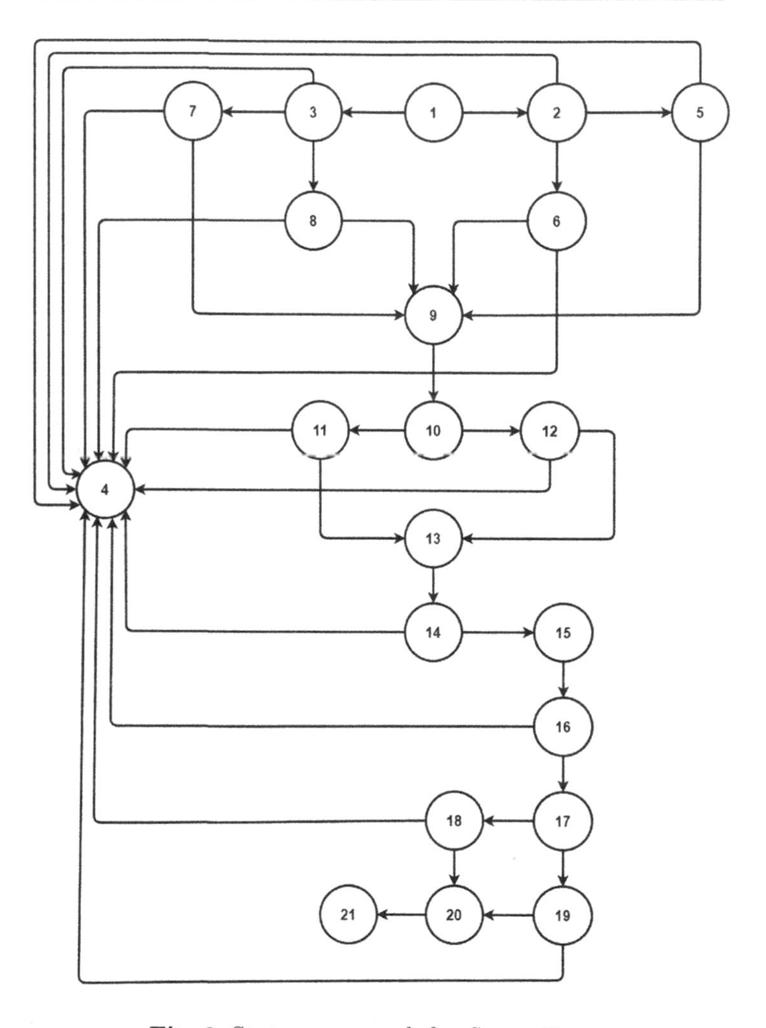

Fig. 6. State space graph for *Soccer Team.*

4 Conclusions

This paper introduces a new approach for testing MAS containing a formal method for assessing the testability of the model under analysis and for generating test cases. Focusing on testing MAS using a specific organizational model called $\mathcal{M}$oise$^+$, the system was successfully mapped to a CPN model here named *CPN4M*, and test cases could be calculated using the *state transition path* adequacy criterion. In the end, test cases were generated, which can be used to guide the execution of system tests.

The contributions presented in this paper are: (i) a formal definition for a Colored Petri Net mapping for $\mathcal{M}$oise$^+$; (ii) a new approach for assessing the testability of the system using the adequacy criterion *state transition path*; (iii) the development of a method to generate and specify test cases.

Future works include the development of a tool to automatically generate the flow control graph, counting the paths, and automatically generating the test case tables. Another possibility is to perform simulations on MAS where there is interference from the environment. We can also mention the realization of a study on different types of specifications $\mathcal{M}$oise$^+$ identifying the best test method for each type of specification.

References

1. Abran, A., Moore, J.W., Bourque, P., Dupuis, R., Tripp, L.: Software Engineering Body of Knowledge. IEEE Computer Society, Angela Burgess (2004)
2. Ammann, P., Offutt, J.: Introduction to Software Testing. Cambridge University Press, Cambridge (2016)
3. Barnier, C., Mercier, A., Jamont, J.P., et al.: Toward an embedded multi-agent system methodology and positioning on testing. In: 2017 IEEE International Symposium on Software Reliability Engineering Workshops (ISSREW), pp. 239–244. IEEE (2017)
4. Dignum, V., Dignum, F.: Modelling agent societies: co-ordination frameworks and institutions. In: Brazdil, P., Jorge, A. (eds.) EPIA 2001. LNCS (LNAI), vol. 2258, pp. 191–204. Springer, Heidelberg (2001). https://doi.org/10.1007/3-540-45329-6_21
5. Girault, C., Valk, R.: Petri Nets for Systems Engineering: A Guide to Modelling, Verification, and Applications. Springer, Heidelberg (2013). https://doi.org/10.1007/978-3-662-05324-9
6. Gonçalves, E.M., Rodrigues, B.C., Machado, R.A.: Assessment of testability on multiagent systems developed with organizational model $\mathcal{M}$oise. In: Moura Oliveira, P., Novais, P., Reis, L.P. (eds.) EPIA 2019. LNCS (LNAI), vol. 11805, pp. 581–592. Springer, Cham (2019). https://doi.org/10.1007/978-3-030-30244-3_48
7. Gonçalves, E.M., Quadros, C., Saldanha, J.: Uma análise comparativa da especificaçao formal em sistemas multi-agente: os desafios e as exigências uma década depois. In: Anais do X Workshop-Escola de Sistemas de Agentes, seus Ambientes e apliCaçoes, Maceio, Brazil, vol. 1 (2016)
8. Hannoun, M., Boissier, O., Sichman, J.S., Sayettat, C.: MOISE: an organizational model for multi-agent systems. In: Monard, M.C., Sichman, J.S. (eds.) IBERAMIA/SBIA -2000. LNCS (LNAI), vol. 1952, pp. 156–165. Springer, Heidelberg (2000). https://doi.org/10.1007/3-540-44399-1_17
9. Horling, B., Lesser, V.: A survey of multi-agent organizational paradigms. Knowl. Eng. Rev. **19**(4), 281–316 (2004)
10. Houhamdi, Z.: Multi-agent system testing: a survey. Int. J. Adv. Comput. (2011)

11. Hübner, J.F., Boissier, O., Kitio, R., Ricci, A.: Instrumenting multi-agent organisations with organisational artifacts and agents. Auton. Agent. Multi-Agent Syst. **20**(3), 369–400 (2010). https://doi.org/10.1007/s10458-009-9084-y
12. Hübner, J.F., Sichman, J.S., Boissier, O.: A model for the structural, functional, and deontic specification of organizations in multiagent systems. In: Bittencourt, G., Ramalho, G.L. (eds.) SBIA 2002. LNCS (LNAI), vol. 2507, pp. 118–128. Springer, Heidelberg (2002). https://doi.org/10.1007/3-540-36127-8_12
13. Jensen, K.: Coloured Petri Nets: Basic Concepts, Analysis Methods and Practical Use, vol. 1. Springer, Heidelberg (2013)
14. Jensen, K., Kristensen, L.M., Wells, L.: Coloured Petri Nets and CPN tools for modelling and validation of concurrent systems. Int. J. Softw. Tools Technol. Transfer **9**(3–4), 213–254 (2007). https://doi.org/10.1007/s10009-007-0038-x
15. Kerraoui, S., Kissoum, Y., Redjimi, M., Saker, M.: MATT: multi agents testing tool based nets within nets. J. Inf. Organ. Sci. **40**(2), 165–184 (2016)
16. do Nascimento, N.M., Viana, C.J.M., von Staa, A., Lucena, C.: A publish-subscribe based architecture for testing multiagent systems. In: SEKE, pp. 521–526 (2017)
17. Nguyen, C.D., Perini, A., Bernon, C., Pavón, J., Thangarajah, J.: Testing in multi-agent systems. In: Gleizes, M.-P., Gomez-Sanz, J.J. (eds.) AOSE 2009. LNCS, vol. 6038, pp. 180–190. Springer, Heidelberg (2011). https://doi.org/10.1007/978-3-642-19208-1_13
18. Padgham, L., Winikoff, M.: Developing Intelligent Agent Systems: A Practical Guide, vol. 13. Wiley, Chichester (2005)
19. van den Broek, E.L., Jonker, C.M., Sharpanskykh, A., Treur, J., Yolum, I.: Formal modeling and analysis of organizations. In: Boissier, O., et al. (eds.) AAMAS 2005. LNCS (LNAI), vol. 3913, pp. 18–34. Springer, Heidelberg (2006). https://doi.org/10.1007/11775331_2
20. Winikoff, M.: BDI agent testability revisited. Auton. Agent. Multi-Agent Syst. **31**(5), 1094–1132 (2017). https://doi.org/10.1007/s10458-016-9356-2
21. Zhu, H.: A formal analysis of the subsume relation between software test adequacy criteria. IEEE Trans. Softw. Eng. **22**(4), 248–255 (1996)
22. Zhu, H., He, X.: A methodology of testing high-level Petri nets. Inf. Softw. Technol. **44**(8), 473–489 (2002)

Knowledge Representation, Logic and Fuzzy Systems

A Fuzzy Approach for Classification and Novelty Detection in Data Streams Under Intermediate Latency

André Luis Cristiani[1](✉), Tiago Pinho da Silva[2], and Heloisa de Arruda Camargo[1]

[1] Federal University of São Carlos, São Carlos, SP 13565-905, Brazil
andre.cristiani@estudante.ufscar.br, heloisa@dc.ufscar.br
[2] University of São Paulo, São Carlos, SP 13566-590, Brazil
tpinho@usp.br

Abstract. Novelty detection is an important topic in data stream classification, as it is responsible for identifying the emergence of new concepts, new patterns, and outliers. It becomes necessary when the true label of an instance is not available right after its classification. The time between the classification of an instance and the arrival of its true label is called latency. This is a common scenario in data streams applications. However, most classification algorithms do not consider such a problem and assume that there will be no latency. On the other hand, a few methods in the literature cope with the existence of infinite latency and novelty detection in data streams. In this work, however, we focus on the scenario where the true labels will be available to the system after a certain time, called intermediate latency. Such a scenario is present in the stock market and weather datasets. Moreover, aiming for more flexible learning to deal with the uncertainties inherent in data streams, we consider the use of fuzzy set theory concepts. Therefore, we propose a method for classification and novelty detection in data streams called Fuzzy Classifier with Novelty Detection for data streams (FuzzCND). Our method uses an ensemble of fuzzy decision trees to perform the classification of new instances and applies the concepts of fuzzy set theory to detect possible novelties. The experiments showed that our approach is promising in dealing with the emergence of new concepts in data streams and inaccuracies in the data.

Keywords: Data streams · Classification · Novelty detection · Fuzzy

1 Introduction

Data streams are characterized by a sequence of infinite examples that arrives in a continuous and high-speed way, making it impossible for storing data in

This study was financed in part by the Coordenação de Aperfeiçoamento de Pessoal de Nível Superior - Brasil (CAPES) - Finance Code 001.

R. Cerri and R. C. Prati (Eds.): BRACIS 2020, LNAI 12320, pp. 171–186, 2020.
https://doi.org/10.1007/978-3-030-61380-8_12

memory. Because of this, they must be analyzed and discarded [1]. Moreover, due to their non-stationary nature, these data may change in their distribution or present new concepts over time. These characteristics are called concept-drift and concept evolution [2]. Thus, an important aspect of algorithms that deal with data streams is to identify such characteristics and adapt their decision model to eventual changes in the data.

Novelty Detection is a sub-task of classification in data streams whose objective is to identify new patterns in the examples that arrive from the data stream, while classifying the examples belonging to known concepts. In this way, we can identify the emergence of new concepts (concept evolution) and changes in known concepts (concept-drift) in order to adapt the classification model.

The availability of real labels and the use of external feedback is essential to define how to update the decision model. The time between the classification of an example and the arrive of its real label is called latency. According to the literature, there are three types of latencies: when the label is never available for the system, called extreme latency; when the label is available right after the example classification, called null latency; when the real label is only available after a certain time after the example classification, called intermediate latency [3]. In this work we focus on the intermediate latency scenario, which is a more realistic assumption than considering the availability of the real labels right after the classification. Such scenario, for instance, is presented in stock market [4] and weather datasets [5].

The changes in the data caused by concept-drifts and concept evolution along with the intermediate latency scenario can lead to uncertain situations. To cope with such problem the use of fuzzy set theory is a way to make the learning process more flexible, enabling the algorithm to adapt better to this situations. Recently, many methods that apply the concepts of fuzzy set theory have been developed [6]. Following this same line, looking for a more flexible learning experience, we propose a method called FuzzCND (Fuzzy Classifier with Novelty Detection for data streams).

Our proposal is based on an ensemble that uses fuzzy decision trees, called FuzzyDT [7]. First, FuzzCND trains the initial decision model based on the data available for training. After, new examples that arrive from the data stream are classified. The identification of outliers occurs through the limits of decisions that were traced on the leaves of the trees during training. We applied our method to synthetic and real data and obtained promising results compared to other algorithms that deal with the classification and novelty detection tasks.

The rest of this article is organized as follows: Sect. 2 presents related works. In Sect. 3, we describe the proposed method in detail. In Sect. 4, we present the experiments and their settings, together with the results. Finally, the conclusions and future work are presented in Sect. 5.

2 Related Work

Classification in data streams is a growing area of research with many approaches available [8,11]. Despite this increase in the literature, a not well explored issue in the data streams classification is the presence of latency regarding the availability of true labels. It plays a fundamental role in algorithms that deal with data streams, especially when it comes to non-stationary data, directly impacting the results [3]. In data streams classification, most studies consider null latency for obtaining the true labels [8,11], which can be unrealistic in most applications, and only a few consider extreme or intermediate latency [12,13].

The existence of latency in a data stream application generates the need to detect novelty, due to unknown patterns or new concepts that may emerge through time, that can not be monitored by true label information. To deal with this scenario the existing approaches usually divide the task into two steps, offline and online [2,9,10,14]. In the first step, offline, a set of labeled data is used to train the initial decision model. In the second step, online, new unlabeled examples from the data stream are classified, new patterns of novelties are detected and the decision model is updated with the most current data available.

In this scenario, Multiclass Novelty Detector for Data Streams (MINAS) [2] is one of the most popular. This algorithm does not use real labels and updates its decision model incrementally. MINAS classifies new instances as known classes, known patterns or unknown news patterns. Examples are considered unknown when they are beyond the reach of the decision boundaries of all micro-clusters that make up the decision model. Unknown examples are stored in a temporary memory. Whenever this memory has a number of examples, a clustering technique is applied to find cohesive groups that can represent a novelty pattern, a known class extension or an extension of a novelty pattern. This method considers the appearance of multiple novelty patterns. Despite being great at novelty detection, MINAS does not adapt well to inaccuracies in the data and classifies many examples as unknown, which can harm the results in several scenarios.

Following a similar idea, however, considering a more flexible learning, the Possibilistic Fuzzy Novelty Detector (PFuzzND) [9] is an algorithm for multiclass novelty detection based on the Possibilistic-Fuzzy C-Means (PFCM) [15] method, which also considers extreme latency. In the offline phase, the algorithm clusters the training data using the Fuzzy C-Means (FCM) [16] method. Each cluster is summarized in a data structure called Supervised Possibilistic Fuzzy Micro-Cluster (SPFMiC). In the online phase, incoming examples are classified using the same calculation as the typicality of PFCM. Unknown examples are stored in a temporary memory. Whenever this memory has a certain number of examples, the FCM algorithm is used to find cohesive groups that represent a novelty pattern.

In the current data streams classification literature, ensemble of classifiers has being widely used due to the results achieved [17]. However, few of the ensemble-based classification techniques can detect novelty. In this context, Enhanced Classifier for Data Streams with novel class Miner (ECSMiner) [10] is based on an ensemble of classifiers for classification and novelty detection in data streams under time constraints. In the same way, Class-based Micro Classifier Ensemble (CLAM) [14] performs classification and novelty detection in data streams with recurring classes through a class-based ensemble, where each known class has a micro-cluster ensemble. Both methods consider intermediate latency to obtain the labels.

Finally, motivated by the current results of ensemble methods in the classification task, and the use of fuzzy set theory to deal with the changes in the data stream, we propose the FuzzCND method that uses the concepts of fuzzy set theory and has adaptability to different latency values.

3 A Fuzzy Classifier for Data Streams with Novel Class Detector

Our proposed method, Fuzzy Classifier for Data Streams with Novel Class Detector (FuzzCND), uses the FuzzyDT classification algorithm in combination with the PFCM clustering method, to create models with flexible decision boundaries, which can help the identification of outliers, and the detection of novel patterns. Moreover, similarly to most of the novelty detection algorithms in the literature, our algorithm is divided in two phases, offline and online.

3.1 Offline Phase

The offline phase is responsible for training the initial classification model. Algorithm 1 details this phase, which receives as input parameters the labeled data, divided into chunks *initChunks*, which will be used for training, the fuzzification parameter m and the number of clusters per classifier K.

In Algorithm 1, a Fuzzy Decision Tree (FuzzyDT) [7] is trained for each available labeled data chunk (step 3), which are built with the examples that arrive sequentially, in the order they appear. Next, to enable the identification of outliers, decision boundaries are created on each of their leaf nodes based on the training data (step 4). This process will be described in detail in the next sub-section. Finally, each trained tree is incorporated into the ensemble of classifiers $\mathcal{E}$ (step 5).

Algorithm 1. Offline phase

Input: $initChunks, m, K$
Output: $\mathcal{E}$
1: $\mathcal{E} \leftarrow \emptyset$
2: **for all** Chunk C_i in $initChunks$ **do**
3: $model \leftarrow$ TRAINFDT(C_i)
4: CREATEDECISIONBOUNDARY($model, m, K, C_i$)
5: $\mathcal{E} \leftarrow \mathcal{E} \cup model$
6: **end for**
7: **return** $\mathcal{E}$

Creating Decision Boundaries During Training: We assume that instances of the same class have similar distributions and are independent of instances of other classes. Therefore, instances that are close to each other, according to a certain distance measure, belong to the same class.

When an example of a new class arrives, if no mechanism is implemented, the decision trees will not be able to correctly classify the new instances belonging to the new class. To prevent this situation, we define decision boundaries for the known classes, so instances out of these boundaries can be further investigate.

To define the decision boundaries for the decision trees in the ensemble, the training instances classified by each of the leaf nodes are clustered using the FCM algorithm and a summary of each resulting cluster is saved. In this way the clusters obtained from each leaf constitute the decision boundaries for the known classes.

The number of clusters built on each leaf node i is proportional to the number of training instances classified by the leaf node i, being defined by:

$$k_i = (t_i/T) * K \tag{1}$$

where t_i represents the number of training instances belonging to the leaf node, T represents the size of the chunk and K the number of clusters per classifier.

Summary of Cluster Information: The information for each cluster generated in the training stage is stored in summary structures called Supervised Possibilistic Fuzzy Micro Cluster (SPFMiC) [9].

An SPFMiC [9] is defined as a vector (M^e, T^e, $\overline{CF1^e_\mu}$, $\overline{CF1^e_T}$, SSD^e, N, t, *class_id*), where M^e represents a linear sum of the membership of examples raised to m; T^e a linear sum of the typicalities of examples raised to n; $\overline{CF1^e_\mu}$ is the linear sum of examples weighted by their membership; $\overline{CF1^e_T}$ is the linear sum of examples weighted for their tipicalities; SSD^e the sum of the distances of the examples for the prototype of the micro-cluster, increases of m and weighted by the example's membership; N the number of examples associated to the micro-cluster and *class_id* a class associated with the micro-cluster.

An SPFMiC includes statistics from a cluster that makes it possible to calculate its centroid and the membership and typicality of new examples. More details on how the calculations are performed can be found at [9]. In FuzzCND, the SFPMiCs are used only to define decision boundaries for the known classes, they are not incrementally updated as in its original work.

3.2 Online Phase

The online phase receives examples from the data stream and is responsible for three operations: classifying unlabeled examples, detecting novelties and updating the decision model through labeled examples. Algorithm 2 presents details of this phase, which requires as input parameters an initial threshold of classification *init_θ*, a parameter for adapting the classification threshold θ_{adapt}, the fuzzification parameter m, values necessary for the calculation of typicality $(\alpha, \beta, K$ and $n)$, the number of clusters for novelty detection k_short, the number of examples T to perform novelty detection process, the size of each chunk C and the maximum size of the ensemble S.

In Algorithm 2, each example that arrives is analyzed to see if it is an instance for classification or training (step 4). If the instance has no label, the algorithm checks whether this instance represents a *Foutlier* (step 6). For this, the typicality of x is calculated for all SPFMiCs belonging to the leaf nodes of all classifiers in the ensemble. If the maximum typicality value obtained is greater than the classification threshold $init\_\theta$ subtracted from θ_{adapt}, it is a sign that x belongs to one of the SPFMiCs of the leaf nodes and it is submitted to the classification process; otherwise, x represents a *Foutlier* and is stored in the temporary memory of unknown examples $unknown\_mem$.

Definition 1 (*Foutlier*). *An unlabeled instance represents a Foutlier (i.e. filtered outlier) if it is outside the decision boundaries of all classifiers.*

The classification process begins with the fuzzification of the continuous attributes of x (step 7). After, the instance is classified using the majority vote of the classifiers belonging to the ensemble (step 8). Whenever an unlabeled instance is marked as *Foutlier* and added to the unknown memory, the algorithm checks whether there is a minimum number of examples in the temporary memory to start the execution of the novelty detection process (step 11).

When a labeled instance arrives, it is added to the temporary memory of labeled examples $labeled\_mem$ (step 16) and the algorithm checks if there is a minimum number of examples to train a new fuzzy decision tree with more recent data (step 17). If so, a new FuzzyDT is trained with the instances belonging to the temporary memory of labeled examples (step 18). Whenever a new classifier is trained, the algorithm checks whether the ensemble has the maximum number of classifiers (step 19), if so, the classifier with the worst performance, based on the most recent labeled data stored in $labeled\_mem$, is removed (step 20) and the new FuzzyDT is incorporated into the ensemble (step 22).

Novelty Detection: Whenever the temporary memory of unknown examples ($unknown\_mem$) reaches a predefined number of examples, the novelty detection process, described in Algorithm 3 is initiated. The algorithm receives as input parameters the examples marked as unknown ($unknown\_mem$), the fuzzification parameter m, the number of clusters k_short, the minimum number of examples for a cluster to be valid min_weight and the ensemble $\mathcal{E}$.

In Algorithm 3, first all SPFMiCs belonging to the leaf nodes are collected (step 1). After, the instances present in the ($unknown\_mem$) are submitted to the clustering process through the FCM algorithm, resulting in k_short clusters (step 2). Each resulting cluster is analyzed to verify if it is valid. For this, the algorithm evaluates whether the fuzzy silhouette coefficient [18] is greater than 0 and whether the cluster has a representative number of examples (step 4).

Algorithm 2. Online phase

Input: $DS, init\_\theta, \theta_{adapt}, m, \alpha, \beta, n, K, k\_short, T, C, S$

```
1: unknown_mem ← ∅
2: labeled_mem ← ∅
3: while !IsEmpty(DS) do
4:     if !IsNull(x.class) then
5:         x ← Next(DS)
6:         if !IsFOutlier(ℰ, x, α, β, n, K, init_θ, θ_adapt) then
7:             x_fuzz ← FuzzifiesAttributes(x)
8:             x.class ← MajorityVoting(ℰ, x_fuzz)
9:         else
10:            unknown_mem ← unknown_mem ∪ x
11:            if |unknown_mem| ≥ T then
12:                unknown_mem ← NoveltyDetection(unknown_mem, m, k_short, ℰ)
13:            end if
14:        end if
15:    else
16:        labeled_mem ← labeled_mem ∪ x
17:        if |labeled_mem| ≥ C then
18:            model ← TrainFDT(labeled_mem)
19:            if |ℰ| >= S then
20:                removesWorstClassifier(ℰ, labeled_mem)
21:            end if
22:            ℰ ← ℰ ∪ model
23:        end if
24:    end if
25: end while
```

To assign a label to a new valid cluster, the algorithm checks the fuzzy similarity FR [19] between the valid cluster and each of the SPFMiCs belonging to the leaf nodes (steps 6 and 7). If the maximum FR between the new cluster and the SPFMiCs belonging to the leaf nodes is greater than a user-defined threshold σ, the new cluster represents an extension of an existing SPFMiC. Therefore, the new valid cluster receives the same label as the SPFMiC that has the greatest similarity to it. Otherwise, a temporary label "novelty" is associated with the new valid cluster. After, each instance belonging to that valid cluster receives the cluster label (step 14) and is removed from the memory of unknown examples (step 15). When the labels of these instances are available, a new classifier will be trained and incorporated into the ensemble of classifiers. Instances belonging to invalid clusters remain in the temporary memory of unknown examples until they become part of a valid cluster.

Impact of the Emergence of New Classes on Classifier Performance: Since new classes can appear and older classes can disappear, classifiers in the ensemble can recognize different sets of labels, which can negatively impact the classifier's performance. There are two situations in which an ensemble of classifiers facing concept evolution may experience reduced performance [10]:

Algorithm 3. Novelty Detection

Input: $unknown\_mem, m, k\_short, min\_weight, \mathcal{E}$

1: $all\_clustes\_leafs \leftarrow$ GETALLLEAFNODESCLUSTERS($\mathcal{E}$)
2: $temp\_clustes \leftarrow$ FCM($unknown\_mem, m, k\_short$)
3: **for all** Cluster $temp\_ci$ in $temp\_clusters$ **do**
4:     **if** VALIDATE($temp\_ci, min\_weight$) **then**
5:         **for all** Cluster $cluster\_li$ in $all\_clusters\_leafs$ **do**
6:             $FR \leftarrow$ SIMILARITY($temp\_ci, cluster\_li$)
7:             $all\_FR \leftarrow all\_FR \cup FR$
8:         **end for**
9:         $(label_{max}, max\_fr) \leftarrow$ MAX(all_FR)
10:        **if** $max\_fr < \sigma$ **then**
11:            $label_{max} \leftarrow$ "*novelty*"
12:        **end if**
13:        **for all** Instance x_i in $temp\_ci$ **do**
14:            $x_i.class \leftarrow label_{max}$
15:            $unknown\_mem \leftarrow$ REMOVE($x_i, unknown\_mem$)
16:        **end for**
17:    **end if**
18: **end for**
19: **return** $\mathcal{E}$

- *Situation 1*: Older classifiers have not been trained with data from a new class that has emerged and will erroneously classify instances belonging to that new class.
- *Situation 2*: Newer classifiers have not been trained with data from an older class and will erroneously classify an instance of that older class.

For both situations, solutions were proposed on the ensemble voting scheme to improve the adaptation of the algorithm. To mitigate Situation 1, priority was given to the vote of newer classifiers. If a newer classifier votes for a class that an older classifier does not recognize, voting priority is given to the newer classifier. Uniform voting was also used by other researchers [10,20] and demonstrated good results in our initial observations.

On the other hand, the ensemble dynamics is enough to handle Situation 2. When a class becomes outdated and its instances stop coming from the stream, new classifiers will be trained without that class and old classifiers that recognize it tend to be removed from the ensemble. If this outdated class reappears, it will be treated as new class and a new classifier will be trained with its data.

A new decision model is always trained when a new T sized data chunk is filled. As a result, the ensemble is always up to date, as the distribution of data may change due to the evolution of the concept. Experimental results will be presented in the next section.

4 Experiments

In order to verify the advantages of the approach described in this work, we consider two different metrics and three types of experiments. The datasets and evaluation metrics used in the experiments will be explained in detail below.

4.1 Datasets

To perform the experiments, two synthetic datasets and two real datasets were used. Table 1 shows the details of each dataset.

Table 1. Datasets used in experiments.

Identifier	#Instances	#Instnaces (offline)	#Attributes	#Classes	#Classes (offline)
MOA3 [2]	100,000	10,000	4	4	2
RBF [9]	48,588	2,000	2	5	3
Electricity [4]	45,312	720	9	2	2
NOAA [5]	18,159	80	9	2	2

The synthetic datasets MOA3 and RBF were generated using a function available in the MOA tool. The MOA3 is formed by hyperspherical clusters with concept-drift over time. In addition, for each 30,000 examples, new classes appear and old classes disappear. The RBF is a stationary dataset that shows disappearance and occurrence of new classes for each 10,000 examples.

The Electricity real dataset contains data collected on the New South Wales (AUS) electricity market from 7 May 1996 to 5 December 1998, where the instances represent a collection at each 30 min. In this market, prices are not fixed and are affected by market demand and supply, they are set every 5 min.

The NOAA real dataset contains data for daily weather measurements, collected by Offutt Air Force Base in Bellevue, Nebraska, over 50 years. According to the authors, the capture of measurements related to a specific day becomes available after a few days of collection. Both real datasets present real world problems in which the intermediate latency applies, being ideal for evaluating our proposal.

4.2 Experimental Setup

The parameters used in our algorithm during the experiments are described in Table 2. During the experiments, some parameters demonstrated different sensitivities for each dataset and had their values adapted.

Table 2. Parameters by dataset.

Dataset	init_θ	θ_{adapt}	K	k_short	C	S	T
MOA3	0.95	0.30	12	4	2,000	6	40
RBF	0.95	0.34	12	4	2,000	6	40
Electricity	0.70	0.50	12	4	720	6	40
NOAA	0.70	0.60	12	4	365	6	40

4.3 Evaluation I

Since the ECSMiner and CLAM source code is not available, and as far as we know, these are the only algorithms that deal with intermediate latency and novelty detection available in the literature, we compared the performance of FuzzCND[1] with four other algorithms, that only deal with classification in the data stream: Adaptive Random Forest Hoeffding Tree (ARF) [11]; Hoeffding Adaptative Tree (HAT) [8]; Adaptative Decision Tree Ensemble (ADTE) (See footnote 1) and C4.5 [21]. This type of comparison allows us to better understand the impact of the proposed novelty detection method on the accuracy of the algorithm.

The ARF and HAT algorithms consider null latency and update their decision models incrementally. As the C4.5 algorithm does not have mechanisms to adapt to the evolution of concepts, it was trained with the data available for training and remained so throughout its execution. Finally, for FuzzCND and ADTE, an intermediate latency was considered. The ADTE algorithm is a version of FuzzCND using classic decision trees, which only adapts to the concept-drift over time, without the implementation of the new detection mechanism.

As FuzzCND has mechanisms for detecting novelties, examples that belong to an unknown class classified as novelties were considered correct. Known concepts classified as novelty or novelties classified as a known concept were considered errors. To observe the behavior of the algorithms over time, we present the evaluation results at each interval, defined by a number of examples, according to the characteristics of each dataset. For each evaluation moment, the accuracy of the instances classified during the evaluation moment was calculated.

Figure 1 shows the results of the algorithms. To evaluate the performance of the algorithm in the synthetic datasets, for both, a latency of 2,000 points and evaluation moments each 1,000 points were defined. It is possible to see that the results obtained by FuzzCND in all datasets are comparable to the ARF and HAT algorithms that consider null latency to obtain the true labels. The C4.5 algorithm shows in all cases the worst scenario that can happen when the algorithm is not able to update its decision model.

In Fig. 1a and Fig. 1b, the gray vertical lines represent the moment when a new class emerged. It is notable the proposed novelty detection mechanism was effective. In the MOA3 (Fig. 1a), while the ADTE algorithm took around 10,000

[1] Available at https://github.com/andrecristiani/.

points to adapt to the emergence of a new class, FuzzCND took less than 1,000 points and effectively recognized most of the examples, marking it as a novelty and maintaining accuracy above 96%. The same happens for the RBF (Fig. 1b), when a new class appears, ADTE's performance drops to around 60%, while FuzzCND remains above 90%.

For the real datasets, the values of latency, evaluation time and chunk size were defined according to each scenario presented above. For Electricity (Fig. 1c), a latency of 48 points was considered, which would be equivalent to a day delay to obtain the true label, the chunk size of the 720, which would be equivalent to training a new classifier every 15 days and the evaluation moments every 1,440 points, which would be equivalent to one evaluation per month.

It is possible to notice that FuzzCND managed to adapt better to the concept-drift than ADTE, mainly at the beginning of the execution, where FuzzCND presented an accuracy of 80 % while ADTE started below 40 %. It is also possible to note that in some moments, FuzzCND outperformed HAT.

For NOAA (Fig. 1d), a latency of 60 points was considered, which would be equivalent to 60 days of delay. The chunk size was defined as 365, which would be equivalent to one new classifier per year and the moments of evaluation every 730 points, which would be equivalent to an evaluation every two years. It is possible to note that until the 10th evaluation moment, ADTE had a better performance than FuzzCND, but from then on, our algorithm was better adapted to the concept-drift, having better performance than ADTE and HAT and finished with an accuracy of 65%, while HAT close to 60% and ADTE 50%.

4.4 Evaluation II

In the second experiment, illustrated in Fig. 2, we analyzed the accuracy of FuzzCND for different latency values. It is possible to observe that the latency value directly impacts the performance of the algorithm, the higher the latency value, the lower the accuracy. For the synthetic datasets, latency values of 100, 1K, 2K, 5K and 10K were imposed for evaluation criteria.

In Fig. 2a, which represents the performance of the algorithm for the MOA3 dataset, for latency values 100, 1K and 2K the algorithm obtained an average of 99% in accuracy, while for latency of 5K and 10K, in certain moments of evaluation, the accuracy was around 90% and 60%, respectively.

Figure 2b represents the performance for the RBF dataset. In this case, for the latency of 100 points, the algorithm remained accurate between 95% and 99% during a large part of the data stream. In its worst case, the accuracy dropped to 85%. For the 10K latency, the accuracy was between 90% and 96%. The worst case was with the 5K latency value, which reached an accuracy of 65%.

Figure 2c represents the performance for the Electricity dataset. Being real data, the latency values were divided as follows: 24, representing a delay of 12 h; 48, representing 1 day; 240, representing 5 days; 480, representing 10 days and 1,440, representing 30 days. It is possible to notice in many evaluation moments, the algorithm obtained similar results in all latency values. The main

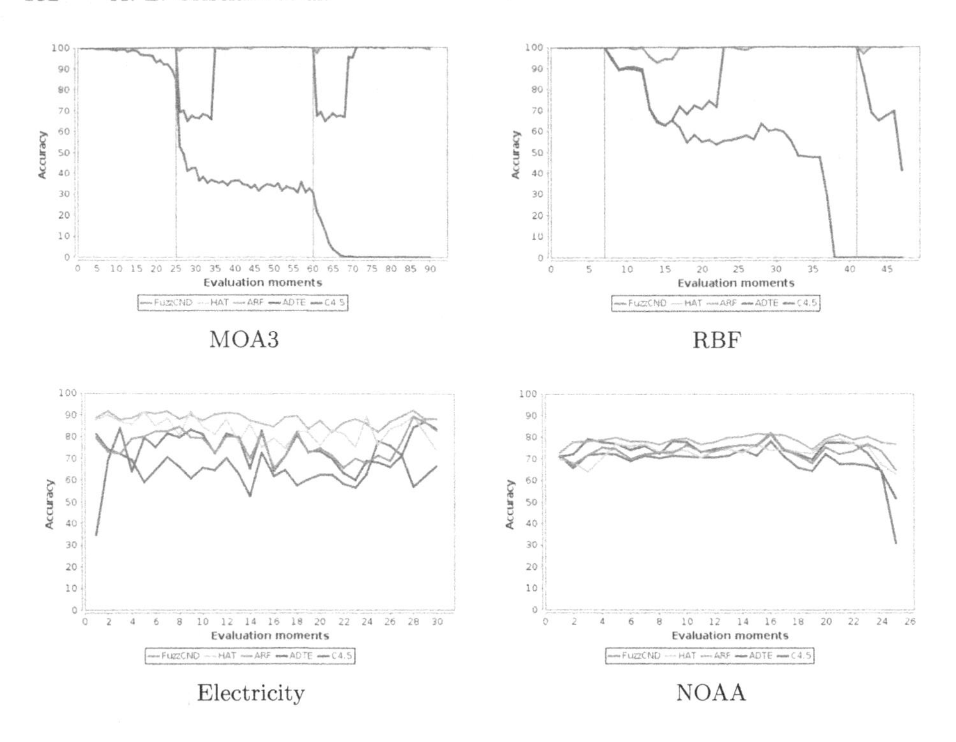

Fig. 1. Comparison of the accuracy of the algorithms.

discrepancy occurs with the highest latency values, 480 and 1,440, which have their performance about 10% lower, when compared to the others, between the evaluation moments: from 3 to 5 and from 23 to 27.

Figure 2d represents the performance for the NOAA dataset. Being real data, the latency values were divided as follows: 10, representing 10 days late; and so on with 30, 60, 180 and 365. It is possible to notice that the algorithm was consistent even with higher latency values. The worst case occurred at the time of evaluation 18, where the latency value 24 was close to 66% while the highest latency value, 1,440, was close to 64%.

In this experiment, the same parameter initializations were used as described in Table 2. These values were defined considering the latency value equal to 2,000. It is important to note that the FuzzCND performance can be superior if the parameters are adjusted to their respective latency value.

4.5 Evaluation III

In the third and last experiment, we analyzed the rate of unknown examples classified by the algorithm, using the measure *UnkR* [2], defined by:

$$Unkr = \frac{1}{\#C}(\sum_{i=1}^{\#C} \frac{\#UnkR_i}{\#Exc_i}) \tag{2}$$

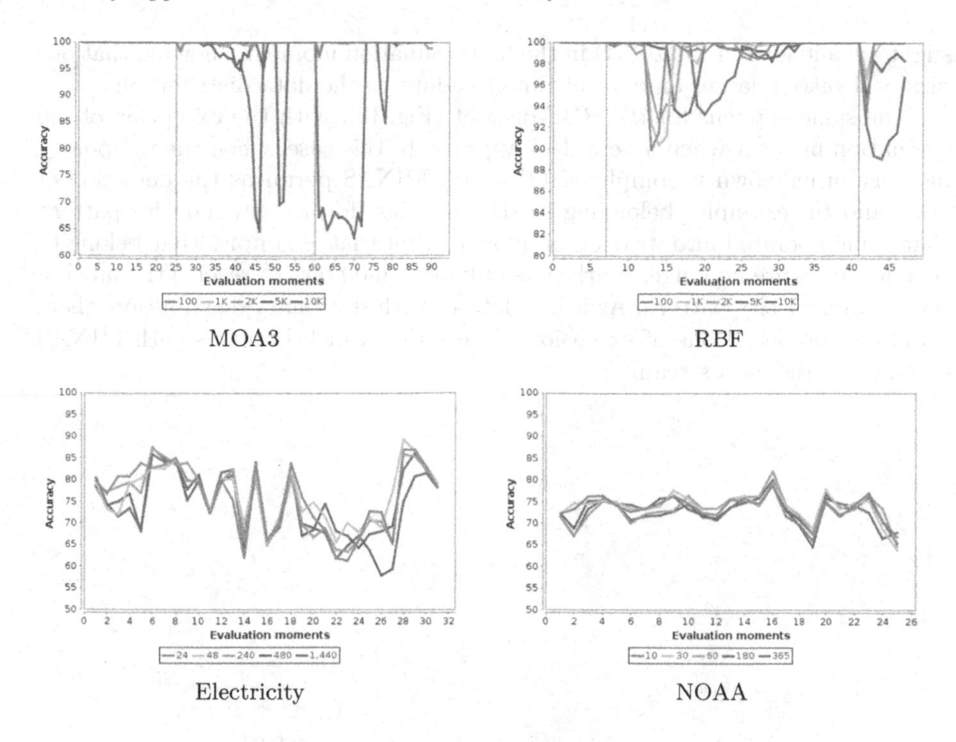

Fig. 2. Algorithm performance for different latency values.

where $\#C$ represents the total number of classes, $\#UnkR_i$ the number of examples belonging to class C_i classified as unknown and Exc_i the total number of examples belonging the class C_i. This metric helps to understand the behavior of the algorithm throughout the data stream. The *UnkR* increases the presence of examples of new classes or the presence of a change of concept.

To this metric, it is necessary the algorithms perform the classification with an unknown label option. Thus, was compared with two other algorithms that have this same characteristic: MINAS [2] and PFuzzND [9].

To make a fair comparison, for the three algorithms we initialized the parameters $K = 12$ and $T = 40$; for MINAS and PFuzzND the parameter ts was initialized with a value of 200. In addition, to avoid the influence of true labels on the results, we applied extreme latency in the FuzzCND algorithm. Thus, unlike the other algorithms, FuzzCND did not update its decision model.

For this experiment, we consider only the datasets that have the emergence of a new class, which would be MOA3 and RBF. As it was considered extreme latency for FuzzCND, the rate of unknowns was analyzed until the appearance of the first instance of a new class. Figure 3 shows the results obtained.

For the MOA3 dataset (Fig. 3a), it is possible to notice that the concepts of fuzzy set theory applied in FuzzCND and PFuzzND made it better adapted to concept drif than MINAS, classifying as unknown a much smaller number of examples during all moments of evaluation. However, PFuzzND obtained a slight advantage over FuzzCND in the first evaluation moment, showing that our proposal takes a larger amount of time to adapt to the data distribution.

The same happens for the RBF dataset (Fig. 3b), with the exception of the evaluation moment when a new class appears. In this case, when the temporary memory of unknown is completed ($T = 40$), MINAS performs the detection of news and the examples belonging to the new class form a new novelty pattern that is incorporated into its decision model. After that, examples that belong to this new class are no longer marked as unknown and start to receive the label of the created novelty pattern. As it is a dataset with stationary distribution, there should be no detections of extension of novelties, which happens with MINAS throughout the data stream.

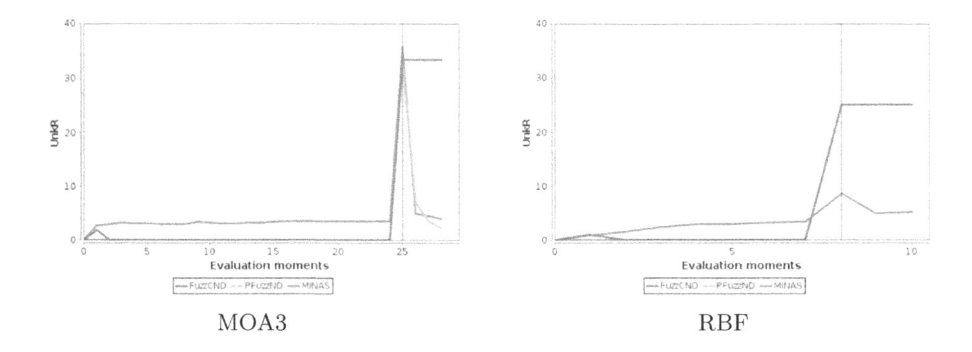

Fig. 3. UnkR obtained by the algorithms

The experiments showed that FuzzCND maintained a good performance in all experiments, with results comparable to algorithms that consider null latency. It was possible to see that the proposed novelty detection method increased the accuracy rate of the algorithm. In addition, when analyzing the rate of examples classified as unknown, our approach showed a performance similar to PFuzzND and superior to MINAS in both datasets. While MINAS identifies examples as unknown even in stationary datasets, FuzzCND identifies a smaller amount and only at the first moment of evaluation.

5 Conclusions

This article proposed FuzzCND, an ensemble fuzzy method for classification and novelty detection in data streams with labels with intermediate delay.

The results showed that FuzzCND adapts well to the concept-drift and identifies new classes efficiently, maintaining its predictive power. Our method demonstrated that the use of fuzzy techniques makes the algorithms better adapt to possible inaccuracies in the data, maintaining a good accuracy in different latency values in real-world scenarios. In addition, our method maintain a lower rate of unknowns than algorithms that do not use fuzzy techniques.

This article opens several perspectives for future work, especially considering novelty detection and adaptation to different types of latency. As future steps, we intend to evolve the novelty detection method to detect different patterns of novelty, and adapt our algorithm to deal with different types of label delay, including extreme latency. In addition, we intend to compare our approach with works that consider intermediate latency and perform the novelty detection.

References

1. Babcock, B., Babu, S., Datar, M., Motwani, R., Widom, J.: Models and issues in data stream systems. In: Proceedings of the Twenty-First ACM SIGMOD-SIGACT-SIGART Symposium on Principles of Database Systems (2002)
2. de Faria, E.R., Ponce de Leon Ferreira Carvalho, A.C., Gama, J.: MINAS: multiclass learning algorithm for novelty detection in data streams. Data Min. Knowl. Discov. **30**(3), 640–680 (2015). https://doi.org/10.1007/s10618-015-0433-y
3. Souza, V., Pinho, T., Batista, G.: Evaluating stream classifiers with delayed labels information. In: 7th Brazilian Conference on Intelligent Systems (BRACIS) (2018)
4. Harries, M.: SPLICE-2 comparative evaluation: electricity pricing. Technical report 1, p. 6, University of New South Wales, Sydney, Australia (1999)
5. Elwell, R., Polikar, R.: Incremental learning of concept drift in nonstationary environments. IEEE Trans. Neural Netw. **22**(10), 1517–1531 (2011). https://doi.org/10.1109/TNN.2011.2160459
6. Škrjanc, I., Iglesias, J.A., Sanchis, A., Leite, D., Lughofer, E., Gomide, F.: Evolving fuzzy and neuro-fuzzy approaches in clustering, regression, identification, and classification: a survey. Inf. Sci. **490**, 344–368 (2019)
7. Cintra, M.E., Monard, M.C., Camargo, H.A.: FuzzyDT – a fuzzy decision tree algorithm based on C4.5. In: Proceedings of the Brazilian Congress on Fuzzy Systems (2012)
8. Bifet, A., Gavaldà, R.: Adaptive learning from evolving data streams. In: Adams, N.M., Robardet, C., Siebes, A., Boulicaut, J.-F. (eds.) IDA 2009. LNCS, vol. 5772, pp. 249–260. Springer, Heidelberg (2009). https://doi.org/10.1007/978-3-642-03915-7_22
9. da Silva, T.P.: Abordagem Fuzzy para Detecção de Novidade em Fluxo Contínuo de Dados. Dissertação de mestrado. Universidade Federal de São Carlos (2018)
10. Masud, M., Gao, J., Khan, L., Han, J., Thuraisingham, B.M.: Classification and novel class detection in concept-drifting data streams under time constraints. IEEE Trans. Knowl. Data Eng. **23**(6), 859–874 (2011)
11. Gomes, H.M., et al.: Adaptive random forests for evolving data stream classification. Mach. Learn. **106**, 1469–1495 (2017). https://doi.org/10.1007/s10994-017-5642-8
12. Souza, V.M.A., Silva, D.F., Gama, J., Batista, G.E.: Data stream classification guided by clustering on nonstationary environments and extreme verification latency. In: Proceedings of the 2015 SIAM International Conference on Data Mining. Society for Industrial and Applied Mathematics (2015)
13. da Silva, T.P., Souza, V.M.A., Batista, G.E.A.P.A., de Arruda Camargo, H.: A fuzzy classifier for data streams with infinitely delayed labels. In: Vera-Rodriguez, R., Fierrez, J., Morales, A. (eds.) CIARP 2018. LNCS, vol. 11401, pp. 287–295. Springer, Cham (2019). https://doi.org/10.1007/978-3-030-13469-3_34

14. Al-Khateeb, T., Masud, M.M., Khan, L., Aggarwal, C., Han, J., Thuraisingham, B.: Stream classification with recurring and novel class detection using class-based ensemble. In: 12th International Conference on Data Mining, Brussels, pp. 31–40 (2012). https://doi.org/10.1109/ICDM.2012.125
15. Pal, N.R., Pal, K., Keller, J.M., Bezdek, J.C.: A possibilistic fuzzy c-means clustering algorithm. IEEE Trans. Fuzzy Syst. **13**(4), 517–530 (2005). https://doi.org/10.1109/TFUZZ.2004.840099
16. Bezdek, J.C., Ehrlich, R., Full, W.: FCM: the fuzzy c-means clustering algorithm. Comput. Geosci. **10**(2–3), 191–203 (1984)
17. Krawczyk, B., Minku, L.L., Gama, J., Stefanowski, J., Wožniak, M.: Ensemble learning for data stream analysis: a survey. Inf. Fusion **37**, 132–156 (2017)
18. Campello, R.J.G.B., Hruschka, E.R.: A fuzzy extension of the silhouette width criterion for cluster analysis. Fuzzy Sets Syst. **157**(21), 2858–2875 (2006)
19. Xiong, X., Chan, K.L., Tan, K.L.: Similarity-driven cluster merging method for unsupervised fuzzy clustering. In: Proceedings of the 20th Conference on Uncertainty in Artificial Intelligence. AUAI Press (2004)
20. Gao, J., Fan, W., Han, J.: On appropriate assumptions to mine data streams: analysis and practice. In: Seventh IEEE International Conference on Data Mining (ICDM 2007), Omaha, NE, pp. 143–152 (2007)
21. Quinlan, J.R.: C4.5: Programs for Machine Learning. Morgan Kaufmann Series in Machine Learning, 1st edn. Morgan Kaufmann, Burlington (1993)

A Fuzzy Reasoning Method Based on Ensembles of Generalizations of the Choquet Integral

Giancarlo Lucca[1(✉)], Eduardo N. Borges[1], Helida Santos[1], Graçaliz P. Dimuro[1,2], Tiago C. Asmus[1,2], José A. Sanz[2], and Humberto Bustince[2]

[1] Universidade Federal do Rio Grande, Rio Grande, RS, Brazil
{giancarlo.lucca,eduardoborges,helida, gracalizdimuro,tiagoasmus}@furg.br

[2] Universidad Publica de Navarra, Pamplona, Navarra, Spain
{joseantonio.sanz,bustince}@unavarra.es

Abstract. An efficient way to deal with classification problems is by using Fuzzy Rule-Based Classification Systems. A key point in this kind of classifier is the Fuzzy Reasoning Method (FRM). This mechanism is responsible for performing the classification of examples into predefined classes. There are different generalizations of the Choquet integral in the literature that are applied in the FRM. This paper presents an initial study of a new FRM that is ensemble-based, which combines different generalizations in a more effective final classifier. We have constructed two distinct ensemble decision-making methods considering the majority and weighted voting approaches. The experimental results demonstrate that the performance of the proposed methods is statistically equivalent compared to the state-of-art generalizations.

Keywords: Classification · Ensemble · Fuzzy Reasoning Method · Generalizations of the Choquet integral

1 Introduction

Data classification is the process of finding, using supervised machine learning, a model or function that describes different data classes [18]. The purpose of classification is to automatically label new examples or observations by applying the learned model or function. This model is based on the characteristics of the training data.

The scientific literature presents a broad set of supervised machine learning algorithms [4], which can be organized according to the characteristics they use in learning. Multilayer Perception (MLP) Neural Networks [22], k-Nearest Neighbor (kNN) [1], Naïve Bayes [28], Logistic Regression [23], C4.5 [38], and Support Vector Machines (SVM) [15] are examples of well-known classification

R. Cerri and R. C. Prati (Eds.): BRACIS 2020, LNAI 12320, pp. 187–200, 2020.
https://doi.org/10.1007/978-3-030-61380-8_13

algorithms. Although some classifiers individually provide solutions which are very accurate in particular scenarios, they may not effectively recognize data classes in complex problems, where there are large sets of patterns and/or a significant number of incomplete data samples or irrelevant features.

In order to improve the classification results, techniques for combining classifiers have been proposed, aiming to take advantage of several classification schemes. The outputs of each classifier can be combined in a final decision that improves the generalization ability. These techniques are called ensemble methods [37]. For instance, Random Forest [10] consolidates the results of multiple decision trees. AdaBoost [20] manipulates the examples, training each classifier sequentially with different samples, trying to label the errors of the previous classifiers correctly. Stacking [19] combines multiple base classifiers trained by using different learning algorithms employing a meta-classifier.

Ensembles aggregate different labeling data strategies to make a more robust classifier, aiming to reach higher accuracy. There should be different solutions for the problem to be solved, i.e., it is essential to get a diversity among the results found by these algorithms [30]. Also, ensemble methods can be adapted to deal with imbalanced class datasets [21].

A more flexible way to deal with complex real-life classification problems is by using Fuzzy Rule-Based Classification Systems (FRBCSs) [27]. These systems are tolerant of imprecision, uncertainty, partial truth, and approximations. They make usage of the interpretability provided by de fuzzy rules to produce a system intelligible. An FRBCS is composed of an inference mechanism called the Fuzzy Reasoning Method (FRM) [13,14], that is responsible for predicting the class in which the examples belong. This FRM uses a database that contains information about the membership functions associated with the linguistic labels considered by the fuzzy rules, and a rule base composed by the rules learned. Chi-RW [26], FURIA [24], and IVTURS [39] are examples of fuzzy rule-based classification algorithms.

The main FRMs used by FRBCSs are based in the aggregation function [9] maximum. That is, the class predicted is the one that achieves the maximum compatibility with the example. This FRM is also known as the Winning Rule (WR). There is also an FRM that makes usage of the combination of different rules, by using an Additive Combination (AC). Barrenechea et al. [7] proposed a FRM that uses the Choquet integral [12] as aggregation operator, achieving satisfactory results. After that, generalizations of the Choquet integral were used as aggregation operators in the FRM of the fuzzy classifier FARC-HD [2]. Precisely, introducing the C_T [32], CC [35], C_F [34] and C_{F1F2} [31] integrals.

Each generalization of the Choquet integral used in the FRM produces a different classifier. Having this in mind, in this paper, we intend to produce a unique FRM that could make the final decision based on the outputs of different generalizations. We have selected for building an ensemble functions that best generalize the Choquet integral in each of the studies presented before. The experimental results show that the performance of the proposed methods is statistically superior to several baselines, and the classification quality was statistically equivalent compared to the state-of-art generalizations [16].

This paper is organized as follows. In Sect. 2, the preliminaries concepts are introduced. After that, in Sect. 4, we present the new FRM that is based on the ensemble technique. The experimental framework is detailed in Sect. 5. Conclusion and some future directions are given in the last section.

2 Preliminaries

This section introduce the basic concepts used in this paper, considering $N = \{1, \ldots, n\}$, for an arbitrary $n > 0$.

Definition 1 [9]. *A function $A : [0,1]^n \rightarrow [0,1]$ is an aggregation function if:*

- ***(A1)*** *A is increasing in each argument:*
  $\forall i \in \{1, \ldots, n\} : x_i \leq y \Rightarrow A(x_1, \ldots, x_n) \leq A(x_1, \ldots, x_{i-1}, y, x_{i+1}, \ldots, x_n)$;
- ***(A2)*** *A satisfies boundary conditions: (i) $A(0, \ldots, 0) = 0$; (ii) $A(1, \ldots, 1) = 1$.*

An aggregation function $A : [0,1]^n \rightarrow [0,1]$ is said to be averaging if and only if: **(AV)** $\forall (x_1, \ldots, x_n) \in [0,1]^n : \min\{x_1, \ldots, x_n\} \leq A(x_1, \ldots, x_n) \leq \max\{x_1, \ldots, x_n\}$.

Definition 2 [29]. *An aggregation function $T : [0,1]^2 \rightarrow [0,1]$ is a t-norm if, for all $x, y, z \in [0,1]$, it satisfies the following properties:*

(T1) *Commutativity: $T(x,y) = T(y,x)$;*
(T2) *Associativity: $T(x, T(y,z)) = T(T(x,y), z)$;*
(T3) *Boundary condition: $T(x,1) = x$.*

Definition 3 [8,11]. *A function $O : [0,1]^2 \rightarrow [0,1]$ is said to be an overlap function if it satisfies the following conditions:*

(O1) *O is commutative;*
(O2) *$O(x,y) = 0$ if and only if $xy = 0$;*
(O3) *$O(x,y) = 1$ if and only if $xy = 1$;*
(O4) *O is increasing;*
(O5) *O is continuous.*

Definition 4 [5]. *A bivariate function $C : [0,1]^2 \rightarrow [0,1]$ is a copula if it satisfies the following conditions, for all $x, x', y, y' \in [0,1]$ with $x \leq x'$ and $y \leq y'$:*

(C1) *$C(x,y) + C(x',y') \geq C(x,y') + C(x',y)$;*
(C2) *$C(x,0) = C(0,x) = 0$;*
(C3) *$C(x,1) = C(1,x) = x$.*

Definition 5 [36]. *A function $\mathfrak{m} : 2^N \rightarrow [0,1]$ is said to be a fuzzy measure if, for all $X, Y \subseteq N$, the following conditions hold:*

- ($\mathfrak{m}1$) *Increasingness: if* $X \subseteq Y$, *then* $\mathfrak{m}(X) \leq \mathfrak{m}(Y)$;
- ($\mathfrak{m}2$) *Boundary conditions:* $\mathfrak{m}(\emptyset) = 0$ *and* $\mathfrak{m}(N) = 1$.

In this paper we consider as fuzzy measure the Power Measure (PM), defined, for each $A \subseteq N$, by:

$$\mathfrak{m}(A) = \left(\frac{|A|}{n}\right)^q, \text{ with } q > 0. \tag{1}$$

In Barrenechea et al. [7], the exponent q was genetically adapted for each data class. A study of the performance of different fuzzy measures is presented in [33].

Definition 6. *Let* $\mathfrak{m} : 2^N \rightarrow [0,1]$ *be a fuzzy measure. The discrete Choquet integral is the function* $\mathfrak{C}_{\mathfrak{m}} : [0,1]^n \rightarrow [0,1]$, *defined, for all* $\boldsymbol{x} = (x_1, \ldots, x_n) \in [0,1]^n$, *by:*

$$\mathfrak{C}_{\mathfrak{m}}(\boldsymbol{x}) = \sum_{i=1}^{n} \left(x_{(i)} - x_{(i-1)}\right) \cdot \mathfrak{m}\left(A_{(i)}\right), \tag{2}$$

where $\left(x_{(1)}, \ldots, x_{(n)}\right)$ *is an increasing permutation on the input* $\boldsymbol{x}$, *that is,* $0 \leq x_{(1)} \leq \ldots \leq x_{(n)}$, *where* $x_{(0)} = 0$ *and* $A_{(i)} = \{(i), \ldots, (n)\}$ *is the subset of indices corresponding to the* $n - i + 1$ *largest components of* $\boldsymbol{x}$.

Observe that by using the distributivity property of the product, the standard Choquet integral, Eq. (2), can be also written in its expanded form in Eq. (3).

$$\mathfrak{C}_{\mathfrak{m}}(\boldsymbol{x}) = \sum_{i=1}^{n} \left(x_{(i)} \cdot \mathfrak{m}\left(A_{(i)}\right) - x_{(i-1)} \cdot \mathfrak{m}\left(A_{(i)}\right)\right). \tag{3}$$

3 Generalizations of the Choquet Integral

The usage of the Choquet integral as the aggregation function in the FRM of an FRBCS was proposed in [7]. After that, different generalizations of this integral were also proposed for applications in FRBCS [16].

Firstly, in [32], it was introduced the concept of pre-aggregation function: a function that satisfies the boundary conditions but it may be not increasing in the whole domain, just in some specific directions, showing that a way to construct such aggregation-like functions is by replacing the product operator of the standard Choquet integral (Eq. 2) by a t-norm, introducing the concept of C_T-integral, which is defined as follows:

Definition 7 [32]**.** *Let* $T : [0,1]^2 \rightarrow [0,1]$ *be a t-norm. Taking as basis the Choquet integral, we define the function* $\mathfrak{C}_{\mathfrak{m}}^{\mathbb{T}} : [0,1]^n \rightarrow [0,n]$, *for all* $\boldsymbol{x} = (x_1, \ldots, x_n) \in [0,1]^n$, *by*

$$\mathfrak{C}_{\mathfrak{m}}^{T}(\boldsymbol{x}) = \sum_{i=1}^{n} T\left(x_{(i)} - x_{(i-1)}, \mathfrak{m}\left(A_{(i)}\right)\right). \tag{4}$$

Also considering the standard Choquet integral, in [34], the product operator was replaced by different functions F satisfying some specific conditions. This generalization is known as C_F-integral. Observe that these functions can be either averaging or non-averaging.

Definition 8 [34]. *Let $F : [0,1]^2 \to [0,1]$ be a left 0-absorbent function (i.e., $\forall y \in [0,1] : F(0,y) = 0$) with right neutral element (i.e., $\forall x \in [0,1] : F(x,1) = x$). The Choquet-like integral based on F with respect to $\mathfrak{m}$, called C_F-integral, is the function $\mathfrak{C}_{\mathfrak{m}}^F : [0,1]^n \to [0,1]$, defined, for all $\boldsymbol{x} \in [0,1]^n$, by*

$$\mathfrak{C}_{\mathfrak{m}}^F(\boldsymbol{x}) = \min\left\{1, \sum_{i=1}^{n} F\left(x_{(i)} - x_{(i-1)}, \mathfrak{m}\left(A_{(i)}\right)\right)\right\} \tag{5}$$

The expanded form of the Choquet integral, defined in Eq. (3), where used as basis of the concept of CC-integrals [35] and C_{F1F2}-integrals [17,31]. We highlight that the first generalization always presents averaging characteristics, and the second may not. A CC-integral is a generalization that replaces the product operator by copulas.

Definition 9 [35]. *Let $C : [0,1]^2 \to [0,1]$ be a bivariate copula. The Choquet-like copula-based integral (CC-integral) with respect to $\mathfrak{m}$ is defined as a function $\mathfrak{C}_{\mathfrak{m}}^C : [0,1]^n \to [0,1]$, given, for all $\boldsymbol{x} \in [0,1]^n$, by*

$$\mathfrak{C}_{\mathfrak{m}}^C(\boldsymbol{x}) = \sum_{i=1}^{n} C\left(x_{(i)}, \mathfrak{m}\left(A_{(i)}\right)\right) - C\left(x_{(i-1)}, \mathfrak{m}\left(A_{(i)}\right)\right). \tag{6}$$

Finally, considering the C_{F1F2}-integrals [31] this generalization replace the product operator of the expanded Choquet integral by two functions $F_1, F_2 : [0,1]^2 \to [0,1]$ satisfying some specific condition, obtaining:

Definition 10. *Let $\mathfrak{m} : 2^N \to [0,1]$ be a symmetric fuzzy measure and $F_1, F_2 : [0,1]^2 \to [0,1]$ be two functions fulfilling: (i) F_1-dominance, that is, $F_1 \geq F_2$; (ii) F_1 is increasing in the direction $(1,0)$. A $C_{F_1F_2}$-integral is defined as a function $\mathfrak{C}_{\mathfrak{m}}^{(F_1,F_2)} : [0,1]^n \to [0,1]$, given, for all $x \in [0,1]^n$, by*

$$\mathfrak{C}_{\mathfrak{m}}^{(F_1,F_2)}(\boldsymbol{x}) = \min\left\{1, x_{(1)} + \sum_{i=2}^{n} F_1\left(x_{(i)}, \mathfrak{m}\left(A_{(i)}\right)\right) - F_2\left(x_{(i-1)}, \mathfrak{m}\left(A_{(i)}\right)\right)\right\}. \tag{7}$$

This paper proposes the usage of all these generalizations in an ensemble method. Table 1 present some instances of these generalizations that achieved the best classification results.

4 A New Ensemble-Based FRM

In this section, we describe the new reasoning method that makes usage of an ensemble approach. To do so, consider a classification problem consisting of m

Table 1. Generalizations of the Choquet integral that are used in this study.

Generalization	Equation
Standard Choquet Integral	$\mathfrak{C}_{\mathfrak{m}}^{T_P}(\boldsymbol{x}) = \sum_{i=1}^{n} (x_{(i)} - x_{(i-1)})\mathfrak{m}(A_{(i)})$
C_T-integral	$\mathfrak{C}_{\mathfrak{m}}^{T_{HP}}(\boldsymbol{x}) = \sum_{i=1}^{n} \begin{cases} 0 & \text{if } x_{(i)} - x_{(i-1)} = \mathfrak{m}(A_{(i)}) = 0 \\ \frac{(x_{(i)}-x_{(i-1)})\mathfrak{m}(A_{(i)})}{x_{(i)}-x_{(i-1)}+\mathfrak{m}(A_{(i)})-(x_{(i)}-x_{(i-1)})\mathfrak{m}(A_{(i)})} & \text{otherwise} \end{cases}$
CC-integral	$CC_{\mathfrak{m}}^{T_M}(\boldsymbol{x}) = \sum_{i=1}^{n} \min(x_i, \mathfrak{m}(A_{(i)})) - \min(x_{i-1}, \mathfrak{m}(A_{(i)}))$
C_F-integral$_{AVG}$	$C_{\mathfrak{m}}^{F_{NA}}(\boldsymbol{x}) = \sum_{i=1}^{n} \begin{cases} x_{(i)} - x_{(i-1)} & \text{if } x_{(i)} - x_{(i-1)} \leq \mathfrak{m}(A_{(i)}) \\ \min\{\frac{x_{(i)}-x_{(i-1)}}{2}, \mathfrak{m}(A_{(i)})\} & \text{otherwise} \end{cases}$
C_F-integral$_{N-AVG}$	$C_{\mathfrak{m}}^{F_{NA2}}(\boldsymbol{x}) = \sum_{i=1}^{n} \begin{cases} 0 & \text{if } x_{(i)} - x_{(i-1)} = 0 \\ \frac{x_{(i)}-x_{(i-1)}+\mathfrak{m}(A_{(i)})}{2} & \text{if } 0 < x_{(i)} - x_{(i-1)} \leq \mathfrak{m}(A_{(i)}) \\ \min\{\frac{x_{(i)}-x_{(i-1)}}{2}, \mathfrak{m}(A_{(i)})\} & \text{otherwise.} \end{cases}$
C_{F1F2}-integral$_{(GM,F_{BPC})}$	$C_{\mathfrak{m}}^{(GM,F_{BPC})}(\boldsymbol{x}) = \sum_{i=1}^{n} \sqrt{x_i \mathfrak{m}(A_{(i)})} - x_{i-1}\mathfrak{m}(A_{(i)})^2$
$C_{F_1F_2}$-integral$_{(GM,T_{Luk})}$	$C_{\mathfrak{m}}^{(GM,T_{Luk})}(\boldsymbol{x}) = \sum_{i=1}^{n} \sqrt{x_i \mathfrak{m}(A_{(i)})} - \max(0, x_{i-1} + \mathfrak{m}(A_{(i)}) - 1)$
$C_{F_1F_2}$-integral$_{(F_{GL},T_M)}$	$C_{\mathfrak{m}}^{(F_{GL},T_M)}(\boldsymbol{x}) = \sum_{i=1}^{n} \sqrt{\frac{x_i(\mathfrak{m}(A_{(i)})+1)}{2}} - \min(x_{i-1}, \mathfrak{m}(A_{(i)}))$
$C_{F_1F_2}$-integral$_{(F_{GL},F_{BPC})}$	$C_{\mathfrak{m}}^{(F_{GL},F_{BPC})}(\boldsymbol{x}) = \sum_{i=1}^{n} \sqrt{\frac{x_i(\mathfrak{m}(A_{(i)})+1)}{2}} - x_{i-1}\mathfrak{m}(A_{(i)})^2$

AVG=averaging; N-AVG = non-averaging

training examples $\mathbf{x}_p = (x_{p1}, \ldots, x_{pn}, y_p)$, with $p = 1, \ldots, m$, where x_{pi}, with $i = 1, \ldots, n$, is the value of the ith attribute variable and $y_p \in \mathbb{C} = \{C_1, C_2, ..., C_M\}$ is the label of the class of the pth training example.

Moreover, considering $x_p = (x_{p1}, \ldots, x_{pn})$, a n-dimensional vector of attribute corresponding to an example $\mathbf{x}_p$. The fuzzy rules that are used in this paper are of the form:

$$\text{Rule } R_j : \text{ If } x_{p1} \text{ is } A_{j1} \text{ and } \ldots \text{ and } x_{pn} \text{ is } A_{jn} \text{ then } x_p \text{ in } C_j^k \text{ with } RW_j, \quad (8)$$

where R_j is the label of the jth rule, A_{ji} is an antecedent fuzzy set shaping a linguistic term, C_j^k is the label of the consequent fuzzy set C^k modeling the class associated to the rule R_j, with $k \in \{1, \ldots, M\}$, and $RW_j \in [0, 1]$ is the rule weight [25].

Then, considering $x_p = (x_{p1}, \ldots, x_{pn})$ as a new example to be classified, L the number of rules in the rule base, M the number of classes of the problem, and $\mathcal{A} = \{\mathbb{A}_1, \ldots, \mathbb{A}_H\}$ the set of H adopted generalizations of the Choquet integral. The five following steps form the new FRM ensemble-based:

1. *Matching degree*: the intensity of the activation of the if-part of the rules with the example x_p. It is calculated using a t-norm, $T' : [0, 1]^n \rightarrow [0, 1]$, in our case, the product.

$$\mu_{A_j}(x_p) = T'(\mu_{A_{j1}}(x_{p1}), \ldots, \mu_{A_{jn}}(x_{pn})), \text{ with } j = 1, \ldots, L. \quad (9)$$

2. *Association degree*: it is the association degree of the example x_p with the class of each rule in the rule base, given by:

$$b_j^k(x_p) = \mu_{A_j}(x_p) \cdot RW_j^k, \text{ with } k = Class(R_j), j = 1, \ldots, L. \quad (10)$$

3. *Example classification soundness degree for all classes*: at this point, for each class, the positive information given by the fired rules in the previous step is aggregated:

$$Y_k^{\mathbb{A}_h}(x_p) = \mathbb{A}_h\left(b_1^k(x_p), \ldots, b_L^k(x_p)\right), \text{ with } k = 1, \ldots, M, \ h = 1 \ldots H \quad (11)$$

where $\mathbb{A}_h \in \mathcal{A}$ are different generalizations of the Choquet integral (Table 1).

4. *Classification*: This step defines the predicted data class for each classifier. To do so, we used the label that maximizes the aggregated value.

$$F^{\mathbb{A}_h}(Y_1^{\mathbb{A}_h}, \ldots, Y_M^{\mathbb{A}_h}) = \arg \max_{k=1,\ldots,M} (Y_k^{\mathbb{A}_h}). \quad (12)$$

5. *Ensemble decision making (EDM)*: The final decision is made in this step. We propose two ensemble decision-making methods:

   Majority Voting: By this EDM, the final classification is obtained by using the Mode [6], which is a statistical operator that returns the element that occurs most often in the vector composed by the labels of the classes obtaining in Step 4 for each generalization of the Choquet Integral $\mathbb{A}_h \in \mathcal{A}$. Notice that the Mode is not a function, in the sense that for the same input vector, it may occur a tie. In this case, we randomly choose one of those elements that tied.

   Then, for simplicity, denote each class label obtained in Step 4 by:

$$c^{\mathbb{A}_h} = F^{\mathbb{A}_h}(Y_1^{\mathbb{A}_h}, \ldots, Y_M^{\mathbb{A}_h}),$$

   and consider the vector $\boldsymbol{R}$ of ordered pairs composed by such class labels and their respective winner aggregated fuzzy membership values of Step 4, given by:

$$\boldsymbol{R} = \left[(c^{\mathbb{A}_1}, Y_c^{\mathbb{A}_1}), \ldots, (c^{\mathbb{A}_H}, Y_c^{\mathbb{A}_H})\right] = \left[(c^{\mathbb{A}_h}, Y_c^{\mathbb{A}_h})\right]_{h=1\ldots H}.$$

   Now let $\boldsymbol{R}[1]$ be the vector of the first components of the ordered pairs of $\boldsymbol{R}$, namely:

$$\boldsymbol{R}[1] = \left[c^{\mathbb{A}_1}, \ldots, c^{\mathbb{A}_H}\right] = \left[c^{\mathbb{A}_h}\right]_{h=1\ldots H}.$$

   Then, the final classification result is given by:

$$c_{final} = \text{Mode}(\boldsymbol{R}[1]). \quad (13)$$

   Weighted Voting: By this EDM, the final classification is given by the maximum of the sums of the winner-aggregated fuzzy membership values with respect of each class obtained in Step 4. In the case, whenever there is a tie, the result is chosen randomly among the values that produce the tie.

Table 2. An example of the proposed EDM with two classes and five generalizations of the Choquet integral.

$\boldsymbol{R} = \left[(c^{\mathbb{A}_h}, Y_c^{\mathbb{A}_h})\right]_{h=1\dots5}$	$[(1,0.5),(2,0.8),(1,0.1),(2,0.7),(1,0.6)]$
$\boldsymbol{R}[1] = \left[c^{\mathbb{A}_h}\right]_{h=1\dots5}$	$[1,2,1,2,1]$
$\boldsymbol{R}[2] = \left[Y_c^{\mathbb{A}_h}\right]_{h=1\dots5}$	$[0.5,0.8,0.1,0.7,0.6]$
Majority Voting (Mode)	1
$S(C_m)_{m=1\dots2}$	$S(C_1) = 1.2$, $S(C_2) = 1.5$
Weighted Voting (max)	2

Then let $\boldsymbol{R}[2]$ be the vector of the second components of the ordered pairs of $\boldsymbol{R}$, namely:

$$\boldsymbol{R}[2] = \left[Y_c^{\mathbb{A}_1}, \dots, Y_c^{\mathbb{A}_H}\right] = \left[Y_c^{\mathbb{A}_h}\right]_{h=1\dots H},$$

and define $R[1][h]$ the element of $R[1]$ in the position $h = 1 \dots H$, and similarly define $R[2][h]$. The sum of the winner aggregated fuzzy membership values with respect of each class in $\{C_1, \dots, C_M\}$, as obtained in Step 4, is given, for each $m \in \{1, \dots, M\}$, by:

$$S(C_m) = \sum_{h=1}^{H} \begin{cases} R[2][h] & \text{if } R[1][h] = m \\ 0 & \text{otherwise.} \end{cases}$$

Then, the final classification result is given by:

$$c_{final}(C_1, \dots, C_M) = \arg \max_{m=1\dots M} S(C_m). \tag{14}$$

Example 1. Consider a fictitious example with two classes and five generalizations of the Choquet integral. Possible results obtained with EDM Majority and Weighted Voting are given in Table 2.

5 Experimental Framework

In this section, we describe the experiments we conducted in order to empirically validate and check the quality of our ensemble-based FRM. Firstly, we describe the 27 real datasets used in the experiments. Then we present the configuration of the proposal and the achieved results. The effectiveness of our approach is compared to the currently state-of-the-art aggregation method C_{F1F2}-integral [16], showing statistically equivalent results.

5.1 Datasets

In order to facilitate the comparison of results, we have picked the same 27 datasets used in the first application of the Choquet integral in the literature [32]. The used datasets are available in KEEL[1] dataset repository [3]. Table 3

[1] http://keel.es.

Table 3. Datasets used in this study

Id.	Dataset	#Inst	#Att	#Class	#DataType
App	Appendiciticis	106	7	2	R
Bal	Balance	625	4	3	R
Ban	Banana	5,300	2	2	R
Bnd	Bands	365	19	2	R, I
Bup	Bupa	345	6	2	R
Cle	Cleveland	297	13	5	R, I
Eco	Ecoli	336	7	8	R
Gla	Glass	214	9	6	R
Hab	Haberman	306	3	2	I
Hay	Hayes-Roth	160	4	3	I
Iri	Iris	150	4	3	R
Led	Led7digit	500	7	10	R
Mag	Magic	1,902	10	2	R
New	Newthyroid	215	5	3	R, I
Pag	Pageblocks	5,472	10	5	R,I
Pho	Phoneme	5,404	5	2	R
Pim	Pima	768	8	2	R
Rin	Ring	740	20	2	R
Sah	Saheart	462	9	2	R, I, N
Sat	Satimage	6,435	36	7	I
Seg	Segment	2,310	19	7	R
Tit	Titanic	2,201	3	2	R
Two	Twonorm	740	20	2	R
Veh	Vehicle	846	18	4	R
Win	Wine	178	13	3	R
Wis	Wisconsin	683	11	2	I
Yea	Yeast	1,484	8	10	R

describes each dataset, showing an identifier (Id.), the number of instances (#Inst), attributes (#Att), classes (#Class) and type of data (#DataType), which can be Real (R), Integer (I) or Nominal (N).

We used the same evaluation scheme presented in [7,32], that consist in a 5-fold cross-validation. This scheme randomly splits a dataset into five subsets (folds) with 20% of the examples. Each fold is used as a test set on which a distinct model is applied, trained using the remaining four folds. The classification performance is computed as the average of the accuracy (percentage of examples classified correctly) in the five folds. We highlight we have used the 5-fold partition available in the dataset repository.

In order to reduce the size of the training sets and make the results comparable with the baselines, the following datasets were stratified sampled at 10%: Magic, Pageblocks, Penbased, Ring, Satimage, and Twonorm. Some examples containing missing information were removed, e.g., in the Wisconsin dataset.

5.2 Configuration of the Proposal

In this subsection we present the configuration of the methods. We stress out that it is the same used in different generalizations of the Choquet integral presented in the literature [32,34,35]. We set the following parameters of the FRBCS:

- Conjunction operator = Product t-norm;
- Number of linguistic labels per variable = 5;
- Shape of membership functions = triangular;
- Minimum support = 0.05;
- Threshold for the confidence = 0.8;
- Maximum depth of the search tree = 3.

With respect to the parameters of the genetic algorithm, we consider:

- Number of individuals in the population = 50;
- Number of evaluations = 20,000;
- Bits for gene in the gray codification = 30.

5.3 Results

The classification results are summarized Table 4, which presents the accuracy for each FRBCS and dataset. The achieved results considering different datasets are available in the rows, while distinct aggregation functions Choquet, CC-integral, CT-integral, F_{NA}, GM–F_{BPC}, GM–T_{Luk}, F_{GL}–T_M, F_{GL}–F_{BPC}, and F_{NA2}, previously introduced in Table 1, are presented per columns. Last two columns show the performance of the proposed ensemble-based FRMs.

For each dataset, we highlight in **boldface** the highest accuracy among all the different methods. Similarly, the lowest value is underlined. Moreover, the mean result, considering the 27 datasets, is available for each method.

Having into consideration the results obtained in Table 4, we notice that F_{GL}-T_M achieved the best mean accuracy (80.63%), closely followed by our approach, Weighted (80.36%) and Majority Voting (80.22%). Moreover, both methods proposed in this paper achieved superior results than the remainder methods, which ranged from 78.85 up to 80.08%.

Considering the performance of the considered approaches, the Majority Voting achieved the highest accuracy mean in one dataset, and the Weighted Voting achieve it twice. Although they do not show the best result in most datasets, our ensemble-based FRMs can be considered as stable, since they showed the lowest results only in the dataset Titanic, which have presented a similar behavior in many different approaches. Moreover, all other FRMs achieved the worst result on multiple data sets.

Table 4. Obtained Results in testing by different approaches: accuracy in %

Dataset	Choquet	CC-integral	CT-integral	F_{NA}	$GM\text{–}F_{BPC}$	$GM\text{–}T_{Luk}$	$F_{GL}\text{–}T_M$	$F_{GL}\text{–}F_{BPC}$	F_{NA2}	Majority	W-Voting
App	80.13	85.84	82.99	82.99	**86.80**	84.89	84.89	85.84	85.84	84.89	83.94
Bal	82.40	81.60	82.72	82.56	89.12	**89.76**	87.84	88.32	88.64	87.84	88.00
Ban	**86.32**	84.30	85.96	86.09	84.79	85.23	85.00	85.13	84.60	85.79	85.42
Bnd	68.56	71.06	**72.13**	69.40	66.96	66.67	70.55	65.80	70.48	70.21	70.50
Bup	66.96	61.45	65.80	**67.83**	56.22	58.57	**67.83**	58.91	64.64	64.35	64.64
Cle	55.58	54.88	55.58	57.92	81.86	**84.53**	58.57	81.56	56.55	56.89	56.89
Eco	76.51	77.09	80.07	78.88	72.53	73.18	82.14	72.53	80.08	**84.24**	81.27
Gla	64.02	69.17	63.10	64.51	78.66	79.43	65.44	**79.46**	66.83	64.02	66.82
Hab	72.52	74.17	72.21	73.51	94.00	94.67	69.59	**94.67**	71.87	71.22	72.52
Hay	79.49	**81.74**	79.49	78.72	80.86	80.18	81.00	79.97	79.43	78.75	81.00
Iri	91.33	92.67	93.33	93.33	95.25	95.98	94.67	94.52	94.00	94.67	**96.00**
Led	68.20	68.40	68.60	68.60	81.42	**82.44**	69.40	81.74	69.80	69.00	70.00
Mag	78.86	79.81	79.76	80.02	75.38	74.74	**80.97**	75.13	79.70	80.55	80.39
New	94.88	93.95	95.35	93.49	78.87	78.87	**97.67**	78.87	96.28	95.35	95.81
Pag	94.16	93.97	94.34	93.97	95.48	**96.03**	94.52	**96.03**	94.15	94.71	94.89
Pho	82.98	82.94	**83.83**	82.86	58.56	58.96	81.38	58.02	81.44	82.35	82.37
Pim	73.95	74.21	74.87	**75.64**	70.00	69.60	75.51	69.60	75.52	74.08	74.22
Rin	90.95	87.97	88.78	90.27	**91.89**	90.41	90.95	91.35	89.86	91.35	89.73
Sah	69.69	70.78	70.77	68.61	**93.29**	92.86	70.56	92.81	70.12	69.27	69.48
Sat	79.47	79.01	80.40	78.54	**92.30**	91.76	80.72	91.89	80.41	79.78	79.78
Seg	93.46	92.25	93.33	92.55	68.20	68.67	**93.77**	70.69	92.42	93.68	93.33
Tit	78.87	78.87	78.87	78.87	**79.47**	79.47	78.87	79.32	78.87	78.87	78.87
Two	84.46	85.14	85.27	83.92	71.43	70.56	91.89	70.77	92.57	91.89	91.89
Veh	68.44	69.86	68.20	67.97	79.39	80.12	70.93	**81.26**	68.08	69.86	68.91
Win	93.79	93.83	96.63	96.03	96.83	**97.33**	97.19	96.14	96.08	97.19	97.17
Wis	**97.22**	95.90	96.78	96.34	85.15	83.23	96.63	81.28	96.78	96.20	96.93
Yea	55.73	57.01	56.53	56.40	54.72	53.50	58.63	54.38	57.08	58.90	**58.96**
Mean	78.85	79.18	79.47	79.25	79.98	80.06	**80.63**	79.85	80.08	80.22	80.36

To check whether our improvements are statistically significant in fact, we performed several paired Wilcoxon tests [40,41] comparing the proposed ensemble-based FRMs with the other approaches. We used the non-parametric Wilcoxon signed-rank test because the samples are not normally distributed.

Table 5 shows the results, where each row is related to our approaches. For each comparison performed by a distinct column, we present the obtained 1-tailed p-value and ranks, where R^+ is the rank related with our approach and R^- the rank related with different FRMs. Moreover, we underline the obtained p-values lower than 0.1 (level of confidence = 90%).

Analyzing the results presented in Table 5, our approaches have a performance that is statistically superior or equivalent to all other FRMs. Considering the Majority Voting, the results are statistically different from standard Choquet, CC-integral, F_{NA}. In addition to these FRMs, Weighted Voting presented statistically higher accuracy than CT-integral. When a state-of-the-art generalization C_{F1F2} is applied, our new approaches are equivalent, reinforcing the idea that an ensemble-based FRM is a promising path to be followed.

Table 5. Statistical study comparing our different approaches

FRM		Choquet	CC-integral	CT-integral	F_{NA}	GM–F_{BPC}	GM–T_{Luk}	F_{GL}–T_M	F_{GL}–F_{BPC}	F_{NA2}
Majority	p	0.01	0.06	0.17	0.02	0.90	0.88	0.07	0.77	0.62
	R^+	288.5	267.5	244.5	281.5	194	194.5	116.5	201.5	207.5
	R^-	89.5	110	133.5	96.5	184	183.5	261.5	176.5	170.5
W-Voting	p	0.00	0.01	0.03	0.00	0.76	0.81	0.24	0.70	0.06
	R^+	311.5	288.5	276.5	296.5	201.5	199	142	205	263.5
	R^-	66.5	89.5	101.5	81.5	176.5	179	236	173	114.5

6 Conclusions

Since the Choquet integral was proposed as an aggregation operator to aggregate the information in the Fuzzy Reasoning Method, many generalizations these functions where proposed in the literature. Each generalization introduces a different FRM and consequently, a different classifier each time. This study aimed to combine these generalizations in an ensemble-based FRM. Precisely, we proposed using two different ensemble approaches, Majority and Weighted Voting, to combine the information provided by these different generalizations.

From the obtained results, it is noticeable that our new method achieved satisfactory results, achieving a superior performance compared to almost all individuals generalizations. Moreover, the experiments show that this method can be considered stable since it performs poorly for only one dataset.

Finally, this approach is an effective way to aggregate information on the FRM. Future work will improve the method by inserting more diversity in the generalizations, considering different fuzzy measures, and introducing a new level of classification through the staking algorithm [19,42]. Besides, we intend to compare this method versus classical ensemble algorithms such as Random Forest [10].

Acknowledgments. This study was supported by PNPD/CAPES (464880/2019-00) and CAPES Financial Code 001, CNPq (301618/2019-4), FAPERGS (17/2551-0000872-3, 19/2551-0001279-9, 19/2551-0001660), and AEI/UE, FEDER (PID2019-108392GB-I00).

References

1. Aha, D., Kibler, D., Albert, M.: Instance-based learning algorithms. Mach. Learn. **6**(1), 37–66 (1991). https://doi.org/10.1007/BF00153759
2. Alcalá-Fdez, J., Alcalá, R., Herrera, F.: A fuzzy association rule-based classification model for high-dimensional problems with genetic rule selection and lateral tuning. IEEE Trans. Fuzzy Syst. **19**(5), 857–872 (2011)
3. Alcalá-Fdez, J., et al.: KEEL: a software tool to assess evolutionary algorithms for data mining problems. Soft Comput. **13**(3), 307–318 (2009)
4. Alpaydin, E.: Introduction to Machine Learning, 4th edn. The MIT Press, Cambridge (2020)

5. Alsina, C., Frank, M.J., Schweizer, B.: Associative Functions: Triangular Norms and Copulas. World Scientific Publishing Company, Singapore (2006)
6. Altman, D.G.: Practical Statistics for Medical Research. CRC Press (1990)
7. Barrenechea, E., Bustince, H., Fernandez, J., Paternain, D., Sanz, J.A.: Using the Choquet integral in the fuzzy reasoning method of fuzzy rule-based classification systems. Axioms **2**(2), 208–223 (2013)
8. Bedregal, B.C., Dimuro, G.P., Bustince, H., Barrenechea, E.: New results on overlap and grouping functions. Inf. Sci. **249**, 148–170 (2013)
9. Beliakov, G., Pradera, A., Calvo, T.: Aggregation Functions: A Guide for Practitioners. Springer, Berlin (2007). https://doi.org/10.1007/978-3-540-73721-6
10. Breiman, L.: Random forests. Mach. Learn. **45**, 5–32 (2001). https://doi.org/10.1023/A:1010933404324
11. Bustince, H., Fernandez, J., Mesiar, R., Montero, J., Orduna, R.: Overlap functions. Nonlinear Anal.: Theory Methods Appl. **72**(3–4), 1488–1499 (2010)
12. Choquet, G.: Theory of capacities. Annales de l'Institut Fourier **5**, 131–295 (1953–1954)
13. Cordon, O., del Jesus, M.J., Herrera, F.: Analyzing the reasoning mechanisms in fuzzy rule based classification systems. Mathware Soft Comput. **5**(2–3), 321–332 (1998)
14. Cordón, O., del Jesus, M.J., Herrera, F.: A proposal on reasoning methods in fuzzy rule-based classification systems. Int. J. Approximate Reason. **20**(1), 21–45 (1999)
15. Cortes, C., Vapnik, V.: Support vector networks. Mach. Learn. **20**, 273–297 (1995)
16. Dimuro, G.P., et al.: The state-of-art of the generalizations of the choquet integral: from aggregation and pre-aggregation to ordered directionally monotone functions. Inf. Fusion **57**, 27–43 (2020). https://doi.org/10.1016/j.inffus.2019.10.005
17. Dimuro, G.P., et al.: Generalized CF1F2-integrals: from choquet-like aggregation to ordered directionally monotone functions. Fuzzy Sets Syst. **378**, 44–67 (2020). https://doi.org/10.1016/j.fss.2019.01.009
18. Duda, R.O., Hart, P.E., Stork, D.G.: Pattern Classification (2nd Edition). Wiley-Interscience (2000)
19. Dzeroski, S., Zenko, B.: Is combining classifiers with stacking better than selecting the best one? Mach. Learn. **54**(3), 255–273 (2004). https://doi.org/10.1023/B:MACH.0000015881.36452.6e
20. Freund, Y., Schapire, R.E.: Experiments with a new boosting algorithm. In: Proceedings of the International Conference on Machine Learning, pp. 148–156 (1996)
21. Galar, M., Fernandez, A., Barrenechea, E., Bustince, H., Herrera, F.: A review on ensembles for the class imbalance problem: bagging-, boosting-, and hybrid-based approaches. IEEE Trans. Syst. Man Cybern. Part C (Appl. Rev.) **42**(4), 463–484 (2012)
22. Haykin, S.: Neural Networks: A Comprehensive Foundation. Prentice-Hall Inc., Upper Saddle River (2007)
23. Hosmer, D.W., Lemeshow, S.: Applied Logistic Regression, vol. 354. Wiley-Interscience (2004)
24. Hühn, J., Hüllermeier, E.: FURIA: an algorithm for unordered fuzzy rule induction. Data Mining Knowl. Discov. **19**(3) (2009)
25. Ishibuchi, H., Nakashima, T.: Effect of rule weights in fuzzy rule-based classification systems. IEEE Trans. Fuzzy Syst. **9**(4), 506–515 (2001)
26. Ishibuchi, H., Yamamoto, T.: Rule weight specification in fuzzy rule-based classification systems. IEEE Trans. Fuzzy Syst. **13**(4), 428–435 (2005)

27. Ishibuchi, H., Nakashima, T., Nii, M.: Classification and Modeling with Linguistic Information Granules. Advanced Approaches to Linguistic Data Mining. Advanced Information Processing. Springer, Berlin (2005). https://doi.org/10.1007/b138232
28. John, G., Langley, P.: Estimating continuous distributions in Bayesian classifiers. In: Proceedings of the Conference in Uncertainty in Artificial Intelligence, pp. 338–345 (1995)
29. Klement, E.P., Mesiar, R., Pap, E.: Triangular Norms. Kluwer Academic Publisher, Dordrecht (2000)
30. Kuncheva, L.I.: Diversity in multiple classifier systems. Inf. Fusion **6**(1), 3–4 (2005). Diversity in Multiple Classifier Systems
31. Lucca, G., Dimuro, G.P., Fernandez, J., Bustince, H., Bedregal, B., Sanz, J.A.: Improving the performance of fuzzy rule-based classification systems based on a nonaveraging generalization of CC-integrals named $C_{F_1F_2}$-integrals. IEEE Trans. Fuzzy Syst. **27**(1), 124–134 (2019). https://doi.org/10.1109/TFUZZ.2018.2871000
32. Lucca, G., et al.: Pre-aggregation functions: construction and an application. IEEE Trans. Fuzzy Syst. **24**(2), 260–272 (2016)
33. Lucca, G., Sanz, J.A., Dimuro, G.P., Borges, E.N., Santos, H., Bustince, H.: Analyzing the performance of different fuzzy measures with generalizations of the choquet integral in classification problems. In: 2019 IEEE International Conference on Fuzzy Systems (FUZZ-IEEE), pp. 1–6, June 2019. https://doi.org/10.1109/FUZZ-IEEE.2019.8858815
34. Lucca, G., Sanz, J.A., Dimuro, G.P., Bedregal, B., Bustince, H., Mesiar, R.: CF-integrals: A new family of pre-aggregation functions with application to fuzzy rule-based classification systems. Inf. Sci. **435**, 94–110 (2018)
35. Lucca, G., et al.: CC-integrals: Choquet-like copula-based aggregation functions and its application in fuzzy rule-based classification systems. Knowl.-Based Syst. **119**, 32–43 (2017)
36. Murofushi, T., Sugeno, M., Machida, M.: Non-monotonic fuzzy measures and the Choquet integral. Fuzzy Sets Syst. **64**(1), 73–86 (1994)
37. Opitz, D., Maclin, R.: Popular ensemble methods: an empirical study. J. Artif. Intell. Res. **11**, 169–198 (1999)
38. Quinlan, J.: C4.5: Programs for Machine Learning. Morgan Kauffman, San Francisco (1993)
39. Sanz, J., Fernández, A., Bustince, H., Herrera, F.: IVTURS: a linguistic fuzzy rule-based classification system based on a new interval-valued fuzzy reasoning method with tuning and rule selection. IEEE Trans. Fuzzy Syst. **21**(3), 399–411 (2013)
40. Siegel, S.: Nonparametric Statistics for the Behavioral Sciences. McGraw-Hill, New York (1956)
41. Wilcoxon, F.: Individual comparisons by ranking methods. Biometrics **1**, 80–83 (1945)
42. Wolpert, D.H.: Stacked generalization. Neural Netw. **5**(2), 241–259 (1992). https://doi.org/10.1016/S0893-6080(05)80023-1

A Useful Tool to Support the Ontology Alignment Repair

Miriam Oliveira dos Santos(✉), Carlos Eduardo Ribeiro de Mello(✉), and Tadeu Moreira de Classe(✉)

Graduate Program in Informatics (PPGI), Federal University of State of Rio de Janeiro (UNIRIO), Rio de Janeiro, Brazil
{miriam.santos,mello,tadeu.classe}@uniriotec.br

Abstract. Ontologies are used in Information Systems to manage and share knowledge pertaining to different domains, considering it is common that one particular piece of knowledge is modeled individually into different ontologies. Alignment processes are carried out towards interoperability; however, the results are not effective in most cases, requiring manual repair and review by domain experts. Thus, this work introduces a repair and review tool for the alignment of ontologies with the goal of assisting experts in the alignment repair tasks. Using the technology acceptance model, a proof of concept was made to analyze the perception by domain experts regarding the tool's usefulness, ease, and intention of use. From statistical analyses in terms of reliability, default path, inference, and correlation, it has been found that study participants had a positive perception of the tool's usefulness and ease in the tasks of ontology alignment repair. Therefore, the tool and its functionalities proved to be useful towards assisting domain experts in their tasks of reducing the semantic heterogeneity among ontologies pertaining to the same domain through alignment repair.

Keywords: Ontology matching · Ontology alignment repair · Data visualization · Technology acceptance model · Proof of concept

1 Introduction

The use of ontologies is frequently associated with Information Systems for knowledge management and sharing [1]. Overall, they describe a domain of interest or any piece of knowledge related to a specific area. They specify the meanings of terms and concepts comprising the domain, establishing a relationship among them, and having as one of its main characteristics the hierarchical organization among such concepts [2].

Ontologies are used for data integration, search, and analysis from different information sources. In most cases, it is common that the ontologies of one particular domain are modeled differently. For example, in the Biomedicine domain it is easy to find ontologies that address the concepts of one particular disease modeled in different ways [3].

R. Cerri and R. C. Prati (Eds.): BRACIS 2020, LNAI 12320, pp. 201–215, 2020.
https://doi.org/10.1007/978-3-030-61380-8_14

In Biomedicine there are recognized laboratories, such as the *European Molecular Biology Laboratory (EMBL-EBI)*[1], the *Alliance of Genome Resources*[2] and *Monarch Initiative*[3]; which aim at the integration of clinic and research data in different parts of the world. Such initiatives aim at the collaboration and diffusion of information in the discovery of new diseases and medications, applying and using ontologies to address real problems [3,4].

The interoperability among such data relies on the fact that different ontologies can relate to each other. Aiming at decreasing the distance and aligning the concepts of different ontologies within a particular domain, ontology matching processes are carried out. Such processes consist of highlighting relationships and/or correspondences among entities pertaining to different ontologies, directed to reduce their semantic heterogeneity [5]. The result is an alignment comprised of a set of correspondences or mappings among the concepts of two or more ontologies [6].

Some researches indicate that completely automated alignment methods are still not very effective [7]. Even when applying automatic strategies, such as text comparison, linguistic, semantic and structural alignment, as well as other strategies, the concepts among ontologies may not be related to each other correctly [8,9].

Thus, the ontology alignment processes require manual intervention by the domain expert (expert of a particular area). The expert's task is to repair the alignment created by automatic means, confirming and removing existing correspondences, or including new associations among concepts. Therefore, the alignment repair task aims at ensuring higher precision among executed mappings [10,11].

Researches in the literature investigate strategies of automatic or semiautomatic alignment. However, such strategies do not provide the appropriate support to the domain expert in the alignment repair stage, basically limited to displayed mappings generated through complex interfaces [11,12].

In such context, this work has the purpose to introduce and analyze the RAOSystem tool, through a proof of concept using the technology acceptance model (TAM [13]), investigating its usefulness and usability. This tool has been developed to give support to the domain expert in the process of ontology alignment repair and review, by means of functionalities based on collaboration and data visualization features, such as chats, comments, graphs, trees, heatmaps, statistics etc.

The remaining of the article is organized as follows: Sect. 2 shows the theoretical bases on ontology matching and ontology alignment repair. Section 3 presents some related works. Section 4 presents the RAOSystem and its main modules. Section 5 presents the proof of concept and how it was made. Section 6 presents data analyses based on TAM, with a discussion about the results. Finally, Sect. 7 presents the final considerations.

[1] https://www.ebi.ac.uk/spot/oxo/search.
[2] https://www.alliancegenome.org/.
[3] https://monarchinitiative.org.

2 Theoretical Bases

2.1 Ontology Matching

Even within the same domain or area, it is possible to find different ways of representing knowledge, in different ontologies, with the purpose to promote interoperability between representations and the systems that use them, it is necessary to make an alignment among the concepts of ontologies [5]. Thus, it is possible that concepts, such as the ontology matching, are processes that relies on the problem of finding the semantic mappings between two given ontologies, in the attempt to make association among domain concepts [14].

This alignment process among ontologies can be illustrated in Fig. 1 In this simple example, two ontologies are relating (aligning) their terms and concepts towards information interoperability (e.g., Person with Human; Volume with Book; Id with ISBN etc.). As a result of this process an alignment of ontologies is generated. Such alignment is comprised of a set of correspondences including the representation of relationship, equivalence, disjunction or association, and one value that represents the strength of similarity among concepts [4].

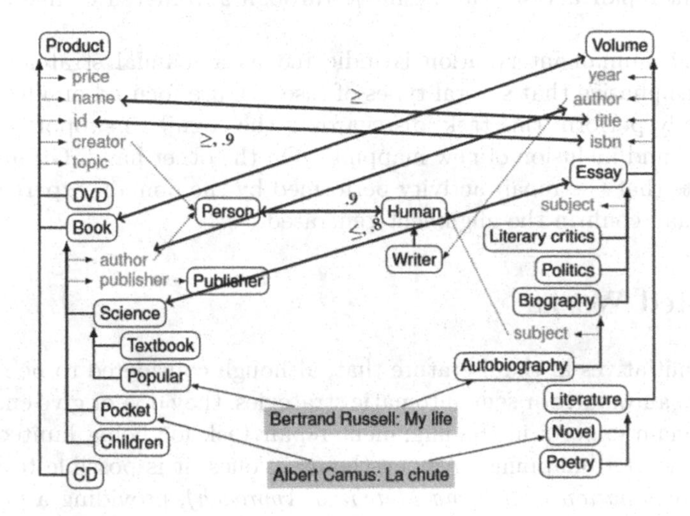

Fig. 1. Ontology matching example between two domain ontologies [4].

Different techniques can be applied in the generation of alignments among ontologies, such as techniques based on the comparisons of strings, linguistic, structure, and semantic. Techniques based on the comparison of strings take into consideration the text-similarity among terms from their chains of characters. However, some words that produce similar spellings can have distinct semantic (e.g., "Parente" in Portuguese X "Parent" in English) [8]. Linguistic-based techniques use external resources, such as thesaurus and dictionaries in the search for synonyms and meaning among terms [8]. As for techniques based on structural alignment, the analysis is made regarding entity composition among ontologies (entities, properties, relationships, superclass, etc.) [9]. Semantic-based techniques are built on the inference of terms to confirm correspondences [4].

2.2 Ontology Alignment Repair

Although there are different techniques to run an automatic ontology alignment, they are neither accurate nor completely effective [7]. Due to the complexity of the ontology alignment task, we consider that human intervention by domain experts is necessary to verify and adjust the mappings produced, leading to more accurate results [15].

Taking the above into consideration, a subsequent process is required, known as *Ontology Alignment Repair* (OAR or RAO – Reparação de Alinhamento de Ontologias - in Portuguese) [16]. This task consists of refining the alignment generated, allowing the domain expert to include, remove, or confirm correspondences among terms [4]. The task also allows identifying logical inconsistencies towards making the final alignment more consistent [16]. The implementation of alignment repair actions can be made through automated or manual strategies [10].

Although human intervention is indicated as a manual strategy. Falconer et al. [17] emphasize that several types of research are focused on algorithms to automatically perform this task, disregarding the cognitive support of users in the analysis and inclusion of new mappings. On the other hand, Granitzer et al. [18] indicate that the repair activity performed by the domain expert is the key to correct and confirm the alignment generated.

3 Related Works

There are initiatives in the literature that, although considered to be referenced in the use of automatic or semiautomatic strategies, they do not give enough support to domain experts in the alignment repair task for being limited to show only the generated mappings. Among the main ones, it is possible to highlight: *COMA (Combination of Schema Matching Approach)*, providing a platform to analyze mappings, which are represented by trees of concepts [12]. *ASMOV (Automated Semantic Matching of Ontologies With Semantic Validation)*, presenting a web page with ontology terms and its respective relationships [10]. And *AML (Agreement Maker Light)*, about which Faria et al. [11] present an interface representing ontologies as a graph, allowing the visualization by the expert.

The RAOSystem tool proposed in this work combines the main visualization functionalities indicated in related works, such as trees, graphs, heatmaps, tables, etc. However, the main difference among them is the fact that RAOSystem allows domain experts to work simultaneously in the alignment repair. Moreover, collaborative characteristics have been implemented, for example, sending messages to other domain experts and the possibility to discuss relationships through chats and comments. Thus, it makes it possible for experts to discuss and agree on a particular mapping directly within the system. Finally, the proposed tool makes it possible to keep a history of ontology alignment repair, producing a collective memory of the actions performed, through a gradual repair.

4 RAOSystem - A Tool to Support the Ontology Alignment Repair

RAOSystem is a tool that has the purpose to support domain experts in the task of ontology alignment repair. The tool architecture (Fig. 2) encompasses its main modules, and each one has specific characteristics for the task of alignment repair and review. This tool allows the repair and alignment of ontologies through data visualization features to facilitate the alignment review, inclusion, and exclusion; collaboration elements, such as messages, comments and communications among domain experts, which allow exchanging knowledge and validation of the work under execution; apart from functionalities, such as graphs and statistics that allow users to see how their work is progressing.

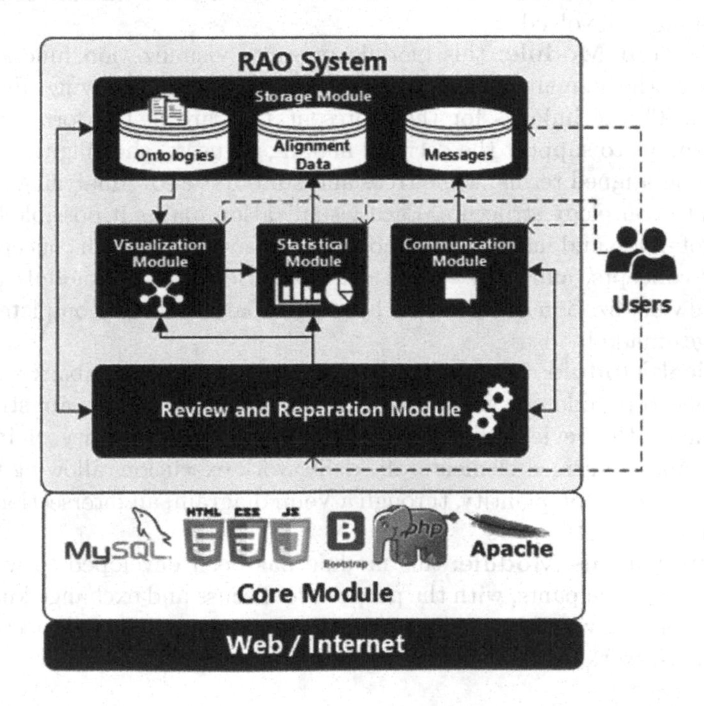

Fig. 2. RAOSystem architecture.

The tool is a web-based system. The core was built with PHP language supported by native HTML, CSS, and JavaScript. The user interface was developed using the Bootstrap framework. The database used was the MySQL 8.0.16. All of these technologies were run at an Apache server. The main RAOSystem's modules detailed in Fig. 2 are presented below:

- **Review and Repair Module**: this one can be considered the main system's module. The main user interface gives access to all system options. In this module, expert users can have access to ontology alignments available in the storage module. Thus, the user can visualize alignments already mapped (Fig. 3A), being able to perform actions, such as: remove, insert, confirm, comment, disagree, and agree with existing alignments (Fig. 3B). This module gives the user access to visualizations of information on ontologies (visualization module), statistical information related to the alignment repair process (statistical module), and the message and communications history among users (communications module).
- **Storage Module**: as the name indicates, this module is responsible for storing system information. Ontology alignment for specific domains are loaded and pursued by administrator users to make them available to the domain experts. This module is responsible for storing in the database the information related to the alignment of ontologies, alignment repair, the progress of tasks, and communications by exerts, as well as all other existing concepts of every ontology involved.
- **Visualization Module**: this module presents visualization functionalities to support the domain expert both in repairing and reviewing alignments (Fig. 3C). The techniques for the representation are in the form of graphs and heatmaps to support the decision making by users, showing the relationship among aligned terms, while trees and sunburst allow analyzing concepts through the ontology structure. Each visualization makes it possible to select options of union and intersection among terms, such as view all concepts, view mapped concepts, and view visited concepts. Besides, this module allows a complete visualization of ontologies by users, apart from the complete history of changes made by users.
- **Statistical Module**: this module lets the users to have feedback about the work done (or undone), as well as how much alignment repair still needs to be made, the review of automatic mappings, the quantity of included, removed and confirmed elements, the daily work execution, allowing the user to follow, in terms of quantity, through a Venn diagram, an intersection among ontologies (Fig. 3D).
- **Communications Module**: this module has been developed to work as a chat among participants, with the purpose to discuss and exchange knowledge on alignments. Every alignment on the platform presents its respective chat (discussions, work distribution, and general comments).

5 Proof of Concept

This section presents a Proof of Concept (evidence that a potential product or service may be useful and successful [19]), on the RAOSystem associated to how

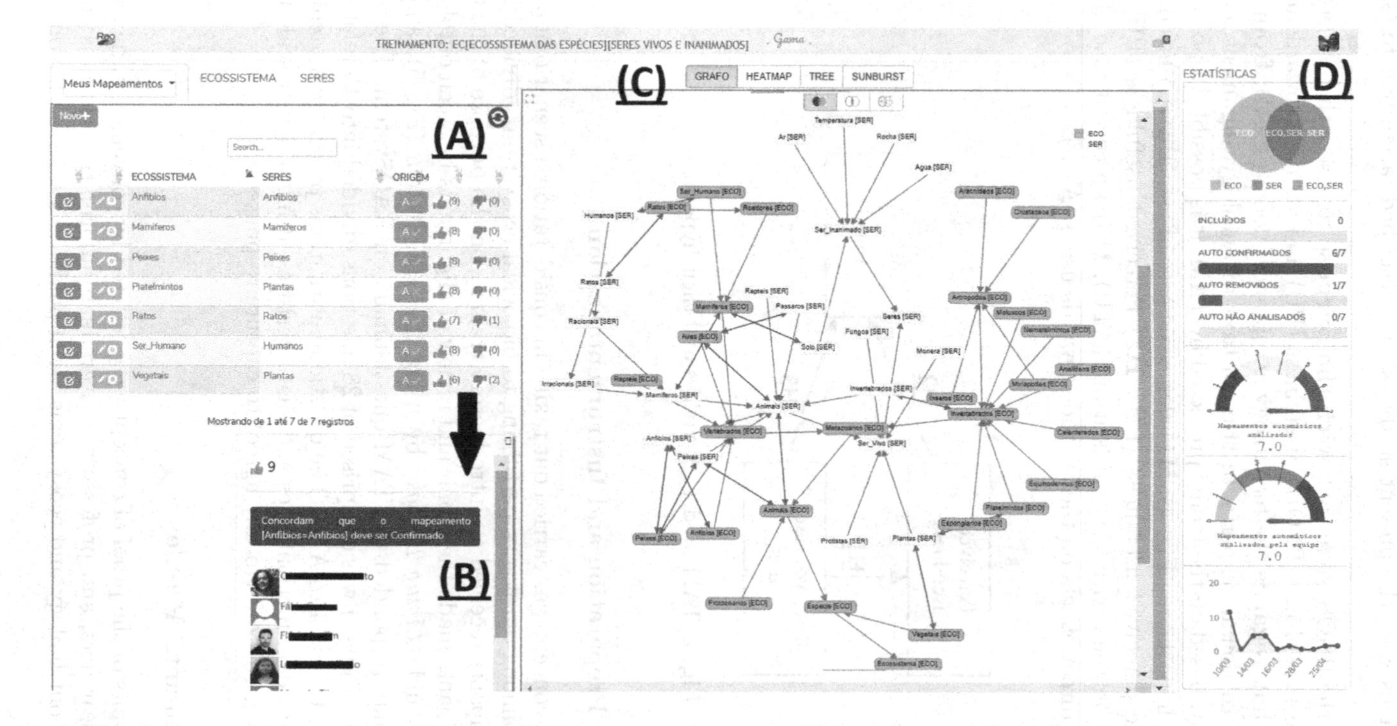

Fig. 3. A) Ontology matching concepts. B) Alignment acceptance by users. C) Ontology matching visualization. D) Ontology matching statistics.

domain experts perceive its usefulness and ease of use, evaluated using TAM Model [13]. This way, the present study can be described, according to GQM [20], as:

Analyze the RAOSystem tool; **with the purpose** to assess the perception related to the ease of use (PEOU), usefulness (PU) and attitude towards using (ATU); **in what regards** the technology acceptance model (TAM); **from the perspective of** domain experts; **in the context of** ontology alignment repair.

In this way, based on the model projected by TAM, it is possible to postulate the following hypotheses (Fig. 4): **H1**) Perceived ease of use (PEOU) has direct effects on the perceived usefulness (PU). **H2**) Perceived usefulness (PU) has direct effects on the attitude towards using (ATU). **H3**) Perceived ease of use (PEOU) has direct effects on the attitude towards using (ATU).

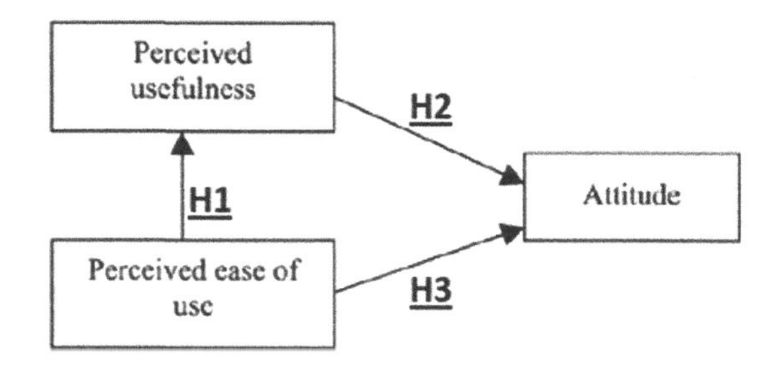

Fig. 4. TAM Model - Variables and their hypotheses.

5.1 Study Preparation and Instrumentalization

This proof of concept was carried out taking in consideration a scenario of alignment repair among ontologies pertaining to biological ecosystems, divided into: 1) the participants received an introductory training on the purpose of the tool; 2) the participants made the repair and review of alignments between ontologies *"Ecosystem"* and *"Living Beings"* by using the RAOSystem (Fig. 5); and, 3) the participants responded the TAM questionnaire on RAOSystem. The TAM questionnaire (Table 1) was comprised of 28 questions, divided into 15 related to PEOU, 9 to PU, and 4 to ATU. The questionnaire's items presented statements about RAOSystem functionalities and characteristics, using a *Likert* scale from 1 (totally disagree) to 5 (totally agree) to capture responses[4].

5.2 Participants' Profile

The participants of the proof of concept were selected by convenience, including students, researchers, and professors of a graduate information systems course who know about biology and ecosystems. Such public was selected for ontology

[4] Questionnaire and Data available at https://bit.ly/RAOSys-TAMData.

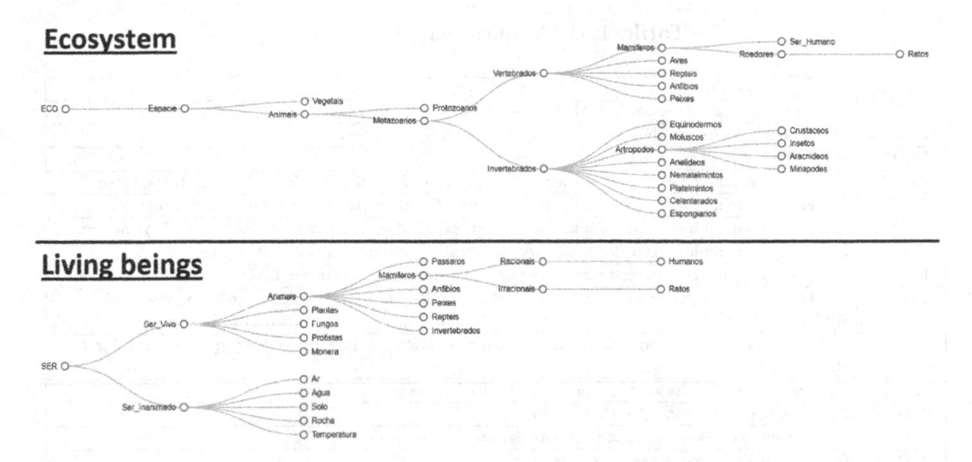

Fig. 5. RAOSystem's tree visualization example, ontologies "Ecosystem" and "Living Beings".

domain reasons, however, any domain experts could have participated, since the purpose is to verify the tool's usefulness, ease, and usability.

5.3 Data Collection and Analysis

This proof of concept was carried out in March 2020, involving 11 participants (according to Nielsen [21], technological usability tests require no more than 5 participants). Data collection took place individually, without communications among participants. All data were analyzed through the *R Statistics (3.2.2)* software and its libraries.

5.4 Threats of Conclusion

The main **threat of the conclusion** of the study can be attributed to the statistical power of the analysis methods, due to the existence of different statistical models and several ways to employ them. To mitigate such threat, normality and inference tests have been used, according to the behaviors observed in datasets gained from statistical data analysis.

Highlights of internal validity include the following: **threats of the construction**, considering the researcher expectations, such threat was carried out so that researcher had no contact with participants; **the threat of the training**, considering that participants had no knowledge about the tool, and this risk was mitigated through training and explanation about it.

Table 1. TAM questionnaire items.

Code	Item (Question)
PEOU1	The information available about RAOSystem are easy to understand.
PEOU2	It is necessary to frequently refer to tutorials to use the RAOSystem.
PEOU3	RAOSystem's icons and labels are easy to understand.
PEOU4	It is easy to remind how to perform the task "Insert Ontology Alignment Repair" in the RAOSystem.
PEOU5	It is easy to remind how to perform the task "Remove Ontology Alignment Repair" in the RAOSystem.
PEOU6	It is easy to remind how to perform the task "Confirm Ontology Alignment Repair" in the RAOSystem.
PEOU7	It is easy to remind how to perform the task "Mapping Annotation" in the RAOSystem.
PEOU8	It is easy to remind how to send a message to another domain expert in the RAOSystem.
PEOU9	It is easy to navigate and to analyze concepts of an ontology from the visual features available in the RAOSystem.
PEOU10	It is easy to analyze the correspondences between two ontologies through available visual resources in the RAOSystem.
PEOU11	I can easily identify who are the other users involved in the project (if they exist).
PEOU12	I get confused when using the RAOSystem
PEOU13	I make many mistakes when I use the RAOSystem
PEOU14	It is easy to learn how to use the RAOSystem.
PEOU15	In general, I consider the RAOSystem easy to use.
PU1	The use of the RAOSystem makes it possible to better organize the tasks of the ontology aligning repair.
PU2	The use of the RAOSystem makes it possible to improve the results of the tasks of the ontology aligning repair.
PU3	The use of the RAOSystem makes possible the teamwork.
PU4	The use of the RAOSystem allows us to save time-related to the tasks of the ontology aligning repair.
PU5	The communication features available in the RAOSystem allow users to learn about the domain concepts based on the messages from other domain experts.
PU6	The communication features available in the RAOSystem allow users to take decisions (insert, remove and confirm ontology alignment) based on the ontology concepts
PU7	The RAOSystem provides useful resources to support the collaborative work without depend on the physical participation of the domain experts.
PU8	The RAOSystem provides useful resources to support the collaborative work without depend on the domain experts who are logged into the system at the same time.
PU9	Overall, I consider the RAOSystem useful to do the ontology aligning repair task.
ATU1	I intend to use the RAOSystem in tasks of the ontology aligning repair in case of I have chances.
ATU2	I recommend to the other domain experts the use of the RAOSytem in the ontology aligning repair task.
ATU3	My attitude is favorable to the use of the RAOSystem in ontology aligning repair projects.
ATU4	I like the idea of participating in tasks of the ontology aligning repair using the RAOSystem.

6 Results and Discussion

Table 2 presents absolute and relative frequencies (score), descriptive statistics, correlation of each questionnaire's item, and the reliability measure for scales, *Cronbach's alpha*. The present study took into consideration positive relationships about user perception of the items with a score above 4. However, it is possible to notice that some questions, such as PEOU2 ("It is necessary to frequently refer to tutorials to use the RAOSystem."), PEOU12 ("I get confused when using the RAOSystem") and PEOU13 ("I make many mistakes when I use the RAOSystem"), present most of the answers with perceptions of disagreement. Those are reverse (or negative) questions, which have been analyzed inversely compared to the others, so that disagreeing with statements represent positive perceptions.

The measures of *corrected item-total correlations* (**Corr. column**) and *Cronbach's alpha coefficients* (**Alpha column**) were used to measure the

Table 2. TAM variables and question results (Descriptive statistics, correlation and reliability)

Variable	Question	Answers Frequency					Tot.	Score	Mode	S.D.	Corr.	Alpha
		1	2	3	4	5						
PEOU	PEOU1	0	0	0	3	8	11	4.73	5.00	0.467	0.392	0.614
	PEOU2	7	1	0	2	1	11	2.00	1.00	1.549	0.719	
	PEOU3	0	1	0	7	3	11	4.09	4.00	0.831	0.456	
	PEOU4	0	0	2	2	7	11	4.45	5.00	0.820	0.334	
	PEOU5	0	0	0	1	10	11	4.91	5.00	0.302	0.645	
	PEOU6	0	0	0	2	9	11	4.82	5.00	0.405	0.377	
	PEOU7	0	0	1	2	8	11	4.64	5.00	0.674	0.539	
	PEOU8	0	0	1	3	7	11	4.55	5.00	0.688	0.722	
	PEOU9	0	0	1	8	2	11	4.09	4.00	0.539	0.430	
	PEOU10	0	0	2	5	4	11	4.18	4.00	0.751	0.263	
	PEOU11	0	0	2	2	7	11	4.45	5.00	0.820	0.281	
	PEOU12	7	4	0	0	0	11	1.36	1.00	0.505	0.701	
	PEOU13	6	3	0	2	0	11	1.82	1.00	1.168	0.832	
	PEOU14	1	0	0	1	9	11	4.55	5.00	1.214	0.245	
	PEOU15	0	0	0	2	9	11	4.82	5.00	0.405	0.316	
PU	PU1	0	0	0	3	8	11	4.73	5.00	0.467	0.568	0.749
	PU2	0	0	0	2	9	11	4.82	5.00	0.405	0.435	
	PU3	0	0	1	2	8	11	4.64	5.00	0.674	0.464	
	PU4	0	0	0	5	6	11	4.55	5.00	0.522	0.467	
	PU5	0	1	0	3	7	11	4.45	5.00	0.934	0.497	
	PU6	0	0	1	3	7	11	4.55	5.00	0.688	0.783	
	PU7	0	0	0	5	6	11	4.55	5.00	0.522	0.866	
	PU8	0	0	1	4	6	11	4.45	5.00	0.688	0.680	
	PU9	0	0	0	1	10	11	4.91	5.00	0.302	0.266	
ATU	ATU1	0	0	0	3	8	11	4.73	5.00	0.467	0.875	0.809
	ATU2	0	0	0	1	10	11	4.91	5.00	0.302	0.093	
	ATU3	0	0	0	2	9	11	4.82	5.00	0.405	0.690	
	ATU4	0	0	0	2	9	11	4.82	5.00	0.405	0.690	
Total								4.15				0.786

questionnaire's reliability and correlation among items and variables. The corrected item-total correlation corresponds to the individual correlation of the questionnaire's items and variables (sets of answers), calculated through the *Pearson's correlation coefficient* [22]. It is important to measure the validity of every questionnaire's question as a whole, based on the following values: **<0.1, no correlation; ≥0.1, low correlation; ≥0.3, moderate correlation; and, ≥0.5, high correlation.** *Cronbach's alpha* is an instrument towards the questionnaire's reliability, considering it is also possible to measure subscales (total reliability of the questionnaire and PEOU, PU, and ATU variables individually). In the analysis of alpha coefficient, the following values must be observed [23]: **>0.9, excellent; >0.8, good; >0.7, acceptable; >0.6, questionable; >0.5, poor; and, ≤0.5, unacceptable.**

In this proof of concept, it was observed values for reliability (*alpha*) following the values: PEOU, **questionable** (*alpha* = 0.614); PU, **acceptable** (*alpha* = 0.749); **good** for ATU (*alpha* = 0.809); and, in general, the total questionnaire had ***acceptable*** reliability (*alpha* = 0.786). One technique that makes it possible to improve the questionnaire's reliability is *"scale purification"* [24], which means eliminating the items with the purpose to analyze the value of the alpha coefficient. If alpha goes up, the removed item is not highly correlated with scale, becoming irrelevant, otherwise, it must not be removed.

To remove items, this study considered the item removal strategy the items with item-total correlation in the ranges of low correlation and non-existent correlation, should be eliminated. Thus, questions PEOU10 ("It is easy to analyze the correspondences between two ontologies through available visual resources in the RAOSystem"), PEOU11 ("I can easily identify who are the other users involved in the project (if they exist)"), PEOU14 ("It is easy to learn how to use the RAOSystem"), PU9 ("Overall, I consider the RAOSystem useful to do the ontology alignment repair task.") and ATU2 ("I recommend to the other domain experts the use of the RAOSytem in the ontology alignment repair task.") were removed from analysis for a second round of analysis (Table 3–R1). As a result of the first purification (R1), there was an increase in the questionnaire's reliability (*alpha*), considering that PEOU recorded **acceptable** reliability (*alpha* = 0.755), ATU an **excellent** reliability (*alpha* = 0.93), and the total questionnaire **good** reliability (*alpha* = 0.843). However, when analyzing the item-total correlation, question PEOU15 ("In general, I consider the RAOSystem easy to use") had recorded value *alpha* = −0.195 (**no correlation**), thus being removed for another round of purification (Table 3–R2). Although alpha recorded a slight improvement in the second round of purification (R2; PEOU's *alpha* = 0.782 – still **acceptable**; and, total questionnaire *alpha* = 0,855 – still **good**), it was considered irrelevant regarding the questionnaire's reliability.

Table 3. Cronbach's alpha during the scale purification

Turn	PEOU	PU	ATU	Total
Before Purification	0.614	0.749	0.809	0.786
Purification 1 (R1)	0.755	0.763	0.938	0.843
Purification 2 (R2)	0.782	0.763	0.938	0.855

Upon the analysis of the questionnaire's reliability, hypotheses H1, H2, and H3 were analyzed. Table 4 and Fig. 6 show the results of hypothesis tests from the analysis of the *default path*, which sought to identify the relationship between variables PEOU, PU, and ATU. The statistical confidence interval adopted in those tests was 95% ($\alpha = 0.05$). Every hypothesis was analyzed according to statistical tests that made it possible to assess the model (Fig. 4), such as an analysis of the *standard coefficient* (β), indicating the *interference degree* of one variable on another one; the validity of the hypothesis based on **T-test** and the *effect size (Cohen's D)*, thus allowing to refute or accept those hypotheses.

Table 4. Hypothesis analysis results.

Hypothesis	Correlation	β	T-test p-value	Effect size	Conclusion
H1: (PEOU→PU)	−0.385	−0.249	2.12e−06	2.8001	Accept
H2: (PU→ATU)	0.478	0.117	0.2559	0.4987	Reject
H3: (PEOU→ATU)	−0.085	0.433	1.45e−07	3.3631	Accept

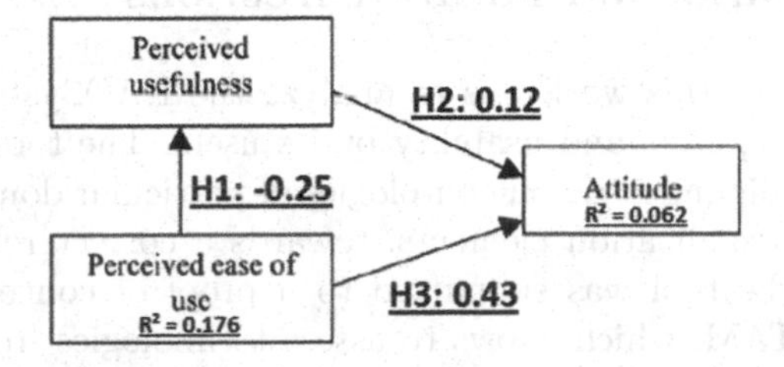

Fig. 6. Default path analysis among TAM variables.

Overall, based on the results, **it is possible to accept** with at least 95% ($\alpha = 0.05$) certainty that PEOU **had significant effects** on PU (H1: *p-value* = 2.12e−06) with **large *effect size*** and that PEOU caused a significant effect on ATU (H3: *p-value* = 1.45e−07) recording a **large *effect size*** as well. In contrast to H2 (*p-value* = 0.2559), considering that **it is not possible to state** that PU **had significant effects** on ATU, the **hypothesis was rejected**.

Upon the analysis of Table 2 it is easy to notice that, in most questions, users had a positive perception of the RAOSystem concerning PEOU, PU, and ATU. To confirm this positive perception, the *"score"* values of every TAM variable were submitted to statistical tests of inference (Table 5): i) to verify the behavior of normality regarding data distributions, which could make it possible to, ii) choose the most appropriate statistical test of hypothesis, comparing all score data and the score mean of every TAM variable.

Table 5. Inferred statistical analysis of the *"score"* value from each TAM variable.

TAM variable	Shapiro-Wilk	Normality result	Inferential test	p-value
PU	0.0692	Normal	T-test	2.73e−12
PEOU	0.5422	Normal	T-test	1.38e−15
ATU	1.37e−05	Not Normal	Wilcoxon test	0.0025

Thus, all data samples were submitted to the *Shapiro-Wilk* test of normality (indicated to small data samples), considering *alpha* = 0.05. In PU (*p-value* = 0.0692) and PEOU (*p-value* = 0.5422), the *T-test* was applied, since a normal distribution was followed. As for ATU (*p-value* = 1.37e−05), since it not followed a normal distribution, the *Wilcoxon* test was used. After such analysis, it was possible to state with at least 95% ($\alpha = 0.05$) certainty, that the RAOSystem: **was perceived as useful** (PU: *p-value* = 2.73e−12), **easy to use** (PEOU: *p-value* = 1.38e−15) and that **there is the intention of use by participants** (ATU: *p-value* = 0.0025).

7 Final Remarks and Future Directions

The main purpose of this work was to analyze the RAOSystem tool based on perceived usefulness, ease, and usability by its users. The tool enables users to review and repair alignment among ontologies of particular domains from collaborative and data visualization elements, towards a correct relationship among concepts. To so, the tool was submitted to a proof of concept, involving the analysis through TAM, which allows to assess technologies from the aspects of perceived ease of use, usefulness, and attitude towards using. As a result, the tool was perceived by users as useful, easy to use, with those users reporting their intentions towards using.

This work is part of a bigger project aimed at solutions of ontology alignment repair. The approach in this study was only related to the tool, however, this research also seeks models and metrics to enhance coverage (inclusion of new relationships), improvements in terms of accuracy, success and error rates (based on [25]) to bring better results to the ontologies alignment repair, as well as significance and usefulness of the visualization resources used.

In this way, the expectation is to also use the tool in an actual assessment scenario where domain experts of areas like Biomedicine, for example, can use the tool for the repair of ontologies that are used in the interoperability among their systems, ensuring that terms are correctly related.

References

1. Nguyen, A., Gardner, L., Sheridan, D.: Towards ontology-based design science research for knowledge accumulation and evolution. In: Proceedings of the 52nd Hawaii International Conference on System Sciences (2019)
2. Shvaiko, P., Euzenat, J.: Ontology matching: state of the art and future challenges. IEEE Trans. Knowl. Data Eng. **25**(1), 158–176 (2011)
3. Harrow, I., et al.: Ontology mapping for semantically enabled applications. Drug Discovery Today **24**, 2068–2075 (2019)
4. Banouar, O., Raghay, S.: Interoperability of information systems through ontologies: state of art. Int. J. Comput. Sci. Inf. Secur. **14**(8), 392 (2016)
5. Silva, J., Baiao, F.A., Revoredo, K.: Alinhamento interativo de ontologias usando anti-padroes de alinhamento: um primeiro experimento. In: Anais do XII Simposio Brasileiro de Sistemas de Informação, pp. 208–215 (2016)
6. Euzenat, J.: First experiments in cultural alignment repair. In: Presutti, V., Blomqvist, E., Troncy, R., Sack, H., Papadakis, I., Tordai, A. (eds.) ESWC 2014. LNCS, vol. 8798, pp. 115–130. Springer, Cham (2014). https://doi.org/10.1007/978-3-319-11955-7_10

7. Li, H., et al.: User validation in ontology alignment: functional assessment and impact. Knowl. Eng. Rev. **34**, e15 (2019)
8. Cheatham, M., Hitzler, P.: String similarity metrics for ontology alignment. In: Alani, H., et al. (eds.) ISWC 2013. LNCS, vol. 8219, pp. 294–309. Springer, Heidelberg (2013). https://doi.org/10.1007/978-3-642-41338-4_19
9. Duan, S., Fokoue, A., Srinivas, K.: One size does not fit all: customizing ontology alignment using user feedback. In: Patel-Schneider, P.F., et al. (eds.) ISWC 2010. LNCS, vol. 6496, pp. 177–192. Springer, Heidelberg (2010). https://doi.org/10.1007/978-3-642-17746-0_12
10. Meilicke, C.: Alignment incoherence in ontology matching. Ph.D. Thesis, Universitat Mannheim (2011)
11. Faria, D., et al.: The agreement makerlight ontology matching system. In: Meersman, R., et al. (eds.) OTM 2013. LNCS, vol. 8185, pp. 527–541. Springer, Heidelberg (2013). https://doi.org/10.1007/978-3-642-41030-7_38
12. Massmann, S., Raunich, S., Aumuller, D., Arnold, P., Rahm, E.: Evolution of the COMA match system. In: International Conference on Ontology Matching, vol. 814, pp. 49–60 (2011)
13. Davis, F.D.: Perceived usefulness, perceived ease of use, and user acceptance of information technology. MIS Quart. **13**(3), 319 (1989)
14. Gargouri, F., Jaziri, W.: Ontology Theory, Management, and Design: Advanced Tools and Models. Information Science Reference (2010)
15. Li, Y., Stroe, C., Cruz, I.F.: Interactive visualization of large ontology matching results. In: VOILA@ ISWC, p. 37 (2015)
16. Pesquita, C., Faria, D., Santos, E., Couto, F.M.: To repair or not to repair: reconciling correctness and coherence in ontology reference alignments. In: Proceedings of the ISWC Ontology Matching Workshop (2013)
17. Falconer, S.M., Noy, N.F., Storey, M.A.D.: Ontology mapping-a user survey. In: OM. Citeseer (2007)
18. Granitzer, M., Sabol, V., Onn, K.W., Lukose, D., Tochtermann, K.: Ontology alignment - a survey with focus on visually supported semi-automatic techniques. Future Internet **2**(3), 238–258 (2010)
19. Sensinum: what is proof of concept in software development? Likely something else than you think (2010). https://sensinum.com/proof-of-concept-in-software-development/. Accessed 14 May 2020
20. Basili, V.R.: Software modeling and measurement: the Goal/Question/Metric paradigm (1992)
21. Nielsen, J.: Why you need to test with 5 users. In: Nielsen Norman Group: Evidence-Based User Experience Research, Training, and Consulting (2000)
22. Gasparin, M., Isabela, H.M., Cristine, S.C.: Psychometric properties of the international outcome inventory for hearing AIDS. Braz. J. Otorhinolaryngol. **76**(1), 85–90 (2010)
23. George, D., Mallery, M.: IBM SPSS Statistics 19 Step by Step: A Simple Guide and Reference, 12th edn. Pearson, Boston (2012)
24. Parasuraman, A.P., Zeithaml, V.A., Berry, L.L.: SERVQUAL: a multiple-item scale for measuring consumer perceptions of service quality. J. Retail. **64**(1), 12–40 (1988)
25. Dragisic, Z., Ivanova, V., Li, H., Lambrix, P.: Experiences from the anatomy track in the ontology alignment evaluation initiative. J. Biomed. Seman. **8**(1), 56 (2017)

Aggregation with Weak, Axiological and Strong Sufficientarian Functions

Henrique Viana[1(✉)] and João Alcântara[2]

[1] Instituto Federal de Educação, Ciência e Tecnologia do Ceará (IFCE), Rodovia CE-040, Km 137,1 s/n, Aracati, CE, Brazil
henrique.viana@ifce.edu.br

[2] Departamento de Ciência da Computação, Universidade Federal do Ceará (UFC), P.O.Box 12166, Fortaleza, CE 60455-760, Brazil
jnando@lia.ufc.br

Abstract. Aggregation functions are employed to combine inputs that are typically interpreted as degrees of membership in fuzzy sets, degrees of preference, strength of evidence or support of a hypothesis. The behavior of each aggregation function can be associated to the area of distributive justice, which refers to fairness in the way data are distributed. Two important theories of distributive justice commonly considered are the utilitarianism and egalitarianism, but some alternative theories, as the sufficientarianism, has gained attention recently. This paper presents three different versions of sufficientarianism as aggregation functions: the weak, axiological and strong sufficientarianism. Additionally, we introduce new logical properties related to the sufficientarianism and show which of these new aggregation functions satisfy them. In particular, we prove the axiological aggregation functions introduced here satisfy an important logical property whereas the weak sufficientarian functions do not. On the other hand, we also show the strong sufficientarian aggregation functions may lead to the loss of logical properties.

Keywords: Aggregation functions · Decision making · Sufficientarianism · Utility theory

1 Introduction

Aggregation of information is a basic concern for all kinds of knowledge based systems, from image processing to decision making, from pattern recognition to machine learning. From a general point of view we can say that aggregation has as purpose the simultaneous use of different pieces of information (provided by several sources) in order to come to a conclusion or a decision [5]. Aggregation functions (operators) are employed to combine inputs that are typically interpreted as degrees of membership in fuzzy sets, degrees of preference, strength of evidence, or support of a hypothesis, and so on. There exists a large number of different aggregation functions that differ on the assumptions related to their

R. Cerri and R. C. Prati (Eds.): BRACIS 2020, LNAI 12320, pp. 216–230, 2020.
https://doi.org/10.1007/978-3-030-61380-8_15

data types and on the properties of their results [1]. An aggregation function can also consider how the data are distributed, caring more about some kind of data, or treating every input equally, for example. The way of each aggregation function behaves can be associated to the area of distributive justice [4,11].

The term distributive justice refers to how fair data are distributed. In modern society, for instance, this is an important principle, as it is generally expected that all goods will be distributed throughout society in some manner. In a society with a limited amount of resources, the question of fair allocation is often a source of debate and contention. Distributive principles vary in numerous dimensions. They vary in what is considered relevant to distributive justice (utility, income, wealth, opportunities, jobs, welfare, etc.); in the nature of the recipients of the distribution (individuality or groups of people, reference classes, etc); and on what basis the distribution should be made (equality, maximization, according to individual characteristics, according to free transactions, etc). In this paper, the focus is primarily on principles designed to cover the distribution of benefits and burdens of economic activity among individuals in a society.

Two important theories of distributive justice commonly considered are the utilitarianism [8,13,15,17] and egalitarianism [9,10]. Utilitarianism is the view that the moral value of a distribution of utility is the non-weighed sum of each individual's utility. The basic Utilitarian approach to justice is to maintain that when we act to maximize utility, we are also acting justly (and vice versa). Egalitarianism comes in many different versions. Basically, an egalitarian favors equality of some sort: People should get the same, or be treated the same, or be treated as equals in some respect. An alternative view expands on this last-mentioned option: People should be treated as equals, should treat one another as equals, should relate as equals, or enjoy an equality of social status. Egalitarianism is a versatile doctrine because there are several different types of equality, or ways in which people might be treated the same, or might relate as equals, that might be thought desirable. However, all of the egalitarians approaches have something in common, that is the objective of decreasing inequality.

Besides these two theories of distributive justice, we have the Sufficientarianism [6,14], which aims at ensuring that each person has an adequate amount of benefits. For instance, we recognize the instrumental importance of having enough sleep, enough money and setting aside enough time. Obviously, this requires a criterion for how much is adequate. Typically, the criterion of adequacy is something like enough to meet basic needs, avoid poverty, or have a minimally decent life, which we refer commonly as the poverty line or sufficiency line. The principle accommodates the concern we normally have for people who are badly off in absolute terms. According to most versions, Sufficiency rejects others theories of distributive justice, such as utilitarianism and egalitarianism.

This work aims at exploring the logical properties of sufficientarian aggregation functions. We will consider three approaches of sufficientarianism: weak [6], axiological [7] and strong [12]. Weak sufficientarianism states that benefiting individuals below the poverty line matters more the worse off those people are; and above the poverty line no priority is to be assigned. Axiological sufficientarianism focus on the maximization of well-being of the individuals when two distributions have the same amount of benefits. Strong sufficientarianism states that benefits that lift individuals above some poverty line matter more than equally large benefits that do not.

The paper is structured as follows. In Sect. 2, we will present some basic notions of the framework of aggregation and the basic logical properties of the framework. In Sect. 3, we will present two weak sufficientarian aggregation functions, the *headcount* and *shortfall*, and their justice relations, together with their logical properties. In Sect. 4, we will introduce the axiological sufficientarian aggregation functions and justice relations and will prove their logical properties. In Sect. 5, we will introduce strong sufficientarian aggregation functions and justice relations and analyze their logical properties. Finally, in Sect. 6 we will conclude the paper.

2 The Framework

2.1 Preliminaries

In this section, we present some fundamental notions about frameworks of aggregation and their logical properties. We assume a fixed population of agents $A = \{1, \ldots, n\}$, and a set of outcomes $\Omega = \{\omega_1, \ldots, \omega_m\}$, where each outcome $\omega_i \in \Omega$ is represented by a n-dimensional utility vector. Each outcome ω_i can be viewed as a possible world or an alternative which contains the utility levels of all agents. For $\omega_i = (\omega_i^1, \ldots, \omega_i^n)$, we will refer to ω_i^k as the utility value of the agent k in the outcome ω_i. For any ω_i^k, we will assume that $\omega_i^k \in [0, 1]$. We will use the binary relation $\leq$ to rank these utility levels. We define $<$ as follows: $\omega_i^k < \omega_j^k$ iff $\omega_i^k \leq \omega_j^k$ and $\omega_j^k \neq \omega_i^k$. Hence, the ranking of the outcomes will only depend on these utility values contained in each vector.

We assume that $\leq_f$ over Ω is a reflexive and transitive binary relation (i.e., a pre-order), where $f : [0, 1]^n \rightarrow [0, 1]$ is an aggregation function. We will refer to $\leq_f$ as an f justice relation. A pre-order $\leq_f$ is total if $\forall \omega_i, \omega_j \in \Omega$, $\omega_i \leq_f \omega_j$ or $\omega_j \leq_f \omega_i$. We define $<_f$ as follows: $\omega_i <_f \omega_j$ iff $\omega_i \leq_f \omega_j$ and $\omega_j \not\leq_f \omega_i$, and $\approx_f$ as $\omega_i \approx_f \omega_j$ iff $\omega_i \leq_f \omega_j$ and $\omega_j \leq_f \omega_i$. When $\omega_i <_f \omega_j$, we say ω_j is more just (or preferable) than ω_i with respect to f; when $\omega_i \leq_f \omega_j$, we say ω_j is at least as just as ω_i with respect to f; and $\omega_i \approx_f \omega_j$ denotes ω_i is as just as ω_j with respect to f. For instance, the *arithmetic mean* justice relation can be defined as follows.

Definition 1 (*Arithmetic mean* Justice Relation [1]). *Let $A = \{1, \ldots, n\}$ be a set of agents, $\Omega = \{\omega_1, \ldots, \omega_m\}$ be a set of outcomes, where each $\omega_i = (\omega_i^1, \ldots, \omega_i^n)$ and $\omega_i^k \in [0, 1]$, for $k \in \{1, \ldots, n\}$. We define $\leq_{mean}$ over Ω as $\omega_i \leq_{mean} \omega_j$ iff $\frac{\omega_i^1 + \cdots + \omega_i^n}{n} \leq \frac{\omega_j^1 + \cdots + \omega_j^n}{n}$.*

Example 1. Consider $A = \{1, 2, 3\}$ and $\Omega = \{\omega_1, \omega_2, \omega_3, \omega_4, \omega_5\}$, where $\omega_1 = (0.1, 0.1, 0.1), \omega_2 = (0.1, 0.1, 0.3), \omega_3 = (0.2, 0, 0.2), \omega_4 = (0.1, 0.2, 0)$ and $\omega_5 = (0, 0.1, 0.1)$. The *arithmetic mean* justice relation for Ω is equivalent to $\omega_5 <_{mean}$

$\omega_1 \approx_{mean} \omega_4 <_{mean} \omega_3 <_{mean} \omega_2$. We can say that ω_2 is more just than each other outcome with respect to *arithmetic mean*; ω_3 is more just than ω_4, which is as just as ω_1; and ω_1 is more just than ω_5.

2.2 Logical Properties

The idea of sufficientarianism is commonly traced back to Harry Frankfurt's doctrine of sufficiency [6]. Frankfurt claims that the doctrine of sufficiency aims at maximizing the number of individuals at or above sufficiency (also denoted as the poverty line in the literature). In the context of our framework of aggregation, we translate it as

Definition 2 (Frankfurt Sufficientarianism). *An outcome ω is at least as good as another ω' if and only if the number of agents at or above sufficiency in ω is at least as large as that in ω'.*

For our framework, we will denote the sufficiency (or sufficiency line) as s, where $s \in [0, 1]$. Sufficientarian justice relations show additional distinguishing behaviors when compared to other justice relations. These behaviors are viewed as representative of a humanitarian principle [16], characterized by the following important property:

Definition 3 (Weak Povertymin for s (WPM-s) [16]). *Let $A = \{1, \ldots, n\}$ be a set of agents, $\Omega = \{\omega_1, \ldots, \omega_m\}$ be a set of outcomes, where each $\omega_i = (\omega_i^1, \ldots, \omega_i^n)$ and $s \geq 0$. We say an aggregation function f satisfies the Weak Povertymin for s if for all $\omega_i, \omega_j \in \Omega$, if (1) there exists a $k \leq n$ such that $\omega_i^k < \omega_j^k$ and $\omega_i^k < s$; (2) for every $l \leq n$ such that $\omega_j^l < s$, we have $\omega_i^l \leq \omega_j^l$; then $\omega_i <_f \omega_j$.*

Weak Povertymin gives priority to the agents below the sufficiency s, while it is silent to the cases where the agents are above s [16]. We argue that **(WPM-s)** can be seen as a humanitarian condition, since it tries to favor a group of agents instead of prioritizing a unique agent (as it happens with the egalitarianism, where it gives the absolute priority to the worst off agent [16]). The agents above s are not considered essential for the group's choice. Furthermore, **(WPM-s)** can be followed by these three properties:

Definition 4 (Weak Absolute Priority of those Below s (WAPA-s) [16]). *Let $A = \{1, \ldots, n\}$ be a set of agents, $\Omega = \{\omega_1, \ldots, \omega_m\}$ be a set of outcomes, where each $\omega_i = (\omega_i^1, \ldots, \omega_i^n)$ and $s \geq 0$. We say an aggregation function f satisfies Weak Absolute Priority of those Below s if for all $\omega_i, \omega_j \in \Omega$, if (1) there exist k, k' such that $s \leq \omega_j^k < \omega_i^k$; (2) $\omega_i^{k'} < \omega_j^{k'} \leq s$; (3) for $l \neq k, k'$, $\omega_i^l = \omega_j^l$; then $\omega_i \leq_f \omega_j$.*

(WAPA-s) solves the conflicts for the agents below s prioritizing those agents with higher utility value below s, whilst it ignores the utility values of those agents above or at s, and also those where the utility values are the same for both agents.

Definition 5 (Strong Pareto (SP) [16]). *Let $A = \{1, \ldots, n\}$ be a set of agents, $\Omega = \{\omega_1, \ldots, \omega_m\}$ be a set of outcomes, where each $\omega_i = (\omega_i^1, \ldots, \omega_i^n)$. We say an aggregation function f satisfies Strong Pareto if for all $\omega_i, \omega_j \in \Omega$, if there exists k, $\omega_i^k < \omega_j^k$ and for all $l \neq k$, $\omega_i^l \leq \omega_j^l$, then $\omega_i <_f \omega_j$.*

Strong Pareto might be interpreted as the principle of personal good, where the utility values refers to the good of the agents. An outcome where all utility values are higher or equal than other outcome (with at least one utility value higher), it might be considered more just.

Definition 6 (Anonymity (A) [16]). *Let $A = \{1, \ldots, n\}$ be a set of agents, $\Omega = \{\omega_1, \ldots, \omega_m\}$ be a set of outcomes, where each $\omega_i = (\omega_i^1, \ldots, \omega_i^n)$. We say an aggregation function f satisfies Anonymity if for all $\omega_i, \omega_j \in \Omega$, if ω_i is a permutation of ω_j, then $\omega_i \approx_f \omega_j$.*

Anonymity is a condition of impartiality, which states that the identity of the agents should not matter in a justice relation. We have the following result:

Theorem 1. [16] *If a justice relation $\leq_f$ satisfies **(WAPA-s)**, **(SP)** and **(A)**, then it satisfies **(WPM-s)**.*

3 Weak Sufficientarian Aggregation Functions

In this section, we present two approaches on the idea of weak sufficientarianism. Recalling [12], there are two central views of sufficientarianism:

- **Weak Sufficientarianism:** Any benefit above s, no matter how small, and no matter to how few agents, outweighs any benefit below s, no matter how large, and no matter to how many agents. Above s, equally large benefits matter more the worse off the agent is;
- **Strong Sufficientarianism:** Benefits that lift agents below some threshold level s matter more than equally large benefits that do not, whether they occur above or below s.

One of the simplest principles of weak sufficientarianism is the *headcount* claim [14], which aims at maximizing the number of agents above sufficiency.

Definition 7 (*headcount* Justice Relation). *Let $A = \{1, \ldots, n\}$ be a set of agents, $\Omega = \{\omega_1, \ldots, \omega_m\}$ be a set of outcomes, where each $\omega_i = (\omega_i^1, \ldots, \omega_i^n)$, $\omega_i^k \in [0, 1]$ for $k \in \{1, \ldots, n\}$. We define the number of agents above $s \in [0, 1]$ in ω_i as $hc(\omega_i, s) = \#(\{\omega_i^k \mid \omega_i^k \geq s\})$, where $\#(A)$ is the cardinal of the set A. We define $\leq_{hc_s}$ over Ω as $\omega_i \leq_{hc_s} \omega_j$ iff $hc(\omega_i, s) \leq hc(\omega_j, s)$.*

Besides *headcount*, the *shortfall* [7] is another sufficientarian measure of aggregation. It simply adds up each agent's total gap from their utility values (where an agent's *shortfall* is zero if his/her utility value is at or above s). The total *shortfall* operator adds up the *shortfall* from s across agents below s, and takes the unweighted sum to be the measure of the disvalue of the group [7]. Differently from *headcount* justice relation, where an outcome ω_j is more preferable than ω_i if the number of agents above or at s in ω_j is greater than in ω_i, *shortfall* justice relation is concerned with the minimization of the total amount of deficit of the agents below s.

Definition 8 (*shortfall* Justice Relation). *Let $A = \{1, \dots, n\}$ be a set of agents, $\Omega = \{\omega_1, \dots, \omega_m\}$ be a set of outcomes, where each $\omega_i = (\omega_i^1, \dots, \omega_i^n)$, $\omega_i^k \in [0,1]$ for $k \in \{1, \dots, n\}$. We define the shortfall of the agents below $s \in [0,1]$ in ω_i as $sh(\omega_i, s) = \sum_{\omega_i^k < s} s - \omega_i^k$. We define $\leq_{sh_s}$ over Ω as $\omega_i \leq_{sh_s} \omega_j$ iff $sh(\omega_i, s) \geq sh(\omega_j, s)$.*

Shortfall aggregation function is sensible for the variations of utility values.

Example 2. Let us consider the following outcomes and sufficiency line: $\omega_1 = (0.1, 0.1, 0.1), \omega_2 = (0.1, 0.1, 0.3), \omega_3 = (0.2, 0, 0.2), \omega_4 = (0.1, 0.2, 0), \omega_5 = (0, 0.1, 0.1)$ and $s = 0.2$. The resulting justice relation for *headcount* and *shortfall* are $\omega_1 \approx_{hc_s} \omega_5 <_{hc_s} \omega_2 \approx_{hc_s} \omega_4 <_{hc_s} \omega_3$ and $\omega_5 <_{sh_s} \omega_1 \approx_{sh_s} \omega_4 <_{sh_s} \omega_2 \approx_{sh_s} \omega_3$.

Note that ω_1 is as just as ω_5 with respect to *headcount* for $s = 0.2$, since they have three agents below sufficiency. However, ω_1 is more just than ω_5 with respect to *shortfall*, because the shortfall of the agents below $s = 0.2$ for ω_1 is equal to 0.3 and for ω_5 is equal to 0.4. The *headcount* and *shortfall* justice relations satisfy the results below:

Theorem 2. *Let $\leq_{hc_s}$ and $\leq_{sh_s}$ be the headcount and shortfall justice relations, respectively. (1) $\leq_{hc_s}$ satisfies **(WAPA-s)** and **(A)**, but **(SP)** and **(WPM-s)** are not satisfied in general; (2) $\leq_{sh_s}$ satisfies **(WAPA-s)**, **(A)** and **(WPM-s)**, but **(SP)** is not satisfied in general.*

Proof. **(WAPA-s)** Let $A = \{1, \dots, n\}$ be a set of agents, $\Omega = \{\omega_1, \dots, \omega_m\}$ be a set of outcomes, where each $\omega_i = (\omega_i^1, \dots, \omega_i^n)$ and $s \geq 0$. Suppose we have $\omega_i, \omega_j \in \Omega$, such that (i) there exist k, k' such that $s \leq \omega_j^k < \omega_i^k$; (ii) $\omega_i^{k'} < \omega_j^{k'} \leq s$; (iii) for $l \neq k, k'$, $\omega_i^l = \omega_j^l$. Then we have two cases: (1) by (ii), if $\omega_i^{k'} < \omega_j^{k'} = s$, then $hc(\omega_i, s) < hc(\omega_j, s)$ and $sh(\omega_i, s) < sh(\omega_j, s)$ ((i) does not add any value to *headcount* or $shortfall$ and (iii) adds the same value for them since for $l \neq k, k'$, $\omega_i^l = \omega_j^l$); (2) by (ii), if $\omega_i^{k'} < \omega_j^{k'} < s$, then $hc(\omega_i, s) = hc(\omega_j, s)$ and $sh(\omega_i, s) < sh(\omega_j, s)$. Therefore, $\omega_i \leq_{hc_s} \omega_j$ and $\omega_i <_{sh_s} \omega_j$.

(A) It is easy to see that if an utility value is below s (above or at s), any permutation of it shall also be below s (above or at s). Therefore, for all $\omega_i, \omega_j \in \Omega$, if ω_i is a permutation of ω_j, then $\omega_i \approx_{hc_s} \omega_j$ and $\omega_i \approx_{sh_s} \omega_j$.

(SP) Suppose $\omega_i, \omega_j \in \Omega$, if there exists k, $\omega_i^k < \omega_j^k$ and for all $l \neq k$, $\omega_i^l \leq \omega_j^l$. If for all $\omega_i, \omega_j \in \Omega$, $s \leq \omega_i^l \leq \omega_j^l$ and $s \leq \omega_i^k < \omega_j^k$, then $\omega_i \approx_{hc_s} \omega_j$ and $\omega_i \approx_{sh_s} \omega_j$. Therefore, $\omega_i \not<_{hc_s} \omega_j$ and $\omega_i \not<_{sh_s} \omega_j$.

(WPM-s) Let $A = \{1, \dots, n\}$ be a set of agents, $\Omega = \{\omega_1, \dots, \omega_m\}$ be a set of outcomes, where each $\omega_i = (\omega_i^1, \dots, \omega_i^n)$ and $s \geq 0$. Suppose we have $\omega_i, \omega_j \in \Omega$, such that (i) there exists a $k \leq n$ such that $\omega_i^k < \omega_j^k$ and $\omega_i^k < s$; (ii) every position l that $\omega_j^l < s$ implies $\omega_i^l \leq \omega_j^l$. In the worst case we have (ii) every position l that $\omega_j^l < s$ implies $\omega_i^l = \omega_j^l$. Then the *shortfall* of each l is equivalent for ω_i and ω_j. By (i) we have $\omega_i^k < \omega_j^k$ and $\omega_i^k < s$, so the *shortfall* of ω_j^k is lower than the shortfall of ω_i^k. Then, $sh(\omega_i, s) < sh(\omega_j, s)$. Therefore, $\omega_i <_{sh_s} \omega_j$. □

Both justice relation do not satisfy **(SP)** because they are silent to the cases of agents above s. **(WPM-s)** is not satisfied by $\leq_{hc_s}$ since the conclusion $\omega <_f \omega'$ is too strong for this operator. However, we may consider a weaker version of **(WPM-s)**.

Definition 9 (Weaker Povertymin for s (wPM-s)). *Let $A = \{1, \dots, n\}$ be a set of agents, $\Omega = \{\omega_1, \dots, \omega_m\}$ be a set of outcomes, where each $\omega_i = (\omega_i^1, \dots, \omega_i^n)$ and $s \geq 0$. We say an aggregation function f satisfies Weaker Povertymin for s if for all $\omega_i, \omega_j \in \Omega$, if (1) there exists a $k \leq n$ such that $\omega_i^k < \omega_j^k$ and $\omega_i^k < s$; (2) every position l that $\omega_j^l < s$ implies $\omega_i^l \leq \omega_j^l$; then $\omega_i \leq_f \omega_j$.*

Besides, we also need consider the condition a Weaker version of Strong Pareto, called Weak Pareto.

Definition 10 (Weak Pareto (WP)). *Let $A = \{1, \dots, n\}$ be a set of agents, $\Omega = \{\omega_1, \dots, \omega_m\}$ be a set of outcomes, where each $\omega_i = (\omega_i^1, \dots, \omega_i^n)$. We say an aggregation function f satisfies Weak Pareto if for all $\omega_i, \omega_j \in \Omega$, if $\omega_i^l \leq \omega_j^l$, then $\omega_i \leq_f \omega_j$.*

Differently from Strong Pareto, in Weak Pareto an outcome where all utility values are higher or equal than other outcome might be considered at least equally just (not necessarily with at least one utility value higher). Then the following result is obtained:

Corollary 1. *If a justice relation $\leq_f$ satisfies **(WAPA-s)**, **(WP)** and **(A)**, then it satisfies **(WPM-s)**.*

Proof. It follows directly from Theorem 1.

Therefore, *headcount* justice relation satisfies the following properties:

Theorem 3. *Let $\leq_{hc_s}$ be the headcount justice relation, then $\leq_{hc_s}$ satisfies **(WAPA-s)**, **(WP)**, **(A)** and **(WPM-s)**.*

Proof. Similar to the proof of Theorem 1. □

Headcount justice relation satisfies **(WPM-s)** because when $\omega_j^k < s$, the conclusion is that $\omega_i \approx_f \omega_j$; otherwise, it concludes that $\omega_i <_f \omega_j$. Weak Pareto is achieved because the utility values above s are equivalent w.r.t to the pre-order $\leq$ and do not bring any impact to the conclusion of the property (note that *shortfall* also satisfies Weak Pareto and the same results hold to *shortfall* justice relation).

4 Axiological Weak Sufficientarian Aggregation Functions

In the previous section, it was showed a first access to weak sufficientarian aggregation functions and justice relations. However, none of them, in fact, satisfies **(SP)**. In order to overcome such weakness, we give now a further step by presenting another class of weak sufficientarian aggregation functions: they are the axiological weak sufficientarian aggregation functions.

4.1 Axiological Weak Sufficientarianism

As said earlier, the doctrine of sufficiency has established itself as a distinctive position among many theories of distributive justice. It has attracted many proponents and they have proposed several variants of the doctrine of sufficiency that retain its general spirit [3]. For instance, in [2], it was pointed out that the doctrine of sufficiency (or sufficientarianism) is committed to both the "positive thesis" and the "negative thesis":

- **The Positive Thesis:** It is morally important for agents to have enough resources;
- **The Negative Thesis:** Once everybody has enough resources, whether somebody has more or less than others has absolutely no moral significance.

In general terms, proponents and opponents of sufficientarianism find the positive thesis plausible. Thus, the usual target of criticisms is focused on sufficientarianism's commitment to the negative thesis. According to the negative thesis, the moral insignificance of some agents having more than others is limited to situations in which everybody has enough resources. This means that there is room for sufficientarians to give priority to those who are below such threshold if we discard the negative thesis condition.

There are different ways to approach the violation of the negative thesis. One of them is to depict the sufficientarianism as an axiological principle, i.e., a criterion for ranking the interpretation in terms of goodness. In [7], it was proposed the Axiological Sufficientarianism, which can be translated into the following principle

Definition 11 (Axiological Sufficientarianism Principle [7]**).** *An outcome ω_j is strictly better than another outcome ω_j if and only if either (1)* $\sum_{\omega_j^k < s} s - \omega_j^k < \sum_{\omega_i^k < s} s - \omega_i^k$ *or (2)* $\sum_{\omega_j^k < s} s - \omega_j^k = \sum_{\omega_i^k < s} s - \omega_i^k$ *and* $\sum_{\omega_i^k \geq s} \omega_i^k < \sum_{\omega_j^k \geq s} \omega_j^k$.

In plain words, an outcome ω_j is strictly better than ω_i if and only if either (1) the weighted sum of shortfall in ω_i is strictly higher than that in ω_j or (2) the weighted sum of shortfall in ω_j is the same as that in ω_i and the unweighted sum of utility values above or at s in ω_i is strictly smaller than that in ω_j. Inspired by this definition, we devise a version of *headcount* and *shortfall* as axiological aggregation functions, dubbed respectively as *ahc* and *ash*.

Definition 12 (Axiological Justice Relations) *Let $A = \{1, \ldots, n\}$ be a set of agents, $\Omega = \{\omega_1, \ldots, \omega_m\}$ be a set of outcomes, where each $\omega_i = (\omega_i^1, \ldots, \omega_i^n)$, $\omega_i^k \in [0,1]$ for $k \in \{1, \ldots, n\}$ and $s \in [0,1]$. (1) We define the number of agents above s in ω_i as $hc(\omega_i, s) = \#(\{\omega_i^k \mid \omega_i^k \geq s\})$, where $\#(A)$ is the cardinal of the set A. We define $\leq_{ahc_s}$ over Ω as $\omega_i \leq_{ahc_s} \omega_j$ iff (if $hc(\omega_i, s) = hc(\omega_j, s)$ then $\sum_{\omega_i^k \geq s} \omega_i^k \leq \sum_{\omega_j^k \geq s} \omega_j^k$; otherwise $hc(\omega_i, s) \leq hc(\omega_j, s)$. (2) We define the shortfall of the agents below s in ω_i as $sh(\omega_i, s) = \sum_{\omega_i^k < s} s - \omega_i^k$. We define $\leq_{ash_s}$ over Ω as $\omega_i \leq_{ash_s} \omega_j$ iff if $sh(\omega_i, s) = sh(\omega_j, s)$ then $\sum_{\omega_i^k \geq s} \omega_i^k \leq \sum_{\omega_j^k \geq s} \omega_j^k$; otherwise $sh(\omega_i, s) \geq sh(\omega_j, s)$.*

When the outcomes are not equally sufficient, these functions behave as the original *headcount* and *shortfall* aggregation functions. Otherwise, they behave as a *sum* function of the utility values above or at s.

Example 3. Let us consider the following outcomes and sufficiency line: $\omega_1 = (0.1, 0.1, 0.1), \omega_2 = (0.1, 0.1, 0.3), \omega_3 = (0.2, 0, 0.2), \omega_4 = (0.1, 0.2, 0), \omega_5 = (0, 0.1, 0.1)$ and $s = 0.2$. The results of *headcount*, *shortfall* and their axiological versions are: (i) The resulting pre-order $\leq_{hc_s}$ is $\omega_1 \approx_{hc_s} \omega_5 <_{hc_s} \omega_2 \approx_{hc_s} \omega_4 <_{hc_s} \omega_3$; (ii) The resulting pre-order $\leq_{ahc_s}$ is $\omega_1 \approx_{ahc_s} \omega_5 <_{ahc_s} \omega_4 <_{ahc_s} \omega_2 <_{ahc_s} \omega_3$; (iii) The resulting pre-order $\leq_{sh_s}$ is $\omega_5 <_{sh_s} \omega_1 \approx_{sh_s} \omega_4 <_{sh_s} \omega_2 \approx_{sh_s} \omega_3$; (iv) The resulting pre-order $\leq_{ash_s}$ is $\omega_5 <_{ash_s} \omega_1 <_{ash_s} \omega_4 <_{ash_s} \omega_2 <_{ash_s} \omega_3$.

Axiological weak justice relations come as a tie-breaker for the original relation. For instance, $\omega_1 \approx_{hc_s} \omega_5$ and $\omega_5 <_{ahc_s} \omega_1$; and $\omega_2 \approx_{sh_s} \omega_3$ and $\omega_3 <_{ash_s} \omega_2$.

Theorem 4. *$\leq_{ahc_s}$ satisfies **(WAPA-s)**, **(A)** and **(WPM-s)**, but **(SP)** and **(WPM-s)** are not satisfied in general. $\leq_{ash_s}$ satisfies **(WAPA-s)**, **(A)**, **(SP)** and **(WPM-s)**.*

Proof. The proofs are similar to Theorems 2 and 3. For **(SP)**, Suppose $\omega_i, \omega_j \in \Omega$, if there exists k, $\omega_i^k < \omega_j^k$ and for all $l \neq k$, $\omega_i^l \leq \omega_j^l$. We have three cases to analyze: (1) If for all $\omega_i, \omega_j \in \Omega$, $\omega_i^l \leq \omega_j^l < s$ and $\omega_i^k < \omega_j^k < s$, then $\omega_i \approx_{hc_s} \omega_j$ and $sh(\omega_i, s) < sh(\omega_j, s)$. Therefore, $\omega_i \not<_{ahc_s} \omega_j$ $\omega_i <_{ash_s} \omega_j$; (2) for all $k' \in \{1, \ldots, n\}$, $\omega_i^{k'} \geq s$. We have that $\sum_{k''} \omega_i^{k''} < \sum_{k''} \omega_j^{k''}$. Then $\omega_i <_{ahc_s} \omega_j$ and $\omega_i <_{ash_s} \omega_j$; (3) there is a $k' \in \{1, \ldots, n\}$ such that $\omega_i^{k'} < s$ and for all $k'' \neq k'$, $\omega_i^{k''} \geq s$. In this case if $\omega_j^{k'} \geq s$, it is easy to conclude that $hc(\omega_i, s) < hc(\omega_j, s)$ and $sh(\omega_i, s) < sh(\omega_j, s)$. Therefore, $\omega_i \not<_{ahc_s} \omega_j$ and $\omega_i <_{ash_s} \omega_j$. □

Although the axiological *shortfall* justice relation is different from its original version, i.e., $\leq_{sh_s} \not\equiv \leq_{ash_s}$, in terms of the logical properties exposed, they are equivalent. However, we can adjust the Axiological Sufficientarianism Principle in order to satisfy the missing logical properties for *headcount* justice relation.

Definition 13 (Axiological Justice Relations (2)). *Let $A = \{1, \ldots, n\}$ be a set of agents, $\Omega = \{\omega_1, \ldots, \omega_m\}$ be a set of outcomes, where each $\omega_i = (\omega_i^1, \ldots, \omega_i^n)$, $\omega_i^k \in [0,1]$ for $k \in \{1, \ldots, n\}$ and $s \in [0,1]$. (1) We define the number of agents above s in ω_i as $hc(\omega_i, s) = \#(\{\omega_i^k \mid \omega_i^k \geq s\})$, where $\#(A)$ is the cardinal of the set A. We define $\leq_{Ahc_s}$ over Ω as $\omega_i \leq_{Ahc_s} \omega_j$ iff (if $hc(\omega_i, s) = hc(\omega_j, s)$ then $\sum_k \omega_i^k \leq \sum_k \omega_j^k$; otherwise $hc(\omega_i, s) \leq hc(\omega_j, s)$. (2) We define the shortfall of the agents below s in ω_i as $sh(\omega_i, s) = \sum_{\omega_i^k < s} s - \omega_i^k$. We define $\leq_{Ash_s}$ over Ω as $\omega_i \leq_{Ash_s} \omega_j$ iff if $sh(\omega_i, s) = sh(\omega_j, s)$ then $\sum_k \omega_i^k \leq \sum_k \omega_j^k$; otherwise $sh(\omega_i, s) \geq sh(\omega_j, s)$.*

When the outcomes are not equally sufficient, these functions behave as the original *headcount* and *shortfall* aggregation functions. Otherwise, they behave as a *sum* function of all utility values.

Example 4. Let us consider the following outcomes and sufficiency line: $\omega_1 = (0.1, 0.1, 0.1), \omega_2 = (0.1, 0.1, 0.3), \omega_3 = (0.2, 0, 0.2), \omega_4 = (0.1, 0.2, 0), \omega_5 = (0, 0.1, 0.1)$ and $s = 0.2$. The results of *headcount* and *shortfall* in their axiological versions are: (i) The pre-order $\leq_{ahc_s}$ is $\omega_1 \approx_{ahc_s} \omega_5 <_{ahc_s} \omega_4 <_{ahc_s} \omega_2 <_{ahc_s} \omega_3$; (ii) The pre-order $\leq_{Ahc_s}$ is $\omega_5 <_{Ahc_s} \omega_1 <_{Ahc_s} \omega_4 <_{Ahc_s} \omega_2 <_{Ahc_s} \omega_3$; (iii) The pre-order $\leq_{ash_s}$ is $\omega_5 <_{ash_s} \omega_1 <_{ash_s} \omega_4 <_{ash_s} \omega_2 <_{ash_s} \omega_3$; (iv) The pre-order $\leq_{Ash_s}$ is $\omega_5 <_{Ash_s} \omega_1 \approx_{Ash_s} \omega_4 <_{Ash_s} \omega_3 <_{Ash_s} \omega_2$.

Note two outcomes can be equivalent for their original and axiological justice relations, as it happens in $\omega_1 \approx_{sh_2} \omega_4$ and $\omega_1 \approx_{Ash_2} \omega_4$. Or the case where the axiological justice relations produce opposite results: for *shortfall*, we have $\omega_2 \approx_{sh_2} \omega_3, \omega_2 <_{ash_2} \omega_3$ and $\omega_3 <_{Ash_2} \omega_2$. An axiological justice relation (2) is also robust enough to satisfy Strong Pareto for both *headcount* and *shortfall* aggregation functions.

Theorem 5. *Let $\leq_{Ahc_s}$ and $\leq_{Ash_s}$ be the axiological headcount and shortfall justice relations (2), respectively. $\leq_{Ahc_s}$ and $\leq_{Ash_s}$ satisfy **(WAPA-s)**, **(SP)**, **(A)** and **(WPM-s)**.*

Proof. We will show that they satisfy **(SP)** (the proof of the other properties follows a similar reasoning). We have three cases to analyze: (1) If for all $\omega_i, \omega_j \in \Omega$, $\omega_i^l \leq \omega_j^l < s$ and $\omega_i^k < \omega_j^k < s$, then $\sum_k \omega_i^k \leq \sum_k \omega_j^k$ and $sh(\omega_i, s) < sh(\omega_j, s)$. Therefore, $\omega_i <_{Ahc_s} \omega_j$ and $\omega_i <_{Ash_s} \omega_j$; (2) For all $k' \in \{1, \ldots, n\}$, $\omega_i^{k'} \geq s$. We have that $\sum_{k''} \omega_i^{k''} < \sum_{k''} \omega_j^{k''}$. Then, $\omega_i <_{Ahc_s} \omega_j$ and $\omega_i <_{Ash_s} \omega_j$; (3) There is a $k' \in \{1, \ldots, n\}$ such that $\omega_i^{k'} < s$ and for all $k'' \neq k'$, $\omega_i^{k''} \geq s$. In this case if $\omega_j^{k'} \geq s$, it is easy to conclude that $hc(\omega_i, s) < hc(\omega_j, s)$ and $sh(\omega_i, s) < sh(\omega_j, s)$. Therefore, $\omega_i <_{Ahc_s} \omega_j$ and $\omega_i <_{Ash_s} \omega_j$. $\square$

5 Strong Sufficientarianism

In Sects. 2 and 3, weak and axiological weak sufficientarian justice relations were described, respectively. Now we will show an approach to strong sufficientarian justice relations. For this, we will define strong versions of the *headcount* and *shortfall* operators based on the statement above:

Definition 14 (Strong *headcount* Justice Relation). *Let* $A = \{1, \ldots, n\}$ *be a set of agents,* $\Omega = \{\omega_1, \ldots, \omega_m\}$ *be a set of outcomes, where each* $\omega_i = (\omega_i^1, \ldots, \omega_i^n)$*,* $\omega_i^k \in \mathbb{N}$ *for* $k \in \{1, \ldots, n\}$*. We define the number of agents across* $s \in [0, 1]$ *w.r.t.* ω_i *and* ω_j *as* $shc(\omega_i, \omega_j, s) = \#(\{\omega_i^k \mid \omega_i^k \geq s > \omega_j^k\})$*, where* $\#(A)$ *is the cardinal of the set* A*. We define* $\leq_{shc_s}$ *over* Ω *as* $\omega_i \leq_{shc_s} \omega_j$ *iff* $shc(\omega_i, \omega_j, s) \leq shc(\omega_j, \omega_i, s)$*.*

This definition comes directly from strong sufficientarianism: an outcome ω_j is better than ω_i if the number of agents above s in ω_j and below s in ω_i is greater than in ω_i. Furthermore, it is silent for the case where an agent is below s in both outcomes.

Example 5. Let us consider the following outcomes and sufficiency line: $\omega_1 = (0.1, 0.1, 0.1), \omega_2 = (0.1, 0.1, 0.3), \omega_3 = (0.2, 0, 0.2), \omega_4 = (0.1, 0.2, 0), \omega_5 = (0, 0.1, 0.1)$ and $s = 0.2$. The results of *headcount* and its strong version are: (i) The resulting pre-order $\leq_{hc_s}$ is $\omega_1 \approx_{hc_s} \omega_5 <_{hc_s} \omega_2 \approx_{hc_s} \omega_4 <_{hc_s} \omega_3$; (ii) The resulting pre-order $\leq_{shc_s}$ is $\omega_1 \approx_{shc_s} \omega_5 <_{shc_s} \omega_2 \approx_{shc_s} \omega_4 <_{shc_s} \omega_3$.

Note that, although we have slightly different definitions, *headcount* and strong *headcount* operators produce the same justice relation.

Theorem 6. $\omega_i \leq_{hc_s} \omega_j \Leftrightarrow \omega_i \leq_{shc_s} \omega_j$.

Proof. ($\Rightarrow$) Suppose $\omega_i \leq_{hc_s} \omega_j$. by definition, $hc(\omega_i, s) \leq hc(\omega_j, s)$. We need to show that $shc(\omega_i, \omega_j, s) \leq shc(\omega_j, \omega_i, s)$. For any k, we have the following cases: (1) $s \leq \omega_i^k \leq \omega_j^k$ or $s \leq \omega_j^k \leq \omega_i^k$. This case is ignored in the *headcount* and the strong *headcount*; (2) $\omega_i^k \leq \omega_j^k < s$ or $\omega_j^k \leq \omega_i^k < s$. In this we count +1 in the *headcount* for both outcomes and for the strong *headcount* no value is added for the outcomes. So, there is no difference for *headcount* and strong *headcount* in this case; (3) $\omega_i^k < s \leq \omega_j^k$ or $\omega_j^k < s \leq \omega_i^k$. In this case, we add +1 in the *headcount* and strong *headcount* for one outcome and +0 to the other outcome. *Headcount* and strong *headcount* behave equivalently in this case. Therefore, $shc(\omega_i, \omega_j, s) \leq shc(\omega_j, \omega_i, s)$ and $\omega_i \leq_{shc_s} \omega_j$.

($\Leftarrow$) Analogous to the previous case. □

Now, let us analyze how the strong *shortfall* operator works:

Definition 15 (Strong *shortfall* Justice Relation). *Let* $A = \{1, \ldots, n\}$ *be a set of agents,* $\Omega = \{\omega_1, \ldots, \omega_m\}$ *be a set of outcomes, where each* $\omega_i = (\omega_i^1, \ldots, \omega_i^n)$*,* $\omega_i^k \in [0, 1]$ *for* $k \in \{1, \ldots, n\}$*. We define the shortfall of the agents across* s *w.r.t.* ω_i *and* ω_j *as* $ssh(\omega_i, \omega_j, s) = \sum_{\omega_i^k < s \leq \omega_j^k} s - \omega_i^k$*. We define* $\leq_{ssh_s}$ *over* Ω *as* $\omega_i \leq_{sh_s} \omega_j$ *iff* $ssh(\omega_i, \omega_j, s) \geq ssh(\omega_j, \omega_i, s)$*.*

The same scenario that occurred for the strong *headcount* will not occur with the Strong *shortfall* operator: the strong *shortfall* justice relation is not equivalent to the original *shortfall* justice relation. Now, for the humanitarian principle related to the strong sufficientarianism, we can state below a strong version of (**WPM-**s):

Definition 16 (Strong Povertymin for s **(SPM-**s**)).** *Let* $A = \{1, \ldots, n\}$ *be a set of agents,* $\Omega = \{\omega_1, \ldots, \omega_m\}$ *be a set of outcomes, where each* $\omega_i = (\omega_i^1, \ldots, \omega_i^n)$ *and* $s \geq 0$. *We say an aggregation function* f *satisfies Strong Povertymin for* s *if for all* $\omega_i, \omega_j \in \Omega$, *if (1) there exists a* $k \leq n$ *such that* $\omega_i^k < \omega_j^k$ *and* $\omega_i^k < s$; *(2) every position* l *that* $\omega_j^l \leq \omega_i^l$ *implies* $\omega_j^l \geq s$ *or* $\omega_i^l < s$, *then* $\omega_i \leq_f \omega_j$.

As in (**WPM-**s), this principle is followed by some other properties. One of them we introduce below:

Definition 17 (Indifference for those Above or Below s **(IAB-**s**)).** *Let* $A = \{1, \ldots, n\}$ *be a set of agents,* $\Omega = \{\omega_1, \ldots, \omega_m\}$ *be a set of outcomes, where each* $\omega_i = (\omega_i^1, \ldots, \omega_i^n)$ *and* $s \geq 0$. *We say an aggregation function* f *satisfies Indifference for those Above or Below* s *if for all* $\omega_i, \omega_j \in \Omega$, *if (1) there exist* k, k' *such that* $s \leq \omega_j^k < \omega_i^k$; *(2)* $\omega_i^{k'} < \omega_j^{k'} < s$; *(3) for* $l \neq k, k'$, $\omega_i^l = \omega_j^l$; *then* $\omega_i \approx_f \omega_j$.

The addition of (**IAB-**s) comes from the definition of the strong sufficientarianism: benefits that lift agents above some threshold level s matter more than equally large benefits that do not, whether they occur above or below s. For the strong aggregation functions that we introduced, we considered null any benefit occurring above or below s.

Theorem 7. *If a justice relation* $\leq_f$ *satisfies (**WAPA-**s), (**IAB-**s), (**WP**) and (**A**), then it satisfies (**SPM-**s).*

Proof. It is similar to the proof of Theorem 1. □

We have the following results for the strong *headcount* and *shortfall*.

Theorem 8. *Let* $\leq_{shc_s}$ *and* $\leq_{ssh_s}$ *be the strong headcount and shortfall justice relations, respectively.* $\leq_{shc_s}$ *satisfies (**WAPA-**s), (**IAB-**s), (**WP**), (**A**) and (**SPM-**s).* $\leq_{ssh_s}$ *satisfies (**WAPA-**s), (**IAB-**s), (**WP**) and (**SPM-**s), but (**A**) is not satisfied in general.*

Proof. We will show the proofs for (**IAB-**s) and (**A**). The rest is similar to the previous Theorems. (**IAB-**s) Let $A = \{1, \ldots, n\}$ be a set of agents, $\Omega = \{\omega_1, \ldots, \omega_m\}$ be a set of outcomes, where each $\omega_i = (\omega_i^1, \ldots, \omega_i^n)$ and $s \geq 0$. Suppose that for $\omega_i, \omega_j \in \Omega$, we have (i) there exist k, k' such that $s \leq \omega_j^k < \omega_i^k$; (ii) $\omega_i^{k'} < \omega_j^{k'} < s$; and (iii) for $l \neq k, k'$, $\omega_i^l = \omega_j^l$. Then it follows that $shc(\omega_i, \omega_j, s) = shc(\omega_j, \omega_i, s)$ and $ssh(\omega_i, \omega_j, s) = shc(\omega_j, \omega_i, s)$, since (i) and (ii) do not alter in the strong *headcount* and *shortfall*, and (iii) adds the same value in both outcomes. Therefore, $\omega_i \approx_{shc_s} \omega_j$ and $\omega_i \approx_{ssh_s} \omega_j$. **(A)** $\leq_{ssh_s}$ does not satisfy **(A)**. As a counterexample, suppose that we have the outcome $\omega = (0.4, 0.3, 0, 0)$, its permutation $\omega' = (0, 0.4, 0.3, 0)$ and $s = 0.2$. We have that $ssh(\omega, \omega', 0.2) = 0.2 > 0.1 = ssh(\omega', \omega, 0.2)$. Therefore, $\omega >_{ssh_s} \omega'$. □

It is possible to strengthen the **(SPM-s)** property as follows:

Definition 18 (Stronger Povertymin for s (S_1PM-s)). *Let $A = \{1, \ldots, n\}$ be a set of agents, $\Omega = \{\omega_1, \ldots, \omega_m\}$ be a set of outcomes, where each $\omega_i = (\omega_i^1, \ldots, \omega_i^n)$ and $s \geq 0$. We say an aggregation function f satisfies Stronger Povertymin for s if for all $\omega_i, \omega_j \in \Omega$, if (1) there exists a $k \leq n$ such that $\omega_i^k < \omega_j^k$ and $\omega_i^k < s$; (2) every position l that $\omega_j^l \leq \omega_i^l$ implies $\omega_j^l \geq s$ or $\omega_i^l < s$; then $\omega_i <_f \omega_j$.*

This condition can be achieved similarly as shown in Theorem 7, but replacing the Weak Pareto with the Strong Pareto property.

Theorem 9. *If a pre-order $\leq_f$ satisfies **(WAPA-s)**, **(IAB-s)**, **(SP)** and **(A)**, then it satisfies **(S_1PM-s)**.*

Proof. It follows straightforwardly from Theorem 7. □

For the two justice relations described in this section, we have this result.

Theorem 10. *$\leq_{shc_s}$ and $\leq_{ssh_s}$ satisfy neither **(S_1PM-s)** nor **(SP)**.*

Proof. **(S_1PM-s)** Let $A = \{1, \ldots, n\}$ be a set of agents, $\Omega = \{\omega_1, \ldots, \omega_m\}$ be a set of outcomes, where each $\omega_i = (\omega_i^1, \ldots, \omega_i^n)$ and $s \geq 0$. Suppose we have $\omega_i, \omega_j \in \Omega$, such that (i) there exists a $k \leq n$ such that $\omega_i^k < \omega_j^k$ and $\omega_i^k < s$; and (ii) for every position l such that $\omega_j^l \leq \omega_i^l$, we obtain $\omega_j^l \geq s$ or $\omega_i^l < s$; By (i), suppose that $\omega_i^k < \omega_j^k < s$. In this case, we will have $shc(\omega_i, \omega_j, s) = ssh(\omega_j, \omega_i, s)$ and $ssh(\omega_i, \omega_j, s) = ssh(\omega_j, \omega_i, s)$. Therefore, $\omega_i \not<_{shc_s} \omega_j$ and $\omega_i \not<_{ssh_s} \omega_j$. **(SP)** It is similar for $\leq_{hc_s}$ and $\leq_{sh_s}$. □

In order to satisfy **(SP)**, we can also define an axiological version of the strong sufficientarian justice relations.

Definition 19 (Axiological Strong Justice Relations). *Let $A = \{1, \ldots, n\}$ be a set of agents, $\Omega = \{\omega_1, \ldots, \omega_m\}$ be a set of outcomes, where each $\omega_i = (\omega_i^1, \ldots, \omega_i^n)$, $\omega_i^k \in [0, 1]$ for $k \in \{1, \ldots, n\}$ and $s \geq 0$. (1) We define the number of agents across $s \in \mathbb{N}$ w.r.t. ω_i and ω_j as $shc(\omega_i, \omega_j, s) = \#(\{\omega_i^k \mid \omega_i^k \geq s > \omega_j^k\})$, where $\#(A)$ is the cardinal of the set A. We define $\leq_{sahc_s}$ over Ω as $\omega_i \leq_{sahc_s} \omega_j$ iff (if $shc(\omega_i, \omega_j, s) = shc(\omega_j, \omega_i, s)$ then $\sum_k \omega_i^k \leq \sum_k \omega_j^k$; else $shc(\omega_i, \omega_j, s) \leq shc(\omega_j, \omega_i, s)$. (2) We define the shortfall of the agents across s w.r.t. ω_i and ω_j as $ssh(\omega_i, \omega_j, s) = \sum_{\omega_i^k < s \leq \omega_j^k} s - \omega_i^k$. We define $\leq_{sash_s}$ over Ω as $\omega_i \leq_{sash_s} \omega_j$ iff if $ssh(\omega_i, \omega_j, s) = ssh(\omega_j, \omega_i, s)$ then $\sum_k \omega_i^k \leq \sum_k \omega_j^k$; else $ssh(\omega_i, \omega_j, s) \geq ssh(\omega_j, \omega_i, s)$.*

Finally, for the axiological strong justice relations we have

Theorem 11. *$\leq_{sahc_s}$ satisfies **(WAPA-s)**, **(IAB-s)**, **(SP)**, **(A)** and **(S_1PM-s)**. $\leq_{sash_s}$ satisfies **(WAPA-s)**, **(IAB-s)**, **(SP)** and **(S_1PM-s)**, but does not satisfy **(A)** in general.*

Proof. These results follows from a similar argument of Theorems 2, 5 and 8. □

6 Conclusions

In this paper, we introduced some sufficientarian aggregation functions based on the *headcount* and *shortfall* operators. We made it by using the approaches of the weak, axiological and strong sufficientarianism. Besides the definition of the aggregation functions and their justice relations, we also considered some logical properties related to the sufficientarianism: they are the Weak and Strong Povertymin, which are implied by some other additional logical properties.

We showed that the original idea of sufficientarianism proposed by Frankfurt [6] describes (in part) a weak Povertymin justice relation, but it does not satisfy an important property in the area of aggregation, which is the Strong Pareto **(SP)** property. By extending to other different notions of sufficientarianism, we searched for the satisfaction of **(SP)** and also the satisfaction of new logical properties.

The axiological sufficientarianism [7] extends the classical sufficientarianism with a tie breaker condition for those cases when the outcomes have exactly the same number (or amount of utility value) of agents below sufficiency. We used two different kinds of tie breaker functions: an axiological *headcount* and an axiological *shortfall*. In practice, the axiological sufficientarianism is an approach that joins the classical sufficientarianism with the utilitarianism. It behaves as a sufficientarian aggregation function when the results are different; otherwise, it behaves as an utilitarian (*sum*) aggregation function for the tie breaker condition. The advantage of using the axiological sufficientarianism in the aggregation is that it satisfies **(SP)**.

Lastly, we began to analyze another form of sufficientarianism called strong sufficientarianism (we assume the classical one is treated as the weak sufficientarianism). We glimpsed two different kinds of operators: the strong *headcount* and strong *shortfall*. As a result, we showed that the strong version of *headcount* does not bring anything new to the results of the aggregation, since they are equivalent to its classical version. However, the strong *shortfall* loses the Anonymity **(A)** logical property. Consequently, we proved these strong sufficientarian justice relations satisfy the Strong Povertymin property. Finally, we considered a axiological version of the strong sufficientarian justice relations to overcome again the loss of **(SP)** in these justice relations.

Some work still needs to be done. For instance, what other characterizations of sufficientarianism can be proposed, along with the axiological and strong sufficientarianism? For the Strong sufficientarianism, a deep research in its logical properties is still missing. Besides, a future work about other possible strong operators that satisfy all logical properties is envisaged (in special, a version of strong *shortfall* which satisfies **(A)**).

References

1. Beliakov, G., Pradera, A., Calvo, T.: Aggregation Functions: A Guide for Practitioners. Studies in Fuzziness and Soft Computing. Springer, Heidelberg (2009). https://doi.org/10.1007/978-3-540-73721-6
2. Casal, P.: Why sufficiency is not enough. Ethics **117**(2), 296–326 (2007)
3. Chung, H.: Prospect utilitarianism: a better alternative to sufficientarianism. Philos. Stud. **174**(8), 1911–1933 (2016). https://doi.org/10.1007/s11098-016-0775-3
4. Cook, K.S., Hegtvedt, K.A.: Distributive justice, equity, and equality. Annu. Rev. Sociol. **9**(1), 217–241 (1983)
5. Detyniecki, M.: Fundamentals on aggregation operators. This manuscript is based on Detyniecki's doctoral thesis and can be downloaded from (2001). http://www.cs.berkeley.edu/~marcin/agop.pdf
6. Frankfurt, H.: Equality as a moral ideal. Ethics **98**(1), 21–43 (1987)
7. Hirose, I.: Axiological sufficientarianism. In: Fourie, C., Annette, R. (eds.) How Much Is Enough? Sufficiency and Thresholds in Health Care (2014)
8. Mill, J.S.: Utilitarianism. In: Seven Masterpieces of Philosophy, pp. 337–383. Routledge (2016)
9. Rawls, J.: Justice as Fairness: A Restatement. Harvard University Press, Cambridge (2001)
10. Rawls, J.: A Theory of Justice. Harvard University Press, Cambridge (2009)
11. Rescher, N.: Distributive Justice. Information and Interdisciplinary Subjects Series. University Press of America, G - Reference (1982)
12. Segall, S.: What is the point of sufficiency? J. Appl. Philos. **33**(1), 36–52 (2016)
13. Sen, A.: Utilitarianism and welfarism. J. Philos. **76**(9), 463–489 (1979)
14. Shields, L.: The prospects for sufficientarianism. Utilitas **24**, 101–117 (2012)
15. Smart, J.J.C., Williams, B.: Utilitarianism: For and Against. Cambridge University Press, Cambridge (1973)
16. Tungodden, B.: Egalitarianism: Is Leximin the Only Option? Working papers, Norwegian School of Economics and Business Administration (1999)
17. Williams, B.: A Critique of Utilitarianism. Cambridge University Press, Cambridge (1973)

An Alternative to Power Measure for Fuzzy Rule-Based Classification Systems

Frederico B. Tiggemann[1], Bryan G. Pernambuco[1(✉)], Giancarlo Lucca[1], Eduardo N. Borges[1], Helida Santos[1], Graçaliz P. Dimuro[1,2], José A. Sanz[2], and Humberto Bustince[2]

[1] Universidade Federal do Rio Grande, Rio Grande, RS, Brazil
{frederico.bender99,bryansgalanip,giancarlo.lucca,eduardoborges,helida, gracalizdimuro}@furg.br

[2] Universidad Publica de Navarra, Pamplona, Navarra, Spain
{joseantonio.sanz,bustince}@unavarra.es

Abstract. An effective way to deal with classification problems, among other approaches, is using Fuzzy Rule-Based Classification Systems (FRBCSs). These classification systems are mainly composed of two modules, the Knowledge Base (KB) and the Fuzzy Reasoning Method (FRM). The KB is responsible for storing information related to the problem, while the FRM performs the classification of new examples based on the KB. A key point in the FRM is how the information given by the triggered fuzzy rules is aggregated. Several FRMs have been proposed in the literature employing generalizations of the Choquet Integral as the aggregation function. The usage of this function to perform the aggregation is efficient because it uses fuzzy measures to model the interrelationships between the data. Indeed, the performance of the classification system is strongly related to the underlying fuzzy measure. However, many of those generalizations have used fuzzy measures that do not properly model the interaction between the data. In this context, we intend to enhance the way how the relationships between the rules are modeled in the FRM of an FRBCS. In order to accomplish that, we propose the use of a well known fuzzy measure, the Sugeno lambda, as an alternative to the Power Measure, which is widely used in many generalizations of the Choquet Integral. The experimental evaluation shows statistical equivalent results comparing our method with the state-of-the-art fuzzy classifier.

Keywords: Fuzzy measure · Choquet Integral · Sugeno Lambda

1 Introduction

Many researchers have focused on acquiring data from the significant growth of information in several research fields [14,23]. These data in the raw format may be unorganized, disordered, and chaotic, which does not reflect any useful content about the analyzed field. Researchers in the field of data science [37],

R. Cerri and R. C. Prati (Eds.): BRACIS 2020, LNAI 12320, pp. 231–244, 2020.
https://doi.org/10.1007/978-3-030-61380-8_16

machine learning [33], and data mining [22] have contributed to the development of technologies and techniques that enabled the use of these data to improve the detection of trends [7] and patterns [16] that can be used for future benefit.

Classification is a process in which the input data is classified (labeled) based on predetermined knowledge. Classification problems may vary from the simplest one, like classifying animals breed based on each animal characteristics [8], to complex ones, such as classifying human diseases based on medical tests [27].

One approach widely used to deal with classification problems is the Fuzzy Rule-Based Classification System (FRBCS) [25], since it commonly provides an interpretable model supported by linguistic terms, also achieving good performances. The FRBCS has two main components, the Knowledge Base (KB), which stores the information about the problem, and the Fuzzy Reasoning Method (FRM), which performs an inference procedure using the information stored in the KB to classify new examples. A key point of the FRM is that one can achieve a cohesive result even without a perfect match between the antecedents of the rules and the system observation [12].

The most used FRMs in the literature are the Winning Rule (WR) and the Additive Combination (AC) [6]. The WR uses maximum matching as aggregation function, classifying a new example based on the fuzzy rule with the greatest association degree [1,41]. However, considering just the rule with the highest compatibility leads to the loss of information, since the information provided by other fuzzy rules, with lesser compatibility, are ignored. Nowadays, the most accurate algorithms use AC, which aggregates all the triggered rules, for each class, by using a normalized sum [12].

Barranechea et al. proposed in [5] an FRM that aggregates all information using the Choquet Integral. This aggregation function was first introduced in the theory of capacities by Gustave Choquet [11]. The Choquet Integral is an averaging aggregation function defined with respect to a fuzzy measure, that considers not only the importance of each attribute to be aggregated, but it also models the interactions between the variables [6].

One of the most successful classes of fuzzy measures is the Sugeno Lambda measure [9,38]. Besides, this measure has been suggested for computing the Sugeno fuzzy integral [28]. The Sugeno Lambda measure models the relationship between the data, once it takes into account the given association degrees to calculate the λ and the fuzzy value to be aggregated. Therefore, in this paper, we propose the usage of the Choquet Integral [11] aggregation function combined with Sugeno Lambda fuzzy measure [36].

Our approach is applied in the FRM of the Fuzzy Association Rule-Based Classification method for High-Dimensional problems (FARC-HD) [2], a state-of-the-art fuzzy classifier. Moreover, this classifier is the same one used as basis of the application of different generalizations of the Choquet Integral [15].

The paper is organized as follows. Section 2 presents some preliminary concepts necessary for the development of our work. Section 3 exposes some literature review related to the concepts used in our work. Section 4 defines both fuzzy measures compared in this paper and shows an example of the application of both fuzzy measures in the aggregation function of the FRM. Section 5 defines the specifications of our experiments, starting with the datasets used, and finally, our FARC-HD configuration. Section 6 display the results obtained with our methods and compare them with the state-of-the-art fuzzy classifier. Finally, Sect. 7 concludes our work, resuming the proposal of our article and our contribution to the literature.

2 Preliminary Concepts

This section introduces some concepts that are necessary to better understand the remainder of the paper.

Definition 1. *A function $g \in \mathbb{R}$, where $\mathbb{R}$ is the set of real values, is called monotonic if satisfies the following two conditions:*

(i.) g is entirely non-increasing;
(ii.) g is entirely non-decreasing.

A function that monotonically increases does not have to increase entirely; it just must not decrease. A function is called monotonically increasing (or non-decreasing), if for all x and y such that $x \leq y$, has $g(x) \leq g(y)$, so g preserves the order. Otherwise, a function is called monotonically decreasing (or non-increasing) whenever $x \leq y$, has $g(x) \geq g(y)$, so g reverses the order.

2.1 Fuzzy Measure

In this subsection, we recall the notion of fuzzy measure, which is going to be a differential purpose of our work. In the following, consider the set $N = \{1, \dots, n\}$ for an arbitrary positive integer n.

Definition 2. *A function $m \colon 2^N \rightarrow [0, 1]$ is a fuzzy measure if, for all $X, Y \subseteq N$, it satisfies the following properties:*

(m1) Increasingness: if $X \subseteq Y$, then $m(X) \leq m(Y)$;
(m2) Boundary conditions: $m(\phi) = 0$ and $m(N) = 1$.

In the context of aggregation functions, fuzzy measures are used for evaluating the relationship among the elements to be aggregated.

2.2 Aggregation Function

A relevant class of operators used in this paper are aggregation functions [6].

Definition 3. *A function $A \colon [0, 1]^n \rightarrow [0, 1]$ is said to be an n-ary aggregation function if the following conditions hold:*

(A1) A is increasing in each argument: for each $i \in (1, \dots, n)$, if $x_i \leq y$, then $A(x_1, \dots, x_n) \leq A(x_1, \dots, x_{i-1}, y, x_{i+1}, \dots, x_n)$;

(A2) A satisfies the boundary conditions: $A(0,\ldots,0) = 0$ *and* $A(1,\ldots,1) = 1$.

In this paper, we consider the Choquet integral related to fuzzy measures, defined as:

Definition 4. *Let* $m\colon 2^N \rightarrow [0,1]$ *be a fuzzy measure. The discrete Choquet integral of* $\vec{x} = (x_1,\ldots,x_n) \in [0,1]^n$ *with respect to fuzzy measure* m *is defined as a function* $C_m\colon [0,1]^n \rightarrow [0,1]$*, given by:*

$$C_m(\vec{x}) = \sum_{i=1}^{n} \left(x_{(i)} - x_{(i-1)}\right) \cdot m\left(A_{(i)}\right) , \tag{1}$$

where, $(x_{(1)},\ldots,x_{(n)})$ *is an increasing permutation on* x*, i.e.,* $0 \leq x_{(1)} \leq \ldots \leq x_{(n)}$*, with the convention that* $x_{(0)} = 0$*,* $A_{(i)} = ((i),\ldots,(n))$ *is the subset of indices of* $n-i+1$ *largest components of* x*, and* $m(A_{(i)})$ *is the fuzzy measure.*

The Choquet Integral aggregates the inputs considering the importance of all available distinct groups of criteria. This function offers greater adaptability in the modeling process, representing a larger class of aggregation function [31].

3 Related Work

Through a bibliographic review, we are able to identify the popularity of FRBCSs and the importance of aggregation functions and the associated fuzzy measures. This approach has been used in several research fields such as pattern classification [24], medical problems [19,39], image processing [4], among others. In order to achieve the best results, it is important to use an adequate fuzzy measure in the aggregation function of the FRM.

Murofushi et al. [34] studied fuzzy measures and aggregation functions, specifically the Choquet Integral, becoming well known in this search field. The authors developed the concept that fuzzy measures could be non-monotonic once the former is not necessarily additive. This work spread along further researches concerning fuzzy measures and Choquet integral, corroborating several models.

Distinct search fields use Choquet Integral as the aggregation function, due to the importance of groups of criteria. Magadum et al. [32] use Choquet Integral combined with Sugeno Lambda fuzzy measure to aggregate marks of the students in different subjects with respect to the weights of each subject. The subject's importance is measured based on its weights. Comparing the indices provided by the Choquet integral, they rank the students for an admission process.

Elkano et al. [17] analyze the inaccuracy when using Fuzzy Rule-Based Classification Systems (FRBCSs) based on the Chi algorithm [10] for Big Data problems. This problem occurs due to the variation of the configuration of the cluster, generating different models. Based on the Chi algorithm, Elkano et al. designed a new FRBCS for BigData classification problems (CHI-BD) using the MapReduce paradigm algorithm [13]. This approach achieves the same model regardless of the number of computing nodes considered.

Recent work, such as González et al. [20], studied how the massive use of data affects the reasoning methods for Fuzzy Rule-Based Classification Systems applied to big data problems. In some cases, the standard reasoning model proved to be inefficient to perform some calculations. In order to solve this, the authors propose a new model that eliminates the need to review all the rules in every inference process, generating the rule that best adapts to the particular example.

4 Application of Fuzzy Measures in FRBCS

The fuzzy measure is a subjective evaluation introduced by Choquet in 1953 and defined by Sugeno in 1974 for fuzzy integrals [42]. M. Sugeno, one of the pioneers on fuzzy measures, defined them as a monotonic set function [40]. This concept extends the definition of probability measure, meaning that the fuzzy measure $m(A)$ of a subset A, of the referential set X, expresses the degree of confidence of $x_0 \in A$, where x_0 is an unknown element of X. In this interpretation, the monotonicity of the fuzzy measure is essential [40].

However, posterior studies [34,35] describe that monotonicity is inessential. T. Murofushi et al. [35] define a fuzzy measure as: "*A (monotonic) fuzzy measure on X is a monotone set function defined on* 2^X *which vanishes at the empty set. A non-monotonic (or signed) fuzzy measure is a set function defined on* 2^X *which vanishes at the empty set*". Therefore the authors claim that a fuzzy measure is not necessarily a measure. The difference between a fuzzy measure and a measure (or a non-monotonic fuzzy measure and a signed measure) is that the former is not necessarily additive.

The performance of a fuzzy measure, in the context of aggregation functions, can be described as the fidelity modeling the relationship (association degree) of the elements that are going to be aggregated [5]. When comparing standard fuzzy measures, the Power Measure (See Eq. 2) was the one that achieved a statistical superior performance in [31]. Taking this into account, we applied a new fuzzy measure in the context of FRBCSs known as Sugeno Lambda, which models the relationship between the data, to compare with the state-of-the-art Power Measure.

In this paper, we analyze the usage of two different fuzzy measures combined with the Choquet Integral to perform the aggregation inside the FRM of a fuzzy classifier. Therefore, in this section, we present the definitions of Power Measure, the Sugeno Lambda, and the FRM where both measures are applied. In order to make the process more comprehensible, cohesive and complete, we also provide an example of how the FRM performs its calculations.

Power Measure. This measure was introduced by Barrenechea et al. [5], in the same study that introduced the FRM that uses the Choquet integral as aggregation operator. The power measure is an improvement of the cardinality or uniform measure, which has an exponent q, that is genetically learned, Eq. (2). Consequently, for each class of the problem, this formula behaves like a different fuzzy measure by learning the most appropriate exponent. Note that if $q = 1$, this measure behaves as the cardinality or uniform measure.

$$m_{PM}(A) = \left(\frac{|A|}{n}\right)^q, \text{ with } q > 0. \tag{2}$$

Sugeno Lambda. One of the most widely and successfully used classes of fuzzy measures is the class of Sugeno Lambda measures [29]. This measure solves a polynomial, Eq. (4), and we use Flanagan's library to accomplish this step [18]. The Sugeno Lambda measure models the relationship between the values, once it regards the given association degrees to calculate the λ and, so, the fuzzy value to be aggregated. Besides, it is not static for all classes.

A fuzzy measure g is called a Sugeno Lambda measure if it, additionally, satisfies the following property, for all $A, B \subseteq \mathrm{X}$ with $A \cap B = \emptyset$:

$$g(A \cup B) = g(A) + g(B) + \lambda \cdot g(A) \cdot g(B), \text{ for } \lambda > -1 \tag{3}$$

The value of λ is found by using Eq. (4), which is a polynomial with a $n - 1$ degree [26].

$$\lambda + 1 = \prod_{i=1}^{n}(1 + \lambda g_{(i)}) \tag{4}$$

In Eq. (3), the λ value describes how close the given Sugeno measure is to a probability measure: the smaller $|\lambda|$, the closer these measures are. Whenever $-1 < \lambda < 0$, it is subadditive, but if $\lambda = 0$, the formula corresponding to the Sugeno measure transforms into an additivity formula. Finally, if $\lambda > 0$, then it is superadditive.

The lattice representation, shown in Fig. 1, is a handy way to represent fuzzy measure coefficients. For the example shown with $n = 4$, the lattice shows all the 2^n possible aggregations between the coefficients, μ_{12} represents the aggregation $\mu(\{x_1, x_2\})$ as well as all coefficients. The coefficient value starts at 0 and tends to be updated until it approaches 1.

Whenever the Power Measure is used, it will model the general levels of the lattice, and not exactly the relationship between the data. Considering that the exponent q is equal to 1, the power measure would be $\frac{4}{4}$ for the first level $(\mu_1, \mu_2, \mu_3, \mu_4)$, $\frac{3}{4}$ for the second, $\frac{2}{4}$ for the third, and, finally, $\frac{1}{4}$ for the last one. This value does not change depending on the μ value. On the other hand, Sugeno Lambda measure models the relationship between the data and not the general levels. So, the μ value impacts directly into the Sugeno Lambda measure.

4.1 The Usage of Fuzzy Measure in FRM

As mentioned previously, the FRM is an inference procedure that derives conclusions from a set of fuzzy if-then rules stored in the Knowledge Base (KB) [12]. Once the KB has been learned, and a new example is ready to be classified, the FRM is responsible for performing this task. The four steps of this process are described as follows:

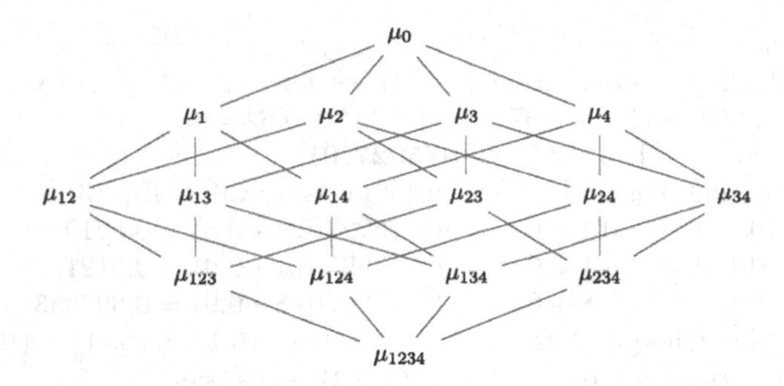

Fig. 1. Behavior of the coefficients of a fuzzy measure forming a lattice structure. This figure illustrates the lattice of the coefficients of a fuzzy measure considering $n = 4$. (Adapted from M. Grabisch [21]).

Step 1 *Matching degree*, calculates the activation strength between the pattern and the if-part for the rules in the Rule Base (RB) by applying a t-norm [20].

Step 2 *Association degree*, calculates the association degree of the pattern with the classes according to each rule in the RB. Then it incorporates the weight of the rule using a function, making possible to increase higher values and penalize the lower ones [12].

Step 3 *Example classification soundness degree for all classes.* Uses an aggregation function [30] for each class that combines the positive association degree given by the triggered fuzzy rules of the previous step.

Step 4 *Classification*, applies a decision function over Step 3 to determine the class label corresponding to the maximum soundness degree.

In order to explain the operations of the FRM using the Choquet Integral with both presented fuzzy measures, we will introduce a short example using random rule association degrees (obtained after **Steps 1 and 2**), that is, we demonstrate the third and fourth steps. We compare the Sugeno Lambda measure with the Power Measure, once the Power Measure is an improvement using GAs of the cardinality, and it is worthy for most cases. To do so, consider the following example:

Example 1. For **Step 3** of the FRM, in which the classification is performed, consider three different classes with n = 3, $C_1 = [0.1, 0.8, 0.9], C_2 = [0.82, 0.83, 0.86]$, and $C_3 = [0.2, 0.35, 0.4]$. For these classes, we compare them using the Choquet Integral as the aggregation function, with distinct fuzzy measures. We assume that for the Power Measure (PM), exponent q is equal to 1 (to make the calculations simple). Regarding the boundary condition presented in Sect. 2.1, $m(N) = 1$. The values computed for each one are the following:

C_1 • Association degrees: $[0.1, 0.8, 0.9]$
- * PM:
  - · Using Choquet Integral, Eq. (1):
    $\left((0.1 - 0.0) \cdot \frac{3}{3}\right) + \left((0.8 - 0.1) \cdot \frac{2}{3}\right) + \left((0.9 - 0.8) \cdot \frac{1}{3}\right) = 0.600.$

* Sugeno:
  · Finding λ value, from Eq. (4) we have: $\lambda + 1 = (0.1\lambda + 1)(0.8\lambda + 1)(0.9\lambda + 1) = -0.072\lambda^3 - 0.89\lambda^2 - 0.8\lambda = 0$.
  So, $\lambda = \{-11.3852, \mathbf{-0.975927}, 0\}$.
  · Finding Sugeno Lambda fuzzy measure value, Eq. (3):
  $g(0.1, 0.8) = 0.1 + 0.8 + -0.975927(0.1 * 0.8) = 0.821926$;
  $g(0.1, 0.9) = 0.1 + 0.9 + -0.975927(0.1 * 0.9) = 0.912167$;
  $g(0.8, 0.9) = 0.8 + 0.9 + -0.975927(0.8 * 0.9) = 0.997333$.
  · Using Choquet Integral, from Eq. (1): $\big((0.1 - 0.0) \cdot 1\big) + \big((0.8 - 0.1) \cdot 0.912167\big) + \big((0.9 - 0.8) \cdot 0.997333\big) = 0.83825$.

C_2 • Association degrees: [0.82, 0.83, 0.86]
* PM:
  · Using Choquet Integral, Eq. (1)
  $\big((0.82 - 0.0) \cdot \frac{3}{3}\big) + \big((0.83 - 0.82) \cdot \frac{2}{3}\big) + \big((0.86 - 0.83) \cdot \frac{1}{3}\big) = 0.837$.
* Sugeno:
  · Finding λ value, from Eq. (4) we have: $\lambda + 1 = (0.82\lambda + 1)(0.83\lambda + 1)(0.86\lambda + 1) = -0.585316\lambda^3 - 2.0996\lambda^2 - 1.51\lambda = 0$.
  So, $\lambda = \{-2.59172, \mathbf{-0.995402}, 0\}$.
  · Finding Sugeno Lambda fuzzy measure value, Eq. (3):
  $g(0.82, 0.83) = 0.82 + 0.83 - 0.995402(0.82 * 0.83) = 0.972529$;
  $g(0.82, 0.86) = 0.82 + 0.86 - 0.995402(0.82 * 0.86) = 0.978043$;
  $g(0.83, 0.86) = 0.83 + 0.86 - 0.995402(0.83 * 0.86) = 0.979482$.
  · Using Choquet Integral, Eq. (1): $\big((0.82 - 0.0) \cdot 1\big) + \big((0.83 - 0.82) \cdot 0.978043\big) + \big((0.86 - 0.83) \cdot 0.979482\big) = 0.859165$.

C_3 • Association degrees: [0.2, 0.35, 0.4]
* PM:
  · Using Choquet Integral, Eq. (1)
  $\big((0.2 - 0.0) \cdot \frac{3}{3}\big) + \big((0.35 - 0.2) \cdot \frac{2}{3}\big) + \big((0.4 - 0.35) \cdot \frac{1}{3}\big) = 0.317$.
* Sugeno:
  · Finding λ, from Eq. (4) we have: $\lambda + 1 = (0.2\lambda + 1)(0.35\lambda + 1)(0.4\lambda + 1) = -0.027\lambda^3 - 0.285\lambda^2 + 0.05\lambda = 0$.
  So, $\lambda = \{-10.7282, \mathbf{0.172616}, 0\}$.
  · Finding Sugeno Lambda fuzzy measure value, Eq. (3):
  $g(0.2, 0.35) = 0.2 + 0.35 + 0.172616(0.82 * 0.83) = 0.562083$;
  $g(0.2, 0.4) = 0.2 + 0.4 + 0.172616(0.82 * 0.86) = 0.613809$;
  $g(0.35, 0.4) = 0.35 + 0.4 + 0.172616(0.83 * 0.86) = 0.774166$.
  · Using Choquet Integral, Eq. (1): $\big((0.2 - 0.0) \cdot 1\big) + \big((0.35 - 0.2) \cdot 0.613809\big) + \big((0.4 - 0.35) \cdot 0.774166\big) = 0.33078$.

Once **Steps 2 and 3** have been executed, and the soundness degree for each class has been fully computed, **Step 4** predicts the class based on the largest value triggered (highlighted in **bold-face**) by the aggregation function, as presented in Table 1.

Table 1. Computed values by the third step of the FRM, using Choquet Integral as the aggregation function with PM and Sugeno Lambda measures.

Class	PM	Sugeno
C_1	0.6	0.83825
C_2	**0.837**	**0.859165**
C_3	0.317	0.33078

Analyzing Table 1, for both fuzzy measures, the class C_2 would be the predicted class. However, for the class C_1, the Sugeno Lambda measure computed a higher value than the Power Measure, which means that for Sugeno Lambda measure the relationship of that class for Power Measure is stronger.

5 Experiment Specifications

In this section, we present the difference between the fuzzy measures used in this work. Then we provide real-world classification problems used for training and testing. Besides, we also give the configuration for the FARC-HD classifier.

5.1 Dataset

We selected 27 real-world datasets from the KEEL dataset repository [3]. We have picked these datasets due to the fact that they are the same ones, which were considered in the first study that applied the Choquet integral in the FARC-HD fuzzy classifier [31]. Since this is an initial study, we intend to perform a fair comparison among these datasets and two different fuzzy measures.

Table 2. Summary of the dataset properties considered in this study.

Id.	Dataset	#Inst.	#Att.	#Class	Id.	Dataset	#Inst.	#Att.	#Class
App	Appendiciticis	106	7	2	Pag	Pageblocks	5472	10	5
Bal	Balance	625	4	3	Pho	Phoneme	5404	5	2
Ban	Banana	5300	2	2	Pim	Pima	768	8	2
Bnd	Bands	365	19	2	Rin	Ring	740	20	2
Bup	Bupa	345	6	2	Sah	Saheart	462	9	2
Cle	Cleveland	297	13	5	Sat	Satimage	6435	36	7
Eco	Ecoli	336	7	8	Seg	Segment	2310	19	7
Gla	Glass	214	9	6	Tit	Titanic	2201	3	2
Hab	Haberman	306	3	2	Two	Twonorm	740	20	2
Hay	Hayes-Roth	160	4	3	Veh	Vehicle	846	18	4
Iri	Iris	150	4	3	Win	Wine	178	13	3
Led	led7digit	500	7	10	Wis	Wisconsin	683	11	2
Mag	Magic	1902	10	2	Yea	Yeast	1484	8	10
New	Newthyroid	215	5	3					

Table 2 presents the characteristics of these datasets. In the columns, we have the identifier (Id.), the (Dataset) name, the number of instances (#Inst), the number of attributes (#Att), and the number of data classes (#Class).

In order to achieve the best result and avoid overfitting, we applied a 5-fold cross-validation procedure, splitting the datasets into five partitions containing 20% of the examples. This process is repeated five times to complete all iterations, and in each of the 5 steps, one different partition is used for testing and the other four are used for training. The general performance of the model is evaluated according to each testing partition, based on the accuracy rate (the number of correctly classified examples divided by the total number of examples). After calculating each partition performance, we use the average result of the five testing partitions to generate the output.

5.2 Classifier Configuration

As mentioned before, this paper uses FARC-HD [2] as the fuzzy classifier. Thus, in this subsection we present its main steps, as well as the parameters' configuration. To ease the comprehension of the classifier, we show in Fig. 2, the main steps that are performed, followed by a brief explanation of each one.

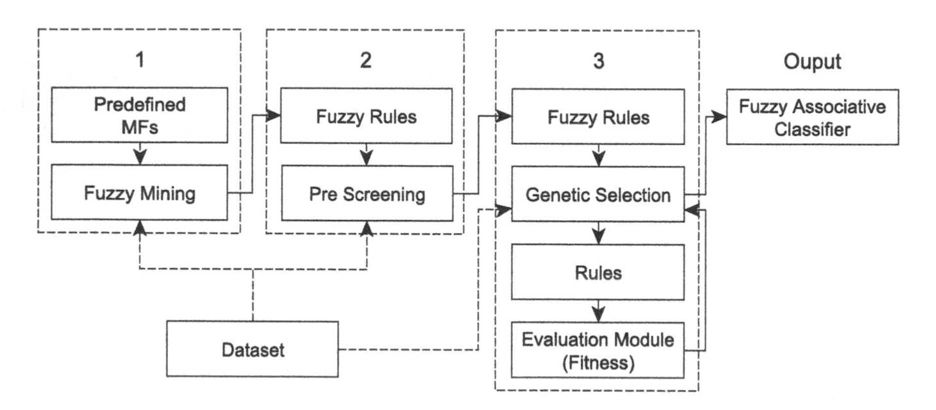

Fig. 2. The steps of the FARC-HD considered in this paper, proposed by Alcála et al. [2]. The architecture of the FARC-HD is composed of three main modules that are individually fed by a dataset and a set of fuzzy rules generated by the previous step, with an exception for the first step that uses predetermined MFs (Membership Functions).

Step 1 Aims at limiting the order of the associations in the association rule extraction,

Step 2 Uses a subgroup discovery based on an improved Weighted Relative Accuracy measure (wWRAcc') to preselect the most interesting rules,

Step 3 Uses Genetic Algorithms (GAs) to select and tune a compact set of fuzzy association rules.

The parameters set-up for the FARC-HD algorithm are the following:

(i) The conjunction operator T is the product t-norm;
(ii) The rule weight RW j is the certainty factor;
(iii) There are 5 linguistic labels per variable;
(iv) The minimum support is 0.05;
(v) The threshold for the confidence is 0.8;
(vi) The depth of the search trees is limited to 3;
(vii) k_t is the parameter that determines the number of fuzzy rules that covers each example, and it equals to 2.

The features considered for the evolutionary process are the following:

- The populations are composed by 50 individuals;
- 30 bits per gen are considered for the Gray codification;
- The maximum number of evaluations is 20,000.

6 Results

In this section, we present, in Table 3, the accuracy achieved using different fuzzy measures applied to the 27 real world datasets. Each row represents the average accuracy rate obtained in the 5-fold datasets, and each column represents the Choquet Integral combined with distinct fuzzy measures, Power Measure (PM) and Sugeno Lambda (Sugeno). We highlight the most accurate result for each dataset in **bold-face**. Finally, we apply Wilcoxon's test to conduct such pairwise comparisons.

Table 3. Accuracy achieved using different fuzzy measures.

Dataset	PM (%)	Sugeno	Dataset	PM (%)	Sugeno
App	80.13	**83.98**	Pag	94.16	**94.34**
Bal	**82.40**	79.36	Pho	**82.98**	82.40
Ban	**86.32**	82.83	Pim	73.95	**75.13**
Bnd	**68.56**	66.07	Rin	**90.95**	86.35
Bup	**66.96**	62.03	Sah	69.69	**69.9**
Cle	**55.58**	55.21	Sat	79.47	**79.93**
Eco	76.51	**77.99**	Seg	**94.46**	91.39
Gla	**64.02**	62.15	Tit	78.87	78.87
Hab	72.52	**74.82**	Two	**84.46**	80.00
Hay	79.49	**79.69**	Veh	**68.44**	66.31
Iri	91.33	**94.00**	Win	93.79	**94.35**
Led	68.20	**68.40**	Wis	**97.22**	96.05
Mag	78.86	**80.34**	Yea	**55.73**	54.85
New	**94.88**	93.95			

*Mean PM **79.20** Sugeno 78.46

Analyzing Table 3 we can see that for all of the 27 datasets used, the Power Measure excelled in 14 datasets when compared to Sugeno Lambda. However, the final accuracy average is close for both fuzzy measures, being 79.20% for Power Measure, and 78.46% for Sugeno Lambda. In order to have a more precise and conclusive result, we also performed a pairwise statistical test. Precisely, we have applied the non-parametric Wilcoxon rank test [43]. The results related with this test are available in Table 4. In this table, R^+ represents the rank obtained by the Sugeno Lambda fuzzy measure, and R^- is the ranking related with the Power Measure.

Table 4. Wilcoxon Statistical Test to compare Sugeno Lambda (R^+) versus Power Measure (R^-).

Comparison	R^+	R^-	p-value
Sugeno vs PM	179.5	381.5	0.067

Analyzing the results obtained in Table 4, one can observe that the Power Measure achieves a superior rank. However, considering a level of confidence of 5%, the null hypothesis is accepted, thus indicating that there are no statistical differences between the approaches.

7 Conclusion

In this paper, we apply the Sugeno Lambda fuzzy measure combined with the Choquet Integral in the FRM of a fuzzy classifier. The approach in our proposal is compared to the Power Measure (PM), which is widely used in generalizations of the Choquet Integral.

In this initial experimental study, 27 real-world datasets were used. We have shown that, in most of the cases, Power Measure still performs better than the proposed Sugeno Lambda measure, despite reaching close results. To corroborate our findings, we made use of Wilcoxon's test, supporting that Power Measure achieves a superior rank. Notwithstanding, considering a level of confidence of 5%, the Sugeno Lambda measure can be equivalent to PM.

We highlight that this methodology appears to be a good strategy to improve the classification system quality, once the results obtained using Sugeno Lambda measure models the relationship between the data. The Power Measure does not allow that, which motivates us to tackle this drawback. As mentioned previously, this is an initial study using Sugeno Lambda measure. In future work, we aim to use wider real-world datasets, better optimized libraries, and we also plan to improve the fuzzy classifier and upgrade the interaction between the rules.

Acknowledgment. This study was supported by PNPD/CAPES (464880/2019-00) and CAPES Financial Code 001, CNPq (301618/2019-4), FAPERGS (17/2551-0000872- 3, 19/2551-0001279-9, 19/2551-0001660), and AEI/UE, FEDER (PID2019-108392GB-I00).

References

1. Abe, S., Thawonmas, R.: A fuzzy classifier with ellipsoidal regions. IEEE Trans. Fuzzy Syst. **5**(3), 358–368 (1997)
2. Alcala-Fdez, J., Alcala, R., Herrera, F.: A fuzzy association rule-based classification model for high-dimensional problems with genetic rule selection and lateral tuning. IEEE Trans. Fuzzy Syst. **19**(5), 857–872 (2011)
3. Alcalá-Fdez, J., et al.: KEEL: a software tool to assess evolutionary algorithms for data mining problems. Soft Comput. **13**(3), 307–318 (2009)
4. Bárdossy, A., Samaniego, L.: Fuzzy rule-based classification of remotely sensed imagery. IEEE Trans. Geosci. Remote Sens. **40**(2), 362–374 (2002)
5. Barrenechea, E., Bustince, H., Fernandez, J., Paternain, D., Sanz, J.A.: Using the choquet integral in the fuzzy reasoning method of fuzzy rule-based classification systems. Axioms **2**(2), 208–223 (2013)
6. Beliakov, G., Pradera, A., Calvo, T., et al.: Aggregation Functions: A Guide for Practitioners, vol. 221. Springer, Heidelberg (2007). https://doi.org/10.1007/978-3-540-73721-6
7. Bicego, M., Grosso, E., Otranto, E.: A hidden Markov model approach to classify and predict the sign of financial local trends. In: da Vitoria Lobo, N., et al. (eds.) SSPR /SPR 2008. LNCS, vol. 5342, pp. 852–861. Springer, Heidelberg (2008). https://doi.org/10.1007/978-3-540-89689-0_89
8. Borneman, J.: Race, ethnicity, species, breed: Totemism and horse-breed classification in America. Comparative Stud. Soc. History **30**(1), 25–51 (1988)
9. Cao, Y.: Aggregating multiple classification results using choquet integral for financial distress early warning. Expert Syst. Appl. **39**(2), 1830–1836 (2012)
10. Chi, Z., Yan, H., Pham, T.: Fuzzy algorithms: with applications to image processing and pattern recognition, vol. 10. World Scientific (1996)
11. Choquet, G.: 54, “theory of capacities”. Ann. Inst. Fourier **5**(131), 295 (1953)
12. Cordón, O., Del Jesus, M.J., Herrera, F.: A proposal on reasoning methods in fuzzy rule-based classification systems. Int. J. Approximate Reason. **20**(1), 21–45 (1999)
13. Dean, J., Ghemawat, S.: Mapreduce: simplified data processing on large clusters. Commun. ACM **51**(1), 107–113 (2008)
14. Deshpande, A., Guestrin, C., Madden, S.R., Hellerstein, J.M., Hong, W.: Model-driven data acquisition in sensor networks. In: Proceedings of the Thirtieth International Conference on Very Large Data Bases, vol. **30**, pp. 588–599 (2004)
15. Dimuro, G.P., et al.: The state-of-art of the generalizations of the choquet integral: from aggregation and pre-aggregation to ordered directionally monotone functions. Inf. Fusion **57**, 27–43 (2020)
16. Duda, R.O., Hart, P.E., Stork, D.G.: Pattern Classification. Wiley, Hoboken (2012)
17. Elkano, M., Galar, M., Sanz, J., Bustince, H.: CHI-BD: a fuzzy rule-based classification system for big data classification problems. Fuzzy Sets Syst. **348**, 75–101 (2018)
18. Flanagan, M.T.: Michael Thomas Flanagan's Java scientific library (2007)
19. Giannoglou, V.G., Stavrakoudis, D.G., Theocharis, J.B., Petridis, V.: Genetic fuzzy rule-based classification systems for tissue characterization of intravascular ultrasound images. In: 2012 IEEE International Conference on Fuzzy Systems, pp. 1–8. IEEE (2012)
20. González, A., Pérez, R., Romero-Záliz, R.: Reasoning methods in fuzzy rule-based classification systems for big data problems. In: IoTBDS (2019)
21. Grabisch, M.: A new algorithm for identifying fuzzy measures and its application to pattern recognition. In: Proceedings of 1995 IEEE International Conference on Fuzzy Systems, vol. 1, pp. 145–150. IEEE (1995)

22. Han, J., Pei, J., Kamber, M.: Data Mining: Concepts and Techniques. Elsevier (2011)
23. Holmes, I.J.S., Anderson, J.E.: Glucose test data acquisition and management system (Dec 6 1994), uS Patent 5,371,687
24. Ishibuchi, H., Nakashima, T., Morisawa, T.: Voting in fuzzy rule-based systems for pattern classification problems. Fuzzy Sets Syst. **103**(2), 223–238 (1999)
25. Ishibuchi, H., Nakashima, T., Nii, M.: Classification and Modeling with Linguistic Information Granules: Advanced Approaches to Linguistic Data Mining. Springer, Heidelberg (2004). https://doi.org/10.1007/b138232
26. Keller, J.M., Osborn, J.: Training the fuzzy integral. Int. J. Approximate Reason. **15**(1), 1–24 (1996)
27. Kim, S.Y., et al.: Classification of usual interstitial pneumonia in patients with interstitial lung disease: assessment of a machine learning approach using high-dimensional transcriptional data. Lancet Resp. Med. **3**(6), 473–482 (2015)
28. Kuncheva, L.: Fuzzy Classifier Design, vol. 49. Springer, Heidelberg (2000). https://doi.org/10.1007/978-3-7908-1850-5
29. Leszczyński, K., Penczek, P., Grochulski, W.: Sugeno's fuzzy measure and fuzzy clustering. Fuzzy Sets Syst. **15**(2), 147–158 (1985)
30. Lucca, G.: Aggregation and pre-aggregation functions in fuzzy rule-based classification systems (2018)
31. Lucca, G., et al.: Preaggregation functions: construction and an application. IEEE Trans. Fuzzy Syst. **24**(2), 260–272 (2015)
32. Magadum, C., Bapat, M.: Ranking of students for admission process by using choquet integral. Int. J. Fuzzy Math. Arch. **15**(2), 105–113 (2018)
33. Michie, D., Spiegelhalter, D.J., Taylor, C., et al.: Machine learning. Neural Stat. Classif. **13**(1994), 1–298 (1994)
34. Murofushi, T., Sugeno, M., Machida, M.: Non-monotonic fuzzy measures and the choquet integral. Fuzzy Sets Syst. **64**(1), 73–86 (1994)
35. Murofushi, T., Sugeno, M., et al.: Fuzzy measures and fuzzy integrals. Fuzzy measures and integrals: theory and applications, pp. 3–41 (2000)
36. Nguyen, H.T., Kreinovich, V., Lorkowski, J., Abu, S.: Why Sugeno lambda-Measures. Departmental Technical Reports (CS) (2015)
37. Provost, F., Fawcett, T.: Data science and its relationship to big data and data-driven decision making. Big Data **1**(1), 51–59 (2013)
38. He, Q., Chen, J.-F., Yuan, X.-Q., Li, J.: Choquet fuzzy integral aggregation based on g-lambda fuzzy measure. In: 2007 International Conference on Wavelet Analysis and Pattern Recognition, vol. 1, pp. 98–102 (2007)
39. Sanz, J.A., Galar, M., Jurio, A., Brugos, A., Pagola, M., Bustince, H.: Medical diagnosis of cardiovascular diseases using an interval-valued fuzzy rule-based classification system. Appl. Soft Comput. **20**, 103–111 (2014)
40. Sugeno, M.: Theory of fuzzy integrals and its applications. Doctoral thesis, Tokyo Institute of technology (1974)
41. Uebele, V., Abe, S., Lan, M.S.: A neural-network-based fuzzy classifier. IEEE Trans. Syst. Man Cybern. **25**(2), 353–361 (1995)
42. Verma, N.K.: Estimation of fuzzy measures using covariance matrices in gaussian mixtures. Appl. Comput. Intell. Soft Comput. **2012** (2012)
43. Wilcoxon, F.: Individual comparisons by ranking methods. In: Kotz, S., Johnson, N.L. (eds.) Breakthroughs in Statistics, pp. 196–202. Springer, Heidelberg (1992). https://doi.org/10.1007/978-1-4612-4380-9_16

FT-BlinGui: A Fuzzy-Based Wearable Device System to Avoid Visually Impaired Collision in Real Time

Elidiane Pereira do Nascimento and Tatiane Nogueira(✉)

DCC, Federal University of Bahia, Salvador, BA, Brazil
ellydiani@gmail.com, tatiane.nogueira@ufba.br

Abstract. There are several technological applications and computational techniques for detection and prevention of obstacles that aim at assisting the locomotion of robots, autonomous vehicles, people with visual impairments, unmanned aerial vehicles, detection of anomalies in runways, among others. However, the imprecision and uncertainty in the perception of obstacles in real time are not detected in advance to avoid them. This is due to the dynamism of the environment, where obstacles can present different behavioral scenarios and are constantly changing. To provide a more efficient perception of obstacles in dynamic scenarios, the fuzzy systems embedded in technological applications have been presented as an efficient approach for the detection and prevention of obstacles more effectively and intuitively. In this paper, therefore, we will investigate the applicability of the fuzzy time series for the forecasting collision with obstacles. The time series analyzed was obtained through a wearable device that detects, in real time, obstacles for the visually impaired. Simulation results demonstrate that FT-BlinGui can provide a collision-avoidance alerting system for wearable devices as a result of Fuzzy Time Series forecasting.

Keywords: Fuzzy time series · Forecasting · Collisions · Obstacle avoidance

1 Introduction

Collision with obstacles is a problem that can occur in different ways and different contexts. Most common collisions are suffered by the visually impaired, mobile robots, unmanned aerial vehicles and ground vehicles. The risks of collisions become high with a constant movement of people, vehicles, animals, besides the existence of objects obstructing the passage, as well as the velocity of them.

Technological advances have enabled the development of several computer systems for preventing collisions. These systems take advantage of the ubiquitous technology with miniaturized wearable, mobile, and portable prototypes, which for the most part make use of sensors to capture information from the

R. Cerri and R. C. Prati (Eds.): BRACIS 2020, LNAI 12320, pp. 245–258, 2020.
https://doi.org/10.1007/978-3-030-61380-8_17

environment. Research on collisions with obstacles in different application contexts advances over time and is addressed in different parts of the world. Initially, in 1975 in France, Thourel [22] launched research on the detection of obstacles on railway lines; already in 1990 in Germany, Storjohann *et al.* [21] developed an obstacle detector system for automatically guided vehicles; in 1991 in Taiwan, Pan *et al.* [13] proposed a new approach to solve the collision detection problem for mobile robots in the presence of moving obstacles; in 1994, in the United States, Shoval *et al.* [16] described in his research the use of an obstacle prevention system with a mobile robot, as a guidance device for blind and visually impaired people. This research on collisions with obstacles has expanded and very recent research has been generated [2,9,10,23–25] that has advanced with most modern technologies and more accurate equipment to facilitate detection.

There are several researches that explore obstacle detection methods to avoid collisions. However, many of these methods disregard the uncertainties caused by the dynamics of the environment, as for example, type of walk of visually impaired people. Besides, data collected along time is more suitable for best prediction. Therefore, based on the works of Song and Chissom [18–20] it is possible to study this dynamic process through the Time Series (TS). To improve TS with the uncertainty inherent to the data, SONG and Chissom [18,20] introduced for the first time the definition of a Fuzzy Time Series (FTS) to consider inaccurate knowledge of the real world. According to Lima et al. (2015) [11], fuzzy approaches have several advantages over crisp ones. Fuzzy approaches have more flexible decision boundaries, and thus are characterized by their higher ability to adjust to a specific domain of application and more accurately reflect its particularities. In general, fuzzy models are expressed by a set of linguistic fuzzy rules, which are derived from the experience of skilled operators, or by a set of fuzzy implications that locally represents the input–output relationship in the process [11].

Collision-avoidance for visually impaired is a real-life problem with inherent uncertainty and imprecision data. As already pointed out in some similar researches that have used FTS for problems of obstacles detection [5,6,8,12], the scope of FTS forecasting [1,14] is proper to our proposal: FT-BlinGui, a fuzzy-based wearable device system to avoid visually impaired collision in real time.

Considering the improvements obtained by the use of FTS to predict obstacles, the purpose of FT-BlinGui is to apply well-known FTS forecasting models to predict risk of collisions by means of an wearable device, which needs to be capable to detect, in real time, obstacles for the visually impaired. Therefore, for this work, we have conducted experimental studies to compare the accuracy of four well-known FTS forecasting model: Chen's model [3], Singh 's model [17], Huarng's model [7] and Chen-Hsu' model [4].

The sections of this paper are organized as follows: in Sect. 2, we present FTS definitions; in Sect. 3, we present FTS forecasting models used in this study; in Sect. 4, we present the results and discussion; and, finally, in Sect. 5, we conclude our work by pointing out future directions.

2 Fuzzy Time Series

The concept of Fuzzy Time Series (FTS) was introduced by SONG and CHISSOM (1991)[18,19] by applying the concepts of Fuzzy Sets Theory [26] to model a special dynamic process, whose observations are represented by fuzzy sets.

In the context of this work, the fuzzy sets are suitable to represent the distances from a visually impaired person to an obstacle, since its perception can vary considerably according to context or conditions.

Therefore, to make use of a FTS, the observations of a TS, which are the distances from a visually impaired person to an obstacle, are translated to fuzzy sets. According to [18], in usual time series, historical data is used to establish the relationship between interest values at different times, but the relationship in the FTS is applied to the knowledge of past experience in the model and the knowledge has the form of fuzzy relationships.

FTS is a dynamic process, with the following Definitions [1]:

Definition 1. *Fuzzy Time Series (FTS)*
Let $Y(t)(t = \ldots, 0, 1, 2, \ldots)$, a subset of R (real time), be the universe of discourse on which fuzzy sets $f_i(t)(i = 1, 2, \ldots)$ are defined and let $F(t)$ be a collection of $f_1(t), f_2(t), \ldots$. Then $F(t)$ is called a FTS defined on $Y(t)(t = \ldots, 0, 1, 2 \ldots)$.

In the above definition, $F(t)$ can be understood as a linguistic variable and $f_i(t)(i = 1, 2, ...)$ as the possible linguistic values of $F(t)$. A FTS is, then, seen as a linguistic variable, whose fuzzy sets (linguistic values) describe each series observation of the universe of discourse.

Definition 2. *Model order*
The order of a FTS model is related to the number of past observations considered to represent a future one. Suppose an observation fuzzified to $f_i(t)$ is caused by $f_i(t-1)$ only, i.e., $f_i(t-1) \rightarrow f_i(t)$. Then this relation can be expressed as $f_i(t) = f_i(t-1) \circ R(t, t-1)$, where "$\circ$", is an arithmetic operator. $f_i(t-1) \rightarrow f_i(t)$ is called a Fuzzy Logical Relationship (FLR), where $f_i(t-1)$ and $f_i(t)$ are called the current observation and the next observation, respectively; and $R(t, t-1)$ is the fuzzy relationship between $f_i(t-1)$ and $f_i(t)$. Such relation is called the first-order model of $F(t)$.

If $f_i(t)$ is caused by $f_i(t-1), f_i(t-2), \ldots, f_i(t-n)$, then this FLR is represented by $f_i(t-n), \ldots, f_i(t-2), f_i(t-1) \rightarrow f_i(t)$, and called n-th order FTS.

There are two main categories of FTS. The classification is determined by their models in terms of their relationships with time t as follows.

Definition 3. *Time-invariant and Time-variant Models*
The temporal dependence of FTS is associated to the FLR, which are based on time-invariant and time-variant definitions. Suppose $R(t, t-1)$ is a first-order model of $F(t)$. If for any time t, $R(t, t-1)$ is independent of t, i.e., for any t, $R(t, t-1) = R(t-1, t-2)$, then $F(t)$ is called a time-invariant FTS, otherwise it is called a time-variant fuzzy time series.

Definition 4. *Fuzzy Time Series forecasting*
The general procedure for FTS forecasting, considering time-invariant FTS, are defined in 5 *steps as follows* [1,14,18].

1. *Determination of the universe of discourse* $Y(t)$ *for a real valued time series and decompose* $Y(t)$ *into equal or unequal length intervals.*
2. *Once the universe of discourse and the length intervals are identified, the time series is split into such intervals. Then the fuzzification is done by replacing a data point (time series observation) with its correspondent fuzzy set.*
3. *Establishment of the FLRs, which are denoted as* $A_l \rightarrow A_r$*:* A_l *represents the fuzzy set of the k-number of lagged observations of fuzzy time series, and* A_r *represents the fuzzy set of the* $k+1$*th observation. The value of k denotes the order of the FTS model. When* $k = 1$*, it is called first order model, and when* $k > 1$*, it is called high order model.*
4. *Once the order of the model is determined, the FLRs are used to relate previous and current observation, and then forecast the fuzzy set of the next observation.*
5. *Defuzzification of the predicted values, i.e., convert* A_r *to a numeric value.*

As illustration of the process, suppose a wearable device in the time $t = 10\,\text{s}$ has detected the distance from a visually impaired person to an obstacle as $x(t) = 200\,\text{cm}$, describing a TS observation. Fuzzyfying $x(t)$, we can, for example, take A_2 as its most pertinent fuzzy set, so $f_i(t) = A_2$. After fuzzifying all the observations of the TS, we get the FLRs to find the compatible fuzzy set with a new observation. Supposing we have an FLR $A_2 \rightarrow A_4$, as the last distance detected by the device is compatible with A_2, the next distance to be detected will be related to A_4, i.e., $f_i(t+1) = A4$ is a fuzzy prediction of a new distance $x(t)$. Finally, it is necessary defuzzify A_4 to find the real value of such prediction: $x(t+1)$.

It is important to highlight each stage previously defined has very significant contribution in forecasting performance of FTS. Therefore, different techniques have been applied in these stages providing several FTS forecasting models. We have used four well-known FTS forecasting models, which are described in the following section.

3 Models for Fuzzy Time Series Forecasting

There are several models for FTS forecasting, and most of them have emerged from the general model defined in the previous section, which followed the concepts described by SONG and Chissom (1993) in [18]. Next, we described the models used in this work.

3.1 Chen's Model [3]

Chen's model was proposed to simplify the step 4 of FTS forecasting proposed by Song and Chissom (1993) in [18]. Therefore, CHEN (1996) [3] suggested to group

Table 1. Fuzzy logical relationships

$A_1 \to A_1$	$A_1 \to A_2$	$A_2 \to A_3$
$A_3 \to A_2$	$A_4 \to A_2$	$A_3 \to A_4$
$A_2 \to A_5$	$A_2 \to A_4$	$A_5 \to A_1$

Table 2. Fuzzy logical relationships groups

Group 1:	$A_1 \to A_1$	$A_1 \to A_2$	
Group 2:	$A_2 \to A_3$	$A_2 \to A_4$	$A_2 \to A_5$
Group 3:	$A_3 \to A_2$	$A_3 \to A_4$	
Group 4:	$A_4 \to A_2$		
Group 5:	$A_5 \to A_1$		

the derived FLRs into groups based on the current observation of the FLRs. For example, based on Table 1, we can obtain FLR groups shown in Table 2.

Therefore, to predict the next value of a FTS, Chen (1996) [3] follows the rules:

(i) If the fuzzification of a TS observation is compatible with a fuzzy set of an antecedent rule of a FLRs group composed by just one rule, the forecasted value is the consequent of such a rule. For example, suppose the last distance from a visually impaired person to an obstacle was $x(t) = 200$ cm, describing a TS observation, and it was fuzzified to A_4. From Table 2, we can predict that the fuzzy set of $x(t+1)$ is A_2;

(ii) If the fuzzification of a TS observation is compatible with a fuzzy set of an antecedent of a rule of FLRs group composed by more than one FLR, the forecasted value is obtained by considering the midpoints of each consequent of the rules of such a group. Following the same example, let $x(t)$ be fuzzified to A_3. From Table 2, we can predict that the fuzzy set of $x(t+1)$ is calculated by the midpoints of A_2 and A_4.

Therefore, Chen (1996) [3] performs the step 4 of FTS forecasting using arithmetic operations, which greatly reduced the computation complexity in Song and Chissom's models [18], which uses the max-min composition for the derivation of a fuzzy relation.

3.2 Huarng's Model [7]

Huarng's model integrates domain-specific knowledge (or heuristic) with Chen's model to improve forecasting. Therefore, in step 4 of FTS forecasting, besides the group of FLRs proposed by Chen (1996) [3], He adds some heuristic knowledge to establish the fuzzy logical relationship groups.

Suppose $x(t) = A_j$ and the fuzzy logical relationship group for A_j is $A_j \to A_q, A_r, \ldots$. The proper fuzzy sets, $A_{p1}, A_{p2}, \ldots, A_{pk}$, for $x(t+1)$ can be selected by heuristic function $h(x_1; x_2, \ldots, A_q, A_r, \ldots) = A_{p1}, A_{p2}, \ldots, A_{pk}$; where $x_1, x_2, \ldots$ are heuristic variables. Then, a heuristic fuzzy logical relationship group is obtained: $A_j \rightarrow A_{p1}, A_{p2}, \ldots, A_{pk}$.

Let $x(t)$ be fuzzified to A_3. From Table 2, we can predict that the fuzzy set of $x(t+1)$ is calculated by using some knowledge about A_2 and A_4. In the context of this work, the heuristic knowledge, h, can show the increase or decrease of the distance from a visually impaired person to an obstacle. In that sense, the fuzzy sets A_2 and A_4 could be ordered, which facilitates the selection of a proper fuzzy set, by the heuristic function for $x(t+1)$.

According HUARNG [7], regardless if F(t) is a time-invariant or time-variant fuzzy time series, the heuristic models can be applied according to the algorithm.

3.3 Chen-Hsu's Model [4]

CHEN and HSU (2003) also present a new approach for step 4 of FTS forecasting. In that way it uses a set of guidelines to determine whether the trend of the forecasting goes up or down. In the context of this work, assume that we want to forecast the distance from $x(t)$, then the *difference of differences* is defined as $(x(t-1) - x(t-2)) - (x(t-2) - x(t-3))$, meaning the distances between the time $t-1$ and $t-2$ and between the time $t-2$ and $t-3$.

Considering a fuzzy logical relationship $A_i \rightarrow A_j$, where A_i denotes the fuzzified distance $x(t-1)$ and A_j denotes the fuzzified distance $x(t)$, then:

1. If $j > i$ or $j < i$ or $j = i$, and the *difference of differences* of the distances between the time $t-1$ and $t-2$ and between the time $t-2$ and $t-3$ is positive, then the trend of the forecasting will go up, and some guideline representing such tendency is used to forecast the distance $x(t+1)$;
2. If $j > i$ or $j < i$ or $j = i$, and the *difference of differences* of the distances between the time $t-1$ and $t-2$ and between the time $t-2$ and $t-3$ is negative, then the trend of the forecasting will go down, and some guideline representing such tendency is used to forecast the distance $x(t+1)$;

In [4], CHEN and HSU (2003) suggest three specifics guidelines to forecast the enrollments of the University of Alabama, a well-known TS. In our work, we considered as guidelines, the number of observations that composes each fuzzy set by which the universe of discourse was partitioned.

3.4 Singh's Model [17]

Differently from the previous, SINGH (2008) [17] proposed a model of order three, i.e., $x(t+1)$ is caused by $x(t-2)$, $x(t-1)$ and $x(t)$. Therefore, $f_i(t+1)$ is computed as $f_i(t+1) = f_i(t-1) * R(t, t-1, t-2)$, where the fuzzy relation R is considered a numeric value rather than a fuzzy relational matrix and is being computed as difference between differences in the consecutive values of time t with $t-1$ and of values of time $t-1$ with $t-2$. To compute the fuzzy relation R, the author suggests some computational algorithms that extensively utilizes a time variant parameter being obtained by the differences of the past 3 observations as current observation to forecast the values for the next one.

The next section will present our proposal, which has used the models previously presented.

4 FT-BlinGui

BlinGui [15] is a device with a system coupled to a bag with eight ultrasonic sensors fixed in its front: four sensors directed from top and four sensors directed to bottom, for the perception of obstacles in different directions. Such a bag, as observed in Fig. 1, is a low-cost wearable device developed to be attached to the chest of a visually impaired person in order to emit vibrating alerts, voice alerts and beeps for collision-avoidance.

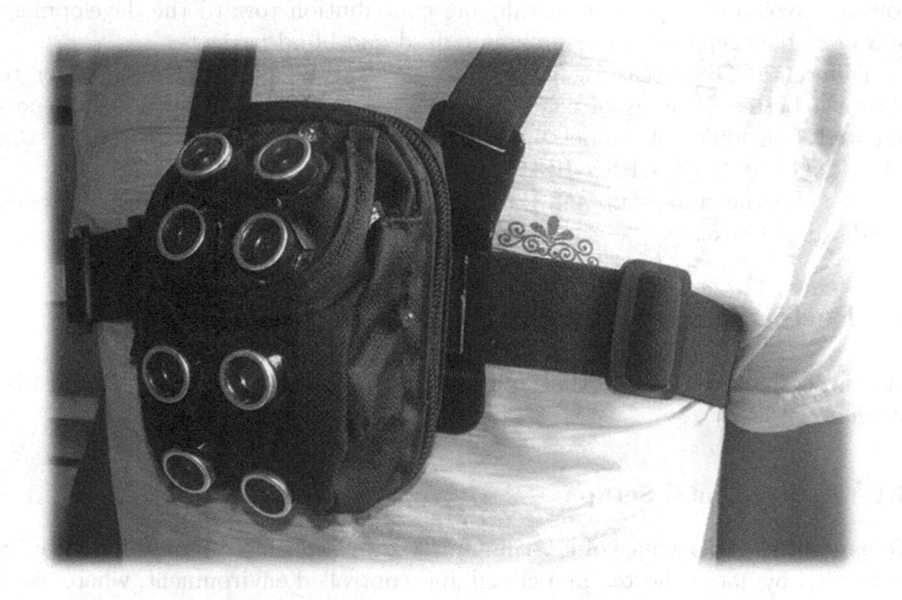

Fig. 1. BlinGui: electronic device to detect obstacles.

We are now proposing FT-BlinGui, an approach capable of improving BlinGui for collision-avoidance by means of fuzzy time series forecasting. With this approach, we can detect the risk of collision with more precision, since we detect how far is the obstacle according to the type of walk of the visually impaired person when using BlinGui.

To apply FT-BlinGui, time series data is composed by the distances from a human to an obstacle detected by BlinGui collected along time. Each observation in such a time series corresponds to the distance from a human to an obstacle during a period of a type of walk. Therefore, five different time series can be obtained from BlinGui, depending on the type of walking: 1) Slow walk, which describes a walk made in 10 s; 2) Light walk, which describes a walk made in 5 s; 3) Sport walk, which describes a walk made in 4 seconds; 4) Vigorous walk, which describes a walk made in 3 s; and 5) Athletic walk, which describes a walk made in 1 s.

Once the time series has been obtained, FT-BlinGui predicts the next distance from a human to the obstacles, detecting the risk of collision and enabling the visually impaired to avoid collision with them.

The prediction of a collision is made by FT-BlinGui using Fuzzy Time Series forecasting models. According to [1], FTS forecasting is an emergent research area to deal with the problems associated with uncertainty, vagueness and imprecision. The authors argue there are two reasons why FTS is more suitable than the conventional forecasting systems: i) It can process both crisp and fuzzy values, and ii) It does not require large datasets as in statistical forecasting models. Therefore, Fuzzy Time Series forecasting on embedded system for collision-avoidance represents a valuable contribution toward the development of a portable vision aid for visually impaired and blind patients.

Different FTS forecasting model can be used by FT-BlinGui in order to obtain a better accuracy of collision-avoidance. In this paper, we have experimented four different models, as previously mentioned: CHEN (1996) [3], HUARNG (2001) [7], CHEN-HSU (2004) [4] and SINGH (2008) [17].

The experimental setup and the results obtained by the proposed approach are discussed next.

5 Performance Evaluation

In this section, we describe the experimental setup, simulation results and discussion of them.

5.1 Experimental Setup

To assess the performance of FT-BlinGui, we have considered five different FTS composed by data collected in a closed and controlled environment, where users have moved towards and obstacle using BlinGui.

For our experiment, a wall was used as a static obstacle, and an user was positioned 5 m distant from this wall. Then, the person using BlinGui moved toward this obstacle 5 times: each one with a different type of walk.

Therefore, five time series were obtained: 1) Slow walk, by which 110 observations were collected; 2) Light walk, by which 74 observations were collected; 3) Sports walk, by which 59 observations were collected; 4) Vigorous walk, by which 37 observations were collected; and 5) Athletic walk, by which 30 observations were collected.

For better visualization, Fig. 2 displays a time series with five peaks, each peak corresponds to the beginning of a time series obtained with the different type of walk previously mentioned. When a type of walk is started, a person has distanced again 5 meters (500 cm) from the obstacle. As can be observed, when the speed of the walk increases, less observations are collected because the obstacle is reached faster.

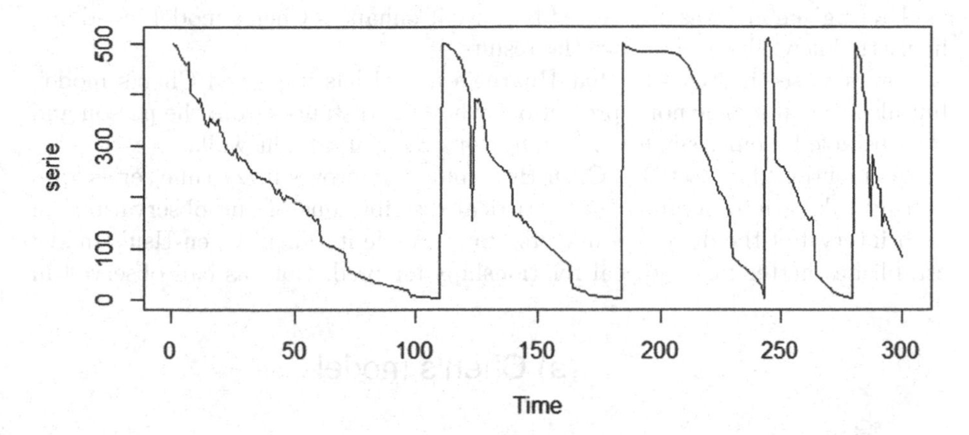

Fig. 2. Time series obtained for the experiments. The time is represented in seconds and the distances are represented in centimeters

Once obtained this series, FT-BlinGui is used to predict how far a person is to an obstacle. For that, different fuzzy time series forecasting models were applied over the time series collected to check which one is more feasible to detect the next distance of the obstacle. For example, considering a person using BlinGui has moved toward an obstacle with a Slow walk, FT-BlinGui needs to be capable to detect how distant is him/her to the obstacle.

To compare the results of different FTS forecasting methods used by FT-BlinGui, for each model we have follow the four steps: 1) Partitioning of the universe of discourse in five equal-length intervals. 2) Fuzzification of the time series observations translating them in fuzzy sets; 3) Establishment of the fuzzy logical relationships; and 4) Forecasting and measure of models' accuracy by means of the mean square error (MSE).

All the FTS forecasting methods used in our approach are available in the AnalyzeTS[1] R package. Next, we present the results obtained by each model.

5.2 Results and Discussion

Using the defined experimental setup, we have used four forecasting models to check which one is the most suitable for collision-avoidance by means of FT-BlinGui.

The classical and well-known Chen's model does not present a satisfactory result for FT-BlinGui. As can be observed in Fig. 3(a), Chen's model has predicted the next distance simplifying its arithmetic operations reducing the dis-

[1] https://CRAN.R-project.org/package=AnalyzeTS.

tances to the average. Therefore, small distances from the person and the obstacle implies in high error in the prediction. As conclusion, such method can not be used by BlinGui because high errors in prediction lead to high risk of collision.

Huarng's model was also tested because it enhances Chen's model by adding heuristic knowledge to improve the results.

One can see in Fig. 3(b) that Huarng's model has improved Chen's model, but also presents some noise predictions when the distances from the person and the obstacle becomes smaller, specially with Slow and Light walk.

As mentioned in Sect. 3.3, Chen-Hsu's model improves fuzzy time series forecasting taking into account the statistical distributions of the observations in each interval of the discourse universe to re-divide it. Then, Chen-Hsu's model establishes better fuzzy logical relationships for prediction, as can observed in

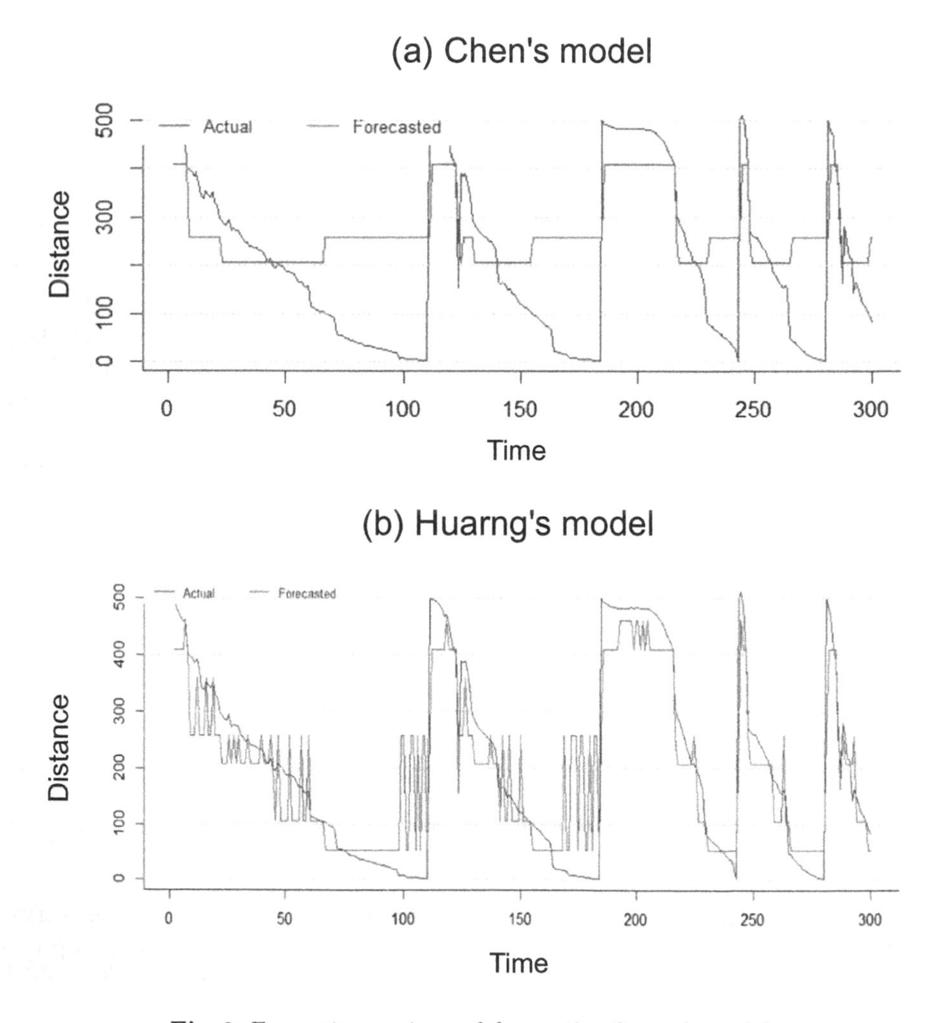

Fig. 3. Fuzzy time series and forecasting for each model.

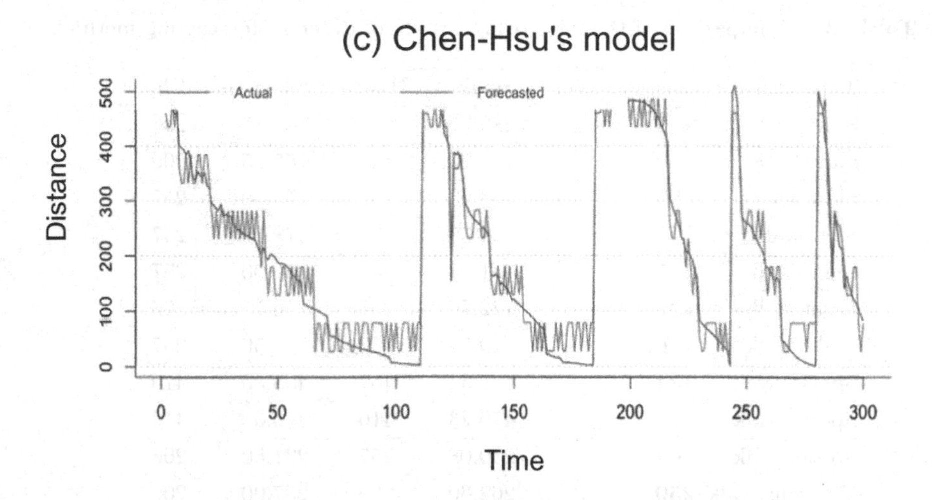

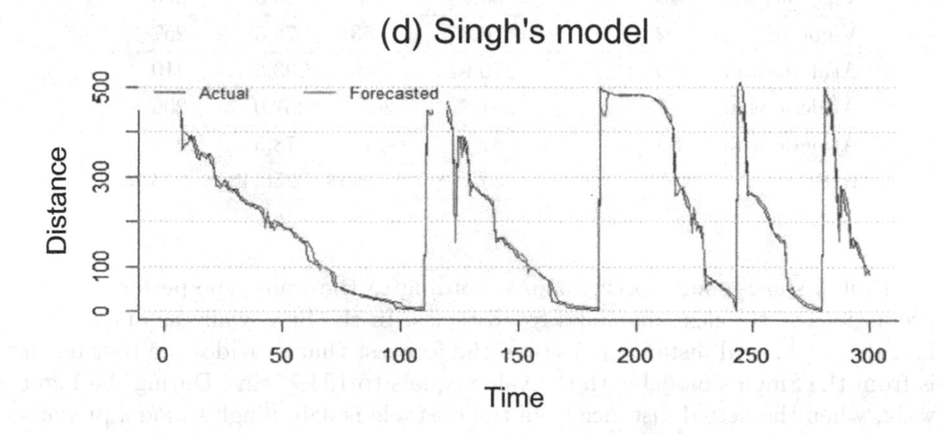

Fig. 3. (*continued*)

Fig. 3(c). Although this model has improved the forecasting, some predictions with noise still remains in every type of walk.

Finally, Singh's model obtained the best result for FT-BlinGui. As expected, It gives improved forecasting with few data because it utilizes a time variant parameter that considers the past 3 observations as current observation to forecast the values for the next observation, being a model of order three.

One can notice in Fig. 3(d) that Singh's model obtained predictions values very close to the real one.

Besides the results presented in Fig. 3, Table 3 presents the MSE obtained by each model and the forecast values obtained of distances in centimeters for some real values of the series. Therefore, Singh's method is the one with the greatest accuracy for the prediction of collisions with obstacles.

Table 3. A comparison of the forecasting results of different forecasting methods.

Type of walk	Distances (cm)	Singh	Huarng	Chen-Hsu	Chen
Slow walk	**283**	283.77	206	288.87	206
Slow walk	**189**	193.77	104	167.75	206
Slow walk	**52**	53.93	53	78.50	257
Slow walk	**7**	20.05	257	78.50	257
Light walk	**385**	381.73	257	359.00	257
Light walk	**121**	122.70	257	116.75	206
Light walk	**6**	10.54	53	78.50	257
Sports walk	**484**	481.31	461	435.50	410
Sports walk	**465**	470.23	410	486.5	410
Sports walk	**208**	230.08	257	231.50	206
Vigorous walk	**259**	262.60	206	257.00	206
Vigorous walk	**49**	53.00	104	53.0	206
Vigorous walk	**2**	6.85	53	78.5	257
Athletic walk	**389**	270.64	410	333.5	410
Athletic walk	**217**	251.86	206	257.0	206
Athletic walk	**83**	89.92	53	78.5	257
MSE		*255.48*	6728.53	2256.40	21332.61

Table 3 shows some observations according to the walk type performed and predictions of the next distances to obstacles. In the Slow walk, as one notices in Table 3, the real distance is 189 cm, the forecast that provides the best result is from the Singh's model with the value equals to 193.77 cm . During the Light walk, when the actual distance from the obstacle is 385, Singh's model predicts the distance equals to 381.73. During Sports walk, when the distance value is 208, only Chen's model approaches the real value providing the distance 206. However, for Vigorous walk, Chen's model provides the same predicted value 206 when the expected on is 259, 49 and 2, respectively, as shown in Table 3. Finally, in the Athletic walk, the closest predictions to the real value (83) are those ones estimated by Singh's model (89.92) and Chen-Hsu's model (78.5).

6 Conclusion

We have presented FT-BlinGui, a fuzzy-based wearable device system to avoid a visually impaired collision. The proposed system can provide collision warnings in real time by predicting the next distance between a visually impaired and an obstacle considering his/her type of walk. The accuracy of the proposal was shown for a variety of scenarios that involve walk conditions and different forecasting models.

We conclude, from the comparative performance of four well-known Fuzzy Time Series forecasting models, FT-BlinGui can be effective for collision detection handling data uncertainty inherent to different types of walk. Simulation results demonstrate that FT-BlinGui can provide a collision-avoidance alerting system to BlinGui as a result of Singh's model forecasting.

Finally, although the Singh's model performed well in general, it is evidenced by Table 3 that it also fails at times (slow walk, 189; vigorous walk, 2; among others). Such results, can be a limitation of the proposed approach, as a prediction that accuses a distance higher than the real one can cause a collision.

As future work, besides modifications on the proposal to improve the forecasting accuracy, we also intend to test FT-BlinGui embedded in BlinGui for visually impaired users when walking through not controlled spaces, such as streets and avenues, by notifying the user about the risks of collision. For such a real scenario, it is important to highlight the FT-Blingui needs to be calibrated with more examples to construct a previous time series to become capable to predict the next observation, i.e. the distance from the visually impaired and the obstacle.

As an improved evaluation, the use of FTS forecasting will be compared with other obstacle detection methods in the literature in a future work.

Acknowledgments. This work was supported by CAPES (Coordination for the Improvement of Higher Education Personnel - Brazilian federal government agency). Any opinions, findings, and conclusions or recommendations expressed in this material are those of the authors and do not necessarily reflect the views of CAPES.

References

1. Bose, M., Mali, K.: Designing fuzzy time series forecasting models: a survey. Int. J. Approximate Reason. **111**, 78–99 (2019)
2. Chavan, S., Bhitale, P.P., Devrukhakar, S.S., Jadhav, S.V., Paradkar, K.R.: Real-time collision detection for visually impaired people. J. Microcontroller Eng. Appl. **5** (2018)
3. Chen, S.M.: Forecasting enrollments based on fuzzy time series. Fuzzy Sets Syst. **81**, 311–319 (1996)
4. Chen, S.M., Hsu, C.C.: A new method to forecast enrollments using fuzzy time series. Int. J. Appl. Sci. Eng. **3**, 234–244 (2004)
5. Dimeas, F., Avendaño-Valencia, L.D., Aspragathos, N.A.: Human-robot collision detection and identification based on fuzzy and time series modelling. Robotica **33**(9), 1886–1898 (2015)
6. Hu, Y., Meng, X., Zhang, Q., Park, G.K.: A real-time collision avoidance system for autonomous surface vessel using fuzzy logic. IEEE Access **8**, 108835–108846 (2020)
7. Huarng, K.: Heuristic models of fuzzy time series for forecasting. Fuzzy Sets Syst. **123**(3), 369–386 (2001)
8. Kao, S.L., Lee, K.T., Chang, K.Y., Ko, M.D., et al.: A fuzzy logic method for collision avoidance in vessel traffic service. J. Navigation **60**(1), 17–31 (2007)
9. Karakaya, S., küçükyıdız, G., Ocak, H.: One dimensional moving obstacle detection method for mobile robotics applications, pp. 1–4 (2018)

10. Krishnakumar, S., Mridha, B., Naves, M.J.N., Kowsalya, K.: Intelligent walker with obstacle detection technology for visually challenged people. In: 2017 IEEE International Conference on Power, Control, Signals and Instrumentation Engineering (ICPCSI), pp. 2863–2867, September 2017
11. Lima, N.N., et al.: Nonlinear fuzzy identification of batch polymerization processes. In: Gernaey, K.V., Huusom, J.K., Gani, R. (eds.) 12th International Symposium on Process Systems Engineering and 25th European Symposium on Computer Aided Process Engineering, Computer Aided Chemical Engineering, vol. 37, pp. 599–604. Elsevier (2015). https://doi.org/10.1016/B978-0-444-63578-5.50095-5
12. Mohammadian, M.: Modelling, control and prediction using hierarchical fuzzy logic systems. Robotic Systems: Concepts, Methodologies, Tools, and Applications, p. 187 (2020)
13. Pan, T., Luo, R.C.: A feasible collision-detection algorithm for mobile robot motion planning with moving obstacles. In: Proceedings IECON 1991: 1991 International Conference on Industrial Electronics, Control and Instrumentation, vol. 2, pp. 1011–1016 (1991)
14. Panigrahi, S., Behera, H.: A study on leading machine learning techniques for high order fuzzy time series forecasting. Eng. Appl. Artif. Intell. **87**, 103245 (2020)
15. Pereira, E., Nascimento, F.M., Vieira, V.: Blingui: uma solução vestivel de apoio a pessoas portadoras de deficiência visual na detecção de obstaculos estáticos em ambientes internos. XIII Simpósio Brasileiro de ComputaÇão Ubiqua e Pervarsiva (2016)
16. Shoval, S., Borenstein, J., Koren, Y.: Mobile robot obstacle avoidance in a computerized travel aid for the blind. In: Proceedings of the 1994 IEEE International Conference on Robotics and Automation, vol. 3, pp. 2023–2028 (1994)
17. Singh, S.: A computational method of forecasting based on fuzzy time series. Math. Comput. Simul. **79**(3), 539–554 (2008)
18. Song, Q., Chissom, B.S.: Forecasting enrollments with fuzzy time series—part i. Fuzzy Sets Syst. **54**(1), 1–9 (1993)
19. Song, Q., Chissom, B.S.: Fuzzy time series and its models. Fuzzy Sets Syst. **54**(3), 269–277 (1993)
20. Song, Q., Chissom, B.S.: Forecasting enrollments with fuzzy time series—part ii. Fuzzy Sets Syst. **62**(1), 1–8 (1994)
21. Storjohann, K., Zielke, T., Mallot, H.A., von Seelen, W.: Visual obstacle detection for automatically guided vehicles. In: Proceedings of the IEEE International Conference on Robotics and Automation, vol. 2, pp. 761–766, May 1990
22. Thourel, B.: Detection of obstacles on railway level-crossings. In: 1975 5th European Microwave Conference, p. 322 (1975)
23. Utaminingrum, F., et al.: A laser-vision based obstacle detection and distance estimation for smart wheelchair navigation. In: 2016 IEEE International Conference on Signal and Image Processing (ICSIP), pp. 123–127 (2016)
24. Valencia, D., Kim, D.: Quadrotor obstacle detection and avoidance system using a monocular camera. In: 2018 3rd Asia-Pacific Conference on Intelligent Robot Systems (ACIRS), pp. 78–81 (2018)
25. Xungu, S., Notununu, L., Mbizeni, A., Dickens, J.: Design of railway obstacle detection prototype. In: 2017 Pattern Recognition Association of South Africa and Robotics and Mechatronics (PRASA-RobMech), pp. 56–61 (2017)
26. Zadeh, L.: Fuzzy sets. Inf. Control **8**(3), 338–353 (1965)

Genetic Learning Analysis of Fuzzy Rule-Based Classification Systems Considering Data Reduction

Allen Hichard Marques dos Santos, Matheus Giovanni Pires(✉), and Fabiana Cristina Bertoni

Department of Exact Sicences, State University of Feira de Santana, Feira de Santana, Brazil
allenhichard21@gmail.com, {mgpires,fcbertoni}@uefs.br

Abstract. In this paper, we evaluated the Knowledge Base building of Fuzzy Rule-Based Classification Systems (FRBCS) with the purpose of find a balance between the accuracy and interpretability objectives. Regarding to build, we compared two well-known algorithms: Wang-Mendel, to generate the rule base, and NSGA-II, to learn the rules and tuning the membership functions. The Wang-Mendel algorithm was also used to introduce a seed in NSGA-II initial population, in order to increase its quality and improve convergence speed. Taking into account that the automatic building of the fuzzy systems knowledge base is challenger, because of the amount of data available in several real problems, we analysed the impact of data reduction on it. The experiments were carried out with 23 datasets divided into small and medium-large size, and the results showed that the use of genetic learning is suitable to large datasets as well as data reduction, improving the accuracy and interpretability of the FRBCS arising.

Keywords: Fuzzy classification systems · Genetic rule learning · Tuning membership functions · Wang-Mendel algorithm

1 Introduction

Several real world problems deal with classification tasks and several approaches have been proposed to manage them. Fuzzy Rule-Based Systems (FRBS) have proved to be very effective as classifiers [15]. With the increasing amount of available data, the automatic generation of the fuzzy system knowledge base is an active and promising research area, since the proposal automatic method should be able to learn from this large amount of data, optimizing the fuzzy systems accuracy and interpretability. In the last decades, Multiobjective Genetic Algorithms (MOGA) have been so extensively used to this purpose, presenting effective findings [1,4–6,10,15,22,24,25]. However, the computational resources and time required by genetic algorithms for generating FRBS increases considerably with the increasing of dataset size and its dimensionality. These problems

R. Cerri and R. C. Prati (Eds.): BRACIS 2020, LNAI 12320, pp. 259–271, 2020.
https://doi.org/10.1007/978-3-030-61380-8_18

can be tackled by data reducing techniques, which consists of removing redundant, information-poor data and/or erroneous data from the dataset to obtain a subset which maintain the semantic integrity of the original dataset [18]. Among the most important and frequently used techniques are Feature Selection (FS) and Instance Selection (IS) and the Genetic Algorithms (GA) are one of the most widely used methods for these purposes [16,18,19,26].

Therefore, in this paper, we focus on the use of MOGA for learning FRBS knowledge base from the reduced datasets. We investigate if data reduction techniques can help improving the interpretability of the generated FRBS, preserving or even improving their accuracy in comparison to models generated by using the original set.

2 Literature Review

Many studies can be found in the literature which use MOGA for FRBS learning, considering IS and FS. In [20], the application of IS methods to a Genetic Fuzzy System for classification is described, in order to discover which reduction methods outperforms the others. In the experiments, a set of 20 small size datasets was considered and the results highlight that a specific family of IS methods was effective in decrease the Genetic Fuzzy System complexity while accuracy was maintained.

In [6], the authors investigated the use of Training Set Selection to reduce the set of instances required by a MOGA to generate FRBS for regression problems. The Training Set Selection is integrated in a co-evolutionary framework: cyclically, a single-objective GA selects a subset of instances which are used by the MOGA for generating the FRBS. The GA maximizes an index that measures the quality of the reduced set of instances.

The work presented in [4] exploits a MOGA to generate FRBS with different trade-offs between classification accuracy and rule base complexity. In order to learn the rule base, the authors employ a rule and condition selection approach which aims to select a reduced number of rules from a heuristically generated rule base and concurrently reduce the number of conditions for each selected rule. The approach was tested on fifteen classification benchmarks, achieving considerable accuracies and a low complexity rule base.

Furthermore, [5] focus on MOGA applied to learn concurrently the rule base and the data base of Mamdani FRBS and propose to tackle the issue by exploiting the synergy between two different techniques. The first technique is based on a novel method which reduces the search space by learning rules from a heuristically generated rule base using the Wang-Mendel Algorithm. The second technique performs instance selection by exploiting a co-evolutionary approach where cyclically a genetic algorithm evolves a reduced training set which is used in the evolution of the MOGA. The approach was tested on twelve datasets and the results show a reduction in FRBS execution time.

The work in [15], performed a set of experiments that are related to the training of a FRBS with different versions of the same dataset, reducing its size.

The authors evaluated the effectiveness of 36 training set selection methods when combined with genetic fuzzy rule-based classification systems. Their findings showed that the performance in test was affected by the usage of different reduced training sets, resulting in decrease of computational time and complexity of the fuzzy rule-based models, maintaining accuracy when compared to the models generated by using the overall training set.

In [25], the authors propose a new Genetic Fuzzy System, the Linguistic TSK, for obtaining accurate and simple linguistic fuzzy rule base models to solve regression problems. In order to reduce the complexity of the learned models while keeping a high accuracy, they propose a Genetic Fuzzy System which consists of three stages: instance selection, multi-granularity fuzzy discretization of the input variables and the evolutionary learning of the rule base. This proposal was validated using 28 datasets and the results show the approach obtains the simplest models while achieving a similar accuracy to the best approximated models. Recent approaches can also be found in [1,13,21,24], which apply Genetic Fuzzy Systems associated to IS and/or FS, presenting the benefits of data reduction.

Based on the presented review, in this paper we aim to verify if reduced datasets affect the building of the FRBS knowledge base, when two well-known algorithms, Wang-Mendel and NSGA-II, are used to do it.

3 Fuzzy Rule-Based Classification Systems

Classification is an important task found in various fields such as pattern recognition, decision making, data mining and modeling. The classification task consist of assign a class c_i from a set of classes $C = (c_1, c_2, ..., c_j)$ to an object e_p, which belongs to a set of objects $E = (e_1, e_2, ..., e_n)$, also named patterns or instances, described by the values of its attributes or features $F_p = (a_{p1}, a_{p2}, ..., a_{pm})$.

A Fuzzy Rule-Based Classification System (FRBCS) is a Fuzzy Rule-Based System where the rules were designed to solve a problem of classification, which contain linguistic variables defining the features in their antecedent, and a class in their consequent part. A classic fuzzy rule has the following structure:

$$R_k : \textbf{IF } X_1 \ is \ A_1 l_1 \textbf{ AND } X_2 \ is \ A_2 l_2 \textbf{ AND } ... \textbf{ AND } X_m \ is \ A_m l_T$$
$$\textbf{THEN } Class = c_i$$

where R_k is the rule identifier, $\mathbf{X} = (X_1, ..., X_m)$ are the features of the patterns set, $P_m = (A_{ml_1}, ..., A_{ml_T})$ is a fuzzy partition of T fuzzy sets on variable X_m, and $c_i \in C$ is the class.

In a FRBCS, the inference mechanism applies the set of fuzzy rules to an input example in order to determine the class it belongs to. The Classic Reasoning Method (CRM), proposed by [11], was used in this work.

4 Proposed Method

The aim of this work is evaluate a MOGA for learning FRBCS knowledge base as from the original and reduced datasets and its performance in comparison to the simple e frequently used Wang-Mendel algorithm. We intend to investigate whether the reduction data can improve accuracy and interpretability of the FRBCS in relation to models generated by using the original dataset. Therefore, we considered the construction of fuzzy system knowledge base in two situations: the first, using the Wang-Mendel algorithm to generate the rule base, considering the membership functions uniformly distributed inside the variable domain; and the second one, applying the MOGA NSGA-II for learning rule base and tuning of membership functions, simultaneously. The analysis is going to be done considering six approaches, three using Wang-Mendel algorithm and three others using MOGA, running them with overall and reduced datasets.

4.1 Chromosome Codification

The chromosomes codification is defined in two parts. The first one is related to Rule Base (RB), and the second part is related to membership functions (MF). So, the representation of one chromosome is defined by $C = C^{RB}C^{MF}$.

The codification of C^{RB} is composed of $s.(m + 1)$ natural numbers, where s is the number of rules and m is the number of antecedents of the rules. Each gene of a chromosome represents the fuzzy set index that is representing a feature, and the last gene represents the class of the problem. As an example, let a classification problem with two input variables, X_1 and X_2, and a output variables $c_i \in C = (c_1, c_2)$. Let $P_1 = (A_{1l_1}, A_{1l_2}, A_{1l_3})$ and $P_2 = (A_{2l_1}, A_{2l_2}, A_{2l_3})$ the partitions fuzzy of the variables X_1 and X_2, respectively, and the following rules:

$$R_1 = \textbf{IF } X_1 \textit{ is } A_{1l_3} \textbf{ AND } X_2 \textit{ is } A_{2l_1} \textbf{ THEN } Class = c_1$$
$$R_2 = \textbf{IF } X_1 \textit{ is } A_{1l_2} \textbf{ AND } X_2 \textit{ is } A_{2l_3} \textbf{ THEN } Class = c_1$$
$$R_3 = \textbf{IF } X_1 \textit{ is } A_{1l_1} \textbf{ THEN } Class = c_2$$

The Fig. 1 illustrates the codification of the C^{RB} part. The number zero represents the "don't care" condition.

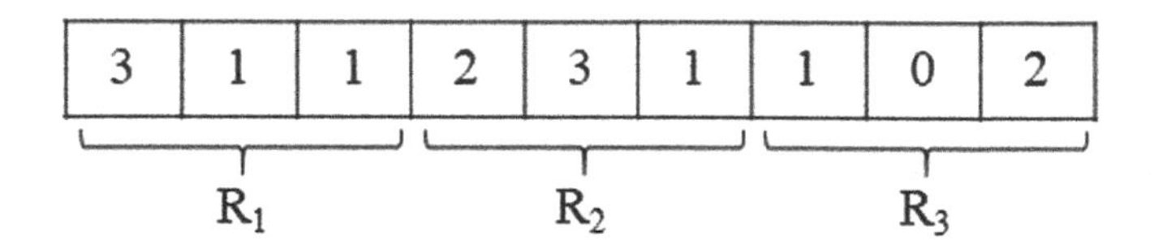

Fig. 1. Codification of the C^{RB}.

In relation to codification of the membership function parameters, there is one simple and widely used way, which uses all of them. The drawback is the greater the number of input variables, and consequently fuzzy sets, the greater the chromosome size. In other words, when dealing with high-dimensional datasets, the search space will be quite complex.

To deal with this problem, based on [5,6], we decided to represent each fuzzy set by only a value, more precisely, the core of fuzzy set. In this work, we adopt triangular fuzzy sets A_{ml_T} defined by the tuple $(a_{m,j}, b_{m,j}, c_{m,j})$, $j \in [1..T]$, where $a_{m,j}$ and $c_{m,j}$ correspond to the left and right extremes of the support of A_{ml_T}, and $b_{m,j}$ to the core. Further, we assume that $a_{m,1} = b_{m,1}$ and $b_{m,T} = c_{m,T}$, and for $j = 2...T-1, b_{m,j} = c_{m,j-1}$ and $b_{m,j} = a_{m,j+1}$. As $b_{m,1}$ and $b_{m,T}$ coincide with the extremes of the universe, the partition of each linguistic variable X_m is completely defined by $T-2$ real numbers. The Fig. 2 shows an uniform partition fuzzy of a variable with five fuzzy sets, and the Fig. 3 illustrates its respective chromosomal codification.

To take the "don't care" condition into account, a new fuzzy set A_{ml_0} is added to all the partitions $P_m, m = 1...M$. This fuzzy set is defined by a membership function equal to 1 on the overall universe.

To ensure a good integrity level of membership functions, in terms of order, coverage and distinguishability [17], $\forall j \in [2, T-1]$, we force $b_{m,j}$ to fluctuate in the interval $\left[b_{m,j} - \frac{b_{m,j}-b_{m,j-1}}{2}, b_{m,j} + \frac{b_{m,j+1}-b_{m,j}}{2}\right]$, as proposed in [5].

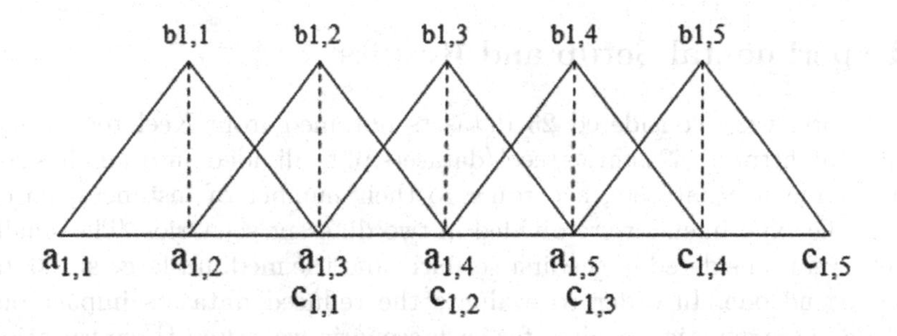

Fig. 2. Example of an uniform partition fuzzy.

$b_{1,2}$	$b_{1,3}$	$b_{1,4}$

Fig. 3. Codification of the C^{MF}.

4.2 Initial Population

The initial population is generated by a random process considering the interval mentioned in Sect. 4.1. Besides that, we use the Wang-Mendel algorithm [27] to generate a seed, that is, one chromosome generated by this algorithm is introduced into the initial population. The idea of using a seed is increase initial population quality and improve convergence speed [23].

4.3 Genetic Operators, Objective Functions and Stop Condition

In this work we use binary Tournament selection, Simulated Binary crossover (SBX) and Polynomial Mutation. In relation to objective functions, the NSGA-II has two objectives: the accuracy rate maximization (ACC) and the interpretability rate maximization (ITP), which are calculated for each fuzzy rule or chromosome, defined by (1) and (2), respectively:

$$ACC = \frac{ECC}{Total_{examples}} \tag{1}$$

$$ITP = \frac{Total_{rules} - Rules_{generated}}{Total_{rules}} \tag{2}$$

where ECC is the number of examples correctly classified, $Total_{examples}$ is the quantity of the examples, $Total_{rules}$ is the maximum number of rules, $Rules_{generated}$ is the number of rules generated by NSGA-II. The number of examples correctly classified is measured using the Classic Reasoning Method, proposed by [11], which is widely used in the literature.

5 Experimental Setup and Results

In this work, were considered 23 datasets obtained from Keel repository [2], available at http://sci2s.ugr.es/keel/datasets.php, divided into small size and medium-large size datasets, according to their number of instances, based on [15]. So, the experiments were divided in two different scenarios. The small size datasets were considered in the first scenario and the medium-large size datasets in the second one. In order to evaluate the reduced datasets impact on the knowledge base genetic learning, for each scenario we tested three variations of the datasets: overall datasets and both, instances and features reduced datasets, which were obtained from [7] and [12], respectively.

In addition, as the Wang-Mendel (WM) algorithm has been widely used to generate the rules of fuzzy systems [3,8,9,14], we evaluated the knowledge bases with rules generated by Wang-Mendel, and with rules and membership function parameters generated by NSGA-II. To facilitate understanding of the different combinations between learning algorithm and datasets, the abbreviations for each approach evaluated is described in Table 1.

The NSGA-II parameters were defined empirically, combining different values for each parameter. The values tested for the population size were 50, 100 and 150. The crossover rate was evaluated with the values 0.7, 0.8 and 0.9, and the mutation rate values were 0.5, 1.0 and 2.0. The stop condition was 100 generations, since the best chromosome does not improve its quality after that. Following the combination of the parameters, the best values obtained were population size = 100, crossover rate = 0.9, mutation rate = 0.5. The *ten-fold cross validation* approach was used. In addition, each fold was tested three times with different random seeds. Then, 30 runs were performed for each dataset and the results presented in next sections express the average of these 30 runs.

Table 1. Abbreviations of the approaches.

Abbreviation	Approach: rules generated by
WM-TOTAL	Wang-Mendel with overall datasets
NSGA2-TOTAL	NSGA-II with overall datasets
WS-IS	Wang-Mendel with instances reduced datasets
NSGA2-IS	NSGA-II with instances reduced datasets
WM-FS	Wang-Mendel with features reduced datasets
NSGA2-FS	NSGA-II with features reduced datasets

5.1 Small Datasets Results

In Tables 2 and 3 are shown average and standard deviation from accuracy and interpretability, respectively. The values in boldface means the best for each dataset. In according to Table 2, the NSGA2-TOTAL approach had the best results in seven datasets, NSGA2-IS had the best results in five datasets and NSGA2-FS was the best in only one. Taking into account the average over all datasets, the NSGA2-TOTAL was the best. In the other side, the results from Table 3 shows that NSGA2-FS was the best in all datasets, excepting in two datasets, which tied with WM-FS.

Table 2. Accuracy results from small datasets.

Dataset	WM-TOTAL	NSGA2-TOTAL	WM-IS	NSGA2-IS	WM-FS	NSGA2-FS
australian	0.8159 ± 0.0361	0.8159 ± 0.0361	0.8159 ± 0.0361	0.8101 ± 0.0557	0.5884 ± 0.0693	$\mathbf{0.8464} \pm 0.0354$
balance	0.6909 ± 0.0565	$\mathbf{0.8485} \pm 0.0423$	0.7022 ± 0.0475	0.8202 ± 0.0466	0.5330 ± 0.0606	0.7190 ± 0.0589
bupa	0.5316 ± 0.0615	0.6651 ± 0.0416	0.4988 ± 0.0684	$\mathbf{0.6738} \pm 0.0639$	0.5358 ± 0.0768	0.5498 ± 0.0729
ecoli	0.5624 ± 0.0784	$\mathbf{0.7410} \pm 0.0645$	0.5716 ± 0.0862	0.7176 ± 0.0524	0.4693 ± 0.1490	0.5563 ± 0.1445
glass	0.5342 ± 0.0983	$\mathbf{0.6592} \pm 0.0785$	0.5408 ± 0.1015	0.6347 ± 0.0881	0.3848 ± 0.0745	0.5025 ± 0.1138
haberman	0.7316 ± 0.0580	0.7303 ± 0.0621	0.6895 ± 0.0743	$\mathbf{0.7466} \pm 0.0593$	0.3868 ± 0.2197	0.7205 ± 0.0544
heart	0.7852 ± 0.0593	0.7852 ± 0.0593	0.7519 ± 0.0861	$\mathbf{0.8148} \pm 0.0370$	0.6741 ± 0.1109	0.7543 ± 0.0920
hepatitis	0.8274 ± 0.1380	$\mathbf{0.8771} \pm 0.0935$	0.8017 ± 0.1425	0.8554 ± 0.0927	0.5458 ± 0.3054	0.7984 ± 0.1134
iris	0.9133 ± 0.0521	$\mathbf{0.9511} \pm 0.0453$	0.9000 ± 0.0537	0.9289 ± 0.0708	0.8519 ± 0.0421	0.9351 ± 0.0404
newthyroid	0.8695 ± 0.0969	$\mathbf{0.9258} \pm 0.0426$	0.8742 ± 0.0824	0.8726 ± 0.0768	0.8193 ± 0.0768	0.8563 ± 0.0582
pima	0.6057 ± 0.0586	$\mathbf{0.7484} \pm 0.0395$	0.6200 ± 0.0669	0.7211 ± 0.0607	0.7267 ± 0.0357	0.7319 ± 0.0497
wine	0.8925 ± 0.0608	0.9059 ± 0.0520	0.9147 ± 0.0797	$\mathbf{0.9490} \pm 0.0400$	0.6513 ± 0.1170	0.7343 ± 0.1269
wisconsin	0.9430 ± 0.0311	0.9430 ± 0.0311	0.9430 ± 0.0311	$\mathbf{0.9487} \pm 0.0258$	0.8540 ± 0.0629	0.9395 ± 0.0222
average	0.7464 ± 0.0681	$\mathbf{0.8151} \pm 0.0530$	0.7403 ± 0.0736	0.8072 ± 0.0592	0.6170 ± 0.1077	0.7419 ± 0.0756

In order to verify if there is statistical difference among the approaches results, we first have applied the Friedman test. If it exists, then we applied a post-hoc procedure, namely the Bonferroni test. For both tests, we let level of significance = 0.05.

Applying the Friedman test on the accuracy and interpretability respectively, we achieved $[X^2(5) = 35.762; p < 0.001]$ and $[X^2(5) = 58.489; p < 0.001]$. As p-values are lower than the level of significance, we can reject the null hypothesis and affirm that there exist statistical differences. The Table 4 shows the adjusted p-values obtained by applying the Bonferroni tests on accuracy and interpretability. The results in boldface means there are not statistical differences.

Table 3. Interpretabiliy results from small datasets.

Dataset	WM-TOTAL	NSGA2-TOTAL	WM-IS	NSGA2-IS	WM-FS	NSGA2-FS
australian	0.4824 ± 0.0054	0.4824 ± 0.0054	0.3370 ± 0.0163	0.3491 ± 0.0370	0.9815 ± 0.0115	$\mathbf{0.9963} \pm 0.0008$
balance	0.8562 ± 0.0006	0.9834 ± 0.0176	0.7078 ± 0.0178	0.9707 ± 0.0138	0.9584 ± 0.0128	$\mathbf{0.9941} \pm 0.0012$
bupa	0.8473 ± 0.0038	0.9696 ± 0.0177	0.7617 ± 0.0260	0.9392 ± 0.0365	0.9810 ± 0.0064	$\mathbf{0.9933} \pm 0.0008$
ecoli	0.7536 ± 0.0049	0.8362 ± 0.0446	0.6306 ± 0.0274	0.7108 ± 0.0356	0.9128 ± 0.0193	$\mathbf{0.9709} \pm 0.0025$
glass	0.7793 ± 0.0080	0.9227 ± 0.0309	0.6527 ± 0.0459	0.8619 ± 0.0331	0.9554 ± 0.0132	$\mathbf{0.9875} \pm 0.0034$
haberman	0.9361 ± 0.0018	0.9887 ± 0.0034	0.8796 ± 0.0112	0.9775 ± 0.0074	0.9840 ± 0.0078	$\mathbf{0.9921} \pm 0.0016$
heart	0.1058 ± 0.0074	0.1058 ± 0.0074	0.0498 ± 0.0174	0.1721 ± 0.1626	0.9288 ± 0.0481	$\mathbf{0.9826} \pm 0.0271$
hepatitis	0.0806 ± 0.0148	0.8319 ± 0.0873	0.0234 ± 0.0258	0.8754 ± 0.1230	0.9167 ± 0.0394	$\mathbf{0.9695} \pm 0.0099$
iris	0.8919 ± 0.0036	0.9719 ± 0.0114	0.7419 ± 0.0264	0.9234 ± 0.0114	$\mathbf{0.9778} \pm 0.0000$	$\mathbf{0.9778} \pm 0.0000$
newthyroid	0.8941 ± 0.0025	0.9793 ± 0.0076	0.8146 ± 0.0233	0.9506 ± 0.0211	$\mathbf{0.9845} \pm 0.0000$	$\mathbf{0.9845} \pm 0.0000$
pima	0.8397 ± 0.0027	0.9782 ± 0.0297	0.7554 ± 0.0116	0.9346 ± 0.0599	0.9841 ± 0.0065	$\mathbf{0.9953} \pm 0.0013$
wine	0.2178 ± 0.0098	0.4128 ± 0.1988	0.0890 ± 0.0428	0.8508 ± 0.0581	0.9657 ± 0.0135	$\mathbf{0.9800} \pm 0.0037$
wisconsin	0.6231 ± 0.0037	0.6231 ± 0.0037	0.5838 ± 0.0236	0.8042 ± 0.0635	0.9873 ± 0.0039	$\mathbf{0.9957} \pm 0.0010$
average	0.6391 ± 0.0053	0.7758 ± 0.0358	0.5406 ± 0.0243	0.7939 ± 0.0510	0.9652 ± 0.0140	$\mathbf{0.9861} \pm 0.0041$

Table 4. Statistical results from small datasets.

Approach 1 vs Approach 2	ACC	ITP
WM-FS vs WM-IS	0.239	0.000
WM-FS vs WM-TOTAL	0.206	0.007
WM-FS vs **NSGA2-FS**	0.178	**1.000**
WM-FS vs NSGA2-IS	0.000	0.416
WM-FS vs NSGA2-TOTAL	0.000	1.000
WM-IS vs WM-TOTAL	1.000	1.000
WM-IS vs NSGA2-FS	1.000	0.000
WM-IS vs NSGA2-IS	0.316	0.035
WM-IS vs **NSGA2-TOTAL**	**0.082**	0.003
WM-TOTAL vs NSGA2-FS	1.000	0.000
WM-TOTAL vs NSGA2-IS	0.363	1.000
WM-TOTAL vs **NSGA2-TOTAL**	**0.096**	0.696
NSGA2-FS vs NSGA2-IS	0.416	0.004
NSGA2-FS vs **NSGA2-TOTAL**	**0.113**	0.042
NSGA2-IS vs **NSGA2-TOTAL**	**1.000**	1.000

Taking into account the results of the Tables 2 and 4, we can see that the best approach is the NSGA2-TOTAL, with a value on average 0.8151, and the worst is the WM-FS. However, analyzing the statistical difference among the approaches, we found that NSGA2-TOTAL only differs from WM-FS. As the NSGA2-TOTAL is the best one (considering the Table 2), but it doesn't have statistical difference to other approaches, and the Wang-Mendel is simpler than NSGA-II to build the rules, we can conclude that when it comes to the accuracy for small datasets, the WM-TOTAL approach is the most suitable.

Regarding interpretability, we found that the best average is given by the NSGA2-FS approach (0.9861), and the worst by the WM-IS. These values can be seen in Table 3. However, observing the Table 4, the NSGA2-FS approach has no significant statistical difference to the WM-FS. Once more, as the Wang-Mendel is simpler than NSGA-II to build the rules, the WM-FS approach is the most appropriate.

5.2 Medium-Large Datasets Results

To medium-large datasets, the results are presented in Tables 5 and 6. The values in boldface means the best results for each dataset. In according to Table 5, NSGA2-FS was the best approach in accuracy, with a value on average 0.8510. In case of interpretability, NSGA2-IS was the best approach in nine datasets with a value on average 0.9950, how we can see in Table 6.

Considering medium and large datasets, Friedman test found $[X^2(5) = 31.724;\ p < 0.001]$ on the accuracy and $[X^2(5) = 39.483; p < 0.001]$ on the interpretability, for significance = 0.05. The p-values are lower than the level of significance, so we also reject the null hypothesis and apply Bonferroni tests. Table 7 shows these results, highlighting in boldface those that have no statistical difference.

Table 5. Accuracy results from medium-large datasets.

Dataset	WM-TOTAL	NSGA2-TOTAL	WM-IS	NSGA2-IS	WM-FS	NSGA2-FS
banana	0.5719 ± 0.0175	0.6762 ± 0.0402	0.5151 ± 0.0615	0.6389 ± 0.0528	0.5662 ± 0.0207	**0.6970** ± 0.0454
magic	0.6845 ± 0.0124	**0.8189** ± 0.0094	0.7131 ± 0.0180	0.7865 ± 0.0107	0.7171 ± 0.0563	0.8169 ± 0.0094
page-blocks	0.6813 ± 0.0433	**0.9454** ± 0.0038	0.7852 ± 0.2617	0.9273 ± 0.0149	0.7825 ± 0.0333	0.9394 ± 0.0094
penbased	**0.9552** ± 0.0053	**0.9552** ± 0.0053	0.6214 ± 0.1136	0.6409 ± 0.0954	0.9533 ± 0.0070	0.9533 ± 0.0070
phoneme	0.6976 ± 0.0162	**0.7805** ± 0.0205	0.6344 ± 0.1338	0.7472 ± 0.0298	0.6743 ± 0.0750	0.7772 ± 0.0170
ring	0.8066 ± 0.0085	0.8073 ± 0.0145	0.6103 ± 0.0615	0.7641 ± 0.0289	0.8073 ± 0.0154	**0.8095** ± 0.0145
segment	0.8056 ± 0.0148	0.8398 ± 0.0238	0.4983 ± 0.0907	0.6182 ± 0.1567	0.7844 ± 0.0231	**0.8667** ± 0.0293
thyroid	0.4806 ± 0.0239	0.9247 ± 0.0072	0.9299 ± 0.0057	0.9371 ± 0.0060	0.5715 ± 0.0741	**0.9381** ± 0.0065
titanic	0.5770 ± 0.0814	**0.7783** ± 0.0289	0.6770 ± 0.0009	0.7760 ± 0.0306	0.7460 ± 0.0369	0.7671 ± 0.0318
twonorm	0.7518 ± 0.0188	0.9438 ± 0.0155	0.6596 ± 0.0303	0.8250 ± 0.0230	0.7870 ± 0.0273	**0.9445** ± 0.0155
average	0.7012 ± 0.0242	0.8470 ± 0.0169	0.6644 ± 0.0778	0.7661 ± 0.0449	0.7390 ± 0.0369	**0.8510** ± 0.0186

Table 6. Interpretability results from medium-large datasets.

Dataset	WM-TOTAL	NSGA2-TOTAL	WM-IS	NSGA2-IS	WM-FS	NSGA2-FS
banana	0.9983 ± 0.0001	0.9994 ± 0.0001	0.9988 ± 0.0005	**0.9995** ± 0.0001	0.9846 ± 0.0006	0.9947 ± 0.0011
magic	0.9793 ± 0.0002	0.9997 ± 0.0002	0.9992 ± 0.0003	**0.9998** ± 0.0001	0.9368 ± 0.0020	0.9979 ± 0.0010
page-blocks	0.9886 ± 0.0003	0.9988 ± 0.0000	0.9983 ± 0.0004	**0.9990** ± 0.0000	0.9477 ± 0.0056	0.9854 ± 0.0070
penbased	0.6624 ± 0.0013	0.6624 ± 0.0013	0.9835 ± 0.0141	**0.9850** ± 0.0135	0.3241 ± 0.0146	0.3241 ± 0.0146
phoneme	0.9880 ± 0.0002	0.9992 ± 0.0003	0.9971 ± 0.0010	**0.9995** ± 0.0001	0.9450 ± 0.0058	0.9919 ± 0.0025
ring	0.8378 ± 0.0012	0.9991 ± 0.0000	0.9929 ± 0.0023	**0.9988** ± 0.0005	0.7266 ± 0.0151	0.9932 ± 0.0001
segment	0.8452 ± 0.0018	0.8870 ± 0.0145	0.9890 ± 0.0079	**0.9961** ± 0.0004	0.4019 ± 0.0347	0.9043 ± 0.0152
thyroid	0.9282 ± 0.0006	0.9636 ± 0.0120	0.9979 ± 0.0006	**0.9995** ± 0.0000	0.7960 ± 0.0076	0.9893 ± 0.0121
titanic	0.9950 ± 0.0002	0.9985 ± 0.0005	0.9988 ± 0.0006	**0.9990** ± 0.0001	0.8908 ± 0.0065	0.9641 ± 0.0082
twonorm	0.7688 ± 0.0010	**0.9995** ± 0.0001	0.9646 ± 0.0096	0.9743 ± 0.0143	0.5861 ± 0.0149	0.9950 ± 0.0007
average	0.8992 ± 0.0007	0.9507 ± 0.0029	0.9920 ± 0.0037	**0.9950** ± 0.0029	0.7539 ± 0.0107	0.9140 ± 0.0063

Assessing this statistical evaluation, looking at the accuracy, we found that the NSGA2-FS, the best approach according Table 5, is statistically different from all approaches that use the Wang-Mendel, but it is statistically similar to the other genetic approaches. Considering that the average number of generations was 421 for the NSGA2-TOTAL, 727 for the NSGA2-FS and 206 for the NSGA2-IS, we can conclude that the NSGA2-IS approach is the most recommended. Finally, analyzing interpretability, the p-values for the NSGA2-IS (the approach that presented the best numerical results), NSGA2-TOTAL and WM-IS approaches allow us to observe that there is not statistically significant difference among them. So, by the same principle of previous evaluations, the WM-IS approach is the most suitable.

Table 7. Statistical results from medium-large datasets.

Approach 1 vs Approach 2	ACC	ITP
WM-IS vs WM-TOTAL	1.000	0.547
WM-IS vs WM-FS	1.000	0.002
WM-IS vs **NSGA2-IS**	0.347	**1.000**
WM-IS vs NSGA2-TOTAL	0.000	1.000
WM-IS vs NSGA2-FS	0.000	1.000
WM-TOTAL vs WM-FS	1.000	1.000
WM-TOTAL vs NSGA2-IS	1.000	0.002
WM-TOTAL vs NSGA2-TOTAL	0.012	0.181
WM-TOTAL vs NSGA2-FS	0.012	1.000
WM-FS vs NSGA2-IS	1.000	0.000
WM-FS vs NSGA2-TOTAL	0.042	0.000
WM-FS vs NSGA2-FS	0.042	0.252
NSGA2-IS vs NSGA2-TOTAL	0.729	**1.000**
NSGA2-IS vs **NSGA2-FS**	**0.729**	0.023
NSGA2-TOTAL vs **NSGA2-FS**	**1.000**	1.000

6 Conclusion

In this paper we have performed a study of the use of MOGA for learning FRBCS knowledge base as from the original and reduced datasets and we compared its performance to Wang-Mendel algorithm. We investigated if the data reduction can improve accuracy and interpretability of the generated FRBCS. The analysis was carried out by considering six approaches, three using Wang-Mendel algorithm to generate the rule base and three using NSGA-II to learn the knowledge base (membership functions and rule base). Wang-Mendel algorithm and NSGA-II were tested considering original datasets and reduced datasets, by instances and features.

To this aim, 23 datasets of different sizes have been used and the numerical and statistical results showed that, to small datasets, the Wang-Mendel algorithm is the most suitable, given that is simpler, less costly and present similar results to NSGA-II. Related to accuracy, the results pointed to the use of original dataset. IS and FS are not necessary. In case of interpretability, the best results were found using FS for both, Wang-Mendel and NSGA-II. For medium and large datasets, regarding accuracy, the results demonstrated the best approaches were those which used NSGA-II, highlighting IS as a promising technique to data reducing. Considering interpretability, the results suggested the use of Wang-Mendel algorithm also associated to IS technique. Therefore, we can conclude that data reduce techniques improve the accuracy and interpretability results for medium and large datasets. In general, IS and FS allow to obtain a smaller dataset, but similar in performance classification comparing to the original dataset. We also observed that MOGA for FRBCS learning is more suitable for large datasets. For small datasets, the Wang-Mendel algorithm is sufficient.

References

1. Alamaniotis, M., Jevremovic, T.: Hybrid fuzzy-genetic approach integrating peak identification and spectrum fitting for complex gamma-ray spectra analysis. IEEE Trans. Nucl. Sci. **62** (2015). https://doi.org/10.1109/TNS.2015.2432098
2. Alcalá-Fdez, J., et al.: KEEL data-mining software tool: data set repository, integration of algorithms and experimental analysis framework. J. Multiple-Valued Logic Soft Comput. **17**(2–3), 255–287 (2011)
3. Antonelli, M., Ducange, P., Lazzerini, B., Marcelloni, F.: Multi-objective evolutionary generation of Mamdani fuzzy rule-based systems based on rule and condition selection. In: 2011 IEEE 5th International Workshop on Genetic and Evolutionary Fuzzy Systems (GEFS), pp. 47–53 (2011). https://doi.org/10.1109/GEFS.2011.5949489
4. Antonelli, M., Ducange, P., Marcelloni, F.: Multi-objective evolutionary rule and condition selection for designing fuzzy rule-based classifiers. In: 2012 IEEE International Conference on Fuzzy Systems, pp. 1–7 (2012). https://doi.org/10.1109/FUZZ-IEEE.2012.6251174
5. Antonelli, M., Ducange, P., Marcelloni, F.: An efficient multi-objective evolutionary fuzzy system for regression problems. Int. J. Approx. Reason. **54**(9), 1434–1451 (2013). https://doi.org/10.1016/j.ijar.2013.06.005
6. Antonelli, M., Ducange, P., Marcelloni, F.: Genetic training instance selection in multiobjective evolutionary fuzzy systems: a coevolutionary approach. IEEE Trans. Fuzzy Syst. **20**(2), 276–290 (2012). https://doi.org/10.1109/TFUZZ.2011.2173582
7. Bertoni, F.C., Pires, M.G.: Aplicação de algoritmos evolutivos multiobjetivo na seleção de instâncias. In: Simpósio Brasileiro de Sistemas de Informação, Lavras, MG, pp. 261–268 (2017)
8. Cardoso, M., Loula, A., Pires, M.G.: Automated fuzzy system based on feature extraction and selection for opinion classification across different domains. Int. J. Uncertain. Fuzziness Knowl.-Based Syst. **24**, 93–122 (2016)
9. Cintra, M.E., Monard, M.C., Camargo, H.A.: Data base definition and feature selection for the genetic generation of fuzzy rule bases. Evol. Syst. **1**(4), 241–252 (2010). https://doi.org/10.1007/s12530-010-9018-6

10. Cordon, O., Herrera, F., Hoffmann, F., Magdalena, L.: Genetic Fuzzy Systems: Evolutionary Tuning and Learning of Fuzzy Knowledge Bases, vol. 19 (2001). https://doi.org/10.1142/4177
11. Cordón, O., del Jesus, M.J., Herrera, F.: A proposal on reasoning methods in fuzzy rule-based classification systems. Int. J. Approx. Reason. **20**(1), 21–45 (1999). https://doi.org/10.1016/S0888-613X(00)88942-2
12. Correia, M.G., Bertoni, F.C.: Seleção de características utilizando um algoritmo genético multiobjetivo. In: IV Workshop de Iniciação Científica em Sistemas de Informação, Lavras, MG, pp. 37–40 (2017)
13. Darwish, S.: Uncertain measurement for student performance evaluation based on selection of boosted fuzzy rules. IET Sci. Meas. Technol. **11** (2016). https://doi.org/10.1049/iet-smt.2016.0265
14. de Castro Ribeiro, M.G., et al.: Detection and classification of faults in aeronautical gas turbine engine: a comparison between two fuzzy logic systems. In: 2018 IEEE International Conference on Fuzzy Systems (FUZZ-IEEE), pp. 1–7 (2018). https://doi.org/10.1109/FUZZ-IEEE.2018.8491444
15. Fazzolari, M., Giglio, B., Alcalá, R., Marcelloni, F., Herrera, F.: A study on the application of instance selection techniques in genetic fuzzy rule-based classification systems: accuracy-complexity trade-off. Knowl.-Based Syst. **54**(C), 32–41 (2013). https://doi.org/10.1016/j.knosys.2013.07.011
16. Fernández, A., López, V., José del Jesus, M., Herrera, F.: Revisiting evolutionary fuzzy systems: taxonomy, applications, new trends and challenges. Knowl.-Based Syst. **80**, 109–121 (2015). https://doi.org/10.1016/j.knosys.2015.01.013
17. Gacto, M.J., Alcalá, R., Herrera, F.: Interpretability of linguistic fuzzy rule-based systems: an overview of interpretability measures. Inf. Sci. **181**(20), 4340–4360 (2011). https://doi.org/10.1016/j.ins.2011.02.021
18. García-Pedrajas, N., de Haro-García, A., Pérez-Rodríguez, J.: A scalable approach to simultaneous evolutionary instance and feature selection. Inf. Sci. **228**, 150–174 (2013). https://doi.org/10.1016/j.ins.2012.10.006
19. García-Pedrajas, N., Pérez-Rodríguez, J.: Multi-selection of instances: a straightforward way to improve evolutionary instance selection. Appl. Soft Comput. **12**(11), 3590–3602 (2012). https://doi.org/10.1016/j.asoc.2012.06.013
20. Giglio, B., Marcelloni, F., Fazzolari, M., Alcala, R., Herrera, F.: A case study on the application of instance selection techniques for genetic fuzzy rule-based classifiers. In: IEEE International Conference on Fuzzy Systems, pp. 1–8 (2012). https://doi.org/10.1109/FUZZ-IEEE.2012.6251191
21. Gorbunov, I., Subhankulova, S.R., Hodashinsky, I., Yankovskaya, A.: Comparative analysis of feature selection algorithms in construction of fuzzy classifiers. In: IEEE 10th International Conference on Application of Information and Communication Technologies (AICT), pp. 1–3 (2016). https://doi.org/10.1109/ICAICT.2016.7991669
22. Herrera, F.: Genetic fuzzy systems: taxonomy, current research trends and prospects. Evol. Intell. **1**, 27–46 (2008)
23. Mirshekarian, S., Süer, G.A.: Experimental study of seeding in genetic algorithms with non-binary genetic representation. J. Intell. Manuf. **29**(7), 1637–1646 (2018). https://doi.org/10.1007/s10845-016-1204-3
24. Ravindranath, V., Ra, S., Ramasubbareddy, S., Remya, S., Nalluri, S.: Genetic algorithm based feature selection and MOE Fuzzy classification algorithm on Pima Indians Diabetes dataset, pp. 1–5 (2017). https://doi.org/10.1109/ICCNI.2017.8123815

25. Rodriguez-Fdez, I., Mucientes, M., Bugarín, A.: Reducing the complexity in genetic learning of accurate regression TSK rule-based systems, pp. 1–8 (2015). https://doi.org/10.1109/FUZZ-IEEE.2015.7337930
26. Tsai, C.F., Chen, Z.Y., Ke, S.W.: Evolutionary instance selection for text classification. J. Syst. Softw. **90**, 104–113 (2014). https://doi.org/10.1016/j.jss.2013.12.034
27. Wang, L.X., Mendel, J.M.: Generating fuzzy rules by learning from examples. IEEE Trans. Fuzzy Syst. Man Cybern. **22**(6), 1414–1427 (1992)

Towards a Theory of Hyperintensional Belief Change

Marlo Souza(✉)

Institute of Mathematics and Statistics, Federal University of Bahia - UFBA, Av. Adhemar de Barros, S/N, Ondina, Salvador, BA, Brazil
msouza1@ufba.br

Abstract. AGM's belief revision is one of the main paradigms in the study of belief change operations. Despite its popularity and importance to the area, it is well recognized that AGM's work relies on a strong idealisation of the agent's capabilities and on the nature of beliefs themselves. Particularly, it is well-recognized in the area of Epistemology that Belief and Knowledge are hyperintensional notions, but to our knowledge only a few works have explicitly considered how hyperintensionality affects belief change. In this work, we investigate belief change operations based on Berto's topic-sensitive framework and provide three different pseudo-contraction operations to account for the hyperintensional behaviour of beliefs. Our work highlights the connection of a foundational hyperintensional theory of belief with the results of AGM Belief Change. Also we propose and characterise different possible contraction-like operations in this framework.

1 Introduction

Belief Change is the area that studies how doxastic agents change their minds after acquiring new information. Currently, the most influential approach to Belief Change in the literature is the AGM paradigm [1]. Although the AGM approach and hypotheses have been questioned in the literature [11,32], it has led to profound developments for the problem of belief dynamics, influencing areas such as Computer Science, Artificial Intelligence, and Philosophy.

One particular criticism against the AGM approach which has received a great deal of attention concerns the idealised nature of doxastic agents in their work [21,23,33]. Namely, given the AGM postulates which govern rational belief change, the belief state of an agent is characterised by a consequentially-closed[1] set of beliefs, and that belief revision is an intensional operator, i.e. based on the language semantics/proof theory. In fact, Gärdenfors [16, p. 9], one of the original authors of the seminal AGM work, acknowledges that AGM's notion of belief is but merely an idealisation "judged in relation to the rationality criteria for the epistemological theory".

[1] While AGM [1] does not adhere to any specific logical language, the authors assume the logic is supraclassical, meaning it proves all truths of classical logic.

R. Cerri and R. C. Prati (Eds.): BRACIS 2020, LNAI 12320, pp. 272–287, 2020.
https://doi.org/10.1007/978-3-030-61380-8_19

This becomes particularly clear, in our opinion, in Hansson's [22] criticism of the use of consequentially-closed belief sets as models of the agent's belief state. Hansson argues that on a dynamic level, the agent's belief state depends not only on the meaning of her beliefs but also on something else, which the author identifies with its syntactic structure. More yet, it has been argued in the literature [7,20,24,38] that resource-bounded agents are not required to believe all consequences of their currently held beliefs, even if they are logically capable of rational inquiry, since an agent can fail to reach the conclusion of a reasoning process due to a lack of cognitive resources. The aim of this work is to investigate belief change operations that take into consideration these cognitive characteristics of resource-bounded agents.

We call attitudes, such as belief, that depend not only on the sentential intension but also on its hyperintensional contexts of *hyperintensional attitudes* [9]. By hyperintensional context, we mean some component of its contents (or meaning), other than its sentential intensions, which influences the truth-conditions of such hyperintensional attitudes [35], i.e. these attitudes can draw distinctions between necessarily equivalent contents. For example, while the sentences "*3 is a prime number*" and "*3068 is divisible by 13*" have the same intension, as logical necessities, they certainly cannot be transparently substituted for the other in the sentence "Alice believes that *3 is a prime number.*"

In the literature, different treatments for the hyperintensional nature of beliefs have been proposed [9,27,38]. One particular approach of interest in this work aims to explicitly represent the structure of the agent's beliefs as an important component of its meaning, e.g. [5,10,26]. In special, Berto [5] proposes a hyperintensional notion of belief based not directly on the structure of formulas, but on what formulas *talk about* - by means of a mereological theory of topics and an *aboutness* relation between formulas and topics. Thus a topic-sensitive theory of beliefs.

In this work, we will investigate different proposals for a hyperintensional belief contraction operations, and their axiomatic characterisation by postulates, based on Berto's [5] notion of topic-sensitive belief. With that, we wish to establish connections between the work of hyperintensional models for beliefs, such as that of Berto, with the work on AGM Belief Change. As such, we introduce in Sect. 4 topic-sensitive belief contractions and later, in Sect. 5, we generalise this treatment for more general hyperintensional notions.

2 Related Work

Extensive work has been published on general notions of belief change not constrained by the laws of classical logic, such as the work on belief change for non-classical [12,15,30,31], paraconsistent [18] or substructural logics [3]. We will focus on work which are connected to, or can be used to study, hyperintensional notions of belief change.

Girard and Tanaka [18] have investigated dynamic belief change operators for many-valued logics, which in principle could be used to model some hyperintensional notions, as classical consequences need not to be valid in such logics.

Interestingly, the authors show that for the logic of paradoxes investigated in the work, the behaviour of two epistemic actions over the agent's belief state coincide with the their behaviour for the classical logic of conditional beliefs, as investigated by Van Benthem [37].

While we believe this work proposes an interesting and general framework for non-classical notions of belief and change, including *some* notion of hyperintensionality, it is not clear how hyperintensional contexts can be encoded in it. Particularly, since their models are based on intensional interpretations of connectives, it is not directly clear how to construct (a class of) models that can differentiate two (classically) equivalent formulas φ and ψ without degenerating the interpretation of these connectives. More yet, it seems clear that given a support logic $\mathcal{L}$, their notion of belief is not hyperintensional within their logic, i.e. if formulas φ and ψ are equivalent in $\mathcal{L}$, then the agent believes in φ iff they believe in ψ.

Berto [5] proposes a topic-sensitive hyperintensional logic of conditional beliefs, in which belief revision is interpreted as conditionalisation. In this logic, the author shows that the notion of conditional belief satisfies minimal desiderata for logics of belief change [8]. To our knowledge, this is the first work explicitly proposing the integration of hyperintensional phenomena within a theory of belief change and, in fact, it constitutes the main inspiration for our work.

To define the notion of topic-sensitive conditional belief, Berto employs a possible-worlds model extended with a mereological structure of topics or contents, to which the elements of the language refer by means of a content-describing function. The hyperintensional nature of beliefs in his theory - and of belief change, encoded by conditional beliefs - comes thus from the *aboutness* relation between formulas and topics, and how this can be explored to define which beliefs are achievable by means of some information. It is not clear in his work, however, how this theory is connected with abstract Belief Change Theory results, as the characterisation of constructibility of such operators for non-classical logics [30]. Our work investigates precisely how this topic-sensitive notion of belief change is connected to the well-known results in the literature.

It is interesting to notice that, since impossible-world semantics has been a popular approach to model for hyperintensional beliefs [9,27], work on belief change based on impossible-world semantics can be connected to our work and, in fact, may present tools and properties with which we can evaluate our contribution.

Badura and Berto [4] propose the use an impossible-world semantics framework to overcome several limitation of Lewis' modal analyses of *Truth in Fiction*. Their modelling is interesting because it can be used to represent paraconsistent and hyperintensional notions of belief revision, based on counterfactual reasoning. While their interpretation of the notion of *truth in fiction* statements is conceptually connected to belief revision, the authors also do not establish the connection with Belief Chang theory, as this was not their goal in the first place.

Fermé and Wassermann [13] have also proposed the use of impossible worlds semantics as a framework to study iterated belief expansion. The authors extend

Grove's models [19] for classical propositional logic with one impossible world $w_\bot$ satisfying all formulas of the language. This world is, however, included in the model merely as a technical device in order to represent the belief state in which the agent has inconsistent belief, and not to model any kind of hyperintensional property of beliefs. If their approach was to be extended, however, we believe such a framework can be used to define hyperintensional notions of revision, contraction and expansion, based on the authors encoding of such operations.

Perhaps the work closest to ours is that of Santos et al. [34] on pseudo-contractions. Given a logic $\mathcal{L}$, the authors investigate how these operations operators can be defined for sublogics of $\mathcal{L}$, thus studying how contraction-like operators can be defined for consequence operators which can take into consideration (possibly) hyperintentional differences between formulas in the reasoning process. While their work does investigate the connections between hyperintensional beliefs and belief change, it does not do so explicitly. Our work highlights exactly the connection to a foundational hyperintensional theory of belief - based on the notion of *aboutness*. Also we propose and characterise different possible contraction-like operations in this framework.

3 Preliminaries

Let us consider a logic $\mathcal{L} = \langle L, Cn \rangle$ where L is the logical language and $Cn : 2^L \to 2^L$ is a consequence operator. In AGM's approach, the belief state of an agent is represented by a belief set, i.e. a consequentially closed set $K \subseteq L$ of $\mathcal{L}$-formulas, i.e. $K = Cn(K)$.

On a belief set, the AGM authors investigated three basic belief change operators: expansion, contraction and revision. Belief expansion blindly integrates a new piece of information into the agent's beliefs. Belief contraction removes a currently believed sentence from the agent's set of beliefs, with minimal alterations. Finally, belief revision is the operation of integrating new information into an agent's beliefs while maintaining consistency.

Among these basic operations, only expansion can be univocally defined. The other two operations are defined by a set of rational constraints or postulates, usually referred to as the AGM postulates or the Gärdenfors postulates. These postulates define a class of suitable change operators which represent different rational ways in which an agent can change her beliefs. Let $K \subseteq L$ be a belief set and $A, B \subseteq \mathcal{L}$ finitely definable sets of formulas. An operation $- : 2^L \times 2^L \to 2^L$ is called a contraction operation if it satisfies the following postulates[2]:

(closure) $K - A = Cn(B - A)$
(success) $A \nsubseteq K - A$
(inclusion) If $K - A \subseteq K$
(vacuity) If $A \nsubseteq K$ then $K - A = K$
(recovery) $K \subseteq Cn(K - A \cup A)$

[2] In the following, we will adopt Ribeiro and Wassermann's [29] generalisation of the AGM postulates, which correspond to a choice multiple contraction.

(extensionality) If $Cn(A) = Cn(B)$ then $K - A = K - B$

AGM require the foundational logic $\mathcal{L}$ to satisfy some properties in order for belief change operations to be definable. These properties are the following:

- **inclusion**: $\Gamma \subseteq Cn(\Gamma)$.
- **idempotence**: $Cn(\Gamma) - Cn(Cn(\Gamma))$.
- **monotonicity**: If $\Gamma \subseteq \Gamma'$ then $Cn(\Gamma) \subseteq Cn(\Gamma')$.
- **tarskianicity**: If Cn satisfies inclusion, idempotence and monotonicity.
- **compactness**: for any $\varphi \in Cn(\Gamma)$, there is some finite $\Gamma' \subseteq \Gamma$ s.t. $\varphi \in Cn(\Gamma')$.
- **supraclassicality**: Let Cn_0 be the classical logic consequence operator, $Cn_0(\Gamma) \subseteq Cn(\Gamma)$ for any $\Gamma \subseteq L$.
- **deduction theorem**: If $\beta \in Cn(\Gamma \cup \{\alpha\})$ then $\alpha \rightarrow \beta \in Cn(\Gamma)$.

To construct contraction operations, AGM [1] introduce the notion of partial meet contractions. Let $K, A \subseteq L$ be sets of formulas, the remainder set $K \bot A$ is the set of sets K' satisfying:

- $K' \subseteq K$
- $A \nsubseteq Cn(K')$
- $K' \subset K'' \subseteq K$ implies $A \subseteq Cn(K'')$

A partial meet contraction $-$ is an operation for which there is a function γ s.t. for any K and A

$$K - A = \bigcap \gamma(K \bot A)$$

and γ satisfies: $\gamma(K \bot A) \subseteq K \bot A$ and $\gamma(K \bot A) = \varnothing$ iff $K \bot A = \varnothing$. AGM [1] show that an operation $-$ is a partial meet contraction if, and only if, it is a contraction operation. Further, Hansson and Wasserman [23] have shown that for any monotonic and compact logic, an operation $-$ is a partial meet contraction if and only if it satisfies

(**success**) $A \nsubseteq K - A$
(**inclusion**) If $K - A \subseteq K$
(**relevance**) If $\beta \notin K \backslash K - A$, then there is some $K - a \subseteq B' \subseteq K$ s.t. $A \nsubseteq Cn(B')$ but $A \subseteq Cn(B' \cup \{\beta\})$
(**uniformity**) If for any $K' \subseteq K$ $A \subseteq Cn(K')$ iff $B \subseteq Cn(K')$, then $K - A = K - B$

This study has spammed several investigations on the generalizability of AGM belief change to other logics, such as Horn Logics [12], Description Logics [14,29], non-compact logics [28]. In this work we will explore these characterisability results which correlate properties of the foundational logics and the relation between postulates and constructions. In the following sections, we will investigate how the previous postulates and constructions need to be altered in order to account for the hyperintensional nature of beliefs.

4 Topic-Sensitive Hyperintensional Pseudo-Contraction

In this Section, we will investigate different ways to define hyperintensional belief contractions, based on Berto's [5] topic-sensitive models for beliefs, and provide axiomatic characterisations for the proposed operations, as commonly done in the work following AGM. Here we point out that our operations are not contraction operations in the usual sense, but pseudo-contractions [34]. This is because our operations formally do not satisfy the usual *(inclusion)* and *(relevance)* postulates, in regards to the foundational logic but weaker versions of these postulates.

Berto [5] introduces his topic-sensitive intensional models aiming to investigate hyperintensional belief revision. He proposes a logic with hyperintensional conditional beliefs $B^{\varphi}\psi$ and interpret these conditional/counterfactual formulas as belief revisions. We will not present Berto's language in this work and will focus on the use of his topic structures to define our belief change operations.

In the remainder of this section, we will suppose our foundational logic to be the classical propositional logic $\mathcal{L}_0 = \langle L_0, Cn \rangle$ defined over some fixed countable propositional symbol set P.

Definition 1. *We call a topic structure a complete join-semilattice $\mathcal{T} = \langle T, \leqslant \rangle$, i.e. a structure such that T is a non-empty set of topics, and $\leqslant \subseteq T \times T$ satisfies reflexivity, transitivity and antisymmetry and for any set $X \subseteq T$ there is a least upper bound $\bigvee X \in T$.*

Any such semilattice $\mathcal{T} = \langle T, \leqslant \rangle$ defines an join operation $\oplus : T \times T \to T$ s.t. $x \oplus y = \bigvee \{x, y\}$ that satisfies idempotence, associativity and commutativity. For consistency of notation, for any set $X \subseteq T$, we will denote the element $\bigvee X$ as $\oplus X$. We will also denote the topic structure $\langle T, \leqslant \rangle$ as $\langle T, \oplus \rangle$ since these structures are inter-definable [6]. We define topic selection functions as functions that map each propositional formula to a (complex) topic in a certain topic structure. As required by Berto, our topic selection function must satisfy the following consistency criterion: the topic of a formula consists of the amalgamation of the topics of each atomic formula that composes it.

Definition 2. *Let $\langle T, \oplus \rangle$ be a topic structure, we call a function $\tau : L_0 \to T$ a topic-selection function if for any formula $\varphi \in L_0$ it holds that $\tau(\varphi) = \oplus \{\tau(p) \mid p \in At(\varphi)\}$, where $At(\varphi)$ is the set of atomic propositions occurring in φ.*

Berto [5] employs such structures to define a conditional model, with which he defines his conditional belief modalities $B^{\varphi}\psi$. His proposal of conditional belief amounts to, after revising the beliefs of the agent by a conditionalisation formula φ, removing from the agent's belief set any belief that does not concern a specific topic - defined as the same topic of φ. We will re-define this notion appealing only to the concept of consequence operators to highlight the connection between Berto's topic-sensitive conditional beliefs and AGM's belief change.

Topic structures and topic selection functions give rise to restricted consequence operations which are conditioned on a particular topic. This consequence operator encoded all the information that an agent can deductively achieve regarding a specific topic. In the following, we denote by $L(t) = \{\varphi \in L \mid \tau(\varphi) \leqslant t\}$ the language of a given topic t, i.e. all the formulas expressing some information about t - for some specified logical language L, topic structure and topic selection function.

Definition 3. *Let $\mathcal{T} = \langle T, \oplus \rangle$ be a topic structure, $t \in T$ be a topic and $\tau : L_0 \to T$ be a topic-selection function. We define the consequence operator sensitive to t (relative to $\mathcal{T}$ and τ), the operation $C_{\downarrow t} : 2^{L_0} \to 2^{L_0}$ s.t.*

$$C_{\downarrow t}(\Gamma) = Cn(\Gamma) \cap L(t)$$

It is easy to see that topic-sensitive consequence operators do not satisfy the inclusion property, as any formula φ in Γ which is not included in the topic of interest is discarded. More yet, these operators satisfy the following properties.

Proposition 4. *Let $\langle T, \oplus \rangle$ be a topic structure, $t \in T$ be a topic, and τ be a topic selection function on L_0, then C_t satisfies monotonicity, compactness and strong iteration below.*

- *Strong iteration: $C_{\downarrow t}(Cn(\Gamma)) = C_{\downarrow t}(\Gamma)$*

With topic-sensitive consequence operators, we can extend Berto's definition of topic-restricted conditional beliefs - equivalent to topic-restricted revision- to all three basic AGM operators. In this work we will focus on contraction - since we are interested in establishing connections with general results on definability of contraction operators in abstract logics in Sect. 5.

Definition 5. *Let $\langle T, \oplus \rangle$ be a topic structure, $t \in T$ be a topic, τ be a topic selection function on L_0 and $- : 2^{L_0} \times 2^{L_0} \to 2^{L_0}$ and AGM contraction on L_0. We define the t-sensitive contraction operation $-_{\downarrow t} : 2^{L_0} \times 2^{L_0} \to 2^{L_0}$ as:*

$$K -_{\downarrow t} A = C_{\downarrow t}(K - A)$$

It is easy to see that these operations can be similarly defined in Berto's [5] framework by means of his conditional beliefs[3]. We can now characterise topic-sensitive contractions by the following postulates.

Theorem 6. *Let $\langle T, \oplus \rangle$ be a topic structure, $t \in T$ be a topic, τ be a topic selection function on L_0 and $- : 2^{L_0} \times 2^{L_0} \to 2^{L_0}$. An operator $-$ is a t-sensitive contraction operation iff for any $K, A \subseteq L_0$, s.t. $K = Cn(K)$ and A is finitely-definable, it satisfies:*

(t-closure) $K - A = Cn(K - A) \cap L(t)$

[3] Namely, we can equivalently define $\psi \in K -_{\downarrow \tau(\varphi)} \varphi$ by the formula $B^{\psi \vee \neg \psi} \psi \wedge B^{\neg \varphi} \psi$ in their language.

(success) If $A \not\subseteq Cn(\varnothing)$, *then* $A \not\subseteq Cn(K - A)$
(t-inclusion) $K - A \subseteq K \cap L(t)$
(t-vacuity) If $A \not\subseteq K$, *then* $K \cap L(t) \subseteq K - A$
(t-relevance) $\beta \in K \cap L(t) \setminus K - A$ *then there is some* $K' \subseteq K$ *s.t.* $K - A \subseteq K'$, $A \not\subseteq K'$ *but* $A \subseteq Cn(K' \cup \{\beta\})$.
(extensionality) If $Cn(A) = Cn(B)$ *then* $K - A = K - B$

Proof (Sketch of the proof). The satisfaction of the postulates is immediate from the fact that it is a t-sensitive contraction, and thus defined as a restriction of the result of an AGM contraction which satisfies AGM postulates, and that $C_{\downarrow t}$ satisfied monotonicity, compactness and strong iteration. To show that such an operation is a t-sensitive contraction, we have to show that there is some AGM contraction $\dot{-}$ s.t. $K - A = (K \dot{-} A) \cap L(t)$. It is easy to see that, since $-$ satisfies (success), and $\mathcal{L}_0$ is monotonic and compact, then $K - A$ can be extended to a maximal element K' s.t. $K - A \subseteq K' \subseteq K$ s.t. $A \not\subseteq Cn(K')$ [23]. As such, it suffices to define a selection function

$$f(K, A) = \begin{cases} K & \text{if } A \subseteq Cn(\varnothing) \\ \{K' \in K \perp A \mid K - A \subseteq K'\} & \text{otherwise} \end{cases}$$

It is easy to see $K - A = (\bigcap f(K, A)) \cap L(t)$ based on the original proof of AGM [1] of constructibility of contraction operations by partial meet. □

While topic-restrained contractions can be neatly characterised, they can present some unintuitive behaviour. Consider the following example.

Example 1. Lets consider an agent that beliefs both in p : *The moon is made of cheese* and q : *all swans are white*, which relate to different/disjoint topics. Thus the agent's belief state can be characterised by the belief set $K = Cn(\{p, q\})$. Now consider the case in which the agent comes to discover that not all swans are white, it is admissible by AGM postulates a contraction operation $-$ s.t. $K - q = Cn(q \to p)$. By definition, $-$ induces a topic-sensitive contraction $-_{\downarrow\tau(p)}$ s.t. $K -_{\downarrow t(p)} q = C_{\downarrow t}(\varnothing)$. Well, it is clear that removing the information q should have no influence on p, when considering the topic at hand of $\tau(p)$.

The reason for this is that when removing and information q we consider all available information in K that may lead to q, not only the information restricted to the topic at hand. In fact, it is easy to see that the consequence operator $C_{\downarrow t}$ behaves similarly, and may use information outside the realm of the topic discussed at hand t to make conclusions, as in $C_{\downarrow t(p)}(\{q \to p, q\}) = C_{\downarrow t(p)}(\{p\})$, which may not always be desirable[4]. As such, we propose the notion of topic-specific consequence. Notice that topic-specific consequences differ from topic-sensitive consequences (Definition 3) in that they select by topic both the arguments used in reasoning and also in the conclusions.

[4] Consider, for example, the case of searching for an algebraic proof for the Fundamental Theorem of Algebra. While we can prove this theorem by analytic methods, such a proof would not be desirable as it employs non-algebraic axioms and language.

Definition 7. *Let $\mathcal{T} = \langle T, \oplus \rangle$ be a topic structure, $t \in T$ be a topic and $\tau : L_0 \to T$ be a topic-selecting function. We define the topic-specific consequence operator restricted to t, the operation $C_t : 2^{L_0} \to 2^{L_0}$ s.t.*

$$C_t(\Gamma) = Cn(\Gamma \cap L(t)) \cap L(t)$$

Topic-specific consequence operators are in some sense similar to Hansson and Wasserman's [23] local consequence operators. As those operators, our topic-specific consequences constrain the amount of information the agent considers when evaluating some information. Differently then local consequences, however, our operators pose constraints based on the subject of a proposition - which is directly related to its structure in our framework - not on the arguments supporting this belief in the agent's belief state.

Proposition 8. *Let $\langle T, \oplus \rangle$ be a topic structure, $t \in T$ be a topic, and τ a topic selection function on L_0, then C_t satisfies monotonicity, idempotence and compactness.*

To define our operations, we will need to change our notion of remainder set to also consider topics in the selection of relevant subsets of a belief set.

Definition 9. *Let $\mathcal{L} = \langle L, Cn \rangle$ be a logic and C_t a t-specific consequence, for some topic structure and topic selection function. Let $K, A \subseteq L$ be sets of formulas, we define the t-selective remainder set $K \perp_t A$ as the set of K' s.t.*

1. $K' \subseteq C_t(K)$
2. $A \not\subseteq Cn(K')$
3. $K' \subset K'' \subseteq C_t(K)$ *then* $A \subseteq Cn(K'')$

With these topic-sensitive remainder sets we will define two notions of topic-sensitive contractions. The first operation here presented are based on Santos et al.'s [34] pseudo-partial meet contractions, which, as we will see, fail to select the topic-relevant information to be removed from the belief set. Based on that analysis, we propose a revised version which can better capture the influence of topic (or the *aboutness*) of an information in the process of belief change.

Definition 10. *Let $\mathcal{T} = \langle T, \oplus \rangle$ be a topic structure, $t \in T$ be a topic and $t : L_0 \to T$ be a topic-selecting function. Let yet $-_t : 2^{L_0} \times 2^{L_0} \to 2^{L_0}$ be an operator and $K \subseteq 2^{L_0}$. We say $-_t$ is a t-dependent contraction operator on K if there is a selection function γ s.t. for any finitely definable $A \subseteq L_0$:*

$$K -_t A = \bigcap \gamma(K \perp_t A)$$

Based on a generalisation of Santos et al.'s [34] characterisation of pseudo-partial meet contractions (Cn^*-partial meet contractions in their terminology) for tarskian logics, we can provide the following characterisation of topic-dependent contractions.

Theorem 11. *Let $\mathcal{T} = \langle T, \oplus \rangle$ be a topic structure, $\tau : L_0 \to T$ be a topic-dependent function and $K \subseteq L_0$ be a set of formulas. An operator $-_t$ is a topic-specific contraction operator on K iff for any finitely definable $A \subseteq L_0$, $-_t$ satisfies:*

(t-logical inclusion) $K - A \subseteq C_t(K)$
(success) If $A \not\subseteq Cn(\varnothing)$ *then* $A \not\subseteq Cn(K - A)$.
(t-uniformity) If for all $K' \subseteq C_t(K)$, $A \subseteq Cn(K')$ *iff* $B \subseteq Cn(K')$, *then* $K - A = K - B$
(t-logical relevance) If $\beta \in C_t(K) \backslash K - A$ *then there is some* $K' \subseteq C_t(K)$ *s.t.* $K -_t A \subseteq K'$, $A \not\subseteq Cn(K')$ *but* $A \subseteq Cn(K' \cup \{\beta\})$.

Proof. The proof of this representation result is exactly the same of that of Santos et al. [34] (Theorem 25) by observing that their proof does not requires C to satisfy inclusion, as the uses of this property can be removed by using their alternative postulates. Namely, substituting in Part 1 the use of uniformity for uniformity* and, in Part 3 Case 1, the use of logical relevance for relevance*. □

While t-dependent contractions exclude contractions that employ irrelevant information in the reasoning, i.e. outside of the topic in consideration, as described in Example 1, it is too weak, in the sense that it may not be felicitous in excluding relevant information. Consider the following example.

Example 2. Lets consider the same agent of Example 1, coming to discover that it is not the case that p: *The moon is made of cheese* and q:*all swas are white.* It is admissible that, regarding whether the moon being made of cheese, the agent comes to believe $K -_{\downarrow t(p)} \{p, q\} = C_{\downarrow t}(\{p\})$. Since q is not satisfied in the resulting belief state, success is satisfied and thus the agent does not need to give up the believe that p, even though the information q may not be true is irrelevant to the topic at hand.

Since the topic-dependent contractions consider all input information in its reasoning process, it may simply remove beliefs that are irrelevant to the topic at hand. As such, for a topic-sensitive reconsideration of currently held beliefs, the agent must evaluate what is relevant in the incoming information which must be taken into account in order change her beliefs. To model this reasoning process, we propose the following operation.

Definition 12. *Let $\mathcal{T} = \langle T, \oplus \rangle$ be a topic structure, $t \in T$ be a topic and $t : L_0 \to T$ be a topic-selecting function. Let yet $-_t : 2^{L_0} \times 2^{L_0} \to 2^{L_0}$ be an operator and $K \subseteq 2^{L_0}$. We say $-_t$ is a t-specific contraction operator on K if there is a selection function γ s.t.*

$$K -_t A = \bigcap \gamma(K \perp_t C_t(A))$$

Notice that, compared topic-dependent contraction (Definition 10), topic-specific contraction (Definition 12) selects formulas of the relevant topic both on the belief set and in the input information - thus limiting the reasoning of the agent within the topic of interest. From previous results, it is easy to provide the following characterisation.

Theorem 13. *Let $\mathcal{T} = \langle T, \oplus \rangle$ be a topic structure, $\tau : L_0 \to T$ be a topic-selecting function and $K \subseteq L_0$ be a set of formulas. An operator $-_t$ is a topic-specific contraction operator on K iff for any finitely definable $A \subseteq L_0$, $-_t$ satifies (t-logical inclusion), (t-uniformity) and*

(strong success) If $C_t(A) \not\subseteq Cn(\varnothing)$ then $C_t(A) \not\subseteq Cn(K - A)$.
(logical t-relevance) If $\beta \in C_t(K) \backslash K - A$ then there is some $K' \subseteq C_t(K)$ s.t. $C_t(A) \not\subseteq Cn(K')$ but $C_t(A) \subseteq Cn(K' \cup \{\beta\})$.

Proof. The satisfaction of postulates follows immediately from the fact that there is a partial meet contraction $-$ on $\mathcal{L}_0$ s.t. $K -_t A = C_t(K) - C_t(A)$. The construction follows similarly to the case of Theorem 11. □

5 Hyperintensional Belief Pseudo-Contraction

The definition and characterisation of the belief change operations presented in Sect. 4 depend on the properties of the consequence operators $C_{\downarrow t}$ and C_t, not on the topic structures in any sense. As such, we can expect that these operations can be generalized to any consequence operation C that defines a sublogic to a specific foundational logic $\mathcal{L}$. In this Section, we generalise the operations defined in the previous section to any monotonic and compact[5] foundational logic $\mathcal{L}$ and a hyperintensional operator C on this logic.

In the remainder, we will represent hyperintensional reasoning - i.e. reasoning that considers hyperintensional contexts - in an abstract form by means of consequence operators in a given foundational logic, which may take into consideration aspects such as structure, topics, etc. Let's define this notion formally.

Definition 14. *Let $\mathcal{L} = \langle L, Cn \rangle$ be a logic, and $C : 2^L \to 2^L$ be a consequence operator. We say that C is $\mathcal{L}$-consistent, if for every $\Gamma \subseteq 2^L$, $C(\Gamma) \subseteq Cn(\Gamma)$*

Notice that our topic-sensitive and specific consequences are $\mathcal{L}_0$-consistent consequence operators (or subclassical consequences). Based on this notion, we can generalise the operations presented previously.

Definition 15. *Let $\mathcal{L}$ be a monotonic and compact logic, C be a $\mathcal{L}$-consistent consequence operator. We say an operator $- : 2^L \times 2^L \to 2^L$ is a C-sensitive pseudo-partial meet contraction (or C-sensitive contraction) if there is some selection function γ s.t. for any sets of formulas $K, A \subseteq L$, with A finitely definable, it holds that*

$$K - A = C(\bigcap \gamma(K \perp A))$$

[5] Notice that the operations considered here are all defined as partial meet operations of some kind. As such, can can only guarantee the existence of such operations for foundational logics that are monotonic and compact, since otherwise the upper bound property [2] may not hold.

The characterisation of topic-sensitive contraction relies on the fact that the foundational logic is classical propositional logic, as in this case all AGM contraction is a partial meet contraction. As such, we can generalise the provided for any monotonic and compact foundational logic with the following postulates - which are variations of the postulates for pseudo-contractions, c.f. [34].

(C-logical inclusion) If $K - A \subseteq C(K)$.
(C-enforced closure) $K - A = C(K - A)$
(Cn-success) $A \not\subseteq Cn(\varnothing)$ then $A \not\subseteq Cn(K - A)$
(Cn-uniformity) If for all $K' \subseteq K$, $A \subseteq Cn(K')$ iff $B \subseteq Cn(K')$, then $K - A = K - B$
(C-sensitive logical relevance) If $\beta \in C(K) \backslash C(K - A)$ then there is some $K - A \subseteq K' \subseteq K$ s.t. $A \not\subseteq Cn(K')$ but $A \subseteq Cn(K' \cup \{\beta\})$.

It is clear for similar studies in belief change in non-classical logics [14,23,29,34] that connection between postulates and constructions of operators depends on which properties the consequence operators satisfies. As such, we can characterise the following connections.

Proposition 16. *Let $\mathcal{L}$ be a monotonic and compact logic, C be a $\mathcal{L}$-consistent consequence operator and $-_C$ be a C-sensitive partial meet contraction. It holds that:*

1. *$-_C$ satisfies (Cn-success)*
2. *If C satisfies idempotence, $-_C$ satisfies (C-enforced closure).*
3. *If C satisfies monotonicity, $-_C$ satisfies (Cn-uniformity), (C-sensitive relevance), and (C-logical inclusion).*
4. *If C satisfies idempotence and monotonicity, $-_C$ satisfies (C-sensitive logical relevance)*

With that, we can provide the following representation result for C-sensitive contractions.

Theorem 17. *Let $\mathcal{L}$ be a monotonic and compact logic, C be a $\mathcal{L}$-consistent consequence operator satisfying monotonicity and idempotence. An operator $-_C$ is a C-sensitive partial meet contraction iff $-_C$ satisfies (Cn-success), (C-enforced closure), (Cn-uniformity) (C-logical inclusion) and (C-sensitive relevance).*

Proof. Satisfaction of postulates is consequence of Prposition 16. The proof of construction follows similar strategy than that of Theorem 11, observing that $K -_C A = C(K - A)$ for some partial meet contraction $-$ in $\mathcal{L}$. □

Similarly, we consider that general notion of topic-dependent contractions, which corresponds to Santos et al.'s [34] Cn^*-pseudo-partial meet contractions.

Definition 18. *Let $\mathcal{L}$ be a monotonic and compact logic, C be a $\mathcal{L}$-consistent consequence operator. We say an operator $- : 2^L \times 2^L \rightarrow 2^L$ is a C-dependent pseudo-contraction (or C-dependent contraction) if there is some selection function γ s.t. for any sets of formulas $K, A \subseteq L$, with A finitely definable, it holds that*

$$K - A = \bigcap \gamma(C(K) \perp A)$$

To characterise this operations, Santos et al. [34] employ the following additional postulates.

(C-uniformity) If for all $K' \subseteq C(K)$, $A \subseteq Cn(K')$ iff $B \subseteq Cn(K')$, then $K - A = K - B$
(C-logical relevance) If $\beta \in C(K) \backslash (K - A)$ then there is some $K - A \subseteq K' \subseteq C(K)$ s.t. $A \nsubseteq Cn(K')$ but $A \subseteq Cn(K' \cup \{\beta\})$.

Based on our previous generalisation of the Santos et al.'s [34] characterisation of C-dependent contractions to monotonic and compact foundational logics. We can provide the following characterisation.

Theorem 19. *Let $\mathcal{L}$ be a monotonic and compact logic, C be a $\mathcal{L}$-consistent consequence operator satisfying monotonicity and idempotence. An operator $-_C$ is a C-dependent contraction iff $-_C$ satisfies (Cn-success), (C-logical inclusion), (C-logical relevance) and (C-uniformity)*

Finally, we generalise topic-specific contractions into C-specific contractions.

Definition 20. *Let $\mathcal{L}$ be a monotonic and compact logic, C be a $\mathcal{L}$-consistent consequence operator. We say an operator $- : 2^L \times 2^L \to 2^L$ is a C-selective partial meet contraction (or C-selective contraction) if there is some selection function γ s.t. for any sets of formulas $K, A \subseteq L$, with A finitely definable, it holds that*

$$K - A = \bigcap \gamma(C(K) \bot C(A))$$

We can generalise the postulates given in Sect. 4 to the following.

(C-weak success) If $C(A) \nsubseteq Cn(\varnothing)$ then $C(A) \nsubseteq Cn(K - A)$.
(C-selective logical relevance) If $\beta \in C(K) \backslash K - A$ then there is some $K - A \subseteq K' \subseteq C(K)$ s.t. $C(A) \nsubseteq Cn(K')$ but $C(A) \subseteq Cn(K' \cup \{\beta\})$.

Based on previous discussion, it is not difficult to see that the following hold.

Proposition 21. *Let $\mathcal{L}$ be a monotonic and compact logic, C be a $\mathcal{L}$-consistent consequence operator and $-_C$ be a C-selective partial meet contraction. It holds that:*

1. *The operator $-_C$ satisfies (C-uniformity), (C-weak success) and (C-selective logical relevance).*
2. *If C satisfies idempotence and monotonicity, then $-_C$ satisfies (C-logical inclusion).*

Based on Santos et al.'s [34] characterisation of Cn^*-pseudo-contractions and Proposition 21, the following characterisation is immediate.

Theorem 22. *Let $\mathcal{L}e$ be a monotonic and compact logic, C be a $\mathcal{L}$-consistent consequence operator satisfying monotonicity and idempotence. An operator $-_C$ is a C-selective partial meet contraction iff $-_C$ satisfies (C-uniformity), (C-weak success), (C-selective logical relevance) and (C-logical inclusion).*

6 Final Considerations

In this work, we investigated belief change operations arising from hyperintensional treatments of beliefs. Firstly we focused on Berto's [5] topic-sensitive framework for hyperintensionality and generalised the proposed operations to any hyperintensional framework which can be characterised by a (monotonic and idempotent) consequence operator. In doing so, we generalise Santos et al.'s [34] previous results on the characterisation of Cn^*-pseudo contractions (here called C-dependent pseudo-contractions) and propose two related notions of C-sensitive and C-selective pseudo-contractions.

It is important to notice that, differently from Berto [5], we rely on tools from Abstract Logic to define our notion. We believe, however, that the semantic treatment of such notions can provide interesting and insightful intuitions for the construction of different operations. Particularly, since competing approaches represent hyperintensional contexts in different ways, we belief that we can use of impossible-world semantics to investigate different kinds of hyperintensional belief change operations. More yet, we believe semantics-defined operations can be more easily connected to other recent dynamic logic-based theories of belief and mental change [17,36] and to the work on iterated belief change [13,25].

References

1. Alchourrón, C.E., Gärdenfors, P., Makinson, D.: On the logic of theory change: partial meet contraction and revision functions. J. Symb. Logic **50**(2), 510–530 (1985)
2. Alchourrón, C.E., Makinson, D.: Hierarchies of regulations and their logic. In: Hilpinen, R. (ed.) New Studies in Deontic Logic. Studies in Epistemology, Logic, Methodology, and Philosophy of Science, vol. 152, pp. 125–148. Springer, Heidelberg (1981). https://doi.org/10.1007/978-94-009-8484-4_5
3. Aucher, G.: When conditional logic and belief revision meet substructural logics. In: DARe-15 Defeasible and Ampliative Reasoning, Buenos Aires, Argentina (2015)
4. Badura, C., Berto, F.: Truth in fiction, impossible worlds, and belief revision. Aust. J. Philos. **97**(1), 178–193 (2019)
5. Berto, F.: Simple hyperintensional belief revision. Erkenntnis **84**(3), 559–575 (2019). https://doi.org/10.1007/s10670-018-9971-1
6. Birkhoff, G.: Lattice Theory, vol. 25. American Mathematical Society, Providence (1940)
7. Bjerring, J.C.: Impossible worlds and logical omniscience: an impossibility result. Synthese **190**(13), 2505–2524 (2013). https://doi.org/10.1007/s11229-011-0038-y
8. Board, O.: Dynamic interactive epistemology. Games Econ. Behav. **49**(1), 49–80 (2004)
9. Cresswell, M.J.: Hyperintensional logic. Studia Logica: Int. J. Symb. Logic **34**(1), 25–38 (1975)
10. Cresswell, M.J., Von Stechow, A.: "de re" belief generalized. Linguist. Philos. **5**, 503–535 (1982)
11. Darwiche, A., Pearl, J.: On the logic of iterated belief revision. Artif. Intell. **89**(1), 1–29 (1997)

12. Delgrande, J.P.: Horn clause belief change: contraction functions. In: KR, pp. 156–165 (2008)
13. Fermé, E., Wassermann, R.: On the logic of theory change: iteration of expansion. J. Braz. Comput. Soc. **24**(1), 8 (2018). https://doi.org/10.1186/s13173-018-0072-4
14. Flouris, G., Plexousakis, D., Antoniou, G.: On applying the AGM theory to DLs and OWL. In: Gil, Y., Motta, E., Benjamins, V.R., Musen, M.A. (eds.) ISWC 2005. LNCS, vol. 3729, pp. 216–231. Springer, Heidelberg (2005). https://doi.org/10.1007/11574620_18
15. Gabbay, D., Rodrigues, O., Russo, A.: Belief revision in non-classical logics. Rev. Symb. Logic **1**(03), 267–304 (2008)
16. Gärdenfors, P.: Knowledge in Flux: Modeling the Dynamics of Epistemic States. The MIT Press, Cambridge (1988)
17. Girard, P., Rott, H.: Belief revision and dynamic logic. In: Baltag, A., Smets, S. (eds.) Johan van Benthem on Logic and Information Dynamics. OCL, vol. 5, pp. 203–233. Springer, Cham (2014). https://doi.org/10.1007/978-3-319-06025-5_8
18. Girard, P., Tanaka, K.: Paraconsistent dynamics. Synthese **193**(1), 1–14 (2016). https://doi.org/10.1007/s11229-015-0740-2
19. Grove, A.: Two modelings for theory change. J. Philos. Logic **17**(2), 157–170 (1988)
20. Halpern, J.Y., Pucella, R.: Dealing with logical omniscience: expressiveness and pragmatics. Artif. Intell. **175**(1), 220–235 (2011)
21. Hansson, S.O.: Belief contraction without recovery. Studia Logica **50**(2), 251–260 (1991). https://doi.org/10.1007/BF00370186
22. Hansson, S.O.: Defense of base contraction. Synthese **91**(3), 239–245 (1992). https://doi.org/10.1007/BF00413568
23. Hansson, S.O., Wassermann, R.: Local change. Stud. Logica **70**(1), 49–76 (2002)
24. Jago, M.: Logical information and epistemic space. Synthese **167**(2), 327–341 (2009)
25. Jin, Y., Thielscher, M.: Iterated belief revision, revised. Artif. Intell. **171**(1), 1–18 (2007)
26. Levesque, H.J.: A logic of implicit and explicit belief. In: Proceedings of the Fourth National Conference on Artificial Intelligence, pp. 198–202. American Association for Artificial Intelligence, Austin, August 1984
27. Rantala, V.: Impossible worlds semantics and logical omniscience. Acta Philosophica Fennica **35**, 106–115 (1982)
28. Ribeiro, J.S., Nayak, A., Wassermann, R.: Belief change and non-monotonic reasoning sans compactness. In: Proceedings of the AAAI Conference on Artificial Intelligence, vol. 33, pp. 3019–3026 (2019)
29. Ribeiro, M.M., Wassermann, R.: Base revision for ontology debugging. J. Logic Comput. **19**(5), 721–743 (2008). https://doi.org/10.1093/logcom/exn048. http://logcom.oxfordjournals.org/cgi/doi/10.1093/logcom/exn048
30. Ribeiro, M.M., Wassermann, R., Flouris, G., Antoniou, G.: Minimal change: relevance and recovery revisited. Artif. Intell. **201**, 59–80 (2013)
31. Ribeiro, M.M.: Belief Revision in Non-classical Logics. Springer, Heidelberg (2012). https://doi.org/10.1007/978-1-4471-4186-0
32. Rott, H.: Two dogmas of belief revision. J. Philos. **97**(9), 503–522 (2000)
33. Rott, H., Pagnucco, M.: Severe withdrawal (and recovery). J. Philos. Logic **28**(5), 501–547 (1999). https://doi.org/10.1023/A:1004344003217
34. Santos, Y.D., Matos, V.B., Ribeiro, M.M., Wassermann, R.: Partial meet pseudo-contractions. Int. J. Approx. Reason. **103**, 11–27 (2018)

35. Sedlár, I.: Hyperintensional logics for everyone. Synthese, pp. 1–24 (2019)
36. Souza, M., Moreira, Á., Vieira, R.: Dynamic preference logic as a logic of belief change. In: Madeira, A., Benevides, M. (eds.) DALI 2017. LNCS, vol. 10669, pp. 185–200. Springer, Cham (2018). https://doi.org/10.1007/978-3-319-73579-5_12
37. Van Benthem, J.: Dynamic logic for belief revision. J. Appl. Non-Class. Logics **17**(2), 129–155 (2007)
38. Wansing, H.: A general possible worlds framework for reasoning about knowledge and belief. Stud. Logica **49**(4), 523–539 (1990). https://doi.org/10.1007/BF00370163

Machine Learning and Data Mining

An Online Pyramidal Embedding Technique for High Dimensional Big Data Visualization

Adriano Barreto(✉), Igor Moreira, Caio Flexa, Eduardo Cardoso, and Claudomiro Sales

Applied Electromagnetism Laboratory, Federal University of Pará, Belém, PA 66075-110, Brazil
{adrianosb,cssj}@ufpa.br, {igor.moreira,caio.rodrigues}@icen.ufpa.br, eduardo.gil.s.cardoso@gmail.com

Abstract. Visualizing multidimensional Big Data is defying: high dimensionalities hinder or even preclude visual inspections. A means of tackling this issue is to use DR (Dimensionality Reduction) techniques, producing low-dimensional representations of high-dimensional data. Popular DR algorithms (e.g., Principal Component Analysis, t-Distributed Stochastic Neighbor Embedding), albeit helpful, are computationally expensive. Most have $\mathcal{O}(n^2)$ or $\mathcal{O}(n^3)$ ATC (Asymptotic Time Complexity) and/or calculate pairwise distances of the entire data set, exceeding available memory and rendering Big Data DR time-consuming or impracticable. These issues impede the employment of DR for online learning appliances, where recurrent, cumulative model updates are habitual. The stochastic factor of some approaches similarly obstructs any meaningful inspection on how knowledge is spatially disposed. The recently introduced PCS (Polygonal Coordinate System)—an incremental, geometric-based technique with linear ATC—is compelling; however, its restriction to 2-D embeddings amounts to significant information loss. We propose the Big Data ready, incremental PES (Pyramidal Embedding System), which builds on PCS virtues by additionally generating 3-D embeddings through its pyramid-like interspace, mitigating quality degradation. Visual inspections, as well as pairwise distance based statistical analyses, validate the PES ability to retain structural arrangements when embedding high- and low-dimensional data while retaining flexibility in resources consumption.

Keywords: Data visualization · Incremental dimensionality reduction · Online learning · Embedding · Feature extraction

1 Introduction

Multidimensional Big Data visualization is a considerable challenge [9]. Big data sets are sizable enough to render their visualization troublesome or unfeasible for exploratory analyses and the KDD (Knowledge Discovery in Database) processes they aid. Big Data also affects the adoption of ML (Machine Learning) techniques

R. Cerri and R. C. Prati (Eds.): BRACIS 2020, LNAI 12320, pp. 291–306, 2020.
https://doi.org/10.1007/978-3-030-61380-8_20

due to their ATC (Asymptotic Time Complexity): many ML algorithms in the literature possess prohibitive ATCs for their application in massive sets of data. One can employ many techniques to make a data set more intelligible and reduce training times of ML algorithms, among which DR (Dimensionality Reduction).

DR techniques map (i.e., embed) high-dimensional information in a low-dimensional space, describing it in a representative, but streamlined manner. This embedded representation assists in comprehending multidimensional data and in sparing storage space [9]. Whether in detecting outliers [1], understanding how the heart fibers function [6] or reconstructing grayscale images [14], DR algorithms are demonstrably paramount in operating with massive knowledge.

Data sets sometimes contain correlated (and therefore, redundant) information, unnecessarily complicating the feature space. DR mitigates the dimensionality curse [16] and its ramifications (e.g., exponential feature space growth as variables accumulate) by extracting a feature subset of the original data. DR can decrease training times of ML algorithms by simplifying data; however, most DR techniques carry quadratic or cubic ATCs, spending the time otherwise spared from model training and potentially neutralizing their advantages when handling sizable data.

One might regard 3-D representations as a better compromise than 2-D ones concerning information, simplification and visualization. However, DR techniques customarily are used to generate (or limited to) 2-D embeddings, amounting to greater information loss due to the inherent quality degradation of DR [5].

Observing these issues, we propose PES (Pyramidal Embedding System), a DR technique based on the geometric approach of PCS (Polygonal Coordinate System) [3] capable of producing both 3-D and 2-D embeddings. Aside from its linear ATC and its capacity to retain global data structures, PES is incremental. As such, it allows DR without loading the entire data set into memory, thus enabling its employment in Big Data enterprises.

Henceforth, the following organization is adopted: related work is in Sect. 2; in Sect. 3, the PES approach is detailed and its complexity is ascertained; the experiments are described and their results are displayed in Sect. 4; and Sect. 5 contains final considerations on the analyses performed.

2 Related Work

DR approaches are commonly categorized into non-linear and linear. Albeit usually harder to understand and costly to run, non-linear DR better separates otherwise not linearly separable data in detriment of structural fidelity between original and embedded spaces. Linear DR, conversely, preserves global data arrangements (e.g., dissimilarities, cluster shapes and probability distributions).

PCA (Principal Component Analysis) is one of the best known DR techniques for its simplicity and computational efficiency. PCA utilizes orthogonal, linear transformations to maximize variance in the resulting principal components by using eigenvectors of the covariance matrix [13]. Despite being deterministic, it has $\mathcal{O}(n^3)$ ATC [10], making PCA onerous for Big Data.

t-SNE (t-Distributed Stochastic Neighbor Embedding), a non-linear technique, uses t-stochastic distributions to compute similarity among points in a low-dimensional space. t-SNE preserves neighboring structures by transforming point-to-point distances into probabilities [7] and is especially adept with image-based information. It entails a costly, derivative-based optimization process which does not ensure convergence to a solution. Its ATC is $\mathcal{O}(n^2)$, subject to improvement to $\mathcal{O}(n \log n)$ by using tree-like structures to approximate distances [12]. Aside from stochastic and hyperparameter-sensitive, t-SNE usually favors neighboring arrangements at the expense of global ones, thus producing local embeddings.

PCS performs DR on an d-dimensional space by creating a regular, d-sided polygon that interfaces high- and low-dimensional spaces [3]. The edges of this shape represent the original feature vectors projected on a 2-D Cartesian plane. Once the linear transformations to create the interface are applied, the embedded data are obtained by averaging the vectors composing the interspace. The PCS interspace design limits the technique to two-dimensional embeddings, unavoidably incurring significant information loss in the process. PCS has $\mathcal{O}(n)$ ATC.

3 The Pyramidal Embedding System (PES)

A non-parametric DR approach, PES globally embeds information in a quick and scalable manner by applying linear transformations. Since it is incremental, unforeseen observations can be cumulatively embedded *a posteriori.* Allied with its Big Data readiness, one can employ PES on massive sets of data through batch loading, culminating in a controllable memory footprint and rendering the loading or re-embedding of the entire set avoidable. Hence, PES can be utilized in online learning endeavors involving vast amounts of data (e.g., live streaming purposes). Figure 1 captures the composing phases of the PES process.

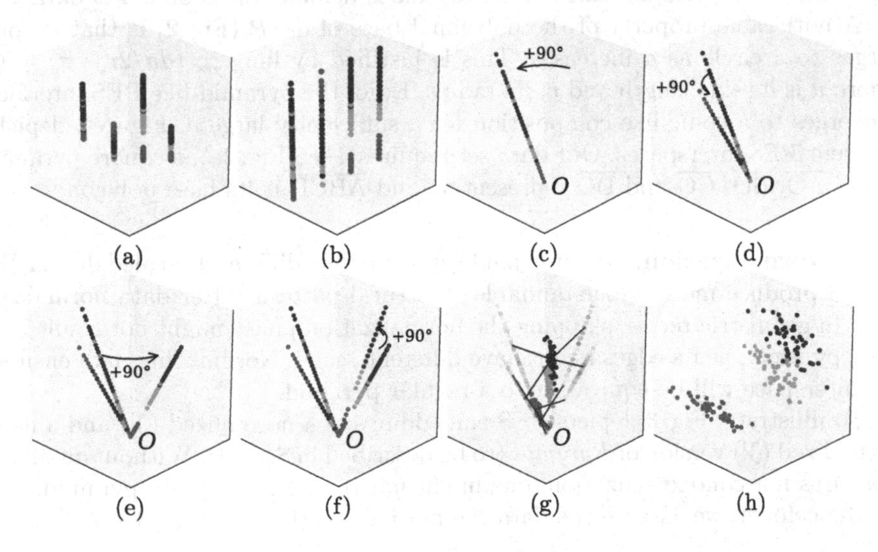

Fig. 1. Iris data set (a) throughout PES—comprised by data normalization (b), interspace conception through feature projection (c to f) and embedding creation through data mapping (g)—to generate its embedding (h).

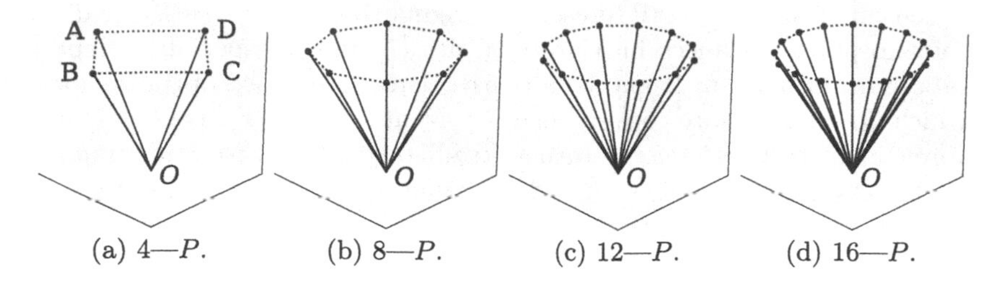

(a) 4—P. (b) 8—P. (c) 12—P. (d) 16—P.

Fig. 2. The interspace of data sets with 4, 8, 12 and 16 dimensions.

PCS resembles PES inasmuch as they are algebraic, their embedding routines entail similar steps (normalization, projection and mapping) and as both conceive an interspace in performing them. Their difference lies on the interspace built: PES employs a pyramid-like interspace capable of building not only two-dimensional embeddings as PCS does, but also three-dimensional ones.

3.1 Describing the Technique

A pyramid is a polyhedron with a polygonal base and triangular sides that meet at a common vertex (the apex). A right pyramid has its apex lying directly above its base centroid. A regular pyramid is right and has a regular polygon base [15].

Assume that $\chi = [x_1, x_2, \ldots, x_n]^\mathsf{T}$ is an n-sized feature vector and that $\mathcal{X} = [\chi_1, \chi_2, \ldots, \chi_i]$, $\mathcal{X} \in \mathbb{R}^d$, $i = [1, 2, \ldots, d]$ is a d-dimensional set containing n observations, such that χ_i is a column of $\mathcal{X}$. Also regard d–P as the interspace bridging original and reduced spaces. Hereinafter, the PES interspace will be abstracted as a pyramid, as the PES routine is demonstrated on a 4-D data set.

A noticeable property of the polygonal base of d—P (Fig. 2) is that it converges to a circle as d increases. This is justified by $\lim_{d\to\infty}(du/2r - \pi) = 0$, where u is its side length and r, its radius. Ergo, the pyramid-like PES interface converges to a conic-like composition for a sufficiently large d. Figure 2 depicts different PES interspaces. Our data set requires 4–P (Fig. 2a), a square pyramid where $\overline{\text{AO}}$, $\overline{\text{BO}}$, $\overline{\text{CO}}$ and $\overline{\text{DO}}$ represent χ_i and $\overline{\text{ABCD}}$ is its base polygon.

Data Normalization. Albeit capable of processing differently scaled data, PES should produce more understandable structural patterns after data normalization. In geometric terms, skipping the normalization phase might not result in a right pyramid, as its edges might have different scales. Normalizing data ensures the interspace will be equivalent to a regular pyramid.

To illustrate, Fig. 3 depicts PES embeddings of a normalized ($\hat{\mathcal{Y}}$) and a non-normalized ($\mathcal{Y}$) version of *Pyramid* (to be described in Sect. 4). Without auxiliary measures nor contextualization, one might not regard Fig. 3c as a pyramid.

Henceforth, we assume the data are normalized (i.e., $\chi_i = \hat{\chi_i}$) through (1).

$$\hat{\chi_i} = \frac{\chi_i - min(\chi_i)}{max(\chi_i) - min(\chi_i)} \tag{1}$$

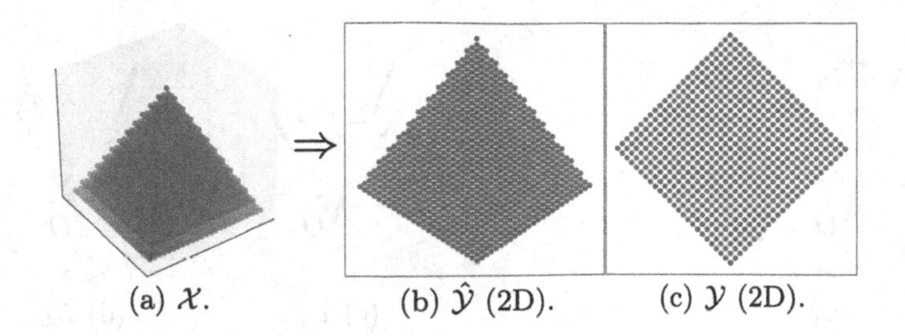

(a) $\mathcal{X}$. (b) $\hat{\mathcal{Y}}$ (2D). (c) $\mathcal{Y}$ (2D).

Fig. 3. The normalization influence on embedding *Pyramid* in 2-D with PES.

Interspace Conception Through Feature Projection. $T1 : \mathbb{R}^1 \Rightarrow \mathbb{R}^3$, $T1(x) = (0, x, x)$—described in (2)—is applied to the transposed vectors in the projection phase, introducing them into the 3-D Cartesian space with congruent angles between x and y axes (*vide* Fig. 4a). The resulting matrices L_i describe the projected vectors to constitute the pyramid edges.

$$L_i = \begin{bmatrix} 0 \\ 1 \\ 1 \end{bmatrix} \times \chi_i^{\mathsf{T}} = \begin{bmatrix} 0 & 0 & \dots & 0 \\ x_1 & x_2 & \dots & x_n \\ x_1 & x_2 & \dots & x_n \end{bmatrix} \tag{2}$$

(a) A projected vector L_i.

Equation (3) defines the rotation of L_i along the z axis. $L_i^{'}$ is obtained by applying $R_z(\theta_i)$ on L_i, where θ_i is the rotation offset in radians calculated in (4), and $R_z(\theta_i)$ is the rotation matrix along the z axis determined in (5).

$$L_i^{'} = R_x(\theta_i) \times L_i, \text{ where} \tag{3}$$

$$\theta_i = \frac{2\pi}{d} \times i, \text{ and} \tag{4}$$

$$R_z(\theta_i) = \begin{bmatrix} \cos(\theta_i) & -\sin(\theta_i) & 0 \\ \sin(\theta_i) & \cos(\theta_i) & 0 \\ 0 & 0 & 1 \end{bmatrix} \tag{5}$$

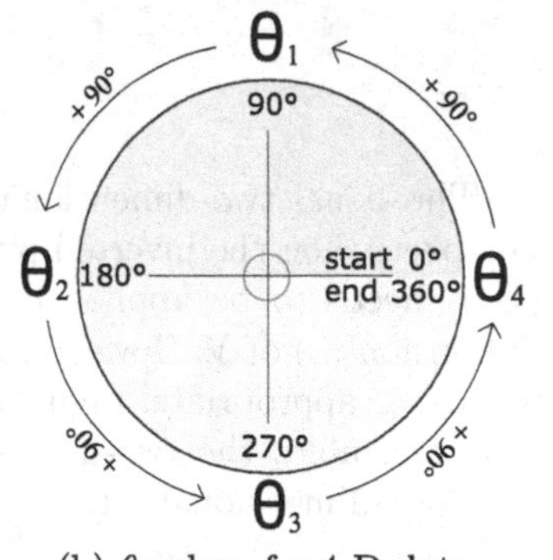

(b) θ values for 4-D data.

Fig. 4. A projected vector L and θ values for 4-D data.

Equation 4 distributes a full trigonometric circle into i congruent θ angles along the z axis based on d. For our 4-D set, we have that $\theta = [90°, 180°, 270°, 360°]$ (*vide* Fig. 4b).

Figure 5 depicts the process for our data set, where observation features of its constituting classes are in red, green and blue. The arrows expose the anticlockwise rotations of L_i after projection. Once $L_i^{'}$ is calculated and assuming the normalization phase was executed, a regular square pyramid is built for our 4-D data set.

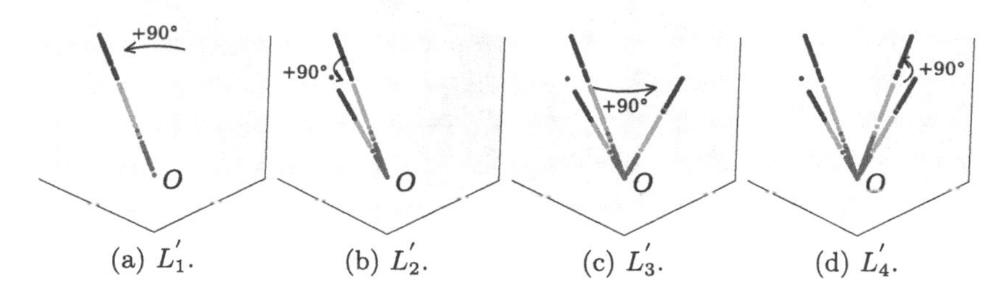

(a) $L_1^{'}$. (b) $L_2^{'}$. (c) $L_3^{'}$. (d) $L_4^{'}$.

Fig. 5. PES projection. (Color figure online)

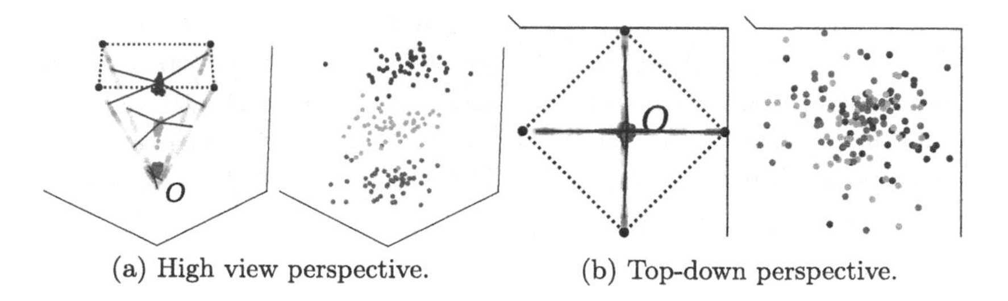

(a) High view perspective. (b) Top-down perspective.

Fig. 6. PES mapping, depicted on 3-D and 2-D perspectives.

Embedding Creation Through Data Mapping. Mapping is the final phase of the process, consisting in representing data in two or three dimensions assisted by the interface created in the previous step. Ultimately, the reduced set of data $\mathcal{Y}$ is obtained by averaging the values of $L^{'}$, as described by (6). Figures 6a and b illustrate the mapping phase in three and two dimensions.

$$\mathcal{Y} = \frac{1}{d} \times \sum_{i=1}^{d} L_i^{'} \tag{6}$$

The usual two-dimensional PES embedding can be defined as a top-down perspective on the inverted pyramid. To obtain it, $T2 : \mathbb{R}^3 \Rightarrow \mathbb{R}^2$, $T2(x, y, z) = (x, y)$ needs to be applied to $\mathcal{Y}$, which is tantamount to discarding the third dimension (z) of $\mathcal{Y}$. However, PES embeddings allow other feature combinations whenever appropriate. Figure 6b shows a top-down perspective of the interspace that resembles the two-dimensional embedding $\text{PES}_{(x,y)}$ to be obtained from this four-dimensional set.

3.2 Asymptotic Time Complexity (ATC)

Algorithm 1 contemplates the essential algorithmic routines of PES, which are applied on a $n \times d$ data set to produce its embedding. In analyzing it, one can see that the asymptotic upper bound $\mathcal{C}$ of PES is $\mathcal{O}(dn)$. In an amortized analysis assuming a significant discrepancy between n and d, we have that $\mathcal{C} = \mathcal{O}(n)$ when $n \gg d$. Conversely, if $d \gg n$, we have that $\mathcal{C} = \mathcal{O}(d)$.

Algorithm 1: PES (Pyramidal Embedding System)

Input: d-dimensional data set $\mathcal{X} = [\chi_1, \chi_2, ..., \chi_d]$ of length n
Output: embedded, 3-dimensional data set $\mathcal{Y}$ of length n

```
1 begin
2   for i ← 1 to d do
3     compute normalized dimension χ̂_i from χ_i (1)                O(n)
4     compute projected edge L_i from χ̂_i (2)                      O(n)
5     compute rotated edge L'_i from L_i with R_z(θ_i) (3)          O(n)
6   end for
7   map data set Y by averaging L'_i, i = [1, 2, ..., d] (6)        O(dn)
8 end
```

4 Performance Analyses

We deemed three perspectives as pertinent to gauge the performance of the selected DR techniques: a visual one that is commonplace in the literature; a numeric one to score the embeddings; and one based on memory and time consumption when embedding highly dimensional knowledge. The visual scrutiny aims to expose the competence of PES under a series of synthetic scenarios, as well as to justify its applicability with real, multidimensional sets of data, against state-of-the-art counterparts. The quantitative evaluation proposed herein is powered by a linear and a non-linear statistic correlation coefficient: Pearson and Spearman, respectively. This approach assumes the pairwise distances distributions reflect the quality of the yielded embeddings [4], which in turn is closely correlated with the visual inspection. The memory and time scrutiny serves to assess the resources invested on embedding massively dimensional data sets.

Table 1. Main characteristics of the data sets under scrutiny.

Type	Name	Observations	Dimensions	Classes	Size	Source
Synthetic	Atom	800	3	2	23.12 KB	[11]
	ChainLink	1,000	3	2	31.49 KB	[11]
	Hepta	212	3	7	6.04 KB	[11]
	Pyramid	2,925	3	1	24.06 KB	[3]
Real	Iris	150	4	3	2.49 KB	[2]
	SmallThyroid	215	5	3	4.90 KB	[2]
	Wine	178	13	3	11.63 KB	[2]
	MNIST	60,000	784	10	303.48 MB	[2]
Big data[a]	Massive-1M	2,000	1,000,000	2	15.36 GB	[3][b]
	1M-G2	2,000	1,000,000	2	15.63 GB	the authors[b]
	1M-G4	2,000	1,000,000	4	15.50 GB	the authors[b]
	1M-G5	2,300	1,000,000	5	18.62 GB	the authors[b]

[a]Composed by Gaussian distributions.
[b]Data set generator script available by accessing https://osf.io/b6qfz.

Table 1 describes the data set collections which the proposed PES, PCS and t-SNE will undergo. Each observation will be colored according either to its class, if there is more than one cluster, or to its index otherwise.

The t-SNE perplexity, exaggeration and learning rate were set to 30, 4 and 500, respectively, adhering to [3,12]. In observance to its stochastic factor, the t-SNE embeddings contained herein were the best of 20 executions with regards to its objective function. This measure was not extended to the remaining techniques, as their routines are entirely deterministic. The experiments were conducted on a server powered by an Intel® Xeon® Silver 4114 CPU @ 2.20 GHz and 32 GB of RAM running a 64-bit copy of Ubuntu 18.04.4 LTS. Python 3.7.3 was the programming language employed; t-SNE was imported from `scikit-learn`. PES and PCS implementations can be attained on the following OSF.io project: https://osf.io/b6qfz. Note that the PCS code, obtained from the author, was not modified for fidelity reasons.

4.1 Synthetic Data Sets

Figure 7 displays the data sets that compose the synthetic collection. *Atom* represents a particular case in which two clusters possess the same center, but differ in density. *ChainLink* illustrates two not linearly separable groupings interposed. *Hepta* represents seven well-separated clusters, and *Pyramid* is a square pyramid representation. All these sets are three-dimensional, enabling us to see how the embedded 3-D representations pose against their original counterparts regarding possible structural changes, despite the absence of DR.

Figure 8 shows three-dimensional embeddings of the synthetic collection. PCS is absent from all 3-D embedding results for reasons clarified in Sect. 3. PES seemingly preserved all global structures; it just rotated the data sets in the space. t-SNE, conversely, obtained mixed results: on one hand, the clusters of *Atom* and *ChainLink* were separated, attesting to its capacity of separating not linearly separable clusters. On the other hand, *Hepta* became a cloud of highly overlapping clusters and *Pyramid* turned into a sliced cone.

Figure 9 depicts two-dimensional embeddings of the same collection. One might remark PCS and PES embeddings resemble: both retained most data structures in detriment of cluster separation, while t-SNE favored cluster segregation over spatial arrangements. This time, t-SNE managed to maintain the

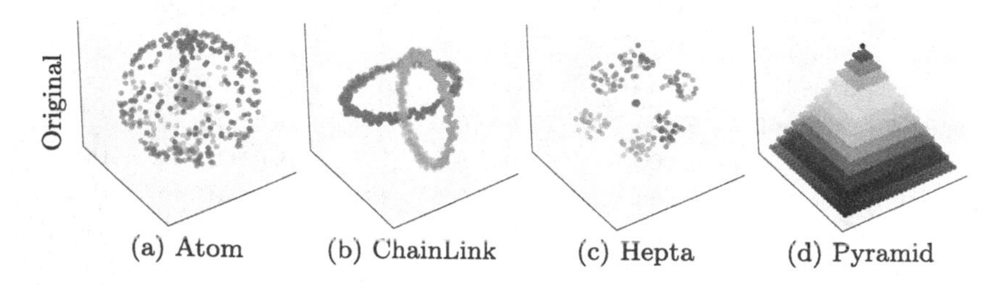

(a) Atom (b) ChainLink (c) Hepta (d) Pyramid

Fig. 7. Original visualizations of four synthetic, 3-D data sets.

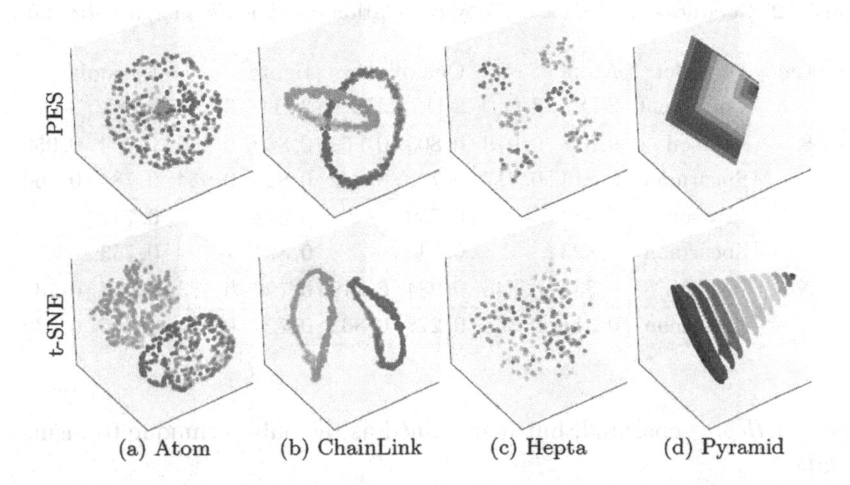

(a) Atom (b) ChainLink (c) Hepta (d) Pyramid

Fig. 8. 3-D embeddings of four synthetic, 3-D data sets.

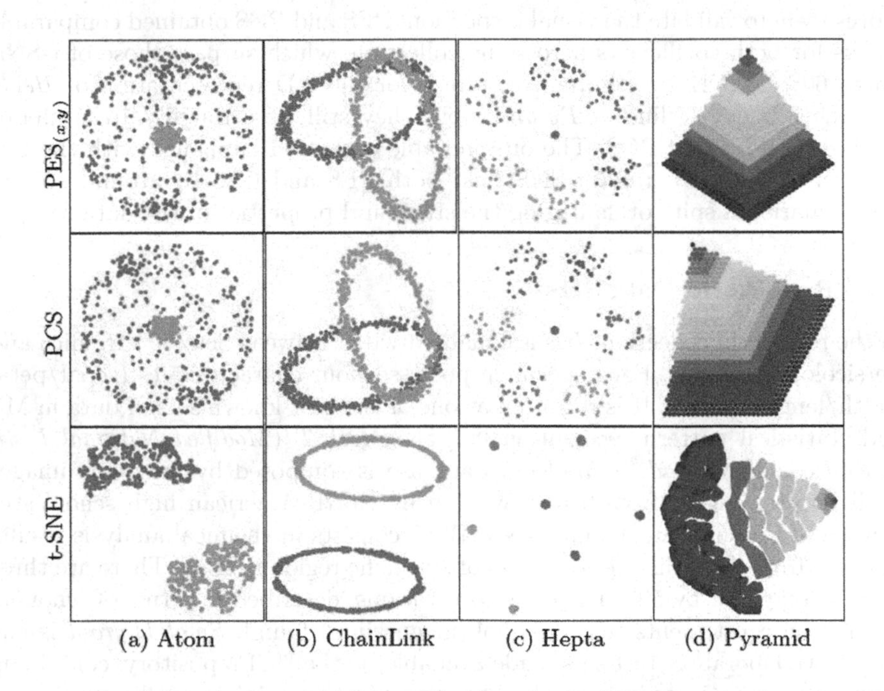

(a) Atom (b) ChainLink (c) Hepta (d) Pyramid

Fig. 9. 2-D embeddings of four synthetic, 3-D data sets.

Table 2. Goodness-of-fit measured by correlation coefficients on synthetic data.

Method	Data Set	Atom		Chainlink		Hepta		Pyramid	
	Coefficient	2-D	3-D	2-D	3-D	2-D	3-D	2-D	3-D
PES	Pearson	0,878	0,979	0,803	0,963	0,869	0,969	0,794	0,959
	Spearman	0,814	0,917	0,795	0,962	0,822	0,954	0,784	0,966
PCS	Pearson	0,882	–	0,759	–	0,871	–	0,741	–
	Spearman	0,823	–	0,748	–	0,824	–	0,733	–
t-SNE	Pearson	0,333	0,348	0,284	0,519	0,796	0,483	0,903	0,730
	Spearman	0,311	0,366	0,278	0,363	0,724	0,434	0,916	0,732

clusters of *Hepta* separated, but it remained as the only technique to dismantle *Pyramid*.

Table 2 exhibits goodness-of-fit results for the synthetic data sets. PES and PCS performed similarly and better than t-SNE in embedding *Atom* and *Hepta*, and PES scored the highest for the *ChainLink* data set. For the most part, the scores seem to validate the visual inspection: PES and PCS obtained comparable marks for both coefficients across the collection, which surpass those of t-SNE up to 60%. t-SNE got competitive scores for the 2-D representation of *Hepta* and the 3-D embedding of *Pyramid*, but they still were roughly 10% inferior to those of PES and PCS. The only notable exception happened with the 2-D representation of *Pyramid*. t-SNE beat both PES and PCS by around 10% in this scenario, in spite of damaging the structural properties of the data set.

4.2 Real-World Data Sets

In the real-world collection, *Iris* is equally divided between Setosa, Virginica and Versicolor specimens. Each specimen possesses four characteristics (sepal/petal width/length in cm.). It is regarded as one of the best known sets of data in ML and statistical pattern recognition [8]. The *MNIST* (*Modified National Institute of Standards and Technology*) database is composed by grayscale images of digit sets ranging from 0 to 9 written by North American high school students and Census Bureau employees. *Wine* consists in chemical analysis results of wines from three distinct cultivars of a specific region in Italy. There are three classes composed by 59, 71 and 48 observations, described in terms of amounts of thirteen constituents (e.g., alcohol, magnesium) found. *SmallThyroid* is one of the five laboratory test sets made available by the UCI repository, containing samples from patients in normal, hypo- and hyperthyroidism conditions.

Figure 10 shows three-dimensional embeddings of the real-world collection. On a positive note, all of them do portray the varying degrees of inter-cluster overlap present across the collection. The PES embedding of *Iris* properly portrays the higher correlation between Virginica and Versicolor while still minimizing inter-cluster overlap. On *SmallThyroid*, PES also displays the three medical conditions in a discernible manner. However, *Wine* and *MNIST* are exposed as clouds of highly overlapping clusters. t-SNE also depicts almost all data sets

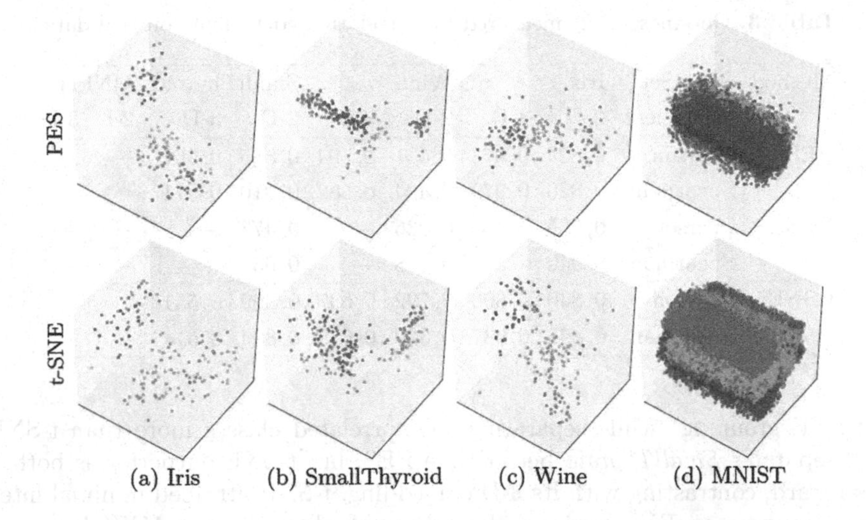

Fig. 10. 3-D embeddings of four real, multidimensional data sets.

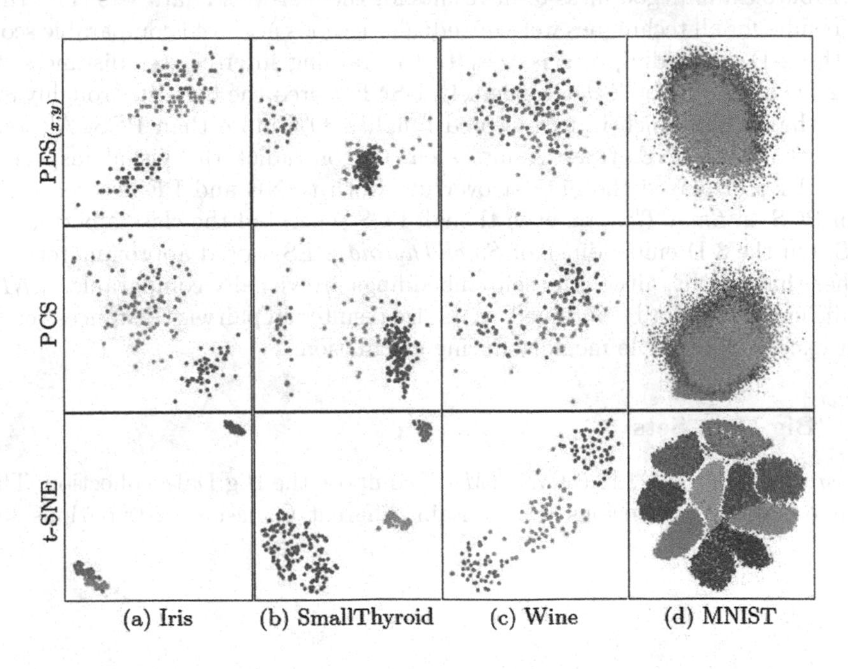

Fig. 11. 2-D embeddings of four real, multidimensional data sets.

in the same manner; however, it separated *MNIST* classes better than PES, befitting the knowledge that t-SNE is specially adequate for embedding image-based data sets.

Figure 11 shows two-dimensional embeddings of the same collection. PES compromised between the closeness of PCS clusters and the excessive separation

Table 3. Goodness-of-fit measured by correlation coefficients on real data.

Method	Data Set	Iris		Wine		SmallThyroid		MNIST	
	Coefficient	2-D	3-D	2-D	3-D	2-D	3-D	2-D	3-D
PES	Pearson	0,824	0,978	0,431	0,594	0,885	0,967	–	–
	Spearman	0,826	0,979	0,404	0,582	0,910	0,971	–	–
PCS	Pearson	0,935	–	0,236	–	0,473	–	–	–
	Spearman	0,940	–	0,228	–	0,636	–	–	–
t-SNE	Pearson	0,840	0,606	0,792	0,637	0,899	0,551	–	–
	Spearman	0,899	0,637	0,829	0,649	0,894	0,614	–	–

of t-SNE groupings while separating the correlated classes more than t-SNE. PCS separates *SmallThyroid* better than PES, but t-SNE outperforms both in this regard, contrasting with its 3-D embedding. t-SNE attained minimal inter-cluster overrun on *Wine* and was the only method to separate *MNIST*.

Table 3 exhibits goodness-of-fit results for the real-world data sets. This time, the results for all techniques were mixed: all methods achieved comparable scores for the 2-D embedding of *Iris*, despite the varying inter-cluster distances. On the 2-D embedding of *Wine*, conversely, t-SNE scored the highest—roughly 80% more than PES, which in turn scored roughly 80% more than PCS. Although coherent with t-SNE, these results seem to contradict the visual inspection, where PES displayed the highest overrun. Both t-SNE and PES scored higher than PCS on *SmallThyroid*, even though PCS separated the classes better than PES. On the 3-D embeddings of *SmallThyroid*, PES scored approximately 75% higher than t-SNE, although their embeddings are visually comparable. *MNIST* coefficients could not be obtained, as its size resulted in pairwise distances vectors that exceeded available memory during calculation.

4.3 Big Data Sets

Massive-1*M*, 1*M*-*G*2, 1*M*-*G*4 and 1*M*-*G*5 compose the Big Data collection. They have 1,000,000 dimensions and contain different Gaussian distributions sets.

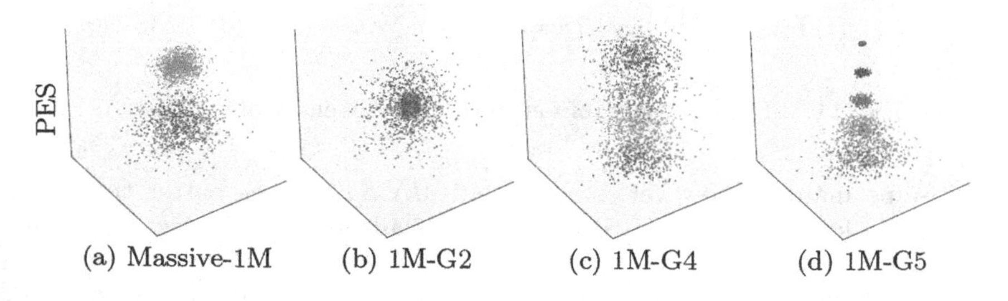

(a) Massive-1M (b) 1M-G2 (c) 1M-G4 (d) 1M-G5

Fig. 12. 3-D embeddings of four massively multidimensional data sets.

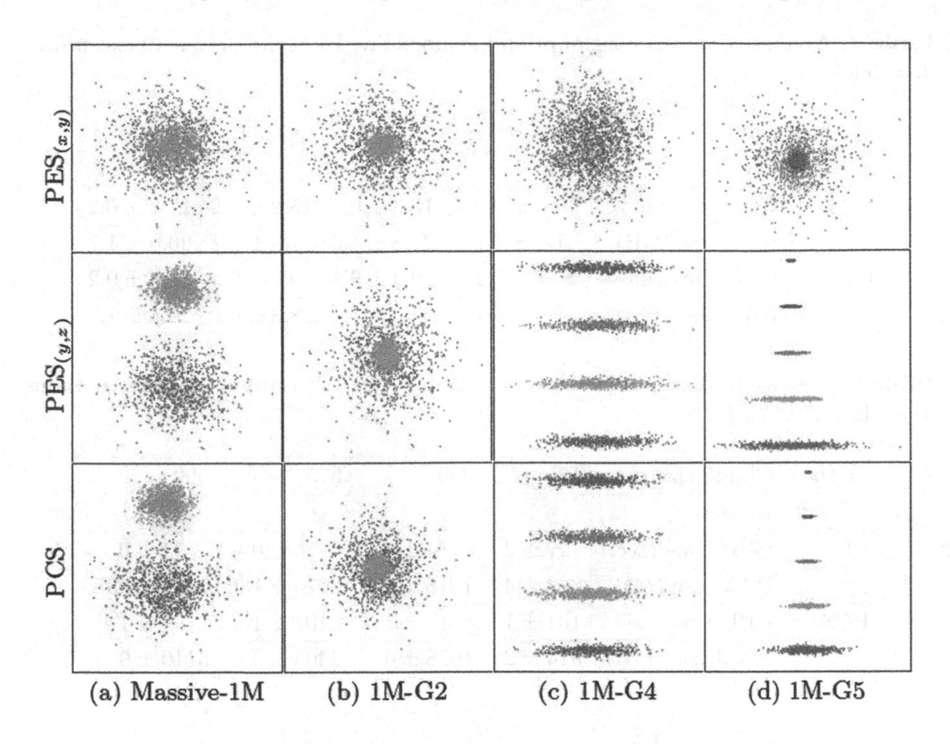

Fig. 13. 2-D embeddings of four massively multidimensional data sets.

Their configurations can be seen and they can be generated by downloading the generator script (*vide* Table 1).

PES and PCS shine in such settings: as incremental techniques, they allow embedding $\mathcal{B}$ observations at a time, controlling the amount of memory used during the embedding process. In contrast, the massive nature of these data sets prevents t-SNE from running and the Pearson and Spearman coefficients from being calculated, since the pairwise distances exceeds available memory during execution. The employment of pairwise distance reliant tools in Big Data scenarios is evidently problematic.

Figure 12 depict the 3-D PES embeddings of the Big Data sets. The entire massive collection were properly depicted by PES in 3-D. The small Gaussian mean differences accumulates as d increases, resulting in linearly separable clusters for a sufficiently large d. No distortion can be seen whatsoever.

The 2-D embeddings of the same collection are exposed in Fig. 13. The top-down approach of $\text{PES}_{(x,y)}$ to produce 2-D embeddings resulted in visualizations with inter-cluster overrun to the point where one might say that, without any contextualization, it is impossible to discern the clusters. However, thanks to its flexibility, PES allows other feature combinations: $\text{PES}_{(y,z)}$, for instance, portrays more meaningful visualizations of *Massive*-1*M*, 1*M*-*G*4 and 1*M*-*G*5. PCS manages to display distinct clusters in *Massive*-1*M*, correctly depicts the

Table 4. Average resources consumption in embedding Big Data across 10 executions ($\mathcal{B} = 500$).

Method	Data Set	Massive-1M	1M-G2	1M-G4	1M-G5
	Resource				
PES	CPU time (min.)	16.0 ± 0.2	15.3 ± 0.1	15.2 ± 0.2	17.2 ± 0.2
	RAM use (GB)	27.2 ± 1.7	27.8 ± 1.3	27.9 ± 1.1	26.0 ± 1.7
PCS	CPU time (min.)	20.7 ± 0.2	19.4 ± 0.1	19.6 ± 0.1	22.5 ± 0.2
	RAM use (GB)	27.9 ± 1.1	28.0 ± 1.1	28.3 ± 0.0	26.8 ± 1.7

Table 5. Average resources consumption in embedding Big Data across 10 executions ($\mathcal{B} = [125, 250, 375, 500]$).

Method	Chunk size $\mathcal{B}$	125	250	375	500
	Resource				
PES	CPU time (sec.)	97 ± 2	84 ± 1	78 ± 0	74 ± 0
	RAM use (MB)	823 ± 44	1518 ± 121	2075 ± 139	2801 ± 235
PCS	CPU time (sec.)	161 ± 1	119 ± 1	107 ± 1	99 ± 1
	RAM use (MB)	974 ± 2	1688 ± 4	2403 ± 3	3110 ± 9

overlapping classes of $1M\text{-}G2$ and separates the classes of $1M\text{-}G4$ and $1M\text{-}G5$. PCS embeddings are remarkably similar to those of $\text{PES}_{(y,z)}$.

For the resource consumption analyses, the average memory and time used in embedding with both techniques were logged across 10 executions with $\mathcal{B} = 500$. Table 4 exposes the results. In spite of sharing the same ATC, PES performed almost 30% faster than PCS overall. In another test, whose results are in Table 5, a set containing $2,000$ random observations of $100,000$ dimensions was embedded by PES and PCS with varying $\mathcal{B}$. PES recurrently took less time than PCS, although the discrepancy diminished as $\mathcal{B}$ increased. PCS used more memory than PES while performing its routine on both analyses.

The data sets of the Big Data collection have, on average, 16.28 GB, while their embeddings have around 91.25 KB—a compression rate greater than 99%. Hence, DR is also an attractive data compression technique, whose application in Big Data is only attainable by incremental techniques such as PES or PCS.

5 Conclusion

Throughout this paper, we introduced the geometry-based Pyramidal Embedding System (PES), showcasing its ability to preserve global data structures when applying DR. Its ATC was proved to be $\mathcal{O}(n)$, attesting its efficiency. Synthetic data set embeddings validated PES in specific scenarios, while real data set embeddings allowed the investigation of its potential application in real data. Its successful embedding of big data sets proved PES capacity in working with

Big Data to produce 2-D and 3-D representations. Divergences between visual and numeric analyses lead to the belief that the Pearson and Spearman coefficients might not properly reflect embedding quality. They are also unusable in Big Data scenarios. Despite that, PES attained the best scores overall.

PES proved to be an auspicious alternative to current DR algorithms. It has many desirable traits, e.g., Big Data readiness, reduced asymptotic time complexity, ability to batch-load massive sets of data and observance to global structures. PES surpasses PCS, as it can resort to 3-D embeddings to retain more information without obstructing visualization.

As prospective work, we intend to explore further the potential of PES by comparing it in a wider selection of data sets with a bigger number of counterparts. Another valuable contribution would be to compare the accuracy of models built on PES embeddings against other DR approaches, aiming to gauge their applicability in pre-processing data. Moreover, a parallel implementation of the algorithm could be developed to enhance its day-to-day performance with massive sets of data whenever memory is widely available.

References

1. Blouvshtein, L., Cohen-Or, D.: Outlier detection for robust multi-dimensional scaling. IEEE Trans. Pattern Anal. Mach. Intell. **41**(9), 2273–2279 (2018)
2. Dua, D., Graff, C.: UCI machine learning repository (2017). http://archive.ics.uci.edu/ml
3. Flexa, C., Gomes, W., Viademonte, S., Junior, C.S., Alves, R.: A geometry-based approach to visualize high-dimensional data. In: 2019 8th Brazilian Conference on Intelligent Systems (BRACIS), pp. 186–191. IEEE (2019)
4. Gracia, A., González, S., Robles, V., Menasalvas, E.: A methodology to compare dimensionality reduction algorithms in terms of loss of quality. Inf. Sci. **270**, 1–27 (2014)
5. Gracia, A., González, S., Robles, V., Menasalvas, E., Von Landesberger, T.: New insights into the suitability of the third dimension for visualizing multivariate/multidimensional data: a study based on loss of quality quantification. Inf. Visual. **15**(1), 3–30 (2016)
6. Li, H., Robini, M.C., Yang, F., Magnin, I., Zhu, Y.: Cardiac fiber unfolding by semidefinite programming. IEEE Trans. Biomed. Eng. **62**(2), 582–592 (2014)
7. Maaten, L., Hinton, G.: Visualizing data using t-SNE. J. Mach. Learn. Res. **9**(Nov), 2579–2605 (2008)
8. Palese, L.L.: A random version of principal component analysis in data clustering. Comput. Biol. Chem. **73**, 57–64 (2018)
9. Praveena, M.A., Bharathi, B.: A survey paper on big data analytics. In: 2017 International Conference on Information Communication and Embedded Systems (ICICES), pp. 1–9. IEEE (2017)
10. Su, Y., Lin, R., Kuo, C.C.J.: Tree-structured multi-stage principal component analysis (TMPCA): theory and applications. Expert Syst. Appl. **118**, 355–364 (2019)
11. Ultsch, A.: Clustering with som: U*c. In: Proceedings of Workshop on Self-Organizing Maps, Paris, France, pp. 75–82 (2005)
12. Van Der Maaten, L.: Accelerating t-SNE using tree-based algorithms. J. Mach. Learn. Res. **15**(1), 3221–3245 (2014)

13. Wang, Y., et al.: A perception-driven approach to supervised dimensionality reduction for visualization. IEEE Trans. Vis. Comput. Graph. **24**(5), 1828–1840 (2017)
14. Wei, X., et al.: Reconstructible nonlinear dimensionality reduction via joint dictionary learning. IEEE Trans. Neural Netw. Learn. Syst. **30**(1), 175–189 (2018)
15. Weisstein, E.W.: Pyramid. Wolfram MathWorld (2002)
16. Yang, L., Song, S., Gong, Y., Gao, H., Wu, C.: Nonparametric dimension reduction via maximizing pairwise separation probability. IEEE Trans. Neural Netw. Learn. Syst. **30**(10), 3205–3210 (2019)

Exceptional Survival Model Mining

Juliana Barcellos Mattos[1](✉), Eraylson G. Silva[1],
Paulo S. G. de Mattos Neto[1], and Renato Vimieiro[2]

[1] Centro de Informática, Universidade Federal de Pernambuco, Recife, PE, Brazil
julianabarcellosmattos@gmail.com, {jbm4,egs,psgmn}@cin.ufpe.br
[2] Universidade Federal de Minas Gerais, Belo Horizonte, MG, Brazil
rvimieiro@dcc.ufmg.br

Abstract. The development of treatments based on the patient's individual characteristics has been an emergent medical approach. The objective is to improve individual responses and overall survival. Thus, there is a need for computational tools able to identify and describe subgroups of patients for which the survival response significantly differs from the overall behaviour. However, there are few algorithms that address this matter. The majority of works of literature aim at building predictive models rather than understanding the characteristics that delineate subgroups with unusual survival. The approaches that provide understanding on factors that interfere in the survival behaviour usually resort to the stratification of the data based on previously known variable's interactions, lacking the ability to shed light into new, possibly unknown, interactions. In contrast to the existent predictive approaches, we propose the use of supervised descriptive pattern mining in order to discover local patterns able to describe subsets of patients that present unusual survival behaviour. In this paper, we present the ESM-AM (Exceptional Survival Model Ant Miner) algorithm, an Exceptional Model Mining approach to the discovery of subgroups with exceptional survival functions that explores the use of ant-colony optimization as search heuristic for the pattern mining task.

Keywords: Exceptional model mining · Survival analysis · Ant-colony optimization · Supervised descriptive pattern mining

1 Introduction

The recent development of large-scale databases and powerful methods for characterising patients is driving medicine towards more personalised (and less aggressive) individual interventions with the ultimate goal of improving the patients' prognosis and survival outcome. Many recent medical works strive for effective methods that can provide a better understanding on factors that interfere in survivability in order to subdivide populations of patients into more specific and uniform subgroups with relation to their survival behaviour. In this sense, many methods for Survival Analysis [10] have been developed over the

R. Cerri and R. C. Prati (Eds.): BRACIS 2020, LNAI 12320, pp. 307–321, 2020.
https://doi.org/10.1007/978-3-030-61380-8_21

years. For most applications, the goal is either to predict the time for the occurrence of a given event or to model the impact of covariates on survival response. Traditional statistical approaches [3] comprise parametric and semi-parametric techniques. The limitations resultant from their distributional and restrictive assumptions have motivated the development of new methods. Machine learning techniques [24] present the advantages of modelling non-linear relationships and deliver high-quality results. These techniques, however, comprise global predictive models usually hard to understand.

Rule induction is a traditional data mining technique that aims at extracting and enumerating patterns by learning sets of rules from a given data. The simplicity of rule models at the representation of understandable patterns gives this technique broad applicability in both predictive and descriptive data mining approaches. We can find in the literature a few rule-based works on survival analysis that strive to extract understandable knowledge from the data. Bazan *et al.* [2] aim at finding groups of patients with different survival estimates by applying rough sets theory to induce a set of decision rules from data. Pattaraintakorn and Cercone [19] employed a rough sets hybrid system to predict the survival time. Liu *et al.* [13] propose the characterisation of high-risk patients by applying the bump hunting method to generate rules. Kronek and Reddy [11] propose a methodology based on Logical Analysis of Data (LAD) for constructing survival patterns used to estimate the survival probability distribution of observations. Wróbel [25] proposes the use of a survival tree to generate a decision list of rules to predict the survival behaviour of new examples. Sikora *et al.* [23] use the survival time to classify patients into *dead* and *alive* classes, and employ a sequential covering strategy for inducing a set of classification rules in order to predict whether the patient would survive longer than a given time. A class of censored observations is used for a post-processing filtration of relevant rules. In [22], the authors handle censoring by proposing a weighting scheme that assigns censored observation to *dead* and *alive* classes with a given probability. Wróbel *et al.* [26] present the LR-Rules, a top-down greedy covering algorithm to induce accurate models for survival time prediction. In order to guide the rule induction process, the authors propose the use of a rule quality measure based on a statistical test that assesses the difference between the survival curve of the subset represented by the rule and its complement. The rules' consequent comprises a survival function fitted to the examples satisfying its premise.

Although the sets of rules deliver explanation over the data, the aforementioned rule-based approaches aim either to predict survival distribution or to classify new observations. Usually, the ones that strive to distinguish subgroups with different survival characteristics resort to the stratification of the time variable or impel the observations to fit predefined classes. This context motivates us to pose the following research question: *is there a more effective approach to characterise subgroups with unusual survival behaviour?*

In this sense, *supervised descriptive rule discovery* [16] is the induction of descriptive rules from labelled data, unifying both descriptive and predictive rule induction approaches. The main goal is, therefore, to understand the underlying phenomena (according to a target) rather than to arbitrarily explain the data or classify new instances. One great advantage of this approach is the ability to discover local patterns. We call *Supervised Descriptive Pattern Mining* (SDPM) the set of tasks that use local patterns to provide any descriptive knowledge about a property of interest.

Subgroup Discovery (SD) [1,8] is one of the earliest SDPM tasks that aims at the discovery of local patterns that describe *interesting* subsets of the data. *Interestingness* is defined as distributional unusualness of a specific target attribute over the subset when compared to its distribution on the whole data (or on the subset's complement). However, one can understand that a deviating distribution of one target attribute does not encompass all forms of interestingness [5] and, thus, SD becomes impracticable in cases where a single target variable cannot express the property of interest. In this context, Exceptional Model Mining (EMM) [5,12] can be considered a multi-target generalisation of SD task, where the concept of the property of interest is extended to a target model. Given a model that best represents the data and thus constitutes the property of interest, the EMM task searches for subgroups of the data for which the model fitted to the subgroup differs substantially from the respective model fitted to the whole data (or to the subgroup's complement).

A perspective from EMM on survival analysis comprehends the data as a potential composition of different data subsets that present distinct survival behaviour. Hence, the EMM rule-based model comprises individual local patterns related to a target function instead of predictive models. Park *et al.* [17] propose an SD approach to analyse survival in breast cancer by presenting a tree-based rule induction approach that uses mean survival time as the target variable. High/low survival groups are derived from subgroups with average survival time significantly higher/smaller than the average for the remaining samples. However, the distributional unusualness of the mean survival time does not encompass all useful insights about the cohort's survival rate provided by a survival function. Although EMM poses a robust computational method for providing comprehensible identification of subgroups with exceptional survival response, there is still little research on this topic.

Finally, the combinatorial nature of rule induction processes poses a great challenge concerning computational cost. When focusing on EMM, these challenges can be even more significant, once the task usually deals with large sets of numerical data and induction of numerical models. Therefore, the strategy employed for traversing large search spaces is an essential issue for a good performance of the method. Apart from exhaustive algorithms, existing approaches on SD and EMM tasks mainly explore the use of greedy heuristic *Beam Search* [7] and of *Evolutionary Computing* [4,14,15,21]. However, to the best of our knowledge, there is no work on EMM exploring the use of bio-inspired meta-heuristic as a search strategy to the pattern discovery process.

In this paper, we address the problem of discovering subgroups of patients with unusual survival behaviour through the perspective of EMM in contrast to the majority of existent predictive approaches. Rather than strive to build accurate models, we aim at building models capable of describing the local exceptionalities existent in the data. Hence, the main goal of this paper is to present the Exceptional Survival Model Ant Miner (ESM-AM) algorithm, an EMM framework designed for discovering subgroups with exceptional survival functions. ESM-AM relies on a measure of exceptionality between survival curves to guide the search for subgroups and returns a ruleset in which each rule describes a discovered subgroup. Differently from most EMM frameworks that employ greedy heuristics, our algorithm employs Ant-Colony Optimisation (ACO) as a search strategy for the rule induction process. We assess the performance of ESM-AM by comparing it to the covering greedy algorithm LR-Rules that, to the best of our knowledge, is the only work in literature for inducing rules based on the difference between survival functions. We evaluate the performance of our ACO-based approach in terms of the characteristics of the resultant rulesets, by comparing ESM-AM and LR-Rules results on 14 survival data sets. We also analyse some individual discovered rules in order to evaluate whether our EMM approach is capable of discovering exceptional subgroups and retrieving essential characteristics from the data. The remainder of this paper is organised as follows. Section 2 gives a brief review of Survival Analysis and introduces the main concepts of EMM framework. Section 3 describes the algorithm proposed in this paper, followed by Sect. 4 that presents the experiments and achieved results. Finally, in Sect. 5, we draw some conclusions and present directions to extend this proposal.

2 Background

2.1 Survival Analysis (SA)

Survival Analysis [10] is the collection of methods and techniques designed to analyse data in which the target variable is the *time* until a given *event* occurs. For *event* one can understand any designated experience of interest that may happen to a subject under study, e.g. a patients' death, relapse from remission, or any other experience. The *time* variable – also referred to as *survival time* – represents the time since the beginning of the study until the occurrence of the event of interest. The main characteristic of survival data that differentiate it from regression problems is the existence of observations for which the survival times are unknown; this phenomenon is called *censoring.* The most common type of censoring happens when an individual's survival time is greater than the time observed during the study – usually because the subject did not suffer the event yet when the study ended or because the subject left the study due to any reason other than the event occurrence. In these cases, we say that the observation is *right-censored.*

Let $\mathbb{S}(A, T, \delta)$ be a survival data set with $|\mathbb{S}|$ observations (instances). Each observation in $\mathbb{S}$ is characterized by a set of descriptive attributes $A = \{A_1, A_2, \ldots, A_{|A|}\}$, a survival time T and a censoring status δ, that indicates whether the subject is censored ($\delta = 0$) or has experienced the event ($\delta = 1$).

Therefore, the ith observation in the data set can be represented as a vector $o_i = (A_{1_i}, A_{2_i}, \ldots, A_{|A|_i}, T_i, \delta_i)$.

In general, survival data is modelled in terms of two functions, namely survival and hazard. The *survival function* $S(t)$ indicates the probability that an individual o_i survives up to a specified future time t, i.e., $S(t) = P(T_i > t)$. The initial value, $S(t = 0) = 1$, represents the fact that no observation has yet suffered the event at the beginning of the study, and thus, the probability of surviving past the initial time is one. The *hazard function* – also referred as conditional failure rate – is usually denoted by $h(t)$ and provides the probability that an individual o_i experiences the event at a certain time t given that it has survived up to that time. In other words, $h(t)$ is a measure of instantaneous failure potential and is the mechanism used at the mathematical modelling of survival data in a large number of survival analysis approaches.

The survival function is usually estimated by the non-parametric Kaplan-Meier (KM) survival estimate [9]. Considering $\mathbb{S}$, we define $\mathcal{T} = \{t_1, t_2, \ldots, t_k | t \in T, k \leq |\mathbb{S}|\}$ the set of unique ordered survival times of $\mathbb{S}$. The estimated probability $\hat{S}(t_j)$ of surviving past a time $t_j \in \mathcal{T}$ is given by Eq. 1:

$$\hat{S}(t_j) = \left(\prod_{\forall t_i \in \mathcal{T}}^{t_{j-1}} \hat{P}(T > t_i | T \geq t_i) \right) \cdot \hat{P}(T > t_j | T \geq t_j) \equiv \hat{S}(t_{j-1}) \left(1 - \frac{d_j}{r_j} \right) \quad (1)$$

where $\hat{S}(t_{j-1})$ is the probability of being alive at the time interval $[t_{j-1}, t_j)$, r_j is the number of patients alive (at risk) just before t_j, and d_j the number of events that happened at the time interval $[t_j, t_{j+1})$; for $t_0 = 0$, $S(t_0) = 1$. The KM survival curve is the plot of the KM survival probabilities against time.

For comparing the survival experience of different groups, the *logrank* [20] statistical test is the most widely used method. It tests the null hypothesis that there is no overall difference between the KM curves of the groups, making use of the events observed within each group versus the number of events that are expected to happen.

Let $G \subseteq \mathbb{S}$ be a group of patients. We define r_j^G the number of patients in G that are at risk just before $t_j \in \mathcal{T}$. The logrank test assumes that the number of events that are expected to happen within a group is proportional to the extent of its risk, i.e. to the proportion r_j^G / r_j. Hence, the number E^G of expected events suffered by G over $\mathcal{T}$ is given by Eq. 2.

$$E^G = \sum_{\forall t_j \in \mathcal{T}} \frac{r_j^G}{r_j} \times d_j \quad (2)$$

For comparing G to its complement $\overline{G} = \mathbb{S} \setminus G$, the logrank test $X^2 \sim \chi_1^2$ is given by Eq. 3, where $O^G(O^{\overline{G}})$ is the number of observed events in $G(\overline{G})$.

$$X^2 = \frac{(O^G - E^G)^2}{E^G} + \frac{(O^{\overline{G}} - E^{\overline{G}})^2}{E^{\overline{G}}} \quad (3)$$

2.2 Exceptional Model Mining (EMM)

The EMM task aims at discovering subgroups that are exceptional in relation to a property of interest. Given a survival data set $\mathbb{S}$(A, T, δ), let A be the set of descriptive attributes used to define subgroups and let $Y = \{T, \delta\}$ be the set of target attributes used to evaluate the described subgroups. Our property of interest is, therefore, defined as the target model $f(Y) = \hat{S}(t)$. The *patterns*, or *descriptions*, are usually taken from a *description language* $\mathcal{D}$ of free choice within EMM design. Here we define a pattern $P \in \mathcal{D}$ as a rule representation in the form

$$P : \quad IF \quad c^1 \ AND \ c^2 \ AND \ \ldots \ AND \ c^L \quad THEN \quad \hat{S}_P(t)$$

where the *conditions* c^l are of the form $a_i = v_{ij}$ for $a_i \in A$ and $v_{ij} \in Domain(a_i)$. We call $L \leq |A|$ the rule's *length*. The consequent $\hat{S}_P(t)$ of the rule comprises the KM model fitted for the observations that satisfy its antecedent.

A *subgroup* corresponding to a pattern P is the set of observations $G_P \subseteq \Omega$ of size $|G_P|$ that are covered by P, i.e. that satisfy its antecedent. The *coverage* of a subgroup is given by the number of observations that it covers. One can understand that *rules* are representations of *patterns/descriptions*, being those three terms equivalent and, thus, implying *subgroups*. The *complement* of a subgroup G_P is the set of observations $\overline{G}_P = \Omega \setminus G_P$ of size $|\Omega| - |G_P|$.

For each subgroup G_P under evaluation, the target model $\hat{S}(t)$ is induced on the set of observations $o_i \in G_P$. Then, the subgroup is evaluated with a *quality measure* $\varphi : \mathcal{D} \rightarrow \mathbb{R}$ that quantifies the difference between the target model fitted on G_P and the target model fitted on $\overline{G}_P$. Here we define the quality measure $\varphi = (1 - p\text{-}value)$ based on the logrank statistical test, assuming values in $(0, 1)$ interval. The smaller the logrank's p-value, the more exceptional is the KM model fitted to the subgroup, and thus the higher is its quality.

3 EMS-AM: Exceptional Survival Model Ant Miner

EMS-AM algorithm is an adaptation of the well-known classification rule induction algorithm Ant-Miner [18]. We adapted the Ant Colony Optimization heuristic to discover subgroups with exceptional KM curves. ESM-AM returns a list of discovered rules of the aforementioned form, and is presented in Algorithm 1.

The algorithm is initialized with an empty list of rules and with a set of uncovered cases comprising all cases in the data set and then it follows a covering-based approach. In each iteration (lines 4–26), a colony of ants constructs a number of rules R_t. Then the best rule R_{best} – according to φ – is selected to be added to the list of discovered rules, and the examples covered by R_{best} are removed from the set of uncovered cases. This process is repeated while the number of remaining uncovered observations do not achieve a maximum threshold or until a maximum number of iterations is reached. Non-significant rules are discarded at a level of significance of α. In case the ant colony is no longer able to discover significant rules, the algorithm is finalized, and the list of rules is returned.

Algorithm 1: High-Level description of ESM-AM algorithm

```
1: uncoveredCases = {all data set cases}
2: DiscoveredRuleList = [ ]
3: it = 0
4: while (uncoveredCases > max_uncovered_cases) or (it < no_of_ants) do
5:    t , j = 1 # ant, convergence indexes
6:    Initialize all trails with the same amount of pheromone
7:    repeat
8:       R_t = ConstructRule(dataset, min_cases_per_rule)
9:       PruneRule(R_t);
10:      PheromoneUpdating(R_t);
11:      if (R_t = R_{t-1}) then
12:         j += 1
13:      else
14:         j = 1;
15:      end if
16:      t += 1;
17:   until (t ≥ no_of_ants) or (j ≥ no_rules_converg)
18:   Choose the best rule R_best;
19:   if quality(R_best) ≥ (1 − α) then
20:      DiscoveredRuleList = DiscoveredRuleList ∪ R_best
21:      uncoveredCases = uncoveredCases \ {examples covered by R_best}
22:   else
23:      break
24:   end if
25:   it += 1
26: end while
27: return: DiscoveredRuleList
```

For the rule induction process (lines 7–17), the ants in the colony generate complete rules following two procedures: rule construction (line 8) and pruning (line 9). Each ant t constructs a rule R_t in a general-to-specific approach by iteratively adding conditions to a initially empty rule until there is no more conditions to be added or until the addition of a condition results in a rule coverage bellow a threshold. The probabilistic choice of a condition $c_{ij} : a_i = v_{ij}$ to be added to the current partial rule depends on both the heuristic function (η) and the pheromone (τ) associated with each condition. It is determined by the probability P_{ij} given in Eq. 4, where $|A|$ is the size of the attribute set, $|D_i|$ is the size of the attribute $a_i \in A$ domain, $x_i = 1$ if the a_i was not yet added to the current rule, or $x_i = 0$ otherwise.

$$P_{ij} = \frac{\eta_{ij} \cdot \tau_{ij}(t)}{\sum_{i=1}^{|A|} x_i \cdot \sum_{j=1}^{|D_i|} (\eta_{ij} \cdot \tau_{ij}(t))} \tag{4}$$

The heuristic function η_{ij} associated with c_{ij} is based on Shannon's entropy (Eq. 5). We considered an initial partition of the observations as those with survival time at least as long as the cohort's average survival time, and those with shorter survival time. The quality of a condition is then the normalized information gain, obtained by further partitioning observations based on it. The class entropy was computed inducing a partition on the observations according to a condition. The heuristic function is given in Eq. 6.

$$H(W|a_i = v_{ij}) = -\sum_{w=1}^{k} P(w|a_i = v_{ij}) \cdot \log_2 P(w|a_i = v_{ij}) \tag{5}$$

$$\eta_{ij} = \frac{\log_2 k - H(W|a_i = v_{ij})}{\sum_{i=1}^{|A|} x_i \cdot \sum_{j=1}^{|D_i|} \log_2 k - H(W|a_i = v_{ij})} \tag{6}$$

After the ant constructs a complete rule R_t, the quality of the discovered subgroup (rule) is computed on the basis of the logrank test as $\varphi = (1 - p\text{-}value)$. Then, a rule pruning procedure iteratively removes conditions from the rule's antecedent, each time eliminating the condition c_{ij} that leads to the largest improvement in the rule quality φ. The pruning stops when no conditions can be removed without decreasing the rule's quality, or when the rule contains only one condition.

Lastly, the pheromone updating process increments the amount of pheromone associated to the conditions used on the pruned R_t according to Eq. 7, where $\tau_{ij}(t+1)$ is the amount of pheromone associated to c_{ij} for the next ant iteration. For the conditions not used on R_t, evaporation process is simulated by the normalization of τ values in $(t+1)$.

$$\tau_{ij}(t+1) = \tau_{ij}(t) + \varphi \cdot \tau_{ij}(t) \tag{7}$$

This ant-based rule induction process is repeated until all ants in the colony have constructed their rules, or until the ants converge to a single rule – according to a threshold of identical sequential rules. For each new algorithm iteration (line 4), a new colony is created and all available conditions receive the same initial amount of pheromone.

Finally, the ESM-AM has five user-defined parameters: (1) *no_of_ants* defines the size of the ant colony; (2) *max_uncovered_cases* defines the maximum number of remaining uncovered cases; (3) *no_rules_converg* defines the number of sequential identical constructed rules R_t in order to consider that the ants have converged to a solution (rule); (4) *min_cases_per_rule* is the minimum required rule coverage; and (5) α defines the level of significance of the logrank test.

4 Experiments and Results

4.1 Experimental Setup

The performance of the ESM-AM algorithm was evaluated on 14 real-world data sets listed in Table 1. All data sets were processed by removing observations containing missing values and by filtering the features. Table 2 presents the

Table 1. Characteristics of 14 datasets used in the experimental study: the number of observations (#obs), the number of descriptive attributes (#att), the number of discretized descriptors (#disc), the percentage of censored observations (%cens), and the survival event description (Event)

Dataset	#obs	#attr	#disc	%cens	Subject of research	Event
actg320	1151	11	3	91.66	HIV-infected patients	AIDS diagnosis/death
breast-cancer	196	80	78	73.98	Node-Negative breast cancer	distant metastasis
cancer	168	7	5	27.98	Advanced lung cancer	death
carcinoma	193	8	1	27.46	Carcinoma of the oropharynx	death
gbsg2	686	8	5	56.41	Breast cancer	recurrence
lung	901	8	0	37.40	Early lung cancer	death
melanoma	205	5	3	72.20	Malignant melanoma	death
mgus	176	8	6	6.25	Monoclonal gammopathy	death
mgus2	1338	7	5	29.90	Monoclonal gammopathy	death
pbc	276	17	10	59.78	Primary biliary cirrhosis	death
ptc	309	18	1	93.53	Papillary thyroid carcinoma	recurrence/progression
uis	575	9	4	19.30	Drug addiction treatment	return to drug use
veteran	137	6	3	6.57	Lung cancer	death
whas500	500	14	6	57.00	Worcester Heart Attack	death

Table 2. Selected descriptive attributes of the data sets with feature selection

Dataset	Descriptive attributes
actg320	tx, txgrp, strat2, sex, raceth, ivdrug, hemophil, karnof, cd4, priorzdv, age
mgus	age, sex, dxyr, pcdx, alb, creat, hgb, mspike
ptc	risk_group, histological_type, age, sex, path_t_stage, path_n_stage, path_m_stage, tumor_status, exome, extrathyroidal_extension, mrna_cluster, mirna_cluster, arm_scna_cluster, methylation_cluster, disease_stage, primary_exome, lowpass, wgs_status
whas500	age, gender, hr, sysbp, diasbp, bmi, cvd, afb, sho, chf, av3, miord, mitype, los

descriptive attributes for the sets that were filtered. As the algorithm only copes with categorical attributes, all numerical variables were discretized with K-Means into five interval categories. The five ESM-AM user-defined parameters were set as follows: $no\_of\_ants = 3000$; $max\_uncovered\_cases = 0$; $no\_rules\_converg = 10$; $min\_cases\_per\_rule = 10$; and $\alpha = 0.05$.

In order to evaluate the models generated by our proposed ACO heuristic approach, ESM-AM algorithm was compared with the LR-Rules [26], a greedy covering algorithm. We used the LR-Rules default parameters defined in the available[1] implementation. The rule models resultant from both algorithms were evaluated according to the following characteristics: number of discovered rules;

[1] LR-Rules algorithm: https://github.com/adaa-polsl/LR-Rules/releases.

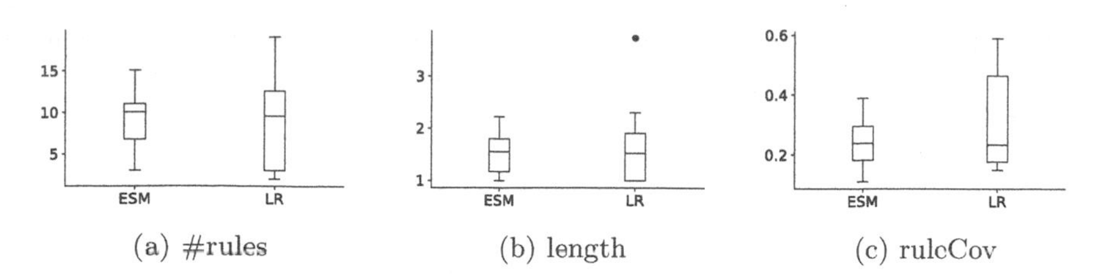

(a) #rules (b) length (c) ruleCov

Fig. 1. Boxplots of the ESM-AM (ESM) and LR-Rules (LR) algorithms results for: (a) the number of discovered rules; (b) the average rule length; and (c) the average relative rule coverage

average rule length; average relative rule coverage; rule set coverage; and integrated Brier score (IBS). The Brier score (BS) [6] measures the square difference between an observation survival status δ_i and its estimated survival probability $\hat{S}(t)$, in a given time T^*. The BS value for an observation o_i (incorporating censoring) is given by Eq. 8. The IBS is the Brier score integrated over all survival times T for all n observations, and is given by Eq. 9, where $\hat{G}(t)$ is the KM estimate of the censoring distribution, obtained from estimating the survival function for $\delta = (1 - \delta)$. The IBS was calculated for each discovered rule and summed over the entire ruleset.

$$BS_i(T^*) = \begin{cases} \frac{1}{\hat{G}(T_i)}[0 - S(T^*)]^2 & \text{if } T_i \leq T^*, \delta_i = 1 \\ \frac{1}{\hat{G}(T^*)}[1 - S(T^*)]^2 & \text{if } T_i > T^* \\ 0 & \text{otherwise} \end{cases} \tag{8}$$

$$IBS = \frac{1}{max(T)} \int_0^{max(T)} \left(\frac{1}{n} \sum_{i=1}^{n} BS_i(T^*) \right) dT^* \tag{9}$$

4.2 Results Analysis

The results for both ESM-AM and LR-Rules algorithms on each data set are presented in Table 3. All figures and additional results, as well as the data sets used in the experiments, are available on ESM-AM website[2].

The ESM-AM algorithm returned rule models with, on average, 9.43 rules of size 1.52 (condition), compared to the LR-Rules' average of 8.93 discovered rules of size 1.63. We notice then that ESM-AM was able of generating compact models concerning both the size of the ruleset and the length of the rules. The coverage of ESM-AM rules was, on average, 25% of the total cases in the data sets, comprising rules that neither cover the majority of the cases nor very small groups. Figure 1 shows the boxplots of the performance of both algorithms with relation to the number of discovered rules, rule length and rule coverage. We

[2] ESM-AM algorithm website: https://github.com/jbmattos/ESM-AM_bracis2020.

also notice that comparing to LR-Rules, ESM-AM results presented smaller variability. When evaluating the coverage of the final rulesets (*setCov*), ESM-AM showed greater variability, presenting in some cases, a higher percentage of observations that remained not covered by any rule. For the IBS results, ESM-AM algorithm presented an average of 0.15 comparing to 0.18 presented by LR-Rules. One could understand the IBS as a measure of the quadratic error between the survival estimates of the observations covered by a rule and their true survival status. Therefore, we notice that the ESM-AM algorithm was able to discover more homogeneous subgroups concerning survival response. Finally, for a level of significance of 5%, the Wilcoxon test showed statistically significant difference between ESM-AM and LR-Rules performances only in terms of the *setCov* criterion (p-value = 0.036).

Besides, to evaluate ESM-AM final models in terms of the subgroups discovered by our EMM framework, we assess whether the induced rules present statistically significant survival functions. Figure 2 presents the KM curves for the rulesets induced on *ptc* and *whas500* data sets. The plots additionally include the cohort's KM curve, given by the *Db* curve. It is possible to observe the significant difference between the survival curve of the study cohort in comparison to the curves induced over the subgroups discovered by ESM-AM, indicating that the algorithm is able to identify local patterns with significant distinct survival response. In a more detailed analysis of the individual discovered rules, we found that the algorithm was able to retrieve information on attributes that stratify the data into different survival experiences.

Table 3. Characteristics on the resultant rule models for ESM-AM (ESM) and LR-Rules (LR) algorithms: the number of discovered rules (#rules), the average rule length (length), the average relative rule coverage (cov±std, ruleCov), the rule set coverage(setCov), and integrated Brier score on the rule set (IBS). Bold values represent the best results.

Metrics	#rules		length		ruleCov		setCov		IBS	
Algorithms	ESM	LR	ESM	LR	ESM	LR	ESM	LR	ESM	LR
actg320	**9**	15	**2.22**	3.73	0.26 ± 0.16	0.15 ± 0.16	1.00	1.00	**0.43**	0.45
breast-cancer	**11**	19	**1.18**	1.95	0.24 ± 0.15	0.17 ± 0.10	0.94	**0.98**	0.01	0.01
cancer	11	**9**	2.00	**1.78**	0.17 ± 0.16	0.25 ± 0.16	0.74	**1.00**	0.08	**0.07**
carcinoma	10	**3**	1.80	**1.00**	0.27 ± 0.20	0.33 ± 0.30	0.99	0.99	0.06	**0.02**
gbsg2	14	**10**	**1.79**	2.30	0.19 ± 0.24	0.22 ± 0.23	1.00	1.00	0.13	**0.11**
lung	9	**7**	**1.00**	1.14	0.35 ± 0.13	0.35 ± 0.14	1.00	1.00	0.12	**0.09**
melanoma	6	**2**	1.00	1.00	0.39 ± 0.17	0.50 ± 0.06	1.00	1.00	0.02	**0.01**
mgus	13	**11**	**1.62**	1.73	0.11 ± 0.08	0.18 ± 0.11	0.71	**1.00**	0.01	0.01
mgus2	**6**	18	**1.17**	1.50	0.18 ± 0.09	0.17 ± 0.09	0.70	**1.00**	**0.38**	1.25
pbc	11	**3**	1.36	**1.00**	0.20 ± 0.28	0.59 ± 0.37	1.00	1.00	0.03	**0.02**
ptc	4	**2**	1.50	**1.00**	0.30 ± 0.36	0.50 ± 0.42	1.00	1.00	0.33	**0.08**
uis	15	**13**	**2.00**	2.08	0.23 ± 0.18	0.22 ± 0.17	0.99	**1.00**	0.36	**0.27**
veteran	**10**	11	1.60	**1.55**	0.18 ± 0.08	0.18 ± 0.08	0.84	**1.00**	0.16	**0.09**
whas500	3	**2**	1.00	1.00	0.38 ± 0.23	0.50 ± 0.19	1.00	1.00	0.04	**0.03**

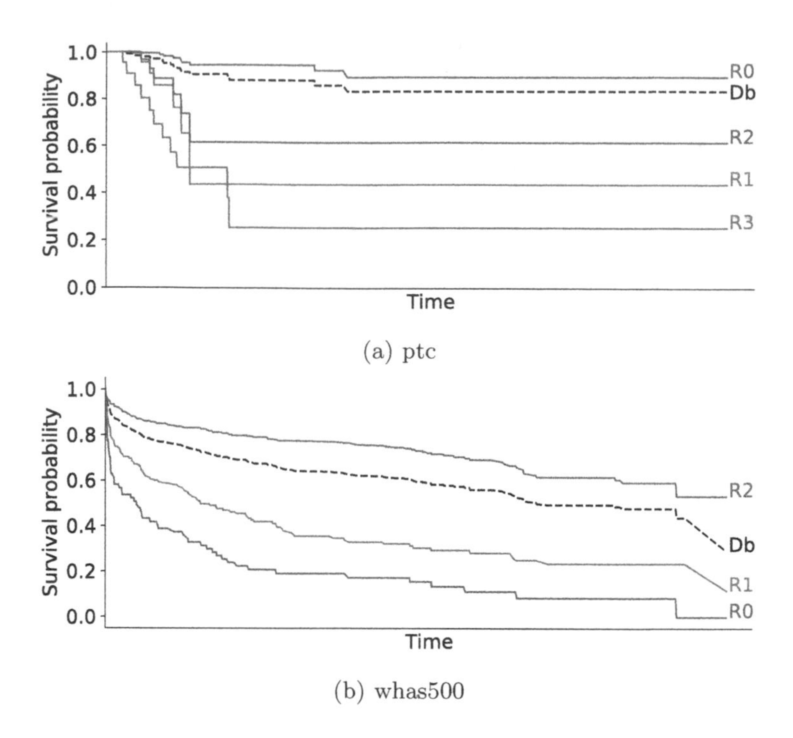

(a) ptc

(b) whas500

Fig. 2. Analysis of the discovered rules for *ptc* (a) and *whas500* (b) data sets; the *Db* curves represent the KM estimates on the entire cohort

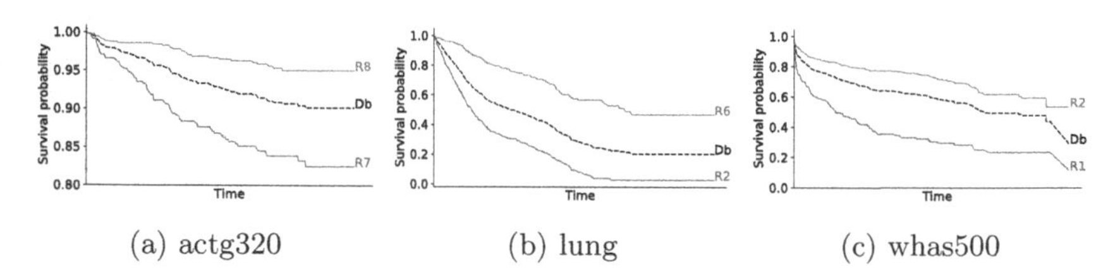

(a) actg320 (b) lung (c) whas500

Fig. 3. Analysis of individual rules induced for *actg320* (a), *lung* (b) and *whas500* (c) datasets; the *Db* curves represent the KM estimates of the study cohort

In the *actg320* data set, the *strat2* variable represents the counting of cells with expression of the CD4 protein, dividing the observations into low/high ($strat2 = 0$ / $strat2 = 1$) counting – where a low counting imply a higher risk for the patient. Among the nine resultant rules induced on this data set, the algorithm recovered such information presenting the following two rules: **R7**: $\{strat2 = 0\}$ and **R8**: $\{strat2 = 1\}$. Figure 3a presents the KM plot of both rules reflecting the expected survival behaviour.

In the *lung* data set, the $stage1 = \{1, 2, 3\}$ variable reflects the overall stage of lung cancer, for $stage1 = 1$ earlier than $stage1 = 3$. For the rule set induced on this data set, the ESM-AM algorithm returned also nine rules, two of them:

R2: $\{stage1 = 3\}$ and **R6**: $\{stage1 = 1\}$. Figure 3b presents the KM curves for both rules, showing that the survivability is better for early lung cancer stage.

In the *whas500* data set, the *chf* variable stands for *congestive heart complications*, dividing the observations into a group of patients that present complications and the ones that do not. The ESM-AM algorithm returned a rule set comprising three rules, two of them: **R1**: $\{chf = True\}$ and **R2**: $\{chf = False\}$. Figure 3c present the plot of both rules, showing that the presence of heart complications decreases the chances of survival.

5 Conclusions

In this paper, we introduce a novel approach to the discovery of subgroups with unusual survival behaviour based on supervised descriptive pattern mining, in contrast to the predictive approaches existent in literature. We presented the ESM-AM (Exceptional Survival Model Ant Miner) algorithm, an EMM framework that uses ACO meta-heuristic for the rule induction process. The algorithm returns a rule set where each rule can be understood as the description of a subgroup that is exceptional with relation to its survival function. Our proposed algorithm is the first approach for EMM task that explores a bio-inspired metaheuristic as search heuristic.

We evaluated our proposal assessing its capability of returning accurate rule models and of discovering interesting subgroups. Therefore, we tested our ACO-based heuristic approach to the discovery of local survival exceptionalities on 14 data sets. The performance of ESM-AM was evaluated in comparison to the LR-Rules algorithm – a greedy covering rule induction algorithm for survival data analysis. Our approach achieved competitive results concerning characteristics of the rulesets, performing similarly to LR-Rules with relation to model size, rule length, rule coverage and IBS. When comparing to LR-Rules, ESM-AM algorithm returned compact rule models and lower IBS, i.e. a smaller difference between a subgroup's survival function and the true survival experience of the observations that it comprises. The low IBS also indicates that the rules discovered by the ESM-AM comprise homogeneous subgroups with respect to survival behaviour. When assessing the algorithm's capability of discovering local survival behaviour exceptionalities, we notice that the ESM-AM was able to discover significant subgroups and to identify data characteristics that interfere in survival experience.

Finally, there are different directions to expand this study: (1) cope with numerical attributes; (2) investigate other quality measures for the subgroups, in other to consider not only its exceptionality but also its coverage, according to the definitions of EMM framework; (3) investigate new heuristic functions and new pheromone updating procedures for the ACO meta-heuristic in order to better capture survival relations on the induction process; (4) tackle problems such as pattern's redundancy, high-dimensionality and false statistical discoveries; and (5) expand the results' analysis with further experimental statistical procedures and more detailed exploratory data analysis.

Acknowledgment. This study was financed in part by the Coordenação de Aperfeiçoamento de Pessoal de Nível Superior - Brasil (CAPES) - Finance Code 001 and by the National Council for Scientific and Technological Development – CNPq.

References

1. Atzmueller, M.: Subgroup discovery. Wiley Interdiscip. Rev.: Data Mining Knowl. Discov. **5**, 35–49 (2015)
2. Bazan, J., Osmólski, A., Skowron, A., Ślçezak, D., Szczuka, M., Wróblewski, J.: Rough set approach to the survival analysis. In: Alpigini, J.J., Peters, J.F., Skowron, A., Zhong, N. (eds.) RSCTC 2002. LNCS (LNAI), vol. 2475, pp. 522–529. Springer, Heidelberg (2002). https://doi.org/10.1007/3-540-45813-1_69
3. Bradburn, M.J., Clark, T.G., Love, S., Altman, D.: Survival analysis part ii: multivariate data analysis-an introduction to concepts and methods. Br. J. Cancer **89**(3), 431 (2003)
4. Carmona, C.J., González, P., del Jesus, M.J., Herrera, F.: Overview on evolutionary subgroup discovery: analysis of the suitability and potential of the search performed by evolutionary algorithms. Wiley Interdiscip. Rev.: Data Min. Knowl. Discov. **4**(2), 87–103 (2014)
5. Duivesteijn, W., Feelders, A.J., Knobbe, A.: Exceptional model mining. Data Min. Knowl. Discov. **30**(1), 47–98 (2016). https://doi.org/10.1007/s10618-015-0403-4
6. Graf, E., Schmoor, C., Sauerbrei, W., Schumacher, M.: Assessment and comparison of prognostic classification schemes for survival data. Stat. Med. **18**(17–18), 2529–2545 (1999)
7. Helal, S.: Subgroup discovery algorithms: a survey and empirical evaluation. J. Comput. Sci. Technol. **31**(3), 561–576 (2016). https://doi.org/10.1007/s11390-016-1647-1
8. Herrera, F., Carmona, C.J., González, P., Del Jesus, M.J.: An overview on subgroup discovery: foundations and applications. Knowl. Inf. Syst. **29**(3), 495–525 (2011). https://doi.org/10.1007/s10115-010-0356-2
9. Kaplan, E.L., Meier, P.: Nonparametric estimation from incomplete observations. J. Am. Stat. Assoc. **53**(282), 457–481 (1958)
10. Kleinbaum, D.G.: Survival analysis, a self-learning text. Biomet. J.: J. Math. Methods Biosci. **40**(1), 107–108 (1998)
11. Kronek, L.P., Reddy, A.: Logical analysis of survival data: prognostic survival models by detecting high-degree interactions in right-censored data. Bioinformatics **24**(16), i248–i253 (2008)
12. Leman, D., Feelders, A., Knobbe, A.: Exceptional model mining. In: Daelemans, W., Goethals, B., Morik, K. (eds.) ECML PKDD 2008. LNCS (LNAI), vol. 5212, pp. 1–16. Springer, Heidelberg (2008). https://doi.org/10.1007/978-3-540-87481-2_1
13. Liu, X., Minin, V., Huang, Y., Seligson, D.B., Horvath, S.: Statistical methods for analyzing tissue microarray data. J. Biopharm. Stat. **14**(3), 671–685 (2004)
14. Lucas, T., Silva, T.C., Vimieiro, R., Ludermir, T.B.: A new evolutionary algorithm for mining top-k discriminative patterns in high dimensional data. Appl. Soft Comput. **59**, 487–499 (2017)
15. Lucas, T., Vimieiro, R., Ludermir, T.: SSDP+: a diverse and more informative subgroup discovery approach for high dimensional data. In: 2018 IEEE Congress on Evolutionary Computation (CEC), pp. 1–8. IEEE (2018)

16. Novak, P.K., Lavrač, N., Webb, G.I.: Supervised descriptive rule discovery: a unifying survey of contrast set, emerging pattern and subgroup mining. J. Mach. Learn. Res. **10**(Feb), 377–403 (2009)
17. Park, J.V., Park, S.J., Yoo, J.S.: Finding characteristics of exceptional breast cancer subpopulations using subgroup mining and statistical test. Expert Syst. Appl. **118**, 553–562 (2019)
18. Parpinelli, R.S., Lopes, H.S., Freitas, A.A.: Data mining with an ant colony optimization algorithm. IEEE Trans. Evol. Comput. **6**(4), 321–332 (2002)
19. Pattaraintakorn, P., Cercone, N.: A foundation of rough sets theoretical and computational hybrid intelligent system for survival analysis. Comput. Math. Appl. **56**(7), 1699–1708 (2008)
20. Peto, R., et al.: Design and analysis of randomized clinical trials requiring prolonged observation of each patient. ii. Analysis and examples. Br. J. Cancer **35**(1), 1 (1977)
21. Pontes, T., Vimieiro, R., Ludermir, T.B.: SSDP: a simple evolutionary approach for top-k discriminative patterns in high dimensional databases. In: 2016 5th Brazilian Conference on Intelligent Systems (BRACIS), pp. 361–366. IEEE (2016)
22. Sikora, M., et al.: Censoring weighted separate-and-conquer rule induction from survival data. Methods Inf. Med. **53**(02), 137–148 (2014)
23. Sikora, M., Mielcarek, M., Kałwak, K., et al.: Application of rule induction to discover survival factors of patients after bone marrow transplantation. J. Med. Inform. Technol. **22**, 35–53 (2013)
24. Wang, P., Li, Y., Reddy, C.K.: Machine learning for survival analysis: a survey. ACM Comput. Surv. (CSUR) **51**(6), 110 (2019)
25. Wróbel, Ł.: Tree-based induction of decision list from survival data. J. Med. Inform. Technol. **20**, 73–78 (2012). http://jmit.us.edu.pl/cms/index.php?page=vol-20-2012
26. Wróbel, Ł., Gudyś, A., Sikora, M.: Learning rule sets from survival data. BMC Bioinform. **18**(1), 285 (2017). https://doi.org/10.1186/s12859-017-1693-x

Particle Competition for Unbalanced Community Detection in Complex Networks

Luan V. C. Martins[1](✉) and Liang Zhao[2]

[1] Institute of Mathematics and Computer Science, University of São Paulo (USP), São Paulo, Brazil
luan.martins@usp.br

[2] Faculty of Philosophy, Science, and Letters at Ribeirão Preto, University of São Paulo (USP), São Paulo, Brazil
zhao@usp.br

Abstract. Unbalanced community structures is a common phenomenon when representing real-world systems using Complex Networks; however, not all community detection model can correctly identify communities of different sizes. In this paper, we propose a community detection technique focusing on the detection of unbalanced communities. The method augments the Particle Competition model: a bio-inspired technique that employs a set of particles (random walkers) into the network to compete for its nodes. However, the detection of unbalanced communities in the original Particle Competition is impossible because every particle shares the same ability to defend and attack nodes. The proposed model introduces a second stage into the system, named Regularization, which explicitly handles the detection of unbalanced communities. This mechanism uses the neighborhood of the nodes to create a custom preference guide for each particle, specifically tailored to the community of the particle. As a result, the model can precisely detect unbalanced community structures. Furthermore, the model achieves higher accuracy and faster computational speed compared to the original Particle Competition model.

1 Introduction

Although humans and machines learn and obtain knowledge differently, that does not stop us from taking inspiration from real-life and nature behavior. Complex Network is a data representation model that has been the focus of much research due to its ability to represent topological and functional relationships in

This work is supported in part by the São Paulo State Research Foundation (FAPESP) under grant numbers 2015/50122-0 and 2013/07375-0, the Brazilian Coordination for Higher Education Development (CAPES), the Pro-Rectory of Research (PRP) of University of São Paulo under grant number 2018.1.1702.59.8, and the Brazilian National Council for Scientific and Technological Development (CNPq) under grant number 303199/2019-9.

R. Cerri and R. C. Prati (Eds.): BRACIS 2020, LNAI 12320, pp. 322–336, 2020.
https://doi.org/10.1007/978-3-030-61380-8_22

data. Research has shown that networks are capable of being used in a variety of situations, ranging from sentiment analysis [18] to 3D DNA folding [4] – a comprehensive review of applications can be found in [6]. One of the salient features of Complex Networks is the presence of *communities*. A community is wildly referred to be a sub-group of nodes more densely connected to themselves than to other nodes of the graph [11].

In graph theory, community detection is akin to graph partition, which is an NP-Complete problem [9]. For this reason, several methods have been developed concerning the efficiency of community detection. However, due to the approximate nature of those methods, not all techniques are appropriate for all community detection tasks. The choice of method must consider the characteristics of the network, such as size, heterogeneity, presence of hierarch in the community structure, and, when dealing with real-life systems, the possibility of communities with different sizes.

Unbalanced community detection is an important topic that has received increasing attention due to the complex nature of data in real-world systems [7], as real-world networks are likely to have community sizes that follow a power-law distribution [12,14]. However, many community detection methods overlooked this aspect: for example, as studied in [10], many community detection algorithms based on modularity optimization may not be able to provide accurate clustering results in unbalanced cases. In order to provide good clustering results, a few techniques have been published targeting such networks. The authors in [21], for instance, propose additional metrics borrowed from graph partition tasks to work around the issues in modularity and employ it to identify unbalanced communities, generating good clustering results. However, the method has a high time complexity order. In the recent work presented in [20], the authors used a set of different measures, each one with their particular strength: because the method identifies communities individually (one by one), it can detect communities that have different sizes or densities in the same network. However, because of the selection of specific network measures, the ability of unbalanced community detection is restricted. In summary, up to now, it still lacks a general and efficient method to detect unbalanced communities in complex networks.

In this paper, we propose a bio-inspired community detection technique with an explicit mechanism to handle unbalanced community. The method is based on the Particle Competition model, and it consists of two stages: Competition and Regularization. The proposed method is more efficient than previous Particle Competition algorithms, making it suitable for larger networks. Moreover, it is able to obtain high accuracy in unbalanced community detection tasks. Competition is a natural behavior that occurs in nature and is usually observed when there is a lack of resources such as food, water, and mates. The proposed method described in this paper applies this concept to the nodes of a graph: random walkers – which are referred to as particles – are placed on the network and must compete with themselves to capture the nodes of the network, which is the limited resource.

The Particle Competition method were originally proposed by [15] and later improved by [17]. Specifically, the original Particle Competition method [15] proposed a model for community detection based on competition through a combination of random and deterministic walks: the random walk acts a novelty finder and helps the particle explore the network and avoid traps. In contrast, the deterministic walk enhances the model by only allowing the particles to move to nodes it already owns; therefore, forcing them to stay in their community and defend it. The work of [17] further improved the model by proposing a combination of the two walking types. Such combination, named preferential walking, allows the particle to choose the next node to navigate freely, but giving preference to node it already owns, i.e., the model has a higher chance of choosing nodes that it visited more frequently in the past. The preferential walking type further improved the model's ability to detect and provide accurate community results.

Previous Particle Competition models rely on knowledge built upon the frequency of visitation each node received to that moment as a guide. The deterministic or preferential walking uses this knowledge to incentivize the particle to continue defending the nodes that are most likely to belong to its' community. As studied in both papers, combining both movement types is essential to obtain good clustering results. However, the models present some shortcomings, which can mainly be summarized as a "lack of vision": 1) because each particle behaves the same, the detection of communities of different sizes is impossible. 2) to achieve good clustering results, a large number of iterations are necessary for the particle to find, own, and defend all members of the community.

The proposed model is a new addition to the Particle Competition family of algorithms. It augments the Particle Competition model by proposing an explicit mechanism to handle unbalanced communities. This mechanism creates an exploration guide generated from a process we named Regularization. In essence, instead of using the history of visits to bias the particle to defend its community, the proposed mechanism builds a guide for each particle based on the neighborhood of each node. The information gained from neighbors of a node provides a strong indication of what community it belongs. Naturally, it is in agreement with most adopted definition of a community, which defines it as a set of nodes densely connect with themselves than to nodes of other communities. By using such information, we create a custom guide for each particle, explicitly tailored to the community it is attempting to dominate. This guide suffers no influence of the particle's equal ability to attack and defend nodes, therefore, allowing for the detection of communities with different sizes.

The contrast from both stages employed in the proposed model provides many advantages compared to the previous Particle Competition models, mainly: 1) shorter running time is required because of the parallel diffusion mechanism introduced by regularization. Consequently, the model requires much fewer iterations in comparison to the original Particle Competition models. 2) The proposed model can better detect the "borders", i.e., nodes from a community that shares links with other communities. 3) More importantly, using the guide from the Regularization step allows each particle to behave differently according to the network structure. Therefore, the new model can detect unbalanced communities.

The remaining of this paper is organized as follows: In Sect. 2, a revision of the original Particle Competition model is presented [15,17]. Section 3 describes the proposed model and its mechanism. Section 4 reports on computer simulations to illustrate how the model works. Still, in this section, we show and discuss the community detection results on artificial networks and real-world data sets. Finally, Sect. 5 concludes this paper.

2 The Particle Competition Model

Consider a graph $\mathcal{G} = \langle \mathcal{V}, \mathcal{E} \rangle$ where $\mathcal{V} = \{v_1, ..., v_V\}$ is a set of nodes (or vertices) and $\mathcal{E}$ is a set of links $\mathcal{E} = \{e_1, ..., e_L\}$. The graph object is represented to a machine as an adjacency matrix A, with $\dim(A) = V \times V$ and a_{ij} denoting whether or not node i and node j are connected: if $a_{ij} = 1$ they are connected, if $a_{ij} = 0$ they do not share a connection.

The Particle Competition uses a set of particles $\mathcal{K} = \{1, ..., K\}$. Each particle has an energy level to guide the exploration process, while each node has a domination-level value. The particle owns the nodes where it has the highest domination-level on it. When a particle visits a node, its domination level on that node strengthens while the domination level of its rival weakens; simultaneously, the energy of a particle decreases when it attacks a node belonging to rival and recharges when visiting a node it owns. On the other hand, the particle's energy level guides the competition by discouraging the particle from wondering around rival communities for too long. When a particle is *exhausted*, i.e., its energy level is too low, the particle is teleported back to a node of its community to recharge the energy and defend its community from rival particles. Otherwise, when the particle is *active*, it is free to explore the network using a combination of two movements: random walking and preferential walking.

The random walking movement type, in which the particle will randomly choose a neighbor node to visit, is responsible for the discovery of new territories and the ability to attack and conquer new nodes. On the other hand, the preferential walking guides the particle to choose the nodes that are most likely to belong to that particle, and as a result, it is responsible for guiding the particles to settle in a community and defend it from rivals. The particle competition can be modeled as a stochastic dynamical system, which is given by

$$p^{(k)}(t+1) = j, j \sim \mathbb{P}^{(k)}_{\text{transition}}(t) \tag{1}$$

$$N_{ik}(t+1) = N_{ik}(t) + \mathbb{1}_{[p^{(k)}(t+1)=i]} \tag{2}$$

$$E^{(k)}(t+1) = \begin{cases} \min\left(w_{max}, E^{(k)}(t) + \Delta\right), \\ \text{if owner(k, i)} \\ \max\left(w_{min}, E^{(k)}(t) - \Delta\right), \\ \text{if } \neg \text{ owner(k, i)} \end{cases} \tag{3}$$

$$S^{(k)}(t) = \mathbb{1}_{[E^{(k)}(t)=w_{min}]} \tag{4}$$

In Eq. 1, $p^{(k)}(t+1)$ represents the network position (node) of particle k at iteration $t+1$. This equation controls which particle is visiting which vertex of the graph. This is done using the time-varying transition matrix $\mathbb{P}^{(k)}_{\text{transition}}(t)$, explained in detail in Eq. 5. Equation 2 counts the amount times that each particle k has visited each node i up to the time t, in other words, it updates the ownership of the particles on the nodes $(N_{ik}(t+1))$. Equation 3 updates the energy level of each particle $E^{(k)}(t)$: it increases by Δ when visiting a node that the particle owns, otherwise it decreases by Δ if the particle is invading a rival's territory. The last equation updates and keeps track of the state of each particle. When the energy level is decreased to the minimal level i.e., $S^{(k)}(t) = 1$, its status is *exhausted* and it should be randomly reset at another node of the network; otherwise, the particle is at *active* status and walks normally in the network.

The transition matrix $\mathbb{P}^{(k)}_{\text{transition}}(t)$, which governs the behavior of the particle, is presented:

$$\begin{aligned} \mathbb{P}^{(k)}_{\text{transition}}(t) = (1 - S^{(k)}(t)) \left[\lambda \mathbb{P}^{(k)}_{\text{pref}} + (1-\lambda)\mathbb{P}^{(k)}_{\text{rand}}\right] \\ + S_{(k)}(t)\mathbb{P}^{(k)}_{\text{rean}}(t) \end{aligned} \tag{5}$$

In summary, Eq. 5 works in the following way: when a given particle's status is *exhausted*, i.e., $S^{(k)}(t) = 1$, the movement policy switches to a defensive strategy using the $\mathbb{P}^{(k)}_{\text{rean}}$ transition matrix, which will return the particle to a node it owns to recharge its' energy and defend its' community. Otherwise, when the particle is *active*, the movement policy will use a combination of random walking, using the transition matrix $\mathbb{P}^{(k)}_{\text{rand}}$, and preferential walking, given by the $\mathbb{P}^{(k)}_{\text{pref}}$ transition matrix. The λ parameter, $0 \leq \lambda \leq 1$, indicates the emphasis on the preferential walking policy, which keeps the particle defending the nodes it is most likely to belong to the particle's community.

3 The Two-Stage Particle Competition Model

The proposed model augments the Particle Competition by using the domination-level information $(N(\tau))$ to create a custom preference guide for each particle, tailored specifically for the community it is trying to detect. By analyzing the neighborhood information of the nodes, the proposed guide can fix the mistakes made by the previous stage and detect when a particle is overstepping the boundaries of its community, thus, allowing for the correct detection of communities with different sizes.

3.1 The Proposed Regularization Mechanism

The Regularization stage's primary goal is to generate the preference matrix $B(\tau)$, which defines the preference-level of each particle upon every node of the network. This preference matrix will act as a guide for the next iteration of Competition and will ultimately contain the network's community structure.

For this purpose, we define the regularization function as:

$$R(i,k,M) = \frac{\sum_{u=1}^{V} (a_{iu} M_{uk})^2}{\sum_{j=1}^{K} \sum_{u=1}^{V} (a_{iu} M_{uj})^2} \tag{6}$$

In short, Eq. 6 uses the neighborhood of a node to create a preference level for each particle. It yields the preference levels of the node i by the particle k, taking into account the preference levels on the input matrix M. M, with $\dim(M) = V \times K$, is a placeholder for either the domination-level matrix N or the preference-level matrix B. a_{ij} represents the link between node i and j. If particle k exerts the most influence on node v neighbors, it will have a higher preference to visit it in the next *epoch*. The Regularization works in the following way:

1. When the Regularization step first starts, Eq. 6 is applied to each entry $B(t_B)$, i.e., $B_{ik}(t_B) = R(i,k,N(\tau))$. As such, the Regularization step initially takes into consideration the domination level of the particles after the Competition step has ended.
$$B_{ik}(t_B) = R(i,k,N(\tau)) \tag{7}$$
2. Additionally, the incremental execution of Eq. 6 on the resulting preference matrix $B(t_B)$ can yield better results in networks with very dense and unbalanced communities. Therefore, it might be desirable to execute Eq. 6, μ more times ($\mu \geq 0$) upon the matrix $B(t_B)$, i.e., $B_{ik}(t_B) = R(i,k,B(t_B - 1))$. μ is the total amount of incremental executions of Eq. 6 applied to $B(t_B)$.

$$B_{ik}(t_B + 1) = R(i,k,B(t_B)) \tag{8}$$

3.2 Transition Matrix for Network Exploration

Both stages of the system are connected in a meaningful way by the walking dynamics of the particles, which governs how they choose the next vertex of the network to navigate, and, therefore, dictates how they compete for the nodes of the network.

In the two-step particle competition model, the preferential rule now employs the preference guide $B(\tau)$, generated by the Regularization:

$$\mathbb{P}^{(k)}_{\text{pref}}(i,j,t) = \frac{a_{ij} B_{jk}(\tau)}{\sum_{u=1}^{V} a_{iu} B_{uk}(\tau)} \tag{9}$$

Equation 9 denotes the likelihood of the particle k visiting the node j from the current node i by taking into consideration the guide $B(\tau)$ matrix, from the Regularization step. The more neighbors owned by the particle k the node j has, the higher the likelihood of it being visited by the particle k is.

3.3 Basic Idea

The algorithm consists of two stages that learn and regulates each other. The Regularization stage learns from the Competition, which sequentially propagates the label information through the network using a random walk. In turn, it gets influenced by the preference guide generated from the Regularization stage, which propagates the neighborhood label in parallel.

The model starts with the Competition step without preferential walking. The Particle Competition runs until convergence and yields the $N(\tau)$ matrix, containing the domination level of the particles over the nodes of the network. Then, the Regularization stage uses the domination matrix $N(\tau)$ to generate the guiding matrix $B(\tau)$. The new preference guide marks the end of the *epoch* τ, and a new one will begin. The Competition stage is restarted, but this time, the particles are influenced by the preference guide. The Competition and Regularization steps are executed alternately until the preference matrix $B(\tau)$ converges, i.e., $||B(\tau) - B(\tau - 1)||_\infty < \epsilon$ where $||.||_\infty$ is the matrix max-norm of the difference between the preference matrix at *epoch* τ and the preference matrix at epoch $\tau - 1$. The Competition must be restarted from scratch each *epoch*, holding no information about the previous execution. Once the model converges, the preference-matrix $B(\tau)$ holds the community structure of the network.

The proposed method is now described in an algorithmic manner (Algorithm 1). It takes as entry the data set to be classified, five user-defined parameters used by each stage. Although the model requires many parameters, only K and μ can influence the performance of the method. All other parameters can remain in their default value, as studied in previous works [15,17].

Algorithm 1. Two-stage Particle Competition Algorithm

1: **procedure** PCR($A, K, \mu, \Delta = 0.2, \lambda = 0.6, \epsilon = 0.05$)
2: $B(0) \leftarrow$ initializeB()
3: $\tau \leftarrow 0$
4: **repeat**
5: $N(\tau) \leftarrow$ particleCompetition(K, A, $B(\tau), \Delta, \lambda$)
6: $\tau \leftarrow \tau + 1$
7: $B(\tau) \leftarrow$ regularization(A, $N(\tau)$, μ)
8: **until** $||B(\tau) - B(\tau - 1)||_\infty < \epsilon$
return $B(\tau)$
9: **end procedure**

The Competition algorithm is already presented in [17] and it will not be repeated here. The Regularization step, presented as a standalone function in Algorithm 2, takes the input graph, the domination-level matrix $N(\tau)$ from the previous Competition step and the user-defined parameter μ.

Algorithm 2. Regularization step

```
1: procedure REGULARIZATION(data, N(τ), μ)
2:     B(0) ← calculateB(data, N(τ))                ▷ Use 6
3:     t_B ← 1
4:     for i to μ do
5:         B(t_B) ← calculateB(data, B(t_B − 1))     ▷ Use 6
6:         t_B ← t_B + 1
7:     end for
       return B(t_B)
8: end procedure
```

4 Numerical Analysis and Simulation Results

This section discusses how the proposed method behaves as well as the impact each parameter has in community detection or data clustering tasks.

4.1 Competition Mechanism Analysis

In this section, the Competition behavior is analyzed, starting by the behavior of the domination-level matrix $N(\tau)$. In order to illustrate its behavior, the algorithm is executed upon the widely used Girvan and Newman benchmark network [8] with $Z_{out}/\langle k \rangle = 0.2$.

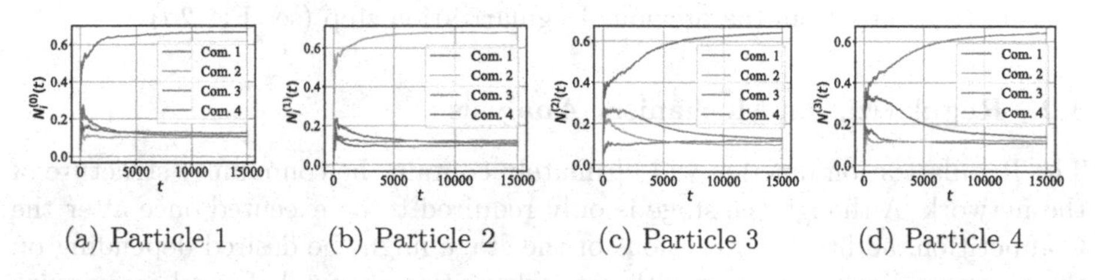

(a) Particle 1 (b) Particle 2 (c) Particle 3 (d) Particle 4

Fig. 1. Example of the average domination levels on the nodes of each $M = 4$ communities of the network, for the $K = 4$ particles in the system. The GN benchmark network was generated with $Z_{out}/\langle k \rangle = 0.2$. Each particle finds and dominates a community of the network.

The normalized $N(t)$ behavior of the Competition is reported in Fig. 1. The analysis shows, for each one of the $K = 4$ particles in the system, the average domination level each particle possesses upon the nodes of $M = 4$ communities in the network. It can be observed that each one of the particles can correctly find, dominate, and defend the members of a community of the network. Furthermore, it can also be noted the particle's ability to expel or remove an enemy particle from its dominated community, such as in Fig. 1d, where particle four is expelled from dominating the members of community 4, which is ultimately owned by particle three.

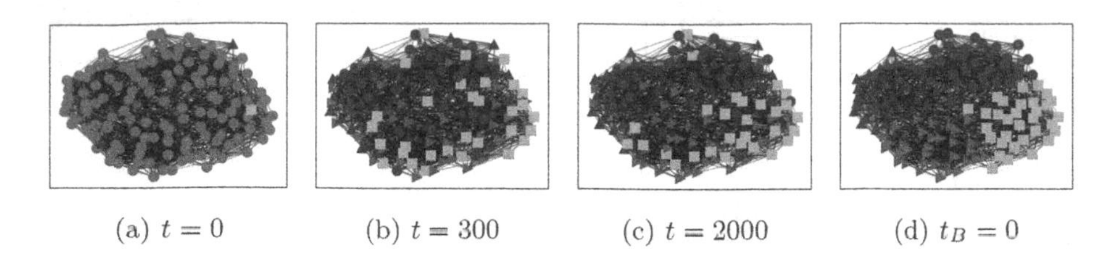

(a) $t = 0$ (b) $t = 300$ (c) $t = 2000$ (d) $t_B = 0$

Fig. 2. Snapshots of a single execution of the proposed method on the GN network with $Z_{out}/\langle k \rangle = 0.3$, in which the color and shape represents a particle's community. (a), (b), (c) shows the $N(t)$ domination-levels of the particles on time 0, 300, and 2000 respectively. It can be noted that at time 2000, each particles roughly settles in a community. (d) shows the first regularization result obtained upon $N(2000)$, which is indeed the expected community result.

In order to illustrate the method's behavior and exemplify how both steps learn with each other, Fig. 2 shows the clustering results during four distinct moments in the process. Figure 2a reflects the domination matrix $N(0)$ on time $t = 0$, the first iteration of the Competition: each particle has been randomly placed in a node. Figure 2b and 2c, shows the Competition at later times when each particle has roughly settled and dominated a community. Finally, Fig. 2d shows the first iteration of the Regularization, which is enough to yield the correct community results in this particular network. Next, the Competition will then be restarted from scratch, but this time each particle takes into preference visiting the nodes from the previous Regularization step (see Fig. 2d).

4.2 Regularization Mechanism Analysis

The Regularization function will ultimately contain the community structure of the network. Although the stage is only required to be executed once after the Competition, additional executions of the stage might be desired depending on the network. That is the case with networks with a very unbalanced community

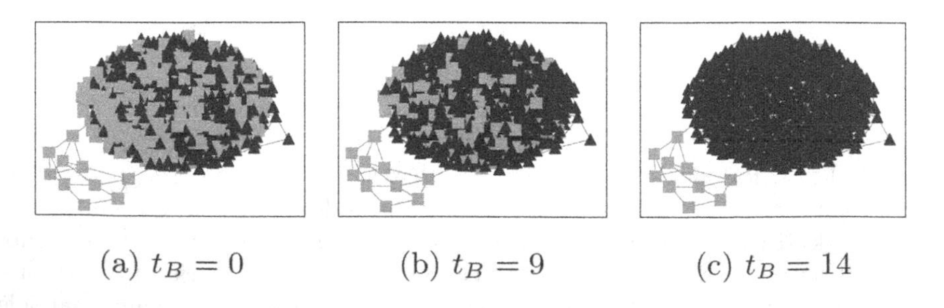

(a) $t_B = 0$ (b) $t_B = 9$ (c) $t_B = 14$

Fig. 3. Additional executions of the regularization function can be beneficial to correctly identify communities in networks with unbalanced or otherwise densely connected communities. The shape and color indicate which particle owns the node.

structure or otherwise dense communities. Figure 3 illustrates the behavior of incremental executions of the Regularization function in extremely unbalanced networks with two communities, one containing 10 nodes and the second having 500: Fig. 3a shows the first execution of the regularization, which is not enough to identify the community in such a degree of unbalance. The incremental executions of the function, is beneficial to community detection, as shown in Fig. 3b and 3c.

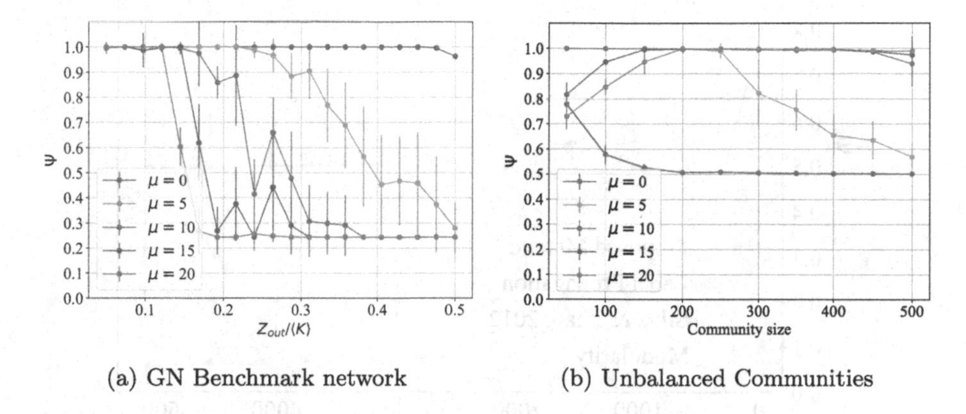

(a) GN Benchmark network (b) Unbalanced Communities

Fig. 4. Impact of different values for μ on different network structures. (a) shows the GN Benchmark network with varying $Z_{out}/\langle k \rangle$, allowing the comparison on different level of community mixture. (b) shows the impact on networks with varying degree of community sizes: the first community has 10 nodes and the second community varies. Averaged over 100 executions.

Finally, we investigate the impact of μ on different networks, the first one is the GN benchmark network, and the second is an unbalanced network. On the GN Network, Fig. 4a shows that higher values of μ are detrimental to the accuracy of the model in networks with poorly defined communities, when the community mixture is too high (given by $Z_{out}/\langle k \rangle$). On the other hand, Fig. 4b shows that higher values for μ help to detect a very unbalanced community structure correctly. Therefore, higher values of μ are only appropriate when the network exhibits well defined or very unbalanced communities.

4.3 Simulation on Artificial Data Sets

To evaluate the ability of the proposed method on unbalanced community detection, a set of artificial networks containing two communities is generated using the following parameters: the size and degree of the first community, the size and degree of the other community, and finally, the number of bridge links, i.e., links connecting the two communities. The nodes of both communities are generated and connected within themselves randomly until the target inner community degree of the community is reached. After both communities are generated, links are created between nodes of both communities at random, until the target number of connecting links is reached. Consequently, a network with two communities of different sizes can be created for an unbalanced community test.

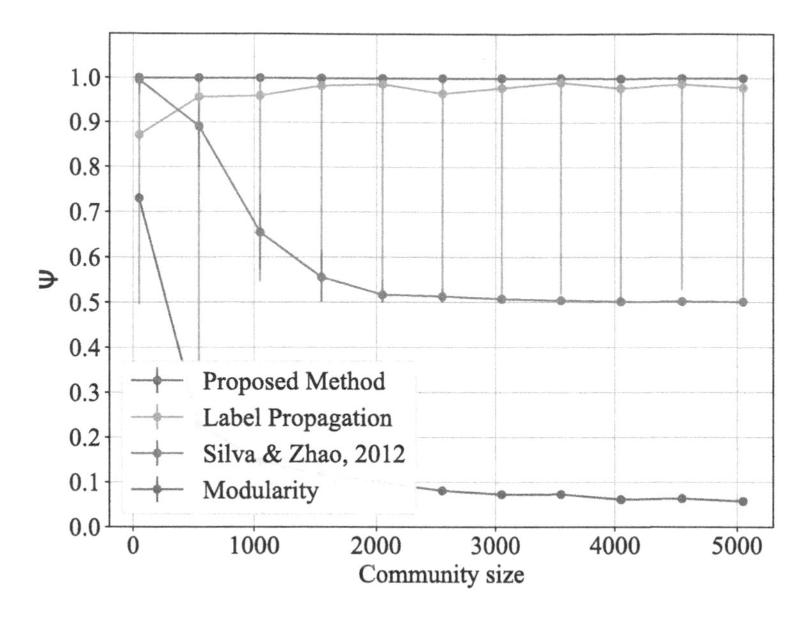

Fig. 5. Unbalanced community detection accuracy averaged over 100 executions with comparison to other three methods. All the methods are applied to a sequence of networks each containing two communities, the first of which containing 50 nodes and the second community varying from 50 to 5050 nodes. Since the first community has fixed size (50 nodes), the networks become much unbalance as the size of the second community increases (shown by x-axis). In the last case, the second community is 100 times larger than the first one, i.e., 50 nodes vs. 5050 nodes.

Figure 5 shows the comparison of the community detection accuracy (Ψ) of the proposed method when faced with networks containing a community of 50 nodes and other community containing nodes ranging between 50 to 5050 elements. The plotted line is the average community detection rate (Ψ), and the error bars represent the best and worst accuracy obtained in 100 attempts. The model was executed with $\Delta = 0.2$, $\lambda = 0.6$, and $\mu = 0$. The figure compares the proposed method against the original Particle Competition method [17], the Fast Greedy Modularity model [5] and the Label Propagation algorithm [16]. The proposed method can correctly classify the network community regardless of the unbalance ratio. On the other hand, Silva & Zhao 2012 starts to misclassify nodes as the unbalance ratio increases. The Modularity based method, as expected [10], also fails to handle the detection of unbalance communities, merely acting as a worst-case scenario. Label Propagation, however, displays an interesting behavior. Although it does have high averages of detection rate, the model is prone sometimes to fail to detect unbalanced classes in the tested network, resulting in accuracy as low as Silva & Zhao, 2012. Although Label Propagation is conceptually similar to the second stage of the proposed method, the initiation procedure is different. Figure 5 indicates the use of both stages (Competition and Regularization) provides a more robust approach to community detection.

4.4 Simulation on Real World Data Sets

Now the proposed method is applied to a set of real-world networks to study its' performance. The first network is the famous Zachery's Karate Club, introduced in [19]. It models the friendship between students of a karate club in the early 1970s by using the social interactions between students inside and outside the club environment. During the data collection, an internal dispute between the teacher and the administration divided the students into two groups. Figure 6 shows the clustering result of the proposed method, which is 100% correct. At the same time, none of the previous versions of the Particle Competition technique [15,17] can correctly identify the communities for all nodes.

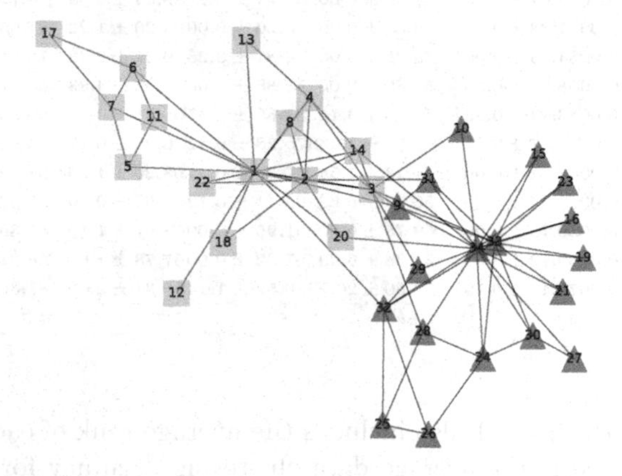

Fig. 6. Community detection result of the proposed method with $K = 2$ on the famous Zachery's karate club network. The proposed method is able to correctly identify the community structure of the network, indicated by the color and shape (community detection result). The proposed method is the first Particle Competition model capable of correctly classifying node 9, which shares many links with both communities.

In order to further test the proposed method as a data clustering technique, the proposed method is applied to 12 well-known real-world data sets from the UCI machine learning repository [2]. It should be noted that for the task at hand (unsupervised learning), the class label of the data is only used to verify each method's result (data clustering accuracy). The method is compared to 7 well-know and established community detection and data-clustering techniques: the Fast Greedy Modularity algorithm [5], Fuzzy C-Means [3], expectation-maximization algorithm [13], K-Means++ with optimized center initialization

[1], the Particle Competition (labelled as Silva & Zhao, 2012) technique from [17] and the Label Propagation algorithm [16]. The Fast Greedy Modularity, Label Propagation, and Particle Competition algorithm [17] are all graph-based techniques and thus have been adapted to generate a graph from the given data set. For this task, the k-nearest neighbor graph formation technique was used with optimized k for each data set.

The parameters of the proposed method are optimized using a simple grid-search algorithm over the following parameters: $0 \leq \mu \leq 10$ and $0 \leq K \leq 30$ to identify the appropriate combination of parameters for each data set, resulting in the best accuracy. The accuracy obtained by the techniques are presented in Table 1, which are the average of 20 independent runs on each data set.

Table 1. Comparison between data classification methods and the proposed algorithm in 20 independent runs

	Modularity	Fuzzy C-Means	Exp-Max	K-Means++	Silva & Zhao, 2012	LP	Proposed method
Iris	85.92 ± 0.00	87.97 ± 0.00	77.58 ± 0.13	83.62 ± 0.60	83.38 ± 0.46	86.95 ± 8.97	90.13 ± 7.60
Breast cancer	60.26 ± 0.00	75.04 ± 0.00	54.02 ± 0.00	82.91 ± 0.27	89.04 ± 0.33	84.72 ± 5.00	88.13 ± 0.0
Wine	81.08 ± 0.00	71.38 ± 0.00	69.83 ± 0.00	96.20 ± 0.00	94.20 ± 1.22	81.05 ± 2.96	96.17 ± 0.15
Glass	77.61 ± 0.00	84.33 ± 0.00	73.35 ± 0.00	74.81 ± 3.28	79.01 ± 2.33	76.51 ± 5.94	79.37 ± 2.03
Ionosphere	55.39 ± 0.00	58.65 ± 0.00	53.85 ± 0.00	58.88 ± 0.05	57.83 ± 4.83	55.67 ± 1.28	59.47 ± 3.28
Vowel	88.37 ± 0.00	83.94 ± 0.12	77.73 ± 3.14	84.65 ± 0.37	90.31 ± 0.16	90.86 ± 0.07	90.32 ± 0.08
Yeast	75.64 ± 0.00	71.87 ± 0.18	51.59 ± 15.39	76.23 ± 0.12	76.26 ± 0.17	75.92 ± 1.14	76.20 ± 0.27
Vertebral	68.82 ± 0.00	67.22 ± 0.00	71.13 ± 1.58	67.32 ± 0.41	68.20 ± 1.23	65.42 ± 4.67	69.96 ± 2.39
Arrhythmia	64.55 ± 0.00	50.24 ± 0.00	55.79 ± 5.62	65.11 ± 0.72	65.92 ± 0.16	65.57 ± 0.32	65.89 ± 0.09
Parkinson	45.97 ± 0.00	59.75 ± 0.00	59.75 ± 0.00	51.96 ± 0.00	58.07 ± 1.10	45.39 ± 1.76	58.16 ± 1.50
Heart disease	56.93 ± 0.00	53.72 ± 0.00	57.38 ± 6.09	57.32 ± 0.60	57.78 ± 1.67	56.01 ± 0.79	57.98 ± 1.67
Ecoli	79.90 ± 0.00	79.12 ± 0.19	57.60 ± 20.75	80.52 ± 1.14	80.31 ± 2.09	81.36 ± 3.76	85.98 ± 0.24
Average rank	4.75	4.66	5.58	3.83	2.75	4.5	1.83

Additionally, the Table 1 also includes the average rank of each technique: it is obtained by ranking the average data clustering accuracy for each data set, i.e., the most accurate algorithm gets rank 1, the second most gets rank two and so on. The average rank is an indicator of the algorithm performance in the selected data sets. The proposed method has the highest rank, indicating that it performs better than others most of the time. The model is followed by the Silva & Zhao, 2012 in second place. However, it should be noted that the technique presented in [17] has some disadvantages, such as higher running times and the inability to detect unbalanced communities, as shown in Fig. 5. Table 2 reports the running times of the proposed method and the original Particle Competition obtained during the tests, which were executed on a personal computer with a Core i7 8550u.

Table 2. Proposed method and Silva & Zhao, 2012 running time in seconds

	Proposed method	Silva & Zhao, 2012
Iris	6.06 s ± 2.12	23.40 s ± 2.43
Breast cancer	8.60 s ± 1.75	63.00 s ± 5.16
Wine	3.82 s ± 1.45	21.05 s ± 1.36
Glass	6.66 s ± 2.42	37.03 s ± 3.57
Ionosphere	7.15 s ± 1.743	39.86 s ± 3.37
Vowel	155.91 s ± 52.84	637.16 s ± 18.25
Yeast	78.62 s ± 24.62	319.17 s ± 50.17
Vertebral	23.68 s ± 8.76	61.01 s ± 4.35
Arrhythmia	71.26 s ± 22.64	113.78 s ± 7.54
Parkinson	2.68 s ± 0.55	20.69 s ± 1.62
Heart disease	11.53 s ± 4.18	51.36 s ± 4.80
Ecoli	9.44 s ± 3.45	57.27 s ± 1.937

5 Conclusion

This paper presents a new stochastic learning model for graph-based data-clustering and community detection consisting of two interactive learning steps: Competition and Regularization. Competition is nature-inspired behavior, which occurs when there is a lack of resources such as food or water. In a network, the particles compete with each other to try to dominate as more as possible nodes. The Regularization step provides a guide for those particles, which introduces a parallel and diffusive mechanism into the model by taking into account the neighbors of each node. The Regularization then helps the next execution of the Competition until a consensus is reached. The first step of the system is stochastic and thrives in exploring the unknown, which is indeed a necessary step for learning new things. The second step, however, builds upon the knowledge and experience obtained over time to guide the learning process. The following *epoch* then combines both visions using a λ parameter, allowing for knowledge to be built.

The computer simulations in this paper show that the method can achieve quite good results in data-clustering and community detection tasks. Furthermore, it shows that the proposed method improves previous Particle Competition models' results in terms of running time and, more importantly, in the ability to cluster unbalanced networks. The proposed method can obtain good community detection accuracy when faced with unbalanced networks, regardless of the unbalanced ratio.

References

1. Arthur, D., Vassilvitskii, S.: k-means++: the advantages of careful seeding. In: Proceedings of the Eighteenth Annual ACM-SIAM Symposium on Discrete Algorithms, pp. 1027–1035. Society for Industrial and Applied Mathematics (2007)
2. Asuncion, A., Newman, D.: UCI machine learning repository (2007)
3. Bezdek, J.C.: Pattern Recognition with Fuzzy Objective Function Algorithms. AAPR. Springer, Boston (1981). https://doi.org/10.1007/978-1-4757-0450-1
4. Cabreros, I., Abbe, E., Tsirigos, A.: Detecting community structures in Hi-C genomic data. In: 2016 Annual Conference on Information Science and Systems (CISS), pp. 584–589. IEEE (2016)
5. Clauset, A., Newman, M.E., Moore, C.: Finding community structure in very large networks. Phys. Rev. E **70**(6), 066111 (2004)
6. Costa, L.F., et al.: Analyzing and modeling real-world phenomena with complex networks: a survey of applications. Adv. Phys. **60**(3), 329–412 (2011)
7. Danon, L., Díaz-Guilera, A., Arenas, A.: The effect of size heterogeneity on community identification in complex networks. J. Stat. Mech.: Theory Exp. **2006**(11), P11010 (2006)
8. Danon, L., Diaz-Guilera, A., Duch, J., Arenas, A.: Comparing community structure identification. J. Stat. Mech.: Theory Exp. **2005**(09), P09008 (2005)
9. Fortunato, S.: Community detection in graphs. Phys. Rep. **486**(3–5), 75–174 (2010)
10. Fortunato, S., Barthelemy, M.: Resolution limit in community detection. Proc. Nat. Acad. Sci. **104**(1), 36–41 (2007)
11. Girvan, M., Newman, M.E.: Community structure in social and biological networks. Proc. Nat. Acad. Sci. **99**(12), 7821–7826 (2002)
12. Guimera, R., Danon, L., Diaz-Guilera, A., Giralt, F., Arenas, A.: Self-similar community structure in a network of human interactions. Phys. Rev. E **68**(6), 065103 (2003)
13. Gupta, M.R., Chen, Y., et al.: Theory and use of the em algorithm. Found. Trends® Signal Process. **4**(3), 223–296 (2011)
14. Palla, G., Derényi, I., Farkas, I., Vicsek, T.: Uncovering the overlapping community structure of complex networks in nature and society. Nature **435**(7043), 814 (2005)
15. Quiles, M.G., Zhao, L., Alonso, R.L., Romero, R.A.: Particle competition for complex network community detection. Chaos: Interdisc. J. Nonlinear Sci. **18**(3), 033107 (2008)
16. Raghavan, U.N., Albert, R., Kumara, S.: Near linear time algorithm to detect community structures in large-scale networks. Phys. Rev. E **76**(3), 036106 (2007)
17. Silva, T.C., Zhao, L.: Stochastic competitive learning in complex networks. IEEE Trans. Neural Netw. Learn. Syst. **23**(3), 385–398 (2012)
18. Wang, H., Zhang, F., Hou, M., Xie, X., Guo, M., Liu, Q.: SHINE: signed heterogeneous information network embedding for sentiment link prediction. In: Proceedings of the Eleventh ACM International Conference on Web Search and Data Mining, pp. 592–600. ACM (2018)
19. Zachary, W.W.: An information flow model for conflict and fission in small groups. J. Anthropol. Res. **33**(4), 452–473 (1977)
20. Žalik, K.R., Žalik, B.: A framework for detecting communities of unbalanced sizes in networks. Phys. A: Stat. Mech. Appl. **490**, 24–37 (2018)
21. Zhang, S., Zhao, H.: Community identification in networks with unbalanced structure. Phys. Rev. E **85**(6), 066114 (2012)

2CS: Correlation-Guided Split Candidate Selection in Hoeffding Tree Regressors

Saulo Martiello Mastelini(✉),
and André Carlos Ponce de Leon Ferreira de Carvalho

Institute of Mathematics and Computer Sciences,
University of São Paulo, São Paulo, Brazil
mastelini@usp.br, andre@icmc.usp.br

Abstract. Incremental machine learning algorithms have been effective alternatives to deal with stream data. The Hoeffding Tree framework is one of the most successful solutions for supervised online prediction tasks. Although online regression tasks are present in several forms, and in many real-life problems, most of the research efforts have been devoted to classification. Existing regression tree solutions have strong limitations, mainly regarding their memory usage and running time. Hence, a new algorithm able to address these aspects in Hoeffding Tree Regressors is a relevant research issue. In this paper, we propose 2CS, a correlation-guided strategy to speed up Hoeffding Tree Regressor training. 2CS is conceptually simple and works by avoiding the exhaustive evaluation of all possible features as split candidates, as occurs in the existing solutions. Moreover, 2CS can be easily merged into existing incremental tree solutions and online tree ensembles algorithms, such as bagging and boosting. Throughout an extensive experimental evaluation, we show that the induction of 2CS-based models can be significantly faster than the traditional Hoeffding Tree Regressor algorithms, whereas retaining similar predictive power and memory use.

Keywords: Hoeffding trees · Online regression · Data stream mining · Split decisions · Incremental learning

1 Introduction

Despite recent advances in technologies for data storage and processing, in many applications, such as big data, the amount of data being produced led to a situation where the computational power available to process all incoming data may not be enough. This occurs because most data is produced continuously, and fast, in the form of potentially unbounded streams [11]. Traditional supervised Machine Learning (ML) algorithms, i.e. *in batch* solutions, were not developed to operate in such circumstances [7,11]. Hence, more efficient solutions must be developed to cope with the requirements of big data [9]. These algorithms must process each incoming datum just once, and they cannot indefinitely store instances [11,12].

R. Cerri and R. C. Prati (Eds.): BRACIS 2020, LNAI 12320, pp. 337–351, 2020.
https://doi.org/10.1007/978-3-030-61380-8_23

In the last years, ML research in data streams has expanded at increasing steps. Among the supervised solutions for data stream mining, the Hoeffding Tree (HT) framework is one of the most explored [11,12,18]. Despite being primarily applied to classification tasks, there are adaptations of this framework for regression [17,21,23]. However, dealing with continuous targets brings additional challenges for incremental algorithms. Differently from classification tasks, there is no well-defined target partitions, i.e., categories. Hence, tree solutions could potentially evaluate infinite data partition possibilities.

Practical HT solutions typically rely on the observed predictive features of the training instances to evaluate split decisions [18,19]. They store input feature values along with necessary statistics to guide these decisions. The required statistics can be incrementally maintained fairly efficiently [18,23]. Nonetheless, the process of seeking for the best split candidates among the features of the problem is computationally expensive. The tree models have to test all features values stored between tree expansions as potential thresholds for new branches creation. The cost becomes even higher as the tree continues processing more instances and gathers more data. Hence, more efficient strategies for split candidate selection are needed for HT regressors (HTRs). More efficient solutions would benefit, for instance, ensemble algorithms, which have received increasing attention from the data stream mining community [13,19,20]. This is particularly true for Boosting ensembles, as their additive nature make a parallel training difficult.

In this work, we investigate how to reduce the processing time needed to perform split decisions, without negatively impacting the prediction error and the required memory. For such, we use a simple yet effective heuristic to speed up split candidate selection in HTRs, named **C**orrelation-guided **S**plit **C**andidate **S**election (2CS). 2CS uses the correlation between numeric features and the target as a heuristic to rank predictive features. Hence, only a reduced subset of features is explored to expand the tree models, avoiding unnecessary computations. This strategy is conceptually simple and easily coupled within the HT framework. In this work, we are focused on stationary data streams, but in the future we intend to extend our analysis to non-stationary environments. Throughout an extensive experimental analysis, we show the capability of the 2CS-based trees of accelerating split decisions, while keeping predictive performance and memory use similar to the traditional HTRs.

The remaining of this text is organized as follows. In Sect. 2 we present background information and important related work. In Sect. 3 we formally introduce 2CS. In Sect. 4 we describe the experimental setup used to compare the 2CS-based tree variants with the traditional HTRs. We show and discuss the obtained results in Sect. 5. We make our final considerations and discuss possible directions for future research in Sect. 6.

2 Background and Related Work

In this paper, we deal with data stream regression tasks, as defined next. We denote as S a possibly unbounded stream of instances in the form

$S = \{(\mathbf{x}^t, y^t)\}_{t=1}^{\infty}$. Each instance $(\mathbf{x}^t, y^t)$ drawn at a timestamp t comes from an input space $\mathbf{X} \subset \mathcal{R}^m$, $m \in \mathbb{Z}^+$, and a target space $Y \subset \mathcal{R}$. The input space can also be referred to as feature space without loss of generality. Formally, a regression task can be formulated as the search for a function $f : \mathbf{X} \rightarrow Y$.

In data stream scenarios, we assume instances arriving continuously over time. Thus, f must be updated in an online fashion. Besides, in some cases we can expect an arbitrarily long delay before instances' labels are available for the incremental learning algorithms [14]. Nonetheless, in this study, we assume the true labels are available immediately after the model predicts the incoming instances.

Tree-based solutions have been extensively applied for data stream mining tasks, from which regression is not an exception [11,19,23]. This comes from the fact that these models are conceptually simple and naturally interpretable [1,11]. As previously mentioned, the HT framework is the most prominent example of a tree-based algorithm family used in online prediction tasks. HTs rely on the Hoeffding Bound (HB) theorem [11,16] to verify whether the model in training gathered enough evidence to enable its growth.

Nevertheless, research efforts on data stream mining were mostly devoted to classification tasks, whereas regression tasks were often overlooked [13,18,19]. Ikonomovska et al. [18] proposed the Fast Incremental Model Tree with Drift Detection (FIMT-DD) algorithm, tackling for the first time regression tasks within the HT framework. In the same work, the authors also presented the Fast Incremental Regression Tree with Drift Detection (FIRT-DD). FIMT-DD uses linear perceptrons as leaf predictors, whereas FIRT-DD uses the target mean [18,19]. When coupled with the drift detection mechanism (indicated by the DD suffix), the tree algorithms use the Page-Hinkley [11] test to detect concept drifts and grow alternate tree branches for the new concepts, very much alike a preceding HT solution for classification tasks [16]. This algorithm is also similar to the Hoeffding Adaptive Tree [2], which also has a regression version [22]. The HTRs were later on applied as base models for ensemble algorithms [13,19] and adapted to multi-target regression tasks [17,21,23].

Both tree algorithms monitor the standard deviation reduction (SDR) in the target space as a measure for split recommendations [18,19]. When deciding whether and how to split, HTRs compare the SDR of the second-best split candidate divided by the SDR of the best one. HTRs verify whether this ratio plus the HB is smaller than 1 [18,21]. In the affirmative case, we can state that the best split candidate is statistically better than the second one and, as a result, the tree creates new branches.

To decrease computational costs, HTs do not attempt to split after each incoming instance. Instead, they wait for $n_{\min}$ instances (also referred to as grace period) between split attempts. Further, in the case where split candidates are equally good, HTs also apply a tie-breaking mechanism to avoid indefinitely waiting for growth [11,18]. If the calculated HB shrinks below a tie-breaking threshold τ, a split is performed with the current best candidate.

In order to evaluate split candidates, HTRs rely on storing the observed feature values along with sufficient statistics related to them. The Extended

Binary Search Tree (E-BST) algorithm is applied for this end [18]. At each of its node, E-BST stores a set of sums related to the elements smaller (at the left side) and larger (at the right side) than each observed feature value. Hence, E-BST stores for the left and right sides the count of elements observed (n), the sum of the target values ($\sum y$), and the sum of squared target values ($\sum y^2$). These statistics are enough to calculate the variance and the standard deviation incrementally [18,19]. Thus, each stored feature value can be evaluated as a potential split point.

When considering datasets with an increased number of features, evaluating all the observed split candidates becomes costly. HTRs calculate the SDR of each observed value for each feature. This action becomes even more impacting when considering ensembles of HTRs. In this work, we hypothesize that the features mostly correlated with the target should provide the best split points. Such idea was previously explored in batch scenarios [15,24], but was not still covered in resource constrained online situations. In our proposal, the HTR traverse just the E-BST of the features most correlated with the target in search for split points. This action ought to decrease the processing time of HTRs without negatively affect their prediction power and memory consumption. This strategy, named 2CS, is detailed in the next section.

3 Correlation-Guided Split Candidate Selection

The split candidate selection in HTRs is guided by estimating the SDR of each partition candidate. In fact, the standard deviation is a measure of the spread of the data. Therefore, HTRs, similarly to traditional batch regression trees [5], aim at reducing the spread of instances lying in each of the created partitions. At the same time, regression trees try to make data partitions in such way that they become as maximally apart from each other as possible, following the "divide-and-conquer" principle [5,18].

In 2CS, we hypothesize that a measure of the relation between the numeric inputs and the target, such as the linear correlation, could give clues of which among them would be the most suited to perform a split decision [15,24]. This conjecture, to the best of our knowledge, was not explored yet in online learning scenarios. As presented in Gama [11], we can easily calculate the linear correlation coefficient using a small set of incrementally maintained statistics. Most interestingly, almost all of them are inherently maintained by the HTRs. The correlation calculation is described in Eq. 1.

$$r = \frac{\sum xy - \frac{(\sum x)(\sum y)}{n}}{\sqrt{\left(\sum x^2 - \frac{(\sum x)^2}{n}\right)\left(\sum y^2 - \frac{(\sum y)^2}{n}\right)}} \tag{1}$$

As previously mentioned in Sect. 2, HTRs already maintain n, $\sum y$, and $\sum y^2$. Moreover, $\sum x$ and $\sum x^2$ are also maintained by HTRs to enable feature standardization [18,21,23]. These measures refer to the sum of observed values for each feature and the sum of their squared values, respectively. Here, for simplicity, we omit the indexing for each j-th feature. Therefore, the only measure missing from Eq. 1 is $\sum xy$, i.e., the sum of the product between the feature values and the corresponding target observations. This additional measure can be easily added to the set of monitored statistics and just adds a constant increment in memory and time processing requirements. Now, we are ready to use the linear correlation as a heuristic to guide split candidate selection. It is important to mention that both positive and negative correlations are an indication of relationships between features and targets. Thus, in 2CS, we take the absolute value of the calculated correlations when ranking the features. Figure 1 summarizes 2CS's operation and how it fits within the HTR split decisions.

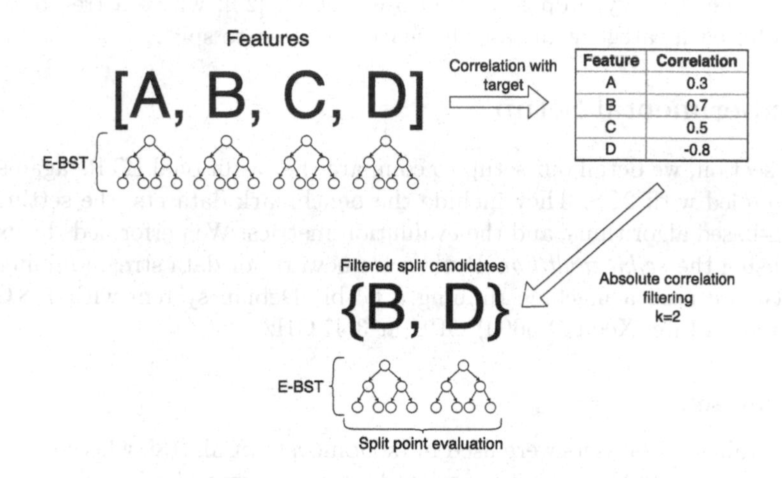

Fig. 1. Overview of 2CS's operation in the HTR framework.

The benefits of using a heuristic of how likely a feature will provide the best split decision are twofold. Firstly, different from batch algorithms, HTRs observe training examples incrementally as they arrive. Hence, at the moment of the splits, only partial information is available to make decisions. Although the HT framework gives us some guarantees that the trained models will perform similarly to batch ones, given enough observations [11,18], shifting the tree growth too much towards what the currently available data describe can lead to overfitting [1,3]. Secondly, by using a heuristic to select a subset of features to evaluate as split candidates, we can avoid performing possibly unnecessary processing efforts. The performance improvements are expected to increase jointly with the number of input features.

The complete functioning of 2CS is straightforward and easily included in the HTR framework. After $n_{\min}$ instances are observed by a leaf node, and the HTR is ready to attempt a split, our proposal performs the following three steps:

1. 2CS calculates the linear correlation between the numeric input features and the target using Eq. 1;
2. The features are ranked according to the absolute value of their linear correlation;
3. 2CS selects the k most correlated features to evaluate as split candidates along with (possibly) existing nominal features.

Here, k is a hyper-parameter that must be adjusted. In our experimental evaluation, we compare different values for k and their impact on predictive performance (refer to Sect. 4 for more information). It is important to note that the cost to calculate the correlations and rank the features accordingly to their values is usually negligible compared with the number of operations commonly performed in split attempts. This claim is supported by our experimental findings. Besides, measuring linear correlations only makes sense when both the features and the target are continuous. Hence, we only consider numerical features when applying the 2CS input feature filtering. Nominal inputs are treated following the strategy proposed by Osojnik et al. [23], where a tree branch is created for each category, in case the feature is used to split.

4 Experimental Setup

In this section, we detail our setup to compare the traditional HTRs against the trees coupled with 2CS. They include the benchmark datasets, the settings for the tree-based algorithms, and the evaluation metrics. We performed the experiments using the *scikit-multiflow* Python framework for data stream mining [22]. For such, we used a machine running a 64-bit Debian system with 128 GB of RAM and an Intel Xeon (X5690) CPU at 3.47 GHz.

4.1 Datasets

All the evaluated datasets were used in Ikonomovska et al. [18], where FIMT-DD was first proposed. All of them are related to stationary tasks. The datasets vary from ≈4000 examples to ≈41000 instances, as shown in Table 1. The majority of the input features in these datasets are numeric. All the datasets are publicly available in platforms such as UCI[1] and OpenML[2].

4.2 Variants of 2CS

We evaluated different settings for 2CS, ranging the correlation rank threshold from $k = 2$ to $k = 5$. For instance, when $k = 2$, only the two numeric features most correlated with the target would be evaluated as split candidates. It is important to note, however, that the HTs include a pre-pruning mechanism when performing splits. They also evaluate the possibility of not performing a split and maintaining the tree as it is, which here we refer as a *null* split option. Thus, HTRs with 2CS considered as split candidates the filtered numerical features, the null split, and the possibly existing categorical input values (as we discussed in Sect. 3).

Here, we want to stress out that some of the evaluated datasets have fewer than 10 input features. This low number of features can reduce the processing time improvement brought by our proposal. We will return to this discussion in the result section.

[1] https://archive.ics.uci.edu/ml/datasets.php.
[2] https://www.openml.org.

Table 1. Datasets used in the experiments.

Dataset	#Examples	#Numeric inputs	#Categorical inputs
Abalone	4177	8	0
Ailerons*	13750	40	0
Elevators*	16599	18	0
House8L	22784	8	0
House16H	22784	16	0
MV*	40768	7	3
Pol	15000	48	0
Wind	6574	14	0
Wine	6497	11	0

* Synthetic dataset.

4.3 Settings Used in the Tree Predictors

During the experiments, we fixed some hyper-parameters for the tree algorithms with values commonly used in the literature [8,18,23]. Split attempts were performed at intervals of $n_{\min} = 200$ examples. We set the significance level of the HB calculation to $\delta = 10^{-7}$, and the tie-break hyper-parameter to $\tau = 0.05$.

Besides, in all cases, we used 200 examples to initiate the tree predictors, providing a "warm" start for the evaluations. Finally, the perceptron weights were started with uniform random values in the range $[-1, 1]$. In case of splits, new nodes inherit their ancestors' weights.

Regarding the decision tree induction algorithms, we considered regression versions of the HT framework, as available in *scikit-multiflow*. They operate very similarly to FIMT-DD/FIRT-DD but do not have concept drift adaptation mechanisms, as at this point we only deal with stationary streams. We denote by HTR_m and HTR_p tree models that use mean and linear perceptron as their prediction strategy, respectively. Although very similar, these tree algorithms have different prediction strategies and might react differently when working along with 2CS.

4.4 Evaluation Strategy

In all the performed experiments, we used the *prequential* strategy for the evaluation of the HTR models [11,20]. In this benchmarking strategy, for each new incoming example, the model first makes a prediction and then learns from it. All tree-based algorithms were applied ten times to each dataset with different random number generator seeds. We report the average results obtained to reduce the effects of randomness and operational system external influences. For all the monitored metrics, we computed their mean value considering all the data seen until each measurement point and also considered windowed measurements. For such, we used a non-overlapping sliding window of size 200 [23].

We chose the Mean Absolute Error (MAE) as the error metric. This metric evaluates the absolute deviations of the tree's predictions compared to the expected target values. Furthermore, we report the amount of time spent by each algorithm (in seconds) and the total of memory resources consumed by the predictors (in MB).

We also performed statistical tests to verify whether the differences in the predictive performance of the models are statistically significant regarding the evaluation metrics. The Friedman test and the Nemenyi post-hoc test were applied with $\alpha = 0.05$, as described by Demšar [6]. We considered the windowed measurements for this end, to take into consideration the different time steps of the stream. We summed all the measurements for the different stream portions to obtain a single measurement per dataset.

5 Results and Discussion

In this section, we present and discuss our experimental results. First, we present the mean measurements after processing all the considered streams. Next, we discuss in details some interesting cases found during our analysis. Finally, we compare the performance of 2CS-based HTRs against traditional solutions, supported by statistical significance tests.

5.1 Overall Results

We start presenting the MAE values for all the compared algorithm variants, in Table 2. The best variants obtained by the HTR algorithm are highlighted in **bold**. The best results per dataset are underlined. We indicate with $k = i$, $i \in \{2, ..., 5\}$, the variants of 2CS. The variant *all* represents the traditional HTR algorithms. The average ranks for the different algorithms are also indicated.

As shown in Table 2, in most cases, the 2CS variants performed very similarly to their traditional counterparts. Despite some ties, MAE differences were observed after the second decimal place, a piece of information here omitted for visual clarity. Nonetheless, in some cases, such as the Pol dataset, the error difference was clear. These differences are a result of distinct tree structures generated by the 2CS strategy and the original HTR framework. In our proposal, correlation is applied as an additional heuristic to guide split selection and avoid excessive computations. Nevertheless, heuristics can sometimes be misleading, as some evaluated cases indicated. Anyhow, $\text{HTR}_p^{k=5}$ obtained the best overall ranking concerning MAE, being closely followed by HTR_m^{all}.

When considering the resulting model sizes, the 2CS-based variants generally originated models smaller than those generated by the original HTR algorithm. This fact is evidenced in Table 3. The best overall solution was $\text{HTR}_m^{k=4}$. Interestingly, the 2CS variants with smaller k hyper-parameter values did not necessarily result in less memory usage. This, again, comes from the fact that the correlation filtering for split candidates leads to different tree structures. In some cases, by evaluating fewer split candidates, the 2CS trees can take longer to split or

Table 2. MAE results. The best results per algorithm are in **bold**, while the best results per dataset are underlined.

Dataset	HTR_m				
	All	$k = 2$	$k = 3$	$k = 4$	$k = 5$
Abalone	2.24 ± 0.00	**2.08 ± 0.00**	2.22 ± 0.00	2.40 ± 0.00	2.40 ± 0.00
Ailerons	**0.00 ± 0.00**	0.00 ± 0.00	0.00 ± 0.00	0.00 ± 0.00	0.00 ± 0.00
Elevators	**0.00 ± 0.00**	0.00 ± 0.00	0.00 ± 0.00	0.00 ± 0.00	0.00 ± 0.00
House8L	**21237.95 ± 0.00**	22466.70 ± 0.00	22010.95 ± 0.00	21965.50 ± 0.00	21933.58 ± 0.00
House16H	**24247.50 ± 0.00**	25984.76 ± 0.00	26058.76 ± 0.00	24711.80 ± 0.00	24877.13 ± 0.00
MV	**1.37 ± 0.00**	4.80 ± 0.00	4.42 ± 0.00	2.97 ± 0.00	2.35 ± 0.00
Pol	**12.62 ± 0.00**	37.31 ± 0.00	37.31 ± 0.00	37.31 ± 0.00	37.31 ± 0.00
Wind	**3.63 ± 0.00**	4.30 ± 0.00	4.18 ± 0.00	3.71 ± 0.00	4.10 ± 0.00
Wine	**0.64 ± 0.00**	0.68 ± 0.00	0.68 ± 0.00	0.67 ± 0.00	0.65 ± 0.00
Average rank	3.33	7.67	7.78	6.44	6.67
Dataset	HTR_p				
	All	$k = 2$	$k = 3$	$k = 4$	$k = 5$
Abalone	1.66 ± 0.09	1.77 ± 0.12	1.67 ± 0.10	**1.58 ± 0.07**	1.59 ± 0.08
Ailerons	0.00 ± 0.00	0.01 ± 0.00	0.01 ± 0.00	**0.00 ± 0.00**	0.00 ± 0.00
Elevators	0.02 ± 0.02	0.02 ± 0.01	0.01 ± 0.00	0.01 ± 0.01	**0.00 ± 0.00**
House8L	21411.20 ± 727.93	21403.59 ± 926.08	21144.45 ± 870.59	**20902.70 ± 862.82**	20917.15 ± 856.20
House16H	24140.92 ± 1358.02	26152.65 ± 1559.79	26447.92 ± 1693.86	24315.09 ± 1368.92	**23987.59 ± 1355.32**
MV	**1.18 ± 0.22**	2.36 ± 0.38	2.12 ± 0.31	1.59 ± 0.26	1.64 ± 0.32
Pol	**14.25 ± 1.38**	27.88 ± 0.27	27.88 ± 0.27	27.88 ± 0.27	27.88 ± 0.27
Wind	3.44 ± 0.33	3.75 ± 0.46	3.45 ± 0.35	3.74 ± 0.45	**3.36 ± 0.31**
Wine	0.62 ± 0.04	0.62 ± 0.04	0.62 ± 0.04	**0.61 ± 0.03**	0.61 ± 0.03
Average rank	4.33	6.56	5.33	3.89	**3.00**

split with increased frequency. We could not find a clear pattern relating the k value and resulting model size. Notwithstanding, excluding $HTR_p^{k=5}$, all the 2CS variants required, in general, less memory than the traditional HTR algorithms.

Table 3. Model size results (in MB). The best results per algorithm are in **bold**, while the best results per dataset are underlined.

Dataset	HTR_m					HTR_p				
	All	$k = 2$	$k = 3$	$k = 4$	$k = 5$	All	$k = 2$	$k = 3$	$k = 4$	$k = 5$
Abalone	1.71	2.79	2.24	0.68	**0.01**	1.71	2.80	2.25	0.69	**0.01**
Ailerons	12.09	**8.88**	10.45	9.88	13.01	12.10	**8.90**	10.47	9.90	13.03
Elevators	11.71	**1.10**	13.14	11.18	12.83	11.72	**1.10**	13.16	11.20	12.86
House8L	**28.30**	32.41	30.03	32.17	30.83	**28.33**	32.44	30.06	32.20	30.86
House16H	50.59	**48.91**	56.99	87.62	60.31	50.62	**48.95**	57.04	87.65	60.34
MV	61.74	77.12	66.99	**57.96**	71.38	61.80	77.17	67.04	**58.00**	71.42
Pol	9.89	**1.87**	1.87	1.87	1.87	9.94	**1.87**	1.87	1.87	1.87
Wind	8.39	10.93	7.82	8.00	**5.56**	8.40	10.94	7.82	8.01	**5.56**
Wine	5.53	4.37	4.25	**3.72**	4.65	5.53	4.38	4.26	**3.73**	4.66
Average rank	5.44	5.00	4.67	**3.89**	5.33	6.44	6.33	6.00	5.22	6.67

Table 4 presents the running times of the algorithms and their variants. This is the characteristic where the 2CS-based variants were clearly superior. Our proposal, as expected, was the fastest method. The improvements, nevertheless, were less pronounced for the datasets with less input features, as expected. When considering HTR_m, the running time increased with the increase of k, as highlighted by the average ranks. The prediction strategy of this tree variant is simple and does not require matrix operations, which reflected in the running time. On the other hand, the same did not occur for HTR_p. The fastest HTR_p variant was the one with $k = 3$. Hence, we did not observe a clear relation between k and the final model runtime, as it occurs with HTR_m. The costs of updating the trees and making predictions ended up overcoming the gains in avoiding the evaluation of all features as split candidates.

Table 4. Running time results. The best results per algorithm are in **bold**, while the best results per dataset are underlined.

Dataset	HTR_m				
	All	$k = 2$	$k = 3$	$k = 4$	$k = 5$
Abalone	2.39 ± 0.16	**1.14 ± 0.02**	2.07 ± 0.13	2.56 ± 0.02	4.74 ± 0.35
Ailerons	12.62 ± 0.43	**6.12 ± 0.07**	9.74 ± 0.29	6.19 ± 0.08	10.79 ± 0.33
Elevators	7.42 ± 0.45	**3.98 ± 0.04**	7.45 ± 0.23	4.97 ± 0.10	7.71 ± 0.33
House8L	13.98 ± 0.46	**7.25 ± 0.12**	11.34 ± 0.31	8.36 ± 0.17	12.90 ± 0.32
House16H	28.08 ± 0.71	**14.31 ± 0.37**	17.33 ± 0.42	21.48 ± 0.42	21.09 ± 0.54
MV	38.09 ± 0.54	**24.56 ± 0.39**	27.67 ± 0.46	30.73 ± 0.59	35.08 ± 0.52
Pol	8.00 ± 0.34	6.35 ± 0.18	6.40 ± 0.25	6.42 ± 0.29	**6.14 ± 0.16**
Wind	5.56 ± 0.24	**3.58 ± 0.11**	3.96 ± 0.21	3.91 ± 0.16	4.35 ± 0.16
Wine	**2.55 ± 0.14**	2.59 ± 0.11	2.76 ± 0.12	2.88 ± 0.20	2.96 ± 0.13
Average rank	4.67	**1.22**	2.89	3.11	4.56
Dataset	HTR_p				
	All	$k = 2$	$k = 3$	$k = 4$	$k = 5$
Abalone	3.49 ± 0.22	**2.71 ± 0.10**	3.03 ± 0.19	4.97 ± 0.23	5.69 ± 0.27
Ailerons	25.12 ± 1.08	22.44 ± 0.55	**22.01 ± 0.57**	22.08 ± 0.82	22.71 ± 0.52
Elevators	14.84 ± 0.55	**13.66 ± 0.63**	14.99 ± 0.52	15.11 ± 0.46	14.94 ± 0.30
House8L	19.62 ± 0.54	**16.61 ± 0.22**	17.05 ± 0.29	17.65 ± 0.31	18.61 ± 0.37
House16H	38.21 ± 0.79	**26.64 ± 0.46**	27.08 ± 0.66	30.79 ± 0.56	30.14 ± 0.54
MV	49.68 ± 1.18	**37.00 ± 0.35**	38.38 ± 0.37	42.19 ± 0.55	45.68 ± 0.43
Pol	28.59 ± 1.24	27.32 ± 0.81	**16.22 ± 0.09**	26.43 ± 0.84	16.26 ± 0.07
Wind	7.34 ± 0.63	5.94 ± 0.24	**3.81 ± 0.06**	5.99 ± 0.36	4.17 ± 0.02
Wine	4.15 ± 0.46	4.37 ± 0.40	**2.89 ± 0.03**	4.25 ± 0.62	3.04 ± 0.04
Average rank	9.11	6.89	6.00	8.56	8.00

5.2 Analysis

As previously mentioned, the use of correlation as a heuristic to guide split feature selection can lead to tree structures different from those induced by the original HTR algorithms. This different growth pattern can have either a positive or a negative impact on the predictive performance of resulting models. According to the experimental results, the use of 2CS usually improved the model size and running time. However, there are cases where 2CS uses more memory or even has a higher runtime than the original solution, e.g., when the 2CS-based trees are much larger than the original HTRs. At first glance, nonetheless, it seems improbable a tree structure built on a reduced amount of information results in lower prediction error. However, we observed this unusual behavior in our experiments. Next, we present our interpretation of this and other interesting cases. For such, we use two of the evaluated datasets.

First, we present the time-varying results for the Pol dataset, which has 48 input features. Due to space limits, we will focus on the HTR_m-based variants. As can be seen on the top chart of Fig. 2a, the MAE values obtained by the 2CS variants were much worse than those obtained by HTR_m^{all}. The analysis of memory usage can give us evidence for this sub-par performance. As showed in the middle chart of the same figure, 2CS variants spent much less memory than the traditional HTRs variants. In fact, they presented a slowly increasing memory behavior. This steady memory increase is probably due to the 2CS-based models performing splits frequently, so the E-BSTs did not gather much data. There are no memory drops, as they occur when a split is performed, and E-BSTs with multiple stored elements are discarded. In fact, E-BSTs are the main source of memory consumption, when compared to the other elements of the trees.

As expected, memory increase and drop occurs for HTR_m^{all}, as depicted in the figure. The bottom chart from Fig. 2a shows that smaller models resulted in lower running times. In this case, we believe that 2CS misguided the tree growth, resulting in undesired performance levels. To overcome this deficiency, we intend to investigate more sophisticated mechanisms for split candidate selection. A potential strategy to pursue would be applying meta-learning for split candidate recommendation [4].

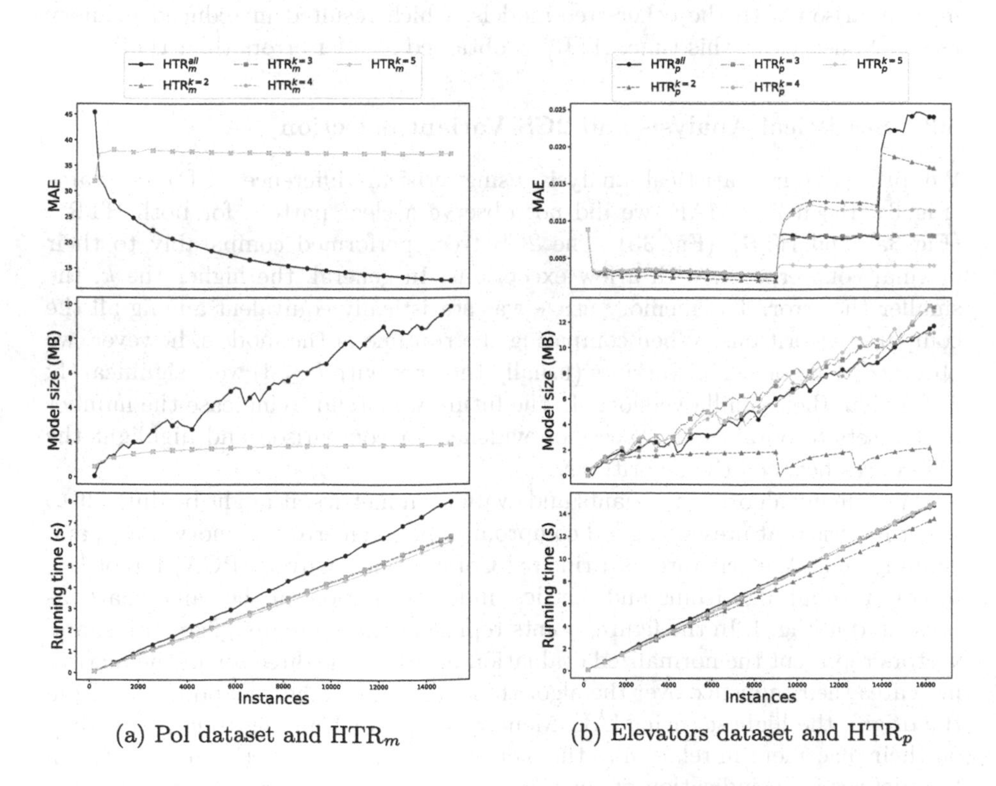

(a) Pol dataset and HTR_m (b) Elevators dataset and HTR_p

Fig. 2. Time varying results for the Pol and Elevators datasets.

The second case discussed refers to the Elevators dataset, whose performance profiles are presented in Fig. 2b. This time, we will only consider the HTR_p variants. Regarding MAE, after ≈8000 instances, the 2CS-based HTR_p become more accurate than HTR_p^{all}. Interestingly, the higher the k, the faster the 2CS-based variants reacted to the sudden error increase. Even the most limited $HTR_p^{k=2}$ was better than HTR_p^{all}. In this context, we observed the opposite of what we saw in the previous case. Our proposal avoided overfitting to the current observed state, as it selected better features for the splits. HTR_p^{all} relies on the currently stored values in the E-BST to decide the splits. However, the current best split decision might be only valid for the data observed so far. The best split for future data might be different from the current estimation. This seems to be the case for the Elevators dataset. Nevertheless, our proposal's variations were able to overcome this problem. This type of situation deserves future investigation. We intend to put special attention to non-stationary problems to evaluate how 2CS might affect the tree algorithms in these scenarios.

Regarding memory and processing time, the 2CS-based variants performed very similarly to HTR_m in the Elevators dataset. 2CS generated faster trees than HTR_p^{all}. They also used varying, but similar memory amounts to HTR_p^{all}. The only clear exception was the memory usage of $HTR_p^{k=2}$. Following the same reasoning used for the Pol dataset, the 2CS variant splits with increased frequency in comparison with the other tree models, which resulted in reduced memory usage. Nonetheless, this time, $HTR_p^{k=2}$ obtained smaller errors than HTR_p^{all}.

5.3 Statistical Analysis and 2CS Variant Selection

We present our statistical analysis using critical difference (CD) diagrams (Fig. 3). Regarding MAE, we did not observe a clear pattern for both HTR_m (Fig. 3a) and HTR_p (Fig. 3b). The 2CS trees performed comparably to their original counterparts, with a few exceptions. In general, the higher the k, the smaller the error. The memory usage was statistically equivalent among all the compared algorithms. When comparing the runtime of the models, however, we observed that some 2CS variants (usually the ones with $k \leq 3$) were significantly faster than their vanilla versions. In the future we intend to increase the number of datasets to obtain more pieces of evidence for comparison and highlight the differences between the algorithms.

With many algorithm variants and evaluation metrics, it might be difficult to select models that present a good compromise between error, memory usage, and running time. We generated a Principal Component Analysis (PCA) biplot [10] to comprise all algorithms and metrics under evaluation on the same chart, as presented in Fig. 4. In the figure, points represent the compared algorithms, and vectors represent the normalized evaluation metrics. The direction of the metrics indicates their influence over the algorithms, i.e., the farther the points are from the origin, the highest their MAE, Memory usage, or Running time (depending on their placement in relation to the metrics' vectors). Lastly, the angle between the metrics is an indication of correlation, in case the vectors have roughly the

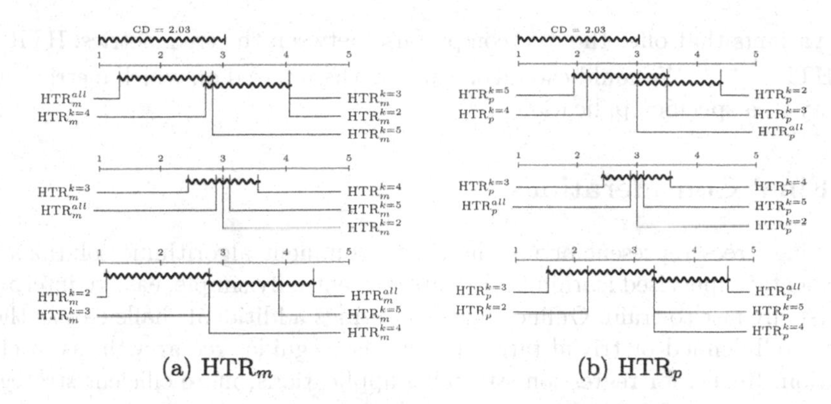

Fig. 3. Statistical tests results: MAE (top), Model size (middle), Running time (bottom). Tree algorithms whose ranks do not differ by at least the critical distance (CD) value are considered statistically equivalent (at $\alpha = 0.05$).

same or opposite directions (which configure positive and negative correlations, respectively). Orthogonality is an indication of no correlation.

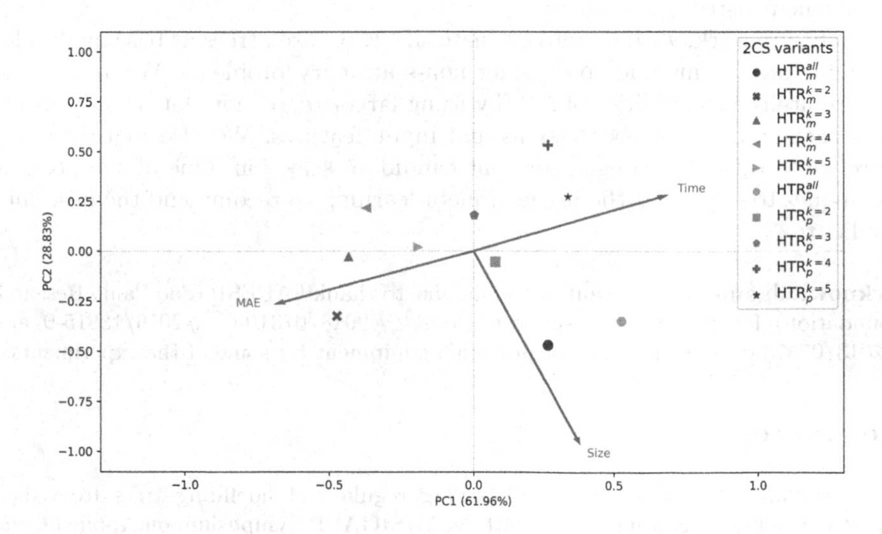

Fig. 4. PCA biplot comprising all the compared algorithms and evaluation metrics.

The obtained biplot enables us to draw interesting observations. Firstly, MAE and the Running time are negatively correlated, i.e., the higher the MAE, the smaller the time spent by the trees, and vice-versa. The resulting tree size does not seem to be related to the error nor the running time of the models. The least accurate variant was $\mathrm{HTR}_m^{k=2}$, the slowest ones were $\mathrm{HTR}_p^{k\in\{4,5,all\}}$, and the biggest ones were the vanilla HTR versions. Towards the origin of the chart we have variants that offer the best compromise between the tree metrics: $\text{HTR}_m^{k=5}$ and $\text{HTR}_p^{k\in\{2,3\}}$. We could choose one among them based on which metric is the priority in a specific application.

6 Final Considerations

Hoeffding Trees represent one of the most prominent algorithmic solutions for incremental supervised learning. They are conceptually simple, easy to interpret, and usually fast to train. Online regression brings additional challenges as there are no well-defined or trivial target partitions to guide tree growth, as in classification. Hence, for regression streaming applications, more efficient strategies for feature split point selection are needed.

In this work, we proposed the use of linear correlation as a complementary heuristic to select a subset of input features to evaluate as split candidates. Our proposal, 2CS, can be easily merged into the HTR framework, has just a single hyper-parameter to tune, and does not increase the amount of the necessary memory. 2CS reduces the processing time by evaluating fewer input features as split candidates. Experimental results showed that 2CS retain the prediction capabilities of vanilla HTRs, whereas using a similar memory footprint and being significantly faster.

As future work, we intend to evaluate the 2CS-based trees as base models for ensemble algorithms and to consider non-stationary problems. We also intend to evaluate the capabilities of 2CS by using larger regression datasets regarding both their number of observations and input features. We also plan to apply more sophisticated strategies for split candidate selection. One of the possible techniques to explore is the usage of meta-learning to recommend the best split candidate.

Acknowledgements. The authors would like to thank FAPESP (São Paulo Research Foundation) for its financial support (grants #2018/07319-6, #2016/18615-0 and #2013/07375-0) and Intel Inc. for providing equipment for some of the experiments.

References

1. Barddal, J.P., Enembreck, F.: Learning regularized hoeffding trees from data streams. In: Proceedings of the 34th ACM/SIGAPP Symposium on Applied Computing, pp. 574–581. ACM (2019)
2. Bifet, A., Gavaldà, R.: Adaptive learning from evolving data streams. In: Adams, N.M., Robardet, C., Siebes, A., Boulicaut, J.-F. (eds.) IDA 2009. LNCS, vol. 5772, pp. 249–260. Springer, Heidelberg (2009). https://doi.org/10.1007/978-3-642-03915-7_22
3. Bishop, C.M.: Pattern Recognition and Machine Learning. Springer, Boston (2006)
4. Brazdil, P., Giraud-Carrier, C., Soares, C., Vilalta, R.: Metalearning: Applications to Data Mining. Springer, Heidelberg (2008)
5. Breiman, L., Friedman, J., Olshen, R., Stone, C.: Classification and Regression Trees. Chapman and Hall, Wadsworth (1984)

6. Demšar, J.: Statistical comparisons of classifiers over multiple data sets. J. Mach. Learn. Res. **7**(Jan), 1–30 (2006)
7. Domingos, P., Hulten, G.: Mining high-speed data streams. In: Proceedings of the sixth ACM SIGKDD International Conference on Knowledge Discovery and Data Mining, pp. 71–80. ACM, Boston (2000)
8. Duarte, J., Gama, J.: Multi-target regression from high-speed data streams with adaptive model rules. In: 2015 IEEE International Conference on Data Science and Advanced Analytics (DSAA), vol. 36678, pp. 1–10. IEEE, Campus des Cordeliers, Paris (2015)
9. Fan, W., Bifet, A.: Mining big data: current status, and forecast to the future. ACM sIGKDD Explor. Newslett. **14**(2), 1–5 (2013)
10. Gabriel, K.R.: The biplot graphic display of matrices with application to principal component analysis. Biometrika **58**(3), 453–467 (1971)
11. Gama, J.: Knowledge Discovery from Data Streams. Chapman and Hall/CRC, London (2010)
12. Gomes, H.M., Barddal, J.P., Enembreck, F., Bifet, A.: A survey on ensemble learning for data stream classification. ACM Comput. Surv. (CSUR) **50**(2), 23 (2017)
13. Gomes, H.M., Barddal, J.P., Ferreira, L.E.B., Bifet, A.: Adaptive random forests for data stream regression. In: 26th European Symposium on Artificial Neural Networks, ESANN 2018, Bruges, Belgium, 25–27 April 2018 (2018). http://www.elen.ucl.ac.be/Proceedings/esann/esannpdf/es2018-183.pdf
14. Grzenda, M., Gomes, H.M., Bifet, A.: Delayed labelling evaluation for data streams. Data Min. Knowl. Disc. **34**, 1237–1266 (2019)
15. Hothorn, T., Hornik, K., Zeileis, A.: Unbiased recursive partitioning: a conditional inference framework. J. Comput. Graph. Stat. **15**(3), 651–674 (2006)
16. Hulten, G., Spencer, L., Domingos, P.: Mining time-changing data streams. In: Proceedings of the Seventh ACM SIGKDD International Conference on Knowledge Discovery and Data Mining, pp. 97–106. ACM (2001)
17. Ikonomovska, E., Gama, J., Džeroski, S.: Incremental multi-target model trees for data streams. In: Proceedings of the 2011 ACM Symposium on Applied Computing, pp. 988–993. ACM (2011)
18. Ikonomovska, E., Gama, J., Džeroski, S.: Learning model trees from evolving data streams. Data Min. Knowl. Disc. **23**(1), 128–168 (2011)
19. Ikonomovska, E., Gama, J., Džeroski, S.: Online tree-based ensembles and option trees for regression on evolving data streams. Neurocomputing **150**, 458–470 (2015)
20. Krawczyk, B., Minku, L.L., Gama, J., Stefanowski, J., Woźniak, M.: Ensemble learning for data stream analysis: a survey. Inf. Fusion **37**, 132–156 (2017)
21. Mastelini, S.M., Barbon Jr., S., de Carvalho, A.C.P.d., Ferreira, L.: Online multi-target regression trees with stacked leaf models. arXiv preprint arXiv:1903.12483 (2019)
22. Montiel, J., Read, J., Bifet, A., Abdessalem, T.: Scikit-multiflow: a multi-output streaming framework. J. Mach. Learn. Res. **19**(1), 2914–2915 (2018)
23. Osojnik, A., Panov, P., Džeroski, S.: Tree-based methods for online multi-target regression. J. Intell. Inf. Syst. **50**(2), 315–339 (2018)
24. Salehi-Moghaddami, N., Yazdi, H.S., Poostchi, H.: Correlation based splitting criterion in multi branch decision tree. Cent. Eur. J. Comp. Sci. **1**(2), 205–220 (2011)

A Distance-Weighted Selection of Unlabelled Instances for Self-training and Co-training Semi-supervised Methods

Cephas A. S. Barreto[1(✉)], Arthur C. Gorgônio[1], Anne M. P. Canuto[1], and João C. Xavier-Júnior[2]

[1] Department of Informatics and Applicated Mathematics (DIMAp), Natal, Brazil
cephas@gmail.com

[2] Digital Metropolis Institute (IMD), Federal University of Rio Grande do Norte (UFRN), Natal, Brazil

Abstract. The use of Semi-supervised Learning (SSL) methods have emerged as an efficient solution to smooth out the problem of availability of labelled instances. Several methods have been proposed in the literature and Self-training and Co-training are two well-known methods. The main aim is to use only a few labelled instances to define a model and to apply this model in a labelling process, in which unlabelled instances are labelled and included in the labelled set. However, the labelling process is always directly dependent on the selection of the unlabelled instances. Moreover, the selection criterion used to select and label new instances has an important effect in the performance of a semi-supervised method. In this paper, we propose a distance-weighted selection of unlabelled instances for Self-training and Co-training semi-supervised methods. In addition, we compare the standard Self-training and Co-training methods against the proposed versions of these two methods over 20 classification datasets.

1 Introduction

In Machine Learning (ML), classification tasks are conceived to analyse an instance of a given problem accordingly to a model (function), aiming to define a label for this instance (also called class). Usually, the performance of a classification system is directly related to the distribution of knowledge spread among the input data points. However, in real world classification problems, the amount of labelled data is usually limited, or in some cases, very hard or expensive to manually label instances [1].

Aiming to overcome the consequences of the lack of labelled data, Semi-Supervised Learning (SSL) methods have been proposed in Literature [2–5]. A SSL method uses only a few labelled data (e.g., 5% or 10%) for iteratively applying training, selecting and labelling new instances. Several methods have been proposed and the two most traditional SSL methods are Co-training [6], and Self-training [7].

R. Cerri and R. C. Prati (Eds.): BRACIS 2020, LNAI 12320, pp. 352–366, 2020.
https://doi.org/10.1007/978-3-030-61380-8_24

Although Self-training and Co-training methods have been efficiently used in several applications, some wrongly classified instances can be included in the labelled set, causing performance deterioration. In fact, these methods use the confidence prediction assigned by a base classifier to select the unlabelled instances to be labelled (selection criterion). In addition, the number of labelled instances used for training a base classifier is small, increasing the chance of selecting wrongly classified instances. In this sense, the selection and labelling of new instances are critical steps for these SSL methods.

Aiming to increase the efficiency of these SSL methods, this paper proposes an approach for selecting unlabelled instances. In order to improve the selection step, our selection approach combines two different strategies, prediction confidence and distance metric, to define a Distance-Weighted Confidence (DWC) selection criterion. In this paper, our selection approach will be applied in two SSL methods, Co-training and Self-training. Nevertheless, it is important to emphasise that this approach can be applied to any semi-supervised method.

An empirical analysis will also be presented to assess the feasibility of our approach as selection criterion. This analysis will compare the standard Co-training (Ct-std) against the Co-training with the Distance-Weighted Confidence (Ct-dwc) as well as the standard Self-training (St-std) against the Self-training-dwc (St-dwc). In this analysis, 20 classification datasets will be used for evaluating the performance of all four aforementioned methods. In addition to the comparison between the standard and DWC-based versions, a brief discussion of the results will be presented.

The remainder of this paper is organised as follows. Section 2 presents the background of Co-training and Self-training methods. Section 3 discusses related work for SSL methods, focusing on these two SSL methods. Section 4 presents the proposed version, focusing on the description of the main differences to the corresponding standard methods. Section 5 describes the experimental methodology while Sect. 6 presents the computational results. Finally, Sect. 7 presents our conclusions and a direction for future work.

2 Background

The next subsections will describe two SSL methods (i.e., Co-training and Self-training). Both methods consider the following names and acronyms, described as: $\boldsymbol{i}$: instance; $\boldsymbol{L}$: labelled set; $\boldsymbol{U}$: unlabelled set; and $\boldsymbol{C}$: classifier.

2.1 Self-training

The main idea of the Self-training SSL technique [7] is simple and efficient (See Algorithm 1). First, a base classifier is trained with the available set of labelled instances (line 2). Then, it uses its own knowledge to label the unlabelled instances (line 3). After that, it selects the best labelled instances based on its confidence prediction (line 4) to be included to the initial labelled set of instances. This process is repeated until the unlabelled set (U) is empty.

Algorithm 1: Self-training

1 **while** *U and is not empty* **do**
2 train C with L;
3 label U using C;
4 select the best labelled-instances of U and join them to L;
5 **end**

2.2 Co-training

Co-training [6] is one traditional SSL technique that iteratively trains two base classifiers on two different views (different attribute subsets), and it uses the predictions of one classifier to select the unlabelled instances to compose the labelled set of the other classifier. Algorithm 2 shows the general functioning of the Co-training method. Its basic idea is similar to the Self-training method regarding the use of base classifiers (the first uses two classifiers while the second uses only one).

Algorithm 2: Co-training

1 create L_1 and L_2 with a vertical split in L;
2 create U_1 and U_2 with a vertical split in U;
3 **while** *U_1 and U_2 are not empty* **do**
4 train C_1 with L_1 and C_2 with L_2;
5 apply C_1 in U_1 and C_2 in U_2;
6 select the best instances from U_1 and include them to L_2;
7 select the best instances from U_2 and include them to L_1;
8 **end**

In the first step, the input dataset is divided into two attribute subsets (vertical split). In doing this, the labelled set (L) has two views, called L_1 and L_2 (line 1). In the same way, the unlabelled set (U) has two views, U_1 and U_2 (line 2). Then, two base classifiers, C_1 and C_2, are trained over L_1 and L_2, respectively (line 4). In the next step, U_1 and U_2 are labelled by its respective classifier, C_1 and C_2 (line 5). Finally (lines 6 and 7), the best labelled instances from U_1 and U_2 are included to L_2 and L_1, respectively. This iterative process is repeated until U_1 and U_2 are empty.

The main objective of including the best labelled instances from one classifier into the labelled set of the other classifier is to allow cooperation between classifiers by crossing of acquired knowledge; and to promote diversity in both labelled sets [6]. For instance, when C_1 classifier selects the best instances to be included to the L_2 labelled set, it cooperates with the model quality of the other classifier (C_2) and these newly included instances are not biased by the prediction of the same view.

2.3 General Discussion

Overall, the functioning described for Self-training (Algorithm 1) and Co-training (Algorithm 2) is similar for both techniques. However, two aspects are important in Co-training. Firstly, the difference among attributes selected for each subset after the vertical split (i.e., Co-training creates two subsets). In these subsets, the attribute selection needs to ensure that no attribute is placed in both subsets (i.e., the intersection must be null), and also that each attribute must be present in one subset (i.e., the union must be total). This aspect guarantees different subsets of the input dataset and it may contribute to the improvement of the classification results. Secondly, as mentioned previously, by allowing the base classifiers to assign labelled instances from crossing subsets introduces cooperation between them.

Considering the standard version of both methods, the selection criterion is performed by the confidence prediction (e.g., one base classifier in St-std or two base classifiers in Ct-std). This criterion is related to the level of confidence that a classifier assigns an instance to a particular class. This confidence can be strongly affected by the distribution of instances within the training set and also by the characteristics of the used classifier. One possible approach to address this issue (the selection of unlabelled instances) is to use the similarity between instances, generally measured by a distance metric. In this sense, the similarity information can be used to select the unlabelled instances.

Another important fact about SSL methods is that if some wrongly classified instances are included to the labelled set, this error will be carried out through each next iterations (snowball case), resulting in weak models (less effective in terms of accuracy). For this reason, it is important to investigate and to build novel approaches for the selection process, aiming to mitigate the selection errors during the execution of SSL methods.

3 Related Work

In the last decades, SSL methods have received special attention due to their strong potential to solve real world problems. In this sense, these SSL methods, such as Co-training and Self-training, have been applied to a wide range of areas, including: image processing and classification [8–10]; bioinformatics [11]; text classification and natural language processing [12,13], among others. The majority of the proposed Co-training and Self-training extensions are based on confidence prediction as selection criterion. However, there are also other versions based on distance metrics. In this sense, we present some SSL studies based on confidence prediction, distance metrics or another approach.

3.1 Confidence-Based Selection

As mentioned previously mentioned, there are studies proposing confidence-based methods to select unlabelled instances in SSL methods, mainly for Co-training and Self-training, such as in [9,14,15]. In [14], for instance,the authors used a Decision Tree combined with a threshold to define the number of labelled instances to be selected. On the other hand, in [15] the authors proposed a Self-training extension based on density peaks of data. This method also used the concept of differential evolution, aiming to discover the data structure for better classifier training and also to optimise the positioning of the newly labelled data. Finally, an empirical analysis was conducted, in which the obtained accuracy results were compared to the standard Self-training version.

Regarding Co-training, in [9] the authors used multiple deep neural networks combined with the general idea of Co-training in order to build a deep multi-view of an image datasets. According to them, this work obtained good results when compared to state-of-the-art methods.

3.2 Distance-Based Selection

In [16] and [17], the authors proposed distance-based selection approaches for Self-training, aiming to use distance metrics in order to select the most similar instances at each iteration of the Self-training method. Particularly, in [17], the authors used the k-NN classifier as noise reduction by aggregating only the nearest labelled instances. On the other hand, the work of [16] used similarity between images for selecting the most similar unlabelled instances in video classification tasks.

A similar idea from previous studies has been used in [18], but it is applied to Co-training. The authors proposed a distance-based selection approach for the Co-training method in combination with the k-means clustering technique. This clustering technique was used to select the most similar (nearest) instances that are able to create a cluster. This idea was then applied to two Co-training steps, data splitting and subsets crossing over. The authors compared their proposal with other state-of-art methods, and it overcame them in all evaluated metrics as purity, normalised mutual information and the unsupervised accuracy.

3.3 Other Selection Approaches and Discussion

Besides the aforementioned studies, there are researches that aim to improve the robustness of SSL methods by changing the selection criterion or using an additional apparatus to improve the selection step, as it plays an important role in SSL methods [15,19–21].

In summary, as indicated by the obtained results, the use of confidence prediction and distance-based metric as criteria for the selection of the best unlabelled instances in Self-training and Co-training methods can be considered as an efficient approach. Although, we believe that there is still need for further improvements, the combination of these two approaches improves the performance of SSL methods.

Based on this, our work proposes a novel selection criterion for SSL methods, more specifically Self-training and Co-training. Our selection criterion uses the confidence prediction of a classifier combined with a distance-based measure to establish the overall confidence of an unlabelled instance to be selected, which has been named as Distance-Weighted Confidence.

4 The Proposed Approach

As Sect. 2 described the functioning of both Self-training and Co-training (i.e., standard versions), this section will discuss the functioning of our proposal. As a matter of fact, this paper proposes a novel approach for selection of unlabelled instances in SSL methods. Our approach uses confidence prediction as the main step for the selection step. Nonetheless, it is important to emphasise that the confidence prediction is combined with a distance measure in order to establish the overall confidence of an unlabelled instance to be selected. Equation 1 formally defines how to compute the Distance-weighted Confidence (DWC).

$$\mathrm{DWC}_i = \max(\forall_{j \in C}\ \mathrm{DWC}_{ij}) \tag{1}$$

where:

$$\mathrm{DWC}_{ij} = \mathrm{conf}_{ij} \times \frac{1}{d_j} \tag{2}$$

where:

- DWC_{ij} is the distance-weighted confidence of instance i to class j;
- $\mathbf{conf_{ij}}$ defines the confidence prediction for an instance i to the j-th class;
- C is the set of classes of a problem;
- $\mathbf{d_j}$ represents the distance between instance i and the centroid of class j in the labelled set.

As illustrated by Eqs. (1) and (2), the confidence prediction to a particular class j is weighed by its inverse distance of the centroid of this class to provide the DWC_{ij} value. The DWC_{ij} value is then calculated to all classes and the highest DWC_{ij} value is used as the overall confidence of instance DWC_i. In this sense, both confidence and distance are being taken into consideration in our selection process.

4.1 Self-training with DWC

Algorithm 3 shows the functioning of Self-training with the DWC approach as selection criterion, which has been named as Self-training with Distance-weighted Confidence (St-dwc). Regarding the differences between St-std and St-dwc, we can describe them as follows:

- the computation of centroids d_j for all classes at each iteration (line 3);
- the computation of DWC_i for each instance of U (line 5);
- the selection of the best instances from U based on DWC_i;

Algorithm 3: Self-training-dwc

1 **while** *U is not empty* **do**
2   train C with L;
3   compute the centroids d of all classes j using L;
4   label U using C;
5   compute DWC_i for each i within U;
6   select the best labelled-instances of U according DWC_i and join them to L;
7 **end**

As an example, lets us consider two instances u_1 and u_2 from U, and d_a being the centroid of class a. In this way, following the flow described in Algorithm 3, suppose that the base classifier assigns class label "a" for both instances u_1 and u_2 (line 4), having a confidence prediction factor ($conf_{ij}$) for u_1 and u_2 equal to 0.80 and 0.90, respectively. In this sense, the base classifier has more confidence in u_2 being labelled as class a than u_1 being labelled in that particular class. Note that, in case of an SSL method being based only on prediction confidence, and on top of that, there is only one instance to be selected, then u_2 instance would be selected, when the standard Self-training (St-std) is used.

However, for our proposed method, the Euclidean distance is also used. Suppose that this distance between d_a and the two instances (u_1 and u_2) are 1.58 and 2.73, respectively. This means that u_1 is much closer to all instances in class a than u_2. Then, the computing of DWC_{ij} for u_1 and class a can be demonstrated as follows:

$$\mathrm{DWC}_{u_1a} = \mathrm{conf\_pred}_i \times \tfrac{1}{d},$$

$$\mathrm{DWC}_{u_1a} = 0.80 \times \tfrac{1}{1.58} = \mathbf{0.5063}$$

On the other hand, DWC_{ij} for u_2 and class a would be 0.3297 (using Eq. (2)). Finally, suppose that DWC_{u_1a} and DWC_{u_2a} are the max values of all DWC_{ij} values for u_1 and u_2, respectively. Then, they are defined as DWC_{u_1} and DWC_{u_2}, respectively. Then, for St-dwc, u_1 would be selected.

The use of DWC intends to benefit instances from U that are close to instances with the same class label in the labelled set. By doing so, we aim to decrease the selection of poorly labelled instances, which often occurs in SSL methods, especially at the first iterations. Hence, instances from U labelled with a certain label but being far from instances within the labelled set, and with the same label, tend to become unavailable for selecting.

4.2 Co-training with DWC

The Co-training version, named Co-training with Distance-weighted Confidence (Ct-dwc) follows the same idea of Self-training. In other words, the DWC values are calculated, using Eq. (1) and (2), for both base classifiers and each one selects the best instances to be labelled. After this selection, Ct-dwc continues with its normal flow, in which the instances selected by one classifier are included in the labelled set of the other classifier.

5 Experimental Methodology

This section describes important details of the used experimental framework, the datasets used in the empirical analysis, the baseline methods for comparative analysis, the predictive accuracy measures and, finally the methods and materials of this empirical analysis.

5.1 The Experimental Framework

The general methodology of this empirical analysis is based on an n-fold cross validation method with $n = 10$ and it can be explained in the following steps:

1. shuffle the dataset;
2. split the dataset into *10* stratified folds;
3. separate fold 1 for validation (Validation set - V);
4. divide the remaining folds (2 to 10) into labelled and unlabelled sets, being 10% for the labelled set (L) and 90% for the unlabelled set (U);
5. build an ML model by performing the SSL algorithms described in this paper using L;
6. validate the built model using V and save the obtained results;
7. repeat steps from 3 (changing the fold used for validation) until all folds have been used as validation.

At the end of this process, it is expected that 10 values have been saved. This process is repeated 10 times, with different data distribution in the folds (step 1). The obtained result is given by the average results of a 10×10-fold cross validation.

5.2 Datasets

In order to evaluate the feasibility of the proposed methods, 20 datasets are selected from well known machine learning repositories[1]. Table 1 presents a brief description of all datasets, including the reference number of the dataset (No), name (Dataset), number of instances (Inst), attributes (Att) and classes, and also the data type of the attributes (categorical - C or Numeric N).

[1] From UC Irvine Machine Learning Repository and available on https://archive.ics.uci.edu/ml/datasets.php.

Table 1. Description of the datasets

No	Dataset	Inst	Att	Class	Type
d1	Abalone	4177	9	28	C, N
d2	Arrhythmia	452	261	13	N
d3	Car	1728	6	4	C, N
d4	Ecoli	336	8	8	C, N
d5	Glass	214	10	6	N
d6	Hill Valley	606	101	2	N
d7	King-Rook vs King Pawn	3196	36	2	C
d8	Leukemia Haslinger	100	50	2	N
d9	Madelon	2600	501	2	N
d10	Multiple Features	2000	64	10	N
d11	Secon	1567	591	2	N
d12	Seeds	210	7	3	N
d13	Semeion	1593	256	10	N
d14	Solar Flare	1389	13	6	C, N
d15	Spectf Heart	267	14	2	N
d16	Tic Tac Toe Endgame	958	9	2	C
d17	Twonorm	7400	21	2	N
d18	Waveform	5000	40	3	N
d19	Wine	4898	12	11	N
d20	Yeast	1484	9	10	N

5.3 Comparative Analysis

As previously mentioned, this paper focuses on evaluating the effectiveness of our proposed *DWC* approach (i.e., an overall confidence for labelling unlabelled instances composed by confidence prediction and weighting distance metric). In order to do this, all SSL methods having the same name are compared, aiming to evaluate the effectiveness of using a DWC-based selection criterion. In other words, the main aim is to perform a pairwise analysis comparing St-std and St-dwc as well as Ct-std and Ct-dwc.

5.4 Predictive Accuracy Measures

All methods are evaluated based on two predictive accuracy measures, which are classification accuracy rate and F-measure. The accuracy rate is well-known and it simply measures the confidence (number of correct predictions) of the analysed model.

On the other hand, F-measure (also called F-score) is the harmonic mean between precision and recall [22] and it is defined as:

$$\text{F-measure} = \frac{(2 * \text{precision} * \text{recall})}{(\text{precision} + \text{recall})} \tag{3}$$

The positive predictive value (precision) is defined as the proportion of the number of true positives and the number of all positive (true and false) labels predicted by the model. Recall, in turn, refers to the proportion of the number of true positives and the summation of true positives with false negatives.

Additionally, the obtained results for both measures will be assessed from a statistical point of view. For this, we will use the Friedman test with a significance of 0.05 and the null hypothesis defines that there is no difference between the average accuracy values of the analysed methods. If the null hypothesis is rejected, the Nemenyi post-hoc test is then performed [23].

5.5 Methods and Materials

In this empirical analysis, the Euclidean distance has been selected for the implementation of the distance metric in St-dwc and Ct-dwc. This distance was selected due to the fact that it is simple and widely used in classification applications.

The implementation of all four methods (e.g., Ct-std, St-std, Ct-dwc and St-dwc) and the development of the experimental framework are based on the Weka API [24]. In these methods, a Decision Tree (J48 - confidence factor = 0.05) was used as base classifier. The percentage of the unlabelled instances that a SSL algorithm aggregates for each iteration is defined to 10% of the total amount of instances in the dataset. Finally, we run the experiments on a desktop PC with Ubuntu 16.04 64 bit operating system driven by an Intel(R) Xeon(R) CPU E5-4610 v4 - 1.80 GHz, 6 core, and RAM with 6 Gb.

6 Experimental Results

This section presents the experimental results, comparing the predictive performance of the proposed approach (Ct-dwc and St-dwc) to two other baseline methods: (a) the Ct-std, and (b) the St-std method, respectively.

6.1 Results for the Predictive Accuracy

Table 2 presents the accuracy values for all four analysed methods (columns) and all 20 datasets (lines). In this table, the bold numbers represent the highest accuracy value between methods with the same methodology (e.g., Ct-std and Ct-dwc), for each dataset. In addition, the highest accuracy value of all evaluated methods is represented by a grey cell, for all datasets. Finally, the last four lines of this table represents: (a) the overall average accuracy over all datasets; (b) the number of wins for each pairwise comparison (Inner_wins); (c) the overall number of wins for each method (General_wins); and (d) the average ranking over all datasets.

As shown in Table 2, we can state that St-dwc obtained the best average accuracy (**66.217%**) among all other methods. Moreover, when analysing each pairwise of methods, we can state that the use of the DWC selection criterion had a positive effect only for the Self-training method since St-dwc outperformed its competing method. However, Ct-dwc was surpassed by Ct-std in the number of wins. Regarding the overall number of wins, once again, St-dwc outperformed all other three methods, obtaining 18 out of 20. Finally, St-dwc obtained the best ranking result (**1.15**), followed by St-std, Ct-std and Ct-dwc, respectively.

In summary, based on Table 2, St-dwc was clearly the best analysed SSL method, which may indicate that our proposal has enhanced the accuracy performance of the Self-training SSL method.

Table 2. Average accuracy for all methods

Dataset	Ct-std	Ct-dwc	St-std	St-dwc
d1	**19.600%**	10.330%	20.038%	**21.809%**
d2	39.660%	**53.520%**	54.889%	**59.304%**
d3	**69.940%**	**69.940%**	74.763%	**76.215%**
d4	**47.810%**	43.910%	74.082%	**74.982%**
d5	35.000%	**42.730%**	**49.048%**	45.736%
d6	50.120%	**50.620%**	46.456%	**50.904%**
d7	**57.360%**	52.190%	94.742%	**95.778%**
d8	64.500%	**69.000%**	60.000%	**73.000%**
d9	**52.330%**	50.810%	53.231%	**55.846%**
d10	**33.350%**	25.500%	63.950%	**69.950%**
d11	**93.590%**	**93.590%**	91.580%	**93.172%**
d12	**78.810%**	71.670%	79.524%	**87.619%**
d13	**36.670%**	23.330%	52.353%	**56.876%**
d14	**32.140%**	29.890%	70.697%	**71.490%**
d15	**63.680%**	**63.680%**	67.622%	**69.622%**
d16	**65.620%**	**65.620%**	66.070%	**68.052%**
d17	76.370%	**77.240%**	79.824%	**82.554%**
d18	**42.110%**	37.640%	69.840%	**74.720%**
d19	**41.970%**	37.610%	43.753%	**44.424%**
d20	28.420%	**31.780%**	49.733%	**52.293%**
Average	**51.453%**	50.030%	63.110%	**66.217%**
Inner_wins	**14**	10	1	**19**
General_wins	1	1	1	**18**
Avg_rank	**3.1**	3.3	2.25	**1.15**

Once a visual analysis was performed, a statistical analysis of the obtained results must be done. In order to do this, a Friedman and Nemenyi post-hoc tests are applied (i.e., applied at the conventional significance level of 5%), as recommended in [23].

Friedman test produced the p-value = 0.0000174. Therefore, the difference in the average accuracy of all four methods is defined as statistically significant. Then, the pairwise comparisons using the Nemenyi post-hoc test was performed. Figure 1a illustrated the Critical difference (CD) diagram for the post-hoc Nemenyi test, as defined by [23]. This diagram provides an interesting visualisation of the statistical significance of the observed paired differences. With this diagram, it is possible to compare all analysed methods against each other and check the results of all these paired comparisons.

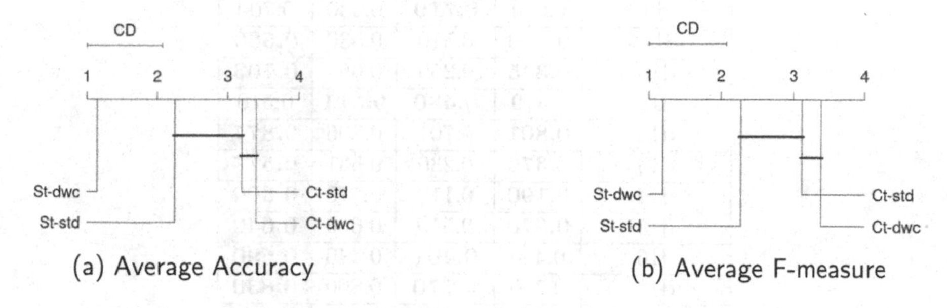

Fig. 1. Critical difference diagram for accuracy and F-measure

According to the values indicated for Fig. 1a, we can state that St-dwc statistically outperformed St-std in. However, the same positive result was not depicted with Co-training. In fact, the St-dwc method surpasses all three SSL methods, from a statistical point of view, since there is no horizontal line connecting St-dwc and any other SSL method.

6.2 Results for the Average F-Measure

Table 3 presents the average values of F-measure for all four methods. As shown in this table, a similar behaviour pattern of the previous section is observed, in which St-dwc obtained the best average F-measure (**0.534**) among all other methods. When analysing each pairwise possibility of methods, we can state that St-dwc and Ct-std outperformed their corresponding methods. In addition, St-dwc obtained the best ranking result (**1.15**), followed by St-std, Ct-std and Ct-dwc, respectively. Finally, in terms of the overall number of wins, St-dwc outperformed all other three methods, obtaining 16 wins, out of 20.

As illustrated by Table 3, St-dwc was clearly the best method. The, the Friedman test was performed and it produced the p-value = 0.00004696. Therefore, the difference between the average F-measure of all four methods is statistically significant. The pairwise comparisons using the Nemenyi post-hoc test produced three statistically significant results and the CD diagram is presented in Fig. 1b. The results depicted in this figure evinced that, once again, St-dwc statistically outperforms all three competing methods.

Table 3. Average F-measure for all methods

Dataset	Ct-std	Ct-dwc	St-std	St-dwc
d1	**0.069**	0.042	0.084	**0.092**
d2	**0.151**	0.150	0.150	**0.190**
d3	0.210	**0.211**	0.350	**0.370**
d4	**0.150**	0.080	0.400	**0.420**
d5	0.160	**0.240**	**0.300**	0.280
d6	0.350	**0.460**	0.340	**0.520**
d7	**0.630**	0.340	0.950	**0.960**
d8	0.640	**0.710**	0.630	0.700
d9	**0.520**	0.510	0.530	**0.560**
d10	**0.345**	0.259	0.651	**0.703**
d11	0.479	**0.480**	**0.511**	0.510
d12	**0.801**	0.701	0.806	**0.877**
d13	**0.379**	0.236	0.531	**0.577**
d14	**0.190**	0.110	0.570	**0.600**
d15	**0.570**	0.510	0.600	**0.640**
d16	0.400	**0.401**	0.440	**0.590**
d17	0.750	**0.770**	0.800	**0.830**
d18	**0.430**	0.380	0.700	**0.750**
d19	**0.194**	0.158	0.142	**0.146**
d20	0.050	**0.070**	0.340	**0.360**
Average	**0.373**	0.341	0.491	**0.534**
Inner_wins	**12**	8	2	**18**
General_wins	1	1	2	**16**
Avg_rank	**3.00**	3.25	2.20	**1.15**

6.3 Discussion of the Results

As presented in Tables 2 and 3 as well as Figs. 1a and 1b, we can conclude that St-dwc statistically outperforms all three competing methods.

One important observation is that the obtained results indicate that our approach has a positive effect only for Self-training. Based on these results, it is believed this is due to the fact that the *DWC* reduces the number of instances being wrongly selected at first iterations more in Self-training. Arguably, this finding may be related to the fact that Co-training uses two base classifiers and a crossing inclusion (unlabelled instances of one classifier is included in the labelled set of the other classifier). Hence, when one classifier defines the confidence of one unlabelled instance, this decision was made using this attribute subset. This confidence may not be the same for the other classifier and this can cause a deterioration in performance, mainly for Ct-dwc since it uses two metrics (confidence prediction and distance).

7 Conclusion and Future Work

This work proposed a novel approach for selecting unlabelled instances in SSL methods by combining two different strategies, prediction confidence and distance metric, to define a Distance-Weighted Confidence selection criterion. In this paper, our selection approach has been applied to Co-training and Self-training. Nevertheless, it is important to emphasise that this approach can be applied to any semi-supervised method.

In order to assess the feasibility of the proposed approach, an empirical analysis was conducted. In this analysis, two versions of Co-training and Self-training, named as St-dwc and Ct-dwc were compared against the standards versions of both methods. In addition, 20 classification datasets and two predictive measures (i.e., accuracy and F-measure) were used for analysis purpose.

In general, our approach applied to Self-training statistically outperformed all three competing methods in both predictive measures used in our analysis as we showed in Figs. 1a and 1b. Moreover, St-dwc obtained the lowest (best raking) values compared to all other methods, and also 18 best results out of 20 for average accuracy, and 16 out of 20 for average F-measure.

On the other hand, Ct-dwc has not achieved good results, on contrary, it was outperformed by all three competing methods. In fact, our DWC selection criterion has not enhanced Co-training's selection step in a way that we would expected. Arguably, this may be related to the fact that Co-training uses two base classifiers, and also two different attribute subsets.

For future work, it would be interesting to extend the experiments in terms of other SSL methods, greater number of datasets, preferring imbalanced ones, and finally different approaches for the formulation of DWC (e.g. linear combination of confidence prediction and distance metric).

Acknowledgements. This work has been financially supported by CAPES (Brazilian Research Council).

References

1. Zhu, X., Goldberg, A.B., Brachman, R., Dietterich, T.: Introduction to Semi-Supervised Learning. Morgan and Claypool Publishers, San Rafel (2009)
2. Chapelle, O., Zien, A.: Semi-supervised classification by low density separation. In: AISTATS, vol. 2005, pp. 57–64 (2005)
3. Wang, W., Zhou, Z.-H.: Analyzing co-training style algorithms. In: Kok, J.N., Koronacki, J., Mantaras, R.L., Matwin, S., Mladenič, D., Skowron, A. (eds.) ECML 2007. LNCS (LNAI), vol. 4701, pp. 454–465. Springer, Heidelberg (2007). https://doi.org/10.1007/978-3-540-74958-5_42
4. Miyato, T., Maeda, S., Koyama, M.: Virtual adversarial training: a regularization method for supervised and semi-supervised learning. IEEE Trans. Pattern Anal. Mach. Intell. **41**(8), 1979–1993 (2018)
5. Zhou, Z.-H., Li, M.: Tri-training: exploiting unlabeled data using three classifiers. IEEE Trans. Knowl. Data Eng. **17**(11), 1529–1541 (2005)

6. Blum, A., Mitchell, T.: Combining labeled and unlabeled data with co-training. In: Proceedings of the Eleventh Annual Conference on Computational Learning Theory, COLT 1998, pp. 92–100. ACM, New York (1998)
7. Yarowsky, D.: Unsupervised word sense disambiguation rivaling supervised methods. In: Proceedings of the 33rd Annual Meeting on Association for Computational Linguistics, pp. 189–196. Association for Computational Linguistics (1995)
8. Jiang, J., Gan, H., Jiang, L., Gao, C., Sang, N.: Semi-supervised discriminant analysis and sparse representation-based self-training for face recognition. Optik **125**(9), 2170–2174 (2014)
9. Qiao, S., Shen, W., Zhang, Z., Wang, B., Yuille, A.L.: Deep co-training for semi-supervised image recognition. CoRR, abs/1803.05984 (2018)
10. Xia, Y., et al.: 3D semi-supervised learning with uncertainty-aware multi-view co-training. In: The IEEE Winter Conference on Applications of Computer Vision (WACV), March 2020
11. Zhe, J., Hong, G.: Predicting pupylation sites in prokaryotic proteins using semi-supervised self-training support vector machine algorithm. Anal. Biochem. **507**, 1–6 (2016)
12. Hajmohammadi, M.S., Ibrahim, R., Selamat, A., Fujita, H.: Combination of active learning and self-training for cross-lingual sentiment classification with density analysis of unlabelled samples. Inf. Sci. **317**, 67–77 (2015)
13. Kim, D., Seo, D., Cho, S., Kang, P.: Multi-co-training for document classification using various document representations: TF-IDF, LDA, and DOC2VEC. Inf. Sci. **477**, 15–29 (2019)
14. Tanha, J., van Someren, M., Afsarmanesh, H.: Semi-supervised self-training for decision tree classifiers. Int. J. Mach. Learn. Cybernet. **8**(1), 355–370 (2017)
15. Di, W., et al.: Self-training semi-supervised classification based on density peaks of data. Neurocomputing **275**, 180–191 (2018)
16. Suzuki, T., Kato, J., Wang, Y., Mase, K.: Domain adaptive action recognition with integrated self-training and feature selection. In 2013 2nd IAPR Asian Conference on Pattern Recognition, pp. 105–109. IEEE, Naha, November 2013
17. Triguero, I., Sáez, J.A., Luengo, J., García, S., Herrera, F.: On the characterization of noise filters for self-training semi-supervised in nearest neighbor classification. Neurocomputing **132**, 30–41 (2014)
18. Bettoumi, S., Jlassi, C., Arous, N.: Collaborative multi-view k-means clustering. Soft. Comput. **23**(3), 937–945 (2019)
19. Vale, K.M.O., et al.: A data stratification process for instances selection in semi-supervised learning. In: International Joint Conference on Neural Networks (IJCNN), pp. 1–8. IEEE (2019)
20. Ma, F., Meng, D., Dong, X., Yang, Y.: Self-paced multi-view co-training. J. Mach. Learn. Res. **21**(57), 1–38 (2020)
21. Karlos, S., Kostopoulos, G., Kotsiantis, S.: A soft-voting ensemble based co-training scheme using static selection for binary classification problems. Algorithms **13**(1), 26 (2020)
22. Powers, D.M.: Evaluation: from precision, recall and f-measure to ROC, informedness, markedness and correlation. J. ML Technol. **2**(1), 37–63 (2011)
23. Demšar, J.: Statistical comparisons of classifiers over multiple data sets. J. Mach. Learn. Res. **7**(Jan), 1–30 (2006)
24. Hall, M., Frank, E., Holmes, G., Pfahringer, B., Reutemann, P., Witten, I.H.: The weka data mining software: an update. ACM SIGKDD Explor. Newslett. **11**(1), 10–18 (2009)

Active Learning Embedded in Incremental Decision Trees

Vinicius Eiji Martins[1](✉), Victor G. Turrisi da Costa[2], and Sylvio Barbon Junior[1]

[1] Computer Science Department, Londrina State University, Londrina, PR, Brazil
{vinicius.martins,barbon}@uel.br
[2] DISI, University of Trento, Trento, Italy
vg.turrisidacosta@unitn.it

Abstract. As technology evolves and electronic devices become widespread, the amount of data produced in the form of stream increases in enormous proportions. Data streams are an online source of data, meaning that it keeps producing data continuously. This creates the need for fast and reliable methods to analyse and extract information from these sources. Stream mining algorithms exist for this purpose, but the use of supervised machine learning is extremely limited in the stream domain since it is unfeasible to label every data instance requested to be processed. Tackling this problem, our paper proposes the use of active learning techniques for stream mining algorithms, specifically incremental Hoeffding trees-based. It is important to mention that the active learning techniques were implemented to match the stream mining constraints regarding low computational cost. We took advantage of the incremental tree original structure to avoid overburdening the original computational cost when selecting a label. In other words, the statistical strategy to grow each incremental tree has supported the execution of active learning. Using techniques of uncertainty sampling, we were able to drastically reduce the number of labels required at the cost of a very small reduction in accuracy. Particularly with *Budget Entropy* there was an average negative impact of accuracy about 4% using only 14% of samples labelled.

Keywords: Stream mining · Active learning · Hoeffding trees

1 Introduction

Data streams are an increasingly common resource that produces a large amount of potentially infinite data in short intervals. Dealing with this type of data,

The authors would like to thank the financial support of the Coordenação de Aperfeiçoamento de Pessoal de Nível Superior - Brasil (CAPES) - Finance Code 001, the National Council for Scientific and Technological Development (CNPq) of Brazil - Grant of Project 420562/2018-4 - and Fundação Araucária.

R. Cerri and R. C. Prati (Eds.): BRACIS 2020, LNAI 12320, pp. 367–381, 2020.
https://doi.org/10.1007/978-3-030-61380-8_25

stream mining algorithms need to face a set of challenges, such as where and how to store data, how to process data in an acceptable time frame and how to deal with its changes in concepts and underlying distributions [11,22].

A common way to extract useful information and patterns from data is through the use of supervised machine learning models, including the decision trees. Incremental decision trees are alternatives from supervised machine learning algorithms for data stream scenarios, in which each single stream sample can be used to update a decision tree. A classic example of incremental decision trees is the Hoeffding Tree (HT) [11], which is based on the Hoeffding Bounds (HB) theory to identify the best split feature during tree growth. HB has gained notoriety in the stream mining scenario for its effectiveness, a fact that impelled its usage in several implementations of incremental decision trees such as Very Fast Decision Tree (VFDT) [11] and Strict Very Fast Decision Tree [6] (SVFDT), the latter focuses on reducing the requirement of computational resources.

However, all these algorithms rely on labelled data, which may be expensive to acquire in the real world and even harder to gather in data stream situations. The challenges of volume and velocity intensify the problem of labelling samples from data streams. Some techniques were developed to overcome this problem for traditional supervised machine learning, such as Semi-Supervised Learning (SSL) [32] and Active Learning (AL) [14]. SSL assumes a small amount of labelled data and a large pool of unlabelled data and uses both to train its model. On the other hand, AL works by intelligently selecting only a subset of data samples to be labelled, allowing the algorithms to train efficiently in more realistic conditions [26].

AL does not require labelled instances before the training begins as it will choose which data instances it will learn from. This is done by asking *queries* containing unlabelled data to an *oracle* that informs it the true label, in several cases the oracle could be a human specialist [26]. With the application of this technique, the model only needs to be trained on a small number of highly informative samples instead of the whole dataset, increasing the efficiency of the training with minimal accuracy losses, following the constraints of reduced access to the complete dataset. AL techniques are broken into various categories, but all are grounded in the same idea: using a sampling technique, the most informative instances are selected or constructed from the input domain and sent to an oracle, human or machine, that will label it and return to the learner, which will then use it to train in a supervised manner.

In recent years, several efforts have been made in the direction to create joint methods with data stream and AL [1,10,19–21,28]. The goal of the major part of the proposed algorithms is to tackle Concept Drift and the detection of Novelty. Concept drift reflects the idea that concepts in the real world are always changing.

Concept drift was addressed using ensemble learning by some works [1,20,28]. Ensemble learning is an important solution used in the stream mining community since they maintain the advantages present in traditional scenarios, such as taking advantage of local competencies from classifiers and robustness to overfitting [7]. Also, ensembles can handle the drifting context ensured by the diversity of committee members.

In [1] the authors propose a framework for use of AL in ensembles with their proposed Query-by-bagging and Query-by-boosting methods, both based upon the paradigm of Query by Committee(QBC) [27]. These methods make the oracle responsible for choosing the data samples which will be labelled and appended to the training dataset that will be broken into various windows that will be used by the ensemble learners to train its models. Besides the fact that concept drifts are not addressed, the framework has a cost of an additional structure.

Shan et al. [28] proposes a framework for ensemble active learning using an ensemble composed of a stable permanent classifier that learns from every data instance that arrives and multiple dynamic ephemeral classifiers that only train with a limited amount of data. A combination of Uncertainty Sampling and Random Sampling is used to determine if a sample inside the data block will be labelled and used to train the ensemble. This combination of a permanent classifier and multiple short lived classifiers make this framework able to adapt to sudden and gradual concept drifts while reducing labeling costs by focusing the queries to the oracle when drift occurs.

In [20] an approach to ensemble active learning is proposed that instead of selecting instances to query based on the amount of disagreement between committee members, it uses a Multi-armed bandit approach, where the most competent member is made responsible for this decision. This approach allows the ensemble to better adapt under concept drifts, specifically when drifts occur in regions of data that regular query sampling techniques register low amounts of uncertainty.

These approaches [1,20,28] present novel ensemble techniques adapted to AL and streaming situations, but they introduce additional complexity and costs to the training procedure.

Alternatively to ensembles, [10] proposed a sequential ID3, grounded on a sequential probability ratio evaluation to reduce the number of samples sufficient to perform a split. They affirm that no theoretical bounds are exposing the extent to which labels can be saved without significantly compromising performance. However, the AL strategy used in [10] has a cost of memory and additional mechanism of control the selection of samples to be labelled. Furthermore, the authors described the implementation of AL with VFDT as a promising approach.

In this work we combine the use of three Hoeffding Tree implementations with AL strategies. The implementations are all variations of the VFDT and SVFDT (SVFDT-I and SVFDT-II). They are robust learners that deal very well with various streaming situations while also performing memory management [6], but they are still supervised machine learning techniques that expect labelled data in the stream. We compared two strategies of AL literature, *Entropy-based* [29] and *Budget Entropy* [34], and proposed a novel mechanism called *Best Budget Entropy*. We took advantage of original implementation from the most memory-friendly HT algorithms to avoid consuming extra memory and reduce the demand for labels with low-cost AL adaptations. Our results exposed quite a few reductions in terms of labels without compromising the predictive performance over a great part of the 26 datasets explored.

In Sect. 2, we introduce the concepts of Active Learning and how it is applied in streaming situations and the incremental trees. In Sect. 3, we explain how the experiments were setup and evaluated. In Sect. 4, the results of the experiments are discussed and analysed and finally, in Sect. 5, we conclude the paper and present directions for future work.

2 Active Learning and Stream Mining

Various challenges permeate data stream mining. The main ones are the volume, the velocity, the volatility of data and constraints of memory consumption. The most efficient solutions demand labels, which can pose difficulties to use the solutions within a real-life scenario. An initial work of Žliobaitė et al. [34] described some theoretical strategies to support mechanisms to control and distribute the labelling over time with balancing capabilities to induce more accurate classifiers and to detect changes. Our proposal arose from strategies and hypothesis related to Žliobaitė et al.'s contributions.

2.1 Active Learning Strategies

The core of AL strategies is composed by sampling strategies and query decision approaches, as shown in Fig. 1. The sampling strategies are different forms of directed search techniques seeking to identify samples from areas of uncertainty. On the other hand, query decision approaches regard methods to decide whether or not to query for the true label, so that the predictive model could train itself with this new instance [34].

Sampling strategies seek for areas in the input domain where the learner believes it will perform incorrect classifications, as opposed to random sampling techniques [3]. There are three main sampling strategies in the literature: membership query synthesis, stream-based and pool-based. Stream and pool-based are part of selective sampling [26].

Membership query synthesis [2] works by querying new synthesized samples based on the underlying distribution of the area of uncertainty in the input domain instead of using the already existing data. This strategy suffers from difficulty in finding methods to synthesize data instances that a human oracle is capable of interpreting. For example, when using an image dataset and interpolation for synthesizing the queried samples, the results may be a mix of different images from the input that does not mean anything to a human, hurdling the job of the oracle [23]. Selective sampling strategies were formulated to solve this problem [3] through the use of several approaches to query data from the input data to the oracle.

Pool-based selective sampling [24] usually assumes the input domain remains unchanged, contains a small number of n labelled instances and a large amount of m unlabelled instances. This method runs for various iterations until a stopping

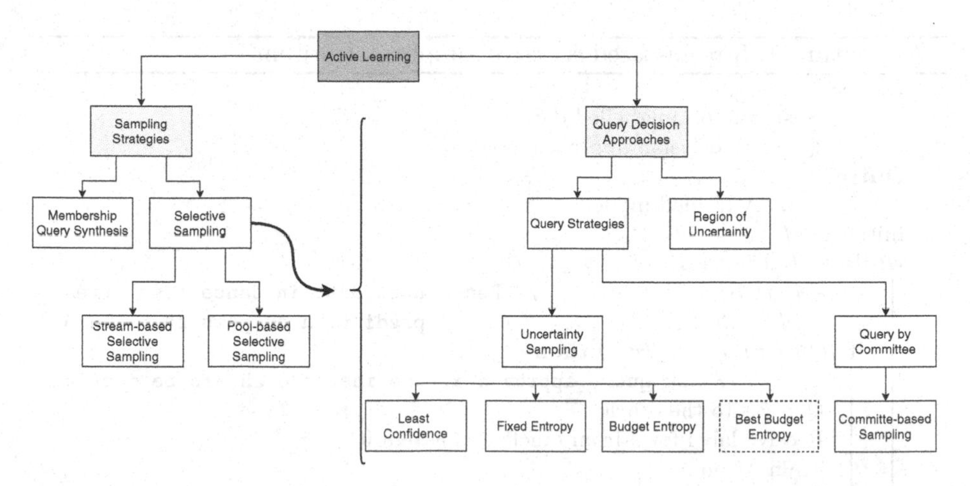

Fig. 1. Overview on the taxonomy of sampling strategies and query decision approaches. Our approach adaptation is highlighted by the dotted border.

criterion is reached, such as the oracle reaches a budget limit. Each iteration ranks the most uncertain instances, queries them to the oracle and adds them to the list of labelled instances, where the model is retrained.

Stream-based selective sampling is the most straightforward method of selective sampling. As data arrives, it will make the decision to query or not the instance to the oracle using the querying approach selected. If the learner decides to not query, that data instance is immediately discarded. This procedure can be seen in Algorithm 1.

Since we are working in a streaming scenario, the use of stream-based selective sampling is the most effective. Due to the velocity in which the data arrives, it is not feasible to use pool-based sampling due to the processes of pooling, ranking and iterations required. This same constraint limits membership query synthesis as the underlying distribution of the data needs to be analysed multiple times due to the changing nature of the data. For that reasons, in this work, we focused on stream-based selective sampling and suitable query decision approaches for a stream scenario.

Query decision approaches regards a method to decide whether or not to query for the true label so that the predictive model can train itself with this new instance [34]. The most traditional approach consists of creating an explicit region of uncertainty $R(S^m)$ where S^m is the set of m instances in the data input domain. The learner first trains on n labelled instances, where $0 < n < m$, to compute $R(S^m)$ and then simply tests each data instance for membership in $R(S^m)$, creating a collection of instances from which it will query the oracle [3]. Each new instance that falls within the region will further reduce the region when recalculated [5].

Another approach is to use query strategies to determine the most informative or uncertain data instances directly and make a decision to query them to

Algorithm 1. Stream-based selective sampling algorithm.

```
Input:
        S: stream of unlabelled data
        Q: query decision approach selected
Output:
        M: A trained model
Initialize M
while (S has next) do
    s ←— next(S)                 // Fetch next data instance from stream
    o ←— M.predict(s)             // Get prediction outputs for s on M
    if (Q.query(o) == True) then
            // Ask query approach if the instance should be queried
        Query s to the oracle
        Receive label for s from oracle and assign it
        Train M on s
    end
end
```

the oracle or not. There are many query strategies. Uncertainty Sampling [24] uses a metric to compute the uncertainty of each data instance and queries it to the oracle if it falls within a certain threshold. Committee-based sampling [9] follows the QBC [27] paradigm, with a committee of k models, where each member classifies a data instance, and the decision to query that instance is based on the classification disagreement of the members of the committee.

Uncertainty Sampling is the most used query strategy due to its simplicity. In a binary problem, for example, it would decide to query in case the probability of the predictions made by the model for either class prediction score is close to 0.5, indicating that the model is unsure as to which class the data instance belongs to. For more classes, *Least Confidence* (LC) [8] may be used. It decides to query data instances where even the class prediction with highest probability is low.

A more general and popular approach to uncertainty sampling is to use *entropy* [29] (H) (Eq. 1), where $\hat{y}_i$ is one of the labels and x^*_H is the instance to be queried to the oracle. H represents the uncertainty over the prediction output distribution with values between 0 (low uncertainty) and $log2(n\_classes)$ (high uncertainty). Technically it is an information-retrieval measure that quantifies the amount of information needed to encode the distribution [26].

$$x^*_H = argmax(H(x_i)) \text{ where } H(x) = -\sum_i P\theta(\hat{y}_i|x)log(P\theta(\hat{y}_i|x)) \qquad (1)$$

A way to use entropy is by first fixing the uncertainty threshold z, and if the entropy value surpasses the specified z value, the data instance being evaluated is queried to the oracle [33]. A variant of this method is by using a budget value that limits the number of queries that can be performed. For example, a value of 0.2 means that only 20% of the instances can be labelled, in a streaming situation we can translate to something like 200 instances every 1000 [34]. We refer to this method as Budget Entropy. Budgets reflect real-life situations where the oracle has limited labeling capability and querying must be kept at a minimum.

L. Korycki et al. [19] proposes a method to decrease the number of queries made under strict budgets by using a hybrid query decision approach that uses both AL and self-labelling techniques. Self-labeling [31] is a semi-supervised learning technique that allows for the learner to label a data instance if it has a high amount of certainty on its class. This can be seen as a direct opposite of AL. This approach allows the learner to increase the number of instances used for its training with no cost. However, concept drifts are not taken in consideration and errors made by the self-labeling mechanism may propagate along the data stream.

B. Krawczyk [21] proposes a framework that is able to deal with concept drifts in limited budget situations by increasing the rate of oracle queries when drift is happening and decreasing in static situations. This framework is simple and effective, but it also has a very large amount of hyperparameters that require tuning, such as the labeling strategy and its own parameters and the adjustable rate of querying.

We proposed another method (highlighted by the dotted box in Fig. 1) grounded on entropy. Instead of a fixed uncertainty threshold, it checks the entropy of the current instance and the instance that came before, if the entropy of the current instance is higher, this instance is queried to the oracle. We call this method *Best Budget Entropy*. It has the advantages of being very simple and no hyperparameters are needed, although concept drifts are not considered directly.

In our work, we compare uncertainly sampling and stream-based selective sampling, since it is fast and effective and matches our main goal of avoiding overburdening the stream mining algorithm, particularly the incremental decision trees with an extra cost when performing AL.

2.2 Incremental Decision Trees

Hoeffding Trees [11] are incremental decision trees optimized for data stream situations. They were designed to deal with infinitely large datasets and each data instance must be read at most once in a small constant time. To achieve that, they use Hoeffding Bounds to assure that the chosen attribute for splitting with n attributes is the same as if it was chosen with infinite attributes by a margin of error ϵ. This process is done based on a function G, for example, Information Gain [25] (Eq. 2, where H is the entropy function, x the attribute and $\hat{y}$ the label), for n examples, let $G(X_1)$ be the highest value and $G(X_2)$ the second highest value among all $G(X_i)$ computed for every attribute in X and that $\Delta G = G(X_1) - G(X_2)$, for a given δ, Hoeffding Bounds guarantee that X_1 is the correct choice for the split with a probability of $1 - \delta$ if n examples were read at the node being trained and $\Delta G > \epsilon^2$.

$$IG = H(\hat{y}) - H(\hat{y}, x) \tag{2}$$

One implementation of a Hoeffding Tree is the Very Fast Decision Tree (VFDT) [11]. First, it allows choosing the G to be either Information Gain or the Gini Index. Additionally, it features a number of optimizations to further speed up the training process:

- **Tiebreak:** Tiebreak happens when two attributes have very similar values from G. Since the decision may require observation of a large number of samples to be made, this mechanism allows the learner to detect when a tie happens and simply split on the current best attribute X_i if $\Delta G < \epsilon < \tau$ for a given τ.
- **G Computation:** Since computing G can be expensive, the VFDT allows accumulation of a minimum number of samples before the G is calculated. This effectively reduces the total amount of time spent calculating G.
- **Memory Management:** In order to limit the amount of memory used, once the maximum memory available is reached, the VFDT deactivates the least promising leaves in order to free memory for new ones.
- **Disabling Poor Attributes:** Removing attributes that do not show potential, memory usage can be further minimized, this is done by dropping attributes that have a value of G with a difference of at least ϵ to the G of the best attribute.
- **Grace Period:** This allows the tree to be initialized with a small subset of data with a conventional learner, allowing the VFDT to reach better accuracies early on with a small number of samples.
- **Rescanning:** If the data arrives slowly enough or is a small finite dataset, previously observed samples can be reexamined.

SVFDT [6] is an optimization made on *VFDT*, it manages to keep a significantly lower memory footprint than the original VFDT while retaining similar predictive performance by enforcing a set of restrictions that ensure a minimum amount of uncertainty, that the leaves observe a similar number of instances and that the attributes used for the splits have relevance to the statistics.

Additionally, there is a mechanism in place that limits unnecessary growth in the tree by checking the Entropy and Information Gain values of the leaves with the other leaves and when the rules for splits in the *VFDT* were met with the Eq. 3, where X represents the observed data instances, $\overline{X}$ their mean, $\sigma(X)$ their standard deviation and x the observation of a new data instance. It is also assumed that X follows a normal distribution.

$$\varphi(x, X) = \begin{cases} True, & \text{if } x \geq \overline{X} - \sigma(X) \\ False, & \text{otherwise} \end{cases} \tag{3}$$

SVFDT is split into two versions: the *SVFDT-I* and *SVFDT-II*. Their difference consists of an additional set of constraints found in the II version that allows the node to skip all the constraints set by the growth mechanism. This set consists of two constraints that check the values of Entropy and Information

Gain for the leaves with their values for when the rules for splits in the *VFDT* were met with the Eq. 4.

$$\varpi(x, X) = \begin{cases} True, & \text{if } x \geq \overline{X} + \sigma(X) \\ False, & \text{otherwise} \end{cases} \tag{4}$$

Our approach to AL allows it to easily plug in any stream mining base learner with minimal cost as it is seamlessly integrated into the learner's input pipeline. This means that our active learning methods work as a separate module to the learner, needing only its prediction statistics to determine what instances should be queried to the oracle and feeding this data to the classifier. This can be seen in Fig. 2.

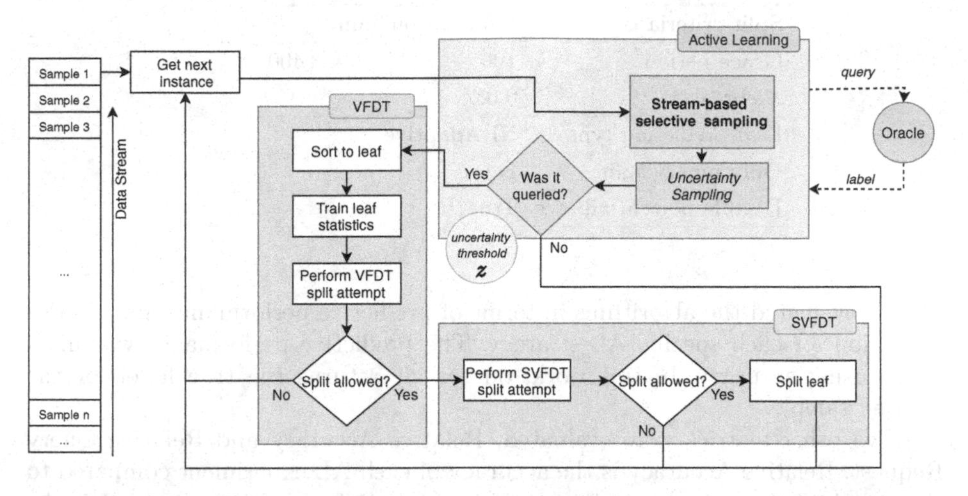

Fig. 2. Overview of stream-based selective sampling coupled to VFDT and SVFDT.

3 Experimental Setup

In this section, we present the experimental definitions to support the proposed AL method embedded into *VFDT*, *SVFDT-I* and *SVFDT-II*. To evaluate the impact of the AL methods in the trees, 26 benchmark datasets, commonly used in data stream mining experiments, were selected: – Datasets from MOA [4]: *Airlines* and *Electricity Normalized*. – Datasets from the UCI [12]: *Poker Hand* and *Covertype*. – Datasets from Weka [18]: *LED24* (with 1M instances and three files with 0%, 10% and 20% noise each) and *RandomRBF* (250k instances and 50 features, 500k instances and 10 features and 1M instances and 10 features). – Datasets from multiple sources: *CTU13* [17] (split into 13 files, one per scenario), *hyperplane* [13], *SEA* [30] and *Usenet* [16].

Prequential evaluation was employed to evaluate the algorithms [15]. Stream-based sampling was used and Uncertainty Sampling was chosen with the three entropy variants (*Entropy*, *Budget Entropy* and *Best Budget Entropy*). This was preferred over calculation of Region of Uncertainty since it is more efficient considering the streaming scenario. The *VFDT* and *SVFDTs* were compared using the parameters seen in Table 1.

Most hyperparameters were chosen with their default value (tiebreaker, split criteria, leaf prediction type and binary splits), while for the grace period we used non-default values and poor attributes are discarded to preserve memory.

Table 1. Parameter values for each incremental tree.

Parameter	*VFDT*	*SVFDT-I*	*SVFDT-II*
Split criteria	Information gain		
Grace period	100		400
Tiebreaker	0.05		
Leaf prediction type	NBAdaptive		
Only binary split	False		
Disable poor attributes	True		

We evaluated the algorithms in terms of predictive performance and queries reduction for each specific AL strategy. The predictive performance was measured using accuracy. In this work, our oracle returns the true label for the queried sample.

Two other metrics were evaluated, Relative Accuracy and Relative Query Request. Relative Accuracy is the accuracy of each AL experiment compared to the standard supervised learning accuracy while Relative Query Request is the percentage of queries made on each experiment related to the total amount of samples in each dataset.

The algorithms and AL strategies were implemented in Python 3.8. The code for this implementation can be seen in https://github.com/Vini7x/pystream-act.

4 Results and Discussions

In this section, first, we present a comparison among *VFDT*, *SVFDT-I* and *SVFDT-II* using the Relative Accuracy and Relative Query Request across several z values. Then, we perform a similar evaluation, but using each AL method across all algorithm to support generalized insights. We observed the queries rate and the impact over accuracy to discuss the trade-off between the reduction of labelling and predictive performance.

Regarding AL relative accuracy from the incremental trees, a very similar performance across four different z values (0.1, 0.2, 0.5 and 0.9) was observed, as Fig. 3 shows. Also, when compared to the usage of all samples, a slight reduction

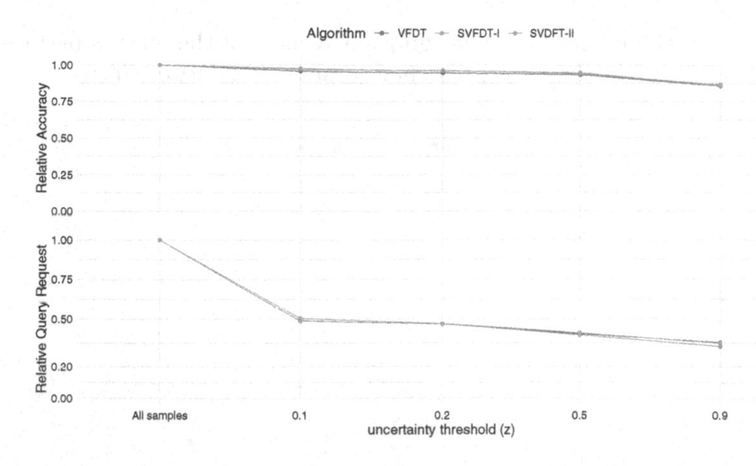

Fig. 3. Performance of different incremental trees based on relative accuracy and relative query reductions over different uncertainty threshold (z) values.

of relative accuracy was observed with the z values of 0.1, 0.2 and 0.5, respectively. A notable reduction was observed when z is equals to 0.9. On the other hand, considering the Relative Query Request, when z equals 0.1, the number of labelled samples was reduced to 49% by *VFDT* and *SVFDT-II* maintaining a low reduction in performance of about 4% and 3%, respectively. If we evaluate a trade-off using a rate of Accuracy per Query Request, *SVFDT-II* was the best combination delivering 13% of accuracy reduction using just 33% of original labelled data, as showed in Table 2.

Table 2. Table of Relative Accuracy and Relative Query request across all uncertainty threshold (z) and incremental trees.

z	Relative Accuracy			Relative Query Request		
Value	*VFDT*	*SVFDT-I*	*SVFDT-II*	*VFDT*	*SVFDT-I*	*SVFDT-II*
All samples	1.00	1.00	1.00	1.00	1.00	1.00
z = 0.1	0.96	0.97	0.97	0.49	0.50	0.49
z = 0.2	0.95	0.96	0.96	0.41	0.41	0.47
z = 0.5	0.93	0.95	0.94	0.41	0.40	0.40
z = 0.9	0.86	0.86	0.87	0.35	0.35	0.33

When evaluated from an AL perspective, we can see that the *Budget Entropy* method was the best performing of the three AL sampling strategies. Although it had the lowest accuracy of all methods, its difference was still minor while resulting in a large reduction in the number of instances queried, as can be seen in Fig. 4. Regardless of the best AL strategy, all of them were very close to the traditional supervised method in accuracy, showing that even though less data was used to train the models, the high informativeness of the queries performed by AL allowed the algorithms to reach very competitive performances.

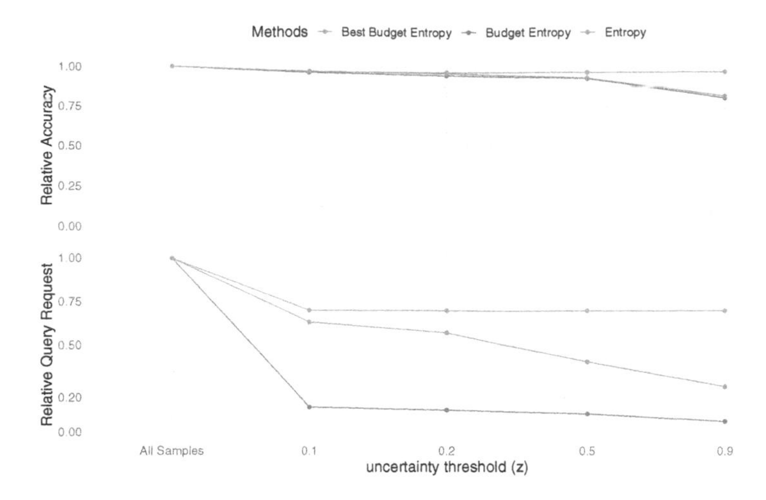

Fig. 4. Performance of AL methods based on relative accuracy and relative query reductions over different z values.

Best Budget Entropy obtained the lowest reduction in number of queries. Since it does not have the z hyperparameter, its results remain stable across all experiments. When observing Relative Query Request, *Entropy* obtained intermediate reductions as Table 3 shows. An impressive reduction was obtained by *Budget Entropy* using $z = 0.9$. It also achieved a relative accuracy of 0.80 using just 6% of available samples. This was the best trade-off between accuracy and number of query request.

Table 3. Table of Relative Accuracy and Relative Query request across all uncertainty threshold (z) and AL methods, *Best Budge Entropy* (BBH), *Budge Entropy* (BH) and *Entropy*.

z	Relative Accuracy			Relative Query Request		
Value	BBH	BH	H	BBH	BH	H
All samples	1.00	1.00	1.00	1.00	1.00	1.00
z = 0.1	0.96	0.96	0.96	0.70	0.14	0.63
z = 0.2	0.96	0.94	0.95	0.70	0.12	0.57
z = 0.5	0.96	0.92	0.92	0.70	0.10	0.40
z = 0.9	0.96	0.80	0.81	0.70	0.06	0.26

Observing the query rates across the relative accuracy intervals of 0.70, 0.75, 0.80, 0.85, 0.90, 0.95 and 1.0 in Fig. 5, it is possible to observe that *Budget*

Entropy kept the number of queries quite reduced for several accuracy intervals. The adapted approach, *Best Budget Entropy*, achieved more stability in comparison to *Entropy* in the highly accurate intervals (0.95 and 1.00).

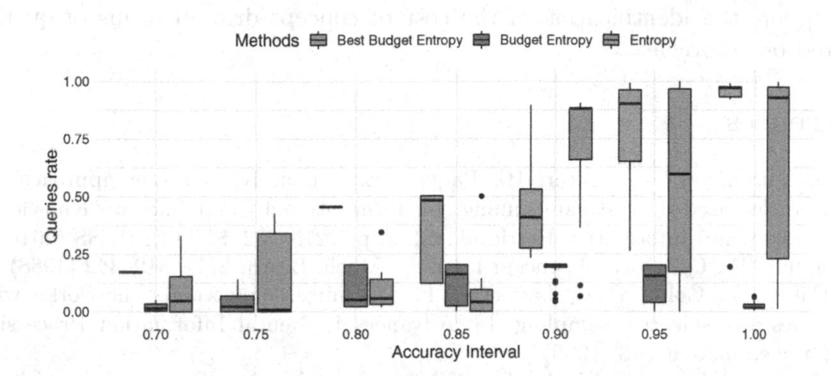

Fig. 5. Percent reduction of queries across relative accuracy interval comparing the AL methods.

Beyond the reduction of label demand, it is important to note that some accuracies obtained with AL methods surpassed the incremental decision tree results with all samples. Precisely, 81 cases distributed among algorithms, methods and some streams ($CTU13-7$, $LED24-10\%$, *Airlines*, *Usenet*, $LED24-20\%$, $CTU13-12$, $CTU13-13$, $CTU13-2$, $CTU13-9$, $CTU13-6$, $RandomRBF-1M$ and *SEA*). Particularly, the best improvement was about 0.8% using 99.8% of samples over $CTU13-13$ with a *SVFDT-I*. The best trade-off was achieved over $RandomRBF-1M$, in which 40.5% of samples were able to induce a model with an improvement of 0.3% over the accuracy of the model created wit all samples. The method used was Entropy. These results indicate that future work investigating alternative strategies to choose the training samples focusing on predictive performance improvements can be viable.

5 Conclusion and Future Work

We evaluated the use of three different AL methods (*Best Budget Entropy*, *Budget Entropy* and *Entropy*) in a streaming scenario for three variations of the Hoeffding Tree (*VFDT*, *SVFDT-I* and *SVFDT-II*). We observed that although regular training has higher accuracy than active learning strategies, the difference was very low in face of the amount of labelled data needed to train the model. *SVFDTs* took more advantage with the use of AL. Furthermore, in some cases *SVFDT-I* coupled with AL was able to improve the results in comparison to the use of all labelled samples. Grounded on the results, it is possible to affirm the *Budget Entropy* is the best Active Learning method to be embedded in the evaluated incremental trees. This method reached the best relation between high accuracy and reduction of query requests, since using just 14% of the total labelled samples result in only 4% of accuracy reduction.

In future work, we will explore the use of AL methods embedded in the studied incremental trees being used as base-learners into ensembles. This study will support the identification of the cost of concept drift in terms of queries required by an oracle.

References

1. Alabdulrahman, R., Viktor, H., Paquet, E.: An active learning approach for ensemble-based data stream mining. In: International Conference on Knowledge Discovery and Information Retrieval, vol. 2, pp. 275–282. SCITEPRESS (2016)
2. Angluin, D.: Queries and concept learning. Mach. Learn. **2**(4), 319–342 (1988)
3. Atlas, L.E., Cohn, D.A., Ladner, R.E.: Training connectionist networks with queries and selective sampling. In: Advances in Neural Information Processing Systems, pp. 566–573 (1990)
4. Bifet, A., Holmes, G., Kirkby, R., Pfahringer, B.: MOA: massive online analysis. J. Mach. Learn. Res. **11**, 1601–1604 (2010)
5. Cohn, D., Atlas, L., Ladner, R.: Improving generalization with active learning. Mach. Learn. **15**(2), 201–221 (1994)
6. da Costa, V.G.T., de Leon Ferreira de Carvalho, A.C.P., Junior, S.B.: Strict very fast decision tree: a memory conservative algorithm for data stream mining. Pattern Recognit. Lett. **116**, 22–28 (2018)
7. da Costa, V.G.T., et al.: Online local boosting: improving performance in online decision trees. In: 2019 8th Brazilian Conference on Intelligent Systems (BRACIS), pp. 132–137. IEEE (2019)
8. Culotta, A., McCallum, A.: Reducing labeling effort for structured prediction tasks. In: AAAI., vol. 5, pp. 746–751 (2005)
9. Dagan, I., Engelson, S.P.: Committee-based sampling for training probabilistic classifiers. In: Machine Learning Proceedings 1995, pp. 150–157. Elsevier (1995)
10. De Rosa, R., Cesa-Bianchi, N.: Confidence decision trees via online and active learning for streaming data. J. Artif. Intell. Res. **60**, 1031–1055 (2017)
11. Domingos, P., Hulten, G.: Mining high-speed data streams. In: Kdd, vol. 2, p. 4 (2000)
12. Dua, D., Graff, C.: UCI machine learning repository (2017). http://archive.ics.uci.edu/ml
13. Fan, W.: Systematic data selection to mine concept-drifting data streams. In: Proceedings of the Tenth ACM SIGKDD International Conference on Knowledge Discovery and Data Mining, pp. 128–137 (2004)
14. Fu, Y., Zhu, X., Li, B.: A survey on instance selection for active learning. Knowl. Inf. Syst. **35**(2), 249–283 (2013)
15. Gama, J.: Knowledge Discovery from Data Streams. Chapman & Hall/CRC, 1st edn. (2010)
16. Gama, J., Kosina, P.: Recurrent concepts in data streams classification. Knowl. Inf. Syst. **40**(3), 489–507 (2013). https://doi.org/10.1007/s10115-013-0654-6
17. Garcia, S., Grill, M., Stiborek, J., Zunino, A.: An empirical comparison of botnet detection methods. Comput. Secur. **45**, 100–123 (2014)
18. Hall, M., Frank, E., Holmes, G., Pfahringer, B., Reutemann, P., Witten, I.H.: The weka data mining software: an update. ACM SIGKDD Explor. Newslett. **11**(1), 10–18 (2009)

19. Korycki, Ł., Krawczyk, B.: Combining active learning and self-labeling for data stream mining. In: Kurzynski, M., Wozniak, M., Burduk, R. (eds.) CORES 2017. AISC, vol. 578, pp. 481–490. Springer, Cham (2018). https://doi.org/10.1007/978-3-319-59162-9_50
20. Krawczyk, B., Cano, A.: Adaptive ensemble active learning for drifting data stream mining. In: International Joint Conference on Artificial Intelligence (Macao), pp. 2763–2771 (2019)
21. Krawczyk, B., Pfahringer, B., Woźniak, M.: Combining active learning with concept drift detection for data stream mining. In: 2018 IEEE International Conference on Big Data (Big Data), pp. 2239–2244. IEEE (2018)
22. Krempl, G., et al.: Open challenges for data stream mining research. SIGKDD Explor. Newsl. **16**(1), 1–10 (2014). https://doi.org/10.1145/2674026.2674028
23. Lang, K., Baum, E.: Query learning can work poorly when a human oracle is used. In: IEEE International Joint Conference on Neural Networks (1992)
24. Lewis, D.D., Gale, W.A.: A sequential algorithm for training text classifiers. In: SIGIR 1994. pp. 3–12. Springer (1994). https://doi.org/10.1007/978-1-4471-2099-5_1
25. Quinlan, J.R.: Induction of decision trees. Mach. Learn. **1**(1), 81–106 (1986)
26. Settles, B.: Active learning literature survey. Computer Sciences Technical Report 1648. University of Wisconsin-Madison (2009)
27. Seung, H.S., Opper, M., Sompolinsky, H.: Query by committee. In: Proceedings of the Fifth Annual Workshop on Computational Learning Theory, p. 287-294. COLT 1992. Association for Computing Machinery, New York (1992). https://doi.org/10.1145/130385.130417
28. Shan, J., Zhang, H., Liu, W., Liu, Q.: Online active learning ensemble framework for drifted data streams. IEEE Trans. Neural Netw. Learn. Syst. **30**(2), 486–498 (2018)
29. Shannon, C.E.: A mathematical theory of communication. Bell Syst. Tech. J. **27**(3), 379–423 (1948)
30. Street, W.N., Kim, Y.: A streaming ensemble algorithm (sea) for large-scale classification. In: Proceedings of the Seventh ACM SIGKDD International Conference on Knowledge Discovery and Data Mining, pp. 377-382. KDD 2001. Association for Computing Machinery, New York (2001)
31. Triguero, I., García, S., Herrera, F.: Self-labeled techniques for semi-supervised learning: taxonomy, software and empirical study. Knowl. Inf. Syst. **42**(2), 245–284 (2013). https://doi.org/10.1007/s10115-013-0706-y
32. Zhu, X.J.: Semi-supervised Learning Literature Survey. University of Wisconsin-Madison Department of Computer Sciences, Technical report (2005)
33. Zliobaite, I., Bifet, A., Holmes, G., Pfahringer, B.: Moa concept drift active learning strategies for streaming data. J. Mach. Learn. Res. - Proc. Track **17**, 48–55 (2011)
34. Žliobaitė, I., Bifet, A., Pfahringer, B., Holmes, G.: Active learning with drifting streaming data. IEEE Trans. Neural Netw. Learn. Syst. **25**(1), 27–39 (2013)

Link Prediction in Social Networks: An Edge Creation History-Retrieval Based Method that Combines Topological and Contextual Data

Argus A. B. Cavalcante(✉), Claudia M. Justel, and Ronaldo R. Goldschmidt

Departamento de Engenharia de Computação, Instituto Militar de Engenharia, Praça Gen. Tibúrcio, 80 - Urca, Rio de Janeiro, RJ 22290-270, Brazil
argus@ime.eb.br

Abstract. Link prediction is an online social network (OSN) analysis task whose objective is to identify pairs of non-connected nodes with a high probability of getting connected in the near future. Recently, proposed link prediction methods consider topological data from OSN past states (i.e., snapshots that depict the network structure at certain moments in the past). Although past states-based methods retrieve information that describes how the network's topology was at the events of link emergence (i.e., moments when the existing edges were created), they do not take into account contextual data concerning those events. Hence, they take the chance to disregard information about the circumstances that may have influenced the appearance of old edges, and that could be useful to predict the creation of new ones. To remedy this issue, this work extends a past states-based method to retrieve both topological and contextual data from the events of edge emergence and combine them to predict links. The extended method presented promising results on experimental data. Overall, it overcame the original method in five different scenarios from five co-authorship OSN frequently used for link prediction method evaluation.

Keywords: Online social network · Data mining · Link mining · Link prediction

1 Introduction

In the last decade, there has been a significant increase in the use of online social networks (OSNs) [13]. An OSN is a structure composed of several participants, such as companies, non-governmental organizations, and government bodies, besides other examples. These OSN participants typically share interactions, like message exchanging and collaborating [18]. In mathematical terms, graphs are the most common representation of OSNs. On these graphs, nodes represent OSN participants while edges, interactions between pairs of participants.

R. Cerri and R. C. Prati (Eds.): BRACIS 2020, LNAI 12320, pp. 382–396, 2020.
https://doi.org/10.1007/978-3-030-61380-8_26

As online social networking became extremely popular, extensive research efforts have been devoted to investigating its behavior and evolution [17]. Typical OSN analysis tasks include: community detection [8], cascade behavior [4], and link prediction [16].

Given the current state of a network (i.e., a network snapshot including information from its participants and the interactions that have already occurred), the link prediction task tries to identify which new links will appear at this network in the future. In other words, taking into account the network state in time t, this task looks after to infer which new interactions between its non-connected elements are more likely to occur in the near future $t'(t < t')$. Link prediction task is recognized as useful in several domains, such as predicting a disease spreading, suggesting new friends, and recommending products and services [11].

One of the main challenges of the link prediction task is to determine the types of information from OSNs that should be considered by the inference process [24]. According to its nature, such information can be classified into two groups: topological and contextual. Topological information is purely structural and only depends on the interactions between network participants (e.g., node's degree and the number of common neighbors between two nodes). On the other hand, the contextual information has explicit semantic related to the subject of OSN. Examples of this type of information include age and place of birth of a user, or his/her profile description, preferred types of music, movies, and others.

Another important aspect of link prediction concerns the treatment of the network states [22]. Most studies consider a single network state (usually the current state) for link prediction. Others consider various past network states, i.e., different past snapshots of the network arranged according to some separation criteria. This past states-based view is fundamentally related to the link prediction task as it characterizes the history of the kind of event one wants to predict: the emergence of new links.

An analysis of the link prediction state-of-the-art shows that the use of past network states has been unexploited. Nevertheless, the few works who did it, like [7] (initially presented at [6]), for example, have used exclusively topological information in link prediction, leaving aside the diversity and richness of contextual information found in OSNs. Therefore, this work's goal is to investigate if combining topological with contextual data may improve link prediction performance when past network states are considered. This investigation was accomplished by extending a past states-based link prediction method to allow the combination of topological and contextual data. The experimental evaluation with five different co-authorship networks led to the superior results of the proposed extension when compared to the original method.

This paper is organized as follows: Section 2 introduces key concepts related to this work. Section 3 presents some articles from state-of-the-art, which have explored similar approaches. Section 4 details the link prediction method used as the baseline of our experiments, followed by Sect. 5, where an extension to the baseline method is introduced. Section 6 describes the experiments and discusses the obtained results. Finally, Sect. 7 concludes the work and points out future research possibilities.

2 Background

The unsupervised approach is one of the most common approaches for solving the link prediction problem [17]. It uses OSN topological and contextual information for calculating similarity metrics on node pairs, based on the assumption that two similar nodes tend to connect when they share common characteristics. In the end, the ranking given by applying these similarity metrics to pairs of non-connected nodes indicates that those pairs have higher chances of getting connected in the future.

Traditional similarity metrics based on topological information include Common Neighbors [23], Jaccard's Coefficient [11], and Adamic-Adar [1]. Time Score [19], Context-based Time Score [3] and Cosine Similarity [27] are samples of similarity metrics that consider contextual data from OSNs.

While topological information is found only as structured data on OSNs, contextual information can be observed as both structured and unstructured data. For example, applying similarity metrics to unstructured data (such as images, videos, and texts) requires an initial step to create an actionable format representation, so data is ready for similarity metrics calculation. Leaving aside images and videos, which require specific treatment and techniques, standard text representations include Bag of Words, TF-IDF, and Doc2Vec.

Bag of words (BoW) is one of the most popular forms of unstructured data representation [9]. The BoW is organized as a vector of words where each word's occurrence is numerically computed. A limitation imposed by the use of the frequency of terms in BoW relates to favoring words extensively used but implicitly common in a set of documents. [30] has developed a method of representation that solves this limitation by penalizing frequent terms in several documents called Term Frequency - Inverted Document Frequency (TF-IDF). In TF-IDF, the score increases proportionally to the number of times a word appears in the document and is offset by the number of documents containing this word.

Although their simplicity, both BoW, and TF-IDF offer a critical disadvantage. The word sequence is not considered, so completely different sentences can result in the same representation if the same words are used. Furthermore, without order, there is a semantic loss of the analyzed text. To address this restriction, [15] proposed a representation model called paragraph vector, or Doc2Vec, which can create a distributed vector representation of variable pieces of text, from a single sentence to a complete document.

To evaluate node pairs similarity with considering their edge attributes, it is previously necessary to consolidate these attributes occurrences. For instance, if one wants to compare the similarity from one user's posts to another, it does not make sense to choose a particular post for each user arbitrarily. The state-of-the-art presents some possibilities to aggregate both structured and unstructured data from OSNs: using the sum as an aggregation function [10], or set theory basic operations (e.g., union and intersection) [20], yet typical relational database aggregation functions (e.g., count, avg, max, min, mode, and empty) [25].

3 Related Work

When analyzing unsupervised link prediction state-of-the-art, one can organize the related works according to two criteria. The first one is regarding the types of information used as input for similarity metrics, while the second is about how the network states were considered.

Notably common in the state-of-the-art, a good number of works have used topological information. This group includes works [7,14,21,33] that, differently from the method proposed in the present work, did not take into account contextual information available in the OSNs discarding important characteristics from network elements. In a similar way to our proposal, [2,20,26,28,29,31] have explored the combination of topological and contextual data.

Concerning the network states, most works have used only a single network state [2,20,26,28,29,31,31,33]. By not analyzing past network states, these works did not consider how the network topology was when news edges were added to it. Hence, those methods take the chance to disregard information about the circumstances that may have influenced the appearance of old edges, and that could be useful to predict the creation of new ones. Few unsupervised link prediction works have explored the use of past network states. For instance, in [7], the authors incorporate past network states and consider at any time t new edges added to the network. After retrieving topological information available on the network, it applies similarity metrics which results are used as input for a clustering algorithm. This clustering process objective is to identify similarity patterns from current network elements that can be used to predict new links. This method's main limitation is using exclusively topological information. To remedy such limitation, the present work extended the method proposed in [7] by taking into account the wealth of contextual information available in a given OSN.

4 An Edge Creation History Based Method

This section describes $\mathbb{H}_T$, a link prediction method based on past network states. Originally proposed by [7], $\mathbb{H}_T$ predicts links in OSN based on the retrieval of information that describes how the network's topology was at the moments when the existing links were introduced in the network. Basically, this method is composed of three stages: Data Preparation, Link Prediction, and Results Evaluation.

4.1 Data Preparation

This stage consists of 6 steps (as shown in Fig. 1) and deals with the preparation of the data for the link prediction task. Here, the network data are obtained, and then a graph is created and partitioned based on user-defined configurations. In sequence, graphs representing the past network states are generated. These graphs contain the history of the network at the time when new links have emerged, including only topological information of the graph structure.

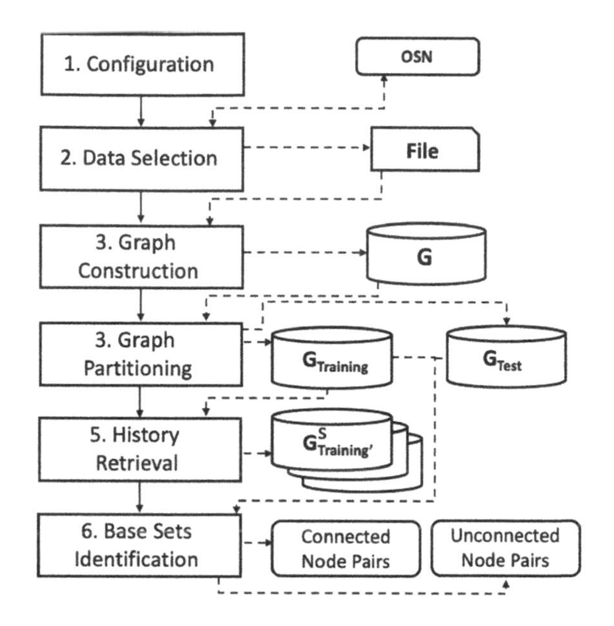

Fig. 1. A. Data Preparation.

Step 1. Configuration: An initial group of settings is required in order to prepare the data that will be used for the next steps and stages. This activity is typically performed by a data analyst with experience in this kind of OSN.

Among these configurations, the following parameters used in the steps *2*, *3*, *4* and *5* should be highlighted: the OSN observation period $[t_i, t_f]$, the training and test partitioning criteria t_c, and the time unit used for separating the past network states. For example, one can only consider OSN data from $t_i = 1997$ to $t_f = 2000$, with $t_c = 1999$. The time unit is described by a timestamp (a year in the last example), which format may vary depending on the OSN. For example, while *yyyy-mm-dd h:mm:ss* timestamp would represent a year, month, day, hour, minute, and seconds, another timestamp, *yyyy*, would represent years only.

Considering the particular dynamics of OSN evolution, the criteria for selecting the observed nodes as the most frequent and active, used in Step *6*, should also be defined. This criterion takes into account a minimum number of edges that indicate the frequency of interactions of a given node, denoted ϕ for training, and ψ for test intervals.

A significant configuration to define is the set D_t with containing only topological similarity metrics used in Step *7*. Another essential choice related to Step *8* corresponds to the clustering algorithm C, besides its specific parameters.

Step 2. Data Selection: Using t_i and t_f defined in Step *1*, OSN data are read and stored in a text file. This process takes place through some facilities offered by OSN (e.g., a data integration service or *API*).

Step 3. Graph Construction: In this step an homogeneous multigraph with attributes[1] is generated, $G = \langle V, E \rangle$, where V is a set of nodes v, E is a set of edges e associated with at least one temporal attribute ($e.t$) indicating the time of appearance of e, such that $t_i \leq e.t \leq t_f$, defined in Step *1*.

Step 4. Graph Partitioning: In this step, $G = \langle V, E \rangle$ is divided into $G_{Training} = \langle V, E_{Training} \rangle$ and $G_{Test} = \langle V, E_{Test} \rangle$. The partitioning takes place based on t_c (defined in Step *1*) where $t_i < t_c < t_f$. $G_{Training}$ and G_{Test} contain the edges that appear in the respective intervals, therefore $t_i \leq e.t \leq t_c$ for training and $t_c < e.t \leq t_f$ for testing.

Step 5. History Retrieval: This step is responsible for obtaining from $G_{Training}\langle V, E_{Training} \rangle$ (i.e. from the training subgraph) the topological information that describes the states of the network at the time when new edges between pairs of nodes never connected before appeared. To be done, the user-defined time unit format (timestamp) is used by applying the time atribute ($e.t$) of each edge to generate a set S of time instants, in which $S = \{t'/\exists e'(e' \in E_{Training} \wedge e' = (u,v), u,v \in V \wedge e'.t = t') \wedge \nexists e(e \in E_{Training} \wedge e = (u,v) \wedge e.t < e'.t)\}$. So for every $s \in S$, this step retrieve the subgraph $G^s_{Training}\langle V, E^s_{Training} \rangle$, where $\forall e \in E_{Training}\ (e.t \leq s) \rightarrow (e \in E^s_{Training})$. These subgraphs portraits the network evolution history $H_T = (G^{s_1}_{Training}, G^{s_2}_{Training}, ..., G^{s_m}_{Training})$, where $s_i \in S$, $|S| = m$, $s_1 < s_2 < ... < s_m$.

Step 6. Base Sets Identification: To continue the process, the necessary supporting sets are identified. Initially, the *Core* set is created. This procedure aims to ensure that spurious nodes, or those with few interactions, are not taken into account in the predictive process, avoiding distortions in the evaluation of results [16]. Then, a filter is applied to the training and test graphs to obtain the nodes that have at least ϕ and ψ edges in $G_{Training}$ and G_{Test}, respectively. The edges that have emerged between elements of the *Core* set in $G_{Training}$ are denoted E_{Old}, while those (still from elements of *Core*) that have emerged in G_{Test}, E_{New}.

The set of connected node pairs to be considered by the next steps is then obtained based on the edges from E_{Old}. Furthermore, $G_{Training}$ unconnected node pairs are identified $((Core \times Core) - E_{old})$. The last set contains the node pairs whose connection potential should be evaluated at the end of the prediction process.

[1] A G graph is said: (a) homogeneous if, and only if, G has only one type of node and one type of edge; and (b) has attributes if, and only if, G contains attributes in its nodes and/or in its edges.

4.2 Link Prediction

In the second stage (Fig. 2), the link prediction is actually performed. First of all, the similarity metrics will be calculated for connected node pairs in order to represent the node pairs in the Euclidean space. These metrics results will be used as input for a clustering algorithm, and its outcome is supposed to represent similarity patterns of the analyzed network. After calculating the same similarity metrics for unconnected node pairs and obtaining their representation in the same Euclidean space, it will be possible to associate them with the clusters obtained before and calculate their final score to identify those pairs most likely to establish a new connection.

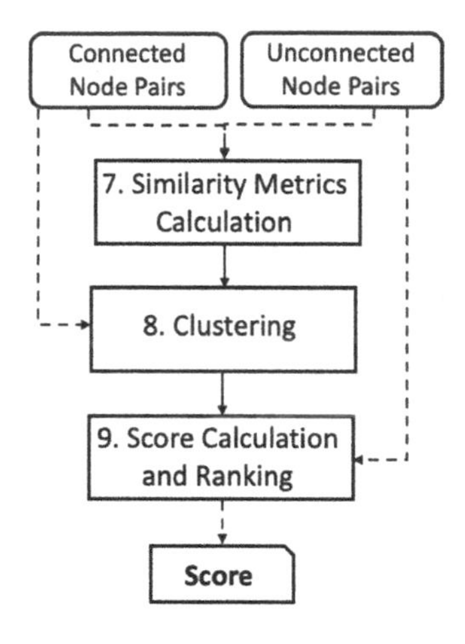

Fig. 2. B. Link Prediction.

Step 7. Similarity Metrics Calculation: For all elements of the connected and unconnected node pairs sets, each one of the topological similarity metrics from $D_t = \{d_{T_1}, d_{T_2}, ..., d_{T_q}\}$ are applied. The similarity metrics values describe a summarized view of network topology at the moment each edge e between node pairs appeared. As a result, is produced a set of elements in q-dimensional space where $q = |D_t|$, or $H_{(d_{T_1}, d_{T_2}, ..., d_{T_q})}$, a simplified representation of the topological evolution history of the network $\mathbb{H}_T$ defined in Step *5*. This set contains the data records extracted from $G^s_{Training}$, $\forall s \in S$ and which portray all the moments when edges appeared during the $[t_i, t_c]$ interval.

On the other hand, these same topological similarity metrics of set D_t are used for the unconnected node pairs set, generating the $P_{(d_{T_1}, d_{T_2}, ..., d_{T_q})}$. This set, however, rather than representing the network history, aims to identify the degree of compatibility between the elements of the unconnected node pairs set, whose connection potential will be calculated in Step 8.

Step 8. Clustering: In this step, the clustering algorithm C is applied on the connected node pair metrics results $H_{(d_{T_1}, d_{T_2}, ..., d_{T_q})}$, creating k *clusters*

$c_1, c_2, ..., c_k$. Each cluster c_j determines a boundary region in q-dimensional space characterized by the appearance of edges during the evolution of $G_{Training}$. The more dense and populated a region is, the more representative the corresponding cluster. This process is illustrated in Fig. 3(a) and (b).

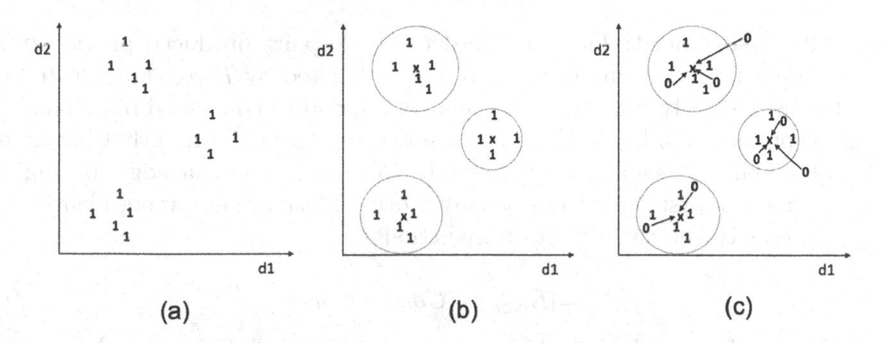

Fig. 3. Clustering process: 1 and 0 represent connected and not connected pairs of nodes respectively, x represent each cluster centroid.

Step 9. Score Calculation and Ranking: Initially, the cluster c to which p, the representation of the unconnected pair of nodes (u, v) in $P_{(d_{T_1}, d_{T_2}, ..., d_{T_q})}$, belongs to is identified by using Eq. 1. Figure 3 (c) illustrates this process.

$$c = \underset{1 \leq j \leq k}{\text{argmin}}\, Cos(p, Cen(c_j)) \tag{1}$$

Next, the frequency with which connected node pairs are found in the set representing the network history $H_{(d_{T_1}, d_{T_2}, ..., d_{T_q})}$ is identified in the *cluster* c, relative to the largest total number of connected node pairs among all the *clusters*. This result is weighted by the number of linked node pairs of H over the number of unconnected node pairs in the set $P_{(d_{T_1}, d_{T_2}, ..., d_{T_q})}$. This process aims to give higher scores to pairs in clusters with a bigger number of connected node pairs and a few unconnected ones.

Finally, the resulting frequency calculation is added to the distance from a particular element p to its cluster, providing an advantage to points closer to this cluster centroid. The score formula is presented on Eq. 2.

$$Score(u, v) = \frac{|H^c|}{\max\limits_{1 \leq j \leq k} (|H^{c_j}|)} \cdot \frac{|H^c|}{|P^c|} + Cos(p, Cen(c)) \tag{2}$$

4.3 Results Evaluation

In this stage (Fig. 4), the results are evaluated to identify which of the unconnected pairs of nodes in $G_{Training}$, pointed out as most likely to create a new link, have indeed been connected in G_{Test}. Here, the evaluation of the link prediction result is performed. This step generates a final report of the whole process.

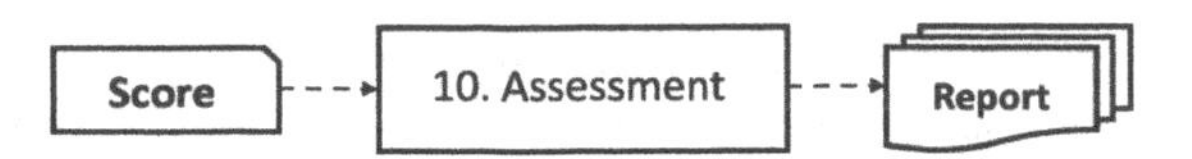

Fig. 4. C. Results Evaluation.

Step 10. Assessment: The list of ordered node pairs produced previously is used as input for this step. The first N pairs, denoted by Top_N, are selected in from the list to verify how many of them correspond to connected node pairs in G_{Test}, as presented in Eq. 3. E_{New} is composed of edges $e = (u, v)$ that belong to G_{Test} and their nodes u and v belong to the $Core$ set. Later, the edges belonging to E_{New} and the first N ordered pairs of $Core \times Core$ are analyzed, identifying $E_{Correct}$, that is, the correct set of predictions.

$$N = |E_{New} \cap (Core \times Core)| \tag{3}$$

Subsequently, we calculate the improvement (Eq. 5) that expresses how many times the performance of the evaluated method is superior to the performance of the random predictor PR (Eq. 4).

$$PR = \frac{|E_{new}|}{|(Core \times Core)| - |E_{old}|} \tag{4}$$

$$Improvement = \frac{|E_{Correct}|/|E_{New}|}{PR} \tag{5}$$

5 Proposed Extension

This section reports how $\mathbb{H}_T$ was extended to take into account contextual information to predict new links in OSNs. To avoid ambiguity, from now on, we will call $\mathbb{H}_{TC}$ to refer to $\mathbb{H}_T$'s extended version. In order to enable $\mathbb{H}_{TC}$ to consider both topological and contextual data, key changes were made on $\mathbb{H}_T$'s steps 1 (Configuration), 5 (History Retrieval), and 7 (Similarity Metrics Calculation). The following paragraphs describe these changes.

Step 1. Configuration: This step was extended so that the data analyst can configure some additional parameters to be considered by $\mathbb{H}_{TC}$. First, the analyst must choose which contextual attributes must be combined with the topological ones. To this end, this step presents a set A containing all contextual attributes available in the OSN. Each contextual attribute $a \in A$ may contain either structured or unstructured data. For simplicity, in this paper, we considered that all unstructured attributes contain textual data. Contextual attributes may be available in nodes (node attributes) or edges (edge attributes). Node attributes contain information that describes the participants, while edge attributes provide data concerning some relationship between them. For example, an OSN could contain attributes like user place of birth and a post's text. Typically, the first is a node attribute that contains structured data, while the former is an edge attribute that stores unstructured data. In formal terms, A can be divided into two subsets, i.e., $A = A_N \cup A_L$, where $A_N \cap A_L = \emptyset$ and A_N and A_L contain node and edge attributes, respectively.

After attribute selection, $\mathbb{H}_{TC}$ creates $A^S = A_N^S \cup A_L^S$, a set with the selected attributes where A_N^S and A_L^S encompass attributes selected from A_N and A_L, respectively. Then, the analyst must choose from a set of available data representations $R = R_S \cup R_U$ the one (denoted by rep) to be considered for each attribute $a \in A^S$. Note that while R_S contains possible data representations for structured data, R_U includes data representations for unstructured data. $R_S = \{numerical, alphanumerical\}$ and $R_U = \{BoW, \text{TF-IDF}, Doc2Vec\}$ contain examples of possible data representations. The analyst must also choose, for each attribute $a \in A^S$, which similarity metric $d_c \in D_c$ will be used for similarity calculation, where D_c is the set of the contextual similarity metrics.

Additionally, for each edge attribute $a_i \in A_L^S$, the user must select a data aggregation operator from a set Op, that contains all aggregation operators provided by $\mathbb{H}_{TC}$. Those operators are responsible for combining the multiple occurrences of a certain attribute. Currently, $\mathbb{H}_{TC}$ provides three aggregation operators: $Op = \{sum, union, average\}$. Based on the analyst's choices, a new set of attributes is defined: $A^C = A_N^R \cup A_L^R$, where $A_N^R = \{(a_i, rep_{a_i}, d_{c,a_i})/a_i \in A_N^S \wedge rep_{a_i} \in R \wedge d_{c,a_i} \in D_c\}$ and $A_L^R = \{(a_i, rep_{a_i}, d_{c,a_i}, op_{a_i})/a_i \in A_N^S \wedge rep_{a_i} \in R \wedge d_{c,a_i} \in D_c \wedge op_{a_i} \in Op\}$.

Step 5. History Retrieval: Besides retrieving topological information that describes the states of the network at the time when new edges between non-connected nodes appeared, this step was extended to retrieve the contextual data concerning each attribute in A^C. Hence, it builds the network evolution history $H_{TC} = (G_{Training}^{s_1}, G_{Training}^{s_2}, ..., G_{Training}^{s_m})$ so that each subgraph $G_{Training}^{s_m}$ also contains in its nodes and edges the values of the contextual attributes in A^C.

Step 7. Similarity Metrics Calculation: This step was extended to consider contextual data in the similarity metrics calculation. Hence, given a pair of nodes u and v, for each $a_i \in A^C$, this step follows one of the alternatives below.

If $a_i \in A_L^R$, nodes u and v incident edges $e_{u_1}, e_{u_2}, ..., e_{u_j}$ and $e_{v_1}, e_{v_2}, ..., e_{v_j}$ will be retrieved and used as input for creating $\overline{U_{a_i}} = rep_{a_i} \otimes op_{a_i(e_{u_1}, e_{u_2}, ..., e_{u_j})}$ and $\overline{V_{a_i}} = rep_{a_i} \otimes op_{a_i(e_{v_1}, e_{v_2}, ..., e_{v_j})}$ based on the application of chosen representation and aggregation function operator.

On the other hand, if $a_i \in A_N^R$, similar depictions $\overline{U_{a_i}} = rep_{a_i}(u, a_i)$ and $\overline{V_{a_i}} = rep_{a_i}(v, a_i)$ will be created with considering applying just the representation defined, as there is no need for combining a node single attribute occurrence.

After that, $\mathbb{H}_{TC}$ applies the chosen similarity metric d_{c,a_i} to $\overline{U_{a_i}}$ and $\overline{V_{a_i}}$ generating $H_{(d_{t_1}, d_{t_2}, ..., d_{t_q}, d_{c,a_1}, d_{c,a_2}, ..., d_{c,a_{|A^C|}})}$ for connected nodes and, similarly, $P_{(d_{t_1}, d_{t_2}, ..., d_{t_q}, d_{c,a_1}, d_{c,a_2}, ..., d_{c,a_{|A^C|}})}$ for non-connected ones.

It is important to note that some data representations, like Doc2Vec, require trained models. This model should learn the representations for all the sentences in a training set (preferably a different one from the dataset used as part of the link prediction process), that will then be used for creating data representations for new data. Another important aspect worth mentioning is that the execution order of data representation and aggregation function may vary according to the chosen data representation. In Doc2Vec, for example, aggregating texts before creating its representation can lead to a semantic loss. In contrast, in the BoW representation, the frequency of terms identification should be made on all words from attribute multiple occurrences.

Although steps 8 (Clustering) and 9 (Score Calculation and Ranking) were not changed as part of the proposed extension $\mathbb{H}_{TC}$, the attributes from set A^C (each one associated with one, and only one, metric from D_c) will affect the number of elements from H and P sets. In this way, the number of dimensions used as input for clustering algorithm C will be $q = |D_t| + |A^C|$. In a similar way, H and P elements will be used for reaching a final score for each unconnected pair of nodes, as discussed in Sect. 4.

6 Experiments and Results

In order to evaluate the performance of the extended method $\mathbb{H}_{TC}$ proposed in this paper, five different co-authorship networks from the ArXiv repository were used. In this kind of network, nodes are scientific papers authors, while edges, the co-authored papers themselves. The networks astro-ph (Astrophysics of Galaxies), cond-mat (Condensed Matter), gr-qc (General Relativity and Quantum Cosmology), hep-th (High Energy Physics - Theory) and hep-ph (High Energy Physics - Phenomenology) were selected because of their popularity in experiments in the area of link prediction [17]. Table 1 introduces the statistical data of these co-authorship networks.

Table 1. Statistical data from co-authorship networks - 1992 a 1998.

$G_{Training}$			*Core*		
Dataset	Nodes	Edges	Authors	$\|E_{New}\|$	$\|E_{Old}\|$
astro-ph	8024	88964	1400	2064	9262
cond-mat	8278	31708	1030	343	2366
gr-qc	2594	9366	294	67	586
hep-th	6692	54592	1382	317	6903
hep-th	6158	24321	1056	421	6903

For the experiments[2], the time frame used was defined by $t_i = 1992$ and $t_f = 1998$, with six years for training and one year for testing, i.e., $t_c = 1997$. In this kind of network, where the available time unit is in the year base, a more extended time frame should better emphasize its historical aspect. In addition

[2] Prototype code is available at: https://gitlab.com/arguscavalcante/link_pred.

to that, the values of parameters $\phi = 6$ and $\psi = 1$ were chosen to filter the most active nodes (i.e., authors with at least one publication by year) in $G_{Training}$ and G_{Test}, respectively.

We used *k-means* [5] as the data clustering algorithm. Its reduced complexity and code availability guided our choice. In order to investigate the influence of the number of clusters, we conducted a sensitivity analysis by applying the cross-validation process [12]. To this end, for each network, we split $G_{Training}$ in two sub-graphs: $G_{Pre-Training}$ and $G_{Validation}$. Then, considering those sub-graphs, we tested the numbers of clusters k from *2* to *10*. This process resulted in the choice of $k = 3$ for the number of *clusters*.

The following link prediction methods were evaluated:

The random predictor (PR): This method deals with the standard comparison proposed by [16], used as a basic reference for the link prediction methods evaluation.

The baseline method [7] ($\mathbb{H}_T$): This method uses exclusively topological information and related similarity metrics *Adamic-Adar*, *Common Neighbours* and *Jaccard Coefficient*, resulting $D_t = \{AA, CN, JC\}$, combined in sets of 2 and 3 elements. The choice of these metrics is due to their popularity in the state-of-the-art [17].

The proposed extension ($\mathbb{H}_{TC}$): This method combines both topological and contextual information. The topological metrics chosen were the same used for $\mathbb{H}_T$, $D_t = \{AA, CN, JC\}$. Based on the set of available edge attributes from $A_L = \{title, abstract, year\}$ the text of paper titles has been selected, resulting $A_L^S = \{title\}$. Additionally, a Doc2Vec representation of dimension 300 from R_U, the Cosine similarity metric from D_c, and finally, an average operator from O_p as the aggregation function were defined. Therefore, $A^C = A_L^R = \{(title, Doc2Vec_{300}, Cosine, Average)\}$.

The Table 2 compiles the experiments results. This table values indicate the improvement factor of $\mathbb{H}_T$ and $\mathbb{H}_{TC}$ with respect to the random predictor PR (see Eq. 5 for improvement factor calculation). The best improvement factor value from both methods comparison is highlighted in bold font.

The first relevant aspect to be mentioned is that excluding two zero improvement results, both $\mathbb{H}_T$ and $\mathbb{H}_{TC}$ performed better than PR (i.e., improvement factor above 1). It means that $\mathbb{H}_T$ and $\mathbb{H}_{TC}$ are more likely to predict new links correctly than an aleatory criterion. Although it is possible to observe gr-qc has a small number of authors and papers, we have not concluded why these above mentioned zero improvement results have occurred for both $\mathbb{H}_T$ and $\mathbb{H}_{TC}$.

Another point worthy of mentioning is that the proposed extension $\mathbb{H}_{TC}$ was superior fourteen times, against three of $\mathbb{H}_T$, besides three draws.

On the other hand, a more meticulous analysis reveals that, in a few cases, the differences in performance among $\mathbb{H}_T$ and $\mathbb{H}_{TC}$ were not high in some situations. Hence, in order to check for statistical significance in those differences, we applied the *Wilcoxon Signed Ranks* [32] with significance level $\alpha = 0.05$ and null hypothesis stated as H_0: *the performance of* $\mathbb{H}_{TC}$ *and* $\mathbb{H}_T$ *are statistically identical.* In the test, we considered the twenty results. The test rejected H_0,

Table 2. Comparing $\mathbb{H}_T$ method [7] and extended method $\mathbb{H}_{TC}$.

Dataset	PR	Method	AA,CN	AA,JC	CN,JC	AA,CN,JC
astro-ph	0,0021	$\mathbb{H}_T$	10,01	7,51	8,88	9,79
		$\mathbb{H}_{TC}$	**26,18**	**44,85**	**31,87**	**25,73**
cond-mat	0,0006	$\mathbb{H}_T$	35,87	22,42	**22,42**	35,87
		$\mathbb{H}_{TC}$	**40,35**	**26,90**	13,45	**40,35**
gr-qc	0,0015	$\mathbb{H}_T$	9,46	**9,46**	0,0	9,46
		$\mathbb{H}_{TC}$	**18,92**	0,0	**9,46**	**18,92**
hep-ph	0,0006	$\mathbb{H}_T$	18,53	9,26	**29,65**	22,23
		$\mathbb{H}_{TC}$	**50,03**	**20,38**	9,26	**50,03**
hep-th	0,0007	$\mathbb{H}_T$	**46,92**	**62,56**	53,17	**43,79**
		$\mathbb{H}_{TC}$	**46,92**	**62,56**	**56,30**	**43,79**

indicating a significant statistical difference between $\mathbb{H}_T$ and $\mathbb{H}_{TC}$. This result reinforces the hypothesis that combining topological and contextual information when using past network states may improve link prediction.

Still considering the close performance values presented by $\mathbb{H}_T$ and $\mathbb{H}_{TC}$ in some situations, it is essential to emphasize that link prediction is a kind of recommendation problem where even low improvements in model accuracy may lead to considerable gains in the domain of the application (e.g., increase the sales of high-priced products).

7 Conclusion

Predicting whether a pair of nodes will connect in the future is a relevant task in OSN analysis known as the link prediction problem. Most of the state-of-the-art studies consider only a static version of the network, leaving aside the possibility of retrieving the network history as made by the past states-based methods. The few works that investigate past network states usage have taken into account exclusively the network topology without considering contextual information. Consequently, the present work aims to investigate the effect of combining topological and contextual information available in the evolutionary network history. Our investigation aimed to improve the performance of the link prediction task by extending the method proposed by [7]. Compared to the traditional method, the proposed extension has reached superior results, confirming the raised hypothesis.

Besides state-of-the-art analysis, this article's main contribution is the extension of the baseline method. The main limitations in our experiments were: (a) we have used only co-authorship networks from Arxiv repository, (b) the only clustering algorithm used was *k-means*, and (c) we have not experimented other attribute and similarity metric combination.

As part of future work, the extended method should be evaluated against other OSNs. This evaluation objective is to test the method in a scenario with other time units than co-authorship networks (yearly basis). Moreover, it is relevant to compare the extended method with results achieved by additional state-of-the-art strategies. Another line of investigation would be to apply different clustering algorithms, checking their performance using the same method and other implications of the complexity of these algorithms' execution time, discussing their feasibility in the analysis of OSNs. Nevertheless, different combinations of unstructured data representations, aggregations functions, and similarity metrics should be assessed.

References

1. Adamic, L.A., Adar, E.: Friends and neighbors on the web. Soc. Netw. **25**(3), 211–230 (2003)
2. Ahmed, N.M., Chen, L.: An efficient algorithm for link prediction in temporal uncertain social networks. Inf. Sci. **331**(C), 120–136 (2016)
3. Cavalcante, A.A., Muniz, C.P., Goldschmidt, R.R.: Context-based time score: an effective similarity function for link prediction in social networks. In: Proceedings of the 24th Brazilian Symposium on Multimedia and the Web, WebMedia 2018, pp. 339–346. Association for Computing Machinery, New York (2018)
4. Cheng, J., Adamic, L., Dow, P.A., Kleinberg, J.M., Leskovec, J.: Can cascades be predicted? In: Proceedings of the 23rd International Conference on World Wide Web, WWW 2014, pp. 925–936. ACM, New York (2014)
5. Damasceno, F.F., Veras, M.B., Mesquita, D.P., Gomes, J.P., de Brito, C.E.: Shrinkage k-means: a clustering algorithm based on the James-Stein estimator. In: 2016 5th Brazilian Conference on Intelligent Systems (BRACIS). IEEE (2016)
6. Florentino, E., Cavalcante, A., Goldschmidt, R.: Um método baseado na evolução dos dados topológicos para a predição de links em redes sociais. In: SBSI 2019 (2019)
7. Florentino, E., Cavalcante, A., Goldschmidt, R.: An edge creation history retrieval based method to predict links in social networks. Knowl.-Based Syst. **205**, 106268 (2020)
8. Fortunato, S.: Community detection in graphs. Phys. Rep. **486**(3) (2010)
9. Harris, Z.S.: Distributional structure. Word **10**(2–3), 146–162 (1954)
10. Hasan, M., Chaoji, V., Salem, S., Zaki, M.: Link prediction using supervised learning. In: Proceedings of SDM 2006 Workshop on Link Analysis, Counterterrorism and Security, January 2006
11. Hasan, M.A., Zaki, M.J.: A Survey of Link Prediction in Social Networks, pp. 243–275. Springer US, Boston (2011)
12. Kohavi, R.: A study of cross-validation and bootstrap for accuracy estimation and model selection. In: Proceedings of the 14th International Joint Conference on Artificial Intelligence, IJCAI 1995, vol. 2, pp. 1137–1143. Morgan Kaufmann Publishers Inc., San Francisco (1995)
13. Kumar, R., Novak, J., Tomkins, A.: Structure and evolution of online social networks, pp. 337–357. Springer, New York (2010)
14. Laishram, R., Mehrotra, K., Mohan, C.K.: Link prediction in social networks with edge aging. In: 2016 IEEE 28th International Conference on Tools with Artificial Intelligence (ICTAI), pp. 606–613, November 2016

15. Le, Q., Mikolov, T.: Distributed representations of sentences and documents. In: Proceedings of the 31st International Conference on International Conference on Machine Learning, ICML 2014, vol. 32, pp. 3111–3119. JMLR.org (2014)
16. Liben-Nowell, D., Kleinberg, J.: The link-prediction problem for social networks. J. Am. Soc. Inf. Sci. Technol. **58**(7), 1019–1031 (2007)
17. Lü, L., Zhou, T.: Link prediction in complex networks: a survey. Phys. A **390**(6), 69:1 69:33 (2011)
18. Mislove, A., Marcon, M., Gummadi, K.P., Druschel, P., Bhattacharjee, B.: Measurement and analysis of online social networks. In: Proceedings of the 7th ACM SIGCOMM Conference on Internet Measurement, IMC 2007, pp. 29–42. ACM, New York (2007)
19. Munasinghe, L., Ichise, R.: Time score: a new feature for link prediction in social networks. IEICE Trans. Inf. Syst. **E95.D**, 821–828 (2012)
20. Muniz, C.P., Goldschmidt, R., Choren, R.: Combining contextual, temporal and topological information for unsupervised link prediction in social networks. Knowl.-Based Syst. **156**, 129–137 (2018)
21. Negi, S., Chaudhury, S.: Link prediction in heterogeneous social networks. In: Proceedings of the 25th ACM International on Conference on Information and Knowledge Management, CIKM 2016, pp. 609–617. ACM, New York (2016)
22. Nettleton, D.F.: Data mining of social networks represented as graphs. Comput. Sci. Rev. **7**(Suppl. C), 1–34 (2013)
23. Newman, M.: Clustering and preferential attachment in growing networks. Phys. Rev. E **64**, 025102 (2001)
24. Pecli, A., Cavalcanti, M.C., Goldschmidt, R.: Automatic feature selection for supervised learning in link prediction applications: a comparative study. Knowl. Inf. Syst. **56**(1), 85–121 (2017). https://doi.org/10.1007/s10115-017-1121-6
25. Popescul, A., Popescul, R., Ungar, L.H.: Statistical relational learning for link prediction (2003)
26. Rummele, N., Ichise, R., Werthner, H.: Exploring supervised methods for temporal link prediction in heterogeneous social networks. In: Proceedings of the 24th International Conference on World Wide Web, WWW 2015 Companion, pp. 1363–1368. ACM, New York (2015)
27. Salton, G.: Automatic Text Processing: The Transformation, Analysis, and Retrieval of Information by Computer. Addison-Wesley Longman Publishing Co., Inc., Boston (1989)
28. Shalforoushan, S.H., Jalali, M.: Link prediction in social networks using Bayesian networks. In: 2015 The International Symposium on Artificial Intelligence and Signal Processing (AISP), pp. 246–250, March 2015
29. Sharma, P.K., Rathore, S., Park, J.H.: Multilevel learning based modeling for link prediction and users' consumption preference in online social networks. Future Gener. Comput. Syst. (2017)
30. Sparck Jones, K.: A statistical interpretation of term specificity and its application in retrieval. In: Document Retrieval Systems, pp. 132–142. Taylor Graham Publishing, London (1988)
31. Tasnadi, E., Berend, G.: Supervised prediction of social network links using implicit sources of information. In: Proceedings of the 24th International Conference on World Wide Web, WWW 2015 Companion, pp. 1117–1122. ACM, New York (2015)
32. Wilcoxon, F.: Individual Comparisons by Ranking Methods. Bobbs-Merrill Reprint Series in the Social Sciences, S541. Bobbs-Merrill Company Incorporated (1945)
33. Wu, J., Zhang, G., Ren, Y.: A balanced modularity maximization link prediction model in social networks. Inf. Process. Manage. **53**(1), 295–307 (2017)

Predicting the Evolution of COVID-19 Cases and Deaths Through a Correlations-Based Temporal Network

Tiago Colliri[1(✉)], Alexandre C. B. Delbem[1], and Liang Zhao[2]

[1] Institute of Mathematics and Computer Science, University of Sao Paulo, São Paulo, Brazil
tcolliri@usp.br, acbd@icmc.usp.br

[2] Faculty of Philosophy, Science, and Letters, University of Sao Paulo, São Paulo, Brazil
zhao@usp.br

Abstract. Given the most recent events involving the fast spreading of COVID-19, policy makers around the world have been challenged with the difficult task of developing efficient strategies to contain the dissemination of the disease among the population, sometimes by taking severe measures to restrict local activities, both socially and economically. Within this context, models which can help on predicting the spread evolution of COVID-19 in a specific region would surely help the authorities on their planning. In this paper, we introduce a semi-supervised regression model which makes use of a correlations-based temporal network, by considering the evolution of COVID-19 in different world regions, in order to predict the evolution of new confirmed cases and deaths in 27 federal units of Brazil. In this approach, each node in the network represents the COVID-19 time series in a specific region, and the edges are created according to the variations similarity between each pair of nodes, at each new time step. The results obtained, by predicting the weekly new confirmed cases and deaths in each region, are promising, with a median and mean absolute percentage error of 21% and 24%, respectively, when predicting new cases, and a median and mean absolute percentage error of 16% and 23%, respectively, when predicting new deaths, for the considered period.

Keywords: Complex networks · Machine learning · Regression · Time series prediction · Temporal networks · COVID-19

This work is supported in part by the São Paulo State Research Foundation (FAPESP) under grant numbers 2015/50122-0 and 2013/07375-0, the Coordenação de Aperfeiçoamento de Pessoal de Nível Superior - Brasil (CAPES) - Finance Code 001, and the Brazilian National Council for Scientific and Technological Development (CNPq) under grant number 303199/2019-9.

R. Cerri and R. C. Prati (Eds.): BRACIS 2020, LNAI 12320, pp. 397–411, 2020.
https://doi.org/10.1007/978-3-030-61380-8_27

1 Introduction

Since the first human populations started to live in groups, the epidemic diseases have been one of the greatest problems faced by humanity. In the modern era, this problem has been aggravated by two combined factors: (1) the fact that human beings have been living more and more in concentrated urban spaces, and (2) that these great concentrations of people, by their turn, are increasingly more interconnected through faster and more efficient worldwide transportation routes [17]. The most recent example regarding this problem involves the COVID-19, which was officially characterized as a pandemic only around four months after its first cases were reported, in China. The fast dissemination of this disease can be particularly challenging for policy makers around the world, which have to face the difficult task of developing efficient strategies for controlling the spread of the disease among the population, oftentimes by taking severe measures to restrict the local activities, both socially and economically. Within this context, models which can help on predicting the spread evolution of the disease in each region would surely help the authorities on their resources planning.

Networks (or *graphs*) are powerful modeling tools for exploring a dataset in terms of the relations between the data instances, both in a static or in a dynamic way. The term *complex network* refers to a graph consisting of a large number of *nodes* (or vertices) joined by *links* (or edges), with a non-trivial topology [3]. Some examples of complex networks include the internet [15], biological neural networks [27], blood distribution networks [31] and power grid distribution networks [2]. There are also several network-based models designed to perform *machine learning* tasks, such as *clustering* [26], *classification* [9,25] and *regression* [16]. Mathematically, a network can be defined as a graph $\mathcal{G} = (\mathcal{V}, \mathcal{E})$, where $\mathcal{V}$ is a set of nodes and $\mathcal{E}$ is a set of tuples representing the edges between pairs of nodes $(i, j) : i, j \in \mathcal{V}$. A *temporal network* is a specific type of *multilayer network* or *multiplex* [12], in the form of $\mathcal{G} = (\mathcal{V}, \mathcal{E}, \mathcal{D})$, in which the additional dimension $\mathcal{D}$ contains an ordered set of temporal indices that represents time [18]. Among the phenomena which have already been modeled through temporal networks are brain connectivity [30], fires events in the Amazon [32], economic trade and social networks [28,29,33], political parties [5,10] and corruption scandals [20].

Although complex networks is a relative new field of study, there are also already well-known network-based models developed specifically to approach problems regarding the spread of epidemic diseases. These models do not necessarily need to make use of very advanced mathematical calculus since, oftentimes, simple models can help to further understand the transmission of infectious agents within human communities [4]. The SI, SIS and SIR models [7,22], for instance, allow one to estimate what would be the critical threshold, in terms of the percentage of infected individuals in a population, for an infectious disease to become endemic. These models can also help on determining which immunization strategies are expected to be more effective, according to the topological characteristics of the network formed by the individuals susceptible to the disease [13,23]. A very interesting survey on this topic was made by Costa et al. [11].

Some recent studies have applied machine learning techniques to predict the spread of epidemics, on a weekly basis. The study made by Al-qaness et al. [1] forecasts the number of weekly new confirmed cases of influenza in China, based on the previously confirmed cases, by using an improved adaptive neuro-fuzzy inference system (ANFIS), achieving a mean absolute percentage error of 32% and 45%, respectively, on the two datasets analyzed. In the work from Arora et al. [6], an artificial recurrent neural network (RNN) was applied on a 7-days testing dataset to predict new confirmed cases of COVID-19 in the states of India, obtaining a mean absolute percentage error of around 6% when predicting the weekly new confirmed cases in 4 states.

Currently, the most common strategy to deal with the COVID-19 pandemic, among policy makers, is to implement isolation policies in the population, through quarantines or lockdowns. These measures have the aim of delaying the peak of the dissemination curve, in order to avoid a suddenly rise on COVID-19 hospitalizations. In this sense, it is more desirable for a given region that its dissemination curve can be more similar to the curves of countries which were less affected by the disease, such as Japan, South Korea and Germany, for instance, than to the curves of countries more severely affected by the disease, such as Italy, Spain and US, for instance. In this paper, we take this type of reasoning one step further, by introducing a semi-supervised regression model which detects the correlations between the COVID-19 curves from different regions, and makes use of these detected correlations for predicting future values for new confirmed cases and deaths on a given region. The model starts by mapping the COVID-19 time series to static networks, where each network represents the current correlations between the time series, at each time step t. Each node in these networks represents the time series for a specific region, and the edges are generated according to the detected correlations between each pair of nodes. Afterwards, these static networks are analyzed in the form of a temporal network, in order to predict the evolution of a specific time series, for the period considered.

The novelty of this model consists in the use of a form of multiple regression analysis, in which it does not take into account the predicted time series' own prior curve for making the predictions. Instead, the model considers, as input attributes, only the curves from other time series in the dataset identified as more correlated to the series to be predicted, and whose lengths are greater or equal to the predicted time step t. Hence, this model is indicated for cases when the time series in the dataset present different initial dates, and one wants to investigate whether it is possible to generate predictions based on the correlations detected between the series. We evaluate the proposed model by applying it on preliminary COVID-19 data from different world regions for predicting the future confirmed cases and deaths in the 27 federal units of Brazil. The obtained results are promising, with the model being able to predict, on a weekly basis, the number of new confirmed cases in each state with a median and mean absolute percentage error of 21% and 24%, respectively, and the new confirmed deaths in each state with a median and mean absolute percentage error of 16% and 23%, respectively.

Regarding the organization of this paper, besides this introduction, we have, in Sect. 2, a description of the methodology and the model, showing how the networks are generated from the input data, and how the predictions are made. In Sect. 3, we present the results obtained by applying the model to real preliminary COVID-19 spreading data, from world regions and from Brazilian federal units. In Sect. 4, we conclude this study with some final remarks.

2 Methodology

The research methodology used in this study is summarized below.

2.1 Database

The database used in this study is built from the preliminary data made publicly available by Dong et al. [14]. This dataset comprises the daily evolution of COVID-19 confirmed cases and deaths in 261 different regions or countries, from 01-22-2020 until 05-26-2020. We also make use of another dataset [8], this one comprising the daily evolution of COVID-19 confirmed cases and deaths in each of the 27 federal units of Brazil, from 02-25-2020 until 05-26-2020. The real values from the later dataset are the ones to be predicted by the model and, for this reason, the future values from each federal unit are suppressed, prior to each prediction, and these values are used later exclusively for performance measuring purposes.

2.2 Description of the Time Series Prediction Model

Regression analysis is a statistical method building a mathematical model that best fits the data for the prediction of the output variable [19]. In simple regression, there is only one independent variable x that affects the value of the dependent variable y, while in multiple regression there are more than one. In this work, for predicting the evolution of the spread of COVID-19 in a given region, we conceive a *semi-supervised* regression model which predicts future values for a time series in a dataset, i.e., the dependent variable, based on the detected similarities between this time series and the other time series in the same dataset, through a correlations *temporal network*. In this case, we are aiming to detect hidden evolution patterns among groups of time series in the dataset, in order to make use of these patterns for prediction purposes. Therefore, this model can be applied in cases when the time series in the dataset present different initial dates, and one wants to investigate the existence of possible correlation patterns between them and, if that is the case, to estimate future values for a given time series based on these detected correlations.

In *machine learning* applied to time series, initially we have an input dataset comprising m instances (or time series) and n time steps t, in the form of $\mathcal{X} =$

$\{X_1, X_2, X_3, ..., X_m\}$, where each instance i consists of n elements, such that $X_i = (x_{i,t=0}, x_{i,t=1}, x_{i,t=2}, x_{i,t=3}, ..., x_{i,t=n})$, as in the following 2d array:

$$\begin{bmatrix} x_{1,1} & x_{1,2} & x_{1,3} & ... & x_{1,n} \\ x_{2,1} & x_{2,2} & x_{2,3} & ... & x_{2,n} \\ ... & ... & ... & ... & ... \\ x_{m,1} & x_{m,2} & x_{m,3} & ... & x_{m,n} \end{bmatrix}. \quad (1)$$

The model proposed in this work starts by bringing all time series in $\mathcal{X}$ to a same starting point $t = 1$. Next, it calculates the variation $\delta X_{i,t}$ for each time series i at each time step t, which is yielded by:

$$\delta X_{i,t} = \frac{X_{i,t} - X_{i,t-1}}{X_{i,t-1}}. \quad (2)$$

This provides us with a 2d variations array $\delta\mathcal{X}$, containing m instances (the time series), and a maximum of $n - 1$ columns (the time steps) for each time series. Afterwards, each row $\delta\mathcal{X}_t$ in this array is mapped as a network G_t, where each node represents a time series and the edges between each pair of nodes are generated according to the similarities between their variations at the time t. The neighbors connected to each vertice i, in each network G_t, are given by:

$$N(i_t) = \epsilon\text{-}radius(i_t), \quad (3)$$

where $\epsilon\text{-}radius(i_t)$ yields a set of instances whose variations $\delta X_{i,t}$ are within the range $[-\epsilon, +\epsilon]$. Following, a community detection algorithm is ran in the networks, in order to group the time series with more similar variations, at each time step t. For this end, one can use, for instance, the *fast greedy* algorithm [21], which detects the community structure based on the greedy optimization of the modularity measure, or also the *walktrap* algorithm [24] which, roughly speaking, is based on the idea that short random walks tend to stay in the same community in the network. At this point, we end up with a set of static networks $\mathcal{G} = \{G_{t=2}, G_{t=3}, G_{t=4}, ..., G_{t=n}\}$, with each of them representing the topological space emerged from the current relations between the variations in δX at the time slice t. Hence, we can also say that this set forms the temporal network $\mathcal{G}$, and each element in the set represents a different time slice t of the temporal network $\mathcal{G}$.

Following, a dictionary D_i is created, for each time series i in the dataset, in the form of a list $\mathcal{D} = \{D_1, D_2, D_3, ...D_m\}$. The set of keys K in D_i are given by all instances j in the dataset which have shared the same community with i, at any time step t, i.e., in any of the networks in $\mathcal{G}$. The set of values V in D_i, for any key j, are given by the respective number of times that the instances i and j have shared the same community in $\mathcal{G}$. Mathematically, we have that a dictionary can be defined as:

$$D \subseteq \{(k, v) \mid k \in K \wedge v \in V\} \wedge \forall(q, w) \in D : k = q \rightarrow v = w, \quad (4)$$

and, in the proposed model, the set of keys K and the set of values V in D_i are yielded by:

$$D_i^K = \{j \mid j \in G_{t,i}^C\} \wedge t \in [2, n] \wedge j \in [1, m] \wedge j \neq i, \quad \text{and} \tag{5}$$

$$D_i^V = u(\{G_t \mid j \in G_{t,i}^C\}) \wedge t \in [2, n] \wedge j \in [1, m] \wedge j \neq i, \tag{6}$$

where $G_{t,i}^C$ provides a set with all instances that share the same community with i in the network G_t.

Finally, the predicted variation for a time series i, at the time step $t > 2$, is equal to the averaged variations of the time series which are in the keys of the dictionary D_i whose length is equal or longer than t, weighted by their respective values in D_i. Thus, the predicted variation $\hat{\delta} X_{i,t}$ is given by:

$$\hat{\delta} X_{i,t} = \begin{cases} \frac{\sum_j D_i^V[k \rightarrow j] \, \delta X_{j,t}}{\sum_j D_i^V[k \rightarrow j]}, \forall j \in D_i^K \mid j \in G_t \wedge u(j) \geq t, \\ \qquad\qquad \text{if } \{j \mid j \in D_i^K \wedge j \in G_t \wedge u(j) \geq t\} \neq \emptyset \\ \emptyset, \text{ otherwise} \end{cases} \tag{7}$$

where $u(A)$ returns the length of array A. In this way, the model is able to predict variations for a time series i, on a time step t, only if at least one of the time series in D_i has a length equal or longer than t. Note that the model, hence, performs a form of multiple regression analysis, in which the prior evolution curve of the time series i (which is the dependent variable) is not taken into account in the predictions, and the independent variables (or predictors) considered in this case are the evolution curves from longer time series in the dataset which are more correlated to i. In this sense, it is worth highlighting the important role played by the ϵ radius threshold parameter in the model, used for generating the edges in the networks. Smaller values of ϵ make the predictions performed by the model more sensitive to local averages in the dataset, with the risk of *overfitting*, while, conversely, higher values of ϵ result in the model considering broader averages when making the predictions, with the risk of *underfitting*. It is also worth noting that, in Eq. 7, by weighting the time series variations more correlated to i by the number of times this correlation has occurred, prior to the time t of the prediction, we are here assuming that the time series i tends to preserve these same correlations in the future.

Model's Demonstration Through a Simple Example. In order to illustrate how the proposed model works, we now present its application on a simple dataset, to be used as example. Let us consider a dataset $\mathcal{X}$, comprising five time series: X_1, X_2, X_3, X_4 and X_5, with different initial dates and different lengths, as shown in Table 1. Since the model makes use of data from the longer time series in the dataset, in order to perform the predictions, then in this case it will be able to predict future values only for X_3, X_4 and X_5. One can observe, from Table 1, that X_1 and X_4 follow an arithmetic progression, while X_2 and X_3 follow a geometric progression, and X_5 – which is the most recent one, having only 2 observations so far – could either follow an arithmetic or a geometric

Table 1. Time series example dataset

	X_1	X_2	X_3	X_4	X_5
1/1/2020	10	2	-	-	-
1/2/2020	15	4	-	-	-
1/3/2020	20	8	3	-	-
1/4/2020	25	16	6	6	-
1/5/2020	30	32	12	9	1
1/6/2020	35	64	24	12	2

progression in the future. However, let us suppose that, in the current context, we are not aware of these evolution patterns for any of these time series, and hence we expect the model to correctly detect these evolution patterns for us, and to also estimate the future values for X_3, X_4 and X_5 accordingly.

Table 2. Daily variations (%) for the example dataset

	X_1	X_2	X_3	X_4	X_5
1st day	50.0	100.0	100.0	50.0	100.0
2nd day	33.3	100.0	100.0	33.3	
3rd day	25.0	100.0	100.0		
4th day	20.0	100.0			
5th day	16.7	100.0			

The first step in the proposed model is to bring all time series to a same starting point $t = 1$, and then to calculate the daily variations, as it is shown in Table 2. Next, the model generates 5 networks, i.e., the maximum length of the series in $\mathcal{X}$ subtracted by 1, where each node represents a time series in $\mathcal{X}$ and the edges between them are created according to the similarities of their daily variations, on each time step t. For accomplishing this task, in this example, we make use of the nearest neighbor technique based on a radius $\epsilon = 10$. Then, the *fast greedy* community detection algorithm is ran in the networks. This results in a temporal network $\mathcal{G}$, formed by the set $\{G_t \mid t \in [1, 5]\}$. Note that, in this step, the model groups the time series with more similar variations, at each time step t, and, as t gets bigger, only the nodes from time series with longer lengths are left in $\mathcal{G}$. The edges evolution among the nodes of $\mathcal{G}$ is shown in Fig. 1, where each row represents one time series in the dataset and each column represents one time slice of the temporal network. The colors denote the community to which each node belongs, at each time slice. In this case, if a node has a white color, it means that this node is not in the temporal network at this time slice.

Following, the model creates a dictionary for each time series i, with a set of keys containing the instances that have shared the same community with i,

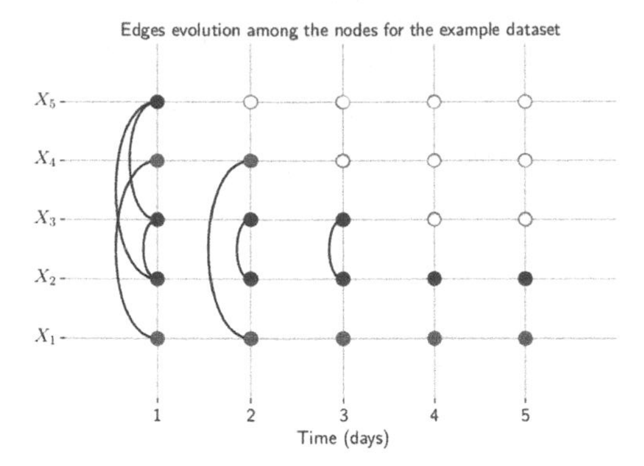

Fig. 1. Time slices showing the edges evolution in the temporal network, for the example dataset. Each row represents one time series and each column represents one time slice of the temporal network. The colors denote the community to which each node belongs, at each time slice. In this case, if a node has a white color, it means that this node is not in the temporal network at this time slice. (Color figure online)

and a set of values equal to the number of times this community sharing has occurred. Hence, in this case, we have that the dictionaries for X_3, X_4 and X_5 are: $D_3 = \{X_2 : 3, X_5 : 1\}, D_4 = \{X_1 : 2\}$ and $D_5 = \{X_2 : 1, X_3 : 1\}$, respectively. Therefore, according to Eq. 7, the variation predicted for X_5 at the time step $t = 2$, is given by: $(1\delta X_{2,2} + 1\delta X_{3,2})/(1 + 1)$. In Fig. 2, we show all predictions made by the model for the example dataset. As one may observe, the model is capable to correctly detect the evolution patterns between X_1 and X_4 and between X_2 and X_3, and to make use of these detected correlations for predicting future values for X_3 and X_4. In the case of X_5, which could either evolve as an arithmetic progression or as a geometric one, the model ends up correlating it to X_2, and therefore the predictions for X_5 follow a geometric progression.

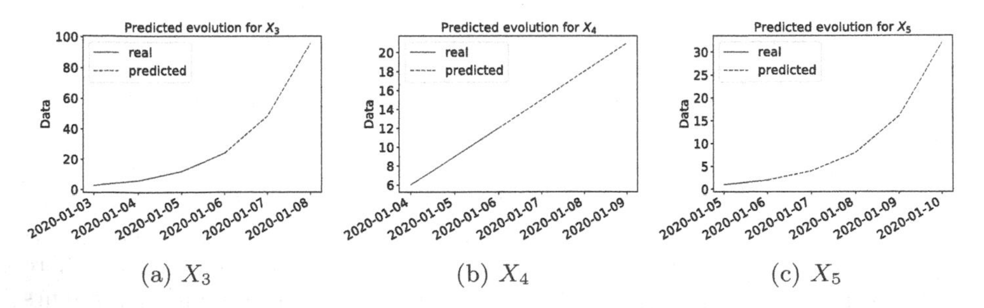

(a) X_3 (b) X_4 (c) X_5

Fig. 2. Predictions performed by the model for the time series (a) X_3, (b) X_4 and (c) X_5, from the example dataset. The blue line shows the series data provided in the dataset, while the blue dashed line indicates the predicted data. (Color figure online)

3 COVID-19 New Cases and Deaths Prediction Results

In this section, we present the obtained results when applying the proposed model to COVID-19 preliminary data, in order to predict the future number of confirmed cases and deaths in the 27 federal units of Brazil.

In order to apply the proposed time series prediction model on the COVID-19 datasets, we start by converting the confirmed cases and deaths daily variations in the database to weekly variations. In this manner, given that the first confirmed case of COVID-19 in Brazil dates from 02-25-2020, in the state of SP, and that the final date in the database is 05-26-2020, we end up with a maximum of 12 weeks for the COVID-19 time series, considering all federal units of Brazil. We predict the number of new confirmed cases and deaths for the next 7 days in each federal unit, for all units which presented at least 20 confirmed cases or deaths in the considered period. Additionally, we opt for not considering the time series from regions located in China from the database as predictors, since these time series may present later corrections in their data, which affects the prediction outputs. We set the value of the ϵ radius threshold parameter, used for generating the edges in the networks, as $\epsilon = Q(\delta X_t, .2)$, where $Q(A, n)$ stands for the n-th quantile of the array A. For detecting communities in the network, we use the *fast greedy community detection* algorithm.

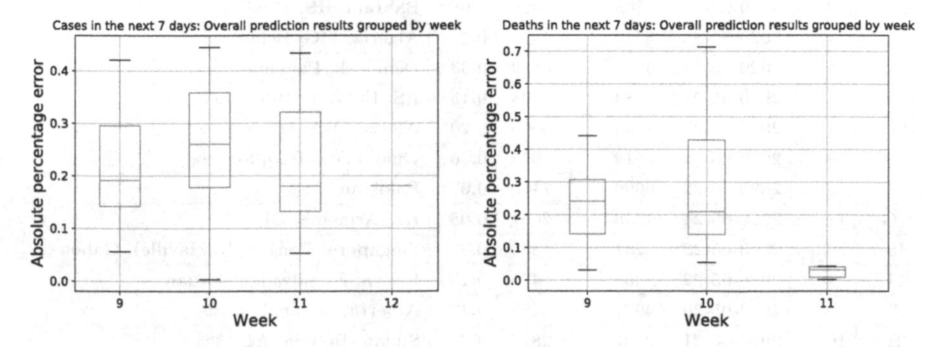

(a) Confirmed new cases in the next 7 days: overall prediction results.

(b) Confirmed new deaths in the next 7 days: overall prediction results.

Fig. 3. Boxplots of the overall prediction absolute percentage errors, grouped by week, for the (a) confirmed new cases in the next 7 days (51 predictions in total) and (b) confirmed new deaths in the next 7 days (25 predictions in total), for each of the 27 federal units of Brazil.

In Fig. 3, we show the boxplots of the absolute percentage errors for the obtained results when predicting the number of COVID-19 confirmed new cases and deaths in the next 7 days, for each of the 27 federal units of Brazil, from weeks 9 to 12. To avoid the inclusion of outliers, we did not consider in this figure predictions made for regions with a current total number of confirmed cases or deaths less than 50, in each week. After applying this filter, we end up with a total of 51 predictions of new confirmed cases and 25 predictions of new confirmed deaths included in the boxplots. The median and mean absolute percentage error for the new cases prediction results, in Fig. 3(a), are 21% and 24%, respectively. This mean absolute percentage error (MAPE) is smaller than the ones obtained by Al-qaness et al. [1], when forecasting the weekly new confirmed cases of influenza in China, which are of 32% and 45%, respectively, on each dataset analyzed. The accuracy obtained by Arora et al. [6], with a MAPE of around 6%, is higher than the one we obtained. However, it is worth mentioning that the testing data used in their study comprised only 4 Indian states and a 7-days time range, while in our case we consider 51 different weekly predictions, for more than 20 states. The median and mean absolute percentage error for the new deaths prediction results, in Fig. 3(b), are 16% and 23%, respectively.

Table 3. Confirmed cases in the next 7 days: most recent predictions performed by the model for the federal units of Brazil.

Label	Week	Date	Predicted	Real	Error	Labels included in the prediction
AC	10	2020-05-26	1121	2019	0.44	Oman, PB, Japan
AL	11	2020-05-24	2003	2398	0.16	Oman, Afghanistan, Kuwait
AM	9	2020-05-15	6256	7665	0.18	Dominican Republic
AP	9	2020-05-22	1865	2025	0.08	Bahrain, RS, Mexico
BA	11	2020-05-22	3840	4429	0.13	Algeria, MG, Belarus
CE	8	2020-05-11	4389	6559	0.33	Denmark, Panama
DF	11	2020-05-23	2384	2108	0.13	RS, Bahrain, Bulgaria
ES	11	2020-05-21	2465	3065	0.20	Algeria, BA, DF
GO	10	2020-05-21	514	695	0.26	Guatemala, Iraq, Senegal
MA	9	2020-05-22	6690	7175	0.07	Pakistan, PE
MG	11	2020-05-24	1948	2057	0.05	RS, Armenia, DF
MS	10	2020-05-23	251	350	0.28	Singapore, Congo (Brazzaville), Gabon
MT	9	2020-05-22	405	479	0.15	Kenya, El Salvador, Jordan
PA	9	2020-05-20	4971	8585	0.42	Alberta, Qatar, Belarus
PB	10	2020-05-21	2187	2877	0.24	Sudan, Bolivia, AC
PE	10	2020-05-21	5341	8323	0.36	Belarus, Pakistan, Qatar
PI	9	2020-05-21	1048	1169	0.10	PB, Sudan, Bolivia
PR	10	2020-05-21	444	764	0.42	Cuba, MG, Bosnia and Herzegovina
RJ	11	2020-05-21	7149	12622	0.43	Qatar, Belarus, Bangladesh
RN	10	2020-05-21	1084	1465	0.26	Cote dIvoire, DF, Bahrain
RO	9	2020-05-22	772	980	0.21	Afghanistan, Bahrain, PB
RR	9	2020-05-23	629	733	0.14	GO, El Salvador, Guatemala
RS	11	2020-05-26	1165	2987	0.61	DF, Bahrain, Armenia
SC	10	2020-05-21	1371	1278	0.07	Ghana, MG, Bahrain
SE	10	2020-05-23	2351	1996	0.18	Bolivia, AL, Afghanistan
SP	12	2020-05-19	14972	18276	0.18	Quebec, Mexico
TO	9	2020-05-20	766	947	0.19	Belarus, Gabon, PI

We believe the prediction accuracy is relatively higher for the number of new confirmed deaths for the reason that these numbers tend to be more reliable than the statistics regarding the new confirmed cases, since the later are subject to some difficulties, such as limited testing capabilities in each region and also the lack of symptoms on some individuals infected by the disease. One can also note, still in Fig. 3, that the boxplot for week 12 is smaller than the ones from previous weeks. This is due to the fact that, up to this date, only the state of SP in Brazil has been infected by COVID-19 for more than 11 weeks, hence this boxplot actually includes only the prediction of new cases performed for the state of SP.

In Tables 3 and 4, we present the most recent predictions performed by the model for the Brazilian federal units, for new confirmed cases and deaths in the next 7 days, respectively, along with their respective real values and absolute percentage errors. The last column in these tables shows the labels with the highest weights considered in each prediction, i.e., the regions in the database whose time series the model identified as being most correlated to the predicted time series and whose lengths are greater or equal to the predicted period. The states of MS, MT, RR and TO do not appear in Table 4 either because they presented less than 20 confirmed deaths caused by COVID-19 in the considered period or because their weekly deaths variations did not match any other in the period, and hence the model did not attempt to perform predictions of new confirmed deaths for these states.

One can note, in Tables 3 and 4, that the model is able to predict both smaller values, as for the new confirmed deaths in GO, PR and SC, and also larger values, such as the new confirmed deaths in SP and RJ. This is because the predictions are made in terms of variations, and not in terms of the actual numbers. It is interesting to analyze these two tables along with Fig. 4, in which we have examples of the correlations networks built by the model for weeks 5, 7 and 9 for the federal units of Brazil and their respective neighbors, i.e., their most correlated world regions in each week. The left column in Fig. 4 shows the correlations networks for new confirmed cases, while the right column shows the correlations networks for the new confirmed deaths. In this sense, one can note, for instance, that the states of RJ and SP presented similar new deaths variations, on week 5, and thus formed the community denoted by the aquamarine color in Fig. 4(b). On weeks 7 and 9, the new deaths variation in these same states did not match the variation from any other region, hence they both appear isolated in Figs. 4(d) and 4(f). In Table 4, the model took into account only the correlation detected in Fig. 4(b) for predicting the new deaths in these two states on week 11, and both predictions achieved a very high accuracy in this case, with an absolute percentage error of less than 1%.

Although the model is able to perform relatively accurate predictions for some regions, as it is shown in Tables 3 and 4 and in the overall performance shown in Fig. 3, it also has some limitations. The most recent predictions made for the state of PA, for instance, listed in the mentioned tables, obtained a high absolute percentage error, of 42% and 84%, for the prediction of new cases and

Table 4. Confirmed deaths in the next 7 days: most recent predictions performed by the model for the federal units of Brazil.

Label	Week	Date	Predicted	Real	Error	Labels included in the prediction
AC	10	2020-05-26	31	24	0.29	Afghanistan, Israel, Belarus
AL	11	2020-05-24	103	106	0.03	Israel, Belarus, BA
AM	10	2020-05-22	579	338	0.71	Peru, PE
AP	9	2020-05-22	41	54	0.24	MG, Luxembourg, PR
BA	11	2020-05-22	123	118	0.04	Egypt, MG, Israel
CE	10	2020-05-25	642	745	0.14	Peru
DF	11	2020-05-23	19	39	0.51	Afghanistan, MG, Belarus
ES	11	2020-05-21	118	114	0.04	South Africa, BA, Moldova
GO	10	2020-05-21	22	18	0.22	North Macedonia, Iraq, Bolivia
MA	9	2020-05-22	156	198	0.21	Romania, Austria
MG	11	2020-05-24	32	70	0.54	Bulgaria, Pakistan, Slovenia
PA	6	2020-04-29	18	113	0.84	PB, Bosnia and Herzegovina
PB	10	2020-05-21	59	85	0.31	Saudi Arabia, MG, Bosnia and Herzegovina
PE	10	2020-05-21	524	627	0.16	AM, Peru, Mexico
PI	9	2020-05-21	28	33	0.15	PR, Afghanistan, Saudi Arabia
PR	10	2020-05-21	17	22	0.23	Saudi Arabia, RS, Slovenia
RJ	11	2020-05-21	1168	1165	0.00	SP
RN	10	2020-05-21	44	59	0.25	Cuba, MG, PB
RO	9	2020-05-22	30	44	0.32	Afghanistan, Israel, Belarus
RS	11	2020-05-26	55	43	0.28	Saudi Arabia, US, Slovenia
SC	10	2020-05-21	16	20	0.20	PR, Tunisia, Saudi Arabia
SE	10	2020-05-23	22	33	0.33	MG, Bulgaria
SP	11	2020-05-12	1095	1098	0.00	RJ

deaths, respectively. This happens when the time series to be predicted presents a drastic shift in its curve, such that this shift does not match the variation from any other time series in the dataset, for that same time step. That is the case for the state of PA, which suffered from a suddenly rise in both of its COVID-19 curves, after the 5th week, causing it to appear isolated in Figs. 4(c), 4(d), 4(e), and 4(f), and which also prevented the model from predicting new deaths for this state after the 6th week, on 2020-04-29. This issue should be addressed in future versions of the model, by allowing it to adapt to such situations, in order to improve its prediction capabilities.

4 Final Remarks

In this work, we introduced a semi-supervised regression model which predicts future values for time series based on a correlations temporal network. The obtained results, by applying the model to predict the new confirmed cases and deaths related to COVID-19, in the 27 federal units of Brazil, are promising, with a mean absolute percentage error of 24%, for the new cases prediction, and a mean absolute percentage error of 23%, for the new deaths prediction. We

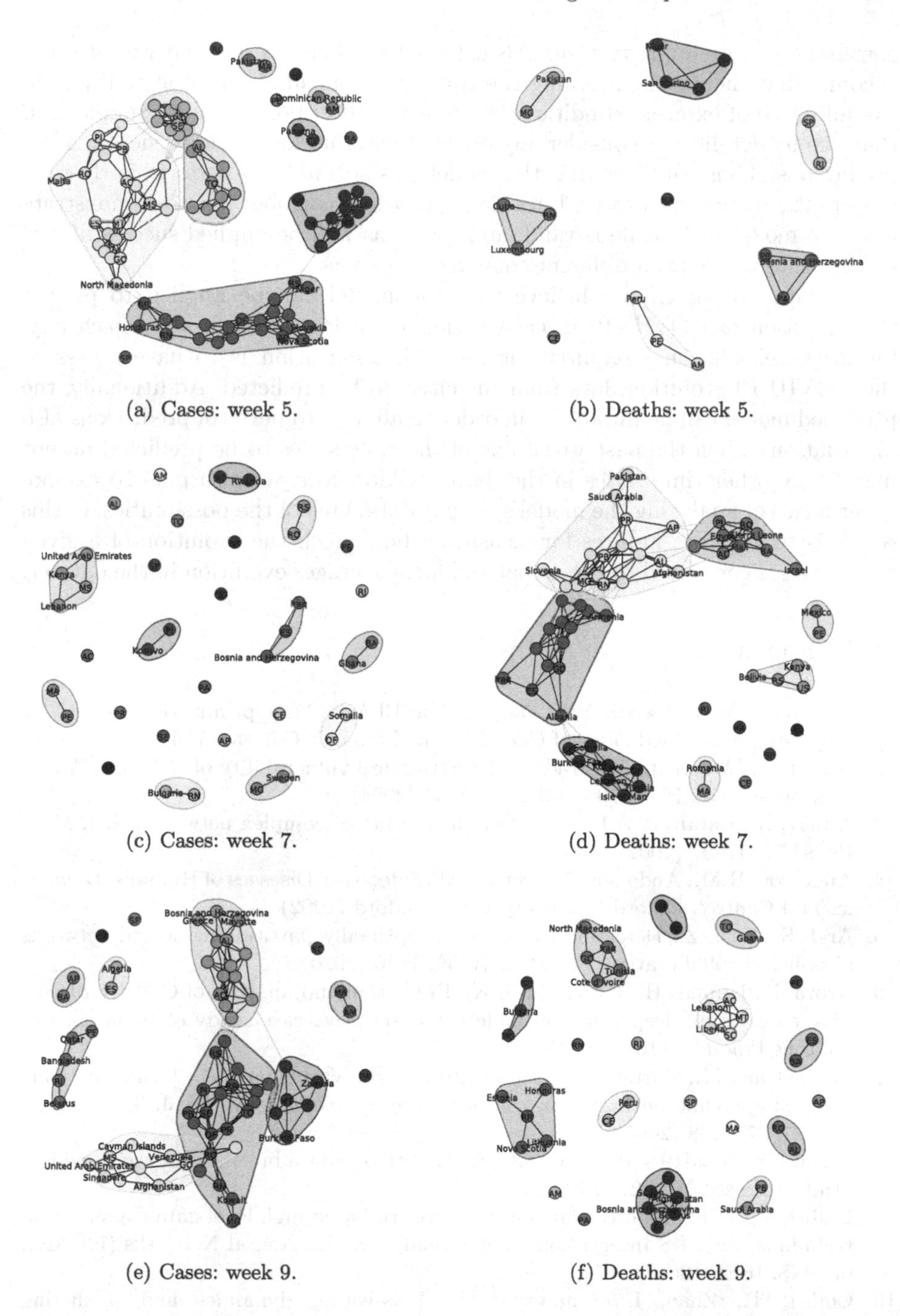

(a) Cases: week 5.

(b) Deaths: week 5.

(c) Cases: week 7.

(d) Deaths: week 7.

(e) Cases: week 9.

(f) Deaths: week 9.

Fig. 4. Examples of correlations networks formed by the federal units of Brazil and their respective neighbors which, in this case, represent their most correlated regions in each week, in terms of COVID-19 confirmed cases (left) and deaths (right) weekly variations. The colors denote the network communities. For the sake of visibility, not all labels are shown in these figures.

consider these preliminary results as satisfactory. Especially when we take into account that the data concerning the context of this application are subject to the influence of external conditions, such as isolation policies, for example, and that the model did not consider any of these external factors for generating the predictions. Thus, the fact that the model was still able to perform fairly well, despite the difficulties involved in this application, corroborates to demonstrate that the model's rationale is valid, and might as well be applied successfully to predict time series from different contexts and areas.

As future research, we believe that the model can be applied to predict the evolution of COVID-19 in each region more locally, such as in each city, for instance. The only requirement for such application is to have access to the COVID-19 evolution data from the cities to be predicted. Additionally, the proposed model can be improved, in order to allow it to perform predictions also in situations when the past variations of the time series to be predicted do not match any other time series in the dataset. Moreover, we also plan to explore other forms of analyzing the model's output data. One of the possibilities, in this sense, is to generate indexes for measuring how much the evolution of a given time series is correlated to the global and local averages evolution in the dataset.

References

1. Al-Qaness, M.A., Ewees, A.A., Fan, H., Abd El Aziz, M.: Optimization method for forecasting confirmed cases of COVID-19 in China. J. Clin. Med. **9**(3), 674 (2020)
2. Albert, R., Albert, I., Nakarado, G.L.: Structural vulnerability of the north American power grid. Phys. Rev. **69**(2), 025103 (2004)
3. Albert, R., Barabási, A.L.: Statistical mechanics of complex networks. Rev. Mod. Phys. **74**, 47–97 (2002)
4. Anderson, R.M., Anderson, B., May, R.M.: Infectious Diseases of Humans: Dynamics and Control. Oxford University Press, Oxford (1992)
5. Aref, S., Neal, Z.: Detecting coalitions by optimally partitioning signed networks of political collaboration. Sci. Rep. **10**(1), 1–10 (2020)
6. Arora, P., Kumar, H., Panigrahi, B.K.: Prediction and analysis of COVID-19 positive cases using deep learning models: a descriptive case study of India. Chaos, Solitons Fractals, 110017 (2020)
7. Barthélemy, M., Barrat, A., Pastor-Satorras, R., Vespignani, A.: Dynamical patterns of epidemic outbreaks in complex heterogeneous networks. J. Theor. Biol. **235**(2), 275–288 (2005)
8. Brasil.IO: Covid19 - dataset - Brasil.IO. https://data.brasil.io/dataset/covid19.html. Accessed May 27, 2020
9. Colliri, T., Ji, D., Pan, H., Zhao, L.: A network-based high level data classification technique. In: 2018 International Joint Conference on Neural Networks (IJCNN), pp. 1–8. IEEE (2018)
10. Colliri, T., Zhao, L.: Analyzing the bills-voting dynamics and predicting corruption-convictions among Brazilian congressmen through temporal networks. Sci. Rep. **9**(1), 1–11 (2019)
11. da F. Costa, L., et al.: Analyzing and modeling real-world phenomena with complex networks: a survey of applications. Adv. Phys. **60**(3), 329–412 (2011)

12. De Domenico, M., Solé-Ribalta, A., Cozzo, E., Kivelä, M., Moreno, Y., Porter, M.A., Gómez, S., Arenas, A.: Mathematical formulation of multilayer networks. Phys. Rev. X **3**(4), 041022 (2013)
13. Dezső, Z., Barabási, A.L.: Halting viruses in scale-free networks. Phys. Rev. E **65**(5), 055103 (2002)
14. Dong, E., Du, H., Gardner, L.: An interactive web-based dashboard to track COVID-19 in real time. Lancet Infect. Dis. (2020)
15. Faloutsos, M., Faloutsos, P., Faloutsos, C.: On power-law relationships of the internet topology. ACM SIGCOMM Comput. Commun. Rev. **29**(4) (1999)
16. Gao, X., et al.: Transmission of linear regression patterns between time series: from relationship in time series to complex networks. Phys. Rev. E **90**(1), 012818 (2014)
17. Harari, Y.N.: Sapiens: A Brief History of Humankind. Random House (2014)
18. Kivelä, M., Arenas, A., Barthelemy, M., Gleeson, J.P., Moreno, Y., Porter, M.A.: Multilayer networks. J. Complex Netw. **2**(3), 203–271 (2014)
19. Kostopoulos, G., Karlos, S., Kotsiantis, S., Ragos, O.: Semi-supervised regression: a recent review. J. Intell. Fuzzy Syst. **35**(2), 1483–1500 (2018)
20. Luna-Pla, I., Nicolás-Carlock, J.R.: Corruption and complexity: a scientific framework for the analysis of corruption networks. Appl. Netw. Sci. **5**(1), 1–18 (2020). https://doi.org/10.1007/s41109-020-00258-2
21. Newman, M.E.: Fast algorithm for detecting community structure in networks. Phys. Rev. E **69**(6), 066133 (2004)
22. Pastor-Satorras, R., Vespignani, A.: Epidemic spreading in scale-free networks. Phys. Rev. Lett. **86**(14), 3200 (2001)
23. Pastor-Satorras, R., Vespignani, A.: Immunization of complex networks. Phys. Rev. E **65**(3), 036104 (2002)
24. Pons, P., Latapy, M.: Computing communities in large networks using random walks. In: Yolum, I., Güngör, T., Gürgen, F., Özturan, C. (eds.) ISCIS 2005. LNCS, vol. 3733, pp. 284–293. Springer, Heidelberg (2005). https://doi.org/10.1007/11569596_31
25. Silva, T.C., Zhao, L.: Network-based high level data classification. IEEE Trans. Neural Netw. Learn. Syst. **23**(6), 954–970 (2012)
26. Silva, T.C., Zhao, L.: Stochastic competitive learning in complex networks. IEEE Trans. Neural Netw. Learn. Syst. **23**(3), 385–398 (2012)
27. Sporns, O.: Network analysis, complexity, and brain function. Complexity **8**(1), 56–60 (2002)
28. Sun, X., Tan, Y., Wu, Q., Chen, B., Shen, C.: TM-Miner: TFS-based algorithm for mining temporal motifs in large temporal network. IEEE Access **7**, 49778–49789 (2019)
29. Tamara, D., Kristijan, P., Ljupcho, K.: Graphlets in multiplex networks. Sci. Rep. **10**(1) (2020)
30. Thompson, W.H., Brantefors, P., Fransson, P.: From static to temporal network theory: applications to functional brain connectivity. Netw. Neurosci. **1**(2), 69–99 (2017). https://doi.org/10.1162/NETN_a_00011
31. West, G.B., Brown, J.H., Enquist, B.J.: A general model for the structure, and allometry of plant vascular systems. Nature **400**, 125–126 (2009)
32. Xubo, G., Qiusheng, Z., Vega-Oliveros, D.A., Leandro, A., Zhao, L.: Temporal network pattern identification by community modelling. Sci. Rep. **10**(1) (2020)
33. Zhang, Y., Li, X., Xu, J., Vasilakos, A.V.: Human interactive patterns in temporal networks. IEEE Trans. Syst. Man Cybern. Syst. **45**(2), 214–222 (2015)

Decoding Machine Learning Benchmarks

Lucas F. F. Cardoso[1(✉)], Vitor C. A. Santos[1], Regiane S. Kawasaki Francês[1], Ricardo B. C. Prudêncio[2], and Ronnie C. O. Alves[3]

[1] Faculdade de Computação, Universidade Federal do Pará, Belém, Brazil
lucas.cardoso@icen.ufpa.br, vitor.cirilo3@gmail.com, kawasaki@ufpa.br
[2] Centro de Informática, Universidade Federal de Pernambuco, Recife, Brazil
rbcp@cin.ufpe.br
[3] Instituto Tecnológico Vale, Belém, Brazil
ronnie.alves@itv.org

Abstract. Despite the availability of benchmark machine learning (ML) repositories (e.g., UCI, OpenML), there is no standard evaluation strategy yet capable of pointing out which is the best set of datasets to serve as gold standard to test different ML algorithms. In recent studies, Item Response Theory (IRT) has emerged as a new approach to elucidate what should be a good ML benchmark. This work applied IRT to explore the well-known OpenML-CC18 benchmark to identify how suitable it is on the evaluation of classifiers. Several classifiers ranging from classical to ensembles ones were evaluated using IRT models, which could simultaneously estimate dataset difficulty and classifiers' ability. The Glicko-2 rating system was applied on the top of IRT to summarize the innate ability and aptitude of classifiers. It was observed that not all datasets from OpenML-CC18 are really useful to evaluate classifiers. Most datasets evaluated in this work (84%) contain easy instances in general (e.g., around 10% of difficult instances only). Also, 80% of the instances in half of this benchmark are very discriminating ones, which can be of great use for pairwise algorithm comparison, but not useful to push classifiers abilities. This paper presents this new evaluation methodology based on IRT as well as the tool decodIRT, developed to guide IRT estimation over ML benchmarks.

Keywords: IRT · Machine learning · Benchmarking · OpenML · Classification

1 Introduction

Machine learning (ML) is a field of artificial intelligence which has been rapidly growing in recent years, partly due to the diversity of applications in different areas of knowledge. There are different types of machine learning algorithms, from unsupervised to supervised ones [3]. In this work, we focus on supervised algorithms, and more specifically on classification algorithms, which are commonly adopted in many applications [1].

R. Cerri and R. C. Prati (Eds.): BRACIS 2020, LNAI 12320, pp. 412–425, 2020.
https://doi.org/10.1007/978-3-030-61380-8_28

Empirically evaluating ML algorithms is crucial for assessing the advantages and limitations of the available techniques. Algorithm evaluation is usually performed by deploying datasets available in repositories [2]. Remarkably, the OpenML repository has been developed as an online platform, allowing researchers to share results, ML strategies and datasets used in benchmarking experiments. Such platform improves reproducibility and minimizes double effort to do the same experiment [2]. Another contribution is the OpenML Curated Classification (OpenML-CC18), a benchmark for classification that has 72 standardized and cured datasets [7].

By relying on benchmark datasets, ML models can be trained and tested using a chosen experimental methodology (e.g., cross-validation) and the performance measures of interest (e.g., success rate). For increasing robustness, different datasets can be adopted in the experiments and the performance measures are averaged across the datasets. This approach however does not allow a deeper analysis of classifier's capacity, since the type of dataset adopted in an experiment may bias the obtained results. For instance, the performance of a ML algorithm can be overestimated if only easy datasets are adopted in the experiments. Also, depending on the datasets, certain ML algorithms may be favored, thus giving the false impression that one classifier is the best in relation to the others [4].

Given the above context, it is important that the performance analysis of a classifier takes into account the complexity of the datasets as well. Previous work [4–6] has addressed this issue by adopting concepts from Item Response Theory (IRT) to provide a more robust approach to devise benchmarks. IRT is commonly used in psychometric tests to assess the performance of individuals on a set of items (e.g., questions) with different levels of difficulty. IRT has been extended to ML evaluation by treating classifiers as individuals and test instances as items. Thus ML algorithms can be ranked by IRT according to their ability to correctly respond the most difficult instances.

The main contributions of this work are: I) an empirical evaluation of IRT to evaluate ML benchmarks, adopting the OpenML-CC18 benchmark as a case study; II) a global assessment of ML algorithms in benchmarks comparison, improving the identification of the strengths and weaknesses among them on the basis of IRT estimators; and III) the decodIRT tool, developed to guide the IRT estimation over ML benchmarks automatically along with a robust rating system to establish proper boundaries between ML algorithms. Regarding the third contribution, rating systems are widely used in other tasks (e.g., chess playing) to indicate the strength of an individual in a competition [8]. Among the existing systems, the Glicko-2 [9] system was used in order to create a ranking that is able to summarize the algorithms' results generated by the IRT. Thus it pointed out the most suitable ML algorithms for the classification tasks proposed by the OpenML-CC18 benchmark.

The remaining of this paper is organized as follows: Sect. 2 contextualizes the main subjects explored in this work, specifically w.r.t. OpenML, Item Response Theory and the Glicko-2 system. Section 3 presents the methodology used, explains how decodIRT tool works and how the Glicko-2 system can be applied to summarize IRT results. Section 4 presents an overall discussion. Section 5 concludes the article and future developments.

2 Background

2.1 OpenML

OpenML is a repository in which machine learning researchers can share data sets, descriptions of experiments and obtained results in as much detail as possible. Hence, a better organization and use of such information is allowed, thus creating a collaborative environment for sharing experiments in a global scale [2]. In OpenML anyone can download a dataset, execute a machine learning method of their choice and share the results obtained with other users, generating discussions and new information about the data sets and the applied algorithms.

In addition, the platform also provides several sets of reference datasets. OpenML-CC18[1] is one of those reference sets that includes 72 existing datasets in OpenML from mid-2018 and that meet several requirements in order to compile a complete reference set. OpenML-CC18 also includes datasets frequently used in benchmarks published in recent years [7]. Several associated metadata is available such as: number of instances, number of features, number of classes and proportion between the minority and majority class of the group.

OpenML actually has a lot to contribute to research in the field of machine learning. The current work aims to apply IRT to evaluate this gold standard to provide greater strength for OpenML-CC18 and ML community. The analysis of the datasets within the IRT perspective allows adding to the reference set new relevant metadata, such as the difficulty of the datasets and discriminatory potential of the data. Such information can be very useful for choosing a benchmarking set and will be better discussed in the following sections.

2.2 Item Response Theory

Andrade, Tavares and Valle (2000) [10] describe IRT as a set of mathematical models that aim to represent the probability that an individual will correctly answer an item according to the item's parameters and the respondent's ability, so that the greater the ability, the greater the chance of a correct response.

The item can be classified according to its associated response. It can be dichotomous, when the responses are just right or wrong, or non-dichotomous, if there are more than two possible answers. Logistical models for dichotomous items are the most used in literature and practice. There are basically three types of models, which differ by the number of item parameters that are estimated. In the 3-parameter logistic model, known as 3PL, the probability of an individual j correctly respond an item i given his ability is defined by the following equation:

[1] Link to access OpenML-CC18: https://www.openml.org/s/99.

$$P(U_{ij} = 1|\theta_j) = c_i + (1 - c_i)\frac{1}{1 + e^{-a_i(\theta_j - b_i)}} \quad (1)$$

where:

- U_{ij} is the dichotomous response that can take the values 1 or 0, being 1 when the individual j hits the item i and 0 when he misses;
- θ_j is the ability of the individual j;
- b_i is the item's difficulty parameter and indicates the location of the logistic curve;
- a_i is the item's discrimination parameter, i.e., how much the item i differentiates between good and bad respondents. This parameter indicates the slope of the logistic curve. The higher its value, the more discriminating the item is;
- c_i is the guessing parameter, representing the probability of a casual hit. It is the probability that a respondent with low ability hits the item.

The 2PL can be derived by simplifying the above model and dropping the guessing parameter, i.e., $c_i = 0$. Finally, the 1PL parameter can be defined dropping the discrimination parameter, i.e., assuming $a_i = 1$.

Although it is possible to obtain negative values of discrimination, they are not expected by IRT. The negative values are not expected because it means that the probability of a hit is higher for individuals with lower ability values, rather than higher than what is normally expected [10].

The logistic curves are estimated from responses collected for a group of items and respondents. According to Martínez-Plumed et al. (2016) [4], there are three possible situations for estimation. In the first situation, the parameters of the items are known, but the ability of the respondents is not known. The second possible situation is just the opposite: the ability of each respondent is defined, but the parameters of the items are not known. In the third case, which is also the most common, both the parameters of the items and the abilities of the respondents are unknown. This work is in the third case and for this situation, the following two-step interactive method proposed by Birnbaum is used [12] for estimation from the group of responses:

- At first, the parameters of each item are calculated with only the answers of each individual. The initial values of respondent's ability can be the number of correct answers obtained. In the case of classifiers, this work used the obtained accuracy as the initial ability;
- Once the parameters of the items are obtained, the ability of the individuals can be estimated. For both item's parameters and respondent's ability, simple estimation techniques can be used, such as the Maximum Likelihood Estimate [4].

IRT is usually applied for educational purposes, in which respondents are students and items are questions. Recently, IRT has been extended to AI, and more specifically to ML, in which respondents are assumed to be AI techniques and items are AI tasks [4–6]. To analyze datasets and learning algorithms through

IRT. In this work, it was used the instances of a dataset as the items and the classifiers were assumed as the respondents. Among the existing IRT models, the 3PL logistic model was used because it is the most complete. The IRT parameters were used to assess the datasets directly, informing the percentage of instances with high difficulty, with great discriminative power and with high chance of casual hit. This allows to have an insight into the complexity of the datasets evaluated and how different classifiers behave in the face of the challenge of classifying different datasets.

2.3 Glicko-2 System

The Glicko-2 system is an extension of Glicko, originally developed by Mark E. Glickman [9] to measure the "strength" of chess players. Today, the system is used worldwide by several organizations such as the Australian Chess Federation [11].

In the Glicko-2 system, each player has three variables used to measure his statistical strength: the R rating, the RD rating deviation and the σ volatility. The R rating is the numerical value itself that measures the strength of a player. As it cannot be said that this value accurately represents the strength of an individual, the RD is used to indicate the reliability of an R rating. The Glicko-2 system estimates the players' strength in a 95% confidence interval, as follows: $[R - 2RD, R + 2RD]$. Thus, there is a 95% chance of the real player's strength is within the calculated range. The lower a player's RD, the more accurate his rating value is. In addition, the σ volatility makes it possible to measure the degree of expected rating fluctuation, i.e., the higher the volatility value, the greater the chances of a player's rating fluctuating within his RD range. Therefore, the lower the value of σ, the more reliable and expected a player's rating is. For example, among players with low volatility, it is possible to state more precisely who is the strongest based on their ratings [11].

To estimate the rating values of the players, the Glicko-2 system uses the so-called classification period. This period is a sequence of games played by the individual in question. At the end of this sequence, the Glicko-2 system can update the player's parameters. To do this, the Glicko-2 system uses the opponents' rating and RD values, along with the score of the game result (e.g., 0 points for defeat and 1 points for victory). If there is no previous player data, the Glicko system uses default initial values for new players, 1500 for rating, 350 for RD and 0.06 for σ [9].

3 Materials and Methods

In order to build the IRT models and analyze ML benchmarks, it was developed a tool called decodIRT[2]. The tool was written mostly in Python together with R language features. The decodIRT has the main objective of assisting the analysis of existing datasets on the OpenML platform as well as the proficiency of

[2] Link to the source code: https://github.com/LucasFerraroCardoso/IRT_OpenML.

different classifiers. For this, it relies on the probability of correctness derived by the IRT model as well as the estimated items' parameters and respondents' ability.

As it can be seen in Fig. 1, the decodIRT tool consists of three scripts designed to be used in sequence. The first script is in charge of downloading the OpenML datasets and generating several ML models and putting them to classify the datasets. It will generate a response matrix, which contains the result of the classification of each test instance. The response matrix will be the input for the second script, which in turn is in charge of calculating the item parameters. The last script will use the data generated by the previous scripts to rank the datasets using the item parameters and use the IRT to analyze the ML models.

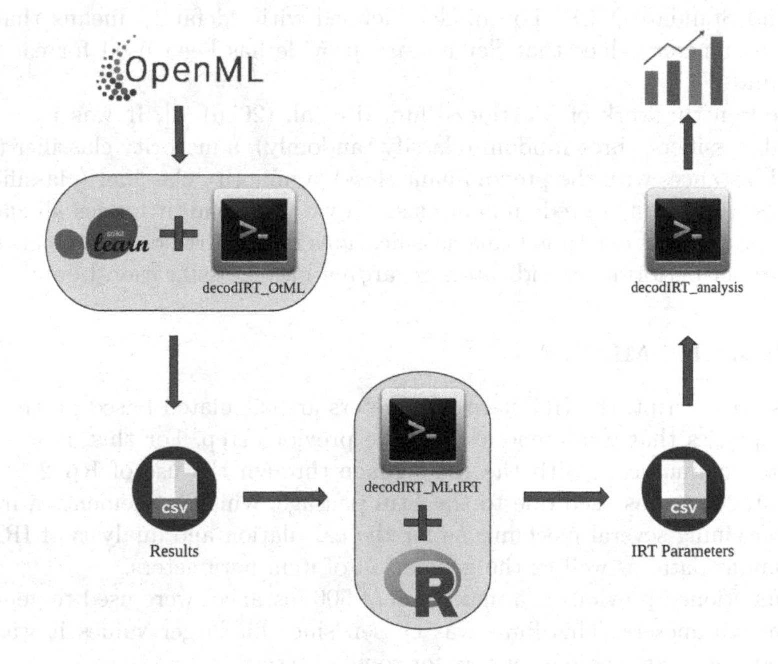

Fig. 1. Flowchart of the decodIRT execution.

3.1 decodIRT_OtML

The first script will download the existing OpenML datasets and execute several classifiers implemented in the Scikit-learn [13] package. These classifiers will generate the response data that will be used to estimate the item parameters. After downloading the dataset, the stratified split for training and testing is performed, to maintain the proportions of the classes. By default, the 70/30 split is used. However, for very large datsets the split is adapted to leave a maximum of 500 instances for testing. Martínez-Plumed et al. (2019) [6] explain that for a very large number of items, packages that estimate IRT values may be stuck at the local minimum or may not converge. Therefore, for very large datasets it is recommended to use less than a thousand instances to generate the parameters.

In order to generate diverse response values, at first, 120 models of Neural Networks (MLP) are executed, varying only the depth of the networks in which the depth starts at 1 and gradually increases to 120. Each of the MLPs is trained using the 10-fold cross-validation.

After that, classifications are performed with a second set of classifiers that will be evaluated by the tool, which are: Naive Bayes Gaussian standard, Naive Bayes Bernoulli standard, KNN of 2 neighbors, KNN of 3 neighbors, KNN of 5 neighbors, KNN of 8 neighbors, Standard Decision Trees, Random Forests (RF) with 3 trees, Random Forests with 5 trees, Standard Random Forests, Standard SVM and Standard MLP. The models defined with "default" means that the default parameter values that Scikit-learn provide has been used for each ML algorithm.

Based on the work of Martínez-Plumed et al. (2016) [4], It was inserted 7 artificial classifiers, three random (classify randomly), a majority classifier (classifies all instances with the predominant class), a minority classifier (classifies all instances with the non-predominant class), a very bad one (it misses all classifications) and a great one (it gets all classifications right). Artificial classifiers serve to set proper performance indicators regarding ideal classification boundaries.

3.2 decodIRT_MLtIRT

In the second script, the IRT item parameters are calculated based on the classifier responses that were generated in the previous step. For this, it was necessary to communicate with the R language through the use of Rpy2 library. The R language was used due to the Ltm package, which implements a framework containing several mechanisms for the calculation and analysis of IRT for dichotomous data, as well as the generation of item parameters.

As mentioned previously, a maximum of 500 instances were used to generate the item parameters. This limit was chosen since for larger values it was not possible to generate the parameters for some datasets.

3.3 decodIRT_analysis

The last step is in charge of doing the analysis on the data generated by the previous steps. By definition, the code will generate in all executions a ranking of the datasets on the percentage of test instances, whose parameter value exceeds the defined limit. For example, 50% of instances have a difficulty value above 1.

If the user does not define any limit for the parameters, standard limit values based on Adedoyin, Mokobi et al. (2013) [14] are used. In Adedoyin's work it is said that to consider an item (instance) as being difficult its value of Difficulty should normally be higher than 1. Items with high discriminative capacity should normally have Discrimination values greater than 0.75, and items with high guessing values are normally greater than 0.2. The analysis of the percentages of each item parameter is one of the interests of this work.

Following the Birnbaum method, to calculate the probability of success for each classifier, the proficiency of each classifier is estimated first. To do so, Python's Catsim package is used, which implements both functions to calculate the probability of success, as well as functions to estimate proficiency using the responses of each classifier and the item parameters.

In order to analyze more generally how the different classifiers perform on the datasets. The concept of True-Score [15] was also implemented. The True-Score is the sum of all the probabilities of a student's (classifier) correct answer to the questions of a test. The purpose of using True-Score is to calculate a final grade, just as we get when we take a test. Thus, it is expected to identify the "aptitude" that the different classifiers have before being presented a dataset. A conceptual view of the innate ability, engineering of the model, through its default hyperparameters settings, it will be put to the test by the Glicko-2 system.

3.4 Ranking of Classifiers by the Glicko-2 System

Glicko-2 [9] has been used to estimate the strength of classifiers and to generate a global performance assessment by simulating a competition among all ML methods. The simulation explores a round-robin tournament, thus all classifiers face each other in the championship and at the end there is a ranking that will reflect how each ML algorithm performed over the competition.

Each dataset is seen as the championship phases and is used as a classification period, so the classifiers face each other in each dataset. The True-Score values obtained by each classifier are used to decide the winner of the match and this is done as follows: if the score is higher than that of your opponent, it is counted as victory, if it is lower it is counted as defeat and if it is the same it is given as a tie. For that, the scoring system used in chess competitions is used, with 1 point for victory, 0 for defeat and 0.5 for a tie. Thus, at the end of each dataset, all classifiers will have their rating, RD and volatility values. These new values are used as the initial values for the next dataset. Once all datasets are finalized, the rating values generated are used to create the ranking of classifiers.

4 Results and Discussion

One of the main objectives of this work is to use the IRT item parameters to give more reliability to the use of the OpenML-CC18 benchmark. The item parameters explored are those of difficulty and discrimination. Although there are 72 datasets available in OpenML-CC18, only 60 were evaluated by decodIRT. This is due to two reasons: (1) the size of datasets, being 11 too large and require a long time for the execution of all ML models; (2) It was not possible to generate the item parameters for the "Pc4" dataset. The latter is still an open issue. Thus, the OpenML-CC18 benchmark evaluated in this work corresponds to 83.34 % of the total reference set. These datasets are then used to performance analysis[3].

4.1 Decoding OpenML-CC18 Benchmark

It was possible to observe that there is a inverse relationship between difficulty and discrimination parameters. The rankings highlights that the most difficult datasets are also the least discriminative and vice versa (Fig. 2). Thus, the difficulty posed by these datasets, despite make them more challenging, they are less suitable to differentiate classifiers. On the other hand, the most discriminatory datasets are not suitable for testing classifiers abilities.

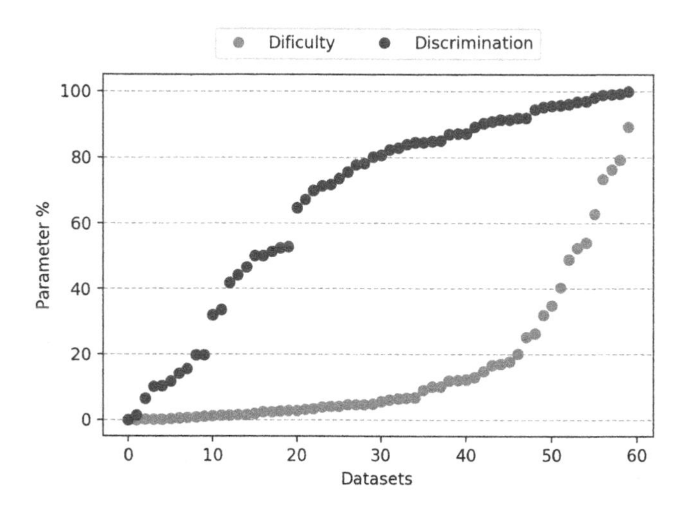

Fig. 2. It shows the percentages of difficult and very discriminative instances arranged in ascending order. There is a certain percentage of discrimination and a percentage of difficulty that are in the same position on the X axis do not necessarily correspond to the same dataset. "tic-tac-toe", "credit-approval" and "optdigits" are respectively the datsets with the most difficult instances. While "banknote-authentication", "analcatdata_authorship" and "texture" are the most discriminative.

Among the analyzed datasets, 49 have less than 27% of instances considered difficult, while only 7 have more than 50% difficult. This means that of the 60 datasets observed, 81.67% of the total have more than 70% of their instances considered easy and only 11.67% are difficult. Therefore, the OpenML-CC18 benchmark, for model comparison purposes, should be adopted with caution. Figure 2 illustrates graphically the classification challenge presented in this benchmark. It is clear that the situation is reversed when compared to the difficulty (blue line) values. In which only 25% of datasets have less than 50% of their instances

[3] All classification results can be obtained at https://github.com/LucasFerraroCardoso/IRT_OpenML/tree/master/benchmarking.

considered to be less discriminatory and 31 of the 60 datasets have at least 80% of very discriminative instances. From this observation one can infer that although the OpenML-CC18 is not considered as difficult as it was expected, it does have appropriate datasets to differentiate good and bad classifiers. In addition, knowing which datasets are more difficult, allows the user to choose his benchmark set more specifically, in case he wants to test the classification power of his algorithm, disregarding the need to test with all datasets. In summary, focusing on datasets that present instances with a higher positive discrimination values.

4.2 Classifiers Performance on OpenML-CC18

Taking into account only the values of True-Score[4] of the classifiers for each dataset, it possible to see a pattern in the overall performance of the classifiers. As expected, the optimal classifier has the highest score and the other artificial classifiers, the worst results. In addition, there are cases where the real classifiers match the score of the optimum and even surpass it. However, in some specific cases the score is reversed and the best classifiers have the worst scores (see Fig. 3).

The score of the optimal classifier be surpassed and the last case mentioned above can happen due to the occurrence of many instances having negative values of discrimination. Therefore, it can be inferred that datasets having many negative instances may not be good for creating benchmarks. A future work would be to carefully analyze how the characteristics of a dataset can cause such situations, in addition to observing how these characteristics can positively or negatively influence the performance of the classifiers. OpenML already has an extensive set of metadata which can be used for this purpose.

Although the True-Score values generated already allow observing the performance of the classifiers, there is still a lot of data to be evaluated individually, towards a natural assessment of the innate ability of these learning algorithms. Thus, the Glicko-2 rating system was used to summarize this information. Table 1 shows the final rating ranking that was obtained.

The positions of the artificial classifiers in the ranking were as expected. The excellent and very bad classifiers assumed the first and last positions, respectively, while the other artificial classifiers performed less than all real classifiers. Among the real classifiers, MLP was the winner, being slightly better than RF classifiers.

When looking at the position of the real classifiers, it is surprising how the MLP got very close to the rating of the optimal classifier, while the other classifiers were more than 100 points below the first place. MLP's "strength" can also be confirmed by looking at the volatility values of the classifiers, in which all have low values. Such values also help to give more confidence with respect to the position of each classifier in the ranking and allow us to infer that the rating

[4] All data generated can be accessed at https://github.com/LucasFerraroCardoso/IRT_OpenML/tree/master/BRACIS.

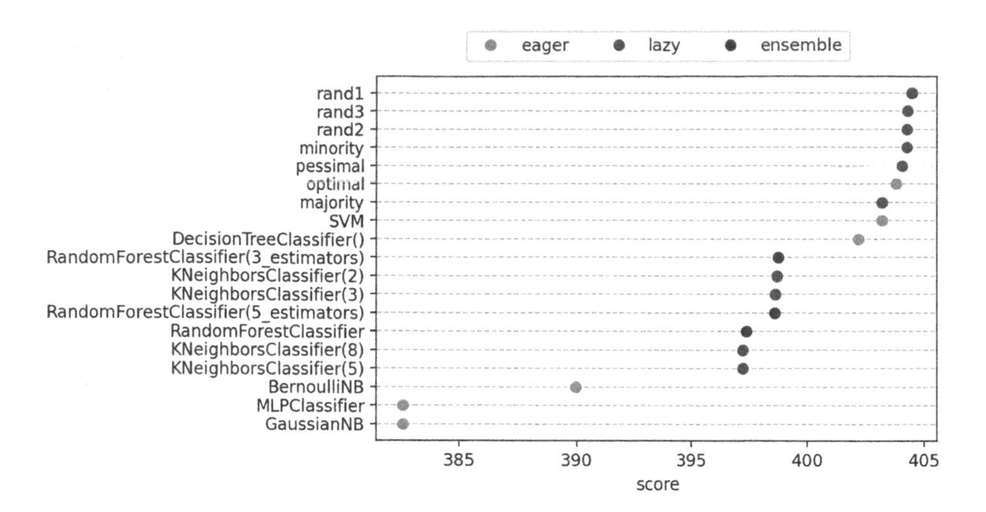

Fig. 3. The True-Score values obtained for the "jm1" dataset.

Table 1. Classifier rating ranking

Classifier	*Rating*	*RD*	*Volatility*
optimal	1732.56	33.25	0.0603
MLPClassifier	1718.65	31.20	0.0617
RandomForestClassifier	1626.60	30.33	0.0606
RandomForestClassifier(5_estimators)	1606.69	30.16	0.0621
RandomForestClassifier(3_estimators)	1575.26	30.41	0.0646
DecisionTreeClassifier()	1571.46	31.16	0.0674
SVM	1569.48	32.76	0.0772
KNeighborsClassifier(3)	1554.15	30.74	0.0646
GaussianNB	1530.86	31.25	0.0683
KNeighborsClassifier(2)	1528.41	30.40	0.0638
KNeighborsClassifier(5)	1526.10	30.27	0.0630
BernoulliNB	1494.87	32.64	0.0770
KNeighborsClassifier(8)	1457.78	30.25	0.0638
minority	1423.01	30.66	0.0631
rand2	1374.78	30.27	0.0605
rand3	1337.27	30.95	0.0600
rand1	1326.38	31.42	0.0610
majority	1301.08	31.74	0.0666
pessimal	1270.46	31.74	0.0603

values in fact represent the "strength" of the classifiers with precision. Considering the rating fluctuation of each classifier within its respective RD range, the final positions may change. The MLP, for example, if it had its "strength" at the lowest value by your RD range, would have a new rating of 1656.25. This would allow classifiers who are 3rd and 4th to overtake if their rating values fluctuate upwards. However, from the 5th position, no classifier can reach even if its rating fluctuates to the maximum of its range.

It can be observed that there are groups of classifiers that have equivalent "strength", in which it is not possible to define, with high confidence, which one of the them present highest aptitude in OpenML-CC18 challenge. This "strength" is translated into the aptitude of these learning algorithms. Since tests were performed with several datasets presenting distinct IRT estimators, and classifiers have same configuration model settings, it is assumed that the all classification results indeed reflects the learning algorithm aptitude, by design. It is important to note that optimization can have an effect of fine-tuning the decision boundaries in the most difficult datasets, and consequently a better performance. On the other hand, it would not allow the assessment of innate, engineering skills of these classifiers. A future work could stress how far one can go in ML optimization, extrapolating the boundaries between ability and aptitude.

In order to bring more credibility to the generated rating values, the Friedman [16] test was carried out. The objective is to identify whether the rating values, in fact, allow to differentiate the "innate ability" of the classifiers. The Friedman test was performed using the rating values of the real classifiers. The execution of the Friedman Test resulted in a p-value of approximately 9.36×10^{-80}. Subsequently, the Nemenyi [17] test was applied to compare and identify which of the distributions differ from each other. Figure 4 presents a Heatmap of the Nemenyi test.

When analyzing the Heatmap, one can observe that the three classifiers having the highest ratings have also high p values, so they do not differ from each other. And even though they are the more performant, they all have high values for at least one different classifier. The other classifiers also have performance similarities, even though they are from different families in some cases. Therefore, it is not evident that there is a clear separation of classifiers into different groups, since some classifiers that are at the top of the ranking do not differ from all classifiers that are in lower positions. This leads us to believe that, although the Friedman Test points out that there are different ML groups, these differences are not statistically significant enough to identify the highest skilled one.

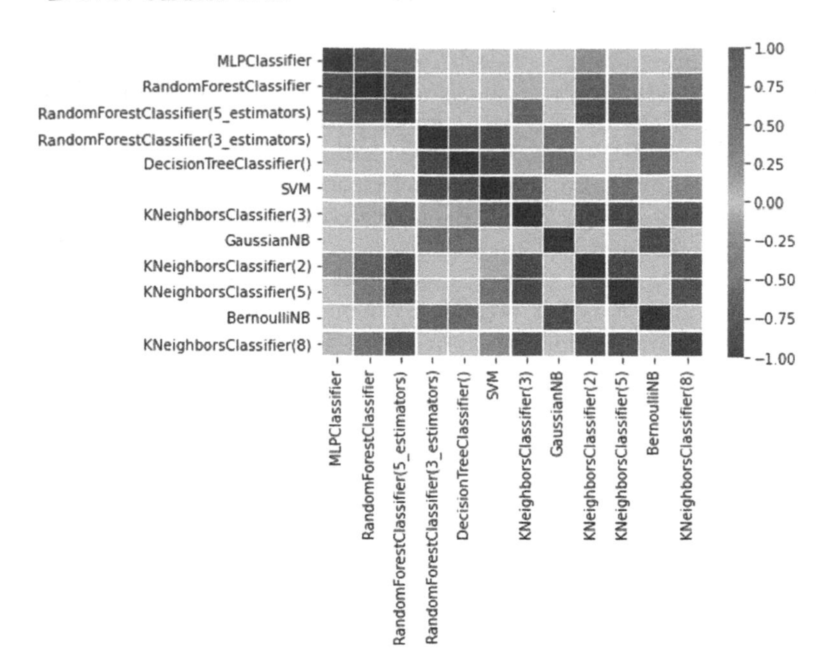

Fig. 4. Heatmap generated from the Nemenyi Test, using only the rating distributions of the real classifiers.

5 Final Considerations

This work explores IRT estimation for the evaluation ML benchmarks. ML benchmarks are commonly used to explore how far ML algorithms can go while handling datasets. The OpenML-CC18 is a gold standard. Although the ML community make broad use of it, such utilization might be taken with caution. From the 60 datasets evaluated in this work, only 12% of their instances are considered difficult; 80% of the instances in half of this benchmark are very discriminating ones which can be of great source for comparisons analysis, but not useful to push classifiers abilities. The benchmark assessment methodology is provided and it can be reproduced using the decodIRT tool. Even though classifier abilities are highlighted by IRT, there were also an issue regarding the innate ability, aptitude, whether one can set the borders between the ML algorithm (by design) and training (optimization). The IRT results were explored by rating systems such as the ones used to evaluate the strength of chess players to establish the ML winner, and consequently providing a glimpse towards a aptitude score of ML algorithms.

References

1. Domingos, P.: A few useful things to know about machine learning. Commun. ACM **55**(10), 78–87 (2012)

2. Vanschoren, J., Van Rijn, J.N., Bischl, B., Torgo, L.: OpenML: networked science in machine learning. ACM SIGKDD Explor. Newslet. **15**(2), 49–60 (2014)
3. Monard, M.C., Baranauskas, J.A.: Conceitos sobre aprendizado de máquina. Sistemas inteligentes-Fundamentos e aplicações **1**(1), 32 (2003)
4. Martínez-Plumed, F., Prudêncio, R. B., Martínez-Usó, A., Hernández-Orallo, J.: Making sense of item response theory in machine learning. In: Proceedings of the Twenty-second European Conference on Artificial Intelligence, pp. 1140–1148. IOS Press (2016)
5. Prudêncio, R.B., Hernández-Orallo, J., Martınez-Usó, A.: Analysis of instance hardness in machine learning using item response theory. In: Second International Workshop on Learning over Multiple Contexts in ECML 2015, Porto, Portugal (2015)
6. Martínez-Plumed, F., Prudêncio, R.B., Martínez-Usó, A., Hernández-Orallo, J.: Item response theory in AI: analysing machine learning classifiers at the instance level. Artif. Intell. **271**, 18–42 (2019)
7. Bischl, B., et al.: OpenML benchmarking suites and the OpenML100. arXiv preprint arXiv:1708.03731 (2017)
8. Samothrakis, S., Perez, D., Lucas, S.M., Rohlfshagen, P.: Predicting dominance rankings for score-based games. IEEE Trans. Comput. Intell. AI Games **8**(1), 1–12 (2014)
9. Glickman, M.E.: Example of the Glicko-2 system, pp. 1–6. Boston University (2012)
10. de Andrade, D.F., Tavares, H.R., da Cunha Valle, R.: Teoria da Resposta ao Item: conceitos e aplicações. ABE, Sao Paulo (2000)
11. Veček, N., Mernik, M., Črepinšek, M.: A chess rating system for evolutionary algorithms: a new method for the comparison and ranking of evolutionary algorithms. Inf. Sci. **277**, 656–679 (2014)
12. Birnbaum, A.L.: Some latent trait models and their use in inferring an examinee's ability. Statistical theories of mental test scores (1968)
13. Pedregosa, F., et al.: Scikit-learn: machine learning in python. J. Mach. Learn. Res. **12**, 2825–2830 (2011)
14. Adedoyin, O.O., Mokobi, T.: Using IRT psychometric analysis in examining the quality of junior certificate mathematics multiple choice examination test items. Int. J. Asian Soc. Sci. **3**(4), 992–1011 (2013)
15. Lord, F.M., Wingersky, M.S.: Comparison of IRT true-score and equipercentile observed-score "equatings". Appl. Psychol. Meas. **8**(4), 453–461 (1984)
16. Pereira, D.G., Afonso, A., Medeiros, F.M.: Overview of Friedman's test and post-hoc analysis. Commun. Stat.-Simul. Comput. **44**(10), 2636–2653 (2015)
17. Nemenyi, P.: Distribution-free multiple comparisons. In: Biometrics, vol. 18, no. 2, p. 263 (1962). 1441 I ST, NW, SUITE 700, WASHINGTON, DC 20005–2210: International Biometric Soc

Towards an Instance-Level Meta-learning-Based Ensemble for Time Series Classification

Caio Luiggy Riyoichi Sawada Ueno[1], Igor Braga[2], and Diego Furtado Silva[1](✉)

[1] Universidade Federal de São Carlos (UFSCar), São Carlos, SP, Brazil
caiol.ueno@gmail.com, diegofs@ufscar.br
[2] Big Data S.A., São Paulo, SP, Brazil
igor.braga@bigdata.com.br

Abstract. State-of-the-art algorithms for classifying time series are based on the combination of classifiers. These ensemble models have the limitation of being extremely costly or depending on a fast algorithm that can damage the accuracy. In this work, we propose using a meta-learning technique that allows choosing, at the instance level, a subset of classifiers to be employed. We explore different approaches to build meta-models based on this idea, including the proposal of an approach based on Long Short-Term Memory trained with a custom loss function. Our results show that this is an approach with great potential to, simultaneously, reduce the number of classifiers needed for each prediction and improve the obtained accuracy.

Keywords: Time series · Classification · Ensemble methods · Meta-learning

1 Introduction

Time series classification (TSC) is a relevant topic nowadays. Due to its relevance, there are a plethora of proposed techniques for this task [1,2,9]. Among all these techniques, ensemble methods have achieved notable results in TSC [3,11,16,26]. While most of these papers seek to combine a large number of classifiers to make the classification more robust, they are computationally expensive. On the other hand, recent algorithms try to circumvent this problem by selecting algorithms to combine, but they make it at random.

In this work, we explore the selection of classifiers to construct ensembles for TSC. However, we look for a way to make this choice more flexible based on the characteristics of the data. This decision leads us to the use of a meta-learning-based strategy. Meta-learning regards a set of techniques technique that learn over another learning process [27].

The views, thoughts, and opinions expressed in this paper belong solely to the authors, and not necessarily to the authors' employer or organization.

R. Cerri and R. C. Prati (Eds.): BRACIS 2020, LNAI 12320, pp. 426–441, 2020.
https://doi.org/10.1007/978-3-030-61380-8_29

For this, we use a meta-learning technique to dynamically select a subset of base classifiers for each instance, the Model Applicability Induction (MAI) [21]. Consider a set of base classifiers $C = \{c_1, c_2, c_3, ..., c_k\}$, where each classifier $c_i, i \in \{1, 2, 3..., k\}$ was trained offline. Our approach considers the output of these classifiers to train a meta-model to predict which base classifiers are good or bad at classifying at the instance level.

It means that, for each time series to be classified, our method selects a subset of base classifiers to consider, aiming to reduce noise on combining them. In other words, it learns when a base classifier has a high probability of correctly classifying a given instance. When a new example is observed, this knowledge is used to decide whether a base classifier will be part of the ensemble. It is essentially a multi-label classification task, in which the decision of whether using or not a classifier is a binary label to be predicted.

Through the exploration of different approaches for the construction of the meta-model based on this technique, we demonstrate that the meta-learning for the instance-level selection of base classifiers has a great potential in TSC. With the correct choice of techniques, we achieve better accuracy rates with fewer classifiers comparing to the naive ensemble. In our experiments, we were able to improve the accuracy, in some cases, by more than 10%. The average number of classifiers used was, in the best case, only 12% of the base classifiers.

The remaining of this paper is organized as follow: Sect. 2 briefly presents recent research efforts on TSC. Section 3 presents the meta-learning approach we explore for TSC in this work. Section 4 defines the experimental setup we used to achieve our objectives. The results obtained in our experiments are presented with a discussion in Sect. 5. Finally, Sect. 6 presents the final remarks of this work.

2 Related Work

For many years, the simple nearest neighbor algorithm was considered difficult to overcome [28]. For this reason, researchers have proposed various TSC approaches using similarity-based algorithms [5,12,13]. Since the decision of distance measure is very relevant, but not straightforward [10], the literature shows that the ensemble of similarity-based methods may benefit the classification [18].

In the last decade, many other ways to deal with the TSC task have been proposed. Some examples are the use of the distances to the training examples as attribute values [14] and reducing the data set to representative substrings for each class, called shapelets [22]. Alternatively, some researchers have proposed transforming time series into an attribute-value space, with techniques based on creating bag-of-patterns [17,23,24]. More recently, there have been several efforts to find deep learning architectures to solve the problem [9].

However, the best results currently found in the literature, for the general context of TSC, are based on ensemble methods. The most straightforward approach in this context is the naive ensemble, where all base classifiers are used to calculate the output, and the majority voting is applied for the final decision. There are different ways to try to improve this approach. For instance, some researchers explored different ways of combining the classifiers and ensuring the diversity between them [3,11]. For example, one of the algorithms considered state-of-the-art in TSC constructs the combination of the base classifiers hierarchically [19]. Specifically, HIVE-COTE first combines classifiers based on similar premises and then aggregates them into a single decision.

However, methods like these are computationally very costly since they depend on all the models for classification. For this reason, researchers try to improve the efficiency of ensemble methods, for instance, by improving the implementations [4]. Another recently explored alternative is through the stochastic selection of training examples and TSC algorithms at the construction stage of decision forests [20,26].

3 Proposed Meta-Learning Approach

As mentioned earlier, our proposal is based on the principle of dynamic ensemble selection. Specifically, we use the simple Model Applicability Induction (MAI) [21]. For a better understanding of this technique, it is interesting to divide the model into levels, as shown in Fig. 1. In level 0, we find the base classifiers, a set of classifiers that predict the class of a given time series. In the next level, we use a meta-model, which is a multi-label classifier. For simplicity, our meta-model is simply a set of independent classifiers, and each of them regards one single base classifier. A level 1 classifier outputs 1 if their respective base classifier has a high probability of predicting the right class of a given instance, or 0 otherwise. We note that we consider that all the time series are segmented in equal lengths, comprising the phenomenon to be classified. Therefore, what we refer to as an instance is such a segmented time series.

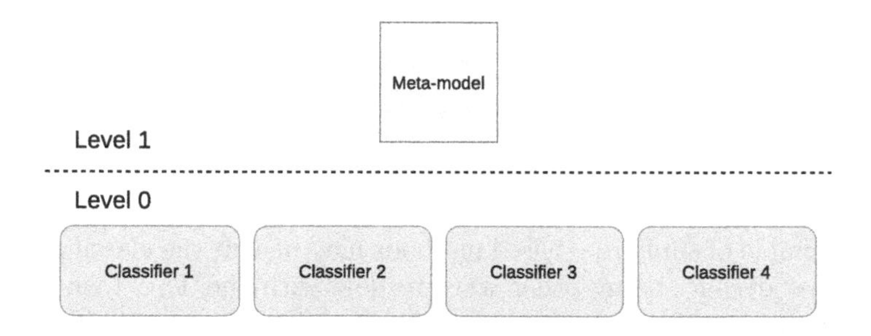

Fig. 1. Meta-model hierarchy levels.

Similar to other ensemble techniques, such as stacking, the training phase of MAI is performed in, at least, two stages. In our proposal, we have three moments of models induction. First, each base classifier is trained individually. These models are saved for later application in the classification stage.

Next, we perform 10-fold cross-validation in the training data to create a new training set for the meta-model. For each fold, 90% of the examples are used to train every level 0 classifier. We use these models to predict the other 10% of instances. The correct classification of each validation example determines the label for each model for each of those examples. If a classifier in level 0 predicts the right label for an example, then it is labeled as 1. Otherwise, its label is set to 0. With this procedure, each example receives a new binary label for each classifier. Finally, the training set for level 1 is ready. Then, we use this data to train the meta-model.

Figure 2 presents an example of the procedure to produce the data set for level 1. In this example, we use three base classifiers, C1, C2, and C3. For each time series, these classifiers create a vector with three values regarding the correct or erroneous classification of each model. This output serves as the label vector for training the meta-model. The time series used as input in this stage is also used as the example that receive these binary labels.

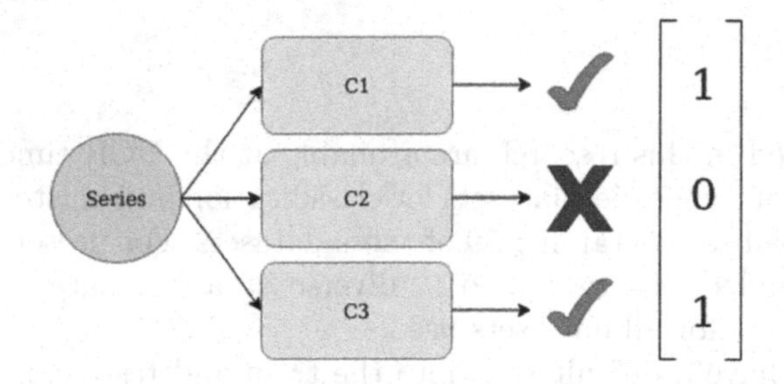

Fig. 2. Example of how we obtain the labels for the level 1 dataset.

Given a new instance, the meta-model must predict which classifiers to use. Then the selected classifiers predict the instance output label, and a majority voting is applied. Notice that only those classifiers that were selected by the meta-model participate in the voting. Figure 3 describe this process.

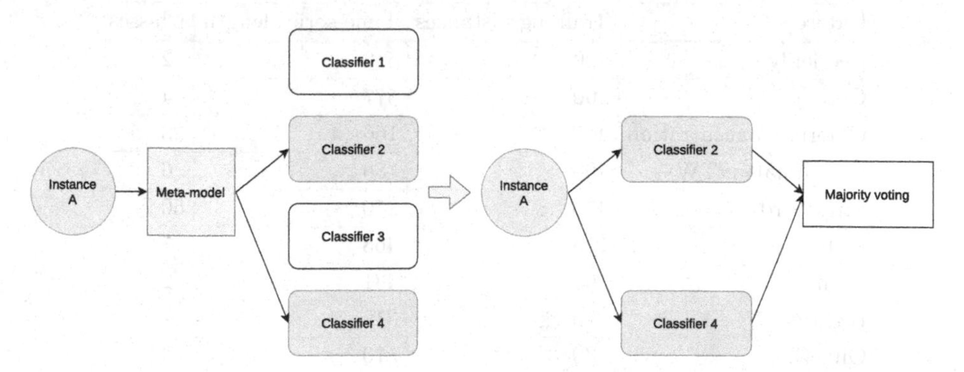

Fig. 3. How a instance is predicted using the meta-learning method.

There are two extreme cases at a prediction: (i) the meta-model can choose all the base classifiers, which means that it is behaving exactly as a naive ensemble;

(ii) the meta-model can choose none of the base classifiers, which means it has no confidence in which classifier to use. In the second case, we used all classifiers, in order not to discard the instance. However, we notice that it may indicate that none of the base classifiers have sufficient information to predict the correct class for the given instance. So, we could consider this instance an outlier or use this information to discover novelties in the streaming scenario. For scope limitation, we do not assess any of these approaches in this paper.

4 Experimental Setup

This section presents the resources and setup of the meta-learning process used in this work. We note that we created a website for this article a supplementary website for reproducible and presentation of detailed results[1].

4.1 Datasets

The datasets used in this research are available at the UCR time series archive [7], a repository of time series datasets for classification and clustering. We experimented with a subset containing 50 of these datasets. For presentation reasons, we limit the detailed exploration to 12 diverse datasets. Later, we present the summarized results for all data sets used.

All datasets have a default split into the train and test sets, and we did no changes on this structure. They are diverse regarding the number of classes, of instances in the train set, and of features - time-series length. Table 1 shows information about the datasets detailed discussed in this paper to demonstrate how the characteristics of the used data widely vary.

Table 1. Information of datasets presented at this paper.

Datasets	Training instances	Time series length	Classes
BeetleFly	20	512	2
Car	60	577	4
ChlorineConcentration	467	166	3
DistalPhalanxTW	400	80	6
FiftyWords	450	270	50
Fish	175	463	7
Ham	109	431	2
Lightning7	70	319	7
OliveOil	30	570	4
ProximalPhalanxTW	400	80	6
ShapeletSim	20	500	2
Wine	57	234	2

[1] https://github.com/diegofurts/mle-tsc.

Also, to use a different representation rather than the raw series as input to the meta-model, a neural network-based autoencoder was implemented to extract new features over the series and reduce the data dimensionality, as illustrated by Fig. 4. The new features, learned by the autoencoder, were used as a different type of input for meta-learning models and compared with raw time series results. Creating new features from the raw time series is important, given it opens a wide range of possibilities for meaningful meta-models. In this case, the extracted features are used as input to train the meta-model, but the base classifiers keep using the raw time series as input.

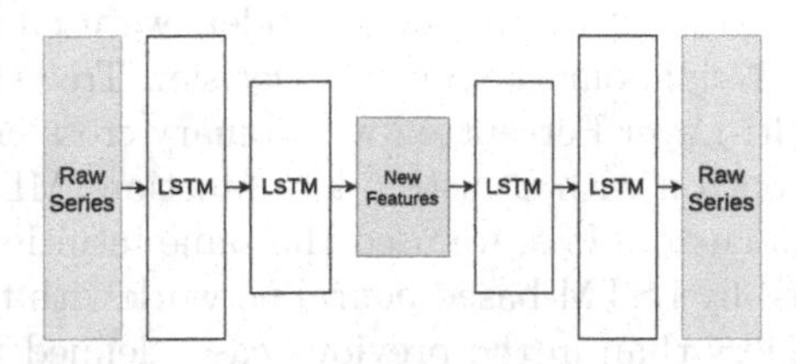

Fig. 4. High-level illustration of the autoencoder used to extract features from the time series.

4.2 Base Classifiers Set

The set of base classifiers comprises well-known but yet simple algorithms for TSC. The goal is to use independent classifiers to introduce a variability at the models' output. In this paper, different techniques and fast training were prioritized. The classifiers used are available at UEA & UCR Time Series Classification Repository [2][2]. Specifically, we used the following methods: Bag Of Patterns [17], BOSS [23], Derivative DTW [12], Derivative Transform Distance [13], Fast Shapelets [22], LPS [6], NN-CID [5], RISE [15], SAX-VSM [25], and TSF [8].

We notice that, once we are not interested in finding the best algorithms configurations, we used the implementations comprised by the Time Series Machine Learning tool[3] with the default setting of parameter values.

4.3 Meta-learning Classifiers

As previously mentioned, we use two approaches to train the meta-model: using the raw time series and applying the features extracted with an autoencoder. For the raw time series, we used the following algorithms in the level 1: *(i)* k-Nearest Neighbour (kNN, where k was chosen by grid search); *(ii)* Long Short-Term Memory (LSTM)-based neural network with binary cross-entropy (LSTM); *(iii)* LSTM-based neural network with a custom loss function (LSTM-cl).

For the last meta-classifier, we approach the two distinct errors with different penalty policies. If it outputs a 0 where it might have been a 1, a false negative, the meta-model suggests us to remove a possibly useful base classifier. Although

[2] http://timeseriesclassification.com/.
[3] https://github.com/uea-machine-learning/tsml/.

this is a misclassification, it implies a lower number of base classifiers, which is interesting from the point-of-view of computational cost. So, we do not penalize this type of error severely. On the other hand, observing a false positive means that the meta-model is adding a useless base classifier to the ensemble, which may damage the classification. Then, we impose a more severe penalty on this error. Given $\hat{y}$ as the predicted label and y as the true label, Eq. 1 defines the custom loss used in this work.

$$Loss = |y - \hat{y}| + |y - \hat{y}|.\hat{y} \tag{1}$$

For the features extracted by the autoencoder, we used the following algorithms: *(i)* k-Nearest Neighbour (kNN); *(ii)* Decision Tree (DT); *(iii)* Random Forest (RF); *(iv)* Multi-Layer Perceptron with binary cross-entropy (MLP); and *(v)* Multi-Layer Perceptron with a custom loss function (MLP-cl).

For the MLP with custom loss, we used the same intuition to define the loss function than the used by LSTM-based neural network with the custom loss. So, we applied the same loss than in the previous case, defined by Eq. 1, but using perceptron-based layers.

5 Results

The results are organized in two subsections: using raw time series as input to meta-learning models and using the features extracted by the autoencoder. We first present the detailed results on the subset presented in Table 1. Later, we present summarize the results obtained in all the assessed datasets.

5.1 Raw Time Series

This subsection presents the results achieved by using the raw time series as input for meta-models. Table 2 shows the accuracy of each method. Those results which are better than naive ensemble at each dataset are highlighted.

The LSTM with custom loss presented better results than other meta-models, obtaining a maximum gain of 9% over naive ensemble accuracy. The nearest neighbor meta-model presented the same accuracy as the naive ensemble in every dataset. This result is an effect of the meta-model quality, that leads the MAI-based model to predict the same label than the naive ensemble for every instance. We will further present the reason behind this effect.

In addition to seeking better accuracy, we use meta-learning to reduce the number of classifiers required for each example. To verify that this result was achieved, we measured the average number of classifiers that the meta-model predicts with the positive label. Table 3 shows the obtained accuracy rates.

We remark that the number of classifiers is notably smaller when applying the custom loss. The numbers obtained by LSTM and kNN are comparable, which does not indicate a reason why the kNN performs equivalently to the naive ensemble. Therefore, we looked at the labels predicted by the meta-model in this case. We noted a large number of examples in which the meta-model

Table 2. Accuracy of all meta-models for each dataset using raw time series.

Datasets	Methods			
	kNN	LSTM	LSTM-cl	NE
BeetleFly	0.75	0.75	0.75	0.75
Car	0.85	0.85	**0.87**	0.85
ChlorineConcentration	0.69	0.69	**0.71**	0.69
DistalPhalanxTW	0.66	0.66	**0.67**	0.66
FiftyWords	0.79	0.79	0.79	0.79
Fish	0.95	0.96	**0.97**	0.95
Ham	0.62	0.63	**0.71**	0.62
Lightning7	0.73	0.75	**0.78**	0.73
OliveOil	0.83	0.83	**0.87**	0.83
ProximalPhalanxTW	0.80	0.80	0.80	0.80
ShapeletSim	0.96	0.97	**0.98**	0.96
Wine	0.70	0.70	**0.78**	0.70

Table 3. Average number of base classifiers used by the meta-model when the raw time series are used as input.

Datasets	Methods		
	kNN	LSTM	LSTM-cl
BeetleFly	**4.50**	7.75	6.75
Car	**8.50**	9.90	9.00
ChlorineConcentration	8.09	9.22	**4.68**
DistalPhalanxTW	7.27	7.55	**3.96**
FiftyWords	7.6	6.11	**5.42**
Fish	9.31	9.93	**8.00**
Ham	6.38	4.30	**1.24**
Lightning7	7.95	7.99	**5.97**
OliveOil	9.33	10.00	**6.00**
ProximalPhalanxTW	8.78	9.22	**8.00**
ShapeletSim	**0.17**	6.18	4.97
Wine	5.19	10.00	**4.00**

suggests discarding all base classifiers. We recall that we deal with this case using the 10 base classifiers, so we do not need to discard the instance. Table 4 shows the average number of base classifiers used when we consider all of them in these exceptional case.

These results show the cause of the poor performance of kNN as a meta-model. Because it is unable to distinguish good base models for classification, it converges to the naive ensemble. Besides, these results show the importance of using a loss function that fits the problem. This fact is observed in both the number of base classifiers and the obtained accuracy.

Table 4. Average number of base classifiers used by the meta-model when the raw time series are used as input, considering all 10 base classifiers when no positive label is predicted by the meta-model.

Datasets	Methods		
	kNN	LSTM	LSTM-cl
BeetleFly	10.00	7.75	**6.75**
Car	10.00	9.90	**9.00**
ChlorineConcentration	10.00	9.22	**4.68**
DistalPhalanxTW	10.00	**7.83**	10.00
FiftyWords	10.00	**7.56**	7.73
Fish	10.00	9.93	**8.00**
Ham	10.00	4.39	**1.24**
Lightning7	10.00	7.99	**5.97**
OliveOil	10.00	10.00	**6.00**
ProximalPhalanxTW	10.00	9.22	**8.00**
ShapeletSim	10.00	6.18	**4.97**
Wine	10.00	10.00	**4.00**

In the next section, we examine whether these conclusions hold if we extract features from the series to use as input for training the meta-model.

5.2 Extracted Features

Table 5 shows the accuracy of each meta-model approach when we apply the autoencoder to extract features from the time series.

Table 5. Accuracy of all meta-models for each dataset using extracted features.

Datasets	Methods					
	kNN	RF	Tree	MLP	MLP-cl	NE
BeetleFly	0.75	0.75	0.75	0.75	**0.8**	0.75
Car	0.85	0.85	0.85	0.85	**0.88**	0.85
ChlorineConcentration	0.69	0.69	0.69	**0.70**	0.68	0.69
DistalPhalanxTW	0.66	0.66	0.66	**0.69**	0.66	0.66
FiftyWords	0.79	0.79	0.79	**0.81**	0.78	0.79
Fish	0.95	0.95	0.95	0.95	0.94	0.95
Ham	0.62	0.62	0.62	**0.65**	0.56	0.62
Lightning7	0.73	0.73	0.73	0.75	**0.78**	0.73
OliveOil	0.83	0.83	0.83	0.83	0.83	0.83
ProximalPhalanxTW	0.80	0.80	0.80	**0.82**	0.81	0.80
ShapeletSim	0.96	0.96	0.96	**0.98**	**0.98**	0.96
Wine	0.70	0.70	0.70	0.70	**0.80**	0.70

Table 6. Average number of base classifiers used by the meta-model when the extracted features are used as input, considering all 10 base classifiers when no positive label is predicted by the meta-model.

Datasets	Methods				
	kNN	RF	Tree	MLP	MLP-cl
BeetleFly	10.00	10.00	10.00	8.50	**6.30**
Car	10.00	10.00	10.00	10.00	**8.00**
ChlorineConcentration	10.00	10.00	10.00	8.94	**4.00**
DistalPhalanxTW	10.00	10.00	10.00	8.14	**8.10**
FiftyWords	10.00	10.00	10.00	**7.82**	8.00
Fish	10.00	10.00	10.00	10.00	**8.00**
Ham	10.00	10.00	10.00	8.00	**2.00**
Lightning7	10.00	10.00	10.00	8.53	**5.56**
OliveOil	10.00	10.00	10.00	10.00	**6.00**
ProximalPhalanxTW	10.00	10.00	10.00	**8.38**	8.83
ShapeletSim	10.00	10.00	10.00	6.02	**5.00**
Wine	10.00	10.00	10.00	10.00	**5.00**

Similar to the noted in the previous section, the meta-models based on neural networks usually achieved results superior to the naive ensemble. On the other hand, neither nearest neighbour, decision tree, or random forest-based models presented any improvement over the naive ensemble.

As in the first experiments, we counted the number of base classifiers used by each meta-model to verify their ability to reduce the number of base classifiers needed to predict the class when trained using the extracted features. But, in this case, we counted all the models applied, including the cases where the meta-models suggest discarding all the base classifiers. Table 6 shows the results.

Again, only neural meta-models reduced the required number of base classifiers, mainly when the custom loss function is applied.

5.3 Results Summarization

As previously stated, presenting all the results in detail is impossible due to space limitations. For this reason, we dedicate this section to graphically summarize all the obtained results, obtained over 50 datasets. The number of training examples in these datasets varies from 16 to 1800. The datasets comprise time series with lengths varying from 15 to 1639, labeled with 2 up to 50 distinct labels.

For this purpose, we focus on the algorithms with customized loss functions. The main reason for this choice is the fact that they usually chose a reduced number of base classifiers and commonly achieved improved accuracy in the previous experiments. Figure 5 illustrates the difference between the different loss functions applied on both raw time series and extracted features cases.

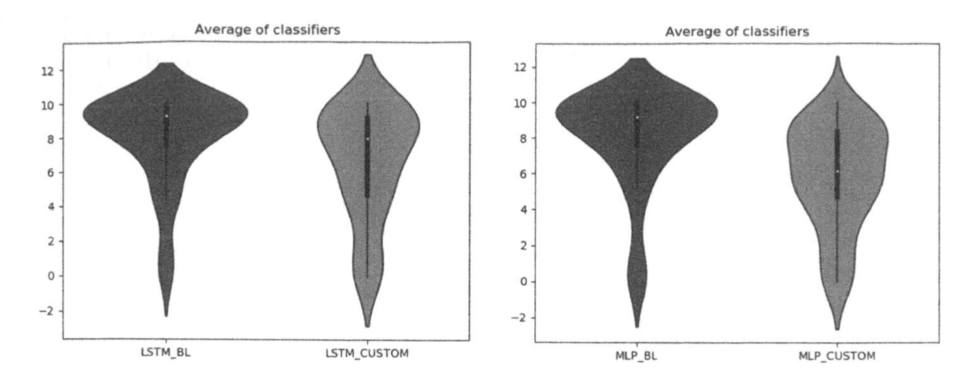

Fig. 5. Estimated distribution of number of selected base classifiers using binary cross-entropy (in blue) and the custom loss function (in orange) when applied on the raw time series (*left*) or the features extracted by the autoencoder (*right*). (Color figure online)

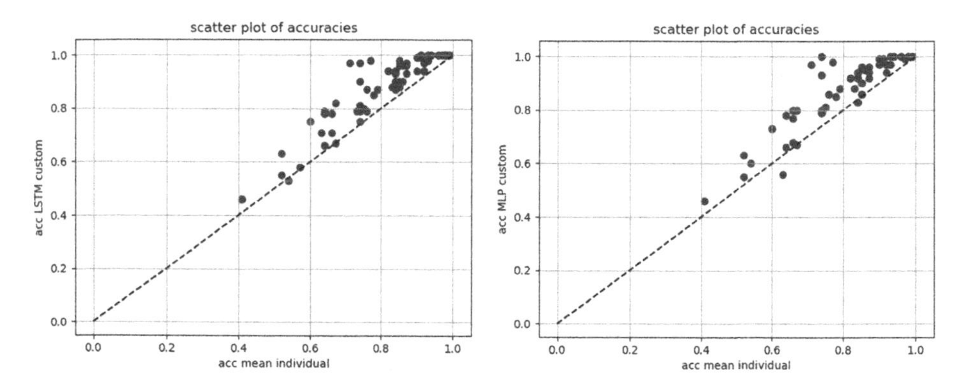

Fig. 6. Comparison of LSTM (*left*) and MLP (both with custom loss function) (*right*) against the mean of accuracy of by individual classifiers.

When we apply binary cross-entropy loss, a standard approach in classification, we can notice a large concentration of the number of models in high values, close to the total number of base classifiers. This effect means that this loss function is not suitable for training the meta-model. On the other hand, the simple proposed custom loss function causes a much better spread. This fact shows that the choice of a suitable loss function has a significant impact on this outcome.

Figure 6 presents a scatter plot comparing the accuracy obtained by LSTM with custom loss and MLP custom loss against the average results of individual classifiers, which simulates the expected accuracy for a random choice of the classifier. As the axes represent the accuracy of each method, the points above the main diagonal indicate a superior result of the proposed meta-learning-based

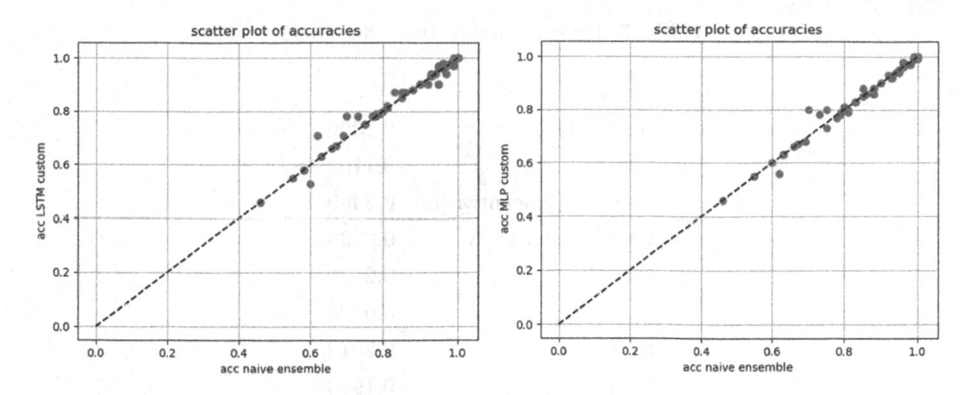

Fig. 7. Results comparison of LSTM with custom loss function (*left*) and MLP with custom loss function (*right*) against the mean of accuracy rates obtained by the naive ensemble.

method. Note that the proposed methods are superior to most accuracy rates obtained by individual models.

These results show a clear superiority of our proposal over individual classifiers. Comparing against the best or worst individual model would be unfair, so we compared against the expected performance of a random choice. As ensemble methods are an alternative to avoid such random choice of the classifiers, we compare our results against those obtained by the naive ensemble method. Figure 7 shows the results.

In this case, we demonstrate that the results obtained by the proposed methods are similar to the naive ensemble. This result shows, at least, one benefit of this work. We demonstrate that it is possible achieving a similar result than naive ensembling with a lower number of classifiers, which allows us to construct faster and more intelligent ensemble methods.

We notice that we did not perform a runtime comparison in our experiments for different reasons. Among them is the fact that the TSC algorithms and the remainder techniques used in this paper are implemented in different programming languages[4]. The results presented in the tables show that the neural networks-based meta-models can significantly reduce the number of base classifiers to predict each instance. That means that our proposal may considerably diminish the classification time, even with the overhead for training the meta-model.

5.4 Further Analysis

When working with models' combination, it is relevant to study the independence between the base classifiers. One way to perform this study is by assessing

[4] The algorithms for TSC are implemented in Java. We used Python to create the necessary code to accomplish the remaining necessary methods.

Table 7. Dependency of base classifiers.

Datasets	ϕ_e
BeetleFly	0.1300
Car	0.1107
ChlorincConcentration	0.2469
DistalPhalanxTW	0.2538
FiftyWords	0.2176
Fish	0.0490
Ham	0.2012
Lightning7	0.1963
OliveOil	0.1074
ProximalPhalanxTW	0.1572
ShapeletSim	0.0567
Wine	0.1839

the dependency measure - or correlated error - between the errors of base classifiers. This shows how much the output labels of each classifier are correlated. Consider a set of classifiers $F = \{\hat{f}_1(x), \hat{f}_2(x), \hat{f}_3(x), ..., \hat{f}_n c(x)\}$, where $\hat{f}_i(x) = \hat{y}$ means that classifier i labeled example x as $\hat{y}$, while y is the true label of x. Equation 2 defines the error correlation between two classifiers.

$$\phi_{i,j} = p(\hat{f}_i(x) = \hat{f}_j(x) | \hat{f}_i(x) \neq y \vee \hat{f}_j(x) \neq y) \tag{2}$$

Given that $\phi_{i,j}$ is symmetric, Eq. 3 defines the mean error correlation between all evaluated classifiers.

$$\phi_e(F) = \frac{2}{n(n-1)} \sum_{i=1}^{n} \sum_{j>i}^{n} \phi_{i,j} \tag{3}$$

Table 7 shows ϕ for the datasets used in this work.

This dependency evaluation shows how much base classifiers are likely to misclassify an instance as the same (wrong) label. It is not concerned about when base classifiers correctly predict an instance's label, once it does not provide any new information for the meta-learning. Usually, when the base classifier errors are different, this favors the construction of an ensemble-based model. This effect occurs because different errors are less relevant in voting than errors in the same class. So, high values of ϕ_e usually implies on a situation where a combined classifier is not expected to improve accuracy.

None of the results obtained has a very high ϕ_e value. However, we can observe some interesting relationships between the accuracy of the LSTM-cl and the correlation of the error. For example, in datasets with higher phi, the naive ensemble does not achieve impressive performance. Even so, the gain accuracy on these datasets was not high. Furthermore, datasets for which there has been no improvement in accuracy have values that are among the highest. An example of this is the FiftyWords dataset. On the other hand, the LSTM-cl performs well on datasets with high ϕ_e. Even with little room for improvement, the obtained results are better than the naive ensemble.

However, we could not find any relationship between the calculated index and the accuracy improvement (or descreasing) when using autoencoder and MLP. For example, using this approach caused a loss of performance in the Fish dataset.

6 Conclusion

In this paper, we explored an instance-level meta-learning approach for time series classification. Our approach is the first in the literature that exploits a data-driven model selection for each new instance to receive a label.

Our results showed that our proposal is promising in time series classification. Specifically, we show that it is possible to reduce the number of required classifiers and improve accuracy compared to a naive ensemble at the same time. Furthermore, we show that this is only possible with the proper construction of the meta-model. We can observe that in the results obtained by the LSTM-based meta-model trained with a custom loss function proposed in this work.

As future work, we intend to extend our experiments to find a loss function even more suitable for the analyzed problem. Besides, we plan to analyze other ways - more efficient and effective - of using features extracted from the time series to create the meta-model. The approach used in this work has not shown exciting results and does not aggregate value to compensate for the overhead of training the autoencoder. We also want to verify the relationship between the proposed techniques and characteristics of the dataset, such as the number of training examples. Finally, once we establish effective guidelines for the construction of meta-models based on the instance-level selection of classifiers for time series classification, we will run runtime experiments on a unified platform, which allows for fair comparisons in this regard.

Acknowledgments. We would like to acknowledge that this research was funded by the Big Data Scholarship Programs and grant #2017/24340-6, São Paulo Research Foundation (FAPESP).

References

1. Abanda, A., Mori, U., Lozano, J.A.: A review on distance based time series classification. Data Min. Knowl. Discov. **33**(2), 378–412 (2018). https://doi.org/10.1007/s10618-018-0596-4
2. Bagnall, A., Lines, J., Bostrom, A., Large, J., Keogh, E.: The great time series classification bake off: a review and experimental evaluation of recent algorithmic advances. Data Min. Knowl. Discov. **31**(3), 606–660 (2016). https://doi.org/10.1007/s10618-016-0483-9

3. Bagnall, A., Lines, J., Hills, J., Bostrom, A.: Time-series classification with COTE: the collective of transformation-based ensembles. IEEE Trans. Knowl. Data Eng. **27**(9), 2522–2535 (2015)
4. Bagnall, A., Flynn, M., Large, J., Lines, J., Middlehurst, M.: On the usage and performance of the hierarchical vote collective of transformation-based ensembles version 1.0 (hive-cote 1.0). arXiv preprint arXiv:2004.06069 (2020)
5. Batista, G.E.A.P.A., Keogh, E.J., Tataw, O.M., De Souza, V.M.A.: CID: an efficient complexity-invariant distance for time series. Data Min. Knowl. Discov. **28**(3), 634–669 (2014)
6. Baydogan, M.G., Runger, G.: Time series representation and similarity based on local autopatterns. Data Min. Knowl. Discov. **30**(2), 476–509 (2015). https://doi.org/10.1007/s10618-015-0425-y
7. Dau, H.A., et al.: The UCR time series archive. IEEE/CAA J. Autom. Sin. **6**(6), 1293–1305 (2019)
8. Deng, H., Runger, G., Tuv, E., Vladimir, M.: A time series forest for classification and feature extraction. Inf. Sci. **239**, 142–153 (2013)
9. Ismail Fawaz, H., Forestier, G., Weber, J., Idoumghar, L., Muller, P.-A.: Deep learning for time series classification: a review. Data Min. Knowl. Discov. **33**(4), 917–963 (2019). https://doi.org/10.1007/s10618-019-00619-1
10. Giusti, R., Batista, G.E.A.P.A.: An empirical comparison of dissimilarity measures for time series classification. In: Brazilian Conference on Intelligent Systems, pp. 82–88. IEEE (2013)
11. Giusti, R., Silva, D.F., Batista, G.E.A.P.A.: Time series classification with representation ensembles. In: Fromont, E., De Bie, T., van Leeuwen, M. (eds.) IDA 2015. LNCS, vol. 9385, pp. 108–119. Springer, Cham (2015). https://doi.org/10.1007/978-3-319-24465-5_10
12. Górecki, T., Łuczak, M.: Using derivatives in time series classification. Data Min. Knowl. Discov. **26**(2), 310–331 (2013)
13. Górecki, T., Łuczak, M.: Non-isometric transforms in time series classification using DTW. Knowl.-Based Syst. **61**, 98–108 (2014)
14. Kate, R.J.: Using dynamic time warping distances as features for improved time series classification. Data Min. Knowl. Discov. **30**(2), 283–312 (2015). https://doi.org/10.1007/s10618-015-0418-x
15. Large, J., Bagnall, A., Malinowski, S., Tavenard, R.: On time series classification with dictionary-based classifiers. Intell. Data Anal. **23**(5), 1073–1089 (2019)
16. Large, J., Lines, J., Bagnall, A.: A probabilistic classifier ensemble weighting scheme based on cross-validated accuracy estimates. Data Min. Knowl. Discov. **33**(6), 1674–1709 (2019). https://doi.org/10.1007/s10618-019-00638-y
17. Lin, J., Khade, R., Li, Y.: Rotation-invariant similarity in time series using bag-of-patterns representation. J. Intell. Inf. Syst. **39**(2), 287–315 (2012)
18. Lines, J., Bagnall, A.: Time series classification with ensembles of elastic distance measures. Data Min. Knowl. Discov. **29**(3), 565–592 (2014). https://doi.org/10.1007/s10618-014-0361-2
19. Lines, J., Taylor, S., Bagnall, A.: Time series classification with HIVE-COTE: the hierarchical vote collective of transformation-based ensembles. ACM Trans. Knowl. Discov. Data (TKDD) **12**(5), 52 (2018)
20. Lucas, B., et al.: Proximity forest: an effective and scalable distance-based classifier for time series. Data Min. Knowl. Disc. **33**(3), 607–635 (2019)
21. Ortega, J.: Exploiting multiple existing models and learning algorithms. In: AAAI-96 Workshop on Integrating Multiple Learned Models for Improving and Scaling Machine Learning Algorithms (1996)

22. Rakthanmanon, T., Keogh, E.: Fast shapelets: a scalable algorithm for discovering time series shapelets. In: SIAM International Conference on Data Mining, pp. 668–676. SIAM (2013)
23. Schäfer, P.: The boss is concerned with time series classification in the presence of noise. Data Min. Knowl. Discov. **29**(6), 1505–1530 (2015)
24. Schäfer, P., Leser, U.: Fast and accurate time series classification with weasel. In: Conference on Information and Knowledge Management, pp. 637–646 (2017)
25. Senin, P., Malinchik, S.: SAX-VSM: interpretable time series classification using SAX and Vector Space Model. In: IEEE 13th International Conference on Data Mining, pp. 1175–1180. IEEE (2013)
26. Shifaz, A., Pelletier, C., Petitjean, F., Webb, G.I.: TS-CHIEF: a scalable and accurate forest algorithm for time series classification. Data Min. Knowl. Discov. **34**, 742–775 (2020)
27. Vilalta, R., Drissi, Y.: A perspective view and survey of meta-learning. Artif. Intell. Rev. **18**(2), 77–95 (2002)
28. Wang, X., Mueen, A., Ding, H., Trajcevski, G., Scheuermann, P., Keogh, E.: Experimental comparison of representation methods and distance measures for time series data. Data Min. Knowl. Discov. **26**(2), 275–309 (2013)

Ensemble of Binary Classifiers Combined Using Recurrent Correlation Associative Memories

Rodolfo Anibal Lobo(✉) and Marcos Eduardo Valle

University of Campinas, Campinas, Brazil
rodolfolobo@ug.uchile.cl, valle@ime.unicamp.br

Abstract. An ensemble method should cleverly combine a group of base classifiers to yield an improved classifier. The majority vote is an example of a methodology used to combine classifiers in an ensemble method. In this paper, we propose to combine classifiers using an associative memory model. Precisely, we introduce ensemble methods based on recurrent correlation associative memories (RCAMs) for binary classification problems. We show that an RCAM-based ensemble classifier can be viewed as a majority vote classifier whose weights depend on the similarity between the base classifiers and the resulting ensemble method. More precisely, the RCAM-based ensemble combines the classifiers using a recurrent consult and vote scheme. Furthermore, computational experiments confirm the potential application of the RCAM-based ensemble method for binary classification problems.

Keywords: Binary classification · Ensemble method · Associative memory · Recurrent neural network · Random forest

1 Introduction

Inspired by the idea that multiple opinions are crucial before making a final decision, ensemble methods make predictions by consulting multiple different predictors [31]. Apart from their similarity with some natural decision-making methodologies, ensemble methods have a strong statistical background. Namely, ensemble methods aim to reduce the variance – thus increasing the accuracy – by combining multiple different predictors. Due to their versatility and effectiveness, ensemble methods have been successfully applied to a wide range of problems including classification, regression, and feature selection. As a preliminary study, this paper only addresses ensemble methods for binary classification problems.

Although there is no rigorous definition of an ensemble classifier [23], they can be conceived as a group of base classifiers, also called weak or base classifiers.

This work was supported in part by CNPq under grant no. 310118/2017-4, FAPESP under grant no. 2019/02278-2, and Coordenação de Aperfeiçoamento de Pessoal de Nível Superior - Brasil (CAPES) - Finance Code 001.

R. Cerri and R. C. Prati (Eds.): BRACIS 2020, LNAI 12320, pp. 442–455, 2020.
https://doi.org/10.1007/978-3-030-61380-8_30

As to the construction of an ensemble classifier, we must take into account the diversity of the base classifiers and the rule used to combine them [23,30]. There are a plethora of ensemble methods in the literature, including bagging, pasting, random subspace, boosting, and stacking [2,10,14,38]. For example, a bagging ensemble classifier is obtained by training copies of a single base classifier using different subsets of the training set [2]. Similarly, a random subspace classifier is obtained by training copies of a classifier using different subsets of features [14]. In both bagging and random subspace ensembles, the base classifiers are then combined using a voting scheme. Random forest is a successful example of an ensemble of decision tree classifiers trained using both bagging and random subspace ensemble ideas [3].

In contrast to the traditional majority voting, in this paper, we propose to combine the base classifiers using an associative memory. Associative memories (AMs) refer to a broad class of mathematical models inspired by the human brain's ability to store and recall information by association [1,13,21]. The Hopfield neural network is a typical example of a recurrent neural network able to implement an associative memory [15]. Despite its many successful applications [16,32–34], the Hopfield neural network suffers from an extremely low storage capacity as an associative memory model [24]. To overcome the low storage capacity of the Hopfield network, many prominent researchers proposed alternative learning schemes [18,27] as well as improved network architectures. In particular, the recurrent correlation associative memories (RCAMs), proposed by Chiueh and Goodman [5], can be viewed as a kernelized version of the Hopfield neural network [8,9,29]. In this paper, we apply the RCAMs to combine binary classifiers in an ensemble method.

At this point, we would like to remark that associative memories have been previously used by Kultur et al. to improve the performance of an ensemble method [22]. Apart from addressing a regression problem, Kultur et al. use an associative memory in parallel to an ensemble of multi-layer perceptrons. The resulting model is called *ensemble of neural networks with associative memory (ENNA)*. Our approach, in contrast, uses an associative memory to combine the base classifiers. Besides, Kultur et al. associate patterns using the k-nearest neighbor algorithm which is formally a non-parametric method used for classification or regression. Differently, we use recurrent correlation associative memories, which are models conceived to implement associative memories.

The paper is organized as follows: The next section reviews the recurrent correlation associative memories. Ensemble methods are presented in Sect. 3. The main contribution of the manuscript, namely the ensemble classifiers based on associative memories, are addressed in Sect. 3.2. Section 4 provides some computational experiments. The paper finishes with some concluding remarks in Sect. 5.

2 A Brief Review on Recurrent Correlation Associative Memories

Recurrent correlation associative memories (RCAMs) has been introduced by Chiueh and Goodman as an improved version of the famous correlation-based Hopfield neural network [5,15].

Briefly, an RCAM is obtained by decomposing the Hopfield network with Hebbian learning into a two-layer recurrent neural network. The first layer computes the inner product (correlation) between the input and the memorized items followed by the evaluation of a non-decreasing continuous activation function. The subsequent layer yields a weighted average of the stored items.

In mathematical terms, a RCAM is defined as follows: Let $\mathbb{B} = \{-1, +1\}$ and $f : [-1, +1] \rightarrow \mathbb{R}$ be a continuous non-decreasing real-valued function. Given a fundamental memory set $\mathcal{U} = \{\mathbf{u}^1, \ldots, \mathbf{u}^P\} \subset \mathbb{B}^N$, the neurons in the first layer of a bipolar RCAM yield

$$w_\xi(t) = f\left(\frac{1}{N}\sum_{i=1}^{N} z_i(t)u_i^\xi\right), \quad \forall \xi \in 1, \ldots, P, \tag{1}$$

where $\boldsymbol{z}(t) = [z_1(t), z_2(t), \ldots, z_N(t)]^T \in \mathbb{B}^N$ denotes the current state of the network and $\mathbf{u}^\xi = [u_1^\xi, \ldots, u_N^\xi]^T$ is the ξth fundamental memory. The activation potential of the output neuron $a_i(t)$ is given by the following weighted sum of the memory items:

$$a_i(t) = \sum_{\xi=1}^{P} w_\xi(t)u_i^\xi, \quad \forall i = 1, \ldots, N. \tag{2}$$

Finally, the state of the ith neuron of the RCAM is updated as follows for all $i = 1, \ldots, N$:

$$z_i(t+1) = \begin{cases} \operatorname{sgn}\left(a_i(t)\right) & a_i(t) \neq 0, \\ z_i(t), & \text{otherwise.} \end{cases} \tag{3}$$

From (2), we refer to $w_\xi(t)$ as the weight associated to the ξth memory item.

In contrast to the Hopfield neural network, the sequence $\{\boldsymbol{z}(t)\}_{t\geq 0}$ produced by an RCAM is convergent in both synchronous and asynchronous update modes independently of the number of fundamental memories and the initial state vector $\boldsymbol{z}(0)$ [5]. In other words, the limit $\boldsymbol{y} = \lim_{t\to\infty} \boldsymbol{z}(t+1)$ of the sequence given by (3) is well defined using either synchronous or asynchronous update.

As an associative memory model, an RCAM designed for the storage and recall of the vectors $\mathbf{u}^1, \ldots, \mathbf{u}^P$ proceeds as follows: Given a stimulus (initial state) $\boldsymbol{z}(0)$, the vector recalled by the RCAM is $\boldsymbol{y} = \lim_{t\to\infty} \boldsymbol{z}(t+1)$.

Finally, the function f defines different RCAM models. For example:

1. The *correlation RCAM* or *identity RCAM* is obtained by considering in (1) the identity function $f_i(x) = x$.

2. The *exponential RCAM*, which is determined by

$$f_e(x;\alpha) = e^{\alpha x}, \quad \alpha > 0. \tag{4}$$

The identity RCAM corresponds to the traditional Hopfield network with Hebbian learning and self-feedback. Different from the Hopfield network and the identity RCAM, the storage capacity of the exponential RCAM scales exponentially with the dimension of the memory space. Apart from the high storage capacity, the exponential RCAM can be easily implemented on very large scale integration (VLSI) devices [5]. Furthermore, the exponential RCAM allows for a Bayesian interpretation [11] and it is closely related to support vector machines and the kernel trick [8,9,29]. In this paper, we focus on the exponential RCAM, formerly known as *exponential correlation associative memory* (ECAM).

3 Ensemble of Binary Classifiers

An ensemble classifier combines a group of single classifiers, also called *weak or base classifiers*, in order to provide better classification accuracy than a single one [23,31,38]. Although this approach is partially inspired by the idea that multiple opinions are crucial before making a final decision, ensemble classifiers have a strong statistical background. Namely, ensemble classifiers reduce the variance combining the base classifiers. Furthermore, when the amount of training data available is too small compared to the size of the hypothesis space, the ensemble classifier "mixes" the base classifiers reducing the risk of choosing the wrong single classifier [19].

Formally, let $\mathcal{T} = \{(\boldsymbol{t}_1, d_1), \ldots, (\boldsymbol{t}_M, d_M)\}$ be a training set where $\boldsymbol{t}_i \in \mathcal{X}$ and $d_i \in \mathcal{C}$ are respectively the feature sample and the class label of the ith training pair. Here, $\mathcal{X}$ denotes the feature space and $\mathcal{C}$ represents the set of all class labels. In a binary classification problem, we can identify $\mathcal{C}$ with $\mathbb{B} = \{-1, +1\}$. Moreover, let $h_1, h_2, \ldots, h_P : \mathcal{X} \to \mathcal{C}$ be base classifiers trained using the whole or part of the training set $\mathcal{T}$.

Usually, the base classifiers are chosen according to their accuracy and diversity. On the one hand, an accurate classifier is one that has an error rate better than random guessing on new instances. On the other hand, two classifiers are diverse if they make different errors on new instances [12,19].

Bagging and random subspace ensembles are examples of techniques that can be used to ensure the diversity of the base classifiers. The idea of bagging, an acronym for *Bootstrap AGGregatING*, is to train copies of a certain classifier h on subsets of the training set $\mathcal{T}$ [2]. The subsets are obtained by sampling the training $\mathcal{T}$ with replacement, a methodology known as *bootstrap sampling* [23]. In a similar fashion, random subspace ensembles are obtained by training copies of a certain classifier h using different subsets of the feature space [14]. Random forest, which is defined as an ensemble of decision tree classifiers, is an example of an ensemble classifier that combines both bagging and random subspace techniques [3].

Another important issue that must be addressed in the design of an ensemble classifier is how to combine the base classifiers. In the following, we review the majority voting methodology – one of the oldest and widely used combination scheme. The methodology based on associative memories is introduced and discussed subsequently.

3.1 Majority Voting Classifier

As remarked by Kuncheva [23], majority voting is one of the oldest strategies for decision making. In a wide sense, a majority voting classifier yields the class label with the highest number of occurrences among the base classifiers [10,35].

Formally, let $h_1, h_2, \ldots, h_P : \mathcal{X} \rightarrow \mathcal{C}$ be the base classifiers. The *majority voting classifier*, also called *hard voting classifier* and denoted by $H_v : \mathcal{X} \rightarrow \mathcal{C}$, is defined by means of the equation

$$H_v(\boldsymbol{x}) = \operatorname*{argmax}_{c \in \mathcal{C}} \sum_{\xi=1}^{P} w_\xi \mathcal{I}[h_\xi(\boldsymbol{x}) = c], \quad \forall \boldsymbol{x} \in \mathcal{X}, \tag{5}$$

where $w_1, \ldots, w_P$ are the weights of the base classifiers and $\mathcal{I}$ is the indicator function, that is,

$$\mathcal{I}[h_\xi(\boldsymbol{x}) = c] = \begin{cases} 1, & h_\xi(\boldsymbol{x}) = c, \\ 0, & \text{otherwise}. \end{cases} \tag{6}$$

When $\mathcal{C} = \{-1, +1\}$, the majority voting ensemble classifier given by (5) can be written alternatively as

$$H_h(\boldsymbol{x}) = \operatorname{sgn}\left(\sum_{\xi=1}^{P} w_\xi h_\xi(\boldsymbol{x}) \right), \quad \forall \boldsymbol{x} \in \mathcal{X}, \tag{7}$$

whenever $\sum_{\xi=1}^{P} w_\xi h_\xi(\boldsymbol{x}) \neq 0$ [7].

3.2 Ensemble Based on Bipolar Associative Memories

Let us now introduce the ensemble classifiers based on the RCAM models. In analogy to the majority voting ensemble classifier, the RCAM-based ensemble classifier is formulated using only the base classifiers $h_1, \ldots, h_P : \mathcal{X} \rightarrow \mathbb{B}$. Precisely, consider a training set $\mathcal{T} = \{(\boldsymbol{t}_i, d_i) : i = 1, \ldots, M\} \subset \mathcal{X} \times \mathbb{B}$ and let $X = \{\boldsymbol{x}_1, \ldots, \boldsymbol{x}_L\} \subset \mathcal{X}$ be a batch of input samples. We first define the fundamental memories as follows for all $\xi = 1, \ldots, P$:

$$\mathbf{u}^\xi = [h_\xi(\boldsymbol{t}_1), \ldots, h_\xi(\boldsymbol{t}_M), h_\xi(\boldsymbol{x}_1), \ldots, h_\xi(\boldsymbol{x}_L)]^T \in \mathbb{B}^{M+L}. \tag{8}$$

In words, the ξth fundamental memory is obtained by concatenating the outputs of the ξth base classifier evaluated at the M training samples and the L input samples. The bipolar RCAM is synthesized using the fundamental memory set $\mathcal{U} = \{\mathbf{u}^1, \ldots, \mathbf{u}^P\}$ and it is initialized at the state vector

$$\boldsymbol{z}(0) = [d_1, d_2, \ldots, d_M, \underbrace{0, 0, \ldots, 0}_{L-\text{times}}]^T. \tag{9}$$

Note that the first M components of initial state $\boldsymbol{z}(0)$ correspond to the targets in the training set $\mathcal{T}$. The last L components of $\boldsymbol{z}(0)$ are zero, a neutral element different from the class labels. The initial state $\boldsymbol{z}(0)$ is presented as input to the associative memory and the last L components of the recalled vector $\boldsymbol{y}$ yield the class label of the batch of input samples $X = \{\boldsymbol{x}_1, \ldots, \boldsymbol{x}_L\}$. In mathematical terms, the RCAM-based ensemble classifier $H_a : \mathcal{X} \to \mathbb{B}$ is defined by means of the equation

$$H_a(\boldsymbol{x}_i) = y_{M+i}, \quad \forall \boldsymbol{x}_i \in X, \tag{10}$$

where $\boldsymbol{y} = [y_1, \ldots, y_M, y_{M+1}, \ldots, y_{M+L}]^T$ is the limit of the sequence $\{\boldsymbol{z}(t)\}_{t \geq 0}$ given by (3).

In the following, we point out the relationship between the bipolar RCAM-based ensemble classifier and the majority voting ensemble described by (7). Let $\boldsymbol{y}$ be the vector recalled by the RCAM fed by the input $\boldsymbol{z}(0)$ given by (9), that is, $\boldsymbol{y}$ is a stationary state of the RCAM. From (2), (3), and (8), the output of the RCAM-based ensemble classifier satisfies

$$H_a(\boldsymbol{x}_i) = \text{sgn}\left(\sum_{\xi=1}^{P} w_\xi h_\xi(\boldsymbol{x}_i)\right), \tag{11}$$

where

$$w_\xi = f\left(\frac{1}{M+L}\sum_{i=1}^{M+L} y_i u_i^\xi\right), \quad \forall \xi = 1, \ldots, P. \tag{12}$$

From (11), the bipolar RCAM-based ensemble classifier can be viewed as a weighted majority voting classifier. Furthermore, the weight w_ξ depends on the similarity between the ξth base classifier h_ξ and the ensemble classifier H_a. Precisely, let us define the similarity between two binary classifiers $H, h_\xi : \mathcal{X} \to \mathbb{B}$ on a set of samples S by means of the equation

$$\texttt{Sim}(H, h) = \frac{1}{\text{Card}(S)} \sum_{\boldsymbol{s} \in S} \mathcal{I}\big[h(\boldsymbol{s}) = H(\boldsymbol{s})\big]. \tag{13}$$

Using (13), we can state the following theorem:

Theorem 1. *The weights of the RCAM-based ensemble classifier given by* (11) *satisfies the following identities for all* $\xi = 1, \ldots, P$*:*

$$w_\xi = f\big(1 - 2 \cdot \texttt{Sim}(H_a, h_\xi)\big), \quad \forall t \geq 1, \tag{14}$$

where the similarity in (14) *is evaluated on the union of all training and input samples, that is, on* $S = X \cup T = \{\boldsymbol{t}_1, \ldots, \boldsymbol{t}_M\} \cup \{\boldsymbol{x}_1, \ldots, \boldsymbol{x}_L\}$.

Proof. Since we are considering a binary classification problem, the similarity between the ensemble H_a and the base classifier h_ξ on $S = X \cup T$, with $N = \text{Card}(S) = M + L$, satisfies the following identities:

$$\texttt{Sim}(H,h) = 1 - \frac{1}{N}\sum_{i=1}^{N} \mathcal{I}[h(\boldsymbol{s}_i) \neq H_a(\boldsymbol{s}_i)] = 1 - \frac{1}{4N}\sum_{i=1}^{N}\left(h(\boldsymbol{s}_i) - H_a(\boldsymbol{s}_i)\right)^2$$
$$= 1 - \frac{1}{2N}\sum_{i=1}^{N}\left(1 - H_a(\boldsymbol{s}_i)h(\boldsymbol{s}_i)\right) = \frac{1}{2}\left(1 - \frac{1}{N}\sum_{i=1}^{N} H_a(\boldsymbol{s}_i)h(\boldsymbol{s}_i)\right)$$

Equivalently, we have

$$\frac{1}{\text{Card}(S)}\sum_{\boldsymbol{s}\in S} H(\boldsymbol{s})h(\boldsymbol{s}) = 1 - 2\cdot\texttt{Sim}(H,h). \tag{15}$$

Now, from (1), (10), and (15), we obtain the following identities:

$$w_\xi = f\left(\frac{1}{N}\sum_{i=1}^{N} y_i u_i^\xi\right) = f\big(1 - 2\cdot\texttt{Sim}(H_a, h_\xi)\big),$$

which concludes the proof.

Theorem 1 shows that the RCAM-based ensemble classifier is a majority voting classifier whose weights depend on the similarity between the base classifiers and the ensemble itself. In fact, in view of the dynamic nature of the RCAM model, H_a is obtained by a recurrent consult and vote scheme. Moreover, at the first step, the weights depend on the accuracy of the base classifiers.

4 Computational Experiments

In this section, we perform some computational experiments to evaluate the performance of the proposed RCAM-based ensemble classifiers for binary classification tasks. Precisely, we considered the RCAM-based ensembles obtained using the identity and the exponential as the activation function f. The parameter α of the exponential activation function has been either set to $\alpha = 1$ or it has been determined using a grid search on the set $\{10^{-2}, 10^{-1}, 0.5, 1, 5, 10, 20, 50\}$ with 5-fold cross-validation on the training set. The RCAM-based ensemble classifiers have been compared with AdaBoost, gradient boosting, and random forest ensemble classifiers, all available at the python's `scikit-learn API` (`sklearn`) [28].

First of all, we trained AdaBoost and gradient boosting ensemble classifiers using the default parameters of `sklearn`. Recall that boosting ensemble classifiers are developed incrementally by adding base classifiers to reduce the number of misclassified samples [23]. Also, we trained the random forest classifier with 30 base classifiers ($P = 30$) [3]. Recall that the base classifiers of the random forest are decision trees obtained using bagging and random subspace techniques [2,14]. Then, we used the base classifiers from the trained random forest ensemble to define the RCAM-based ensemble. In other words, the same base classifiers $h_1, \ldots, h_{30}$ are used in the random forest and the RCAM-based classifiers. The difference between the ensemble classifiers resides in the combining rule. Recall that the random forest combines the base classifiers using majority voting. From the computational point of view, training the random forest and the RCAM-ensemble classifiers required similar resources. Moreover, despite the consult and vote scheme of the RCAM-based ensemble, they have not been significantly more expensive than the random forest classifier. The grid search used to fine-tune the parameter α of the exponential RCAM-based ensemble is the major computational burden in this computational experiment.

For the comparison of the ensemble classifiers, we considered 28 binary classification problems from the OpenML repository [36]. These binary classification problems can be obtained using the command `fetch_openml` from `sklearn`. We would like to point out that missing data has been handled before splitting the data set into training and test sets using the command `SimpleImputer` from `sklearn`. Also, we pre-processed the data using the `StandardScaler` transform. Therefore, each feature is normalized by subtracting the mean and dividing by the standard deviation, both computed using only the training set. Furthermore, since some data sets are unbalanced, we used the F-measure to evaluate quantitatively the performance of a certain classifier. Table 1 shows the mean and the standard deviation of the F-measure obtained from the ensemble classifiers using stratified 10-fold cross-validation. The largest F-measures for each data set have been typed using boldface. Note the exponential RCAM-based ensemble classifier with grid search produced the largest F-measures in 11 of the 28 data sets. In particular, the exponential RCAM with grid search produced outstanding F-measures on the "Monks-2" and "Egg-Eye-State" data sets. For a better comparison of the ensemble classifiers, we followed Demšar's recommendations to compare multiple classifier models using multiple data sets [6]. The Friedman test rejected the hypothesis that there is no difference between the ensemble classifiers. A visual interpretation of the outcome of this computational experiment is provided in Fig. 1 with the Hasse diagram of the non-parametric Wilcoxon signed-rank test with a confidence level at 95% [4,37]. In this diagram, an edge means that the classifier on the top statistically outperformed the classifier on the bottom. The outcome of this analysis confirms that the RCAM-based ensemble classifiers statistically outperformed the other ensemble methods: AdaBoost, gradient boosting, and random forest.

As to the computational effort, Fig. 2 shows the average time required by the ensemble classifiers for the prediction of a batch of testing samples. Note that the most expensive method is identity RCAM-based ensemble classifier while the gradient boosting is the cheapest. The exponential RCAM-based ensemble is less expensive than the AdaBoost and quite comparable to the random forest classifier.

Table 1. Mean and standard deviation of the F-measures produced by ensemble classifiers using stratified 10-fold cross-validation.

Data set	AdaBoost	Gradient boosting	Random forest	Identity RCNN	Exponential RCAM	Exp. RCAM + grid search
Arsene	84.0 ± 5.9	**86.2 ± 7.6**	81.5 ± 8.9	83.8 ± 8.4	83.8 ± 8.4	85.2 ± 10.2
Australian	82.1 ± 3.4	**85.8 ± 3.8**	85.4 ± 3.4	85.3 ± 2.9	85.3 ± 2.9	85.0 ± 2.9
Banana	67.9 ± 2.1	88.1 ± 1.6	88.0 ± 1.3	**88.2 ± 1.2**	88.2 ± 1.2	87.2 ± 1.2
Banknote	**99.6 ± 0.4**	99.5 ± 0.9	99.3 ± 0.7	99.2 ± 0.7	99.2 ± 0.7	98.9 ± 0.9
Blood transfusion	**43.0 ± 13.1**	37.9 ± 11.2	32.3 ± 10.4	33.3 ± 10.6	33.3 ± 10.6	32.5 ± 8.2
Breast cancer Wisconsin	94.7 ± 2.0	95.2 ± 2.4	94.9 ± 3.4	**95.4 ± 2.9**	95.1 ± 3.3	95.2 ± 4.2
Chess	96.5 ± 1.1	97.9 ± 0.8	99.0 ± 0.5	99.0 ± 0.6	99.0 ± 0.6	**99.2 ± 0.4**
Colic	87.1 ± 6.4	86.7 ± 7.4	88.7 ± 5.7	88.6 ± 5.4	88.6 ± 5.4	**88.9 ± 4.6**
Credit approval	86.4 ± 2.9	86.9 ± 3.2	88.4 ± 2.8	**88.4 ± 2.5**	**88.4 ± 2.5**	88.3 ± 2.3
Credit-g	82.3 ± 2.5	84.2 ± 2.8	83.7 ± 2.4	**84.3 ± 2.2**	**84.3 ± 2.2**	83.9 ± 1.8
Cylinder bands	78.3 ± 4.8	84.0 ± 4.8	83.0 ± 6.6	83.3 ± 6.4	83.3 ± 6.4	**87.0 ± 4.2**
Diabetes	63.1 ± 5.2	65.1 ± 6.5	63.9 ± 8.8	**65.6 ± 8.2**	**65.6 ± 8.2**	62.4 ± 7.8
Egg-eye-state	70.1 ± 1.3	78.0 ± 0.9	91.5 ± 0.7	91.8 ± 0.8	91.8 ± 0.8	**92.9 ± 0.8**
Haberman	**35.4 ± 9.5**	30.8 ± 14.2	27.4 ± 13.4	30.6 ± 9.6	30.6 ± 9.6	34.9 ± 12.9
Hill-valley	40.9 ± 5.4	52.9 ± 7.3	54.9 ± 4.6	56.6 ± 3.8	56.6 ± 4.0	**59.1 ± 6.2**
Internet advertisements	98.0 ± 0.3	98.6 ± 0.3	**98.8 ± 0.4**	98.7 ± 0.4	98.7 ± 0.4	98.7 ± 0.5
Ionosphere	94.3 ± 1.7	94.4 ± 2.0	94.2 ± 2.5	94.0 ± 2.5	94.0 ± 2.5	**94.7 ± 2.7**
MOFN-3-7-10	**100.0 ± 0.0**	**100.0 ± 0.0**	99.8 ± 0.2	99.7 ± 0.3	99.7 ± 0.3	99.7 ± 0.5
Monks-2	0.0 ± 0.0	69.3 ± 8.7	93.1 ± 3.3	93.5 ± 3.3	93.5 ± 3.3	**98.5 ± 2.7**
Phoneme	68.3 ± 3.0	75.4 ± 2.4	84.0 ± 3.0	84.1 ± 2.7	84.1 ± 2.7	**85.7 ± 2.0**
Pishing websites	94.4 ± 0.4	95.3 ± 0.5	97.5 ± 0.6	97.4 ± 0.6	97.4 ± 0.6	**97.5 ± 0.5**
Sick	78.3 ± 6.4	88.8 ± 3.9	87.5 ± 3.1	88.6 ± 3.9	88.6 ± 3.9	**89.7 ± 3.6**
Sonar	**83.9 ± 8.0**	81.3 ± 6.2	81.9 ± 11.4	83.3 ± 11.1	83.3 ± 11.1	83.2 ± 11.1
Spambase	91.8 ± 1.5	93.1 ± 1.7	**94.2 ± 1.1**	94.0 ± 1.2	94.1 ± 1.2	94.0 ± 1.2
Steel plates fault	**100.0 ± 0.0**	**100.0 ± 0.0**	99.0 ± 0.8	99.2 ± 0.6	99.2 ± 0.6	99.4 ± 0.7
Tic-Tac-Toe	84.5 ± 2.6	94.8 ± 2.1	95.6 ± 1.2	95.5 ± 1.2	95.5 ± 1.2	**96.5 ± 1.5**
Titanic	**58.8 ± 4.3**	53.8 ± 4.4	53.6 ± 4.2	53.6 ± 4.2	53.6 ± 4.2	53.8 ± 4.4
ilpd	**41.4 ± 11.4**	35.3 ± 15.1	35.1 ± 15.8	37.5 ± 16.6	37.5 ± 16.6	33.5 ± 14.6

Finally, note from Table 1 that some problems such as the "Banknote"' and the "MOFN-3-7-10" data sets are quite easy while others such as the "Haberman" and "Hill Valley" are very hard. In order to circumvent the difficulties imposed by each data set, Fig. 3 shows a box-plot with the normalized F-measure values provided in Table 1. Precisely, for each data set (i.e., each row in Table 1), we subtracted the mean and divided by the standard deviation of the score values. The box-plot in Fig. 3 confirms the good performance of the RCAM-based ensemble classifiers, including the exponential RCAM-based ensemble classifier with a grid search. Concluding, the boxplots shown on Figs. 2 and 3 supports the potential application of the RCAM models as an ensemble of classifiers for binary classification problems.

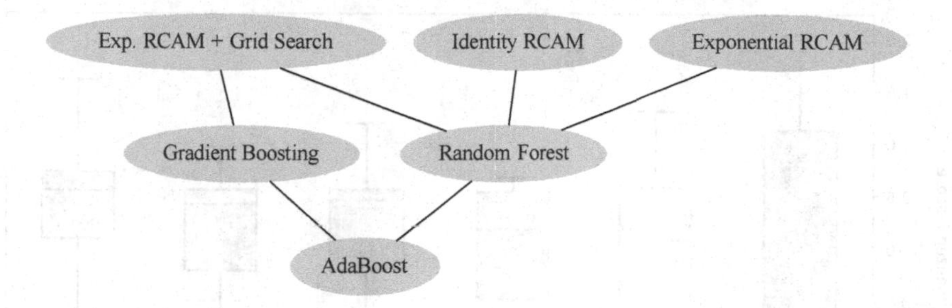

Fig. 1. Hasse diagram of Wilcoxon signed-rank test with a confidence level at 95%.

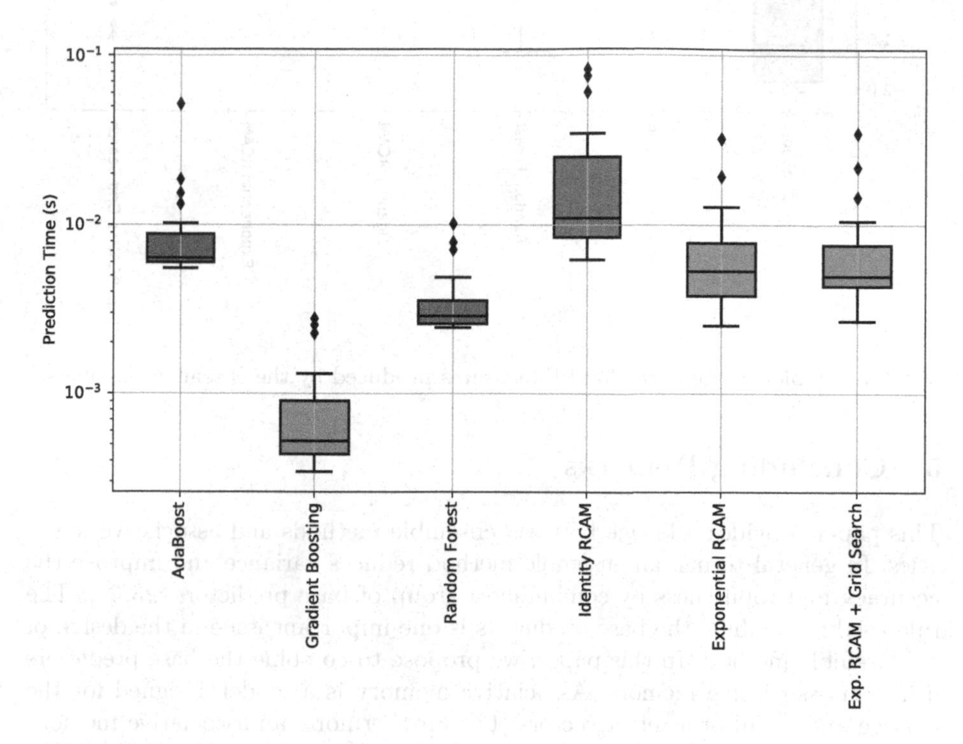

Fig. 2. Box-plot of the average time for prediction of batch of input samples.

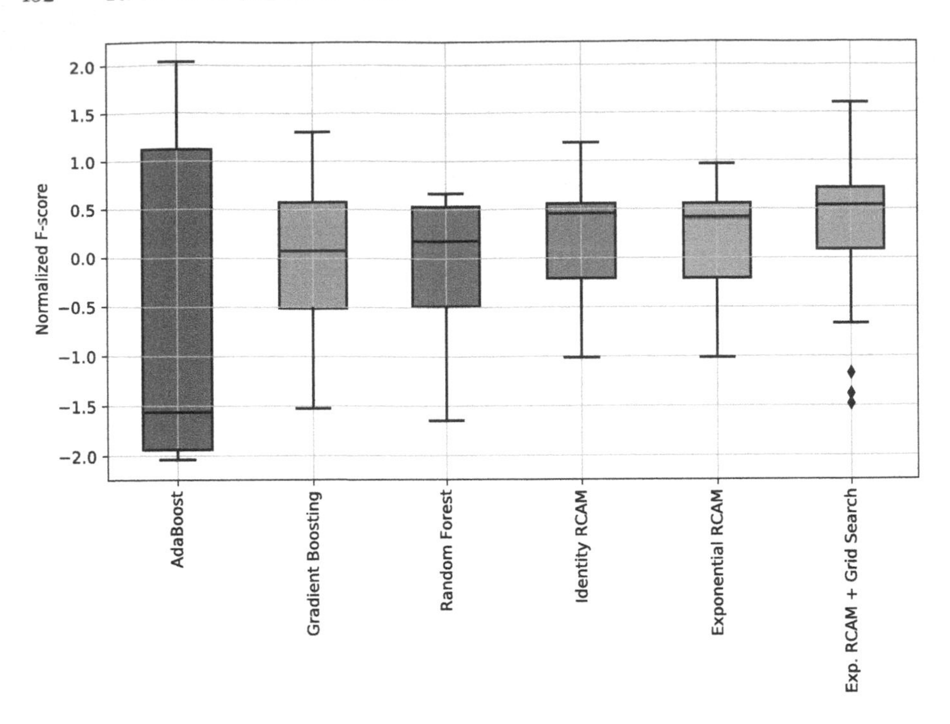

Fig. 3. Box-plot of the normalized F-measures produced by the ensemble classifiers.

5 Concluding Remarks

This paper provides a bridge between ensemble methods and associative memories. In general terms, an ensemble method reduces variance and improve the accuracy and robustness by combining a group of base predictors [23,38]. The rule used to combine the base predictors is one important issue in the design of an ensemble method. In this paper, we propose to combine the base predictors using an associative memory. Associative memory is a model designed for the storage and recall of a set of vectors [13]. Furthermore, an associative memory should be able to retrieve a stored item from a corrupted or partial version of it. In an ensemble method, the memory model is designed for the storage of evaluations of the base classifiers. The associative memory is then fed by a vector with the target of training data as well as the unknown predictions. The output of the ensemble method is obtained from the vector retrieved by the memory.

Specifically, in this paper, we presented ensemble methods based on the recurrent correlation associative memories (RCAMs) for binary classifications. RCAMs, proposed by Chiueh and Goodman [5], are high storage capacity associative memories which, besides Bayesian and kernel trick interpretation, are particularly suited for VLSI implementation [8,9,11,29]. Theorem 1 shows that the RCAM model yields a majority voting classifier whose weights are obtained by a recurrent consult and vote scheme. Moreover, the weights depend on the similarity between the base classifiers and the resulting ensemble. Computational experiments using decision tree as the base classifiers revealed an outstanding performance of the exponential RCAM-based ensemble classifier combined with a grid search strategy to fine-tune its parameter. The exponential RCAM-based ensemble, in particular, outperformed the traditional AdaBoost, gradient boosting, and random forest classifiers.

In the future, we plan to investigate further associative memory-based ensemble methods. In particular, we plan to extend these ensemble methods to multi-class classification problems using, for instance, multistate associative memory models [17,20,25,26].

References

1. Austin, J.: ADAM: a distributed associative memory for scene analysis. In: Proceedings of the IEEE First International Conference on Neural Networks, vol. IV, p. 285. San Diego (1987)
2. Breiman, L.: Bagging predictors. Mach. Learn. **24**(2), 123–140 (1996). https://doi.org/10.1023/A:1018054314350
3. Breiman, L.: Random forests. Mach. Learn. **45**(1), 5–32 (2001). https://doi.org/10.1023/A:1010933404324
4. Burda, M.: Paircompviz: An R Package for Visualization of Multiple Pairwise Comparison Test Results (2013). https://doi.org/10.18129/B9.bioc.paircompviz
5. Chiueh, T., Goodman, R.: Recurrent correlation associative memories. IEEE Trans. Neural Netw. **2**, 275–284 (1991)
6. Demšar, J.: Statistical comparisons of classifiers over multiple data sets. J. Mach. Learn. Res. **7**, 1–30 (2006)
7. Ferreira, A., Figueiredo, M.: Boosting algorithms: a review of methods, theory, and applications. In: Zhang, C., Ma, Y. (eds.) Ensemble Machine Learning: Methods and Applications, pp. 35–85. Springer (2012). https://doi.org/10.1007/978-1-4419-9326-7_2
8. García, C., Moreno, J.A.: The hopfield associative memory network: improving performance with the kernel "Trick". In: Lemaître, C., Reyes, C.A., González, J.A. (eds.) IBERAMIA 2004. LNCS (LNAI), vol. 3315, pp. 871–880. Springer, Heidelberg (2004). https://doi.org/10.1007/978-3-540-30498-2_87
9. García, C., Moreno, J.A.: The kernel hopfield memory network. In: Sloot, P.M.A., Chopard, B., Hoekstra, A.G. (eds.) ACRI 2004. LNCS, vol. 3305, pp. 755–764. Springer, Heidelberg (2004). https://doi.org/10.1007/978-3-540-30479-1_78
10. Géron, A.: Hands-On Machine Learning with Scikit-Learn, Keras, and TensorFlow: Concepts, Tools, and Techniques to Build Intelligent Systems. O'Reilly Media (2019)
11. Hancock, E.R., Pelillo, M.: A Bayesian interpretation for the exponential correlation associative memory. Pattern Recogn. Lett. **19**(2), 149–159 (1998)
12. Hansen, L.K., Salamon, P.: Neural network ensembles. IEEE Trans. Pattern Anal. Mach. Intell. **12**(10), 993–1001 (1990)
13. Du, K.-L., Swamy, M.N.S.: Associative Memory Networks. Neural Networks and Statistical Learning. LNCS, pp. 201–229. Springer, London (2019). https://doi.org/10.1007/978-1-4471-7452-3_8
14. Ho, T.K.: The random subspace method for constructing decision forests. IEEE Trans. Pattern Anal. Mach. Intell. **20**(8), 832–844 (1998)

15. Hopfield, J.J.: Neural networks and physical systems with emergent collective computational abilities. Proc. Nat. Acad. Sci. **79**, 2554–2558 (1982)
16. Hopfield, J., Tank, D.: Neural computation of decisions in optimization problems. Biol. Cybern. **52**, 141–152 (1985)
17. Jankowski, S., Lozowski, A., Zurada, J.: Complex-valued multi-state neural associative memory. IEEE Trans. Neural Netw. **7**, 1491–1496 (1996)
18. Kanter, I., Sompolinsky, H.: Associative recall of memory without errors. Phys. Rev. **35**, 380–392 (1987)
19. Kittler, J., Roli, F.: 2000 Proceedings of the Multiple Classifier Systems: First International Workshop, MCS 2000, Cagliari, Italy, June 21–23. Springer (2003)
20. Kobayashi, M.: Quaternionic Hopfield neural networks with twin-multistate activation function. Neurocomputing **267**, 304–310 (2017). https://doi.org/10.1016/j.neucom.2017.06.013
21. Kohonen, T.: Self-Organization and Associative Memory, 2rd edn. Springer, New York (1987)
22. Kultur, Y., Turhan, B., Bener, A.: Ensemble of neural networks with associative memory (ENNA) for estimating software development costs. Knowl.-Based Syst. **22**(6), 395–402 (2009)
23. Kuncheva, L.: Combining Pattern Classifiers: Methods and Algorithms, 2 edn. Wiley (2014)
24. McEliece, R.J., Posner, E.C., Rodemich, E.R., Venkatesh, S.: The capacity of the Hopfield associative memory. IEEE Trans. Inf. Theory **1**, 33–45 (1987)
25. Minemoto, T., Isokawa, T., Nishimura, H., Matsui, N.: Quaternionic multistate Hopfield neural network with extended projection rule. Artif. Life Robot. **21**(1), 106–111 (2015). https://doi.org/10.1007/s10015-015-0247-4
26. Müezzinoğlu, M., Güzeliş, C., Zurada, J.: A new design method for the complex-valued multistate Hopfield associative memory. IEEE Trans. Neural Netw. **14**(4), 891–899 (2003)
27. Müezzinoğlu, M., Güzelis, C., Zurada, J.: An energy function-based design method for discrete Hopfield associative memory with attractive fixed points. IEEE Trans. Neural Netw. **16**(2), 370–378 (2005)
28. Pedregosa, F., et al.: Scikit-learn: machine learning in Python. J. Mach. Learn. Res. **12**, 2825–2830 (2011)
29. Perfetti, R., Ricci, E.: Recurrent correlation associative memories: a feature space perspective. IEEE Trans. Neural Netw. **19**(2), 333–345 (2008)
30. Polikar, R.: Ensemble learning. In: Zhang, C., Ma, Y. (eds.) Ensemble Machine Learning: Methods and Applications, pp. 1–34. Springer (2012). https://doi.org/10.1007/978-1-4419-9326-7_1
31. Ponti Jr, M.P.: Combining classifiers: from the creation of ensembles to the decision fusion. In: 2011 24th SIBGRAPI Conference on Graphics, Patterns, and Images Tutorials, pp. 1–10. IEEE (2011)
32. Serpen, G.: Hopfield network as static optimizer: learning the weights and eliminating the guesswork. Neural Process. Lett. **27**(1), 1–15 (2008). https://doi.org/10.1007/s11063-007-9055-8
33. Smith, K., Palaniswami, M., Krishnamoorthy, M.: Neural techniques for combinatorial optimization with applications. IEEE Trans. Neural Netw. **9**(6), 1301–1318 (1998)
34. Sun, Y.: Hopfield neural network based algorithms for image restoration and reconstruction II. Perform. Anal. IEEE Trans. Sign. Process. **48**(7), 2119–2131 (2000). https://doi.org/10.1109/78.847795

35. Van Erp, M., Vuurpijl, L., Schomaker, L.: An overview and comparison of voting methods for pattern recognition. In: Proceedings Eighth International Workshop on Frontiers in Handwriting Recognition, pp. 195–200. IEEE (2002)
36. Vanschoren, J., van Rijn, J.N., Bischl, B., Torgo, L.: OpenML: networked science in machine learning. SIGKDD Explor. **15**(2), 49–60 (2013). https://doi.org/10.1145/2641190.2641198
37. Weise, T., Chiong, R.: An alternative way of presenting statistical test results when evaluating the performance of stochastic approaches. Neurocomputing **147**, 235–238 (2015). https://doi.org/10.1016/j.neucom.2014.06.071
38. Zhang, C., Ma, Y. (eds.): Ensemble Machine Learning: Methods and Applications. Springer (2012). https://doi.org/10.1007/978-1-4419-9326-7

Comparative Study of Fast Stacking Ensembles Families Algorithms

Laura Maria Palomino Mariño[1(✉)], Agustín Alejandro Ortiz-Díaz[2], and Germano Crispim Vasconcelos[1]

[1] Centro de Informática, Universidade Federal de Pernambuco, Recife, Brazil
{lmpm,gcv}@cin.ufpe.br

[2] Centro de Ciências Tecnológicas, Universidade do Estado de Santa Catarina, Joinville, Brazil
agaldior@gmail.com

Abstract. One of the main challenges in Machine Learning and Data Mining fields is the treatment of large Data Streams in the presence of Concept Drifts. This paper presents two families of ensemble algorithms designed to adapt to abrupt and gradual concept drifts. The families Fast Stacking of Ensembles boosting the Old (FASEO) and Fast Stacking of Ensembles boosting the Best (FASEB) are adaptations of the Fast Adaptive Stacking of Ensembles (FASE) algorithm to improve run-time, without presenting a significant decrease in terms of accuracy when compared to the original FASE. In order to achieve a more efficient model, adjustments were made in the update strategy and voting procedure of the ensemble. To evaluate the methods, Naïve Bayes (NB) and Hoeffding Tree (HT) are used, as learners, to compare the performance of the algorithms on artificial and real-world data-sets. An experimental investigation with a total of 32 experiments and the application of Friedman and Bonferroni-Dunn statistical tests showed the families FASEO and FASEB are more efficient than FASE with respect to execution time in many experiments, also some methods achieving better accuracy results.

Keywords: Concept drift · Data stream · Ensemble methods

1 Introduction

In recent years, data generated by different sources such as cell phones, sensors, networks, and satellites has increased significantly. Part of these data can be viewed as a sequence of examples that arrive at high rates and can often be read-only once using a small amount of processing time [1]. In the literature, such data are known as data-streams. According to [2], in the streaming scenario,

Supported by Coord. de Aperfeiçoamento de Pessoal de Nível Superior (CAPES) and Fundação de Amparo à Ciência e Tecnologia do Estado de Pernambuco - FACEPE (IBPG-0820-1.03/19).

R. Cerri and R. C. Prati (Eds.): BRACIS 2020, LNAI 12320, pp. 456–470, 2020.
https://doi.org/10.1007/978-3-030-61380-8_31

two primary issues have to be dealt with in the construction of training models: One-pass Constraint, and Concept Drift.

The first aspect, One-pass Constraint, is when a model needs to analyze each of the data only once. This feature is very important in online learning. Second, a concept drift can be described from the change between two different concepts: initial concept (P_I) and final concept (P_F) [3]. Attending to the time it takes to change from P_I to P_F (t_{ch}), this change can be abrupt (sudden) ($t_{ch} \sim 0$) or gradual ($t_{ch} > 0$).

Depending on the taxonomy of the involved aspects in the distribution change, different types of Concept Drift can be analyzed: virtual concept drift (the distribution of instances change but the underlying concept does not), real concept drift (there exist a change in the class boundary), recurrence of the concepts (when previously active concept reappears after some time) [4], and some other types.

The presence of Concept Drift affects the performance of the classification algorithms because models become stale over time. Therefore, it is crucial to adjust the model in an incremental way in order for the algorithm to achieve high accuracy over current unknown instances.

Many of the works published so far have focused mainly on accuracy as a fundamental parameter to establish comparisons between learning algorithms. However, in many real-life scenarios other parameters like run-time should also be taken into account due to their importance. Motivated by previous works [5] that analyzed the performance of several algorithms, it was observed that Fast Adaptive Stacking of Ensembles (FASE) [7] (one of the compared methods) presents good accuracy results, but its run-times indicate there is room for improvement.

The present work aims to introduce the families of methods Fast Stacking of Ensembles boosting the Best (FASEB) and Fast Stacking of Ensembles boosting the Old (FASEO) obtained from the algorithm FASE. All methods present, from both families, are designed to adapt to concept drifts, whether abrupt or gradual and to increase the "efficiency" of their base model. The word "efficiency" has a particular meaning here: it is a "suitable" balance among accuracy and run-time highlighting. Furthermore, "suitable" is associated with the degree to which these aspects have relevance in a given context [6]. In order to obtain the variants of FASE, it was experimentally investigated two main modifications, regarding (i) the update strategy and (ii) the voting procedure of the ensemble. This work is an extension of paper [12].

The paper is organized as follows: Sect. 2 describes related work; Sect. 3 explains the families methods and the explored strategies. Sect. 4 presents the data-set characteristics and provides the experimental results analyzing the main findings. Section 5 provides the performance evaluation of the FASE's family on Sensor data-stream, identify which of the methods had the best performance in the experimental evaluation. Finally, Sect. 6 concludes.

2 Related Works

In this section, a bibliographic study of the Fast Adaptive Stacking of Ensembles (FASE) [7] and Hoeffding-based Drift Detection Methods (HDDM) [8] will be carried out aiming to highlight the methods related.

2.1 FASE

Fast Adaptive Stacking of Ensembles (FASE) [7] is based on the Online Bagging algorithm [9], and uses $HDDM_A$ as a drift detection mechanism to estimate the error. It has a set of adaptive learners to handle Concept Drift explicitly by detecting the changes and updating the model if a Concept Drift is detected. The adaptive learners estimate error rates (by the corresponding change detectors) with a predictive sequential approach (test-then-train). FASE uses weighted voting to combine the predictions of the main and alternative models. It uses a meta-classifier too, combining the predictions of the adaptive learners. For that, it generates a training meta-instance $M = (\widehat{y}_1, \ldots, \widehat{y}_j, \ldots, \widehat{y}_k; y)$ where each $\widehat{y}_j$ is an attribute value and y is its corresponding class label. Each attribute value $\widehat{y}_j$ of the meta-instance M corresponds to the prediction from classifier h_j for the example $\mathbf{z}$. The class label of the meta-instance M is the same label of the original training example [7].

2.2 HDDM

Hoeffding-based Drift Detection Methods (HDDM) authors [8] propose to monitor the performance of the base learner by applying "some probability inequalities that assume only independent, univariate and bounded random variables to obtain theoretical guarantees for the detection of such distributional changes". $HDDM_A$ "involves moving averages and is more suitable to detect abrupt changes" and the second $HDDM_W$ "follows a widespread intuitive idea to deal with gradual changes using weighted moving averages". For both cases, the Hoeffding inequality [10] is used to set an upper bound to the level of difference between averages.

3 FASEO and FASEB Families Methods

This section introduces two families of classifier ensemble methods derived from FASE: FASEO and FASEB. These algorithm families are originated using different change adaptation strategies and methods to combine the predictions of the classifiers that make up the ensemble.

3.1 Overview of the Methods

According to [11], when designing ensemble of classifiers two main points must be considered: (i) how the base classifiers in the ensemble are updated and (ii) how they are combined to make a joint prediction. Taking into account the above assumption, variations were introduced in FASE to search for a better balance between accuracy and other necessary resources (run-time) for its operation.

The original algorithm is composed of a set of adaptive classifiers. Each one is formed by a base classifier and an alternative classifier (both classifiers include a drift detection mechanism) that is generated each time the base classifier issues a *warning* state. Both classifiers, the main and the alternative, process each instance and by weighted voting determines its class. This strategy is followed in the adaptive classifiers that form the ensemble, but also in the level of the meta-classifier that receives as input the predictions of each classifier in the form of a meta-instance [7].

Considering this scenario, the derived methods aim to handle resources more efficiently while keeping accuracy at similar levels. Thus, the main proposed modifications made on FASE are (i) the update strategy and (ii) the voting procedure of the ensemble. As a result of these modifications, two families of ensemble algorithms derived from FASE were devised: FASEO and FASEB.

In general, one of the main modifications made to the algorithm variants derived from FASE was to eliminate the use of alternative adaptive classifiers and create, in the structure of the general model, a parallel ensemble. In this alternative ensemble, a classifier is activated and begins to train once one of the classifiers of the main ensemble reaches the *warning* level. On the other hand, when a concept-drift level is detected, then one of two variants is followed, (i) the oldest classifier, the one that stayed longer in the alternative classifier ensemble is promoted, or (ii) the classifier with the best accuracy is promoted. Based on these strategies, the families of algorithms FASEO and FASEB were created respectively.

Each family of algorithms is based on classifiers that integrate detection mechanisms. So, the associated detector triggers each of the three different drift signals manipulated in the model. The first two methods, (FASEO and FASEB), maintained the meta-classifier proposed in FASE in order to perform class voting while the others combine weighted voting for final decision. To determine the whole weight of each classifier, accuracy, entropy degree and class probabilities are combined in different ways. Moreover, they are also considered two-class voting strategies. The first variant uses combined voting using a meta-classifier, like the FASE algorithm. The second one uses combined voting using weighted majority voting in different ways. The description of each algorithm follows below: The description of each algorithm follows below:

- FASEO: To update the main ensemble, the classifier with more training time in the set of alternative classifiers is promoted. To determine the final class, a meta-classifier is used with its inputs being the meta-instances formed by the predictions of each classifier in the main ensemble.
- FASEO_{wv1}: As in FASEO, to update the main ensemble, the oldest classifier in the set of alternative classifiers is promoted. To vote the final class, the weight of each classifier is computed taking into account accuracy, entropy degree, and the class probability vector.

- FASEO$_{wv2}$: As in FASEO and FASEO$_{wv1}$, to update the main ensemble the classifier with more training time in the set of alternative classifiers is promoted. To vote the final class, the weight of each classifier is computed taking into account only accuracy and the class probability vector.
- FASEO$_{wv3}$: As in the three former cases, to update the main ensemble, the classifier with more training time in the set of alternative classifiers is promoted. To vote the final class, the weight of each classifier is computed taking into account only accuracy and entropy degree.
- FASEB [12]: To update the main ensemble, the classifier with the best accuracy among the alternative classifiers is promoted. The decision on the final class is given by a meta-classifier, whose inputs are the meta-instances formed by the predictions from each classifier in the main ensemble.
- FASEB$_{wv1}$: As in FASEB, to update the main ensemble, the classifier with the best accuracy among the alternative classifiers is promoted. To vote the final class, the weight of each classifier is computed taking into account accuracy, entropy degree, and the class probability vector.
- FASEB$_{wv2}$: As in FASEB and FASEB$_{wv1}$, to update the main ensemble, the classifier with the best accuracy among the alternative classifiers is promoted. To vote the final class, the weight of each classifier is computed taking into account only accuracy and the class probability vector.
- FASEB$_{wv3}$ [12]: As in the three last cases, to update the main ensemble, the classifier with the best accuracy is promoted. To vote the final class, the weight of each classifier is computed taking into account only accuracy and entropy degree.

3.2 The Update Strategy

This section provides a description of update strategy used in the FASEO and FASEB families. In a general way, the derived methods are updated once one of the learners (classifier with change detection mechanism) that compose the main ensemble, experiment any of the following change of states:

(i) A classifier initially *in-control*, triggered a *warning* (through its detection mechanism)
(ii) A classifier suddenly reaches the *drift* level from *in-control* state
(iii) A classifier reaches the *drift* level from state of *warning*
(iv) A classifier, currently in *warning*, return to the (by-default state) *in-control*

In (i), an alternative classifier is activated and placed in a parallel set (ensemble of alternative classifiers). When no *drift* is detected, the learning process is carried out by the learners of the main set.

When one of the classifiers of the main set reached an out-of-control signal (drift), (case (ii) or (iii)) the main set is updated, firstly the drifted classifier is deleted and then is promoted to the main ensemble a) the alternative classifier with the greatest accuracy (FASEB methods) or b) the oldest alternative classifier (FASEO methods). Once the alternative classifier is promoted, it is deleted from the parallel ensemble. Then, in (ii), the model activates a new alternative classifier, since, due to a sudden change, it was not created and, therefore, it was taken "borrowed" from a classifier that triggered warning, therefore, should be "returned to it".

In order to handle the final states there is an arrangement of states: initially, it is assumed that each value corresponding to the status of each classifier that is part of the main ensemble is *in-control* and therefore takes value 0; whenever a classifier of the ensemble enters *warning*, its status has value 1. When a classifier of the main ensemble detects a drift, either by going from *in-control* to *drift* or from *warning* to *drift*, the algorithm quickly updates the ensemble and the state is again *in-control*. Therefore, once the algorithm removes the classifier from the array of alternatives, it updates the state corresponding to the new classifier that became part of the main ensemble to *in-control*.

3.3 Class Voting Strategies

An ensemble of classifiers $H(\mathbf{x})$ are models learned from a set of classifiers $h_1(\mathbf{x}), ..., h_j(\mathbf{x}), ..., h_k(\mathbf{x})$. During the training process, it receives as an input a labeled instance $(\mathbf{x}_i, y_i)$, and the model $H(\mathbf{x})$ aims to predict the class $\widehat{y}_i$ of each instance in unlabeled data-set. To achieve this task, there is different ways of ensemble combination methods like stacking, voting schemes, unweighted voting schemes. Thus, in these methods the classification is done using the stacking and the weighted-voting strategies. The particular way in which these strategies are applied is explained below.

Meta-classifier Strategy: The goal of a meta-learning process is to train a meta-classifier (meta-learner), which will combine the ensemble members' predictions into a single prediction. Thus, the input of the meta-learner are the outputs of the ensemble-member classifiers. In this process, both the ensemble members and the meta-classifier need to be trained. The meta-classifier is the ensemble's combiner, thus it is responsible for producing the final prediction [13]. Similar to the ensemble-members, the meta-classifier of these methods is a one class classifier; it learns a classification model from meta-instances, whose attributes are nominal. As FASE, both methods uses a Prequential methodology to generate meta-instances; the idea of this methodology is to use each instance first to test the model, and then to train the model. Thus, for each original training instance $\mathbf{z} = (\mathbf{x}, y)$ it is generated a training meta-instance $M = (\widehat{y}_1, \ldots, \widehat{y}_j, \ldots, \widehat{y}_k; y)$, where each attribute value $\widehat{y}_j$ of the meta-instance M corresponds to the prediction from the base classifier j in the main ensemble for the original example $\mathbf{z}$. The class label of the meta-instance M is the same label of the original training example.

Weighted Voting Strategies: A weighted voting is a system in which not all learners have the same amount of influence over the outcome because their votes have a different weight. In the classification task, an ensemble classifier can combines the decision of a set of classifiers by weighted voting to classify unknown examples. The weighting methods are best suited for problems where individual classifiers perform the same task [14]. Therefore, that is the reason why in this work was used the weighted majority vote to obtain the final prediction of the class label.

To determine the weight of each classifier, the accuracy (a component of the weight derived from historical performance) and the degree of entropy in its classification (the component of the weight coming from the current behavior of the classifier) are taken into account. A similar idea was previously proposed in [15].

These component are combined in different ways originating three weighting approaches: (1) it is used the classifier accuracy weight, class probability vector and entropy weight; (2) it is used the classifier accuracy weight and class probability vector; and (3) it is used the classifier accuracy weight and entropy weight.

4 Experimental Results and Analysis

4.1 Data-Sets

Table 1 summarizes the main characteristics of the data-sets used in the experiments. A total of 4 synthetic generators and 8 real data-sets were considered. The synthetic data-sets were built with two different sizes: 10000 (10k) and 50000 (50k) instances. Abrupt and gradual controlled concept drifts were introduced. MOA framework allows us to simulate the different types of concept drift using a sigmoid function. Real data-sets employed are available on the MOA website. These data-sets have very diverse characteristics regarding the number of instances, the number of classes, and the presence or absence of different types of concept drift. This diversity allows us to better describe the real problem situations that algorithms may face.

4.2 Experimental Results and Analysis

This section compares the performance between the families of methods and FASE using the synthetic and real data-sets. Both in the synthetic (8) and real (8) data-sets each algorithm is tested and trained using the classifiers HT and NB. In summary, 32 experiments were carried out to evaluate the performance of each method according to the two metrics considered.

Tables 2 and 3 present accuracy rates and run-times achieved by the algorithms in synthetics and real data-sets using both NB and HT as base learners.

In order to improve the visualization, the methods name was exposed as in parentheses: FASEO (FO), $FASEO_{wv1}$ (FOwv1), $FASEO_{wv2}$ (FOwv2) and $FASEO_{wv3}$ (FOwv3). FASEB (FB), $FASEB_{wv1}$ (FBwv1), $FASEB_{wv2}$ (FBwv2), and $FASEB_{wv3}$ (FBwv3).

The first values appearing in each table refer to respective base-learner used, NB or HT. The first rows of each table show the results obtained over the

Table 1. Main characteristics of synthetic and real datasets

Type	Data-sets	Size	# Atributes	# Class
Synthetic	LED	10k & 50k	24	10
	Sine	10k & 50k	2	2
	Waveform	10k & 50k	40	3
	Random-RBF	10k & 50k	40	6
Real	Connect-4	67557	42	3
	Covertype-Sorted	581,012	54	7
	Lung-cancer	32	56	3
	NslKdd99	125,973	41	2
	Pokerhand-1M	1,000,000	10	10
	WineRed	1599	11	9
	Usenet-2	1500	100	2
	Sensor	2,219,803	5	54

synthetic data sets and the last rows show the results over real data sets. Each cell in the tables presents the values reached by the methods. The result indicating improvements with respect to FASE are in highlighted bold (the winner) and italics. Note that higher values in accuracy indicate better performance whereas, for the run-time, the lower values are the better.

Regarding accuracy, the FASE's families methods performs better using NB as base classifier. With this base classifier, FASEB outperformed FASE in 4 out 8 synthetics data-sets. FASE outperformed each one out of all derived methods in 4 data-sets. FASEO, $FASEO_{wv}$ and $FASEB_{wv}$ methods outperformed FASE in 3 data-sets. On the other hand, using HT, FASEB outperformed FASE in 3 out 8 synthetics data-sets. FASE outperformed each one out of all derived methods in 3 data-sets. $FASEO_{wv}$ and $FASEB_{wv}$ methods outperformed FASE in 2 data-sets and FASEO outperformed FASE in 3 data-sets.

In general, FASEB, FASE, $FASEB_{wv2}$ and FASEO presented the best average results. With synthetic data, FASEB had better performance than the other variants followed by FASE, FASEO and $FASEB_{wv2}$. The data-set on which the developed variants performed better were those obtained from the Led generator, where the methods that use weighted voting to perform classification reached better behavior, especially $FASEO_{wv2}$ and $FASEB_{wv2}$. Concerning real data-sets, $FASEO_{wv2}$ and $FASEB_{wv2}$ performed equally or better than FASE in 5 out of 8 real data-sets using NB as base classifier. Similarly, FASEB improved or tied FASE in 5 out of the 8 real data-sets when HT was employed. In general, $FASEO_{wv2}$ and $FASEB_{wv2}$ are the best-ranked methods.

Considering run-time, the implemented variants of FASE had better performance than FASE, in almost all data-sets. In general, $FASEB_{wv3}$, $FASEB_{wv2}$, $FASEB_{wv1}$ and $FASEO_{wv3}$ presented the best average results. In particular, the same result was obtained in synthetic data-sets. FASEB performed slower than

FASE more frequently in real data-sets. FASEO_{wv3}, FASEB_{wv3}, FASEB_{wv1} and FASEB_{wv2} are the best-ranked methods. In conclusion, the variant of FASE with a meta-classifier demanded more run-time to perform the classification than the others with weighted voting. Particularly, FASEB_{wv3} is the fastest among all proposed methods.

Table 2. Mean accuracies in percentage (%) with 95% confidence intervals in scenarios of abrupt and gradual concept drifts with artificial data-sets and real data-sets using NB and HT.

BC	DATA-SET	FO	FOwv1	FOwv2	FOwv3	FB	FBwv1	FBwv2	FBwv3	FASE
NB	LED-10k-gra	67,00	*67,77*	*67,79*	*67,51*	*67,21*	*67,79*	**67,81**	*67,51*	67,10
	Sine-10k-gra	*81,78*	81,52	81,52	81,35	**81,86**	81,60	81,59	81,35	81,71
	Waveform-10k-gra	77,62	77,56	77,53	77,36	77,62	77,58	77,55	77,36	**78,19**
	Random-10k-gra	30,90	30,81	30,79	30,92	30,94	30,84	30,82	30,85	**31,61**
	LED-10k-abr	*69,13*	*69,68*	**69,72**	*69,44*	*69,13*	*69,66*	*69,70*	*69,45*	68,64
	Sine-10k-abr	86,29	86,22	86,23	86,12	86,32	86,25	86,25	86,15	**86,42**
	Waveform-10k-abr	*78,63*	*78,53*	*78,50*	*78,38*	*79,00*	**78,65**	*78,53*	*78,50*	78,38
	Random-10k-abr	30,96	30,84	30,82	30,95	30,87	30,85	30,83	30,95	**31,58**
HT	LED-50k-gra	72,18	*72,52*	*72,57*	*72,49*	72,20	*72,52*	**72,58**	*72,50*	72,41
	Sine-50k-gra	90,80	90,20	90,21	89,84	90,84	90,21	90,22	89,84	**90,91**
	Waveform-50k-gra	81,45	80,54	80,70	80,31	81,42	80,61	80,79	80,38	**81,46**
	Random-50k-gra	*33,54*	32,55	32,64	32,74	**33,57**	32,53	32,67	32,71	33,38
	LED-50k-abr	72,52	*72,79*	*72,85*	*72,75*	72,52	*72,80*	**72,87**	*72,75*	72,73
	Sine-50k-abr	*91,98*	91,35	91,39	91,00	**92,01**	91,36	91,38	91,01	91,98
	Waveform-50k-abr	81,48	80,81	80,97	80,63	81,53	80,82	80,97	80,64	**81,59**
	Random-50k-abr	**33,64**	32,59	32,75	32,71	*33,60*	32,52	32,72	32,67	33,40
NB	Connect-4	74,47	**75,10**	**75,10**	*74,66*	74,48	*74,92*	*74,92*	74,64	74,66
	Covertype-Sorted	68,10	68,61	*69,55*	*69,15*	68,27	68,67	**69,74**	*69,35*	69,06
	Lung-cancer	**77,97**	65,57	66,82	65,57	**77,97**	65,57	66,82	65,57	**77,97**
	NslKdd99	**89,83**	89,79	89,79	89,75	**89,83**	89,79	89,79	89,75	89,81
	Pokerhand-1M	*50,10*	*50,10*	**50,11**	**50,11**	*50,10*	*50,09*	**50,11**	**50,11**	49,87
	WineRed	48,74	48,38	*52,86*	**54,09**	48,74	48,38	*52,86*	**54,09**	50,60
	Usenet-2	70,23	67,31	67,31	66,84	70,11	67,10	67,10	66,27	**72,86**
	Sensor	*86,41*	*89,36*	*88,06*	86,01	*87,15*	**89,86**	*88,30*	*87,00*	86,23
HT	Connect-4	*75,01*	74,89	**75,13**	74,50	*75,06*	74,78	*74,96*	74,44	74,94
	Covertype-Sorted	**74,08**	*72,13*	*72,71*	71,73	*73,70*	*72,46*	*73,28*	71,95	72,04
	Lung-cancer	**74,74**	67,29	66,46	67,29	**74,74**	67,29	66,46	67,29	**74,74**
	NslKdd99	98,62	98,35	98,47	98,41	98,62	98,35	98,47	98,41	**98,67**
	Pokerhand-1M	**54,40**	50,36	52,69	52,81	*54,30*	50,36	52,69	52,81	53,32
	WineRed	48,57	50,13	**54,41**	*54,25*	49,06	50,66	*54,24*	53,79	54,02
	Usenet-2	66,53	*68,07*	67,79	68,10	66,56	*68,07*	67,79	**68,10**	67,95
	Sensor	85,59	*89,39*	*88,43*	*86,71*	*86,06*	**89,55**	*88,84*	*87,71*	86,03

In order to conduct a statistical analysis of the derived methods regarding a control method (the original FASE), the *Friedman* test [16,17] and the Bonferroni-Dunn test [17,18] were applied. The tests were used with a significance level of 5%. The total of the experiment taken into consideration was 32, corresponding to the test performed by the base classifier (NB or HT) in synthetic and real data-sets.

Table 3. Mean of time in percentage (%) with 95% confidence intervals in scenarios of abrupt and gradual concept drifts with artificial data-sets and real data-sets using NB and HT.

BC	DATA-SET	FO	FOwv1	FOwv2	FOwv3	FB	FBwv1	FBwv2	FBwv3	FASE
NB	LED-10k-gra	2,67	2,47	2,34	2,45	2,65	2,30	2,33	**2,27**	3,03
	Sine-10k-gra	1,20	1,13	1,14	1,16	1,24	1,13	1,14	**1,12**	1,42
	Waveform-10k-gra	1,88	1,74	1,69	1,68	1,96	**1,62**	1,65	1,70	2,25
	Random-10k-gra	3,74	3,06	3,20	3,11	3,87	3,09	3,01	**3,00**	4,40
	LED-10k-abr	2,76	2,44	2,38	2,37	2,66	2,31	2,32	**2,30**	2,92
	Sine-10k-abr	1,23	**1,12**	1,15	1,14	1,26	**1,12**	**1,12**	**1,12**	1,39
	Waveform-10k-abr	1,91	1,70	1,70	1,71	2,19	1,94	**1,63**	1,66	1,70
	Random-10k-abr	3,81	3,09	3,26	3,21	3,87	3,16	3,03	**2,98**	4,43
HT	LED-50k-gra	16,39	15,97	15,81	14,87	17,32	**14,41**	15,14	15,16	17,75
	Sine-50k-gra	13,29	12,47	11,97	11,55	12,81	11,64	11,37	**11,12**	13,58
	Waveform-50k-gra	18,43	16,77	16,53	17,42	18,93	**16,51**	17,03	17,04	19,16
	Random-50k-gra	22,79	20,07	21,05	19,92	22,85	19,61	**18,93**	19,19	24,94
	LED-50k-abr	16,48	15,51	15,78	15,17	17,19	**14,37**	14,50	15,02	17,65
	Sine-50k-abr	13,90	12,64	12,45	11,77	12,88	**11,54**	11,67	11,59	13,52
	Waveform-50k-abr	17,52	16,12	16,24	16,80	18,32	16,48	**16,05**	16,12	19,32
	Random-50k-abr	22,86	20,10	19,90	19,68	22,28	19,78	19,32	**18,56**	24,37
NB	Connect-4	11,90	**11,81**	13,03	11,99	12,55	12,45	15,88	12,69	13,58
	Covertype-Sorted	156,96	139,68	132,32	128,72	140,94	124,66	122,60	**120,41**	149,64
	Lung-cancer	**0,05**	0,06	0,06	**0,05**	0,06	0,06	0,06	**0,05**	0,06
	NslKdd99	24,30	24,62	30,45	24,33	26,85	25,42	21,94	**21,60**	27,96
	Pokerhand-1M	183,64	129,78	125,88	126,00	163,13	**123,44**	126,20	130,43	181,06
	WineRed	0,56	0,53	**0,45**	0,51	0,56	0,52	0,54	0,49	0,70
	Usenet-2	0,89	0,92	0,91	0,86	0,86	0,85	0,90	**0,82**	0,92
	Sensor	1028,25	693,5	678,79	695,9	1042,65	**635,06**	644,64	654,65	1153,36
HT	Connect-4	**21,87**	22,60	23,61	22,21	27,94	33,10	26,91	23,55	27,08
	Covertype-Sorted	372,31	269,34	278,12	267,18	333,73	262,40	268,59	275,65	245,78
	Lung-cancer	0,11	0,11	0,14	0,12	0,14	0,17	**0,10**	0,12	0,11
	NslKdd99	1256,79	1218,50	1107,60	1137,01	1052,11	**1013,71**	1155,79	1109,51	1069,80
	Pokerhand-1M	605,85	472,22	427,66	435,17	559,44	**421,31**	433,76	445,22	438,03
	WineRed	0,93	**0,88**	0,92	0,90	1,01	0,94	0,98	1,00	1,09
	Usenet-2	1,51	**1,22**	1,37	1,38	1,70	1,56	1,35	1,40	1,70
	Sensor	4697,27	1085,96	**1059,98**	1076,79	4547,49	1093,21	1063,68	1078,59	1384,75

Regarding accuracy, the best-ranked method was FASEB. FASEB and FASE are significantly better than $FASEB_{wv3}$ and $FASEO_{wv3}$ tacking into consideration all data-sets. With respect to the other methods the observed differences were not statistically significant. The same happens in synthetic data-sets. On the other hand, $FASEO_{wv2}$ is the best ranked in real data-set, but the methods do not present significant differences.

Concerning run-time, the best ranked algorithm was $FASEB_{wv3}$. Tacking into consideration all data-sets, FASE, FASEB and FASEO are the worst ranked and significantly less fast with respect to the others methods. Regarding synthetic data-sets FASE and FASEB are more time consuming methods, significantly less fast with respect to all methods (except FASEO). On the other hand, $FASEO_{wv3}$ and $FASEB_{wv3}$ are the best ranked and only they presents statistical differences respect to the FASEB and FASE in real data-sets.

Figure 1 showed the best measurements located in the first positions. Particularly, Figs. 1(a) show a comparison of the methods accuracies using the Bonferroni-Dunn test in synthetic and real data-sets. Figure 1(b) show a comparison of the methods run-time using the same test tests in all data-sets.

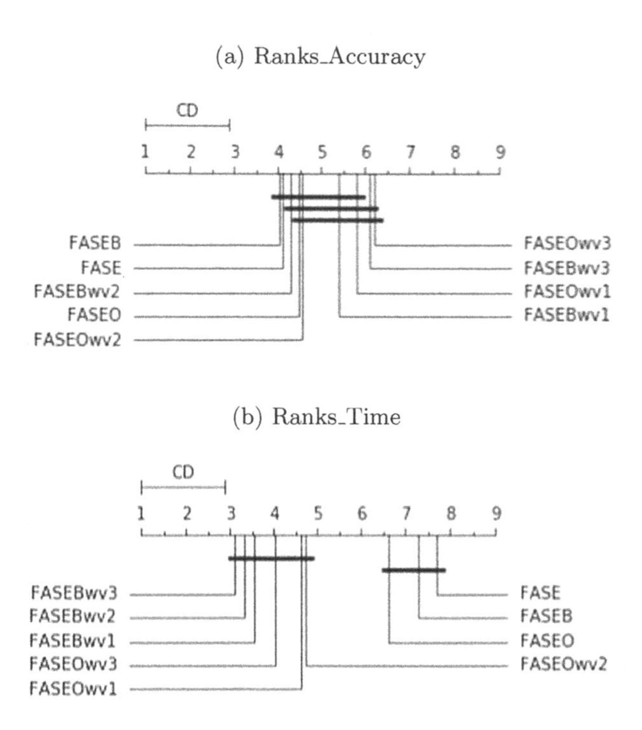

Fig. 1. Comparison of methods accuracies and time using the Bonferroni-Dunn tests with a 5% of significance level.

5 Application of the FASE Family Methods on the Sensor Data-Stream

A sensor, is a device that detects some physical stimulus (such as heat, light, sound, pressure, magnetism, or a particular motion) and responds usually with a transmitted signal resulting of impulse (as for measurement or operating a control) [6]. Normally, it is used to record that something is present or that there are changes in something [19]. Hence the importance of validating the behavior of these methods in data stream from sensors, because it represents a high complexity problem likely to presenting concept drift.

Particularly, the present research compared the performance of the FASEB, FASEO families and FASE method on the Sensor data-set. Sensor Stream [20] contains information collected from 54 sensors deployed in the Intel Berkeley Research Lab (temperature, humidity, light, and sensor voltage). It contains consecutive information recorded over a 2 months period, with one reading every 1–3 minutes. The sensor ID is the target attribute, which must be identified based on the sensor data and the corresponding recording time. This data-set is constituted of 2,219,803 instances, 5 attributes, and 54 classes. This is an interesting and intriguing data-set because, in addition to being much larger than the others, produces considerable variations in the accuracy performance of the methods.

Tables 2 and 3 presented the performance of the evaluated methods on the Sensor data-set among others. In addition, Fig.2 shows the results achievement on Sensor data-set using NB and HT.

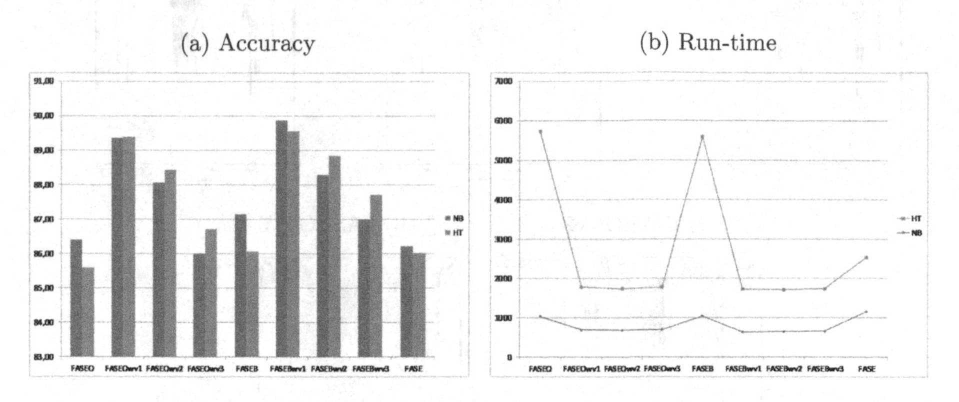

Fig. 2. Comparison of methods accuracies and run-time using NB and HT on Sensor data-stream.

As shown in the Fig. 2, the method that had the best performance on Sensor data-stream using NB and HT was $FASEB_{wv1}$ followed by $FASEO_{wv1}$ regarding the accuracy achievement. The worst result was by FASEO with HT. In general, this method and FASE had a worse rank. On the other hand, $FASEB_{wv2}$ was the best-ranked method regarding run-time. In particular, with NB the best result was achieved by $FASEB_{wv1}$, and $FASEO_{wv2}$ using HT. In general, FASE, FASEO and FASEB were the least fast.

Figure 3 show the performance of FASE derived algorithms with respect to the original algorithm using NB, when processing a fragment of 100,000 instances of the Sensor data-set. As it is possible to see, in all the methods the performance over time during the processing of the instances is more stable both in the FASEB and FASEO families regarding FASE, except in the $FASEO_{wv3}$ method in NB. Frequently, it can be seen that the accuracy values in all cases in the FASE method fall down more than in the methods derived from it.

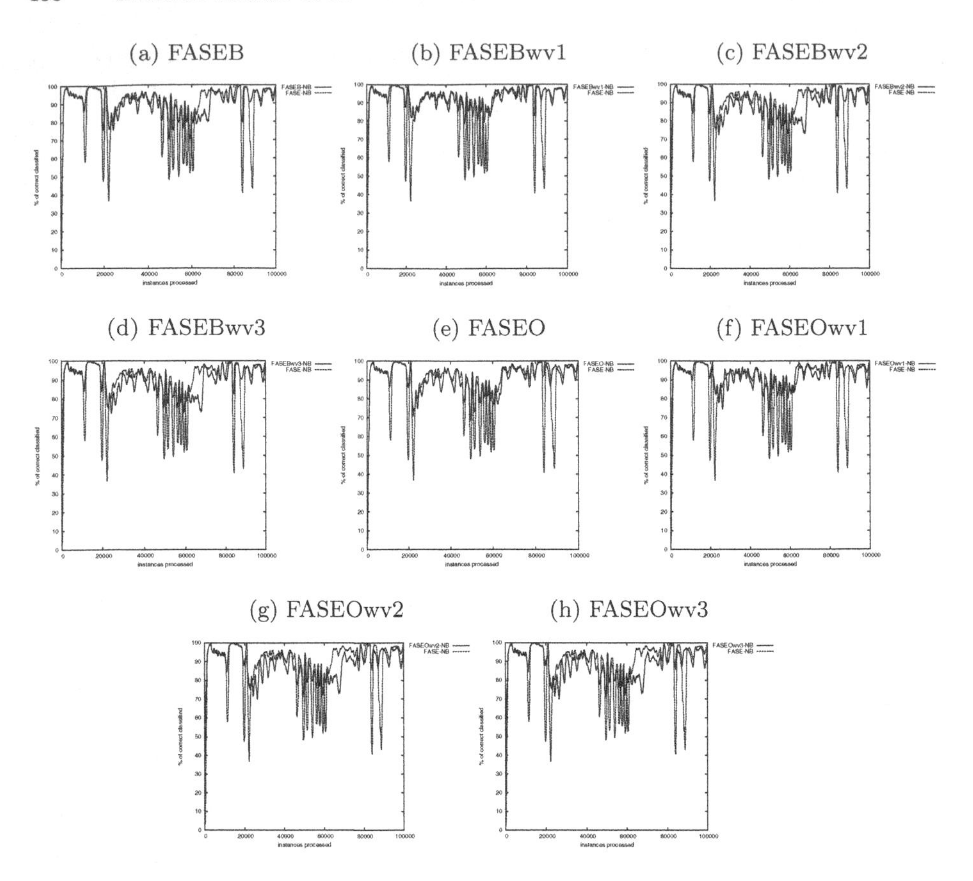

Fig. 3. Performance evaluation of the methods with respect to FASE regarding the accuracy over time on the Sensor data-stream using NB.

6 Conclusions

This work proposed FASEB and FASEO families of algorithms, a total of eight ensemble methods for operation in concept drift scenarios with full access to labeled classes. The algorithms are variants of the FASE ensemble. The main difference of FASEB and FASEO families as compared to FASE is the update strategy employed by the algorithms. While FASE uses adaptive classifiers to keep the ensemble updated, the implemented algorithms have in common a parallel ensemble formed by alternative classifiers, activated and set to be trained when one of the classifiers in the main ensemble issues a warning.

When a Concept Drift is detected, algorithms in the FASEB family boosts the alternative classifier with the greatest accuracy. Algorithms in the FASEO family, instead, promote the oldest active alternative classifier. The proposed variants were compared to FASE through similar parametrization and same testing conditions in order to accordingly evaluate their performance, using HT and NB as base learners. In terms of accuracy, FASEB obtained the best results in most of the tested data-sets using HT and NB, but it was noticed a very close approximation of FASEB_{wv2} as compared to those of FASEB.

In addition, the performance of the methods studied in the Sensor datastream was analyzed because it is a data-set of high complexity, large size, and representative of a real problem prone to presenting concept drift. FASEB_{wv1} achieved the best result regarding accuracy. FASEB_{wv2} was among the first 3 results in accuracy and, at the same time, it was the most rapid.

The statistical significance of the results provided by the experiments were evaluated using the non-parametric Friedman test together with the Bonferroni-Dunn test. Those tests confirmed the proposed algorithms were often significantly better than FASE with respect to run-time. In particular, versions FASEB_{wv2} and FASEO_{wv2}, while not showing significant accuracy losses, were noticeably faster than the original FASE algorithm. This can be very useful in contexts that require quick access to partial information, a high level of accuracy is still needed, but a very fast decision has to be made.

References

1. Gama, J., Gaber, M.: Learning From Data Streams: Processing Techniques in Sensor Network. Springer, Heidelberg (2007). https://doi.org/10.1007/3-540-73679-4
2. Aggarwal, C.: Data Classification: Algorithms and Applications (2014)
3. Frías-Blanco, I.: Nuevos métodos para el aprendizaje en flujos de datos no estacionarios. Granada University (2014)
4. Ortiz-Diaz, A., et al.: Fast adapting ensemble: a new algorithm for mining data streams with concept drift. Sci. World J. **2015**, 1–14 (2015)
5. Barros, R., Silas, G.: An overview and comprehensive comparison of ensembles for concept drift. Inf. Fusion Number C **52**, 213–244 (2019)
6. Stevenson, A.: Oxford Dictionary of English. Oxford University Press, Oxford (2010)
7. Frías-Blanco, I., Verdecia-Cabrera, A., Ortiz-Diaz, A., Carvalho, A.: Fast adaptive stacking of ensembles. In: Proceedings of the 31st Annual ACM Symposium on Applied Computing, pp. 929–934. ACM (2016)
8. Frías-Blanco, I., et al.: Online and non-parametric drift detection methods based on Hoeffding's bounds. Trans. Knowl. Data Eng. **27**(3), 810–823 (2015)
9. Oza, N., Russell, S.: Online bagging and boosting. Artif. Intell. Stat. 105–112 (2001). https://doi.org/10.1109/ICSMC.2005.1571498
10. Hoeffding, W.: Probability inequalities for sums of bounded random variables. J. Am. Stat. Assoc. **58**, 13–30 (1968)
11. Ortiz-Díaz, A.: Algoritmo multiclasificador con aprendizaje incremental al que manipula cambios de conceptos. Universidad de Granada, España (2014)
12. Mariño, L., Hidalgo, J., Barros, R., Vasconcelos, G.: Improving fast adaptive stacking of ensembles. In: International Joint Conference on Neural Networks (IJCNN), pp. 1–8. IEEE (2019)
13. Menahem, E., Rokach, L., Elovici, Y.: Combining one-class classifiers via meta learning. In: Proceedings of the 22nd ACM International Conference on Information & Knowledge Management, pp. 2435–2440. ACM (2013)
14. Shen, H., Lin, Y., Tian, Q., Xu, K., Jiao, J.: A comparison of multiple classifier combinations using different voting-weights for remote sensing image classification. Int. J. Rem. Sens. **39**(11), 3705–3722 (2018)

15. Song, G., Ye, Y., Zhang, H., Xu, X., Lau, R., Liu, F.: Dynamic clustering forest: an ensemble framework to efficiently classify textual data stream with concept drift. Inf. Sci. **357**, 125–143 (2016)
16. Friedman, M.: The use of ranks to avoid the assumption of normality implicit in the analysis of variance. J. Am. Stat. Assoc. **32**(200), 675–701 (1937)
17. Demšar, J.: Statistical comparisons of classifiers over multiple data sets. J. Mach. Learn. Res. **7**, 1–30 (2006). JMLR.org
18. Dunn, O.: Multiple comparisons among means. J. Am. Stat. Assoc. **56**(293), 52–64 (1961)
19. Dictionary Cambridge: Cambridge advanced learner's dictionary. PONS-Worterbucher, Klett Ernst Verlag GmbH (2008)
20. Qun, Z., Xuegang, H., Yuhong, Z., Peipei, L., Xindong, W.: A double-window-based classification algorithm for concept drifting data streams, pp. 639–644. IEEE (2010)

KNN Applied to PDG for Source Code Similarity Classification

Clóvis Daniel Souza Silva(✉), Leonardo Ferreira da Costa, Leonardo Sampaio Rocha, and Gerardo Valdísio Rodrigues Viana

State University of Ceará, Fortaleza, CE, Brazil
{clovis.souza, leonardo.costa}@aluno.uece.br,
{leonardo.sampaio,valdisio.viana}@uece.br

Abstract. The source code similarity problem consists in defining if two given distinct codes are the same program or not. This problem is valuable for polymorphic malware detection, which can generate distinct versions of code applying rules of obfuscation that change the original code. A technique to measure the similarity between source codes is to model the codes as program dependency graphs (PDG) and find an alignment between the graphs that maximizes a similarity function. This work investigates the performance of distinct similarity metrics applied to a genetic algorithm (GA). To distinguish between similar codes and non-similar codes, we use the k-nearest-neighbors (KNN) algorithm based on the similarity of the alignment found by the GA. The experiments are conducted with a database proposed by this work, where the source codes were retrieved from the Codeforces website and obfuscated by the tool CXX-obfuscator.

Keywords: Program dependency graphs · Malware detection · Network alignment problem · Isomorphism problem

1 Introduction

To define if source code is a copy from another is a non-trivial task. There are many forms that a program can be written, and the definition of copy it's a topic to be discussed. After all, two distinct algorithms that solve a given problem does not consist of copies of the same program. In this work, we call two codes similars if one is a copy of the other. To be called a copy, the similar code passes by the obfuscation process to mask its similarity. This problem it's present in malware detection and plagiarism detection [6,7,10].

This work makes use of free source code from a competitive programming database [2], where the copies were generated through tools which applies the obfuscation process.

As strategy to find the similarity between the programs, we transform the programs in program dependency graphs (PDG) and use a genetic algorithm (GA) to find the best alignment between the graphs. The fitness presented by

R. Cerri and R. C. Prati (Eds.): BRACIS 2020, LNAI 12320, pp. 471–482, 2020.
https://doi.org/10.1007/978-3-030-61380-8_32

the alignment it's the similarity found between the source codes. This strategy is also used in [6,7,10].

This work proposes a new method to build the alignment between the graphs through the GA. Also, it proposes a new metric to the GA. And perform the classification of the pairs of source code using the k-nearest-neighbors algorithm.

2 Definitions

Given two source codes C_1 and C_2, the source code similarity problem consists in defining if both are the same program or not. This problem becomes hard when, without loss of generality, C_2 is generated from a set of obfuscation techniques applied to C_1. In this case, C_2 is said to be a version of C_1 and both are similars [6,7,10].

The techniques or operators used for obfuscation aims to hide the identity of a program. Those operators are the same used by polymorphic malwares [10]. The following operators were defined in [7,9]:

- Format Alteration: To erase or add blank lines, or random comments in the code.
- Variable Renaming: Change variable, functions and procedures names.
- Statement Reordering: Change statements ordering without damage the program's logic.
- Statement Replacement: Replace statements with equivalents.
- Control Replacement: Change types of loops.
- Junk Code Insertion: To insert random useless code in the code.

A technique used to measure the similarity between source code is to turn the source code in PDG and find the best alignment between the graphs. Those graphs treat instructions as vertices and variable usage, flow control, and instruction-variable dependence as edges. This work uses two models of PDG. The first, proposed by [7], is called variable dependence graph (VDG). The other model will be referred by PDG*, it's a model created by Frama-C tool [5].

The usage of PDG in the source code similarity problem makes it analogous to the network alignment problem (NAP). The NAP problem is a variant of the graph isomorphism problem (GIP), for which no polynomial-time algorithm is known. The GIP aims to decide if exists a preservative-structural bijection $f : V(G_1) \to V(G_2)$ for given two graphs G_1 and G_2. This bijection f is also called isomorphis between G_1 and G_2. The NAP aims to maximize a function $s(G_1, G_2, f)$ that compute the similarity between G_1 and G_2 for an injective function $f : V(G_1) \to V(G_2)$ representing the found alignment among the PDG [4].

The present papper makes use of two iterative genetic algorithms (IGA) to find an alignment between two PDG. Each IGA presents a distinct method to create an alignment f. The metrics used as fitness for these IGAS are EC, ICE, S^3 and C [6,8,12,13]. Beyond these, it'll be used a new metric proposed by this work, C*. Each IGA will take each metric as fitness and the final alignment will be expressed in function of all the metrics. This work uses the IGA to find the alignment between each pair of codes defined in the proposed database.

To classify the pair of codes as similars or non-similars, we use the KNN algorithm. The KNN takes the fitness of the resulting alignment and classifies the pair of code as similar based on the nearest known nearest neighbors.

The experiments are conducted with a database that we propose, generated from the Codeforces website [2].

3 Related Work

The NAP aims to maximize the total of edges existing in both graphs, namely, find a function $f : V(G_1) \rightarrow V(G_2)$ that maximizes the amount of edges obeying the biconditional $uv \in E(G_1) \leftrightarrow f(u)f(v) \in E(G_2)$ [4]. The related works use the GA as a technique to solve the NAP [6,7,10,13]. In any case, the individual I was defined as a permutation of $V(G_2)$, where G_2 represents the graph with greater size. The building method for the mapping $f : V(G_1) \rightarrow V(G_2)$ was defined as align the first $|V(G_1)|$ elements of I to the ordered elements of $V(G_1)$.

Those works show that in order to measure the similarity between programs as PDG is an efficient method [6,7,10]. However, finding the best alignment is as hard as deciding if there is an isomorphism between the graphs, thus it's ideal to reduce the search space making both graphs retain only the program's crucial information. From the two related models, PDG* and VDG, reduction methods were proposed only for VDG [7,10]. The referred works explore the source code similarity associated with polymorphic malware detection and plagiarism detection. Two of the related works classify source codes as similars and non-similars [6,7].

The VDG model defines each program's instruction as a vertex and assign to each vertex a variable. Starting from the declaration vertex v, when the assigned variable var is modified by another vertex u, the edge vu is built and the variable var is now assigned to u. If var was not modified, the edge vu is built but var still assigned to v. Hence the graph construction follow the code execution flow [7,10]. The PDG* has a building method more complex than VDG, the vertices can represent instructions, loops, and blocks of code, each with different colors. Also, the edges are created by the variable usage and the instruction control dependence. This makes the PDG* with greater size than VDG, but the coloring scheme can be used to improve the search for the best alignment.

The proposed database work with graphs of the same complexity presented in [7] and [6]. The average sizes of the graphs presented in [7] (VDG) and [6] (PDG*) are respectively 21.6 and 41.0 vertices with a standard deviation of 14.57 and 33.8. The total of edges stands for a mean of 37.0 with 26.0 as the standard deviation for the VDG, while for the PDG* 340.0 and 391.3.

4 Applied Metrics

To decide the best alignment for two given graphs, metrics considering the number of edges found in both graphs are used [4]. We call those conserved edges and define the total of conserved edges as $|E_c|$. The metrics are in function of G_1 or $G_{f(G_1)}$ and an alignment f. $G_{f(G_1)}$ is the inducted graph from $V(G_2) \subset f(V(G_1))$. Hence, we describe the following formulas, *edge correctness* EC, *induced conserved substructure* ICS, *symetric substructure score* S^3 and the similarity score C [6,8,12,13]:

$$EC(G_1, G_2, f) = \frac{|E_c|}{|E(G_1)|}. \tag{1}$$

The Eq. 1 emphasizes the conserved edges related to G_1. Which causes this equation to obtain low score if G_1 is denser than G_2 [4].

$$ICS(G_1, G_2, f) = \frac{|E_c|}{|E(G_{f(G_1)})|}. \tag{2}$$

The Eq. 2 emphasizes the conserved edges related to $G_{f(G_1)}$. Likewise EC, ICS obtains low score if $G_{f(G_1)}$ is denser than G_1 [4].

$$S^3(f) = \frac{|E_c|}{|E(G_1)| + |E(G_{f(G_1)})| - |E_c|}. \tag{3}$$

The Eq. 3 focus attention on the conserved edges related to the unique edges of G_1 and $G_{f(G_1)}$. In this case, the score will be low only when few conserved edges exist [13]. This equation scores 1.0 only if the found alignment was an isomorphism between G_1 and $G_{f(G_1)}$.

Another metric proposed by [6] takes advantage of the graph coloring presented in PDG*. Said that the model classifies vertex with colors, [6] presents a color based correctness metric. This dissimilarity metric is represented by Eq. 4, where C_v and C_e are vertices and edge correctness.

$$COR(G_1, G_2) = \alpha * C_v(G_1, G_2) + \beta * C_e(G_1, G_2). \tag{4}$$

The vertices correctness is the total of matched vertices with distinct colors. The edge correctness is the total of edges that should be removed or added to the alignment to become an isomorphism. The author propose 0.9 and 0.1 to the constants α e β, therefore the matching colors weights more than the structural behavior. The similarity between both graphs is represented by C metric showed in Eq. 5 [6].

$$C(G_1, G_2) = \frac{1 - 10 * COR(G_1, G_2)}{|E(G_1)|}. \tag{5}$$

5 Methodology

5.1 The Genetic Algorithm

The experiments were conducted with iterative genetic algorithms. The IGA breaks the problem in lesser problems until reaching an initial subproblem.

Hence, the GA computes the final population for the initial subproblem which is used as the initial population for the next subproblem - next iteration. This method is applied in problems which solution is composed by the solutions of its subproblems [6]. The Algorithm 1 shows the pseudocode of the iterative process.

Data: **Integer** n, **Graph** G_1, G_2
Result: **Individual** P
1 $m \leftarrow Integer(\frac{|V(G_1)|}{n})$;
2 $aux \leftarrow m$;
3 $G_1^0 \leftarrow \{v_0, ..., v_m\} \in V(G_1)$;
4 $P_0 \leftarrow$ random_individuals($V(G_2)$);
5 **for** $i \leftarrow 0$ **to** $n - 1$ **do**
6     $P_{i+1} \leftarrow$ AG(P_i, G_1^i);
7     $G_1^{i+1} \leftarrow \{v_0, ..., v_m, ..., v_{m*i}\} \in V(G_1)$;
8 **end**
9 **return** best_from(P_n);

Algorithm 1: Iterative GA

At each iteration i, the algorithm tries to align the subgraph $G_1^i \subset G_1$ to G_2. Thus, a population P_{i+1} is produced and taken to the next iteration for a new subgraph $G_1^{i+1} \supset G_1^i$. The process is repeated until $G_1^i = G_1$, granting good initial individual to the GA.

The GA has as individual a permutation of $V(G_2)$. However, the fitness is computed by the mapping $f : V(G_1) \rightarrow V(G_2)$, which is built from the individual. This mapping is made by aligning the first $|V(G_1)|$ alleles of the individual to each vertex of $V(G_1)$ [6,7,10,13]. This work proposes a new greedy method to build the alignment from the individual. This method is presented in the Algorithm 2. In the experiments, both methods are used for each GA.

In the IGA it was used cycle crossover and random swap for selection. The population size was 100. The selection took the best individual plus roulette selection. The crossover probability was set to 0.9 and 0.1 for the mutation probability. Only 20 new individuals were generated per generation. The total of generations applied to all the tests was 100.

This work makes use of two IGA, one to each method of building alignment. The GA with the greedy neighborhood alignment will be called IGA-greedy. The other will be called IGA-hybrid. The IGA-hybrid will use the local search at the end of each iteration. The local search algorithm was presented in [6], and it was adapted to work on non-colored graphs.

5.2 Proposed Metric

In this work, it was also proposed a new metric as a variation of C. This variation aims to relate vertices correctness with the number of vertices and the edge correctness with the amount of edges. This approach makes use of Eq. 6.

```
    Data: Sequence V_g1, V_g2, Mapping adj_g1, adj_g2
    Result: Mapping M
 1  M ← ∅;
 2  foreach v ∈ V_g1, v ∉ M do
 3      foreach w ∈ V_g2, w ∉ M do
 4          M[v] ← w;
 5          break;
 6      end
 7      orderedAdj_g1 ← adj_g1[v] ordered by V_g1;
 8      orderedAdj_g2 ← adj_g2[w] ordered by V_g2;
 9      foreach v' ∈ orderedAdj_g1, v' ∉ M do
10          foreach w' ∈ orderedAdj_g2, w' ∉ M do
11              M[v'] ← w';
12              break;
13          end
14      end
15  end
16  return M;
```

Algorithm 2: Greedy neighborhood alignment (mapping)

$$C^*(G_1, G_2) = 1 - \alpha\frac{C_v(G_1, G_2)}{|V(G_1)|} - \beta\frac{C_e(G_1, G_2)}{|E(G_1)|}. \quad (6)$$

With this variation, we hope to relate the correctness with its entity emphasizing the lesser graph.

5.3 Method

The experiments of this work aim to bring a comparison between the presented metrics. The conducted experiment applied both IGA, greedy and hybrid, to PDG* and VDG. From this, it was possible to see which of the models distinguish better between pairs of similar code and pairs of non-similar code.

The PDG were produced from source code samples from the proposed database. For each code, both models were applied. For the PDG*, the metrics EC, ICS, S^3, C, and C^*. To the VDG model, since this model is not colored, only the metrics that didn't consider coloring were used EC, ICS, S^3.

We applied the KNN algorithm to measure the efficiency of each metric for classifying the pairs of code. It was used $k \in 3, 5, 7$. It was computed mean accuracy, specificity, and sensibility from 100 executions of the KNN for each k. The test set was randomly selected at each iteration and consisted of 40.00% of all pairs of samples. At last, we show the best metric for classification applied to each IGA.

The KNN training was performed in two ways, the first consider only the fitness metric to classify the pairs of code. The second is a multi-metric classification, which considers not only the metric used for fitness but all the metrics cited in this work. The usage of multi-metric for classification of source code similarity was proposed in this work. The Fig. 1 shows the proposed schema.

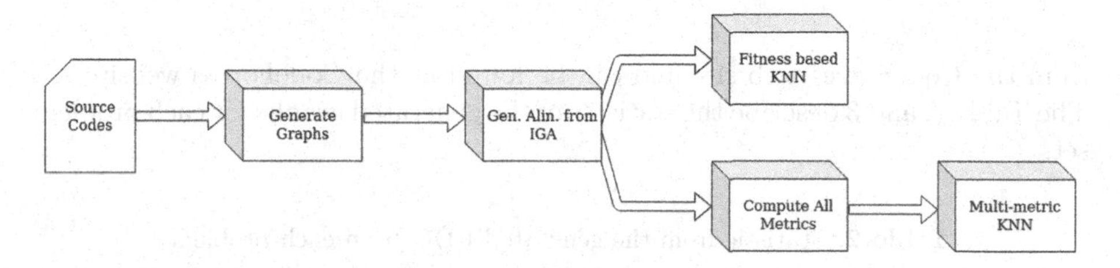

Fig. 1. View of the system

6 Database

To compare the PDG models, instances was collected in C programing language from the web site CodeForces [2]. Five problems were randomly chosen, with 10 unique samples for each problem. Two processes of obfuscation were applied to duplicate unique samples generating the fraud code:

- It was employed the CXX-obfuscator tool [3] to the unique samples, which applies format alteration, variable renaming and statement replacement.
- The second method applies the control replacement and statement replacement operators before to use the CXX-obfuscator tool.

It was produced a dataset with 138 source code samples, where 46 are uniques, 46 obfuscated by the first method and the last 46 are obfuscated by the second method. The Table 1 shows the graphs statistics generated by the samples. In the described table, the graphs represent by VDG^R and VDG are the same, but the VDG^R applies a reduction process on VDG graphs [7].

The reduced size of VDG comparing with PDG* is caused by two reasons. First, the number of instructions in the programs. Each selected problem has a max of 70 lines of code. This makes an upper bound to the number of vertices of VDG since it'll create a vertex for each instruction. The amount of edges is explained by the fact that VDG models only instruction-variable relationship, while PDG* cares about data transformation, data assignment, and flow control. PDG* also presents declaration-vertices, control-vertices, and conditional-vertices, which are always linked to use-vertices or declaration-vertices. All this information makes the PDG* bigger and denser.

Source code with at most 60 instructions was selected to minimize execution cost, also this approach brings graphs with the same complexity as [7] [6]. The choose problem were: 266B, 1196D, 1201C, 1203C e 1213C. All of them retrieved

Table 1. Generated graphs information

	Vertices mean	Vertices standard deviation	Edges mean	Edges standard deviation
PDG*	64.08	30.29	321.22	364.74
VDG	31.56	12.51	57.34	26.88
VDG^R	20.58	10.07	48.58	25.29

from the CodeForces web site and can be found at the CodeForces website [2]. The Tables 2 and 3 describe the statistcs of the generated graphs for each problem set.

Table 2. Statistic from the generated PDG from each problem

	Vertices mean	Vertices std.	Edges mean	Edges std.
266B	67.16	18.45	311.63	118.78
1196D	109.03	29.79	856.25	438.10
1201C	55.82	14.66	191.50	268.36
1203C	33.55	5.40	31.75	19.61
1213C	56.26	10.75	232.33	101.54

Table 3. Statistic from the generated VDG from each problem

	Vertices mean	Vertices std.	Edges mean	Edges std.
266B	17.16	5.04	33.90	12.44
1196D	32.93	11.98	80.10	25.37
1201C	23.10	6.31	63.53	13.42
1203C	16.00	3.82	33.76	9.29
1213C	13.73	6.73	31.63	13.91

The similarity between the generated codes was also measured by the MOSS tool to compare with the malware samples used in [6,7,11]. The MOSS similarity tool, indicates de similarity between a pair of code in two matches. The first match indicate the percentage of the first code is similar to the second code. The second match indicates the percentage of the second code is similar to the first code. The Table 4 shows the similarity distribution to each pair of similar code presented in both databases. Therefore, this database presents codes with the same MOSS similarity level than the malwares used in [6,7].

The conducted experiments makes use of 30 pairs of graphs from each problem set, half similars pairs. The total amount of pairs of code was 150.

Table 4. MOSS similarity from this database

Similarity	This Work		Kim 2016	
80%–100%	6	7	21	9
60%–80%	30	23	10	3
40%–60%	26	17	2	6
20%–40%	28	25	4	9
below 20%	20	38	–	29
Not computed	28	28	–	–

7 Results

Both IGA were applied to the 150 pairs of samples, 30 for each problem set, so graphs of different sizes were considered. For this reason, the similarity pair found was normalized by the problem set which the pair belongs to. The data was normalized by the *Manhatam norm* and the KNN was fed by this normalized data.

The Table 5 shows the accuracy of the classification of the alignments found by both IGA. It is shown that the accuracy of the alignments found by both GA were very low for the GDP* model. Still, for the metrics C e C^*, the IGA-greedy shows better accuracy in comparison to the IGA-hybrid. The best model shown in the table was VDG, this shows that VDG still contains crucial information to distinguish the similarity. The best approach was IGA-greedy(VDG), with $k = 7$. The bold values present the high accuracy found for the specified IGA. In Table 5, The bold numbers present the metric wich obtained a better accuracy for the GA configuration.

Table 5. Fitness based KNN classification - Single metric classification

IGA config.	k = 3					k = 5					k = 7				
	EC	ICS	S^3	C	C*	EC	ICS	S^3	C	C*	EC	ICS	S^3	C	C*
Greedy (GDP)	0.52	0.66	0.65	**0.69**	0.60	0.54	0.67	0.67	**0.69**	0.60	0.53	**0.68**	0.67	0.67	0.62
Greedy (GDV)	0.74	**0.82**	0.78	–	–	0.74	**0.81**	**0.81**	–	–	0.77	0.80	**0.82**	–	–
Hybrid (GDP)	0.54	**0.64**	0.50	0.50	0.51	0.56	**0.67**	0.50	0.51	0.58	0.57	**0.69**	0.50	0.54	0.60
Hybrid (GDV)	0.60	**0.74**	0.63	–	–	0.59	**0.75**	0.63	–	–	0.60	**0.75**	0.65	–	–

The Table 6 present the best results of the KNN multi-metric classification to both IGA. The metric in parenthesis in each row represents the fitness function applied to the GA, however, all the metrics were taken into account for the classification. Therefore, Table 6 shows the result of the pairs aligned by some metric and classified by all the metrics. The table also presents specificity, which is the ratio of non-similars correctly classified, and sensibility, which is the ratio of similars correctly classified. Considering this problem can be applied in malware detection, the sensibility should be the utmost importance. Hence, the configuration found that maximizes the sensibility is illustrated by bold font at Table 6. In this table, the bolds numbers highlights the GA configuration and metric with the max found sensibility.

Table 6. Multi-metric KNN classification - best results

IGA config.	k = 3			k = 5			k = 7		
	Acc	Sen	Esp	Acc	Sen	Esp	Acc	Sen	Esp
Greedy (GDP*, EC)	0.79	0.79	0.79	0.79	0.83	0.75	0.77	0.83	0.72
Greedy (GDV, EC)	0.89	0.89	0.89	**0.90**	**0.91**	**0.90**	0.91	0.90	0.92
Hybrid (GDP*, ICS)	0.72	0.76	0.70	0.75	0.76	0.73	0.75	0.76	0.74
Hybrid (GDV, ICS)	0.72	0.73	0.73	0.74	0.74	0.75	0.75	0.75	0.75

The IGA-hybrid present best individuals for classification with ICS metric as fitness. At IGA-greedy, the same happens to EC metric. It's also possible to see that IGA-greedy generated individuals more clusterized than IGA-hybrid since its accuracy is greater for both PDG* and VDG. The IGA-greedy with VDG model presented greater accuracy, precision, and sensibility.

Besides that, the Table 6 shows a greater overall accuracy than Table 5 for all similarity measures. It's also interesting that the worst metric for classification in Table 5, EC, generated the best alignments for classification with all the metrics. This behavior is explained in Fig. 2, where the same alignment presents better segregation when exposed in function of distinct metrics. It also explains the better performance of the KNN algorithm when taking the multi-metric approach.

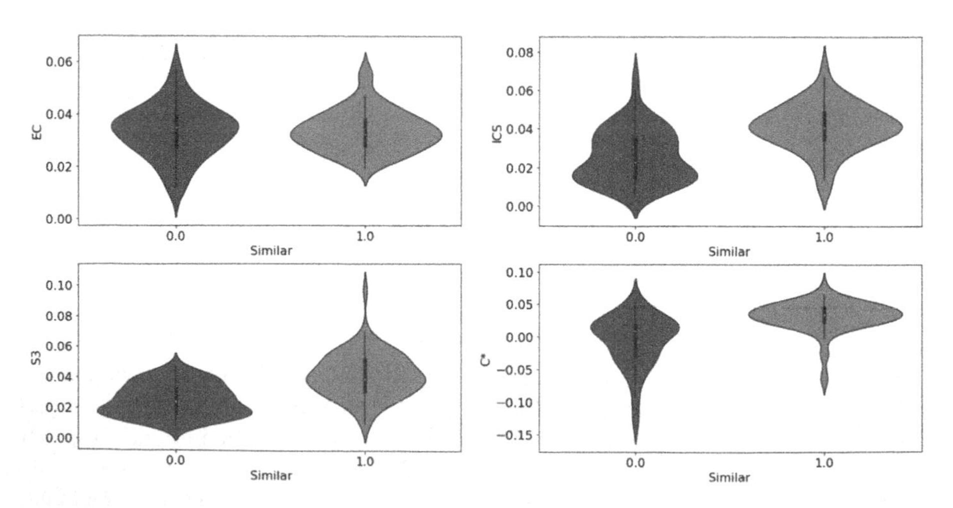

Fig. 2. Segregation from the alignments computed by IGA greedy(GDP,EC) for all metrics

The Fig. 3 shows the multi-metric approach for KNN with $k = 5$, the k which algorithm achieves a greater sensibility. It also shows the missing configurations from Table 6.

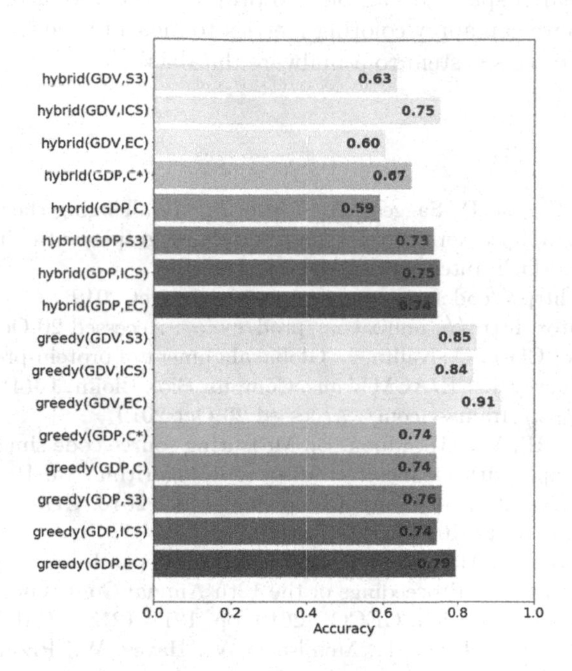

Fig. 3. Multi-metric KNN classification - summary

8 Conclusion

This work presented a study of the metrics associated with the network alignment problem (NAP) applied to source code similarity detection. We applied the iterative genetic algorithm (IGA) for the NAP. Two methods for alignment building was used in IGA, one with the proposed greedy neighborhood algorithm IGA-greedy, another with the method presented in [6,7,10,13]. Applying KNN to the found alignment, it was observed that the IGA-greedy with VDG has high accuracy than any other configuration. Also, the alignments founded by a fitness were more segregated when applied to another metric. With this, we also applied a multi-metric approach with the KNN. In this context, the metrics which better segregates similars codes from non-similars was EC and ICS. Hence, the experimentation shows that IGA-greedy with EC fitness had the highest accuracy with 90%, a sensibility of 91% and specificity of 90%. Also, all the tests were conducted in a database which shows a similarity distribution similar to the malware database presented in [6].

9 Future Works

As future work, it can be proposed methods for PDG* reduction, which graphs have a bigger search space. It can be also proposed a coloring technique for the VDG model, so we can apply coloring metrics to measure the similarity. At last, the application of this system to a malware database.

References

1. Carletti, V., Foggia, P., Saggese, A., Vento, M.: Challenging the time complexity of exact subgraph isomorphism for huge and dense graphs with VF3. IEEE Trans. Pattern Ana. Mach. Intell. **40**(4), 804–818 (2018)
2. CodeForces. http://codeforces.com/. Accessed 20 Oct 2019
3. CXX-Ofuscator. http://stunnix.com/prod/cxxo/. Accessed 20 Oct 2019
4. Elmsallati, A., Clark, C., Kalita, J.: Global alignment of protein-protein interaction networks: a survey. IEEE/ACM Trans. Comput. Biol. Bioinf. **13**(4), 689–705 (2016)
5. Frama-C. https://frama-c.com/. Accessed 20 Oct 2019
6. Kim, J., Choi, H., Yun, H., Moon, B.: Measuring source code similarity by finding similar subgraph with an incremental genetic algorithm. In: Proceedings of the Genetic and Evolutionary Computation Conference 2016. GECCO 2016, pp. 925–932. ACM, Denver (2016)
7. Kim, K., Moon, B.: Malware detection based on dependency graph using hybrid genetic algorithm. In: Proceedings of the 12th Annual Conference on Genetic and Evolutionary Computation. GECCO 2010, pp. 1211–1218. ACM, Portland (2010)
8. Kuchaiev, O., Milenokovic, T., Memisevic, V., Hayes, W., Przulj, N.: Topological network alignment uncovers biological function and phylogeny. J. Roy. Soc. Interface **50**(7), 1341–1354 (2010)
9. Liu, C., Chen, C., Han, J., Philip, S.: GPLAG: detection of software plagiarism by program dependence graph analysis. In: Proceedings of the 12th ACM SIGKDD International Conference on Knowledge Discovery and Data Mining, pp. 872–881. ACM, Philadelphia (2006)
10. Martins, G., Freitas, R. Souto, E.: Virtual structures and heterogeneous nodes in dependency graphs for detecting metamorphic malware. In: 2014 IEEE 33rd International Performance Computing and Communications Conference (IPCCC), pp. 1–8. IEEE, Austin (2014)
11. For a Measure Of Software Similarity. https://theory.stanford.edu/aiken/moss/. Accessed 20 Apr 2020
12. Patro, R., Kingsford, C.: Global network alignment using multiscale spectral signatures. Bioinformatics **28**(23), 28 (2012)
13. Saraph, V., Milenkovic, T.: MAGNA: maximizing accuracy in global network alignment. Bioinformatics **30**(20), 2931–2940 (2014)

Measuring Instance Hardness Using Data Complexity Measures

José L. M. Arruda[1], Ricardo B. C. Prudêncio[2], and Ana C. Lorena[1(✉)]

[1] Instituto Tecnológico de Aeronáutica, São José dos Campos, Brazil
aclorena@ita.br

[2] Centro de Informática, Universidade Federal de Pernambuco, Recife, Brazil
rbcp@cin.ufpe.br

Abstract. Assessing the hardness of each instance in a problem is an important meta-knowledge which may leverage advances in Machine Learning. In classification problems, an instance can be regarded as difficult if it gets systematically misclassified by a diverse set of classification techniques with different biases. The instance hardness measures were proposed with the aim of relating data characteristics to this notion of intrinsic difficulty of the instances. There are also in the literature a large set of measures which are dedicated at describing the difficulty of a classification problem from a dataset-level perspective. In this paper these measures are decomposed at the instance-level, giving a perspective of how each individual example in a dataset contributes to its overall complexity. Experiments on synthetic and benchmark datasets demonstrate the proposed measures can provide a complementary instance hardness perspective when compared to those from related literature.

Keywords: Machine learning · Instance hardness · Data complexity

1 Introduction

Machine Learning (ML) techniques have become largely adopted in real applications from different domains. Some learning problems found in practice can be simple (e.g., linearly separable problems), requiring few training data and the use of low-cost learning algorithms. On the other hand, many (possibly most) learning problems are very complex, possibly with noisy training data and complex decision boundaries, thus requiring more resources and expertise for delivering a successful solution. To understand problem difficulty is crucial not only to assess the potential gains and limits of ML in each application but also to provide explanations and to support algorithm selection and design [13].

Data complexity measures have been proposed along the years to assess the difficulty of learning problems, particularly the complexity of the available training data [1]. Different aspects have been taken into account like class separability and boundary complexity, relevance of input features, data sparsity and dimensionality [6]. These measures have been largely employed in meta-learning studies

R. Cerri and R. C. Prati (Eds.): BRACIS 2020, LNAI 12320, pp. 483–497, 2020.
https://doi.org/10.1007/978-3-030-61380-8_33

for outlining the domain of competence of learning techniques [7,9] and analyzing the diversity of data repositories [8,11], to name a few.

Instance-level analysis of learning difficulty is a more recent perspective in ML [10,12]. The objective is knowing which instances are misclassified and understanding why they are misclassified. Instance hardness measures can be used to understand how each instance (or group of instances) contributes to problem difficulty, which can also improve the learning process. For instance, recent papers have investigated the use of instance hardness measures for creating dynamic ensembles, in which locally competent classifiers are selected for difficult instances [2,14].

Therefore, whilst data complexity measures provide a global perspective of problem difficulty, instance hardness measures provide a local perspective which can be useful for a more fine-grained analysis of a learning problem. Despite sharing common objectives, the connection between the two topics has not been deeply explored yet. On one hand, data complexity measures can be derived from instance hardness measures by simply averaging the hardness values of all instances in a dataset, which was actually done in [12]. On the other hand, instance hardness measures could be derived from data complexity measures, for example, by decomposition. This research direction can be fruitful since the literature of data complexity measures is more consolidated.

In this paper, we follow the approach of measuring instance hardness based on the individual contribution of each instance in the global complexity of a dataset. A set of complexity measures from different categories is then adapted or decomposed to assess how each individual feature contributes to the dataset complexity regarding aspects such as feature overlapping, class separability and dataset topology. Next, these measures are evaluated experimentally in two types of datasets: synthetic 2-dimensional datasets, allowing to visualize the hardness values extracted in the input space; and on a set of benchmark datasets of public repositories, to show how they generalize to datasets with more features. Within these experiments, the log-loss error rates achieved by classifiers from different families are correlated to the values of the hardness measures. Whilst all measures can capture distinct instance hardness aspects, some of them are more successful in that task.

This paper is organized as follows: Sect. 2 describes the instance hardness measures from the literature and the dataset complexity measures. Section 3 discusses how the dataset complexity measures are used in instance hardness assessment. Section 4 presents the materials and methods used in the experiments, whose results are presented in Sect. 5. Section 6 concludes this work.

2 Related Work

The concept of instance hardness was introduced in the work of Smith et al. [12] as a way of pointing out which instances in a dataset are harder to classify and to understand why this happens. This type of knowledge can leverage the design of existent ML techniques, by focusing on instances with a particular set of characteristics. On the other hand, Ho & Basu [3] have tried to characterize if a classification problem is difficult by analyzing the training dataset available for learning. This can be considered a dataset-level analysis.

In this section, we present some relevant related work in both perspectives: dataset-level and instance-level. Initially, dataset complexity measures are introduced, as this topic is more consolidated in literature. Following, related work on instance hardness measures is presented. In this and the following sections, we consider that the measures are extracted from a training dataset T with n labeled data points $\mathbf{x}_i$. Each example is described by m input features and has a discrete class label y_i.

2.1 Data Complexity Measures

Ho & Basu [3] have presented measures which capture distinct aspects related to classification complexity and can be directly extracted from the learning datasets. Their objective is to assess how difficult a dataset is beyond the use of classification error metrics only. These measures are revisited in Lorena et al. [6], which also add some other recent measures. Three distinct categories of measures are common in previous work: (1) feature-based measures try to characterize the overlapping of the feature values in the classes; (2) linearity measures try to quantify whether the dataset is linearly separable; and (3) neighborhood measures characterize the dataset based on information from close examples. All measures are taken at the dataset-level, that is, they are calculated using all data in T.

Table 1 lists the complexity measures adapted in the current work, along with the category they belong to according to the division presented in [6]. Each measure was decomposed into instance hardness measures, as it will be seen in Sect. 3. There are other complexity measures in the literature that were not covered in our work. In fact, whilst some of the complexity measures can be easily decomposed to the instance-level, this is not true for other measures. For instance, the Fisher's discriminant ratio measure (with acronym F1) takes the average and standard deviation of the feature-values for all examples per class, which cannot be estimated from a single individual example. All measures based on plain classification error rates were not adapted either, since they originally only indicate whether an example is misclassified or not, without any confidence level. Although probabilities of classification could be employed instead, we opted to disregard such measures. Finally, other existing categories, like network-based measures which are extracted from a graph built from T, are left for future work.

(a) Feature-Based Measures. Both F2 and F3 measures in Table 1 take into account the overlap of the distributions of input feature values in each class. The size of the overlapping region for a feature f_j can be computed according to Eq. 1 for a problem with two classes.

$$overlap(f_j) = \max\{0, \min\max(f_j) - \max\min(f_j)\}, \tag{1}$$

Table 1. Data complexity measures used in the work.

Category	Name	Acronym
Feature-based	Volume of overlapping region	F2
	Maximum individual feature efficiency	F3
Linearity	Sum of the error distance by linear programming	L1
Neighborhood	Faction of borderline points	N1
	Ratio of intra/extra class NN distance	N2
	Fraction of hyperspheres covering data	T1
	Local set average cardinality	LSC

where:

$$\min\max(f_j) = \min(\max(f_j^{c_1}), \max(f_j^{c_2})),$$
$$\max\min(f_j) = \max(\min(f_j^{c_1}), \min(f_j^{c_2})),$$
$$\max\max(f_j) = \max(\max(f_j^{c_1}), \max(f_j^{c_2})),$$
$$\min\min(f_j) = \min(\min(f_j^{c_1}), \min(f_j^{c_2})).$$

The values $\max(f_j^{y_i})$ and $\min(f_j^{y_i})$ are the maximum and minimum values of each feature in a class $y_i \in \{c_1, c_2\}$. Therefore, based on the maximum and minimum values of a feature assumed by the examples of a given class, it is possible to measure the extent to which they overlap. We take the feature overlapping concept to define instance hardness measures which consider whether the example is in one or more feature overlapping regions.

(b) Linearity-Based Measures. The L1 measure assesses if T is linearly separable by computing the sum of the distances of incorrectly classified examples to a linear boundary (obtained by a linear Support Vector Machine - SVM) used in their classification, as shown in Eq. 2. L1 requires taking the decision values of the SVM predictions for incorrectly classified instances and averaging them. Therefore, it is a measure that can be easily adapted to the instance-level, by taking the individual decision values.

$$L1(T) = \frac{1}{n} \sum_{\mathbf{x}_i : h(\mathbf{x}_i) \neq y_i} \epsilon_i, \tag{2}$$

where ϵ_i corresponds to the distance of $\mathbf{x}_i \in T$ to the correct side of the linear decision border, given by the model $h()$.

(c) Neighborhood-Based Measures. The N1 measure requires building a Minimum Spanning Tree (MST) from the dataset, where each vertex corresponds to an example and the edges are weighted according to their distance. N1 is then given as the percentage of vertices incident to edges connecting examples of opposite classes in the MST (Eq. 3). It is possible to regard on how many instances from other classes one data point is connected to.

$$N1(T) = \frac{1}{n}\sum_{i=1}^{n} I((\mathbf{x}_i, \mathbf{x}_j) \in MST \wedge y_i \neq y_j), \tag{3}$$

where I is the indicator function, which returns 1 if its argument is true and 0 otherwise.

N2 computes the ratio of two sums: intra-class and extra-class, as shown in Eq. 4:

$$N2(T) = \frac{\sum_{i=1}^{n} d(\mathbf{x}_i, nn(\mathbf{x}_i))}{\sum_{i=1}^{n} d(\mathbf{x}_i, ne(\mathbf{x}_i))}, \tag{4}$$

where $nn(\mathbf{x}_i)$ is the nearest neighbor of $\mathbf{x}_i$ belonging to its own class y_i; in turn, $ne(\mathbf{x}_i)$ is the nearest enemy, i.e., the nearest neighbor from a different class; $d()$ is a distance function between two examples. Again, one may easily estimate the contribution of each example to the overall sum.

T1 builds hyperspheres centered at each one of the examples, whose radius are progressively increased until they reach an example of another class [3]. Next, smaller hyperspheres contained in larger hyperspheres are discarded. T1 is then defined as the ratio of the remaining hyperspheres. In this paper we opted to extract a similar view of the dataset by using the Local Set concept described in [4]. The Local-Set (LS) of an example $\mathbf{x}_i$ is defined as the set of points from the dataset T whose distances to $\mathbf{x}_i$ are smaller than the distance between $\mathbf{x}_i$ and $\mathbf{x}_i$'s nearest enemy:

$$LS(\mathbf{x}_i) = \{\mathbf{x}_j | d(\mathbf{x}_i, \mathbf{x}_j) < d(\mathbf{x}_i, ne(\mathbf{x}_i))\}, \tag{5}$$

The Local Set Cardinality (LSC) of an example is then given by the size of $LS(\mathbf{x}_i)$. At a dataset-level, an average of the LSC values for all examples is taken. One may also characterize the hyperspheres formed around each example in $T1$ by extracting indices from their LS.

2.2 Instance Hardness Measures

Smith et al. [12] have defined a way of measuring which instances in a dataset are more difficult to classify. Herewith, instances that are frequently misclassified by a pool of diverse learning algorithms can be considered hard. The work also defines a set of hardness measures intended to understand why these instances are often misclassified. The proposed hardness measures are defined for each instance $\mathbf{x}_i \in T$ as:

- *k-Disagreeing Neighbors* (kDN): percentage of the k nearest neighbors of $\mathbf{x}_i$ which do not share its label.

- *Disjunct Size* (DS): builds a decision tree using the dataset T and, for each instance $\mathbf{x}_i$, verifies the relative size of the disjunct where it is contained.
- *Disjunct Class Percentage* (DCP): for each instance $\mathbf{x}_i$, it returns the percentage of instances in its disjunct which share the same label as $\mathbf{x}_i$.
- *Tree Depth* (TD): given by the depth of the leaf node that classifies the instance in a decision tree. There are two versions of this measure, using pruned (TD_P) and unpruned (TD_U) decision trees.
- *Class Likelihood* (CL): measures the likelihood of an instance belonging to its class assuming the input features are independent.
- *Class Likelihood Difference* (CLD): takes the difference between the likelihood of an example belonging to its class and the maximum likelihood it has to any other class.
- *Minority Value* (MV): it is the ratio of the number of instances sharing the same label of an instance to the number of instances in the majority class.
- *Class Balance* (CB): computes a similar ratio as MV, but taking the size of the complete dataset T instead.

Smith et al. [12] also use their instance hardness measures in two situations: to modify the learning algorithm of a Multilayer Perceptron classifier and to filter difficult instances from a dataset. In the first case, the error function minimized by the MLP training algorithm is modified so that more emphasis is given to non-overlapping instances. In the case of data filtering, the idea is to point instances which are harder than a given threshold as noisy.

3 Complexity-Based Instance Hardness Measures

We now present the definition of the complexity-based instance hardness measures. In [12], the authors report competitive results of their hardness measures when compared to those from the work of Ho & Basu [3], by averaging the values achieved for each of the instances at a dataset-level. Nonetheless, a reverse analysis can be done, that is, it is possible to decompose some of the complexity measures of Ho & Basu [3] to the instance-level, as investigated in this work.

In the following subsections, we present the proposed instances hardness measures, produced from decomposing the dataset complexity measures presented in Sect. 2. Each subsection is dedicated to a category of measure.

3.1 Feature-Based Measures

In the feature-based category, four instance hardness (HD) measures are proposed. The first one takes the number of features for which the instance lies in an overlapping area as:

$$F1_{HD}(\mathbf{x}_i) = \sum_{j=1}^{m} I(x_{ij} > \max\min(f_j) \wedge x_{ij} < \min\max(f_j)). \qquad (6)$$

The other three measures are based on the distance of each instance to the overlapping boundaries. This distance is calculated for each feature, as shown in Eq. 7. A transformation is applied so that a maximum hardness value is obtained for instances lying in the center of an overlapping region, as shown in Eq. 8. Since we have m features for a dataset, we take as suitable measures the minimum ($F2_{HD}$ - Eq. 9), mean ($F3_{HD}$ - Eq. 10) and maximum ($F4_{HD}$ - Eq. 11) of the transformed distance values registered for an example in relation to each feature.

$$d_o(\mathbf{x}_i, f_j) = \frac{\min\max(f_j) - x_{ij}}{\min\max(f_j) - \max\min(f_j)}, \tag{7}$$

$$d_o^t(\mathbf{x}_i, f_j) = \frac{1}{(1 + abs(0.5 - d_o(\mathbf{x}_i, f_j))}. \tag{8}$$

$$F2_{HD}(\mathbf{x}_i) = \min_{i=1}^{m} d_o^t(\mathbf{x}_i, f_j). \tag{9}$$

$$F3_{HD}(\mathbf{x}_i) = \frac{1}{m}\sum_{i=1}^{m} d_o^t(\mathbf{x}_i, f_j). \tag{10}$$

$$F4_{HD}(\mathbf{x}_i) = \max_{i=1}^{m} d_o^t(\mathbf{x}_i, f_j). \tag{11}$$

3.2 Linearity-Based Measures

The modified $L1_{HD}$ measure takes the distance from each instance to the linear SVM decision boundary. This distance is multiplied by the correct label of the example (taking the classes as $y_i \in \{-1, +1\}$), so that instances correctly classified by the linear SVM assume a positive value and instances incorrectly classified show negative values for this measure. In this case, lower values are expected for more difficult instances.

$$L1_{HD}(\mathbf{x}_i) = y_i \epsilon_i. \tag{12}$$

3.3 Neighborhood-Based Measures

The adaptation of the neighborhood data complexity measures to an instance-level analysis is straightforward for the measures N1 and N2. For $N1_{HD}$ one may take, for each point in the MST, the number of nodes from different classes it is connected to, as presented in Eq. 13. Therefore, instances which have more neighbors in the MST from another class(es) are considered harder.

$$N1_{HD}(\mathbf{x}_i) = \sum_{i=1}^{n} I((\mathbf{x}_i, \mathbf{x}_j) \in MST \wedge y_i \neq y_j) \tag{13}$$

In the case of $N2_{HD}$, the ratio of the intra-class and extra-class distances is taken for each individual example:

$$N2_{HD}(\mathbf{x}_i) = \frac{d(\mathbf{x}_i, nn(\mathbf{x}_i))}{d(\mathbf{x}_i, ne(\mathbf{x}_i))} \tag{14}$$

Lower values are expected for easier instances, whose distance to the nearest enemy exceeds the distance to the nearest neighbor from the same class. In order to obtain bounded values for this measure, we take as final output $1 - \frac{1}{1+N2_{HD}}$.

Finally, we added four other measures based on the LS of each example. The first is the ratio of the Local Set Cardinality (LSC) of each example to the total number of examples, adapted from [4]:

$$LCS(\mathbf{x}_i) = \frac{1}{n}|LS(\mathbf{x}_i)|, \tag{15}$$

where $LS()$ is expressed Eq. 5 and $|\cdot|$ represents the cardinality of a set.

The second measure (LS_{radius}) takes the radius of the LS of each example, giving an indicative of the hypersphere that can be formed around each example in the complexity measure T1, as expressed in Eq. 16.

$$LS_{radius}(\mathbf{x}_i) = d(\mathbf{x}_i, ne(\mathbf{x}_i)). \tag{16}$$

The last measures were defined in the work [5] for instance selection in classification problems and are evaluated here as potential measures of instance hardness too: usefulness (Eq. 17) and harmfulness (Eq. 18). The usefulness index of an instance $\mathbf{x}_i$ ($U(\mathbf{x}_i)$) corresponds to the number of instances having $\mathbf{x}_i$ in their local sets. A high usefulness value is expected for instances in dense parts of a given class, so that they are close to many examples from their own class. On the other hand, the harmfulness of an example $\mathbf{x}_i$ ($H(\mathbf{x}_i)$) is given by the number of instances having $\mathbf{x}_i$ as their nearest enemy. A higher harmfulness value is expected for examples in borderline areas or for noisy examples surrounded by examples from a different class, which can be considered harder to classify.

$$U(\mathbf{x}_i) = |\{\mathbf{x}_j|\ d(\mathbf{x}_i, \mathbf{x}_j) < d(\mathbf{x}_j, ne(\mathbf{x}_j))\}|. \tag{17}$$

$$H(\mathbf{x}_i) = |\{\mathbf{x}_j|\ ne(\mathbf{x}_j) = \mathbf{x}_i\}|. \tag{18}$$

4 Experiments

This section presents the experiments performed in this work in an attempt to validate the use of the measures from Sect. 3 for instance hardness assessment.

4.1 Datasets

Two types of datasets are employed in the experiments: synthetic and from benchmarks. The synthetic datasets were generated with the mlbench package from R and are illustrated in Fig. 1. They contain two continuous input features, two classes and a total of 1000 examples each. The dataset with two normals contain some overlap between the classes. Instances in these regions may be considered harder to classify. In the case of datasets circle and xor, while borderline cases can be regarded as harder to separate, examples contained in the center of the classes can be considered easier to classify.

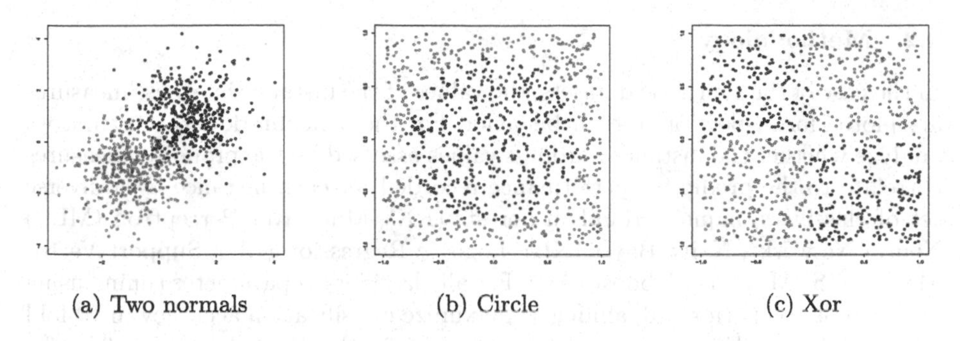

(a) Two normals (b) Circle (c) Xor

Fig. 1. Synthetic datasets used in the experiments

The benchmark datasets were collected from the OpenML repository [15] and present the characteristics summarized in Table 2: number of instances, number of features (nominal/continuous) and majority accuracy rate (MAR). MAR corresponds to the accuracy obtained when predicting all examples as belonging to the majority class. All datasets have two classes, since the feature-based and linearity measures are able to deal with binary problems only. Nonetheless, this does not prevent their use in multiclass datasets, which can be decomposed into multiple binary subproblems and have their hardness values averaged.

All datasets were also normalized in the 0–1 scale, which is required for some of the classifiers used. For classifiers requiring numerical features only, categorical features were encoded according to the one-hot encoding. Feature-based measures disregarded the categorical features in this set of experiments, since they could impose a large overlapping after one-hot encoding. And the neighborhood measures use the Gower distance measure, which is hybrid and deals with both categorical and continuous features.

Table 2. Benchmark datasets employed

Name	# Instances	# Features (cat./cont.)	MAR
Diabetes	768	0/8	0.65
Sonar	208	0/60	0.65
Fertility	100	8/1	0.88
Parkinsons	195	0/22	0.75
Churn	5000	4/16	0.86
Spam	4601	0/57	0.61
Ozone	2534	0/72	0.94
Sa-heart	462	1/8	0.65
Haberman	306	0/3	0.73

4.2 Methodology

In the case of the synthetic datasets, the values of the instance hardness measures are ploted for inspection of their behavior. For the benchmark datasets, first we analyze whether the instance hardness values assessed by the proposed measures correlate to the predictive performance recorded for each instance by a diverse set of classifiers, namely: Random Forests (RF), Multilayer Perceptron (MLP) Neural Networks, Naive Bayes (NB), Logistic Regression (LR), Support Vector Machine (SVM) and Adaboost (AB). For all classifiers, a parameter tuning using grid search was performed, aiming to maximize classification accuracy in 10-fold cross-validation. For each dataset, we keep only the three best classifiers for further analysis (if their performance is better than the MAR), in an attempt to guarantee the best performance in evaluating the difficulty of the instances. Next, the log-loss of the predictions was computed for each instance in a leave-one-out setting. This performance metric takes into account the probabilities of the predictions into each of the classes. As in [12], a high correlation to the error rates computed is expected for measures which capture the intrinsic classification difficulty of the instances.

Next, we verify if the proposed measures capture aspects distinct from those of the instance hardness measures of Smith et al. [12]. The correlation between pairs of measures from each of the previous groups is taken. Low correlation values indicate that the measures may be complementary.

5 Results and Discussion

5.1 Visualization of the Measures Values

Figure 2 presents plots of how the instances from the synthetic datasets were evaluated by the hardness measures proposed in this paper. In this figure, the datasets are shown in the columns and the instance hardness measures are presented in the rows. All plots were normalized so that colder colors are attributed to easier instances, while warmer colors represent harder instances.

It is possible to observe that all the measures present coherent results according to the hardness perspective they measure. For instance, for the feature-based measures, all instances in yellow are in overlapping regions for the two input features, while the blue points are not in any overlapping region. The results get smoother from $F1_{HD}$ to $F4_{HD}$ when the distance to the borders of the overlapping regions are taken into account. But, overall, since the feature-based measures consider overlapping regions orthogonal to each of the feature's axes, they are not able to fully capture the complexity of borderline cases for datasets "two normals" and "circle".

$L1_{HD}$ was not able to assess instance difficulty in non-linear datasets well and can be considered quite successful only for the dataset "two-normals", which shows a linear tendency. Some measures, such as $F1_{HD}$, $N1_{HD}$ and H, assume a lot of tied values for many instances. This may be a problem, since they do not allow to assess the difficulty of the instances smoothly.

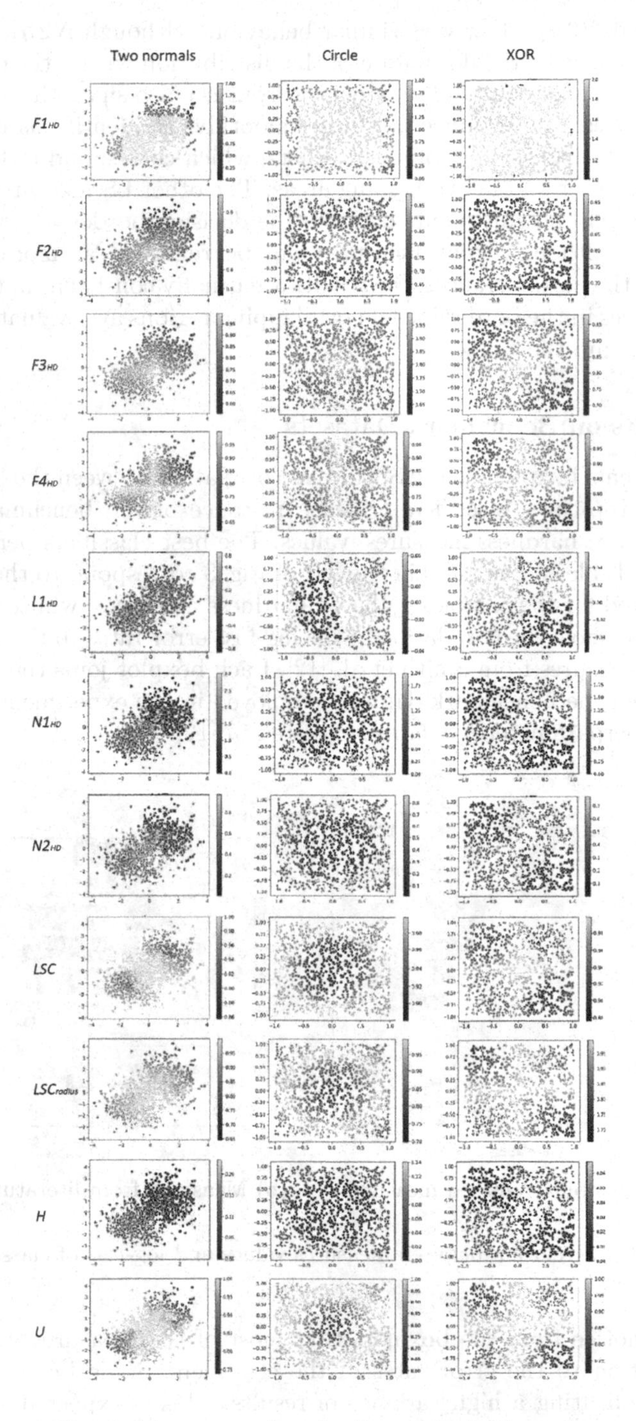

Fig. 2. Measures values for the synthetic datasets

$N1_{HD}$ and $N2_{HD}$ show very similar behaviors, although $N2_{HD}$ is smoother because it also considers information of the distribution within the classes. Most of the measures based on local sets (except from H) also share this property and are able to assess the instance hardness smoothly. H identify as difficult only points that figure as enemies of many others, which correspond to few points in the decision border and in overlapping areas. The other LS measures are able to characterize even difficult points far from the decision border.

Therefore, although some measures can be regarded as more effective in pointing out the hard instances, the results are usually consistent in the datasets, where instances in the borderline and overlapping regions are evaluated as harder than those within the classes.

5.2 Results on Benchmark Datasets

Figure 3 presents boxplots of the Spearman correlation between the log-loss error rates of the top-three classifiers for each instance of the benchmark datasets and the instance hardness measures' values. The best classifiers per dataset are presented in Table 3. The plots in the left of Fig. 3 correspond to the correlation values achieved for the complexity-based hardness measures, whilst the boxplots from the right show the correlation of the log-loss error rates to the values of the complexity measures from Smith et al. [12]. Each boxplot joins the results of all instances from the benchmark datasets employed in the experiments. kDN was run with hyperparameter $k = 5$, as suggested in [12].

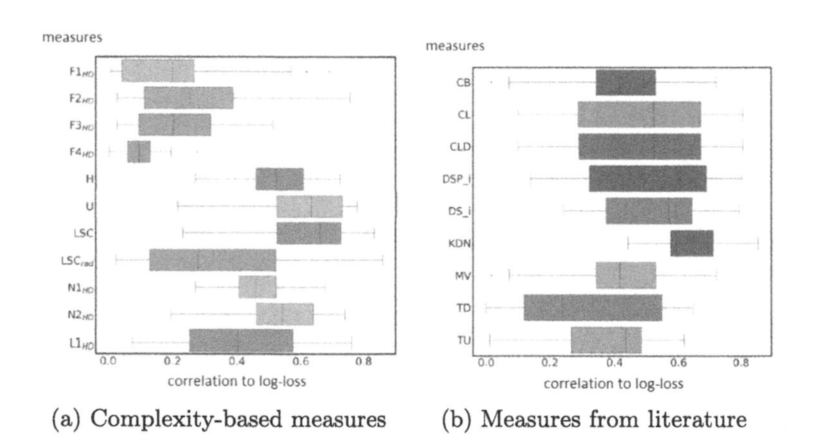

(a) Complexity-based measures (b) Measures from literature

Fig. 3. Correlation between measures values and log-loss of classifiers.

We can notice from the boxplots that most of the measures values show a high correlation to the error rates of the classifiers. Some boxplots are more elongated, indicating a high variance of results. This is expected, since we are joining the results of multiple classifiers and datasets. In fact, different causes of difficulty are captured by each measure, which can be present at different levels

Table 3. Classifiers used in the log-loss computation for each dataset (in dataset fertility, only two classifiers achieved accuracy rates above MAR).

Dataset	Top-performing classifiers
Diabetes	MLP, SVM, RF
Sonar	SVM, RF, AB
Fertility	SVM, RF
Parkinsons	SVM, RF, MLP
Churn	RF, MLP, SVM
Spam	AB, MLP, RF
Ozone	MLP, SVM, RF
Sa-heart	AB, MLP, LR
Haberman	RF, MLP, NB

in the datasets. On the other hand, the $N1_{HD}$ and harmfulness measures show quite stable results, but this is more related to the fact that there are multiple ties in their values. The feature-based measures were the less correlated to the log-loss error rates achieved, which can be attributed to their naive analysis of the overlapping of the classes as orthogonal to the features' axes. They also consider the features as independent and disregard categorical features. In the classification experiments, the NB algorithm, which also regards the features as independent was the worst performing technique in most of the datasets, which corroborates the previous observation that the joint information of the features must be taken into account to capture the information needed to discriminate the classes. The measures based on neighborhood information were in general the best descriptors of instance hardness when related to the classifiers performance, specially U, LSC and $N2_{HD}$, which showed higher correlation values.

The correlation results of the instance hardness measures of the literature (right side of Fig. 3) are usually high for all measures. Nonetheless, one must observe that some of them use classifiers such as decision trees and NB in their computation, which are also used to compute the log-loss values in some datasets and this may have biased the results.

To better evaluate the redundancy between the instance hardness measures proposed in this work and those from the literature, their Spearman correlation was computed. Figure 4 presents the heatmap of these results. The hardness measures of Smith et al. [12] are shown in the columns and the hardness measures from this work are in the rows. This plot joins the results achieved for the multiple datasets used in the experiments. The correlation values are higher for combinations of the kDN and the neighborhood-based measures, since they are all based on nearest neighbor or local information. But there are some measures with low correlation to the those from the literature, particularly the feature-based measures. There are, therefore, interesting complementary results for some pairs of measures that can be explored for better describing instance hardness.

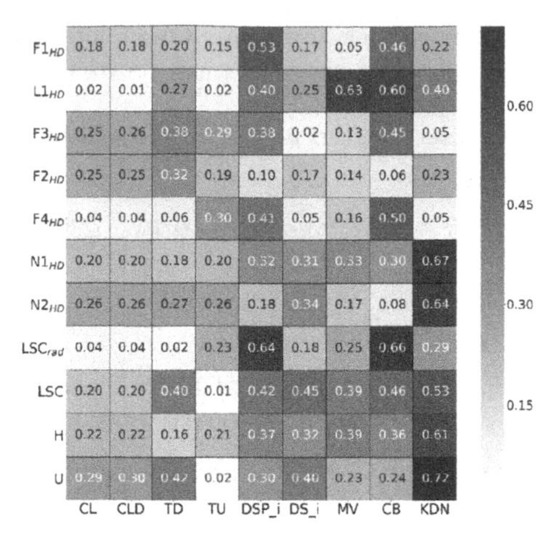

Fig. 4. Correlation between measures

From our experiments, measures such as LSC and $N2_{HD}$ can be regarded as promising alternatives to measure instance hardness and complementing the results from the pre-existing measures. They show high correlation to the log-loss error rates of classifiers and an average to low correlation to previous measures.

6 Conclusion and Future Work

This work adapted a set of measures devoted to estimate the complexity of the classification problems, which are taken at a dataset-level, to an instance-level. Herewith, we aim to quantify how much each instance contributes to the overall complexity of the problem, giving an indicative of instance hardness. We experimentally showed that these measures can be considered indeed good descriptors of instance hardness, allowing to rank the instances according to their difficulty level. This was verified visually in synthetic datasets and by correlating the measures values achieved for a set of benchmark datasets to the error rates of a set of competent classifiers tuned in order to maximize classification accuracy. In particular, all measures based on the concept of local sets, except from harmfulness, present interesting results. They do not have any parameters to be tuned either, an advantage over kDN, one of the most effective instance hardness measures from the literature. We have also shown that the proposed measures may complement those of the related literature.

As future work, we shall investigate the behavior of the metrics for datasets with more classes . We must also evaluate more deeply which measures contribute most for defining the hardness of the instances in the datasets and a comparison to other methods can be expanded. A promising direction can be relating the measures values to the instance hardness as measured by the framework of item response theory [10]. A combination of the measures values can also be evaluated. Finally, it would be worth investigating the effectiveness of such measures for supporting tasks such as dynamic classifier selection, noise detection, active learning, instance selection and self-paced and curriculum learning.

Acknowledgments. The authors would like to thank the research agencies CNPq (grant 305291/2017-3) and FAPESP (grants 2012/22608-8 and 2019/20328-7).

References

1. Basu, M., Ho, T.K.: Data Complexity in Pattern Recognition. Springer, Heidelberg (2006)
2. Dantas, C., Nunes, R., Canuto, A., Xavier-Júnior, J.: Instance hardness as a decision criterion on dynamic ensemble structure. In: 2019 8th Brazilian Conference on Intelligent Systems (BRACIS), pp. 108–113 (2019)
3. Ho, T.K., Basu, M.: Complexity measures of supervised classification problems. IEEE Trans. Pattern Anal. Mach. Intell. **3**, 289–300 (2002)
4. Leyva, E., González, A., Perez, R.: A set of complexity measures designed for applying meta-learning to instance selection. IEEE Trans. Knowl. Data Eng. **27**(2), 354–367 (2014)
5. Leyva, E., González, A., Pérez, R.: Three new instance selection methods based on local sets: a comparative study with several approaches from a bi-objective perspective. Pattern Recogn. **48**(4), 1523–1537 (2015)
6. Lorena, A.C., Garcia, L.P., Lehmann, J., Souto, M.C., Ho, T.K.: How complex is your classification problem? a survey on measuring classification complexity. ACM Comput. Surv. (CSUR) **52**(5), 1–34 (2019)
7. Luengo, J., Herrera, F.: An automatic extraction method of the domains of competence for learning classifiers using data complexity measures. Knowl. Inf. Syst. **42**(1), 147–180 (2013). https://doi.org/10.1007/s10115-013-0700-4
8. Macià, N., Bernadó-Mansilla, E.: Towards UCI+: a mindful repository design. Inf. Sci. **261**, 237–262 (2014)
9. Mansilla, E.B., Ho, T.K.: On classifier domains of competence. In: Proceedings of the 17th International Conference on Pattern Recognition, pp. 136–139 (2004)
10. Martínez-Plumed, F., Prudêncio, R.B., Martínez-Usó, A., Hernández-Orallo, J.: Item response theory in AI: analysing machine learning classifiers at the instance level. Artif. Intell. **271**, 18–42 (2019)
11. Muñoz, M.A., Villanova, L., Baatar, D., Smith-Miles, K.: Instance spaces for machine learning classification. Mach. Learn. **107**(1), 109–147 (2017). https://doi.org/10.1007/s10994-017-5629-5
12. Smith, M.R., Martinez, T., Giraud-Carrier, C.: An instance level analysis of data complexity. Mach. Learn. **95**(2), 225–256 (2013). https://doi.org/10.1007/s10994-013-5422-z
13. Smith-Miles, K.A.: Cross-disciplinary perspectives on meta-learning for algorithm selection. ACM Comput. Surv. (CSUR) **41**(1), 1–25 (2009)
14. Souza, M.A., Cavalcanti, G.D., Cruz, R.M., Sabourin, R.: Online local pool generation for dynamic classifier selection. Pattern Recogn. **85**, 132–148 (2019)
15. Vanschoren, J., van Rijn, J.N., Bischl, B., Torgo, L.: OpenML: networked science in machine learning. SIGKDD Explorations **15**(2), 49–60 (2013)

Simulating Complexity Measures on Imbalanced Datasets

Victor H. Barella[1(✉)], Luís P. F. Garcia[2], and André C. P. L. F. de Carvalho[1]

[1] Institute of Mathematics and Computer Sciences, University of São Paulo, São Carlos, Brazil
{victorhb,andre}@icmc.usp.br
[2] Department of Computer Science, University of Brasília, Brasília, Brazil
luis.garcia@unb.br

Abstract. Classification tasks using imbalanced datasets are not challenging on their own. Classification models perform poorly on the minority class when the datasets present other difficulties, such as class overlap and complex decision border. Data complexity measures can identify such difficulties, better dealing with imbalanced datasets. They can capture information about data overlapping, neighborhood, and linearity. Even though they were recently decomposed by classes to deal with imbalanced datasets, their high computational cost prevents their use on applications with a time restriction, such as recommendation systems or high dimensional datasets. In this paper, we use a Meta-Learning approach to estimate the decomposed data complexity measures. We show that the simulated measures assess the difficulty of the dataset after applying preprocessing techniques to different sample sizes. We also show that this approach is significantly faster than computing the original measures, with a statistically similar estimation error for both classes.

Keywords: Imbalanced dataset · Complexity measures · Meta-Learning

1 Introduction

In Machine Learning (ML), standard classification algorithms tend to perform poorly on classes less represented on the training set. This problem is called the imbalanced data problem [9]. Several approaches have been proposed in the literature to mitigate the effects of such problem, some concerning preprocessing the training data to make it more balanced, others adapting standard classification algorithms to consider the imbalance on the learning or prediction steps, and others may combine both strategies [5,7,11,12]. No technique performs well in all datasets, and their performance will depend on each dataset characteristics.

Data Complexity Measures (CMs) were proposed to assess dataset characteristics, such as data overlapping, neighborhood, linearity, and decision border

R. Cerri and R. C. Prati (Eds.): BRACIS 2020, LNAI 12320, pp. 498–512, 2020.
https://doi.org/10.1007/978-3-030-61380-8_34

complexity [13,16]. Their adaptations for imbalanced datasets are useful for understanding the imbalance problem and the techniques in the literature, as they correlate with the difficulty in imbalanced datasets and sampling sizes of preprocessing techniques [1,2]. One disadvantage is that they have a high computational cost, making them unappropriated in approaches with time restrictions such as Meta-Learning (MtL), genetic algorithms, and iterative ones.

To overcome this challenge, we propose a MtL approach to estimate the data CM for imbalanced datasets. A MtL approach learns from previous experiences, considering, for example, previous applications of techniques on different datasets [4,26]. A meta-dataset is usually created, in which each meta-instance represents a dataset, and each meta-feature represents a dataset characteristic. The approach recommends the target-feature, which can be algorithms, their performance, or a ranking of algorithms [17]. A MtL approach can induce a model to predict the performance of a technique on a dataset based on the dataset characteristics by using a meta-dataset.

In this work, we show that a MtL approach can estimate the CMs with a small predictive error for imbalanced datasets using regressor techniques and standard meta-features. This evaluation considers the CM for both classes, the positive (P) and negative (N). We also show that our approach has a low computational cost, which is faster than calculating the original CMs. We show that the simulated measures are as useful as the original ones on an analysis with real datasets and preprocessing techniques on different balance ratios. We also make available the models in an R package called `SImbCoL`[1].

This paper is separated into five sections. Section 2 describes the CMs used to estimate the difficulty of each class separately. Moreover, we present the main concepts about MtL and describe the standard meta-features used to predict CMs values. Next, Sect. 3 presents the experimental setups designed in this work. The experimental results are shown and discussed in Sect. 4. Section 5 concludes this paper with contributions, limitations and future works.

2 Background

This section presents the background information to describe the proposed approach: Sect. 2.1 describe the main concepts regarding data CMs and Sect. 2.2 introduces the MtL framework, including the process of building a meta-dataset and how to recommend algorithms.

2.1 Data Complexity Measures

The CMs were proposed to assess the difficulty in a training set [13]. They were extended by many studies [14–16,18]. A package called `DCoL` (Data Complexity Library) popularized and proposed generalizations of CMs for multiclass problems [18]. Some limitations of the package were solved [16], and they were

[1] https://github.com/victorhb/SImbCoL.

standardized and implemented in a revised R package called ECoL (Extended Complexity Library) [10]. They were adapted for the imbalance problem by a decomposition strategy measuring the CM for each class separately [2].

The measures can be classified into three different categories: overlapping, neighborhood, and linearity. Such categories are described below.

Feature Overlapping Measures. The feature overlapping measures assess the discrimination power of the predictive attributes. Most of them evaluate the features individually and the most discriminate feature is selected, while others use a combination of the individual feature assessments. The overlapping measures considered in this article are F2, F3, and F4.

- **F2: Volume of Overlap Region.** F2 computes the volume of the classes' overlapping region using the minimum and maximum values of each input attribute per class. If the attribute ranges overlap in a certain region, this region is considered ambiguous for the attribute. Next, a product of the normalized size of the ambiguous regions for all attributes is output. For example, suppose an attribute with values for class 1 between 0 and 1, and the values for class 2 between 0.75 and 1.25. Taking the previous example, F2 for class 1 would be $\frac{0.25}{1} = 0.25$ and F2 for class 2 would be $\frac{0.25}{0.5} = 0.5$.
- **F3: Feature Efficiency.** In F3, one feature is considered efficient, depending on how many examples are not in an ambiguous region. For each attribute, the number of examples from the class of interest out of the ambiguous region is divided by the total number of examples from the class of interest. Then, the maximum of such values among all the input attributes is calculated, which corresponds to the attribute that separates better. F3 is 1− the maximum value calculated.
- **F4: Collective Feature Efficiency.** F4 uses the main concept of F3, but instead of getting the maximum value from all attributes, it combines their discrimination power. First, the most discriminative attribute, according to F3, is found; next, the examples correctly separated by that attribute are removed. The previous steps are repeated until all examples are correctly discriminated or until all attributes are removed. F4 is the proportion of examples not discriminated at the end of the process.

Neighborhood Measures. The neighborhood measures use the concept of Nearest Neighbor (NN) to assess classification difficulty. They use the distance between instances to assess, for example, the shape of decision boundaries and class distributions. In this paper, we considered the measures N1, N2, N3, N4, and T1.

- **N1: The Fraction of Points on the Class Boundary.** N1 builds a minimum spanning tree (MST) that connects all the examples from a dataset based on their distances, despite their classes. Next, it counts the number of examples connected to at least one example from another class. Those examples are considered borderline. The fraction of the number of borderline examples for each class over the size of each class is the final decomposed N1 measure.

- **N2: The Ratio of Average Intra/Inter Class NN Distance.** N2 compares the intraclass and interclass dispersions of the classes. For each example, its distance from the NN of the same class (intraclass) and its distance to the NN of a different class (interclass) are computed. Decomposed N2 is the ratio of the average of the intraclass distances for each class and the average of the interclass distances for each class.
- **N3: Leave-one-out Error Rate of the 1NN Classifier.** N3 gives the leave-one-out training error of a nearest-neighbor classifier, which is easy to be calculated and is a good indicator of the separability of the classes. The decomposed N3 is the error rate per class.
- **N4: Nonlinearity of a 1-NN Classifier.** N4 uses a method that creates a new test set by interpolating two randomly selected examples from the same class multiple times. Then an NN classifier using the training set is used to predict the labels of the examples in the interpolated test set. Decomposed N4 gives the error rate per class achieved in this procedure.
- **T1: Fraction of Maximum Covering Spheres.** T1 tries to explain the training set with hyper-spheres. Suppose that every example in the training set has a hypersphere with radius zero. If we gradually increase the radius of all hyperspheres, some will touch a hypersphere from a different class. When that happens, both hyperspheres stop growing. The method stops when there is no more growing hypersphere. The hyperspheres that are contained in another hypersphere are discarded. Decomposed T1 is the ratio between the number of remaining hyperspheres for each class and the number of examples in each class.

Linear Separability Measures. These measures assess whether the classes can be linearly separable in the attribute space. They assume that a classification problem solved with a hyperplane is simpler than another with a non-linear boundary. The measures from this category considered in this article are L1, L2, and L3.

- **L1: The Minimized Sum of Error Distance of a Linear Classifier.** In L1, one linear model (e.g., a linear SVM) is built using the training dataset and calculating the distances of erroneous instances to the obtained hyperplane. Decomposed L1 is the sum of these distances per class. L1 is equal to 0 for linearly separable problems.
- **L2: The Training Error of a Linear Classifier.** Decomposed L2 is the training error of a linear classifier per class. Higher values are expected for non-linear separable classes.
- **L3: Nonlinearity of the Linear Classifier.** L3 is based on the same method of N4. A test set is interpolated, and instead of an NN classifier, N3 uses a linear classifier to predict the labels of the examples from the test set.

Although the data CMs showed to be useful for different applications, their computational cost may prevent them from being used on applications that have time restriction. To overcome this, we suggest in this paper to estimate them using a MtL approach.

2.2 Meta-learning

Rice, J. (1976) [23] initially addressed the algorithm selection problem. In this study, the author proposed an abstract model to systematize the algorithm selection problem to predict the best algorithm when more than one algorithm is available. The main components in this model are the problem instances space (P) composed by datasets, the instance features space (F) based on the meta-features used to describe the datasets, the algorithms space (A) with the pool of ML algorithms that might be recommended, and the evaluation measures space (Y) responsible for assessing the performance of the ML algorithms in solving the problem instances contained in P. By using the previous sets, the MtL system can obtain an algorithm able to map a dataset x, described by the meta-features f, into one (or more) algorithm α able to solve the problem with an acceptable predictive performance according to Y, i.e., with maximum $y(\alpha(x))$

Smith-Miles, K. (2008) [26] improved this abstract model by proposing generalizations that can also be applied to the algorithm design problem. In this proposal, some components are added: the set of MtL algorithms; the generation of empirical rules or algorithm rankings; the examination of the empirical results, which may guide theoretical support to refine the algorithms.

One crucial component of the previous models is the definition of the set of standard meta-features (F) used to describe the general properties of datasets. These meta-features must be able to provide evidence about the future performance of the algorithms in A [21,27] and to discriminate, with a low computational cost, the performance of a group of algorithms. [24] gathered the most used meta-features in the literature. We consider such meta-features in this paper. Next, we describe the essential categories of meta-features. For further information, please check [24].

The main standard meta-features used in the MtL literature can be divided into:

- **Simple:** meta-features that are easily extracted from data [22], with low computational cost [21]. They are also named *general* measures [6].
- **Statistical:** meta-features that capture statistical properties of the data [22], mainly of localization and distribution, such as average, standard deviation, correlation, and kurtosis. They can only characterize numerical attributes [6].
- **Information-Theoretic:** meta-features based on information theory [6], usually entropy estimates [25], which capture the amount of information in (subsets of) a dataset [26].
- **Model-Based:** meta-features extracted from a model induced from the data [22]. They are often based on properties of decision tree (DT) models [3,19], when they are referred to as *decision-tree-based* meta-features [?].

- **Landmarking:** meta-features that use the performance of simple and fast learning algorithms to characterize the datasets [26]. The algorithms must have different biases and should capture relevant information with a low computational cost.
- **Others:** standalone, time-related, concept and case-based meta-features [17,28], clustering and distance-based measures [20,30], among others. These describe characteristics that do not fit in the other groups.

The definition of the set of problem instances (P) is another concern, when the ideal would be to use a large number of diverse datasets, in order to induce a reliable meta-model. To reduce the bias in this choice, datasets from several data repositories, like UCI[2] [8] and OpenML[3] [29], can be used.

The algorithm space A represents a set of candidate algorithms to be recommended in the algorithm selection process. Ideally, these algorithms should also be sufficiently different from each other and represent all regions in the algorithm space [17]. Different measures can evaluate the models induced by the algorithms. For classification tasks, most of the studies in the MtL use accuracy. However, other indices, like F_β, AUC, and kappa coefficient, can also be used. For regression problems, Mean Squared Error (MSE) or Root MSE (RMSE) (or normalized versions of such measures) are usually employed.

After the extraction of the standard meta-features from the datasets and the evaluation of the performance of a set of algorithms for these datasets, the next step is labeling each meta-example in the meta-base. Brazdil et al. [4] summarize the three main properties frequently used to label the meta-examples in MtL: (*i*) the algorithm that presented the best performance on the dataset (a classification task); (*ii*) the ranking of the algorithms according to their performance on the dataset (a ranking classification task), where the algorithm with the best performance is top-ranked; and (*iii*) the performance value obtained by each evaluated algorithm on the dataset (a regression task).

3 Methods

In this section, we describe the experimental setup performed in this paper. First, we describe how the meta-dataset was built, second, we describe how we evaluated the MtL that estimates the simulated CMs, and third, we explain the computational cost experiment to compare the runtime execution between the groups of measures. Finally, we analyzed the simulated CM on real datasets when preprocessing techniques are used to balance them.

3.1 The Meta-dataset

We used 161 binary datasets, in which 41 datasets have less than 25% of minority class instances, while the remaining 120 ones have more than 25% of minority

[2] https://archive.ics.uci.edu/ml/index.php.
[3] http://www.openml.org/.

class instances. We call these two sets of datasets, respectively, the high imbalanced and the low imbalanced datasets. Table 1 shows the number of examples, features, and percentage of minority class of all 161 datasets considered.

Table 1. Characteristics of the datasets used to build the meta-dataset

Characteristic	Min value	Max value	Mean value
Number of instances	34	5,278	509
Number of features	3	95	16
Percentage of minority class	4%	49%	33%

Both sets combined are used to build the meta-dataset. For each dataset, we extracted the standard meta-features and the decomposed data CMs. The standard meta-feature set corresponds to the meta-features of the meta-base, while the set of CMs corresponds to the target features.

3.2 The Meta-learning

We used regressor models to predict the value of each decomposed CM, induced by the Distance Weighted k-Nearest Neighbor (DWNN), Random Forest (RF) and Support Vector Regressor (SVR). As baselines, we used the Random (RD) and Mean (DF) approaches. The RD approach consists of selecting randomly one value for each CM using the training set. The DF approach consists of using the mean value of each CM on the training set. We performed a leave-one-out sampling to evaluate the strategies. We measured the error of the meta-regressor using Mean Squared Error (MSE). We also analyzed the trade-off between the computational cost of the standard meta-features, the original CMs, and the simulated CMs.

3.3 Preprocessing Techniques Analysis

In order to evaluate whether the simulated CMs can be helpful in practical analysis, we also performed an experiment using two traditional preprocessing techniques, Random Undersampling (RU) and Synthetic Minority Over-sampling Technique (SMOTE) [7]. We randomly selected 19 datasets with less than 25% of the minority class. For each selected dataset, we applied the preprocessing techniques with different sample sizes, up to 100%, in which 0% represents that no instances were sampled and 100% represents a sampled dataset with a proportion of 1 : 1 between the classes. Each selected dataset and its sampled datasets versions are not used in the training phase. For each sampled dataset, we extracted the standard meta-features, the CMs, and the simulated CMs, in which the latter has never seen this dataset nor its original one. In that way, we can track the evolution of both CMs, as the sample size increases. Figure 1 illustrates the experimental pipeline.

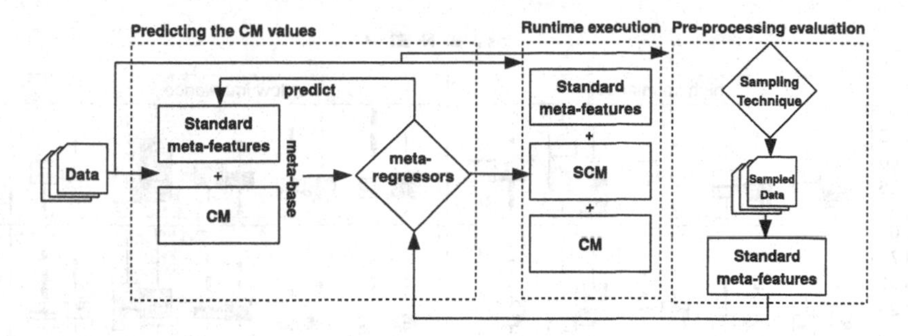

Fig. 1. Evaluation methodology followed in the experiments.

The 161 datasets were selected from the OpenML repository [29]. They represent diverse context datasets, with binary classes and no missing values. The standard meta-features were extracted using the `mfe`[4] package, whereas the CMs were extracted using the `ImbCoL`[5] package. The simulated CMs are available in a R package called `SImbCoL`[6].

4 Results and Discussion

In this paper, we show that a MtL approach is effective in simulating the CMs. For that, first, we evaluated a MtL approach to predict the CMs based on simple and fast meta-features. We show that our approach has a low error rate on estimating them, that it performs better than the baselines. In order to prove the efficiency of that strategy, we also evaluate the time to simulate the CMs. The results indicate that they are faster than the original ones for all datasets. The last analysis shows that the simulated CMs are as helpful as the original ones when estimating the difficulty after applying preprocessing techniques.

Figure 2 shows the MSE for each regression approach for high and low imbalanced datasets. The x-axis shows the regressors, including the baselines in the shadowed area. The y-axis shows the MSE. The colors represent whether the simulation error is related to the positive (P) class, the minority class, or the negative (N), which is the majority class. On the right part of the figure, the name of the CMs in question are displayed.

The MSE analysis indicates that the meta-regressors outperformed the baselines with a better predictive performance for almost all cases. Even on F2 and T1, CMs that the MtL regressors had the highest MSEs, the regressors performed better than the baselines. Compared to the N class, the P class CMs tend to be more difficult to induce, especially on the high imbalanced datasets. Besides, the regressors showed lower MSE for the low imbalanced datasets, compared to those high imbalanced.

[4] https://github.com/rivolli/mfe.
[5] https://github.com/victorhb/ImbCoL.
[6] https://github.com/victorhb/SImbCoL.

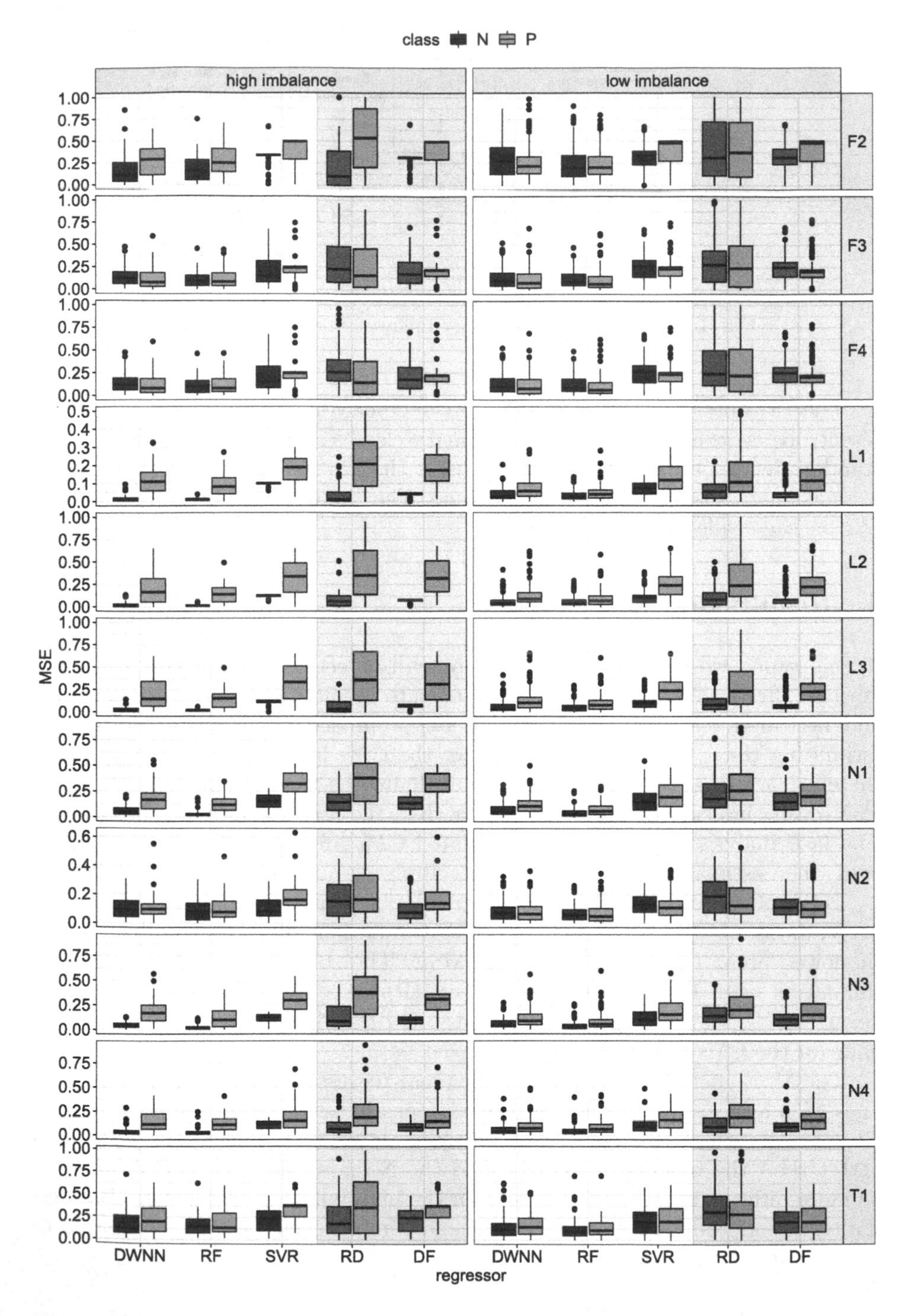

Fig. 2. MSE of the regressors, considering each CM for each class on different levels of imbalance.

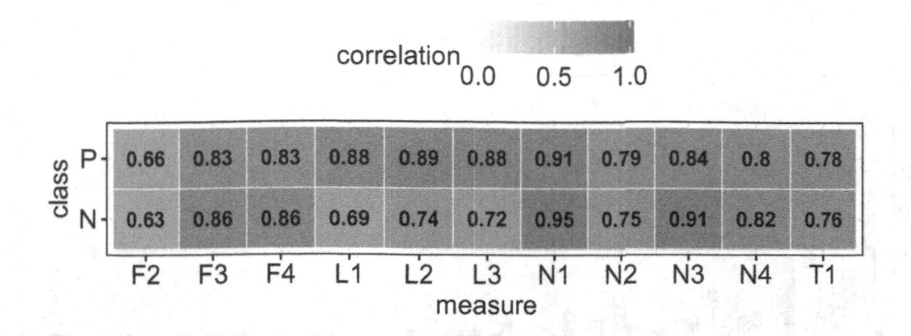

Fig. 3. The correlation between the original CM and the CM simulated by RF.

We performed a paired Friedman-Nemenyi statistical test with a confidence level of 95%. The test confirmed that both DWNN and RF regressors performed better than the baselines and SVR for almost all CMs. Also, the test showed that RF performed better than the DWNN on N2 and N3 CMs. For that reason, in the subsequent analysis, we only consider RF as meta-regressor.

Figure 3 shows a heatmap of the Pearson's correlation between the original and simulated CMs using RF. Each column and row corresponds to the classes and the original CMs, respectively. Each box is colored according to the correlation, from white (lowest correlation) to gray (highest correlation). The correlation values are also shown inside the heatmap's cells.

Most correlations are higher than 70%, and all presented a p-value lower than 0.05. N1 is the CM with the highest correlation for both classes, corroborating with the results on MSE. Although the MSEs of the N class were lower than the P class, the mean values of correlations for the P class is 0.83, and 0.79 for the N class. The linearity measures are responsible for bringing down the mean correlations of the N class. Most of the original linearity CMs values for the N class is grouped close to zero, which made their MSE estimation small but affected negatively their correlation.

Figure 4 illustrates the feature importance of the RF meta-regressor through the increase of MSE considering the top 30 meta-features. The x-axis represents the meta-features sorted, and the y-axis shows the MSE generated by leaving out the meta-feature.

According to the results, the most important meta-features are based on statistical, landmarking, information-theoretic, and model-based. The statistical meta-features are the canonical correlation between the predictive attributes, and the class is present. From landmarking measures, they are related to the performance of simple meta-models induced by the k-NN, the DT algorithm, and the Naive Bayes. The information-theoretic measures highlighted are the mutual information and the concentration coefficient for each pair of attributes. The model-based measures are related to the proportion of training instances to the DT model leaf, the number of nodes of the DT model per number of instances, and the number of nodes per attribute. We observe that there is a difference between the feature importance for the P class and the N class. The

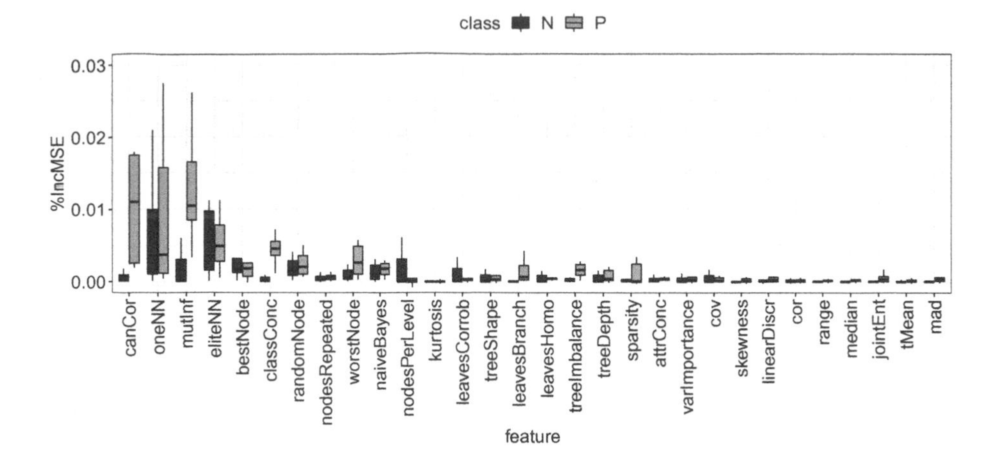

Fig. 4. The feature importance of the meta-dataset using RF.

difficulty of the minority class is related to a group of meta-features that is, according to the results, less relevant to the majority class's difficulty.

Figure 5 compares the time to compute the standard meta-features, the original and simulated CMs. The time is presented on a log-scale to improve visualization. Each point represents a dataset, and those in the diagonal line indicate when the time is similar, the ones above the main diagonal means that y-axis spent more time to be computed than the strategy from the x-axis, while values below that line indicate the opposite.

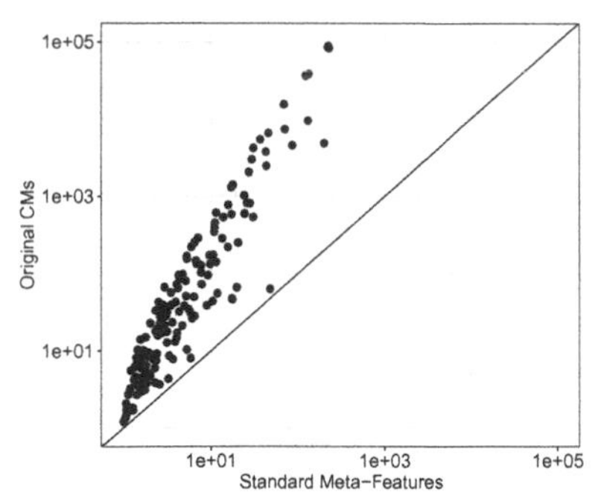

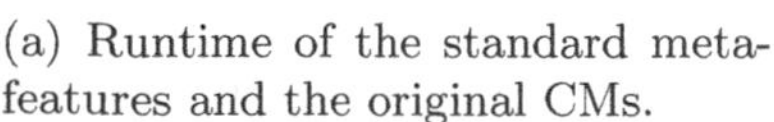
(a) Runtime of the standard meta-features and the original CMs.

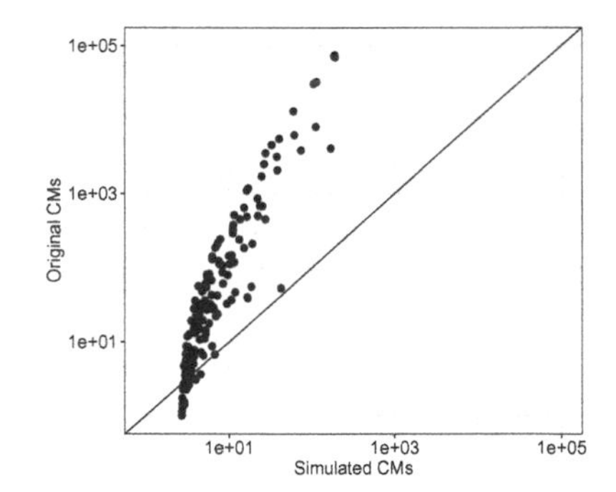

(b) Runtime of the simulated and original CMs.

Fig. 5. Time elapsed to extract the standard meta-features, original and simulated CMs for each dataset.

In Fig. 5a all datasets are above the main diagonal, meaning that, for all datasets, calculating the original CMs took more time than extracting the standard meta-features. The extraction of the standard meta-features is the most time-consuming process of simulating the CMs after the models are built. In Fig. 5b almost all datasets are above the main diagonal, meaning that calculate the original CMs took more time than extracting the simulated CMs. Thus, we show that a MtL approach using such meta-features is faster than calculating the CMs.

In Fig. 6, we can see the mean values of the original and simulated CMs after applying SMOTE and RU with various sample sizes. The selected measures are L2, N1 and N3, the most imformative CMs [1]. The x-axis represents the sample size from 10% to 100%, e.q. how balanced the dataset is, and the y-axis represents the mean values of CMs. The figure shows the results for both classes, P and N, separately.

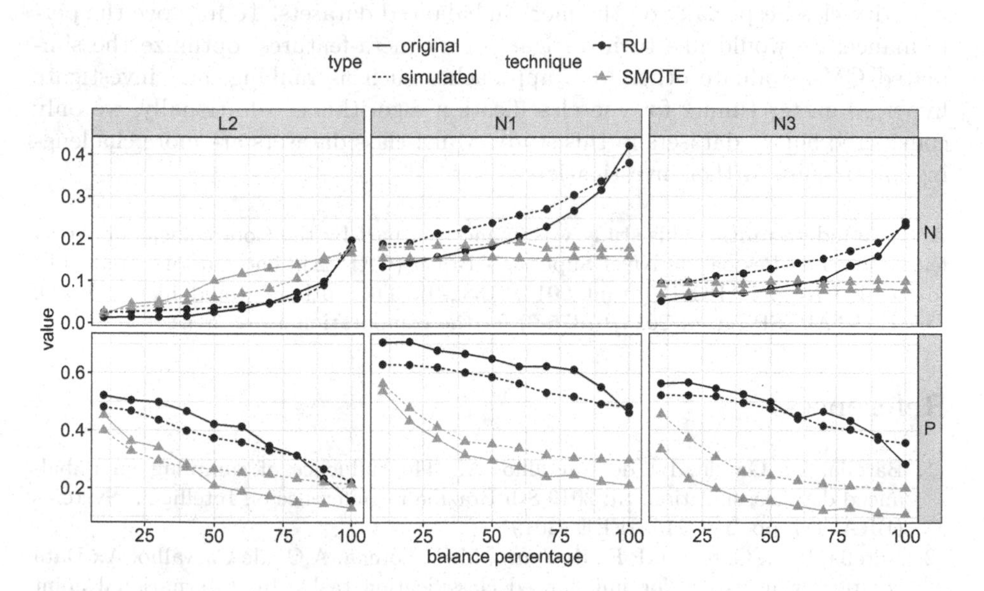

Fig. 6. Mean values of original CMs and simulated CMs after applying SMOTE and RU with various sample sizes.

As the datasets get more balanced, the P class becomes less difficult, and the N class usually gets more difficult. While SMOTE decreases more the complexity than RU in the P class, RU tends to increase more the complexity of the N class. The main difference between original and simulates CMs occurs for N3 measures after applying SMOTE. In all other cases, the simulated CMs are similar to the original ones. Therefore, both original and simulated measures follow this pattern, giving evidence that the simulated CMs are as useful as the original CM to track data complexity when applying data balancing techniques.

5 Conclusions

Measuring data complexity is useful for several ML applications, such as supporting the preprocessing techniques and estimating the expected difficulty of a classification problem. Although CMs are very important in these areas, they have a high computational cost that may prevent their popularization and efficient use. In this paper, we showed that a MtL approach is faster and yet effective to simulate them. For that, meta-models were induced based on standard meta-features, which have a lower computational cost. The main results indicate that the simulated CMs can predict the original CMs with low error and can be obtained at a lower computational cost. Moreover, the simulated CMs also tracks the data complexity when applying preprocessing techniques.

Future work shall look to increase the simulated CMs performance for the minority class, especially on the more imbalanced datasets. To improve the performance, we would like to investigate other meta-features, optimize the simulated CMs, evaluate other MtL approaches such as ranking, and investigate hyperparameter tuning for the classification algorithms. Additionally, we only considered binary datasets in this study. Multi-class datasets are more challenging and require further investigation.

Acknowledgements. This study was financed in part by the Coordenação de Aperfeiçoamento de Pessoal de Nível Superior - Brasil (CAPES). The authors would like to thank CNPq, FAPESP (grant 2015/01382-0). The authors would like to thank CeMEAI-FAPESP (grant 2013/07375-0) for the computational resources.

References

1. Barella, V., Garcia, L., de Carvalho, A.: The influence of sampling on imbalanced data classification. In: 2019 8th Brazilian Conference on Intelligent Systems (BRACIS), pp. 210–215. IEEE (2019)
2. Barella, V.H., Garcia, L.P.F., de Souto, M.P., Lorena, A.C., de Carvalho, A.: Data complexity measures for imbalanced classification tasks. In: International Joint Conference on Neural Networks (IJCNN), pp. 1–8 (2018)
3. Bensusan, H., Giraud-Carrier, C., Kennedy, C.: A higher-order approach to metalearning. In: 10th International Conference Inductive Logic Programming (ILP), pp. 1–10 (2000)
4. Brazdil, P., Giraud-Carrier, C., Soares, C., Vilalta, R.: Metalearning - Applications to Data Mining. Cognitive Technologies, 1st edn. Springer, Heidelberg (2009). https://doi.org/10.1007/978-3-540-73263-1
5. Cano, A., Zafra, A., Ventura, S.: Weighted data gravitation classification for standard and imbalanced data. IEEE Trans. Cybern. **43**(6), 1672–1687 (2013)
6. Castiello, C., Castellano, G., Fanelli, A.M.: Meta-data: characterization of input features for meta-learning. In: Torra, V., Narukawa, Y., Miyamoto, S. (eds.) MDAI 2005. LNCS (LNAI), vol. 3558, pp. 457–468. Springer, Heidelberg (2005). https://doi.org/10.1007/11526018_45
7. Chawla, N.V., Bowyer, K.W., Hall, L.O., Kegelmeyer, W.P.: Smote: synthetic minority over-sampling technique. J. Artif. Intell. Res. **16**, 321–357 (2002)

8. Dua, D., Graff, C.: UCI machine learning repository (2017). http://archive.ics.uci.edu/ml
9. Fernández, A., García, S., Galar, M., Prati, R.C., Krawczyk, B., Herrera, F.: Learning from Imbalanced Data Sets. Springer, Cham (2018). https://doi.org/10.1007/978-3-319-98074-4
10. Garcia, L.P.F., Lorena, A.C.: ECoL: complexity measures for classification problems (2018). https://CRAN.R-project.org/package=ECoL
11. Gonzalez-Abril, L., Nuñez, H., Angulo, C., Velasco, F.: GSVM: An SVM for handling imbalanced accuracy between classes inbi-classification problems. Appl. Soft Comput. **17**, 23–31 (2014)
12. He, H., Bai, Y., Garcia, E.A., Li, S.: ADASYN: adaptive synthetic sampling approach for imbalanced learning. In: International Joint Conference on Neural Networks (IJCNN), pp. 1322–1328 (2008)
13. Ho, T.K., Basu, M.: Complexity measures of supervised classification problems. IEEE Trans. Pattern Anal. Mach. Intell. **24**(3), 289–300 (2002)
14. Ho, T.K., Basu, M., Law, M.H.C.: Measures of geometrical complexity in classification problems. In: Basu, M., Ho, T.K. (eds.) Data Complexity in Pattern Recognition, pp. 1–23. Springer, London (2006). https://doi.org/10.1007/978-1-84628-172-3_1
15. Lorena, A.C., de Souto, M.C.P.: On measuring the complexity of classification problems. In: Arik, S., Huang, T., Lai, W.K., Liu, Q. (eds.) ICONIP 2015. LNCS, vol. 9489, pp. 158–167. Springer, Cham (2015). https://doi.org/10.1007/978-3-319-26532-2_18
16. Lorena, A.C., Garcia, L.P.F., Lehmann, J., de Souto, M.C.P., Ho, T.K.: How complex is your classification problem? A survey on measuring classification complexity. ACM Comput. Surv. (CSUR) **52**(5) (2019)
17. Muñoz, M.A., Villanova, L., Baatar, D., Smith-Miles, K.: Instance spaces for machine learning classification. Mach. Learn. **107**(1), 109–147 (2017). https://doi.org/10.1007/s10994-017-5629-5
18. Orriols-Puig, A., Maciá, N., Ho, T.K.: Documentation for the data complexity library in C++. La Salle - Universitat Ramon Llull, Technical report (2010)
19. Peng, Y., Flach, P.A., Soares, C., Brazdil, P.: Improved dataset characterisation for meta-learning. In: Lange, S., Satoh, K., Smith, C.H. (eds.) DS 2002. LNCS, vol. 2534, pp. 141–152. Springer, Heidelberg (2002). https://doi.org/10.1007/3-540-36182-0_14
20. Pimentel, B.A., de Carvalho, A.C.P.L.F.: A new data characterization for selecting clustering algorithms using meta-learning. Inf. Sci. **477**, 203–219 (2019)
21. Reif, M.: A comprehensive dataset for evaluating approaches of various meta-learning tasks. In: 1st International Conference on Pattern Recognition Applications and Methods, pp. 273–276 (2012)
22. Reif, M., Shafait, F., Goldstein, M., Breuel, T., Dengel, A.: Automatic classifier selection for non-experts. Pattern Anal. Appl. **17**(1), 83–96 (2012). https://doi.org/10.1007/s10044-012-0280-z
23. Rice, J.R.: The algorithm selection problem. Adv. Comput. **15**, 65–118 (1976)
24. Rivolli, A., Garcia, L.P.F., Soares, C., Vanschoren, J., de Carvalho, A.C.P.L.F.: Characterizing classification datasets: a study of meta-features for meta-learning. eprint arXiv (1808.10406), pp. 1–49 (2019)
25. Segrera, S., Pinho, J., Moreno, M.N.: Information-theoretic measures for meta-learning. In: Corchado, E., Abraham, A., Pedrycz, W. (eds.) HAIS 2008. LNCS (LNAI), vol. 5271, pp. 458–465. Springer, Heidelberg (2008). https://doi.org/10.1007/978-3-540-87656-4_57

26. Smith-Miles, K.A.: Cross-disciplinary perspectives on meta-learning for algorithm selection. ACM Comput. Surv. **41**(1), 1–25 (2008)
27. Soares, C., Petrak, J., Brazdil, P.: Sampling-based relative landmarks: systematically test-driving algorithms before choosing. In: Brazdil, P., Jorge, A. (eds.) EPIA 2001. LNCS, vol. 2258, pp. 88–95. Springer, Heidelberg (2001). https://doi.org/10.1007/3-540-45329-6_12
28. Vanschoren, J., Blockeel, H., Pfahringer, B., Holmes, G.: Experiment databases. Mach. Learn. **87**(2), 127–158 (2011). https://doi.org/10.1007/s10994-011-5277-0
29. Vanschoren, J., van Rijn, J.N., Bischl, B., Torgo, L.: OpenML: networked science in machine learning. SIGKDD Explor. **15**(2), 49–60 (2013)
30. Vukicevic, M., Radovanovic, S., Delibasic, B., Suknovic, M.: Extending meta-learning framework for clustering gene expression data with component-based algorithm design and internal evaluation measures. Int. J. Data Min. Bioinfor. (IJDMB) **14**(2), 101–119 (2016)

SSL-C4.5: Implementation of a Classification Algorithm for Semi-supervised Learning Based on C4.5

Agustín Alejandro Ortiz-Díaz(✉), Flavio Roberto Bayer, and Fabiano Baldo

Graduate Program in Applied Computing, Santa Catarina State University, Joinville, Brazil
agaldior@gmail.com, flaviobayer@hotmail.com, fabiano.baldo@udesc.br

Abstract. Classification algorithms have been extensively studied in many of the major scientific investigations in recent decades. Many of these algorithms are designed for supervised learning, which requires labeled instances to achieve effective learning models. However, in many of the real human processes, data labeling is expensive and time-consuming. Because of this, alternative learning paradigms have been proposed to reduce the cost of the labeling process without a significant loss of model performance. This paper presents the Semi-Supervised Learning C4.5 algorithm (SSL-C4.5) designed to work in scenarios where only a small part of the data is labeled. SSL-C4.5 was implemented over the J48 implementation of the C4.5 algorithm available at the WEKA platform. The J48 was modified incorporating a metric for semi-supervised learning. This metric aims at inducing decision tree models able to analyze and extract information from the entire training dataset, including instances of unlabeled data in scenarios where they are the majority. The assessment performed using eight different benchmark datasets showed that the new proposal has achieved promising results compared to the supervised version of C4.5.

Keywords: Classification-algorithms · Semi-supervised-learning · Decision tree

1 Introduction

Data classification tasks have become increasingly difficult due to the complexity and volume of datasets generated in real scenarios. Among these complexities, we can mention a large amount of unlabeled data, the presence of data with multiple labels, the processing of data streams, among others [2]. Additionally,

Supported by the CAPES and FAPESC organizations.

R. Cerri and R. C. Prati (Eds.): BRACIS 2020, LNAI 12320, pp. 513–525, 2020.
https://doi.org/10.1007/978-3-030-61380-8_35

in real-world scenarios, a large amount of data is generated without a specialist to label them beforehand. These datasets with unlabeled instances make many conventional algorithms not suitable for their classification. This is because several inducing algorithms are not designed to extract all possible information from the data provided [4].

In this scenario, labeling the data can be an extremely expensive task [3]. For this reason, it is not practical to previously labeling all instances of data provided to the algorithm, which is usually necessary in the cases of supervised learning [6]. Therefore, semi-supervised learning (SSL) methods have among their objectives to face the challenges of inducing suitable predictive models considering unlabeled instances [1]. SSL techniques allow the algorithms to combine labeled and unlabeled data in order to improve the induced models. Therefore, the aim of the SSL approach is to increase the capacity of the inducing algorithms by extracting significant information from the entire training dataset.

On the other hand, according to [5], decision tree models have desirable properties that should be exploited in SSL. Among these properties, we can highlight the following ones: non-parametric, efficient, easily interpretable, and suitable for predicting in many domains.

In this work, we propose a method for semi-supervised learning based on classification trees. According to the categorization in [13], our proposal is classified as an inductive learning approach since it uses all the information to build the models, both labeled and unlabeled instances. We selected a well-performance supervised learning algorithm that induces decision trees and we introduced on it a semi-supervised metric for creating models using the semi-supervised paradigm.

The supervised learning algorithm selected was the C4.5 [10]. Two of the main benefits of the C4.5 algorithm are that it has relatively quick induction and generally obtains high accuracy values [14]. In addition, we modify it in order to add the impurity metric proposed by Levatić et al. [5]. This metric improves the division of the instances in the tree nodes to enable it to deal with unlabeled data. We take the J48 algorithm as basis for supporting the implementation of the adaptations on C4.5 in order to make it the proposed SSL-C4.5. The J48 is a implementation of C4.5 in Java available at the WEKA platform [16].

The rest of the paper is organized as follows: Sect. 2 references the most relevant related works. Section 3 describes the main characteristics of the well-known C4.5 algorithm, including a subsection to detail some of the used impurity metrics. Section 4 presents in detail the proposed SSL-C4.5 algorithm. Section 5 shows the experiments and assessment results. Finally, Section 6 highlights the main conclusions and future works.

2 Related Works

There are different techniques applied to semi-supervised learning. Among these techniques are Maximization of Expectations (ME), Self-training, or Co-training [13]. Some papers related to these techniques and the present work are described below.

An interesting idea related to the self-training technique is proposed in [11]. This paper introduces a classification algorithm that combines the characteristics of the known methods, Naive Bayes, and C4.5. The authors propose a novel approach to increase the power of semi-supervised methods using a technique called cascade classification. The main feature of this classification technique is the use of two levels of classification. At a first level, a base classifier is used to increase the characteristic space by adding the class predicted by it to the distribution of the probability class of the initial data. On a second level, another classifier is incorporated, which from the new data set extracts the decision for each instance analyzed. The new proposal was compared to other semi-supervised classification methods. Known data sets, with standard references, were used for the experiments. The authors concluded that the new method presented has greater precision according to the results shown.

Another proposal for a semi-supervised classifier, also related to the self-training technique, is presented in [12]. The authors start from the premise of showing that learning with standard decision trees as a base-learner cannot be effective in a self-training algorithm for the semi-supervised learning paradigm. However, through the experiments, the authors showed that improving the probability estimate of the classifiers leads to a better selection metric for the algorithm and produces a better self-training model of classification. Furthermore, they concluded that combining the Laplacian correction, the non-pruning, on an NBTree produced a better probability estimate in the tree classifiers. They also showed that the Mahalanobis distance method for sampling is effective in selecting a set of high-confidence predictions in decision trees.

With a different idea, being the basis of our work, The Semi-Supervised Learning - Predictive Clustering Trees (SSL-PCT) algorithm, described in [5]. This is an adaptation of the Predictive Clustering Trees (PCT) algorithm that allows the use of not only labeled instances, but also the unlabeled instances during the model training. PCT algorithm works under the assumption that data labels can be successively separated using techniques of clustering, allowing the creation of a tree analogously to the operation of the ID3 algorithm. However, PCT uses the Gini statistical dispersion measure to calculate the impurity metric [5].

In their work, Levatić et al. [5] argues that the grouping behavior of PCT can be used not only in the class attribute of an instance but also in the rest of the dataset attributes. Therefore, the idea is to perform a separability analysis considering the values of the instances attributes, rather than assess the separability only considering the class of the instances. By incorporating an unsupervised component in the impurity metric, it is possible to perform data separation even if the instance labeled is not previously known.

Levatić et al. [5] proposes the impurity formula described by the Eq. 1, where a ω parameter is introduced and two impurity metrics are used, one for the supervised learning, $Impurity_l$, based on the Y label of the instances, and another for the unsupervised learning, $Impurity_u$, based on all X_i enumerations of the X attributes of the dataset.

$$ImpuritySSL(E) = \omega \cdot Impurity_l(E_l, Y) + \frac{(1-\omega)}{|X|} \cdot \sum_{i=1}^{|X|} Impurity_u(E, X_i) \quad (1)$$

According to Eq. 1, when the ω value is close to 1 the process tends to be more supervised, while when the value is close to 0 the process tends to be more unsupervised. In their work, Levatić et al. [5] used the standardized Gini measure to calculate the impurity for discrete attributes.

3 C4.5 Algorithm and Impurity Measures

One of the simplest decision tree induction algorithms for classification is the ID3 [9]. The ID3 algorithm has great importance in the classification area and its fundamental ideas are used as the basis for the construction of other algorithms that use the top-down approach. This algorithm is based on the expected information gain formula (Eq. 2). The idea is to reduce the impurity value of the nodes when going deep in the tree, always trying to make the next level of the tree with lower impurity than the last one.

$$Gain(E, X_i) = Impurity(E) - \sum_{E_{X_i,j} \subset E} \frac{|E_{X_i,j}|}{|E|} \cdot Impurity(E_{X_i,j}) \quad (2)$$

Where E is a dataset that has a X set of attributes. The algorithm selects the best attribute X_i of X to divide the set E into subsets $E_{X_i,j}$ which reduces the impurity of the subsets. $E_{X_i,j}$ represents the jth subsets of data in which the data set E is divided after selecting the X_i as the best attribute for the division. The ID3 algorithm has an important drawback, it is only capable of inducing decision trees for attributes with discrete values.

The C4.5 algorithm [10] is an improvement of the ID3 algorithm, created by the same author, which uses the same operating principle. That is, for each leaf node of the tree, the C4.5 algorithm chooses the attribute that most effectively divides its instance set into subsets that tend to one category or another. The C4.5 algorithm has among its objectives to solve the main limitation of the ID3 algorithm. For this reason, it supports induction of trees with numerical attributes of continuous values, allowing such attributes to appear at multiple levels in the same branch of the tree that is being induced. C4.5 uses a disorder measure called entropy to calculate the impurity metric.

3.1 Impurity Measures

This subsection presents a short overview of the metrics used as the basis for the implementation of the proposed algorithm. Actually, this section serves as a compilation of the relevant finds took into account to decide the metrics to be used in the work.

The value of the impurity equation reflects how disorganized or how random the dataset is. A high impurity value is typically found on a random dataset, while a low impurity is typically seen in a well-organized dataset. The impurity measure is a metric that abstracts and quantifies the organization of data in a numerical value. Therefore, various impurity measure formulas can be used according to the characteristics of the data. Below, it is presented some of the most commonly used impurity measures found in the literature.

Entropy. The entropy measure is calculated by the Eq. 3. This equation provides an entropy coefficient between 0 and 1 based on the probability p that an attribute has a nominal class presented in the dataset. Given $p_i = |X_i|/|X|$, where X is the set of classes or possible values for the nominal attribute of an instance, X_i is the subset of instances where the attribute has the ith class of X, and finally n is given by $n = |X|$ [8].

$$S(p) = -\sum_{i=1}^{n} p_i \cdot log_2(p_i) \tag{3}$$

Gini. The Gini metric is represented by the Eq. 4, which generates a continuous value between 0 and 1. The calculated value is based on the probability p that an attribute is a nominal class presented in the dataset. Given $p_i = |X_i|/|X|$, where X is the set of classes or possible values for the nominal attribute of an instance, X_i is the subset of instances where the attribute has the ith class of X, and finally the value of n is given by $n = |X|$ [15].

$$G(p) = 1 - \sum_{i=1}^{n} p_i^2 \tag{4}$$

Variance. To calculate the impurity for a regression model, or when evaluating numerical attributes, the standard deviation is commonly used to estimate the data capacity of a set x, where x_i represents data in x. To calculate the incremental variance, it is possible to use the Eq. 5, so that the sum of x_i, the sum of x_i^2 and n can be easily calculated incrementally. Thus, the method gain efficiency by reducing the runtime.

$$V(x) = \left(\frac{1}{n}\sum_{i=1}^{n} x_i^2\right) - \left(\frac{1}{n}\sum_{i=1}^{n} x_i\right)^2 \tag{5}$$

It is important to keep in mind that the variance of the numerical values generates a non-negative number, but it is not a normalized measure. Taking this issue into account, the variance value of a subset of data is often divided by the variance value of the complete dataset, generating an entropy coefficient between 0 and 1 [5].

4 SSL-C4.5 Algorithm

Many of the supervised classification algorithms that are used to induce decision trees often focusing only to separate the data based on their respective classes. That is, although the other attributes are used for example as a comparison to determine subtrees, these attributes are not considered in the metric to determine the quality of a possible division of a tree node.

On the other hand, in the unsupervised clustering approaches, all instance attributes can be used to subdivide the search space by considering each of them as a dimension. According to [1], for some datasets it can be assumed that separating a data set based on any attribute can be as good as separating a data based on a class attribute of the instances. Therefore, even if a significant number of instances within a dataset are not previously labeled, it is still possible to use the information from their other attributes to group them, so that the information is also used during the processing of the data.

Taking into account these two approaches to different learning paradigms, but with the same objective, to induce a model that represents the dataset, this paper proposes a model capable of joining the beneficial characteristics of both models to obtain a more promising final result. The SSL-C4.5 algorithm is developed over the J48 algorithm, implementation in Weka [16] of the C4.5 [10] algorithm. In this new implementation, changes have been made in the impurity metric used to adapt the original C4.5 to support the semi-supervised learning.

According to Levatić et al. [5], an algorithm that builds a top-down decision tree is based on the use of a heuristic capable of finding the best attribute-value pair in a dataset to reduce the value of the impurity of new partitioned datasets. Different algorithms can use different metrics to achieve this effect and maximize the quality of the result as detailed in Sect. 3.1.

Algorithm 1 shows the pseudocode of the SSL-C4.5 algorithm. In line 2, the empty list called *Analysis_pending_nodes_list* is initialized. In this list, all the nodes of the tree that will be analyzed will be stored. The root node N is added as the first to be analyzed (line 8). Then, inside the while loop (line 9), all nodes obtained as a result of dividing previous nodes (line 14) will be added.

In general, the Algorithm 1 is a variant of any classic algorithm for inducing top-down decision trees. However, a notable difference is present in the function called *Attribute_selection_method* (line 11). This function has as main objective to decide if a node of the tree is going to be divided or not. If this function decides not to divide the node, then this node is labeled with the dominant class in the set of instances associated with that node. However, if this function decides to divide this node, then the attribute that offers the most efficient division is returned. To make this decision, the *Attribute_selection_method* function uses the $information\_gain\_formula$ (Eq. 2). It is necessary to emphasize that this formula (Eq. 2) has been adapted to the semi-supervised paradigm.

Algorithm SSL-C4.5:

S: Set of labeled or unlabeled examples.
$w\epsilon[0,1]$: Weighting parameter to balance the supervised and semi-supervised paradigms.
D : Attribute list.
Attribute_selection_method: It uses impurity measure SSL-C4.5 that calculates the similarity between the examples on the basis of both the class labels and descriptive attributes.

Result: $SSL - C4.5 - Tree$: A decision tree for semi-supervised learning induced according to the S training instances data set.

```
1  begin
2      Analysis_pending_nodes_list ← Ø
3      Created SSL − C4.5 − Tree a tree with a single node N(the root)
4      if tuples in S are all labeled with the same class C then
5          Marking N as a leaf node labeled with the class C
6          return(SSL − C4.5 − Tree)
7      else
8          Analysis_pending_nodes_list ← Analysis_pending_nodes_list ∪ N
9          while Analysis_pending_nodes_list <> Ø do
10             Extract nextNode from Analysis_pending_nodes_list
11             Attribute_selection_method(nextNode, D) Finding the best
               splitting criterion
12             if splitting_criterion = true then
13                 newNodes ← Best_splitting_criterion(nextNode)
14                 Analysis_pending_nodes_list ←
                   Analysis_pending_nodes_list ∪ newNodes
15             else
16                 Marking nextNode as a leaf node labeled with majority
                   class
17     return(SSL − C4.5 − Tree)
```

Algorithm 1: Semi-Supervised Learning C4.5 Algorithm (SSL-C4.5)

Based on the results of a large set of tests, we decided to use the entropy standardized measure of impurities for the supervised learning part of the model. The entropy measure has been tested by several investigations [8] due to its robustness, simplicity, and flexibility. In addition, this measure is compatible with the base supervised implementation, the J48 algorithm. The calculation of the normalized entropy is represented by the Eq. 6, where E_l represents the set of ordered data that is analyzed in a given tree node, ${E_l}^{full}$ represents the set of all the ordered data used in the construction of trees, and S is the heuristic metric (h) of the entropy of impurities.

$$Impurity_l(E_l, Y) = \frac{S(E_l, Y)}{S({E_l}^{full}, Y)} \tag{6}$$

Similarly, we chose to use the standardized Gini impurity metric and the standardized variance impurity for the unsupervised learning part of the model.

This metric has been tested by several investigations [15] due to its easy understanding and implementation, as well as the good results offered. The standardized calculation of impurities without supervision is represented by the Eqs. 7 and 8, where E represents the set of all the data analyzed in a given tree node (whether labeled or not), E^{full} represents the set of all the data used to build the tree (whether labeled or not), G is the Gini heuristic metric of impurities, V is the heuristic metric of impurity variance and X_i is the attribute that is being evaluated.

$$Impurity_u(E, S_i) = \frac{G(E, X_i)}{G(E^{full}, X_i)}, if X_i\_is\_nominal \tag{7}$$

$$Impurity_u(E, S_i) = \frac{V(E, X_i)}{V(E^{full}, X_i)}, if X_i\_is\_numerical \tag{8}$$

These two impurity measures are incorporated into the Eq. 1 of semi-supervised learning proposed by Levatić et al. [5]. It creates a single measure of impurities, which was incorporated into the J48 algorithm replacing its original metric.

Finally, the *Best_splitting_criterion* function (line 13) receives the last node analyzed (*nextNode*), the one decided to be divided. The objective of this function is to divide this node (*nextNode*) using the attribute that guarantees more efficiency. This attribute has already been selected before (line 11). This function returns a list of new nodes (*newNodes*), which are added to the *Analysis_pending_nodes_list* list.

5 Results Assessment

In this section, we evaluate the predictive performance of the SSL-C4.5 algorithm using different datasets. The datasets were mainly obtained from the UCI repository [7]. We compared our proposed algorithm (SSL-C4.5) with the SSL-PCT algorithm proposed by Levatić et al. [5], with two variations of both algorithms respectively, and with the implementation in WEKA of C4.5 algorithm (which is named J48). All the experiments were performed over WEKA [16] framework for data mining. WEKA provides a collection of evaluation tools and a great variety of known algorithms.

5.1 Methodology

We used eight datasets with different features for the experiments to ensure high diversity in the experimentation scenarios. Table 1 shows the features regarding the selected datasets, which are: number of instances (Instances); number of discrete attributes (Nominal); number of continuous attributes (Numeric); and number of classes (Classes). The algorithms and variants of the algorithms involved in the experiments are described below.

- **C4.5 Algorithm:** Supervised algorithm C4.5 [10] implemented in WEKA named as J48.

- **SSL-C4.5-W1.0 Algorithm:** Semi-supervised implementation of SSL-C4.5 over J48 with parameter $\omega = 1$. Equivalent to the C4.5 algorithm. Tested to ensure that the semi-supervised version can be equivalent to the supervised version.
- **SSL-C4.5-WA:** Semi-supervised implementation of SSL-C4.5 over J48 with automatic adaptation of the ω parameter.
- **SSL-PCT-W1.0:** Semi-supervised PCT algorithm with parameter $\omega = 1$. Equivalent to the PCT algorithm [5].
- **SSL-PCT-WA:** Semi-supervised PCT algorithm with automatic adaptation of the ω parameter.

Table 1. General characteristics of the used dataset

Dataset	Instances	Nominal	Numeric	Classes	Missing values
Abalone	4177	0	8	3	No
Adult	32561	9	5	2	Yes
Bank	45211	10	6	2	Yes
Banknote	1372	0	4	2	No
Biodegradation	1055	0	40	2	No
Eyestate	14980	0	14	2	No
Madelon	2000	0	500	2	No
Mushroom	8124	22	0	2	No

During the experiments, the algorithms were set with their default configuration options. The "CollapseTree" and "SubtreeRaising" options were disabled in the C4.5 algorithm (J48 implementation) since these features were not implemented in the SSL-C4.5 algorithm. The confidence factor was set as 0.25 in the C4.5 and SSL-C4.5 algorithms, with a minimum information gain of 0.

All algorithms were evaluated using the 10-fold cross-validation technique. For each experiment, the part of the dataset selected to train the models was fully used. That is, the supervised models were trained with all the training data labeled. However, to assess the learning capability of the semi-supervised algorithms, the class label of part of the instances from the training set was removed, while the rest of the attributes remained intact. Therefore, for the semi-supervised learning, only a certain number of instances were labeled. On the other hand, during the test phase, the class of all instances (including instances whose classes had been removed) is compared with the class predicted by the model, in order to check if the prediction was correct.

For each dataset described in Table 1, the test datasets were created by varying the amounts of labeled instances. That is, each original dataset generated six derived datasets with exactly the same instances but leaving only a small amount of these instances with their labels. The following quantities of labeled instances were used by default: 25, 50, 100, 200, 350 and 500. The remains of the instances were left unlabeled for training each scenario. These datasets, which were kept with the class labels, are randomly selected. The experiments were repeated ten times to leverage the results reliability.

During the tests of the semi-supervised algorithms, it was necessary to define the best value of ω in the automatic adaptation tests. The value of ω was established by performing tests (training/test), using the same test datasets, ranging it from 0.0 (fully unsupervised) to 1.0 (fully supervised) in steps of 0.1. Therefore, 11 values were tested and the value that generates the best model was selected. This can guarantee that the method will never generate a worse model than its supervised version ($\omega = 1.0$).

5.2 Analysis of the Results

All algorithms were evaluated taking into account two fundamental criteria, accuracy and runtime. Figure 1 show the accuracy values of the execution of all the algorithms on each dataset. Analyzing the eight figures, it is possible to observe the behavior of all the algorithms studied during the experiments, including the new proposal.

With a quick analysis, we can perceive that the C4.5 algorithm and the SSL-C4.5-W1.0 algorithm exhibit identical behaviors. In this way, we can confirm that the semi-supervised version implemented (our proposal) is an extension of the original C4.5 algorithm (J48 implementation), whose additional functionality can become inert when $\omega = 1.0$. In addition, we can see that the algorithm versions with the automatic adjustment parameter ω, (SSL-C4.5-WA and SSL-PCT-WA) never have results with accuracies lower than their supervised versions (SSL-C4.5-W1.0 and SSL-PCT-W1.0). In other words, semi-supervised models always even or improve supervised models.

By comparing the accuracy values of the C4.5 and PCT algorithms (SSL-PCT-W1.0), it is not possible to affirm that one algorithm is statistically superior to the other. We can observe that for some datasets such as Abalone, Banknote, Eyestatem, and Mushroom, the accuracy values show alternating behavior. Nevertheless, in other datasets, such as Adult, Bank and Biodegradation datasets, the C4.5 algorithm has superior results. According to our empirical analysis, the C4.5 algorithm is capable of producing more promising results.

On the other hand, we can also observe that for all the algorithms there were improvements in the accuracy values when the semi-supervised component was added to the metrics. That is, when we compare the two variants of the same algorithm, in general, the semi-supervised variant obtained better results (SSL-C4.5-WA compared to SSL-C4.5-W1.0, and SSL-PCT-WA compared to SSL-PCT-W1.0).

Finally, we could observe that between the two algorithms, in their semi-supervised variants (SSL-C4.5-WA compared to SSL-PCT-WA), the new proposal obtained more promising results, although a statistically significant difference could not be established. When evaluating some datasets, such as Banknote, Biodegradation, and Mushroom, it is not possible to notice a marked difference in the accuracy values. However, for other datasets, such as Abalone, Adult, Bank, and Eyestatem, the new proposal has a significant improvement in accuracy values. It is also valid to highlight that when evaluating the Madelon dataset, the opposite result occurred.

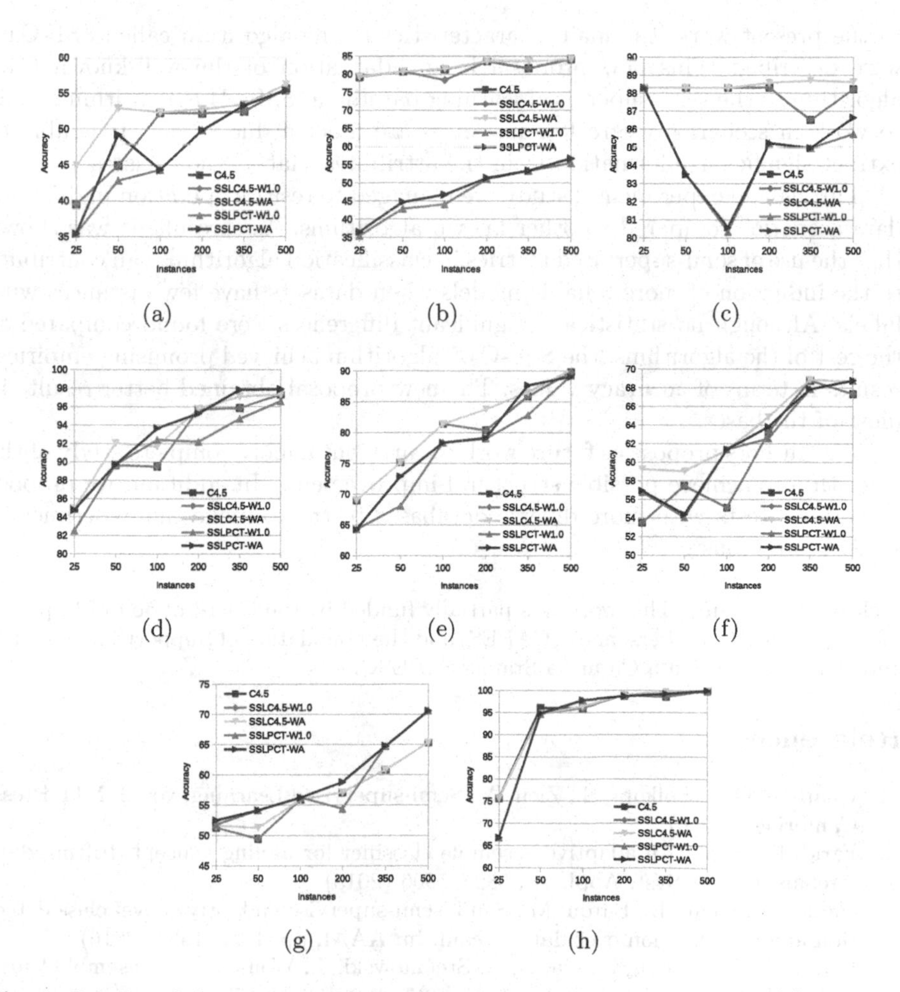

Fig. 1. Results of the accuracy of the algorithms on the eight datasets. **(a)** Abalone dataset, **(b)** Adult dataset, **(c)** Bank dataset, **(d)** Banknote dataset, **(e)** Biodegradation dataset, **(f)** Eyestate dataset, **(g)** Madelon dataset, **(h)** Mushroom dataset.

In general, it can be said that the SSL-C4.5-WA algorithm, offers promising results compared to the other algorithms or variants. In general, the accuracy values of the new model are higher in most experiments.

Although runtime values have not been included in this paper, it is necessary to highlight the following fact. In terms of runtime, semi-supervised algorithms get higher numerical values than supervised algorithms when there is a significant amount of unclassified data. This is mainly because supervised algorithms ignore these instances, while semi-supervised algorithms still need to process this data.

6 Conclusion

In the present work, the main characteristics of the algorithm called SSL-C4.5 were described. This new proposal is an adaptation of the well-known C4.5 algorithm to the semi-supervised learning paradigm. SSL-C4.5 algorithm is able to work in scenarios where few instances are labeled due to also it is able to extract significant information from the attributes that are not classes.

Through an experimental study, we managed to test its operation in different data scenarios compared to other known algorithms. As a result, it was shown that the use of semi-supervised metrics in classification algorithms can contribute to the induction of more reliable models when datasets have few instances with labels. Although no statistically significant differences were found compared to the rest of the algorithms, the SSL-C4.5 algorithm achieved promising empirical results in terms of accuracy values. The new proposal obtained better results in most of the bases.

The authors propose as future work to carry out a more complete study of the algorithm to achieve possible structural improvements. In addition, we propose to perform tests with more diverse databases to try to make the experimental study more robust.

Acknowledgment. This work was partially funded by the Coordination of Improvement of Higher Level Personnel - CAPES, and the Foundation of Support for Research and Innovation of Santa Catarina State - FAPESC.

References

1. Chapelle, O., Scholkopf, B., Zien, A.: Semi-supervised Learning, vol. 2. MIT Press, Cambridge (2006)
2. Farid, D., et al.: An adaptive ensemble classifier for mining concept drifting data streams. Expert Syst. Appl. **15**, 5895–5906 (2013)
3. Haque, A., Khan, L., Baron, M.: Sand: semi-supervised adaptive novel class detection and classification over data stream. In: AAAI, pp. 1652–1658 (2016)
4. Krawczyk, B., Minku, L., Gama, J., Stefanowski, J., Woniak, M.: Ensemble learning for data stream analysis Inf. Fusion **37**, 132–156 (2017)
5. Levatic, J., Ceci, M., Kocev, D., Dzeroski, S.: Semi-supervised classification trees. J. Intell. Inf. Syst. **49**, 461–486 (2017)
6. Li, P., Wu, X., Hu, X.: Mining recurring concept drifts with limited labeled streaming data. In: Sugiyama, Proceedings of 2nd Asian Conference on Machine Learning, vol. 13, pp. 241–252 (2016)
7. Lichman, M.: UCI machine learning repository (2013). http://archive.ics.uci.edu/ml
8. Ortiz-Díaz, A., Baldo, F., Palomino-Mariño, L., Bayer, F., Verdecia-Cabrera, A., Frías-Blanco, I.: Fast adaptive stacking of ensembles adaptation for supporting active learning. A real case application. In: 14th International Conference on Natural Computation, Fuzzy Systems and Knowledge Discovery. ICNC-FSKD (2018)

9. Quinlan, R.: Induction of decision trees. Mach. Learn. **1**, 81–106 (1986)
10. Quinlan, R.: C4.5: Programs for Machine Learning. Morgan Kaufmann Publishers, San Francisco (1993)
11. Stamatis, K., Nikos, F., Sotiris, K., Kyriakos, S.: A semisupervised cascade classification algorithm. Appl. Comput. Intell. Soft Comput. **2016**, 14, Article ID 5919717 (2016)
12. Tanha, J., van Someren, M., Afsarmanesh, H.: Semi-supervised self-training for decision tree classifiers. Int. J. Mach. Learn. Cybern. **8**(1), 355–370 (2015). https://doi.org/10.1007/s13042-015-0328-7
13. Triguero, I., García, S., Herrera, F.: Self-labeled techniques for semi-supervised learning: taxonomy, software and empirical study. Knowl. Inf. Syst. **42**, 245–284 (2013). https://doi.org/10.1007/s10115-013-0706-y
14. Wagh, S., Khati, A., Irani, A., Inamdar, N., Soni, R.: Effective framework of j48 algorithm using semi-supervised approach for intrusion detection. Int. J. Comput. Appl. **94**(12), 23–27 (2014)
15. Wei, Z., Jia, K., Sun, Z.: An automatic detection method for morse signal based on machine learning. In: IIH-MSP-2017: Advances in Intelligent Information Hiding and Multimedia Signal Processing, pp. 185–191 (2017)
16. Witten, I., Frank, E., Trigg, L., Cunningham, M.H.G.H.S.: Weka: practical machine learning tools and techniques with Java implementations. In: Proceedings of the ICONIP/ANZIIS/ANNES 1999 Workshop on Emerging Knowledge Engineering and Connectionist-Based Information Systems, pp. 192–196 (1999)

Multidisciplinary Artificial and Computational Intelligence and Applications

Data Streams Are Time Series: Challenging Assumptions

Jesse Read[1], Ricardo A. Rios[2(✉)], Tatiane Nogueira[2], and Rodrigo F. de Mello[3]

[1] LIX, École Polytechnique, Paris, France
jread@lix.polytechnique.fr
[2] DCC, Federal University of Bahia, Salvador, BA, Brazil
{ricardoar,tatiane.nogueira}@ufba.br
[3] ICMC, University of São Paulo, São Carlos, SP, Brazil
mello@icmc.usp.br

Abstract. The increasingly relevance of data streams in the context of machine learning and artificial intelligence has motivated this paper which discusses and draws necessary relationships between the concepts of *data streams* and *time series* in attempt to build on theoretical foundations to support online learning in such scenarios. We unify the concepts of data streams and time series by assessing their definitions in the literature and discuss the major implications of this claim on the way that data streams research and practice is carried out, showing that many common assumptions are incorrect or unnecessary. We analyzed six data sources typically used in benchmark data-stream classification and found that none of those meet the requirements and assumptions qualifying them for online learning.

Keywords: Data streams · Time series · Statistical Learning Theory

1 Introduction

Data stream analysis is a hot topic in Machine Learning (ML) and Artificial Intelligence (AI), in particular with the increasing ubiquity of sensor networks and the Internet of Things (IoT), as well as the rise of applications involving robotics, reinforcement learning, system monitoring, anomaly detection, social networks and media analysis [1,2,4,8,15,27,30,35,36,40,41] From this perspective, it is worth defining the necessary assumptions associated with streams, as well as their connections with pre-existing concepts from other areas which could support, and even extend, the theoretical foundation of learning in such scenarios. Moreover, it is relevant to study which types of streaming tasks are present in the real world, and which techniques should be used to deal with them.

In this context, this paper discusses and draws a necessary relationship between the concepts of *Time Series* (TS) and *Data Streams* (DS), concluding the latter is a special instance from the former, consequently we could

R. Cerri and R. C. Prati (Eds.): BRACIS 2020, LNAI 12320, pp. 529–543, 2020.
https://doi.org/10.1007/978-3-030-61380-8_36

take advantage of all theoretical foundation and practical experience from time series, most specially from two scientific branches: Statistics and Dynamical Systems [6,9,22,39].

By definition, TS analysis is a research area focused on designing tools to modeling sequences of observations collected over a specific time interval and arranged in chronological order [9,44] as $X = \{x_1, \ldots, x_t, \ldots, x_n\}$, in which t is a given time instant and n is the number of data observations. According to [9], TS analysis is mainly suitable in four application domains: (i) forecasting; (ii) estimation of transfer functions; (iii) effects analysis of unusual intervention events on systems; and (iv) discrete control systems.

DS analysis is here presented as a TS extension, being characterized by an open-ended data flow continuously produced over time and, eventually, at high speed [3,17], i.e. $X = \{x_1, x_2, \ldots, x_t, \ldots\}$. These characteristics meet three important V's usually discussed in the Big Data context: Volume, Velocity, and Volatility. At this point, it is worth mentioning that Big Data does not necessarily considers the continuous collection of observations along time, besides such aspect is commonly assumed in the literature [38]. Given the ever-growing nature of DS, batch-like TS tools cannot be directly applied on such scenarios, however they provide a fundamental background to support new developments in this ML scenario.

The TS background already defines aspects such as the time-uniform and the non-uniform data sampling, the presence or the absence of temporal data dependence, uni or multidimensional data observations, lag operators to formalize relationships between a given observation and past or future ones, in addition to a series of models to represent data characteristics such as the ARIMA models [9], from the Statistics perspective; and the Polynomial and the Radial Basis Function models widely employed by the Dynamical System community [19]. All that theory and practice is obviously useful to give an extra step in DS studies, pointing out useless and useful strategies to be assessed when dealing with endless streams. Simultaneously, the DS scenario brings some new and complementary challenges to the TS area such as: (i) the need of single-pass algorithms to process data as it arrives [26]; (ii) the incremental update of learning models as new data arrives; (iii) the reduction in memory requirements once it is impossible to maintain all data in accessible; (iv) the minimization of time complexity so that the algorithm is capable of updating models without discarding new arriving samples [31].

We observe a complementary connection between the TS and DS areas which have motivated this paper to explore additional relationships so that DS researchers could take advantage of previous TS results, as well as TS researchers could employ their knowledge to help us improving the ML area. From that, we discuss on some aspects that may have major implications on the way DS researches are carried out:

(i) Data streams are time series – see Sect. 3;
(ii) Real-world data streams usually exhibit temporal dependence what contradicts the i.i.d. (identically and independently sampled) principle necessary

to ensure learning guarantees according to the Statistical Learning Theory (SLT) [29,32,48] - see Sect. 4, and

(iii) Real-world data streams do not inherently produce any ground-truth class labels, and this stands against the possibility of building classifiers in general-purpose scenarios. Nevertheless, regression can be performed when temporal dependence is present - see Sect. 5.

The overall message we wish to bring is that data-stream researchers need to pay closer attention to these aspects prior to design, deployment, and evaluation of streaming algorithms. In the remainder of this paper, we elaborate further on these aspects, finishing the paper by making recommendations in this regard.

2 Background

This section introduces the related work and all necessary background concepts used throughout this paper.

2.1 Related Work

There has been some earlier consideration in the literature in regard to temporal dependence in data-stream classification and evaluation. For example, Žliobaitė et al. [49] identified temporal dependence in benchmark data streams and conclude that common measures, such as the classification accuracy and the Kappa statistic can be misleading in the presence of time dependence, and fail to diagnose cases the poor performance when such a dependence is present. Žliobaitė et al. [49] still address the issue of temporal dependence by including labels $y_{t-k}, \ldots, y_{t-1}$ as additional inputs so that a classifier is inferred to predict $_t$. In fact, this is the same as the space embedding techniques from Dynamical Systems employ to unfold time relationships [46], without however estimating the best as possible reconstruction to map the relationship of past class labels to their next values, nor commenting on the availability of such past labels.

In DS concept drift, we must consider that the probabilities of class membership [23] on the feature vectors cannot be always taken as stationary. In fact, this is the same as having a dynamic joint probability distribution (JPD) $P(\mathcal{X} \times \mathcal{Y})$ what stands against the Statistical Learning Theory (please refer to Sect. 2.2), therefore learning bounds cannot be ensured according to such framework. Most specifically the i.i.d. principle and the requirement of a static/fixed JPD to ensure the probabilistic convergence of the empirical risk (sample error) to its expected value [29,32].

Mello et al. [33] also observed temporal dependence but in the context of concept drift detection and concluded that some approach should be considered to unfold time relationships among data stream observations to allow the proper prediction of next values. They do not consider the typical classification scenario, but the regression one instead, which is more general to explain the time dependence effects on learning.

After this brief discussion on the related work, the following section provides details on the framework of the Statistical Learning Theory, which provides the necessary background to discuss supervised learning from a theoretical perspective in light of probabilistic convergence guarantees.

2.2 Statistical Learning Theory

Statistical Learning Theory (SLT) is the formal framework designed by Vapnik [48], which provides convergence guarantees for the supervised learning scenario. This framework relies on an input space $\mathcal{X}$ and an output space $\mathcal{Y}$ to infer some classification or regression function $f : \mathcal{X} \rightarrow \mathcal{Y}$. Training examples are sampled from some JPD $P(\mathcal{X} \times \mathcal{Y})$ and the supervised learning algorithm is responsible for estimating this distribution and then representing the relationship between inputs and their outcomes.

SLT relies on the Law of Large Numbers (LLN) to ensure learning convergence (Eq. 1) by means of the probability of some average estimator $\frac{1}{n}\sum_{i=1}^{n}\xi_i$, computed along the realizations of a random and independent variable, converges to its expected value $\mathbb{E}(\xi)$ [12]. However, this result is only valid if the two following assumptions are satisfied: (i) the data distribution, or the JPD in our case, must be static, i.e., it cannot change along time; and (ii) the random variable ξ must produce independent observations along the collection. It is worth highlighting that these assumptions are in complete contradiction to those typically made in the context of data streams (independently distributed data, and concept drift, respectively). There are other additional assumptions that do not change any of our conclusions in this paper (see more in [29,48]).

$$\lim_{n\rightarrow\infty} P\left(\left|\frac{1}{n}\sum_{i=1}^{n}\xi_i - \mathbb{E}(\xi)\right| > \epsilon\right) = 0 \tag{1}$$

After setting those assumptions, we can take advantage of the LLN to represent learning according to the concept of Generalization in form $|R_{\text{emp}}(f) - R(f)|$, in which $R_{\text{emp}}(f)$ corresponds to the empirical risk, or the error measured in a given sample, and the risk $R(f)$ is the error measured over the whole JPD, as if we had access to all input and output mapping possibilities.

The empirical risk $R_{\text{emp}}(f)$, formulated in Eq. 2(a), relies on some loss function $\ell(.)$ which receives a given input example x_i, its expected output y_i, and the result of the classification or regression function represented by f. If $f(x_i) = y_i$ no error has occurred, so that $\ell(x_i, y_i, f(x_i)) = 0$, otherwise some divergence will be summed up along the assessment of all n input examples. The risk $R(f)$, defined in Eq. 2(b), computed the expected value of the loss function given full access to the JPD $P(\mathcal{X} \times \mathcal{Y})$.

$$(a)\ R_{\text{emp}}(f) = \frac{1}{n}\sum_{i=1}^{n}\ell(x_i, y_i, f(x_i)) \qquad (b)\ R(f) = \mathbb{E}(\ell(x, y, f(x))) \tag{2}$$

From that, Vapnik formulated the Empirical Risk Minimization Principle (EMRP) for a single classification function f as follows $\lim_{n\to\infty} P(|R_{\text{emp}}(f) - R(f)| > \epsilon) = 0$, as a direct result of the LLN [29,32]. However, that is valid if a function f is selected at random and not after training iterations on top of some training set. As consequence, the set of admissible functions $\mathcal{F}$ can be defined as the algorithm bias, i.e., the set of functions a given learning algorithm is capable of using to represent some input space. After that, he assumed any function inside such bias could be selected during the training stage, so that the following automatically holds

$$\lim_{n\to\infty} P(|R_{\text{emp}}(f_1) - R(f_1)| > \epsilon) = 0, \ldots, \lim_{n\to\infty} P(|R_{\text{emp}}(f_m) - R(f_m)| > \epsilon) = 0,$$

if one considers each classification function $f_i \in \mathcal{F}$ is selected independently of the training set and supposing a finite number of functions m composing such bias. Then, Vapnik formulated the such convergence using the Chernoff bound [12] in form

$$P(|R_{\text{emp}}(f_i) - R(f_i)| > \epsilon) \leq 2\exp(-2n\epsilon^2),\ \forall i = 1, \ldots, m,$$

which was a previously proved upper bound for the LLN. Using this bound, the worst-case scenario is that we must consider all such functions; probabilistically as

$$P(\sup_{f\in\mathcal{F}} |R_{\text{emp}}(f) - R(f)| > \epsilon) \leq P(\vee_{i=1}^{m} |R_{\text{emp}}(f_i) - R(f_i)| > \epsilon) \leq \sum_{i=1}^{m} P(|R_{\text{emp}}(f_i) - R(f_i)| > \epsilon) \leq \sum_{i=1}^{m} 2\exp(-2n\epsilon^2),$$

so that a clear upper bound is defined for the probability of converging the empirical to the expected risk, in which the supremum means the greatest as possible divergence between both risks given the space of admissible functions $\mathcal{F}$. The number of functions m inside the algorithm bias is dependent on the sample size; known as the *Shattering coefficient*, referred to as $m(n)$ (as it depends on the sample size), finally obtaining

$$P(\sup_{f\in\mathcal{F}} |R_{\text{emp}}(f) - R(f)| > \epsilon) \leq 2m(n)\exp(-2n\epsilon^2), \quad (3)$$

from which he proved that learning is only possible if $m(n)$ is a polynomial function, otherwise, if exponential, no learning guarantee is ensured according to this theoretical framework. It is very important to mention that $m(n)$ measures the *bias complexity*, for instance, if once considers a supervised learning model composed of k or $k + 1$ hyperplanes, the former is less complex then the latter strategy [32].

3 Data Streams Are Time Series

Next, we look at some of the definitions and assumptions made in the literature regarding data streams and time series.

Data Stream: According to, e.g., [5,16,21,30,41,45] (and those cited therein, as well as others), we can define, without contradiction, a data stream as a continuous flow of data that arrives in chunks, or instances, $[x_1, x_2, \ldots, x_t]$, where x_t is the most recently obtained instance. Each instance is comprised of d attributes, as, e.g., $x_t = [x_{t,1}, \ldots, x_{t,d}]$. There may or may not be a specific time order among these d attributes.

This definition is accepted by a wide range of authors, and other sources beyond data mining and machine learning. For example the Institute for Telecommunication Sciences (ITS)[1], which, according to the Federal Standard 1037C, defines a data stream as a sequence of digitally encoded signals used to represent information in transmission.

Time Series: According to Box and Jenkins in their the landmark book on time series analysis [9] a time series as a set of values sequentially observed over time. Thus, a time series is formally represented as $X = \{x_1, \ldots, x_t, \ldots, x_n\}$, i.e., value x_t at time t. The definition by Keogh et al. [24], in a machine learning context, refers to an ordered set of real-valued variables, and as such is completely compatible with this notation.

Although x_t is often interpreted as a scalar value, in the more general it may indeed be multidimensional, as already dealt with from the perspective of Dynamical Systems [43]. This is linked to the more general concept referred to as *panel* or *longitudinal data* which defines multi-dimensional data measurements collected or make available over time [14,37]. In Statistics and Econometrics, a time series is a particular instance of a panel in which every data record is uniquely identified according to its timestamp. However, if time indices are not considered, such multidimensional data are simply referred to as panel [13].

Therefore, we may consider a time series formally defined as length of panel data indexed over time; and thus, already equivalent to a data stream in terms of notation. Of course, the notation is only one question of similarity between the two fields. Other important aspects are what the data looks and how it is obtained in practice, as this affects particularly how the data is modeled and processed.

Research in data streams typically revolves around several major requirements and assumptions on *how* the data should be dealt with; assumptions which are not typically made when dealing with time series. In other words, data-stream researchers focus on a set of constraints/requirements. Each requirement (R) is associated with a particular assumption (A). We list these concisely as follows (as summarized from [1,4,8,15,27,30,36] among others):

- A1 The end of the stream is indefinite (i.e., possibly infinite);
- R1 The data stream frequency defines the maximum available computational resources (e.g., time and memory) to process single observations;
- A2 Data instance (observation) x_t is observed only at time t (and not before);

[1] ITS provides federal departments and agencies comprehensive definitions of terms used in telecommunications and directly related fields by the U.S. Government and internationally; see https://www.its.bldrdoc.gov/fs-1037/dir-010/_1451.htm.

R2 An action (e.g., prediction) is required at real time t (i.e., immediately, and prior to $t+1$);
A3 Concept drift may eventually occur;
R3 Models must be able to adapt over time;
A4 There is no temporal dependence (each instance of the stream is sampled in an i.i.d. fashion).

We emphasize that unlike assumptions A1–A3, the assumption A4 serves in practice to *remove* a requirement rather than add one. In particular, it allows the learning guarantees discussed above (Sect. 2.2) without further consideration. As we argue in Sect. 4, this assumption may be artificial, being loosely supported in practice.

Let us elaborate by defining more precisely what is a concept, and then a concept drift:

Concept: According to the Dynamical Systems framework, every time series (recall: which may be also a data stream) is produced by a set of functions referred to as *generating process* [6,22]. The sequence of observations produced by some generating process along time compose what we here refer to as a concept. Assuming that data observations x_t follow $x_t \sim P_t(.)$, where P_t is the probability distribution in play at time t, being the same manner as the generating process to Dynamical Systems. The behavior produced by $P_t(.)$, such as time dependencies and trends, is what we refer to as a *concept.*

Let $S_c = \{P_{-\infty}, \ldots, P_t, \ldots, P_0\}$ be a set of probability distributions happening from a distant past, that is why we use this typical notation of $-\infty$ from the Time Series area [9], to the present, defined as time index 0 in P_0. If $P_k = P_t$, the same data distribution is then used and, consequently, the same concept is expressed.

Concept Drift: A concept drift is a change in the behavior of data observations due to a modification in the underlying generating process, if analyzing from the Dynamical System perspective, or in the data distribution, from the point of view of Statistics. In the extreme case of abrupt change, it means a completely different distribution P_{t+1} starts after P_t, so that their corresponding produced observations are very different. A smooth change is equally possible, where P_{t+1} is somehow close to/related to P_t, with smaller differences corresponding to smoother drift. Figure 1 illustrates the connection between concepts c_t and temporal dependence among x_t.

4 Temporal Dependence in Real-World Data Streams

Real-world data streams usually exhibit temporal dependence, contradicting assumption A4 from above. In attempt to illustrate our perspective, consider the growth of bacteria [22] which is modeled using the logistic map in form $x_{t+1} = rx_t(1 - x_t)$ having x_t as the population size at time instant t, x_{t+1} as

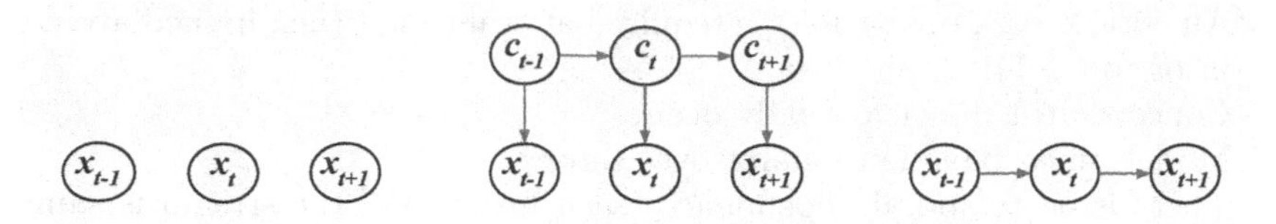

Fig. 1. (a) True independent stream (no drift/single concept). (b) Considering the concept may change over time; represented by c_{t-1}, c_t, c_{t+1}. (c) If we marginalize out the concept variable, we observe a time-dependent stream.

its next size, and $r \in \mathbb{R}_+$ as a complexity factor that implies chaoticity [6,22]. Observe that the current population is dependent of the previous one, illustrating our claim.

Similarly, other real-world phenomena are also modeled using differential equations that make evident the temporal dependence among observations, such as:

(i) the Lorenz system [28] to model the atmospheric convection, which comprises in the formulation of a fluid layer being uniformly warmed from below and cooled from above. This system is also sed to model other phenomena such as lasers, dynamos, thermosyphons, brushless DC motors, electric circuits, chemical reactions and forward osmosis [18,25,34];
(ii) the Sinusoidal function to model temperatures and solar explosions [11,50];
(iii) the Mackey-Glass equation is used to model the qualitative features of physiological dynamics, allowing to represent physiological disorders. This equation has been used to model disorders related to hematology, cardiology, neurology, and psychiatry [7];
(iv) the Rössler attractor to perform topological analysis as well as the modeling of equilibrium in chemical reactions [42].

One can also employ classical tools in the assessment, modeling and forecasting of observations, so that the area of Machine Learning is complemented with fundamental formulations and proofs. In order to exemplify, consider the Auto-Correlation Function (ACF), typically employed to estimate the time lag for ARIMA models [9], which confirms the dependence of data observations along time. Figure 2a illustrates a time series produced with the logistic map. Figure 2b confirms there are relevant time dependences, which result in those spikes above some confidence level (horizontal line above the time-lag axis).

The main issue resultant from such time dependence is that supervised learning has no guarantee according to the SLT [48]. This means any supervised model obtained after training on top of those dependent data produce some inconclusive model, that happens because such type of learning is only ensured if data are sampled in an i.i.d. fashion [29]. Therefore, one need to find some transformation to ensure data independence.

Takens' embedding theorem [46] can be used in conjunction with supervised learning to find the best as possible embedding, unfolding data dependences

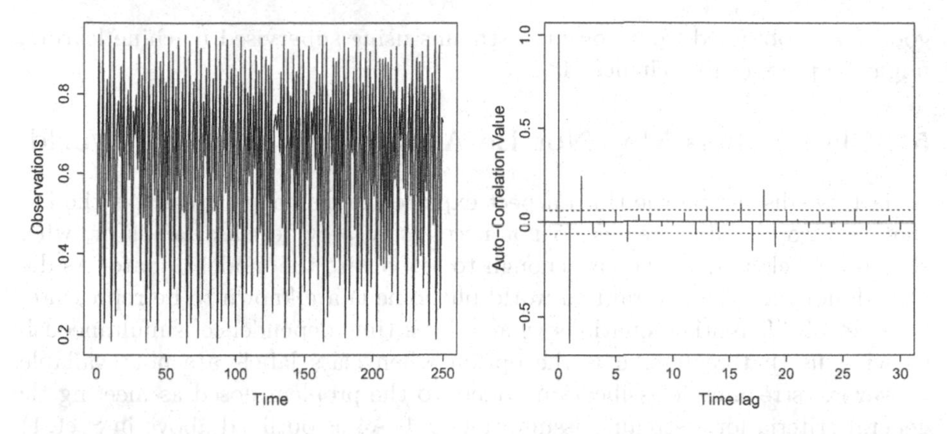

Fig. 2. (a) Time series generated using the Logistic map. (b) Auto-Correlation Function computed on data from the Logistic map.

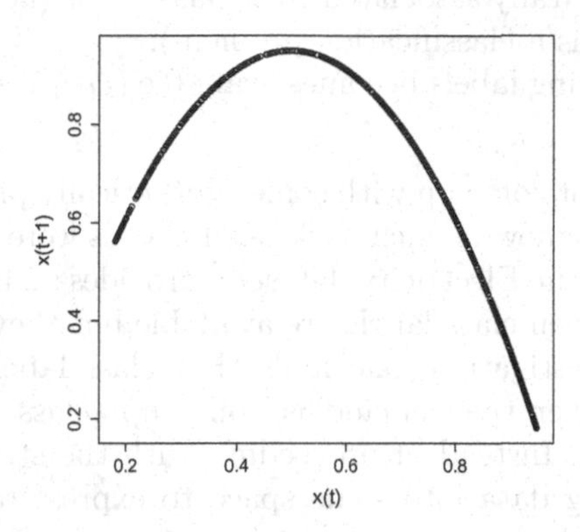

Fig. 3. Phase space obtained for the data illustrated in Fig. 2a.

in another space, a.k.a. phase space, which respects the i.i.d. assumption so supervised learning can be performed. For instance, after applying such approach on the time series shown in Fig. 2a, we obtain the phase space illustrated in Fig. 3 from which one can sample data in an i.i.d. manner.

Observe that Fig. 3 represents a given observation x_t in terms of its previous one x_{t-1}, so that a simple regression function on top of such reconstructed space is enough to model and predict such stream. Therefore, instead of considering some classification task, one can take the data observations to obtain some space containing the regression inputs as well as their outputs, while relying on theoretical guarantees after the SLT (Sect. 2.2).

Of course there are time independent observations forming data streams such as the numbers in a lottery round, numbers in a casino roulette, as well as some background noise that a sensor may read in a given target scenario. However, we state that is unfair to consider supervised learning on any data stream scenario, due to temporal dependence is widely found in real-world streams [22]. Thus, any good result obtained from non-i.i.d. streams using supervised machine learning might be produced by chance [48].

5 Class Labels May Not Be Available in the Real World

At last, we discuss on the third aspect explored in this paper, which is the fact that real-world data streams do not necessarily provide class labels or, when they do, labels may arrive late enough to jeopardize the model update. As discussed along Sect. 4, several real-world phenomena are known to be represented by a set of differential equations what proves time dependence, simultaneously showing us that regression is the option when class labels are not available. However, "streaming classification" refers to the problem posed as meeting the general criteria for a stream (assumptions A1–A4 as outlined above in Sect. 1), in complement to the following assumptions:

- A5 Instances are naturally associated with class labels (i.e., the problem is most naturally posed as a classification problem);
- A6 A stream of training labels becomes available (y_{t-1} is observed at time $t+k$ for some $k > 0$).

Of course one might come up with some synthetically produced stream which has labels, or use well-known benchmarks as if classes were naturally available at some point in future. The Electricity dataset [2] provides an interesting illustration on our perspective, given class labels are available but they were produced from data itself. Upon investigation, one finds that class labels in this benchmark simply indicate whether the demand is going up (class 1) or down (class 2) given the price signal. Instead of proceeding with the streaming classification, one should embedding data into some space to express temporal dependences when needed, as for this dataset, so one can perform some regression to model if prices go up or down instead of assuming such labels. In other words, the class just defines information that is already embedded in the underlying time series, being related to its short-term trend. The labels are in this sense artificial and the problem is in fact a time series analysis problem.

Aside from these examples, we have gathered a list of the most common real-world data sources used in empirical evaluations in data-stream classification papers, and we have enumerated which of the six data-stream assumptions are not met (see Table 1).

An initial observation is that there are relatively few datasets. Another remark, as highlighted in the second column of this table, is that they are either not classification problems (A5) and/or they are time series analysis tasks (A4) and thus not classification tasks – a separate body of literature – as per our discussion above (Fig. 4).

Concretely: Electricity, Airlines and AWS Prices are based on regression problems. These, and also Give-me-some-Credit are time-series analysis problems, where the

[2] This dataset and corresponding sources are available at https://moa.cms.waikato.ac.nz/datasets.

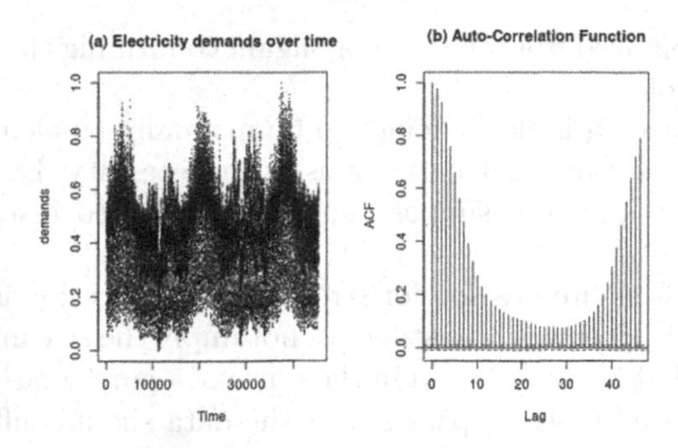

Fig. 4. Snapshot from the energy demands contained in the Electricity dataset (upon which the class label is based) in this well-known data-stream benchmark). We see that (a) Temporal dependence is visible on the feature space, and (b) as confirmed by an Auto-Correlation Function (ACF).

Table 1. Benchmark data-stream classification datasets, and the reason they are not data-stream classification tasks in the real-world scenarios. No indicates the assumption is not met; and ∼ when it is not known or clear.

Dataset	A1	A2	A3	A4	**A5**	**A6**
Electricity				No	No	
Airlines				No	No	
Covertype	∼	No		No		No
Give-me-some-Credit				No		No
KDD Cup 99			∼	∼		No
AWS Prices		∼		No	No	

label comes from the future (a 2-year prediction horizon in the credit case). All those scenarios would take advantage of space embeddings in conjunction with regression models to represent the temporal dependence and yet ensure learning [10]. There is no indication that potential applications relating to Covertype need to meet the constraints of a data-stream problem. KDD Cup 99 has already been criticized as unrealistic (e.g. [47]). Furthermore we note that several of these datasets are over 20 years old.

Of course there are clear classification tasks, such as in the scenario of movie streaming services in which data streams can be built using users' ratings (such as in the AWS Prices dataset[3]). However, such scenarios are actually close to a batch learning system in the sense the modeling cost is high so companies typically retrain classification algorithms only a dozen or less times an year, what is very far from online learning. As an additional observation, as movies are launched, the joint probability distributions representing their learning tasks are likely to change, specially because some movies do not call enough attention

[3] Again, see https://moa.cms.waikato.ac.nz/~datasets/.

and/or they are removed from the catalog, again confirming the time dependence as an important role.

If indeed the stream is slow enough to be manually labeled, this has still no implication that a streaming treatment is even necessary. Even a large crowd of efficient labelers does not suppose an urgent *need* to instance incremental learning.

Similarly, an often cited reason for streaming classification is when data does not fit into memory. However, this still does not imply the streaming classification treatment (indeed, it is not a stream in the sense of a particularly infinite number of instances) – indeed multiple passes over the data should suffice.

6 Recommendations and Conclusions

This paper has challenged various assumptions made implicitly and explicitly in the data stream literature. Many of these assumptions may be unnecessary or even harmful in practice. For example, there is often no need to assume that instances cannot be stored, especially under low frequency arrival. In many cases there is not even a need to consider finite resources (as long as purchase of such resources is balanced against monetary concerns). On the other hand, the assumption of temporal dependence in a stream should be always questioned, or the contrary (of independence) should be justified, prior to committing to a particular approach. Temporal dependence should at least make us question the possibility of space embedding in attempt to proceed with regression models to ensure learning results.

We also recommend to question the feasibility of "streaming classification", given several scenarios could be approached as a time series analysis, a regression, or a very weakly supervised task since the labels will be too scarce for the application of incremental supervised learning. In light of this, a sensible benchmark for streams should always include some time series methods, such as ARIMA or dynamical system models such as partial-differential equations, hidden Markov Models, or recurrent neural networks [20]. An alternative is to chunk the data into time windows, but this again implies a consideration of independence as just discussed above. All this said, many items from the data-stream literature have important applications if considered in the right context, and has much to offer to the evolution of time series analysis, particularly in consideration of the rising number of dynamic data sources and the need for online modeling.

References

1. Tennant, M., Stahl, F., Rana, O., Gomes, J.B.: Scalable real-time classification of data streams with concept drift. Fut. Gener. Comput. Syst. **75**, 187–199 (2017). https://doi.org/10.1016/j.future.2017.03.026
2. Aaij, R.: Tesla: an application for real-time data analysis in high energy physics. Comput. Phys. Commun. **208**, 35–42 (2016). https://doi.org/10.1016/j.cpc.2016.07.022

3. Aggarwal, C.C.: Data Streams: Models and Algorithms. Advances in Database Systems, vol. 31. Springer, Heidelberg (2006). https://doi.org/10.1007/978-0-387-47534-9
4. Aggarwal, C.C.: A survey of stream classification algorithms. In: Data Classification: Algorithms and Applications (2014)
5. Al-Khateeb, T.: Recurring and novel class detection using class-based ensemble for evolving data stream. IEEE Trans. Knowl. Data Eng. **28**(10), 2752–2764 (2016). https://doi.org/10.1109/TKDE.2015.2507123
6. Alligood, K., Sauer, T., Yorke, J.: Chaos: An Introduction to Dynamical Systems. Textbooks in Mathematical Sciences. Springer, New York (2000). https://doi.org/10.1007/b97589
7. Bélair, J., Glass, L., der Heiden, U., Milton, J.: Dynamical disease: identification, temporal aspects and treatment strategies of human illness. Chaos Interdisc. J. Nonlinear Sci. **5**(1), 1–7 (1995)
8. Bifet, A.: Adaptive Stream Mining: Pattern Learning and Mining from Evolving Data Streams. Frontiers in Artificial Intelligence and Applications. IOS Press, Amsterdam (2010)
9. Box, G.E.P., Jenkins, G.M.: Time Series Analysis: Forecasting and Control, 3rd edn. Prentice Hall, Upper Saddle River (1994)
10. de Carvalho Pagliosa, L., de Mello, R.F.: Applying a kernel function on time-dependent data to provide supervised-learning guarantees. Exp. Syst. Appl. **71**, 216–229 (2017). https://doi.org/10.1016/j.eswa.2016.11.028
11. Chen, G., Fang, X., Fan, H.: Estimating hourly water temperatures in rivers using modified sine and sinusoidal wave functions. J. Hydrol. Eng. **21**(10), 05016023 (2016)
12. Devroye, L., Györfi, L., Lugosi, G.: A Probabilistic Theory of Pattern Recognition. Stochastic Modelling and Applied Probability. Springer, New York (1997). https://doi.org/10.1007/978-1-4612-0711-5
13. Diggle, P., et al.: Analysis of Longitudinal Data. Oxford Statistical Science Series. Oxford University Press, Oxford (2002)
14. Frees, E.: Longitudinal and Panel Data: Analysis and Applications in the Social Sciences. Cambridge University Press, Cambridge (2004)
15. Gaber, M.M., Zaslavsky, A., Krishnaswamy, S.: A survey of classification methods in data streams. In: Aggarwal, C.C. (ed.) Data Streams. Advances in Database Systems, vol. 31, pp. 39–59. Springer, Boston (2007). https://doi.org/10.1007/978-0-387-47534-9_3
16. Gama, J., Žliobaitė, I., Bifet, A., Pechenizkiy, M., Bouchachia, A.: A survey on concept drift adaptation. ACM Comput. Surv. **46**(4), 441–4437 (2014). https://doi.org/10.1145/2523813
17. Gama, J.: Knowledge Discovery from Data Streams. CRC Press, Boca Raton (2010)
18. Gorman, M., Widmann, P., Robbins, K.: Nonlinear dynamics of a convection loop: a quantitative comparison of experiment with theory. Physica D **19**(2), 255–267 (1986). https://doi.org/10.1016/0167-2789(86)90022-9
19. Hegger, R., Kantz, H., Schreiber, T.: Practical implementation of nonlinear time series methods: the TISEAN package. Chaos Interdiscip. J. Nonlinear Sci. **9**(2), 413–435 (1999)
20. Hochreiter, S., Schmidhuber, J.: Long short-term memory. Neural Comput. **9**(8), 1735–1780 (1997)

21. Hulten, G., Spencer, L., Domingos, P.: Mining time-changing data streams. In: Proceedings of the 7th ACM SIGKDD International Conference on Knowledge Discovery and Data Mining, KDD 2001, pp. 97–106. ACM, New York (2001). https://doi.org/10.1145/502512.502529
22. Kantz, H., Schreiber, T.: Nonlinear Time Series Analysis, 2nd edn. Cambridge University Press, Cambridge (2003). https://doi.org/10.1017/CBO9780511755798
23. Kelly, M.G., Hand, D.J., Adams, N.M.: The impact of changing populations on classifier performance. In: Proceedings of the 5th ACM SIGKDD International Conference on Knowledge Discovery and Data Mining, pp. 367–371. Citeseer (1999)
24. Keogh, E.J., Chu, S., Hart, D., Pazzani, M.J.: An online algorithm for segmenting time series. In: Proceedings of the 2001 IEEE International Conference on Data Mining, ICDM 2001, pp. 289–296. IEEE Computer Society, Washington, DC, USA (2001)
25. Knobloch, E.: Chaos in the segmented disc dynamo. Phys. Lett. A **82**(9), 439–440 (1981)
26. Krawczyk, B., Minku, L.L., Gama, J., Stefanowski, J., Woźniak, M.: Ensemble learning for data stream analysis: a survey. Inf. Fusion **37**, 132–156 (2017). https://doi.org/10.1016/j.inffus.2017.02.004
27. Krempl, G., et al.: Open challenges for data stream mining research. ACM SIGKDD Explor. Newsl. **16**, 1–10 (2014). https://doi.org/10.1145/2674026.2674028
28. Lorenz, E.N.: Deterministic nonperiodic flow. J. Atmos. Sci. **20**(2), 130–141 (1963). https://doi.org/10.1175/1520-0469(1963)020⟨0130:DNF⟩2.0.CO;2
29. von Luxburg, U., Schölkopf, B.: Statistical Learning Theory: Models, Concepts, and Results, vol. 10, pp. 651–706. Elsevier North Holland, Amsterdam (2011)
30. Masud, M.M.: Facing the reality of data stream classification: coping with scarcity of labeled data. Knowl. Inf. Syst. **33**(1), 213–244 (2012). https://doi.org/10.1007/s10115-011-0447-8
31. McGregor, A., Pavan, A., Tirthapura, S., Woodruff, D.P.: Space-efficient estimation of statistics over sub-sampled streams. Algorithmica **74**(2), 787–811 (2016)
32. Fernandes de Mello, R., Antonelli Ponti, M.: Machine Learning - A Practical Approach on the Statistical Learning Theory. Springer, Cham (2018). https://doi.org/10.1007/978-3-319-94989-5
33. de Mello, R.F., Vaz, Y., Ferreira, C.H.G., Bifet, A.: On learning guarantees to unsupervised concept drift detection on data streams. Exp. Syst. Appl. **117**, 90–102 (2019). https://doi.org/10.1016/j.eswa.2018.08.054
34. Poland, D.: Cooperative catalysis and chemical chaos: a chemical model for the lorenz equations. Physica D **65**(1), 86–99 (1993). https://doi.org/10.1016/0167-2789(93)90006-M
35. Puthal, D.: Lattice-modeled information flow control of big sensing data streams for smart health application. IEEE Internet Things J. **6**(2), 1312–1320 (2019). https://doi.org/10.1109/JIOT.2018.2805896
36. Rajaraman, A., Leskovec, J., Ullman, J.D.: Mining Massive Datasets (2014). http://infolab.stanford.edu/~ullman/mmds/book.pdf
37. Hsiao, C.: Analysis of Panel Data, 2nd edn, p. 382. Cambridge University Press, Cambridge (2003). ISBN: 0-521-81855-9, [uk pound]21.95. Int. J. Forecast. **20**(1), 142–143 (2004)
38. Richards, N.M., King, J.H.: Three paradoxes of big data. Stan. L. Rev. Online **66**, 41 (2013)

39. Rios, R.A., de Mello, R.F.: Applying empirical mode decomposition and mutual information to separate stochastic and deterministic influences embedded in signals. Sig. Process. **118**, 159–176 (2016). https://doi.org/10.1016/j.sigpro.2015.07.003
40. Rios, R.A., Pagliosa, P.A., Ishii, R.P., de Mello, R.F.: TSViz: a data stream architecture to online collect, analyze, and visualize tweets. In: Proceedings of the Symposium on Applied Computing, SAC 2017, Marrakech, Morocco, 3–7 April 2017, pp. 1031–1036 (2017). https://doi.org/10.1145/3019612.3019811
41. Roseberry, M., Cano, A.: Multi-label kNN classifier with self adjusting memory for drifting data streams. In: Torgo, L., Matwin, S., Japkowicz, N., Krawczyk, B., Moniz, N., Branco, P. (eds.) Proceedings of the 2nd International Workshop on Learning with Imbalanced Domains: Theory and Applications, PMLR. Proceedings of Machine Learning Research, ECML-PKDD, Dublin, Ireland, 10 September 2018, vol. 94, pp. 23–37 (2018)
42. Rössler, O.: An equation for continuous chaos. Phys. Lett. A **57**(5), 397–398 (1976). https://doi.org/10.1016/0375-9601(76)90101-8
43. Serrà, J., Gómez, E., Herrera, P.: Audio cover song identification and similarity: background, approaches, evaluation, and beyond. In: Raś, Z.W., Wieczorkowska, A.A. (eds.) Advances in Music Information Retrieval. Studies in Computational Intelligence, vol. 274. Springer, Heidelberg (2010). https://doi.org/10.1007/978-3-642-11674-2_14
44. Shumway, R.H., Stoffer, D.S.: Time Series Analysis and Its Applications: With R Examples. Springer Texts in Statistics, 2nd edn. Springer, Heidelberg (2006). https://doi.org/10.1007/978-3-319-52452-8
45. Silva, J.A., Faria, E.R., Barros, R.C., Hruschka, E.R., de Carvalho, A., Gama, J.: Data stream clustering: a survey. ACM Comput. Surv. **46**(1), 131–1331 (2013). https://doi.org/10.1145/2522968.2522981
46. Takens, F.: Detecting strange attractors in turbulence. In: Rand, D., Young, L.-S. (eds.) Dynamical Systems and Turbulence, Warwick 1980. LNM, vol. 898, pp. 366–381. Springer, Heidelberg (1981). https://doi.org/10.1007/BFb0091924
47. Tavallaee, M., Bagheri, E., Lu, W., Ghorbani, A.A.: A detailed analysis of the KDD CUP 99 data set. In: Proceedings of the 2nd IEEE International Conference on Computational Intelligence for Security and Defense Applications, CISDA 2009, pp. 53–58. IEEE Press, Piscataway (2009)
48. Vapnik, V.N.: The Nature of Statistical Learning Theory. Springer, Heidelberg (1995). https://doi.org/10.1007/978-1-4757-3264-1
49. Žliobaitė, I., Bifet, A., Read, J., Pfahringer, B., Holmes, G.: Evaluation methods and decision theory for classification of streaming data with temporal dependence. Mach. Learn. **98**(3), 455–482 (2014). https://doi.org/10.1007/s10994-014-5441-4
50. Zhang, K., Ng, C.T., Na, M.: Computational explosion in the frequency estimation of sinusoidal data. Commun. Stat. Appl. Meth. **25**(4), 431–442 (2018)

Evaluating a New Approach to Data Fusion in Wearable Physiological Sensors for Stress Monitoring

Clarissa Rodrigues[1], William R. Fröhlich[1], Amanda G. Jabroski[1], Sandro J. Rigo[1(✉)], Andreia Rodrigues[2], and Elisa Kern de Castro[3]

[1] Applied Computing Graduate Program, UNISINOS, São Leopoldo, Brazil
clarissa.ar@gmail.com, william_r_f@hotmail.com, amanda.grams@gmail.com, rigo@unisinos.br

[2] Psychology Graduate Program, University of Vale Do Rio Dos Sinos (UNISINOS), São Leopoldo, Brazil
andreiakrodrigues@yahoo.com.br

[3] Universidade Lusíada, Lisboa, Portugal
elisa.kerndecastro@gmail.com

Abstract. The physiological signs are a reliable source to identify stress states, and wearable sensors provide precise identification of physiological signs associated with the stress occurrence. The literature review shows that the use of physiological signs as a source for stress patterns identification is still a critical investigation subject. Few studies evaluate the effect of combining several different signals and the implications of the data acquisition procedures and details. This article's objective is to investigate the possible integration of data obtained from heart rate variability, electrocardiographic, electrodermal activity, and electromyography to detect stress patterns, considering a new experimental protocol to data acquisition. The data acquisition involved the Trier Social Stress Test, wearable sensor monitoring, and complementary stress perception instruments, resulting in a publicly available dataset. This dataset was evaluated using different machine learning classifiers, considering the obtained annotated data and exploring different physiological features and their combinations.

Keywords: Wearable sensors · Stress · Machine learning

1 Introduction

The stress consists of body response to some situations, and the physiological signs are a source for the identification of this occurrence. Some approaches, such as the Biofeedback [32], consider this aspect to generate effective patient interventions. The biofeedback approach is based on the organism's response and its physiological processes, which are measured through body-driven sensors, further stored and processed by computer applications. Therefore, it allows for awareness of emotional states and training for the voluntary control of physiological and emotional responses.

R. Cerri and R. C. Prati (Eds.): BRACIS 2020, LNAI 12320, pp. 544–557, 2020.
https://doi.org/10.1007/978-3-030-61380-8_37

In recent years, improvements in wearable sensors have presented the possibility of using these devices as sources of data to monitor the user's physiological state. Most of these wearable sensors consist of low-cost devices that provide good quality signals [4, 6, 12]. Consequently, it can generate data utilized as source material to Machine Learning approaches aiming to model and predict these stress states [1, 7, 8]. As a related element, the wide-spread use of mobile devices and their capabilities presents the possibility to collect, process and integrate those physiological signs with more elaborated applications. The physiological data provided by the sensors can be collected on an online basis by mobile devices. In contrast, these devices can support applications to detect specific states and generate interventions to be followed by the users [9–11].

Nevertheless, some critical questions associated with this context are the focus of further research. The physiological wearable sensor data acquisition is a very dynamic field, continually proposing new sensors and improving its capabilities. Some research can be observed in the data analysis and data fusion models, due to the number of possibilities to process the features and integrate the acquired data [9, 04, 11]. Another question of interest is how to ensure the correct identification of some specific data pattern associated with a psychologic state [14]. Finally, there are few works dedicated to evaluating the complete cycle of Biofeedback comprehensively, which comprise using the wearable devices, applying Machine Learning patterns detection algorithms, generating the psychologic intervention, besides monitoring its effects and recording the history of events [9, 15]. Several works were developed considering just one sensor or a few sensors [1–3, 5, 30]. These works, in significant part, do not broadly address the investigation of using a group of different biosignals compositions to identify stress patterns. Papers do not bring enough details on the data acquisition protocol, which is a necessity to make clear the annotation procedures adopted [12, 20, 23].

As outlined above, he literature review shows, that wearables sensors' use to acquire physiological signs as a source to stress patterns identification is still a critical investigation subject. Few studies investigate the effect of combining several different signals and the implications of the data acquisition procedures. Besides, this is a growing area, and there are no standardized and broadly used benchmarking datasets [11, 31]. Our work intends to address these two shortcomings. Therefore, this involves investigating the integration of data obtained from HFV, EDA, EMG to detect stress patterns, considering a new experimental protocol to data acquisition proposed by Psychology researches. The data acquisition involves the Trier Social Stress Test, wearable sensor monitoring, cortisol markers acquisition, and complementary stress perception instruments. Besides that, the generated dataset will be available to broad and open use. It was evaluated using different machine learning classifiers considering the obtained annotated data and exploring different physiological features and their combinations.

The main contributions of this article are: a) Present an experimental protocol to data acquisition regarding stress using the standard TSST protocol and complements with wearable sensors and additional stress perception elements, such as personal questionnaires. b) Investigate a broad set of features and signs to evaluate classification results with a well-known feature set and machine learning classification methods; c) Make available a new dataset, as a publicly available resource. This dataset comprises all the data acquired from the experiments and is designed to promote further research comparison.

2 Background

Stress is a physiological response to internal or external stimuli triggered by the nervous system, particularly by the sympathetic nervous system. The primary physiological known responses are accountable for a broad set of reactions. Some examples are sweat, gastrointestinal discomfort or pains in the stomach, allergies, heart palpitations, altered blood pressure, high cortisol, pupil dilation, the blink of the eye increasing as anxiety levels increase.

Derived from psychophysiology and influenced by different areas, such as behavioral therapy, behavioral medicine, stress intervention research and strategies, biomedical engineering, among others, the Biofeedback [32]. Biofeedback is the organism's response (physiological processes), measured through body-allocated sensors and sent to a base (computer program or application), which allows training for the voluntary control of physiological and emotional responses.

The study of stress is essential to improve understanding of the mechanisms involved and to achieve scientific and technological advances concerning its evaluation and intervention. However, reliable testing, which is capable of generating acute stress in laboratory situations, is necessary to be able to study it, so that there may be experimental control, a safe environment for the participants and generate valid scientific results [34].

Currently, TSST is recognized as the gold standard protocol for stress experiments. There are different adaptations of this protocol, including group and virtual reality options, and adaptations for different age groups [34]. The standard TSST protocol, for adult application in person, consists of three minutes of preparation for a speech, where the participant introduces himself, simulating a job interview. This presentation lasts five minutes, and the last task consists of mental arithmetic exercises, also for five minutes, in front of evaluators (for more details, see the methodology section). The total protocol time is 13 [35].

The responses of elevation of cortisol and HPA axis levels in the TSST application are higher in the morning due to the circadian rhythm. However, both morning and afternoon applications are reliable [35]. TSST can generate robust responses to stress, which are perceptible through psychological, physiological, and biological measures.

3 Related Work

In recent years, many studies have been conducted to detect stress based on wearables measured biosignals and also towards the evaluation of the best psychological intervention to deal with this situation. Among the studies regarding stress detection, the overall focus observed involves choosing a few physiological aspects and the choice of a respective sensor to measure it. When considering the works with the focus on support application for the regulation of anxiety or stress and use of biofeedback, in general, we can observe few studies incorporating a broad set of signs as the source of the stress indication. In the case of using the wearable sensors to online detection of such patterns, then the data preprocessing plays a fundamental role, due to the time spent on this activity. Below are described related works organized by its primary focus. The first part with a focus on data acquisition, preprocessing, and classification. The second with a focus on the online use of wearables as biofeedback support to psychological intervention.

Choi and Osuna [1] describe an approach to detecting mental stress using unobtrusive wearable sensors. It follows the heart rate variability only, using a nonlinear system identification technique known as principal dynamic modes, with a success rate varying from 83% to 69% depending on the experiment carried out. The skin conductance and the possibilities of emotional states were investigated in [26]. Betti et al. [12] describes an experiment using three biosignals (ECG, EDA, EEG) to generate classification models and correlated this with the cortisol level, which is considered an objective and reliable stress marker. The Support Vector Machine (SVM) classification algorithm was used and the results obtained provided 86% accuracy. The data was collected from 15 participants, and 15 data features were analyzed, together with a correlation with cortisol information.

Some works are dedicated to a specific context, such as the truck drivers' work journey, as can be seen in [5], which describes an experiment for data acquisition in real situations. In this work, a deep learning approach was used, compared with a baseline feedforward Artificial neural network. Another specific context, the construction workers' daily routine, is studied in [10], with the selection of EEG signals and Online Multi-Task Learning (OMTL) algorithms. Schmidt et al. [11] describes Wearable Stress and Affect Detection (WESAD), one public dataset integrating several sensors signals and the use of a set of classifiers to identify stress patterns. The data acquisition also integrates the emotional aspect, along with stress. The precision of 93% was obtained with classifiers experiments. The work of Wijman [13] describes the use of EMG signals to identify stress. An experiment with 22 participants was conducted, and both the wearables signals and questionnaires were considered.

In some cases, experiments were made to integrate both sensors acquired signals with other sources, such as smartphone-based activity. Sano [7] performed integration of sensors signals such as EDA while using smartphones to use social networks, read the news, or other related activities. Similarly, [8] approach is dedicated to assessing cognitive problems using the EDA sensor signal and the data obtained with pen movements of the patient while writing with a digital pen able to record some aspects of the writing movements. Paredes [9] explores the design of a smartphone app to interact with users and suggest interventions when detected the necessity, due to a stress situation.

Some works with the main focus on dataset construction. [6] describes the construction of a dataset regarding human movement identification. The wearable approach is one of the possibilities to collect and process the necessary data to identify daily activities. In the case of stress identification, the complementary information regarding human movement can be of interest to support better quality stress patterns identification. Some of the normal daily physical activities will generate impacts on the physiological signals used in most stress patterns recognition approaches. Schmidt et al. [11] describe WESAD, one public dataset integrating several sensors' signals and using a set of classifiers to identify stress patterns. The data acquisition also integrates the emotional aspect, along with stress. The precision of 93% was obtained with classifiers experiments.

Wijsman et al. [3] present a set of wearables sensors in an ambulatory context, dedicated to collecting data from patients for future analysis. Besides that, the experiment provided the opportunity to use a set of previously acquired stress patterns information, used to estimate a stress probability over time, with the wearable data signals online analysis. The primary objective of some works is to provide an integration of components that could support biosignals' online analysis. In work presented by Attaran [4], we can observe an architecture dedicated to integrating the sensor signals acquisition and the processing, even with the dedicated support to execute the classification of the patterns in an online capacity. Therefore, the results can support wearable devices integration the sensors and the online processing, opening the possibility to several applications. Henriques et al. [14] main focus was to evaluate the positive effects of biofeedback software as a mechanism to reduce anxiety in a group of students. The main biosignal monitored, in this case, was the heart rate variability. Two pilot experiments consisting of four weeks each were designed to verify the effectiveness of this computer-based heart rate variability feedback system to help in reducing anxiety and negative mood in college students. Gaglioli et al. [15] describe the main features and preliminary evaluation of a free mobile platform for the self-management of psychological stress. The platform can provide guided relaxation techniques to the users, besides the possibility to show visual information regarding wearables sensors measures from the heart rate. The overall data set obtained during the platform's use is available to the users, as well as the self-reports generated. Dilon et al. [16] describe an experiment using smartphones and games integrated with physiological signs sensors to help in stress reduction. The skin electrical conductance and the TSST test were used as data sources.

Some works are dedicated to analyzing the data generated by the usage of some mobile devices, such as the smartphone. Vildjiounaite et al. [17] described an experiment based on datasets generated by several kinds of use of the smartphone, manually annotated regarding the users' perception of stress during the periods. This approach does not use any additional sensor and can only generate late reports on the identified situation. The opportunities identified in the context of health, with the support of new wearable sensors, communication, and integration possibilities have been described in recent works [24, 25]. Besides the opportunities to data acquisition and monitoring in real-time and with good precision, using Machine Learning approaches to classify these data regarding specific stress, or emotion patterns are promising [27–29]. Particular attention in increasing on the design of systems based on wearables sensors capabilities and in the flexibility and integration aspects [31].

4 Materials and Methods

In this section, we describe the experimental study protocol used to acquire the annotated data from wearables sensors in a session using the TSST protocol.

4.1 Experimental Study Protocol

This is a study with a quasi-experimental, single-group design with interrupted time series. This type of study can present a single group, but in this case, it performs several evaluations, with repeated measures, at different times. The quasi-experimental designs are less controlled than a simple experiment, usually having a control group, but do not present a random distribution of participants to the groups. Participants were selected by convenience. The research was disclosed to undergraduate and graduate students of a university in the metropolitan region. The courses with a high number of students, other than the health area, were selected.

Were excluded from the experiment, people with cardiac disease, psychiatric disorders, or other illness chronic self-reported. Were also excluded people who use psychoactive drugs or beta-blockers, who have consumed caffeine or other stimulants up to three hours before the study, having insomnia or other sleep problems, reported pain at the time of data collection, pregnant or lactating women, or who have passed (in the last 120 days) or are experiencing a severe stressor (e.g., family assault or severe illness). These criteria were stipulated to gain greater control over the experiment due to its influence on physiological stress and cortisol measurements. A total number of 71 participants were selected for the experiment.

The estimated total TSST [36] protocol time, involving pre-tests and post-tests, is 116 min. The experiment consists of the following steps. In step 1, Initial Evaluation, the participant answers the questionnaires to check the inclusion and exclusion criteria. Only participants who meet the requirements for the experiment are selected. In step 2, Habituation, the participant will take a rest time of twenty minutes before the pre-test. This rest helps avoid the influence of events before applying the test and establishes a safe baseline. In step 3, Pre-test, the sensors are allocated, a saliva sample is collected, and the psychological instruments are applied.

In step 4, Explanation of procedure and preparation, the researcher will deliver written and standardized instructions, explaining the activity that the subject will perform. The participant reads the instructions, and the researcher ensures that he understands the task specifications. Then it is sent to the room with the jurors, equipped as if it were a room of a company. The jury is trained to remain neutral during the experiment, not giving positive verbal or non-verbal feedback during the experiment for the subject. The researcher informs that the participant will speak in front of the microphone, with a marked point on the ground, at a distance of one to three meters from the jury table. A camera and a microphone will be used to record the participant. The researcher briefly presents the jurors the objective of the subject in his presentation, remembers the presentation will be recorded, and leaves the room. The participant then begins the preparation for the speech. The committee asks the subject to sit down with paper and pencil and prepare their presentation. The participant will have three minutes to prepare.

In step 5, Free speech presentation, after three minutes of preparation, the participant is requested to go to the marked point and start his speech, being informed that he can not use the notes. If he closes earlier than five minutes, the jurors warn him that he still has time and expect him to talk more. After five minutes, the jurors interrupt the subject and direct it to the next task. In step 6, Arithmetic task, the jurors request an arithmetic task in which the participant must subtract mentally and consecutively the number 17, beginning with 2023. He is asked to perform the calculation as quickly and correctly as possible. At times, the jurors interrupt and warn that the participant has made a mistake, requesting that he begin again. After five minutes, the task is terminated. In step 7, Post-test evaluation, the experimenter receives the subject outside the room for the post-test evaluations. Pre-test, initial response, peak, and stress recovery levels are verified.

In the step 8, Feedback and clarification, the investigator and jurors talk to the subject and clarify what the task was about. They should take this moment to thank for the participation, to resolve any discomfort, and to indicate the return of the data. In step 9, the Relaxation technique, a recording will be used with the guidelines on how to perform a relaxation technique. It will be used only breathing. The inspiratory-expiration time measure will be 10 s per complete cycle, with about six cycles per minute. Participants will be instructed to inhale for four seconds and expire for six seconds. Firstly, they will be guided and trained in the technique and, afterward, will perform the procedure with recording and pacer, in a standardized way for all participants. The physiological measures will be evaluated during the application of the technique, and, afterward, physiological, psychological, and biological measures will be repeated. In step 10, final post-test, some of the psychological instruments will be reapplied, saliva samples will be collected, and the sensors will still be picking up the physiological signals.

In Fig. 1, each of the mentioned steps is indicated regarding the duration of the task and the different kind of data collected.

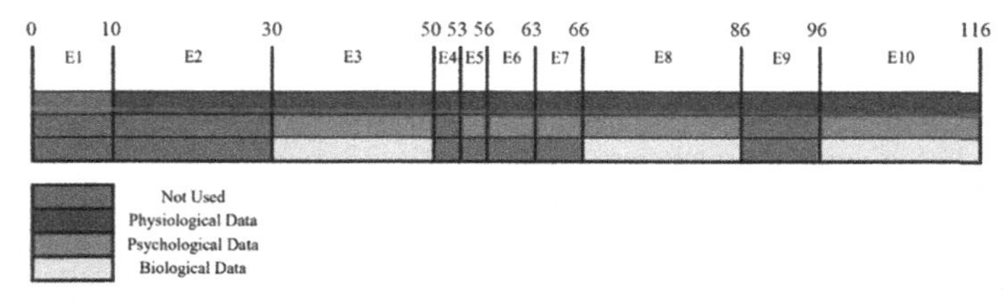

Fig. 1. Overall view of the experimental protocol steps

The instruments applied for the evaluation of the sample are divided among instruments for inclusion or exclusion of the sample in the study; sociodemographic and health questionnaire; instruments for psychological data collection; physiological and; biological. The following is a description of the instruments for collecting additional data.

Some psychological data collection instruments used are commented. The Perceived Stress Scale (PSS) assesses cognitive aspects of stress perception, verifying the indices in which people assess situations in their context. The Inventory of State-Trait Anxiety Inventory (STAI) is used for the verification of anxiety symptoms. It has two scales, one that evaluates anxiety as a state, that is, a temporary situation, and another as a trait, referring to a more stable condition of presence or absence of anxiety during life. The Visual Analogue Scale (EVA) for stress presents in the form of a horizontal line of ten centimeters, enumerated in its extremities with the numbers zero (0) and ten (10), where 0 means "no stress" and 10 means "maximum stress".

The instruments for collecting physiological data are diversified. The BeWell is composed by the following sensors of the BITalino Kit - (PLUX Wireless Biosignals). The Electrocardiogram (ECG) sensor, provides data on heart rate and heart rate variability.

The Electrodermal activity (EDA) sensor allows the capture of the bioelectrical signals sent to the muscular fibers. The electromyography (EMG) sensor allows the data collection of the electrical activity of the sweat glands. Another instrument used was the Polar RS800CX Heart Rate Variability Monitor. This instrument uses a sensor that is attached to the chest by an elastic band. The data collected by the sensor are transmitted simultaneously to the Polar clock, allowing the transfer to a computer for analysis, generally performed in the Kubios HRV Software. The MindField Esense Skin Response: Measures the galvanic responses of the skin (electrodermal activity), and the electrical activity of the sweat glands can be verified. The sensors are placed in the fingers, using an application to verify the response emitted by the sensors.

The salivary dosage of the hormone cortisol is a non-invasive method, which does not require the presence of doctors or nurses for the collection. It is a practical and reliable method for obtaining cortisol analysis. However, it is necessary to take into account the different influences to which this measure is subject, such as the time of collection of the saliva sample, gender, use of stimulants, or medical conditions, which may influence the results obtained. Evaluations of weight and height measurements of each participant were performed for the analysis. Samples of saliva were collected by the participant himself, with the assistance and guidance of the team responsible for the project. The Elisa Kit for Salivary Cortisol from DRG Instruments was used, an enzymatic immunoassay kit to measure active free, solid-phase cortisol, based on the principle of competitive binding.

4.2 Data Acquisition

For the analysis of the physiological data, there was assistance from a specialist in the area. The heart rate variability data was computed through the root mean square of the successive differences (RMSSD) calculation, as it is indicated for HRV evaluations in research contexts.

The cortisol analyzes were carried out in the biology laboratory of the University of Vale do Rio dos Sinos. For this analysis, besides the authors, we counted on the collaboration of the Group of Advanced Studies in Health Psychology, with the support of the technicians of the laboratory of biology and supervision of teachers, who possess the necessary technical skills to carry out these analyses. The sociodemographic and psychological data, together with the cortisol analyzes and physiological responses, were registered to a Statistical Package for the Social Sciences (SPSS), version 25.0. The level of significance considered was 5% ($p < 0.05$).

For the development of the device for biosignal measures, we first verified similar applications developed in the area, performing a systematic review of scientific articles on the subject and searching non-systematically in app stores such as Google Play and Apple Store, not being found devices with the same characteristics. Some applications found in this line evaluate physiological signals and training in biofeedback in a specific and specific time or interventions in Cognitive-Behavior Therapy without the use of sensors for biofeedback.

The vast majority of applications found do not use sensors to obtain physiological responses, but offer intervention through relaxation techniques, for example. There are programs and applications of biofeedback for evaluating physiological signals, with possibilities of allocating sensors in different regions of the body. However, these devices usually use a single sensor, or when they use more, they are allocated separately in different regions of the body. To date, no applications have been found that present continuous, momentary, and automatic measurement characteristics with different sensors in a single wearable, offering empirically supported intervention, which is the final proposal of BeWell. Recently, studies are emerging in this direction, pointing out to be an area with promising results.

After performing these searches, the different types of biofeedback sensors were studied. They selected those that were reliable and were most used, with the possibility of integrating into a single wearable. From these surveys, three sensors were selected to obtain measurements: ECG, EDA, EMG. The ECG sensor is used for the collection of HR and HRV data through the electrical signals emitted by cells in the heart. This sensor allows the capture of these electrical signals and their transformation into numerical values (BITALINO, 2015a). Higher HRV indicates an ideal interaction between the sympathetic and parasympathetic nervous systems. There are different measures of HRV. We will use the RMSSD, because it is more appropriate for our study at the moment and because it is more used in research. The EDA sensor allows the capture of the bioelectrical signals from the motor control neurons in the brain, sent to the muscular fibers. These signals are translated into numerical values, enabling their analysis (BITALINO, 2015b). Electro-dermal activity, also known as galvanic skin response or skin conductance, refers to the ability of the skin to conduct electricity. Skin conductance is associated with the amount of moisture produced by eccrine sweat glands. This activity signals the sympathetic nervous system's activation, which produces more sweat and increases the electrical conductivity of the skin, which can be detected by biofeedback sensors. Stress and SNS activation will be detected by the device, feedback on the functioning of the organism in this regard. Therefore, this kind of biofeedback is a way of measuring the activation of sweat glands in stress situations directly through electrical activity and indirectly. The EMG sensor allows the sweat glands' electrical activity data collection. The transformation of these electrical changes into numerical data, make possible the analysis. Electromyographic biofeedback measures the emitted by the skin during contraction muscular. Motor control neurons signal to the muscle, and that signal is perceived by the biofeedback sensor, that the translates into numerical terms, different applicability to it. This process is related to the skeletal nervous system.

4.3 Methods for Analysis and Evaluation of the Acquired Data

The obtained data was analyzed and evaluated with the well-known data processing chain, consisting of the following main steps:

Preprocessing, segmentation, feature extraction, classification. The biosignals were acquired with BITalino [32]. The data was preprocessed with the support of the available API BioSPPY (https://biosppy.readthedocs.io/). The segmentation of the sensor signals was done considering a sliding window, with a window shift. For the biosignals the window size selected was NN seconds, according to arguments described by Kreibig [18]. The ECG signal was analyzed with Peak detection algorithms. From the peaks, computed the heart rate and statistical features such as mean and standard deviation.

Also, the heart rate variability was obtained from the analysis of the location of the heartbeats. Also were computed the energy in different frequency bands. A detailed description of the HR and HRV analysis can be found in [19].

EDA signal is strongly associated with stressful situations since the Sympathetic Nervous System controls it. Due to its high sensitivity, a lowpass filter is used in several works using this biosignal [20, 21]. In our case, we used a lowpass filter of 5 Hz. With the result of this operation, the following statistical features were calculated: mean, standard deviation, dynamic range. We used skin conductance response (SCR) and skin conductance level (SCL). The first represents a short response for some stimulus. The second represents a baseline conductivity that can slowly vary. These two components were separated, and also additional information such as the number of peaks was computed using the reference provided by [22, 23]. The EMG features were processed, applying different filters. First was applied a lowpass filter (50 Hz) to the raw EMG signal. The result of the processed signal was segmented in 60-second windows. In these windows, the reassures of the different peaks, and the mean amplitude was measured. The second approach used a high pass filter and then segmented in windows of 5 s. The signal in these windows was used to calculate peak frequencies. The spectral energy was computed in bands ranging from 0 to 350 Hz. Details of this approach can be found in [13].

5 Experiments

This section presents details of the experiments conducted and the results obtained during the step of the process and analysis of data. The code developed is written in Python due to the libraries available in this programming language, specific for data analysis, machine learning, and filtering in biosignals. During the verification of data, it is possible to check that there is some absence in part of the signal during some periods. The main hypothesis for it happened is the loss of communication between the BITalino and the computer that stored the signal during the experiments. The reason assessed for this loss of communication must be a function of the signal acquisition rate, as verified in later tests with the wearable.

During the verification of the signs, each participant's data is checked separately, for graphic analysis and signal average to check the plausibility of the data. The developed script reads the files with the raw data of each participant and a CSV file with the annotations of all participants in each step of the experiment. It is stored in dataframes of Pandas library, developed for data analysis. The raw data contains information from all wearable channels. To facilitate the process, the unused channels are discarded in the first step of the data processing. The script verifies the timestamp of the signals and combines all information compared with the data times for each step of the CSV file.

The whole experiment had its steps divided into six categories (Baseline, TSST, Arithmetic, Sensor Post-Test I, Sensor Post-Test II and No Category) and all data is categorized with base in the time for each step. In sequence, the dataframe is stored in other CSV file discarding the category "No Category", because the data in this classification is about steps without relevant information about the experiment. This dataset is generated for participants to be used as training and testing data for the Machine Learning stage. The next step is data filtering, using the BioSSPY library. This library is developed for filtering and frequency analysis of biosignals. For example, Blood Volume Pulse (BVP), Electrocardiogram (ECG), Electromyograph (EMG), Electrodermal Activity (EDA), Electroencephalograph (EEG). This filtering is used the data classified as "Filtered", for having a linear variability rate and all signals there is this kind of filtering.

Each participant has approximately 1.886.000 rows of data for the entire experiment. Due to the computational power limitation, it is not very easy to use a large amount of data to apply some Machine Learning technic. For this reason, a technique called windowing is applied. Windowing, also known as a window function, is a mathematical function with the objective to reduce the amount of data, it has as characteristic of retiring a part of data for each period, the removing of the information is done symmetrically. One way to reduce some eventual distortion, all data that removed is used to calculate the delta and is implemented in place of the data taken. In other words, if there are 1000 rows of data, implementing this technique using windowing with 10 times, the result is 100 rows of data are pure data, and 100 rows of data come from the average of the data removed, 200 lines of data remain. It is a great reducing, and as there is a large amount of data, this reduction is not noticeable.

Lastly, the script creates the dataset to apply the Machine Learning stage. During this step, the dataframe is converted and divided in two lists, the first list is the train dataframe, and this list has data about the signals. The other list is the test dataframe. Both lists have information about the class, which in this context is the "Category" applied in Machine Learning. After processing the data, these lists are stored in NumPy files, to facilitate data handling in any application.

5.1 Data Analysis

The Machine Learning stage used different combinations of the signals with the objective to determine what is the best method. The combinations used are only ECG, only EMG, only EDA, ECG and EMG, ECG and EDA, and EMG and EDA.

The library used to apply Machine Learning is the Scikit-Learn, an open-source library for Python, with modules of different algorithms. For each combination are applied six algorithms, the SVM with Linear Kernel, SVM with Radial Kernel, Decision Tree Classifier, Random Forest Classifier and Gaussian Naive Bayes. For each algorithm, four metrics are used to evaluate the results: accuracy, precision, recall, and F1 score (the combination of recall and precision). Based on this information, it is possible to determine the best method for implementing a system for detecting people's status. During the experiments with the machine learning codes, it was implemented all process using two different windowing approaches, the first using 10 frames and the second with 100 frames.

The process performed with a windowing of 10 times, the code took considerably longer, about 30% more, than windowing 100 times. Most combinations did not return good results when applying most algorithms, and only the ECG signal presented some significant result. Regarding the metrics obtained, both presented results very close to one another. Comparing the metrics resulting from the different windowing performed during the Machine Learning process and all six combinations, the best context in precision was the combination of ECG and EMG.

With the results commented before, it is possible to verify no difference which justifies the use of smaller windowing. Another conclusion is that the EDA signal is not a good option to implement some artificial intelligence to determine the person's status. All the metrics used in combination with this signal showed low accuracy and precision. In contrast, the EMG and ECG signal presented good responses, mainly using the algorithm Gaussian Naïve Bayes.

6 Conclusion

In this paper, we presented a new experimental protocol to acquire physiological data regarding stress situations, based on the well-known TSST protocol, improved with questionnaires for self-reports of the participants and physiological measures obtained with wearables sensors. During the graphical analysis of the participants' signals with the signaled categories, it was evident that the TSST protocol fulfills the objective.

The protocol differentiates from previous works regarding the number of signals, evaluation of the combination, complement with questionnaires annotated with cortisol. This work is part of a broader effort to support online identification of the patterns, which is important to foster biofeedback applications. As a future improvement to this work, new machine learning experiments will be carried out using a larger volume of data and will also be implemented deep learning techniques.

An alternative is the standardization of the volume of data in each category, ensuring that there as much data in one stage as in another. As a suggestion for future works, it is recommended to perform new tests with the wearable reducing the sampling rate and monitoring the stability of the acquisition signals. Reducing the acquisition rate from 1 kHz to 100 Hz, the signal tends to have lower communication losses, but it is necessary to carry out validation.

References

1. Choi, J., Gutierrez-Osuna, R.: Using heart rate monitors to detect mental stress. In: 6th International Workshop on Wearable and Implantable Body Sensor Networks, Berkeley, CA, pp. 219–223 (2009)
2. Wijsman, J., et al.: Towards mental stress detection using wearable physiological sensors. In: 33rd Annual International Conference of the IEEE EMBS, Boston, Massachusetts, USA (2011)
3. Wijsman J., et al.: Towards ambulatory mental stress measurement from physiological parameters. In: Humane Association Conference on Affective Computing and Intelligent Interaction (2013)
4. Attaran, N., Brooks, J., Mohsenin, T.: A low-power multiphysiological monitoring processor for stress detection. In: IEEE SENSORS (2016). https://doi.org/10.1109/ICSENS.2016.7808776
5. Saeed, A. et al.: Deep physiological arousal detection in a driving simulator using wearable sensors. In: IEEE International Conference on Data Mining Workshops (ICDMW), pp. 18–21 (2017). https://doi.org/10.1109/ICDMW.2017.69
6. Saha, S.S., et al.: DU-MD: an open-source human action dataset for ubiquitous wearable sensors. In: 2018 Joint 7th International Conference on Informatics, Electronics and Vision, ICIEV, Kitakyushu, Japan (2018)

7. Sano, A., Picard, R.W.: Stress recognition using wearable sensors and mobile phones. In: Humane Association Conference on Affective Computing and Intelligent Interaction (2013). https://doi.org/10.1109/acii.2013.117
8. Niemann, M., Prange, A., Sonntag, D.: Towards a multimodal multisensory cognitive assessment framework. In: IEEE 31st International Symposium on Computer - Based Medical Systems (2018). https://doi.org/10.1109/cbms.2018.00012
9. Paredes, P., et al.: PopTherapy: coping with stress through pop-culture. In: Proceedings of the 8th International Conference on Pervasive Computing Technologies for Healthcare, 20–23 May (2014)
10. Jebelli, H., Khalili, M.M., Lee, S.A.: Continuously updated, computationally efficient stress recognition framework using Electroencephalogram (EEG) by applying Online Multitask Learning Algorithms - OMTL. IEEE J. Biomed. Health Inform. **23**, 1928–1939 (2018)
11. Schmidt, P., et al.: Introducing WESAD, a multimodal dataset for wearable stress and affect detection. In: Proceedings of the 20th ACM International Conference on Multimodal Interaction, ICMI 2018, pp. 400–408. ACM, New York (2018)
12. Betti, S., et al.: Evaluation of an integrated system of wearable physiological sensors for stress monitoring in working environments by using biological markers. IEEE Trans. Biomed. Eng. **65**(8), 1748–1758 (2018)
13. Wijsman, J., et al.: Trapezius muscle EMG as predictor of mental stress. In: Wireless Health 2010 - WH 2010, pp. 155–163. ACM, New York (2010)
14. Henriques, G., et al.: Exploring the effectiveness of a computer-based heart rate variability biofeedback program in reducing anxiety in college students. Appl. Psychophysiol. Biofeedback **36**(2), 101–112 (2011)
15. Gaggioli, A., et al.: Positive technology: a free mobile platform for the self-management of psychological stress. Ann. Rev. Cyber Therapy Telemed. **12**(May), 25–29 (2014)
16. Dillon, A., et al.: Smartphone applications utilizing biofeedback can aid stress reduction. Front. Psychol. **7**, 832 (2016)
17. Vildjiounaite, A., et al.: Unobtrusive stress detection on the basis of smartphone usage data. Pers. Ubiquit. Comput. **22**(4), 671–688 (2018)
18. Kreibig, S.: Autonomic nervous system activity in emotion: a review. Biol. Psychol. **84**(3), 394–421 (2010)
19. Malik, M.: Heart rate variability. Standards of measurement, physiological interpretation, and clinical use. Eur. Heart J. **17**, 354–381 (1996). Task force of the European society of cardiology and the north American society of pacing and electrophysiology
20. Setz, C., et al.: Discriminating stress from cognitive load using a wearable EDA device. IEEE Trans. Inf. Technol. Biomed. **14**(2), 410–417 (2010)
21. Sun, F.-T., Kuo, C., Cheng, H.-T., Buthpitiya, S., Collins, P., Griss, M.: Activity-aware mental stress detection using physiological sensors. In: Gris, M., Yang, G. (eds.) MobiCASE 2010. LNICST, vol. 76, pp. 211–230. Springer, Heidelberg (2012). https://doi.org/10.1007/978-3-642-29336-8_12
22. Choi, J., Ahmed, B., Gutierrez-Osuna, R.: Development and evaluation of an ambulatory stress monitor based on wearable sensors. IEEE Trans. Inf Technol. Biomed. **16**, 2 (2012)
23. Healey, J., Picard, R.: Detecting stress during real-world driving tasks using physiological sensors. IEEE Trans. Intell. Transp. Syst. **6**(2), 156–166 (2005)
24. Din, S., Paul, A.: Smart health monitoring and management system: toward autonomous wearable sensing for internet of things using big data analytics. Fut. Gener. Comput. Syst. (2018). https://doi.org/10.1016/j.future.2017
25. Tokognon, C.A.: Structural health monitoring framework based on internet of things: a survey. IEEE IoT J. **4**(3), 619–635 (2017)
26. Greco, A., et al.: Skin admittance measurement for emotion recognition: a study over frequency sweep. Electronics **5**, 46 (2016). https://doi.org/10.3390/electronics5030046

27. Scilingo, E.P., Valenza, G.: Recent advances on wearable electronics and embedded computing systems for biomedical applications. Electronics **6**, 12 (2017)
28. Nweke, H.F., Mujtaba, G., Wah, T.Y.: Data fusion and multiple classifier systems for human activity detection and health monitoring: review and open research directions. Inf. Fusion **46**, 147–170 (2018)
29. Verma, P., Sood, S.K.: A comprehensive framework for student stress monitoring in fog-cloud IoT environment: m-health perspective. Med. Biol. Eng. Compu. **57**(1), 231–244 (2018). https://doi.org/10.1007/s11517-018-1877-1
30. Pantelopoulos, A., Bourbakis, N.G.: A survey on wearable sensor based systems for health monitoring and prognosis. IEEE Trans. Syst. Cybern. **40**(1), 1–12 (2010)
31. Mosenia, A., et al.: Wearable medical sensor-based system design: a survey. IEEE Trans. Multiscale Comput. Syst. **3**(2), 124–138 (2017)
32. Guerreiro, J., et al.: BITalino: a multimodal platform for physiological computing. In: ICINCO, pp. 500–506 (2013)
33. Schwartz, M.S.: Biofeedback: A Practioner Guide. The Guilford Press, New York (2016)
34. Allen, A.P., et al.: Biological and psychological markers of stress in humans: focus on the trier social stress test. Neurosci. Biobehav. Rev. **38**, 94–124 (2014)
35. Kudielka, B.M., Hellhammer, H., Kirschbaum, C.: Ten years of research with the trier social stress test. In: Social Neuroscience, pp. 56–83 (2007)

Financial Time Series Forecasting via CEEMDAN-LSTM with Exogenous Features

Renan de Luca Avila(✉) and Glauber De Bona

Computer Engineering Department, Escola Politécnica da Universidade de São Paulo, São Paulo, SP, Brazil
{renan.avila,glauber.bona}@usp.br

Abstract. The most recent successful time series prediction models are a combination of three elements: traditional stochastic models, machine learning models and signal processing techniques. CEEMDAN-LSTM models have combined empirical mode decomposition and long short-term memory neural networks to achieve state-of-the-art results for financial data. In this work, we propose a generalized CEEMDAN-LSTM architecture for time series forecasting capable of dealing with exogenous features as input, and the consequences of input data growth, such as convergence difficulties. Our model was applied to time series from 10 of the most liquid Brazilian stocks, and results show that accuracy is overall improved when compared to the original single feature input CEEMDAN-LSTM architecture..

Keywords: LSTM · CEEMDAN · Neural networks · Time series · Forecasting

1 Introduction

Financial time series analysis and forecasting have had several approaches over time. The traditional models follow a stochastic probabilistic approach, while more recent models are based on machine learning methods.

The traditional models arose in the beginning of the last century with Yule and Walker, who proposed the autorregressive model (AR) [1] describing linear relations between past and future values. After that, Whittle showed better results considering also moving averages [2], extending the AR model into ARMA. Then, Box and Jenkins proposed a method to estimate ARMA coefficients and they also popularized model variations including seasonality and integrality (ARIMA and SARIMA) [3]. More advances on explaining and predicting time series are Engle's results for modelling time series volatility [4] and Hamilton's cointegration theory [5]. These approaches were very succesful to explain time series based on *a priori* hypothesis, generating a single model that can be applied to circumstantial situations.

R. Cerri and R. C. Prati (Eds.): BRACIS 2020, LNAI 12320, pp. 558–572, 2020.
https://doi.org/10.1007/978-3-030-61380-8_38

In contrast to that, more recent models try to focus on learning the behavior of a series from its data, without prior explicit assumptions, such as linearity or stationarity. This approach makes it possible to use machine learning and signal processing techniques with less financial market singularities, for instance market's liquidity, volatility or efficiency. These models are a hybrid combination of neural network models, traditional models and signal processing. This last element is responsible for decomposing or filtering a time series prior to the model fitting.

Models within the machine learning approach have been showing promising results, such as CEEMDAN-LSTM [6], EMD-ARIMA-ANN [7], Wavelet-ARIMA-LSTM [8] and LSTM-SVM [9]. Nevertheless, very little attention has been given to the hidden information of a certain feature's time series present in the considered exogenous historical features: open, high, low, close prices and financial traded volume are taken as input features and the target prediction feature is the next day's closing price. Previous works have shown correlation among these features [10], so accuracy improvement is expected.

Thus, the aim of this work is to improve time series forecasting accuracy of a state-of-art computational model, CEEMDAN-LSTM, by proposing a model architecture and training method capable of using exogenous features. The proposed model is trained and compared considering ten of the top liquidity Brazilian stock market one step-ahead closing price.

2 Materials and Methods

Our work takes Cao's [6] CEEMDAN-LSTM as starting point. The use of CEEMDAN as data preprocessing method is explained and so are the details of data used. Further, we investigate new ways to change the model data and neural architecture in order to use the exogenous features open, high, low and volume. Then finally we propose a generic architecture that might enhance prediction results by using exogenous features.

2.1 CEEMDAN-LSTM

CEEMDAN-LSTM [6] is a time series prediction model that employs signal analysis with complete ensemble empirical mode decomposition with adaptive noise (CEEMDAN) technique [11] and long short-term memory neural networks for learning series' trends and patterns. CEEMDAN decomposes a given signal into different period trend components called intrinsic mode functions (IMF), and one LSTM net for each component is trained to predict the next value. At last, there is a recomposing step for evaluating the next closing price given a past window of days. The general architecture data flow of CEEMDAN-LSTM model is described in Fig. 1.

Here we show that this CEEMDAN-LSTM model can be augmented by proposing a new architecture and training method both capable of dealing with exogenous features such as market metrics of daily open, high and low prices,

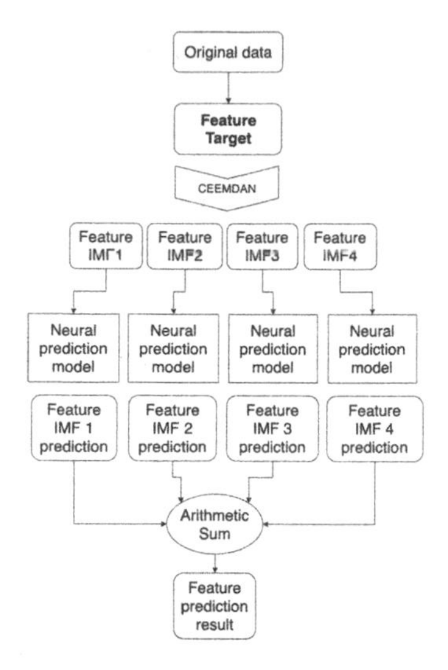

Fig. 1. Architecture of CEEMDAN-LSTM model.

and also daily financial volume in order to predict the closing price. Even though we focus on CEEMDAN, the proposed architecture can be applied to other decomposition methods. In addition, our model proposes a way to deal with the drawbacks of a bigger and multiple input data volume, such as training time, complexity and convergence difficulties.

2.2 Data

The data employed to train and test our model is composed by daily open, high, low and close stock prices, and also daily financial volume, all considering the regular trading hours and corporate events adjustments. Since the closing price series is the one to be predicted, the others are called *exogenous features*. The data was extracted from *Yahoo Finance*, discarding day gaps caused by weekends and holidays.

The time ranges used are from the July 20th of 2018 until December 2nd of 2019[1], for the following Brazilian equities: PETR4, VALE3, BOVA11, ITUB4, BBDC4, B3SA3, BBAS3, ABEV3, MGLU3, VVAR3. These were chosen based on their high liquidity characteristics, which yields more robust predictions [12].

2.3 Data Preprocessing

Our data preprocessing technique consists of three steps, detailed below: firstly, we transform each time series into new series with values in range of 0 to 1;

[1] https://finance.yahoo.com/quote/PETR4.SA/history?period1=1517961600&period2=1549929600&interval=1d&filter=history&frequency=1d.

then we decompose each of these series into different IMFs using CEEMDAN method; finally we tranform each IMF into a series of time windows.

Mathematical Tranformations. For each of the modes obtained from CEEMDAN, to deal with series of values between 0 and 1, we transform the data according to Eq. 1, where x' is the new transformed value, x is the value of a data point in an intrinsic mode function, x_{max} is the greatest value among the whole series and x_{min} is the least.

$$x' = \frac{x - x_{min}}{x_{max} - x_{min}} \tag{1}$$

Signal Processing. The signal processing method we use is the complete ensemble empirical mode decomposition with adaptive noise (CEEMDAN). It is an enhanced version of the empirical mode decomposition (EMD) method. It can be compared to Wavelet decomposition method, which in turn has concepts related to Fourier transform [13].

What distiguishes the EMD based methods is that it extracts variable frequency components based only on numerical data behavior, instead of requiring choices of non-numerical parameters such as the analytical wavelet.

The EMD method [14] gathers components by enveloping the original signal to obtain the average of the upper and bottom envelope, which in turn are obtained interpolating upper and bottom extrema points of the signal with *cubic splines*. Then it discretely subtracts each of the original signal data points from the envelope average. This results into a new signal, also called component or intrinsic mode function (IMF), that undergoes the same process of sifting until the signal's power becomes less than a given threshold or the residue turn into a monotonic function. EMD makes the resulting components able to be point by point arithmetically added to recompose the original signal without data loss.

The Ensemble EMD (EEMD) [15] method performs several EMD decompositions of the original signal with artificially added gaussian white noise, in order to prevent mode mixing. Mode mixing is a phenomenon that occurs in decomposition of specific signal cases when using EMD that produces high frequency modes with low frequencies trends or vice-versa, which happens mostly between adjacent IMFs. Nevertheless, the EEMD still adds new undesired modes into the original signal.

CEEMDAN is a variation of EEMD that tackles this issue by controlling the added noise specifically to each of the modes during the sifting process, providing a better spectral separation [11]. The idea in using CEEMDAN as spectral decomposition of a time series is to allow neural networks to learn specific time frequency patterns, making each mode's trends input for neural network better behaved. An example of decomposition for the time series of the feature closing price is shown in Fig. 2.

In our model's input, each of the features, close price and the 4 exogenous ones (open, high, low and volume), is decomposed with CEEMDAN. Applying

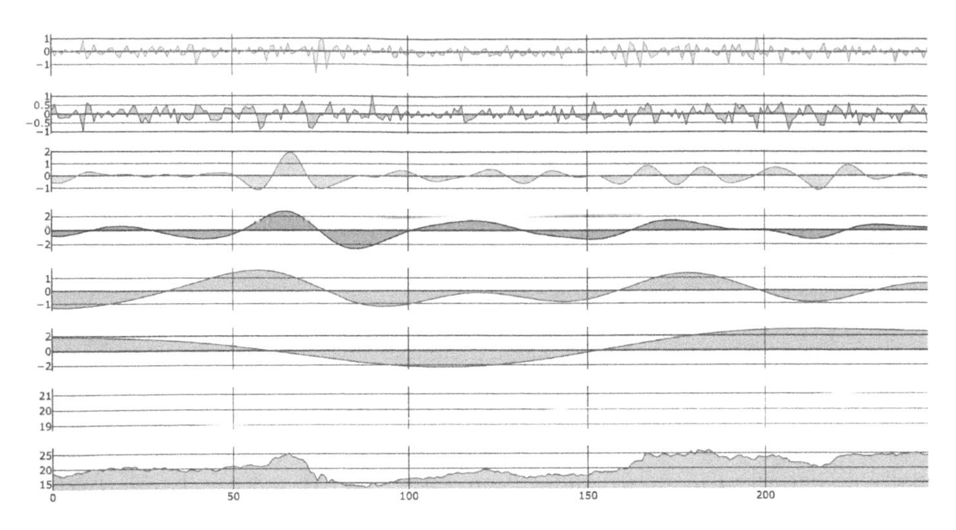

Fig. 2. Example of CEEMDAN decomposition for the closing price signal of PETR4 from February 6th 2018 to February 11th of 2019. Legend from top to bottom: IMF1, IMF2, IMF3, IMF4, IMF5, IMF6, residue and original data. Horizontal axis represents time in days, and vertical axis represents IMF magnitude. Area colored for visualization purposes.

CEEMDAN to an input series yields a set of IMF series whose range is also in the unit interval. We employed a public Python implementation[2] of the CEEMDAN method [16], with all the default settings, except for the scale of added noise ϵ, considered as a model's hyperparameter.

Time Windows. For each of the IMF series resulting from the CEEMDAN decomposition, a series of windows of sequential values is generated. An IMF series of length n yields a series with $n - \Delta t + 1$ windows, where Δt is the window length, and this series of windows is used as input. For example, if the original series is $[1, 2, 3, 4, 5, 6, 7]$ and the window size is 3, the resulting series of windows will be $[[1, 2, 3], [2, 3, 4], [3, 4, 5], [4, 5, 6], [5, 6, 7]]$.

The window length Δt is treated as a hyperparameter and is adjusted according to each series relative frequency. That is, low frequency series have time windows of greater length and high frequency series have lesser window length. The input tensor for the neural network is bidimensional of size $m \cdot \Delta t$, where m is the number of input features.

Data Subsets. We divide the data into training and test sets after preprocessing, with the most recent 10% of data representing the test set. The training data set, in order to allow hyperparameter tuning without affecting test results, has its most recent sequential 20% of data reserved for validation.

[2] https://github.com/laszukdawid/PyEMD.

2.4 Prediction

Financial time series have nonlinear and nonstationary characteristics, and to learn these behaviors we use a double layer of long short-term memory neural networks with hidden states of 128 and 64 units.

Following Cao's neural model [6], we also use two dense network layers of sizes 16 and 4 to improve feature extraction capability. However, we choose LeakyReLu activation function instead of Cao's ReLu, since our larger input data, which includes exogenous features, turns the model more prone to the dying ReLu problem and to loss function instability.

Our neural layer achitecture is shown in Fig. 3, and the hyperparameters related to it are learning rate, maximum number of epochs, dropout Bernoulli's probability and patience for early stopping criterion. The dropout mask is applied only on the tensor input and is kept the same during each epoch.

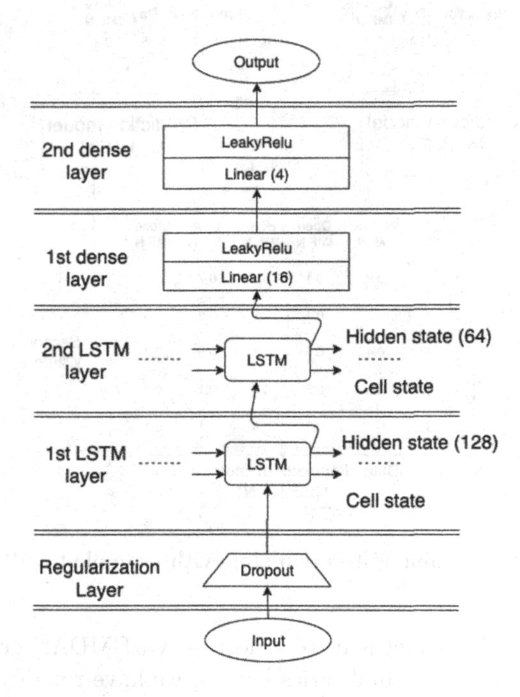

Fig. 3. Neural model with network layer architecture with number of cell units provided in brackets.

The neural network is coded with the help of Keras open deep learning framework[3], with early stopping and regularization self implemented.

The training algorithm used is backpropagation through time with supervised learning, the epoch is defined as a whole succession of predictions over the train set and the metric to optimize with *Adam* after each epoch is the mean squared

[3] https://keras.io/.

error (MSE) measured by the sequence of windows' next day real value against the sequence of windows' next day prediction. The neural model has its initial state set to zeros.

Each neural model's input shown in the training step illustrated in Fig. 4 is a sequence of tensors built by preprocessing each of the intrinsic mode function generated by CEEMDAN decomposition for each of the five features considered. For the different features but same level IMFs to be comparable in frequency behavior, we must use the same length for each feature's original time series and the same time resolution (daily).

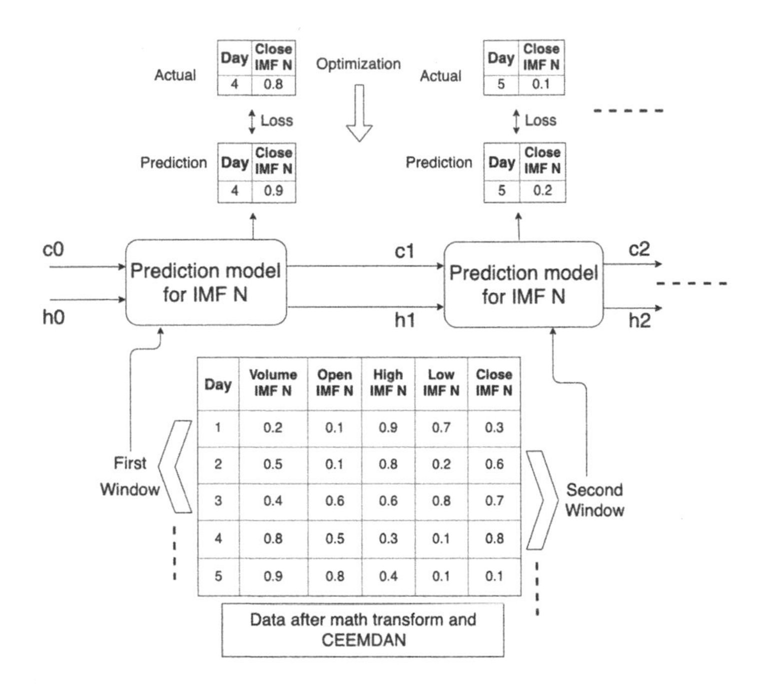

Fig. 4. Training iteration step with example toy data

Considering our 5 distinct features and that CEEMDAN generates around 7 to 9 components for our original series length, we have roughly 40 series to deal with. This implies a significantly larger training time, decreasing the chances for the neural model to converge properly. Nevertheless, long trend intrinsic mode functions are smooth, so the use of simpler and faster methods should result in less errors in the prediction results. A simpler prediction method can prevent errors caused by neural convergence difficulties [17] which, in turn, can be caused by the multiple feature input. Thus, some of the high frequency IMFs, which have more complexity in the time frequency, can be predicted with the neural model while the other ones are left for a simpler prediction method.

In Fig. 2, the low frequencies IMFs, such as IMF4, 5, 6 and residue, clearly share a smooth shape that indicates that the next point prediction can be reasonably performed via polynomial-based methods. Cubic splines are then a natural candidate, as they are employed in the EMD method to generate the envelopes. Thus, we apply cubic splines to predict low frequency IMFs, expecting to benefit both from the neural model's capacity of learning complex patterns in high frequency data and from the spline's stability when it comes to low frequency IMFs.

The cubic spline projection of the next value for a certain target feature IMF is interpolated based on its current window. The use of simpler method for predicting long trend IMFs only from the target feature also solves the problem of different number of IMFs generated by CEEMDAN depending on each input feature because of its random nature, which would occasionally cause the last IMFs neural models not to be fully fed with all of the features information.

At last, the predicted values for the IMFs are summed and the inverse of Eq. 1 is applied to obtain the prediction for the original closing price series.

A generalization of our model architecture is shown in Fig. 5. In our specific case, the inputs are all 5 features (open, high, low, close, volume) and the output predicted feature is the next day's close price only.

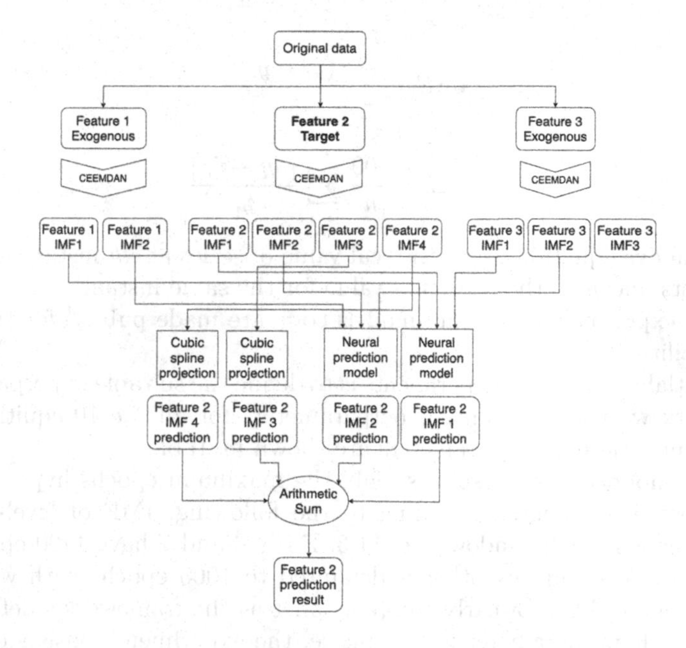

Fig. 5. Model architecture generalization.

The resulting proposed model differs from CEEMDAN-LSTM mainly in the following ways: 1) our model's input considers exogenous features open, high, low and volume as multivariate time series and handles the learning process with the data flow of Fig. 5 and algorithm described on Fig. 4; 2) Our model predicts IMFs of level higher than given threshold (2 or 3) with the help of spline projection instead of neural models, because the higher IMF levels behave more softly and less noisy.

3 Results

The effects of employing cubic splines and considering exogenous inputs are compared relative to the original CEEMDAN-LSTM model, in which all the IMFs were predicted by single feature input neural models with ReLu activation function. In order to reach this comparison, we provide experiments with 4 different models: CEEMDAN-LSTM is the original model; CEEMDAN-LSTM-SPLINE predicts IMFs higher than given threshold (2 or 3) with splines; XCEEMDAN-LSTM considers exogenous features as multivariate time series input; and XCEEMDAN-LSTM-SPLINE, our proposed model, combines both modifications.

The implemented models are evaluated and compared using mean squared error (MSE) 2, the metric minimized during training phase, and mean absolute percentage error (MAPE) 3, a metric that considers the equities' prices magnitude.

$$MSE = \sum_{i=1}^{n} \frac{(y_i - \hat{y_i})^2}{n} \tag{2}$$

$$MAPE = \frac{100}{n} \sum_{i=1}^{n} \frac{\mid y_i - \hat{y_i} \mid}{y_i} \tag{3}$$

In the above equations, y_i is the real value of series at *ith* instant for a series of n instants and $\hat{y_i}$ is the predicted value for the same instant.

All the experiments data and models code are made public[4] for the sake of reproducibility.

The initial phase of the experiments is to define the suitable hyperparameters. In this work we take the same hyperparameters for all the 10 equities we are interested in. These hyperparameters are shown in Table 1.

For the models without splines, only the maximum epochs hyperparameter and window size changes, according to the following: IMF of levels 4 and 5 have 1500 epochs with window sizes of 3, IMFs 6 and 7 have 1200 epochs with window sizes of 4 and any other is defaulted to 1000 epochs with windows of size 5. Minimum delta for early stopping is let as the framework's default.

After the hyperparameter tuning phase, the experiment consists on decomposing, training the model and predicting the test data 10 times for each of the equities considered. It is important to reproduce the experiment several times because of the random nature of CEEMDAN decomposition.

In order to maximize the benefit of employing both LSTM and cubic spline, we must find the appropriate IMF level to switch between neural model and

[4] https://github.com/avilarenan/xlstmceemdan.

Table 1. Hyperparameters

Target feature (close) IMF		1	2	3
CEEMDAN	Noise Scale	0.15	0.15	0.15
Neural Model	Learning Rate	2e−4	8e−4	8e−4
	Max. Epochs	2500	2000	1500
	Window size	2	2	3
	Patience	20	20	20
	Dropout probability	0.2	0.2	0.2

spline methods when predicting. Therefore, we experiment our XCEEMDAN-LSTM-SPLINE model with different thresholds in Fig. 6. The neural/spline threshold at IMF2, for example, refers to predicting the IMF1 and IMF2 with neural model and all the rest with cubic splines.

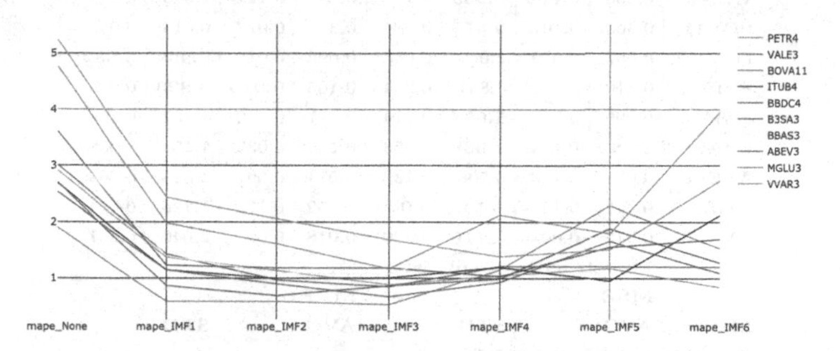

Fig. 6. MAPE (y axis) behavior when moving the IMF level frontier after which we start to predict with splines. From the left to right, starting from using only splines, the model is gradually adding neural nets predictions for the low level (high frequency) IMFs.

Figure 6 shows that the best frontiers for switching from neural model to spline are IMF2 and IMF3. Most of the tickers show better MSE and MAPE as well as less deviation when we use neural nets to predict the IMFs 1, 2 and 3, and use we use splines to predict IMFs 4, 5 and on until the residue. This means that the neural model performs better in the high frequency IMFs prediction and the spline model performs better in the low frequency IMFs.

Table 2 brings the accuracy results for each model and each dataset and different neural/spline thresholds, with best results by symbol in bold. Models with splines have two versions, employing the LSTM only up to IMF2 or IMF3. The highest MAPE error averages are noticed mainly in the equities of B3SA3, VVAR3 and MGLU3. That is because these tickers have followed a stronger monotonic trend when compared to other tickers, making the original time series less stationarity. This behavior makes it more difficult to predict the trend of the ticker, decreasing the corresponding accuracy.

Table 2. Comparison results

	Threshold at IMF3				Threshold at IMF2			
	MSE		MAPE		MSE		MAPE	
Metric	AVG	STD	AVG	STD	AVG	STD	AVG	STD
Symbol	LSTM-CEEMDAN-SPLINE							
PETR4	0.140	**0.006**	**0.930**	**0.055**	0.155	0.015	1.063	0.069
VALE3	**0.347**	0.101	0.996	**0.134**	0.340	0.125	1.098	0.085
BOVA11	**0.302**	**0.046**	0.653	**0.048**	0.331	0.095	0.672	0.038
ITUB4	**0.051**	**0.007**	1.047	**0.038**	0.107	0.019	1.085	0.083
BBDC4	**0.093**	0.015	0.887	0.088	0.151	0.018	0.933	**0.075**
B3SA3	**0.272**	**0.015**	1.217	**0.048**	0.337	0.042	1.343	0.085
BBAS3	0.178	0.048	1.184	0.083	0.222	0.043	1.271	**0.048**
ABEV3	**0.051**	0.006	**0.678**	0.077	0.070	**0.005**	0.928	**0.037**
MGLU3	**0.628**	0.058	1.464	0.065	0.942	**0.040**	1.963	0.062
VVAR3	**0.010**	**0.001**	1.948	**0.076**	0.019	0.003	2.107	0.159
Symbol	XLSTM-CEEMDAN-SPLINE							
PETR4	**0.122**	0.024	0.957	0.109	0.103	0.006	0.972	0.030
VALE3	0.356	**0.099**	**0.995**	0.193	0.297	0.121	1.056	0.083
BOVA11	0.305	0.101	**0.572**	0.064	0.311	0.077	0.610	0.032
ITUB4	0.067	0.019	1.040	0.058	0.098	0.010	**0.969**	**0.032**
BBDC4	0.110	0.032	0.841	0.084	0.105	**0.010**	**0.823**	0.085
B3SA3	0.299	0.050	**1.077**	0.114	0.277	0.032	1.132	0.106
BBAS3	**0.161**	**0.020**	**1.038**	0.053	0.202	0.039	1.156	0.058
ABEV3	0.053	**0.005**	0.708	0.138	0.071	0.007	0.860	0.068
MGLU3	0.717	0.121	**1.193**	**0.049**	1.152	0.178	1.726	0.074
VVAR3	0.015	0.003	**1.710**	0.230	0.018	0.003	2.040	0.337

	Models without splines:			
	MSE		MAPE	
Metric	AVG	STD	AVG	STD
Symbol	LSTM-CEEMDAN			
PETR4	0.247	0.153	1.269	0.165
VALE3	0.567	0.403	1.297	0.280
BOVA11	1.523	1.146	0.910	0.269
ITUB4	0.119	0.063	1.280	0.242
BBDC4	0.201	0.128	1.490	0.418
B3SA3	0.673	0.567	1.569	0.347
BBAS3	0.660	0.478	1.918	0.547
ABEV3	0.106	0.049	0.984	0.197
MGLU3	1.103	0.419	1.850	0.333
VVAR3	0.085	0.097	2.828	1.077
Symbol	XLSTM-CEEMDAN			
PETR4	0.232	0.154	1.480	0.565
VALE3	1.433	1.412	1.290	0.228
BOVA11	2.390	3.336	1.385	0.655
ITUB4	0.280	0.242	1.222	0.170
BBDC4	0.564	0.621	2.304	0.923
B3SA3	6.011	5.410	2.486	1.535
BBAS3	3.973	3.099	3.213	1.436
ABEV3	0.120	0.059	1.458	0.475
MGLU3	2.489	1.841	2.605	0.790
VVAR3	0.123	0.146	2.887	0.904

Also in Table 2, we can see that, for a given ticker, the lowest accuracy can be achieved either by CEEMDAN-LSTM-SPLINE or by XCEEMDAN-LSTM-SPLINE, with thresholds at either IMF2 or IMF3. Hence, in order to compare overall improvements over the top ten liquidity Brazilian equities by architectural modification in the model, we average the results among different tickers and summarized in the Table 3, where the "x" labeling columns refers to the use of exogenous features. The improvements are calculated for each ticker by the percentage of decrease in MSE and MAPE metrics (average and standard deviation) shown in Table 2 relative to original CEEMDAN-LSTM. The overall average in Table 3 is then obtained by averaging the improvement results among tickers for each model. Table 3 shows best results in bold.

Table 3. Average improvements relative to CEEMDAN-LSTM (%) over top ten liquidity Brazilian equities

	Architectural modification improvement relative to vanilla CEEMDAN-LSTM (%)				
Threshold	**IMF3**		**IMF2**		**N/A**
Metric	**spline**	**x+spline**	**spline**	**x+spline**	**x**
mse_avg	60.934	**65.233**	53.686	62.285	−145.209
mse_std	**94.595**	89.910	89.369	92.432	−266.306
mape_avg	26.695	27.054	21.672	**27.234**	−35.755
mape_std	**76.961**	68.464	76.471	76.797	−135.948

Table 3 "spline" columns bring positive results for the adoption of splines for high level IMF predictions. It improves the average standard deviation of the ten experiments for the ten equity symbols by 94% and 76% respectively both in MSE and MAPE metrics when compared to vanilla CEEMDAN-LSTM, and these results are not hardly correlated to the threshold. This improvement comes from the stability of cubic splines because of its low-degree polynomial nature. When compared to a prediction made by a neural net, the cubic spline has indeed more chances to well fit the data when it comes to smooth curves.

Table 3 "x" column shows that the inclusion of exogenous features in the models, by itself, actually decreases the accuracy values. Nevertheless, exogenous features are often beneficial when used only in the low level IMFs, when splines handle the high level ones. This means that the neural models for low frequency IMFs are not taking advantage of the more information available in exogenous data. This is probably caused by the increase of complexity in the problem, as it is harder to find a convergence point when the neural net has more features as inputs.

When we combine both the splines for higher level IMFs and exogenous features with neural networks for the low level IMFs (those with highest frequencies), the model with exogenous variables and splines overcomes, in both accuracy metrics, the results of only adopting splines for both thresholds at IMF2 and IMF3.

Overall, these results indicate that most of the useful correlation among the open, high, low, close and volume features are located in the low level IMFs. These low level IMFs are responsible for the day to week term trends. The high frequency IMFs of the exogenous features do contain relevant information to predict the closing price.

Our XCEEMDAN-LSTM-SPLINE model with threshold at IMF2, presented the greatest average improvement in MAPE accuracy, by 27.23% when compared to the original CEEMDAN-LSTM model. On the other hand, our XCEEMDAN-LSTM-SPLINE with threshold at IMF3 model brings the best MSE accuracy improvement by 65.233%. Considering that MAPE is a better practical measure in this domain, for allowing for different price magnitude, it is worth including exogenous features in the input of the high frequency IMFs, adopting the XCEEMDAN-LSTM-SPLINE model with threshold at IMF2. Table 3 also shows a clear tradeoff between improving average accuracy and worsening standard deviation when using neural instead of spline for predicting IMF3.

Examples of the test predictions for model XCEEMDAN-LSTM-SPLINE with threshold at IMF2 are shown in Fig. 7. Figure 7b was chosen to illustrate stronger prediction deviation from the real data that may happen, that is because VALE3 ticker has shown a greater variance in the train period when compared to test and validation, what caused the higher frequencies IMFs to be more unpredictable, and in this case the neural model had more dificulties following the magnitude of variations.

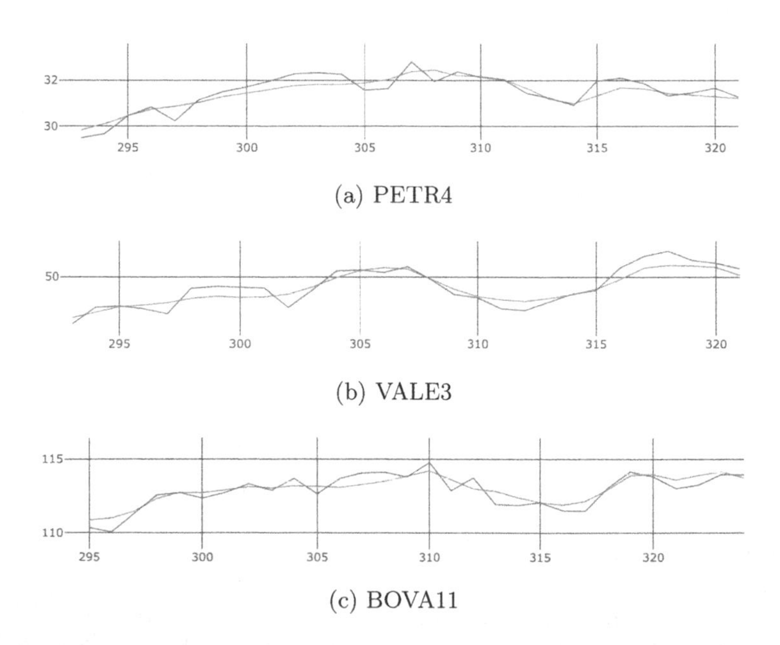

Fig. 7. Example prediction of PETR4, VALE3 and BOVA11 test set. Blue line represents real values and Green line represents predicted values.

4 Conclusion

Our work has taken the CEEMDAN-LSTM model as starting point, proposing methods to deal with multiple feature input as multivariate time series, and a way to deal with convergence difficulties caused by multi-feature input. By our experiments, we have shown that the use of exogenous features by itself does not improve the prediction for every IMF as primarily expected, but only for the ones with highest frequency. When employing splines for high level IMFs and the exogenous features with neural model for low level IMFs, we have obtained significant accuracy average gains for the top ten liquidity Brazilian equity stock symbols in the chosen period as well as decreasing standard deviation among different realizations of the same experiments.

The proposed architecture is designed to be able to support different decomposition methods and different prediction models. There is still room to investigate even more different input features, with data of nature other than financial, and to investigate different predictions models that best suit each CEEMDAN IMFs level range behavior. The key of exploring these types of prediction models relies on experimenting and understanding the behavior of data in each of its perspectives, using the right tools to best extract patterns.

Acknowledgment. The first author thanks BTG Pactual for the support to this work.

Conflict of Interest. Both authors declare they have no conflict of interest.

References

1. Yule, G.U.: On a method of investigating periodicities in disturbed series. Philos. Trans. R. Soc. Lond. Ser. A **226**, 167–298 (1927)
2. Whittle, P.: Hypothesis Testing in Time Series Analysis, vol. 4. Almqvist & Wiksells Boktr, Uppsala (1951)
3. Box, G.E.P., Jenkins, G.M., Reinsel, G.C.: Time Series analysis: forecasting and Control, 4th edn. A John Wiley & Sons, Inc. (2012)
4. Engle, R.F.: Autoregressive conditional heteroscedasticity with estimates of the variance of united kingdom inflation. Econometrica **50**(4), 987–1007 (1982)
5. Hamilton, J.D.: On the interpretation of cointegration in the linear-quadratic inventory model. J. Econ. Dyn. Control **26**(12), 2037–2049 (2002)
6. Cao, J., Li, Z., Li, J.: Financial time series forecasting model based on CEEMDAN and LSTM. Phys. A **519**, 127–139 (2019). https://doi.org/10.1016/j.physa.2018.11.061
7. Büyükşahin, Ü.Ç., Ertekin, Ş.: Improving forecasting accuracy of time series data using a new ARIMA-ANN hybrid method and empirical mode decomposition. Neurocomputing **361**, 151–163 (2019)
8. Wang, Z., Lou, Y.: Hydrological time series forecast model based on wavelet denoising and ARIMA-LSTM. In: Proceedings of 2019 IEEE 3rd Information Technology, Networking, Electronic and Automation Control Conference, ITNEC 2019, no. Itnec, pp. 1697–1701 (2019)

9. Li, X., Long, X., Sun, G., Yang, G., Li, H.: Overdue prediction of bank loans based on LSTM-SVM. In: Proceedings - 2018 IEEE SmartWorld, Ubiquitous Intelligence and Computing, Advanced and Trusted Computing, Scalable Computing and Communications, Cloud and Big Data Computing, Internet of People and Smart City Innovations, SmartWorld/UIC/ATC/ScalCom/CBDCo, pp. 1859–1863 (2018)
10. Rogers, L.C., Zhou, F.: Estimating correlation from high, low, opening and closing prices. Ann. Appl. Probab. **18**(2), 813–823 (2008)
11. Torres, M.E., Colominas, M.A., Schlotthauer, G., Flandrin, P.: A complete ensemble empirical mode decomposition with adaptive noise. In: ICASSP, IEEE International Conference on Acoustics, Speech and Signal Processing-Proceedings, pp. 4144–4147 (2011)
12. Lin, W.T., Tsai, S.C., Sun, D.S.: Price informativeness and predictability: how liquidity can help. Appl. Econ. **43**(17), 2199–2217 (2011)
13. Daubechies, I.: The wavelet transform, time-frequency localization and signal analysis. IEEE Trans. Inf. Theor. **36**(5), 961–1005 (1990)
14. Huang, N.E., et al.: The empirical mode decomposition and the Hilbert spectrum for nonlinear and non-stationary time series analysis. Proc. R. Soc. Lond. A **454**, 903–995 (1998). https://doi.org/10.1098/rspa.1998.0193
15. Wu, Z., Huang, N.E.: Ensemble empirical mode decomposition: a noise-assisted data analysis method. Adv. Adapt. Data Anal. **1**(1), 1–41 (2009)
16. Laszuk, D.: Python Implementation of Empirical Mode Decomposition Algorithm (2017). https://laszukdawid.com/codes/
17. Gers, F.A., Eck, D., Schmidhuber, J.: Applying LSTM to time series predictable through time-window approaches. In: Tagliaferri, R., Marinaro, M. (eds.) Neural Nets WIRN Vietri-01. Perspectives in Neural Computing, pp. 669–676. Springer, London (2001). https://doi.org/10.1007/978-1-4471-0219-9_20

Improved Multilevel Algorithm to Detect Communities in Flight Networks

Camila P. S. Tautenhain(✉), Calvin R. Costa, and Mariá C. V. Nascimento

Instituto de Ciência e Tecnologia, Universidade Federal de São Paulo, São José dos Campos, Brazil
{santos.camila,calvin.costa,mcv.nascimento}@unifesp.br

Abstract. Community detection in networks is a hard to solve combinatorial optimization problem with large-scale applications such as those based on flight networks. To approach this problem, a number of heuristics can be found in the literature, among them we highlight multilevel algorithms which are very efficient in detecting communities in larger networks. In this paper, we propose an improvement of a multilevel community detection algorithm that aims at finding communities so as to maximize the well-known modularity measure. The proposed improvement is the inclusion of a refinement phase to enhance the modularity value in flight networks we compiled. We presented a study case on the communities found for real flight networks. The vertices of the constructed networks correspond to airports in Europe, South America, United States of America and Canada and the arcs represent the flights whose weights indicate the number of flights between pairs of destinations. Experiments with these networks pointed to an improvement in the modularity of the communities found by the proposed algorithm when contrasted with its original version and the Lovain method. We visually identified consistent communities in the flight networks. Besides, in comparison to other reference algorithms on benchmark networks, ours was very competitive.

Keywords: Community detection · Modularity maximization · Multilevel algorithms · Flight networks

1 Introduction

Graphs and digraphs can properly model air traffic, airport and flight networks. Airport networks are often studied in terms of their topology in air transportation problems [8]. Such networks present community structure, that is, the organization of vertices into groups of highly related vertices.

The study performed in [7], for example, was one of the first to observe the delimitation of communities in airport networks. Other previous works have detected communities in networks from geographic regions such as Europe [8],

R. Cerri and R. C. Prati (Eds.): BRACIS 2020, LNAI 12320, pp. 573–587, 2020.
https://doi.org/10.1007/978-3-030-61380-8_39

Brazil [2] and United States of America [3,5]. Gurtner et al. [8] showed how community detection methods applied to airport networks can detect communities composed by airports that are geographically distant. The authors suggested that the design of the European airspace could be possibly improved by the establishment of air control unit zones according to the detected communities.

Besides inferring about air traffic, to detect communities in flight networks can be useful in guiding the solution search of air transportation problems, as e.g. routing-related problems. Jiang et al. [9] observed that in such networks, it was possible to define a solution of the classical Traveling Salesman Problem as a sequence of subpaths. Most of the vertices in each subpath belonged to the same communities.

Detecting communities by modularity maximization is one of the most employed approaches in the literature [12]. On weighted digraphs, the modularity measure is calculated by the difference between the total arc weight within communities and the sum of expected arc weights in a random network with the same in- and out-degree sequence as the original digraph. It is well-known that the modularity maximization problem is computationally prohibitive, even for medium-sized instances. A computationally efficient approach to maximize modularity in large digraphs is by multilevel algorithms, such as those proposed in [15,17]. Multilevel algorithms are solution methods that successively contract arcs to define graph clusterings. The goal of contracting arcs is for the method to assign cluster labels in a reduced graph, therefore, aiming at a more efficient method. In spite of traditional multilevel algorithms having coarsening, partitioning and refinement phases, *ConClus*, which is the multilevel algorithm proposed in [17], does not include the refinement phase.

Both the method proposed in [15] and *ConClus* successively contract arcs of an input digraph until they reach a stop criterion. In both algorithms, the arc contractions process implies that the end vertices of the arcs are merged into a supernode and must belong to the same community. The contractions must stop when it is no longer possible to improve the modularity of the partition resulting from the contractions.

In this paper, we propose *ConClus+* as an improvement of *ConClus*. Different from *ConClus*, *ConClus+* has a refinement phase, after the arc contractions. The refinement phase expands the arc contractions whilst applying a local search algorithm to refine the community attribution of vertices. We performed experiments with *ConClus+* in LFR networks [10] and in real flight networks. We compiled four flight networks whose airports are located in Europe (EU), South America (SA), United States of America (USA) and Canada (CA) and whose arcs represent the number of flights between pairs of airports. *ConClus+* outperformed its original version in all tested LFR networks. Moreover, *ConClus+* found partitions with higher modularity values than those obtained by *ConClus* and Louvain in flight networks. We identified cohesive communities delimited by the countries in SA, communities distributed across geographical regions in EU, CA and in the east coast of the USA, and a single highly connected community on the west coast of the USA.

The rest of this paper is organized as follows: Sect. 2 formally defines the notations used throughout the paper as well as modularity maximization problem; Sect. 3 gives an overview of community detection methods related to this study and some of its applications on flight networks; Sect. 4 describes the proposed algorithm; Sect. 5 presents the computational experiments using artificial LFR networks; Sect. 6 shows an analysis of the results achieved by *ConClus+* in flight networks; and, finally, Sect. 7 wraps up the paper with a discussion of its main contributions giving future work directions.

2 Problem Definition

Let $G = (V, E, \omega)$ be a digraph where V and E are its sets of n vertices and m arcs, respectively, and ω is a function $\omega : E \rightarrow \mathbb{R}_{>0}$ that assigns integer weights to the arcs of the digraph. Each arc is an ordered tuple (i, j). The weighted adjacency matrix of G is defined by $W = [w_{ij}] \in \mathbb{R}_{+}^{n\times n}$, where w_{ij} is the weight of arc (i, j) if $(i, j) \in E$, and 0 otherwise. The total weight of the arcs is given by $\Omega = \sum_{(i,j)\in E} w_{ij}$. The out and in-degree of a vertex v are respectively given by $d_v^+ = \sum_{j\in V} w_{vj}$ and $d_v^- = \sum_{j\in V} w_{jv}$.

Let the communities be identified by integer labels $1, 2, \ldots, k$. A partition of vertices into communities is here given by $P = \{c_1, c_2, \ldots, c_n\}$, where each component c_i indicates the label of the community which vertex i belongs to.

Equation (1) presents a formulation for the modularity in weighted digraphs when considering a partition P.

$$Q(P) = \frac{1}{\Omega}\sum_{i\in V}\sum_{j\in V}\left(w_{ij} - \frac{d_i^+ d_j^-}{\Omega}\right)\delta_{c_i c_j} \tag{1}$$

where $\delta_{c_i c_j}$ is 1 if $c_i = c_j$ and 0, otherwise.

According to Eq. (1), the higher the number of arcs within communities and the lower the expected number of arcs in a random digraph with the same degree sequence as the original vertex sequence, the higher the modularity.

3 Related Works

Because of the difficulty in defining the community detection problem, there is no method widely accepted as the state-of-the-art; instead, there are algorithms that perform better in certain types of graphs than others. Since a complete review is out of the scope of this paper, we shall briefly discuss some reference algorithms. For a comprehensive review of community detection algorithms for directed networks, we refer to [13].

As previously mentioned, multilevel methods successively contract arcs of an input graph. Arc contractions result in the creation of supernodes by merging the end vertices of the arcs.

Noack and Rotta [15] introduced one of the first modularity maximization-based multilevel algorithms to find communities in networks. Starting from a partition in which each vertex is in its own community, their algorithm successively contracts the arc whose end vertices merging into a community results in the largest modularity gain. In addition to contracting arcs, the authors implemented a refinement phase that projects the contracted graphs back to the original and applies a local search procedure to the resulting graphs.

Santos et al. [17] introduced a multilevel algorithm, called *ConClus* which employs a semi-greedy strategy based on the modularity maximization to select the arcs to be contracted. This strategy is based on the construction phase of the Greedy Randomized Adaptive Search Procedure (GRASP) [4] heuristic, which randomly selects arcs amongst those which would result in the largest modularity.

Different from these algorithms, Louvain [1] is a hierarchical method which first finds a partition of vertices into communities, and then merges the vertices belonging to each community into supernodes. Louvain possesses two phases that are repeatedly performed until it is not possible to improve modularity. In the first phase, having singletons as starting partition, vertices are repeatedly moved between communities to maximize the value of the partition. In the second phase, a contracted graph is created by merging all vertices belonging to the same community into a supernode.

Other classical community detection methods worth mentioning are Infomap [16] and OSLOM [11]. Rosvall and Bergstrom [16] explored the duality between the community detection problem and the problem of compressing a message that describes random walks in networks. A random walk describes the process of randomly moving agents through arcs or edges of a network. The authors designed a coding scheme that ensures that repeating movements in network regions, i.e. sets of vertices, result in a higher message compression. Since these regions are likely to represent network communities, it is possible to find network communities by minimizing the message length. Towards this goal, the authors introduced the map equation measure to quantify the description length of such messages.

Lancichinetti et al. [11] designed a measure to evaluate the statistical significance of communities in networks by considering the probability of a community existing in a null random model that represents networks whose degree distributions are equal to the network under study. They proposed an algorithm named Order Statistics Local Optimization Method (OSLOM) to optimize the designed measure.

In all these studies, the methods were applied to detect communities in artificial or benchmark networks. They do not address the air transportation networks specifically. In line with this, the next section shows a brief review of studies addressing community detection in air transportation networks.

3.1 Community Detection in Air Transportation Networks

As mentioned earlier, community detection methods can uncover the community structure of air transportation networks to help designing the airspace, coordinating communications between airports or even guiding the solution of related transportation problems. In this section, we give an overview of studies that provide this type of contribution.

Gurtner et al. [8] constructed an air traffic network whose vertices are airports in Europe and arcs are weighted according to the number of flights between airports. They applied to these networks several community detection methods, among them, Louvain, Infomap and OSLOM. The authors found that, although communities were mostly delimited according to the countries, there are some airports geographically distant from the remaining airports in the community. Gurtner et al. [8] suggested that community detection methods could improve the design of the airspace with, for example, direct communication links between distant airports.

Gegov et al. [5] drew a similar remark for a USA airport network, by stating that airports distant to each other belong to the same community. In [5], the authors conducted an analysis of the evolution of an air network in the USA across different periods of the years in the last three decades. They have also related migration patterns between regions, such as, for example, workers who return home on the weekends, with the identified communities.

Other works have also analyzed communities in air networks in the USA. Among them, we highlight those in [3,6,19]. Ernesto and Naomichi [3] found that most of the airports in a USA network operating in 1997 were grouped according to their geographical location. The authors also quantified the overlapping of certain airports into different communities.

Different from the aforementioned works, Gopalakrishnan et al. [6] studied flight delay in USA networks whose arcs weights model the delay of flights between pairs of airports in certain time windows. By detecting communities in networks representing different time-windows, the authors observed that arcs within communities of flights presented long delays. The analysis of the characteristics of these communities can be used in models to predict flight delays and to guide airline managers' decision-making process.

Wu et al. [19] took into account both the route distances and the volume of passengers in flights to define the network edge weights. They compiled two networks with the flights offered by a traditional airline and a low-cost airline in the USA. The core of the detected communities in these networks coincided with the hub airports. Most of the flight routes on the traditional airline network were between hub airports whereas the low-cost airline operated more distributed routes. On the one hand, despite the low number of routes within communities, they presented a heavy traffic congestion. On the other, there were numerous flights between communities that transported a small quantity of passengers in the low-cost network.

Although, community detection methods are commonly used to analyze air transportation networks, little attention was given to comparing the results achieved by different community detection algorithms in such a domain. In the next section, we introduce a community detection algorithm to investigate the community structure of the compiled networks comparing with algorithms found in the literature.

4 Proposed Algorithm: *ConClus+*

ConClus+ is the name of the proposed algorithm, which is an enhanced version of the two-phase multilevel algorithm *ConClus*: the coarsening phase – when the arcs are contracted – the partitioning phase – which occurs simultaneously to the arc contractions, to define the final partition. Besides the two phases, *ConClus+* has a refinement phase as explained in Algorithm 1.

Algorithm 1. *ConClus+*

```
   Input  : G_0 = (V, E, ω); S ∈ [0; 1]; r ∈ [0; 1]; IT ∈ Z_+
   Output: Partition P
 1 ℰ := ∅
 2 for it = 1 ... IT do
 3     {G_1, G_2, ..., G_Y} := CoarseningPhase(G_0, S)
 4     P_R := Each supernode of G_Y is a community
 5     // Begin Local Search
 6     while ⌊r|V(G_Y)|⌋ ≥ 0 do
 7         P_R := Move one vertex v ∈ V(G_Y) at a time to the community that
           results in the largest modularity gain to P_R until no improvement is
           observed
 8         Expand G_k whose |V_k| is the closest among {G_1, G_2, ..., G_{Y-1}} to
           ⌊r|V(G_Y)|⌋ and assign such a G_k to G_Y
 9         Update P_R by assigning to the expanded vertices the labels of the
           supernodes in P_R
10     end
11     // End Local Search
12     Update(ℰ, P_R )
13     if it > ⌊IT/2⌋ and it  mod  ⌊IT/10⌋ = 0 then
14         G_0 := PermanentCoarsening(G_0, ℰ)
15     end
16 end
17 return P := Highest modularity partition from ℰ
```

The input of Algorithm 1 is a digraph $G_0 = (V, E, \omega)$, a real number $S \in [0; 1]$ used to calculate the stopping criterion of the coarsening phase, a real number $r \in [0; 1]$ to control which of the coarsened digraphs will go through the local search in the refinement phase and the maximum number of iterations, *IT*.

For *IT* iterations – lines 3 to 15 – the algorithm performs a sequence of steps to find the final partition, which is the output of the algorithm. In line 3, the coarsening phase proposed in [17] occurs. It constructs a sequence of successively contracted digraphs $G_1, G_2, \ldots, G_Y$, where $G_k, k \in \{1, 2, \ldots, Y\}$, is the digraph obtained after performing k arc contractions in G_0. Each arc contraction merges the end vertices into a node, here called supernode. Each arc to be contracted is randomly selected among the top 10% largest modularity arc contractions. The stopping criterion for the coarsening phase is to reach $S|V|$ successive arc contractions without improving modularity. To define the partition P_R from the coarsening phase, each supernode is considered a community in the last digraph of the contracted digraphs sequence, i.e. G_Y – line 4.

The novelty of *ConClus+* with respect to *ConClus* in Algorithm 1 is the refinement phase, presented in lines 6 to 10. The refinement phase expands the contracted arcs iteratively until it returns to its original topology. During the expansion, a local search is applied to the partition of the expanded, but still contracted digraph, to refine the partition. For this, starting from G_Y, i.e., the base digraph, each iteration of the refinement phase applies a best improvement local search algorithm to move a vertex at a time to communities that enhance the modularity of the partition. When any improvement is no longer possible, another digraph from the sequence of coarsened digraphs is evaluated. More specifically, the local search is applied to the digraph from the sequence whose number of vertices is the closest to $\lfloor r|V(G_k)|\rfloor$, being G_k the current digraph used in the local search. As such a digraph is expanded with regard to G_k, partition P_R is updated to consider the new vertices. The labels of these vertices are the same as the supernodes in P_R, to which they belonged to.

After the local search, P_R replaces the most similar partition in $\mathcal{E}$, provided that P_R has a higher modularity value. The similarity between two partitions is here evaluated by the number of pairs of vertices which are in the same community in both partitions. The size of $\mathcal{E}$ is at most 5. At every 10 iterations after the $\lfloor T/2\rfloor$-th iteration of the method, a permanent coarsening is applied to the original digraph, to be considered in the next coarsenings of the method. Let the highest valued modularity partition in $\mathcal{E}$ and four other different partitions randomly drawn from $\mathcal{E}$ compose set $r\mathcal{E}$. Vertices from V that belong to the same community in all partitions from $r\mathcal{E}$ are candidates to be permanently contracted in the initial digraph G. Each permanent contraction candidate has a 50% probability to be executed. After all the iterations, the partition with the best modularity is returned.

Figure 1 shows an example of *ConClus+* applied to a small network with 12 vertices. In this figure, each vertex is identified by a different numeric label. The dotted ellipses indicate arcs to be contracted and vertices of the same community are enclosed within an ellipse.

Figure 2 shows an example of the Permanent Coarsening strategy. In this example, $r\mathcal{E} = \mathcal{E}$ since the elite set contains only five partitions, but only two different community structures. Next to the figures representing the partitions, we indicate the number of times they appear in the elite set.

5 Experiments

We generated 40 weighted directed LFR networks using the benchmark software proposed in [10]. Following the widely employed methodology to generate networks in the literature, we produced networks with mixture degree $\mu \in \{0.1; 0.2; \ldots; 0.8\}$ and assigned the same values of μ to the mixture coefficient parameter for the arc weights. For each mixture coefficient, we generated

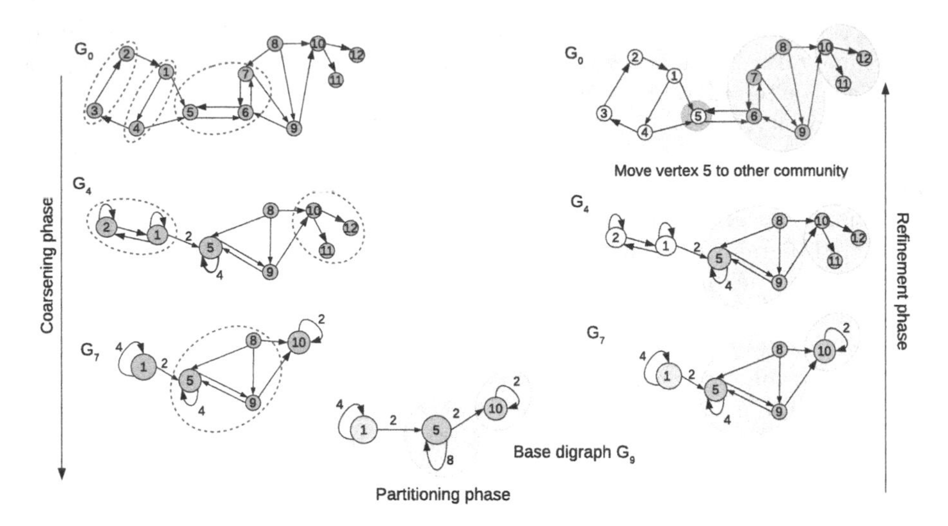

Fig. 1. An example of how *ConClus+* works in a toy network. In the coarsening phase, digraph G_4 is obtained by contracting the vertex pairs $\{1;4\}$, $\{2;3\}$, $\{7;6\}$ and $\{6;5\}$; G_7 is obtained after contracting vertex pairs $\{2;1\}$, $\{10;11\}$ and $\{10;12\}$; and finally the base digraph G_9 is obtained after contracting vertex pairs $\{8;5\}$ and $\{8;9\}$. In the partitioning phase, each supernode is considered a community. In the refinement phase, the communities of the supernodes are expanded producing the respective digraphs: G_7, G_4 and G_0. To obtain a partition with higher modularity value, the local search moves vertex 5 from the community with vertices 6, 7, 8 and 9 in the expanded digraph G_0 to the community with vertices $1, 2, 3$ and 4.

5 different networks with 1000 vertices. Moreover, we set the number of vertices within communities to vary within the interval [10;100]. Besides, to define the networks, we set the average in-degree to 20; the maximum degree to 50; the negative exponent for degree sequence to 2; and the negative exponent for the community size distribution to 1.

In the experiments, the Normalized Mutual Information (NMI) measure [18] assessed the correlation between a partition obtained by a community detection algorithm and the ground-truth partition. The value of NMI is within the range $[0; 1]$ and the closer to 1, the more correlated are the partitions.

The parameters we set for *ConClus+* and *ConClus* are: $S = 0.05$, $r = 0.3$ and $IT = 30$. We decided upon these values after preliminary tests.

Figure 3a shows the average NMI values by mixture coefficient μ between the ground-truth partitions and those obtained by *ConClus+*, *ConClus*, Louvain [1], Infomap [16] and OSLOM [11] over 5 independent runs. Figures 3b and 3c report respectively the corresponding standard deviation from the average NMI values and average running times of the algorithms in seconds. All experiments were carried out on a computer with Intel Xeon E5-1620 3.7-GHz processor and 32GB of main memory.

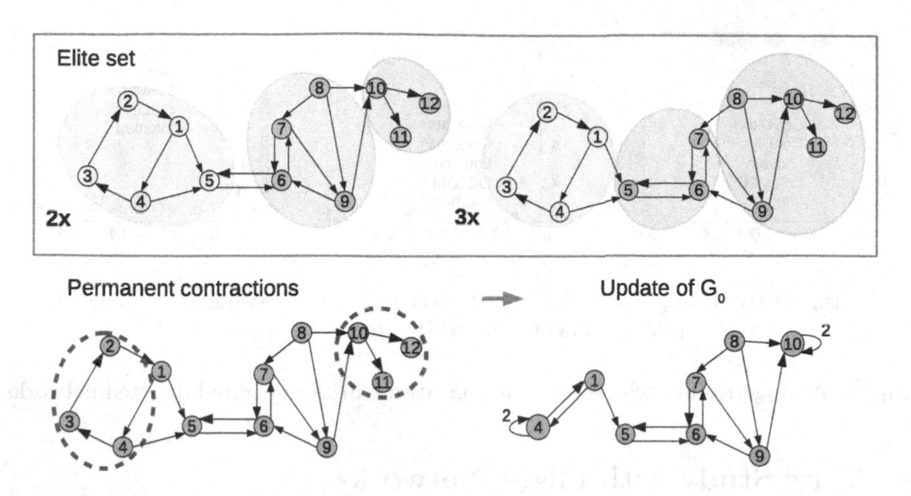

Fig. 2. Example of the Permanent Coarsening algorithm. Vertices 2, 3 and 4 are permanently contracted into a supernode labeled 4 and vertices 10, 11 and 12 into a supernode labeled 10.

ConClus+ obtained better communities than *ConClus* in all the networks. Infomap and *ConClus+* obtained partitions with the highest NMI values in networks with $\mu \leq 0.6$ and $\mu = 0.7$, respectively. *ConClus+* and Louvain, nonetheless, found communities with competitive NMI values when $\mu \leq 0.5$ whereas Infomap failed to find communities when $\mu \geq 0.7$. Although OSLOM found the partitions with the highest NMI values in networks with $\mu = 0.8$, one can observe that the standard deviation values were substantially larger than those presented by the other algorithms. Moreover, *ConClus+* and Louvain obtained better partitions than OSLOM when μ was equal to 0.6 and 0.7. OSLOM was the most time-consuming algorithm in all the networks. The best running times were taken by Louvain, followed by Infomap.

According to the reported results, we can generalize that *ConClus+* achieved satisfactory results on this experiment and outperformed its original version, *ConClus*. The NMI values of the partitions obtained by *ConClus+* were, on average, 1.51% higher than those found by *ConClus*. When considering only the networks with $0.6 \leq \mu \leq 0.7$, however, we observe an average improvement of 5.46% on the NMI values of the partitions achieved by *ConClus+* over *ConClus*. Nonetheless, the results do not allow us to point an algorithm that performed better in all networks. In this context, since we are primarily interested in flight networks, in the next section we present a case study with community detection algorithms on real flight networks we compiled.

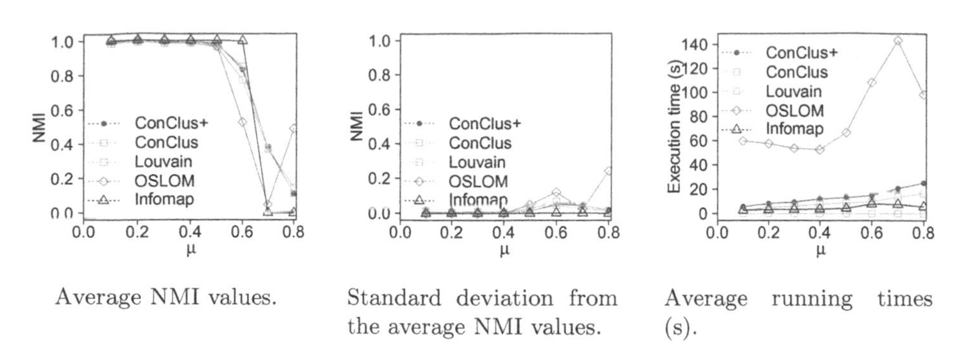

Average NMI values. Standard deviation from the average NMI values. Average running times (s).

Fig. 3. Average results achieved by the algorithms in the weighted directed networks.

6 Case Study with Flight Networks

We compiled flight networks operating in airports located in Europe (EU), United States of America (USA), Canada (CA) and South America (SA) between 13 July 2017 and 13 August 2017, from a database composed of approximately 2.65 million real flights tickets. This is an updated version of the database employed in [14]. The vertices of these networks correspond to airports. There exists an arc between a pair of vertices if there is at least one flight between them. The arc weight is the number of flights between its endpoint airports in the database. Table 1 gives information on the compiled networks regarding number of vertices (n); number of arcs (m); total arc weight (Ω); normalized average (avg), normalized standard deviation (sd) and normalized maximum value for the arc weights (max). The values avg, sd and max were normalized with respect to Ω.

Table 1. Information on the compiled networks.

Network	n	m	Ω	Arc weights		
				avg	sd	max
CA	94	394	39601	0.00254	0.00282	0.01985
SA	210	933	135872	0.00107	0.00159	0.01722
EU	435	7614	415841	0.00013	0.00018	0.00282
USA	389	4528	834373	0.00022	0.0003	0.00321

The SA and CA networks are significantly sparser than EU and USA networks regarding the number of arcs, that is, the number of airport pairs that are connected by at least one flight. The high average number of flights between airports in SA and CA networks along with the high corresponding standard deviation show that some pairs of airports in these networks are connected by numerous flights, whereas others are more sparsely connected. Although there are more flights in the USA than in the EU network, as shown by its Ω values, there are more airports connected by arcs, as indicated by m, in the EU network. In fact, the average number of flights in the EU network shows that flights are more consistently distributed in this network than in the others.

Table 2 reports the average modularity values of the partitions obtained by the modularity maximization algorithms *ConClus+*, *ConClus* and Louvain over 100 independent runs. Infomap was not included in this table because it does not aim at maximizing modularity. In this table, the $\pm$ notation indicates the standard deviation from the reported average values. Louvain is a deterministic algorithm, therefore the standard deviation is zero.

Table 2. Modularity values of the partitions obtained by the algorithms in the compiled networks over 100 independent runs.

Network	Modularity		
	ConClus+	*ConClus*	Louvain
CA	0.44809 ± 0	0.44809 ± 0	0.44487 ± 0
SA	0.72065 ± 0	0.72065 ± 0	0.72065 ± 0
EU	0.345 ± 0.00038	0.34473 ± 0.00059	0.34112 ± 0
USA	0.28212 ± 0.00143	0.2818 ± 0.00141	0.28006 ± 0

All algorithms obtained partitions with the same modularity value in the SA network. Both *ConClus+* and *ConClus* found partitions with higher modularity value than Louvain in the CA network. In the remaining networks, EU and USA, *ConClus+* achieved higher modularity values than *ConClus* and Louvain.

Figures 4, 5, 6 and 7 illustrate the communities found by *ConClus+* and Infomap, respectively, in the networks SA, CA, EU and USA. Infomap was selected for this analysis because it obtained the best partitions in most LFR networks. The proposed *ConClus+* found the highest modularity valued partitions in the flight networks.

On the one hand, flights between countries were less frequent in the SA network and, thereby, grouping airports in the same country resulted in greater improvements in the modularity. Similarly, *ConClus+* and Infomap found communities whose airports are geographically close in the CA network.

On the other hand, the number of flights did not vary much within or between countries in the EU network. Both *ConClus+* and Infomap found cohesive communities in the northern and eastern regions of Europe by grouping together geographically close airports. Infomap, however, detected a major community in the western region of the EU network, whereas *ConClus* was able to distinguish several communities in this region. Even though the communities obtained by *ConClus+* in the western region of the EU network correspond mostly to cohesive regions, they are not necessarily delimited by the countries or geographically close regions. For example, the community primarily composed of airports in the

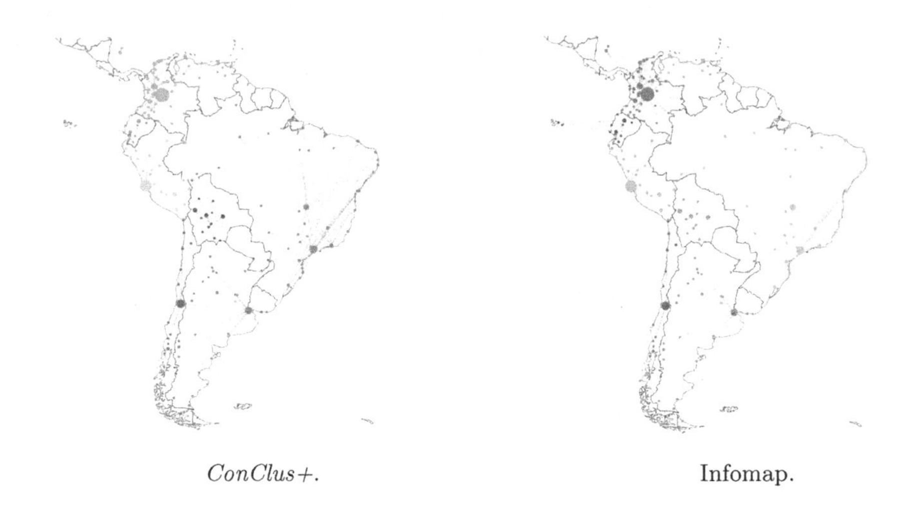

ConClus+. Infomap.

Fig. 4. Communities found in the SA network. The airport in Easter Island, which was omitted from the figure, was classified in the same community as the other airports in Chile.

ConClus+. Infomap.

Fig. 5. Communities found for the CA network.

ConClus+. Infomap.

Fig. 6. Communities found for the EU network.

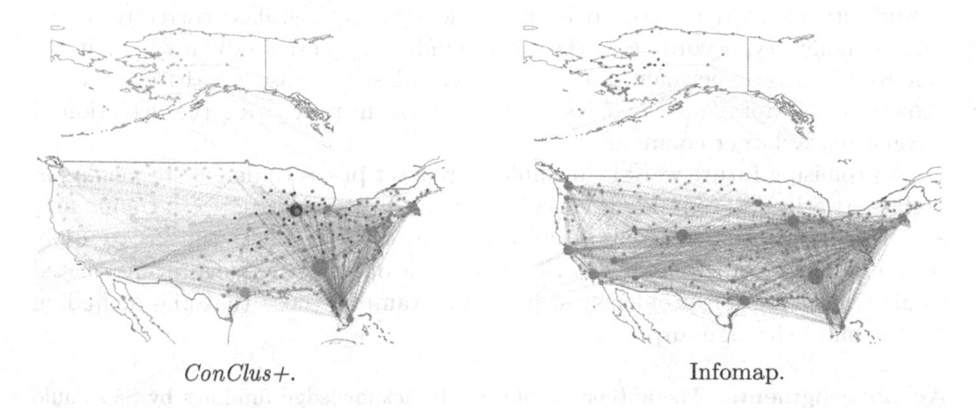

Fig. 7. Communities found for the USA network. The figures on the bottom and top show, respectively, the mainland and the Alaska state of the USA network. All the airports in Hawaii island, which is omitted from the figure, were classified into a single community by both algorithms.

United Kingdom (UK) also includes airports in the south of Portugal, Spain and France. It is worth pointing out that [8] also identified a similar community in their European flight network. In the EU network we compiled, grouping distant airports, for example, in the UK and in the south of Spain, might result in better communities.

A similar observation is valid for the two communities found by *ConClus+* on the east coast of the USA. The west coast and central region of USA, however, were distinctively grouped into two highly connected communities by *ConClus+*. Opposed to *ConClus+*, Infomap failed to distinguish communities in the mainland region of the USA network. *ConClus+* classified all but one airport in Alaska state in the USA network, whereas Infomap distinguished smaller communities.

7 Final Remarks

In this paper, we proposed a refinement phase to a multilevel and algorithm to detect communities in digraphs. The proposed algorithm, called *ConClus+*, aims at maximizing the modularity measure. We have also presented a case study on the communities detected in networks compiled from real flight data in airports located in Europe, South America, United States of America and Canada. To define the arc weights of these networks, we calculated the number of flights between pairs of airports.

We achieved satisfactory results in LFR networks in comparison to a modularity maximization algorithm, Louvain, and two other reference algorithms, Infomap and OSLOM. In the flight networks we compiled, *ConClus+*, achieved communities with slightly higher modularity values than its original version and a classical modularity maximization algorithm, Louvain. In the networks with airports in Europe and United States of America, the slight improvements in modularity resulted in more pairs of vertices being classified correctly in the same community, according to the corresponding highest modularity partitions achieved in the experiments. Besides, a visual study case on these networks showed an advantage of *ConClus+* over Infomap in preventing the detection of inconsistently larger communities.

A promising future work is to employ air ticket prices to define arc weights in flight networks and to study their influence on the resulting network community structure. Another promising future direction of the present study is to apply the communities obtained by *ConClus+* to guide optimization algorithms related to air transportation problems, such as, for example, those that aim at finding travel routes through airports.

Acknowledgments. The authors would like to acknowledge fundings by São Paulo Research Foundation (FAPESP), grant numbers: 2016/22688-2, 2019/22067-6; by Conselho Nacional de Desenvolvimento Científico e Tecnológico (CNPq); and by Coordenação de Aperfeiçoamento de Pessoal de Nível Superior - Brasil (CAPES) - Finance Code 001.

References

1. Blondel, V.D., Guillaume, J.L., Lambiotte, R., Lefebvre, E.: Fast unfolding of communities in large networks. J. Stat. Mech: Theory Exp. **2008**(10), P10008 (2008)
2. Couto, G.S., Silva, A.P.C.D., Ruiz, L.B., Benevenuto, F.: Structural properties of the Brazilian air transportation network. Anais da Academia Brasileira de Ciências **87**(3), 1653–1674 (2015)
3. Estrada, E., Hatano, N.: Communicability graph and community structures in complex networks. Appl. Math. Comput. **214**(2), 500–511 (2009)
4. Feo, T., Resende, M.: Greedy randomized adaptive search procedures. J. Global Optim. **6**, 109–133 (1995)
5. Gegov, E., Postorino, M.N., Atherton, M., Gobet, F.: Community structure detection in the evolution of the United States airport network. Adv. Complex Syst. **16**(01), 1350003 (2013)
6. Gopalakrishnan, K., Balakrishnan, H., Jordan, R.: Clusters and communities in air traffic delay networks. In: 2016 American Control Conference (ACC), pp. 3782–3788. IEEE (2016)
7. Guimerà, R., Mossa, S., Turtschi, A., Amaral, L.A.N.: The worldwide air transportation network: anomalous centrality, community structure, and cities' global roles. Proc. Nat. Acad. Sci. **102**(22), 7794–7799 (2005)
8. Gurtner, G., et al.: Multi-scale analysis of the European airspace using network community detection. PLOS One **9**(5), e94414 (2014)
9. Jiang, Z., Liu, J., Wang, S.: Traveling salesman problems with PageRank distance on complex networks reveal community structure. Physica A **463**, 293–302 (2016)
10. Lancichinetti, A., Fortunato, S., Radicchi, F.: Benchmark graphs for testing community detection algorithms. Phys. Rev. E **78**(4), 046110 (2008)
11. Lancichinetti, A., Radicchi, F., Ramasco, J.J., Fortunato, S.: Finding statistically significant communities in networks. PLOS One **6**(4), 1–18 (2011)

12. Leicht, E.A., Newman, M.E.: Community structure in directed networks. Phys. Rev. Lett. **100**(11), 118703 (2008)
13. Malliaros, F.D., Vazirgiannis, M.: Clustering and community detection in directed networks: a survey. Phys. Rep. **533**(4), 95–142 (2013)
14. Nakamura, K.Y., Coelho, L.C., Renaud, J., Nascimento, M.C.V.: The traveling backpacker problem: a computational comparison of two formulations. J. Oper. Res. Soc. **69**(1), 108–114 (2018)
15. Noack, A., Rotta, R.: Multi-level algorithms for modularity clustering. In: Vahrenhold, J. (ed.) SEA 2009. LNCS, vol. 5526, pp. 257–268. Springer, Heidelberg (2009). https://doi.org/10.1007/978-3-642-02011-7_24
16. Rosvall, M., Bergstrom, C.T.: An information-theoretic framework for resolving community structure in complex networks. Proc. Nat. Acad. Sci. **104**(18), 7327–7331 (2007)
17. Santos, C.P., Carvalho, D.M., Nascimento, M.C.: A consensus graph clustering algorithm for directed networks. Expert Syst. Appl. **54**, 121–135 (2016)
18. Studholme, C., Hill, D.L., Hawkes, D.J.: An overlap invariant entropy measure of 3D medical image alignment. Pattern Recogn. **32**(1), 71–86 (1999)
19. Wu, W., Zhang, H., Zhang, S., Witlox, F.: Community detection in airline networks: an empirical analysis of American vs. southwest airlines. J. Adv. Transp. **2019**, 11 (2019)

Intelligent Classifiers on the Construction of Pollution Biosensors Based on Bivalves Behavior

Bruna V. Guterres[1(✉)], Je N.J. Junior[2], Amanda S. Guerreiro[3], Viviane B. Fonseca[3], Silvia S.C. Botelho[1], and Juliana Z. Sandrini[3]

[1] Science Computer Center, Federal University of Rio Grande, Rio Grande, Brazil
guterres.bruna@furg.br, silviacb.botelho@gmail.com
[2] Institute of Mathematics, Statistics and Physics, Federal University of Rio Grande, Rio Grande, Brazil
jenamjunior@hotmail.com
[3] Institute of Biological Sciences, Federal University of Rio Grande, Rio Grande, Brazil
amandahg@gmail.com, vivibarneche@hotmail.com, juzomer@hotmail.com

Abstract. The aquatic environment is subject to a series of contaminants resulting from anthropogenic activity which may compromise biota health and the quality of water resources. There is an imperative need for cost-effective, accurate and online solutions for monitoring aquatic environments. The present work proposes the construction of aquatic pollution biosensors for detecting both petrochemical and anti-fouling paint compounds based on the behavioral analysis of *Perna perna* mussels through multiple classifier systems. Networks of mussels instrumented with Hall effect sensors and magnets were exposed to Water-Accommodated Fraction of diesel and micro-particles of anti-fouling paint. The hourly behavioral parameters average amplitude, transition frequency and amount of motion reversals were used to infer the contamination status (polluted or not) through voting classifiers. Results presented high accuracy (95.8%) in predicting diesel pollution and non-pollution while lack of data and intrinsic characteristics of anti-fouling paints provided less significant results for detecting its compounds. This paper has demonstrated a promising use of artificial intelligence in the construction of aquatic pollution biosensors using behavioral analysis of bivalves mollusks.

Keywords: Biosensors · Aquatic pollution · Behavioral analysis

1 Introduction

The fouling of aquatic organisms causes deformation at the bottom part of vessel hulls, increasing drag and decreasing their speed [23]. It is considered a major

This study was sponsored by the Coordenação de Aperfeiçoamento de Pessoal de Nível Superior (CAPES) - Brazil.

R. Cerri and R. C. Prati (Eds.): BRACIS 2020, LNAI 12320, pp. 588–603, 2020.
https://doi.org/10.1007/978-3-030-61380-8_40

problem among maritime activities [15] and the use of anti-fouling paints on submerged surfaces is an alternative for minimizing these problems. Many toxic compounds are used in the composition of anti fouling paints and released continuously into the environment [20]. For example, DCOIT (4,5-dicloro-2-n-octil-4-isotiazolin-3-ona) is a broad-spectrum co-biocide intended to act as an anti-fouling agent.

The Environmental Quality Standard (EQS) for DCOIT in marine environments, that is, the maximum concentration for biota protection, is estimated at 0.67 ng.L^{-1} [13]. Nevertheless, DCOIT concentrations above EQS have been reported in Spanish coastal waters (3, 7 μg.L^{-1}) [14] and in the sediment of ports and marinas in Asian countries (281 ng.L^{-1}g) [10]. Since energy and transport are necessary for the production of any goods or services, the availability of oil and its derivatives have paramount importance for assessing the level of economic growth of national economies [2]. Nevertheless, oil exploitation operations are one of potential pollution sources in the marine environment [19]. Oil and its derivatives are perhaps the most complex substances to assess toxicologically.

The aquatic environment is subject to a series of contaminants resulting from anthropogenic activity such as petrochemical compounds and those released by anti fouling paints which may compromise biota health and the quality of water resources. The elaboration of rapid, profitable and reliable aquatic pollution detectors has paramount importance for the protection of aquatic organisms and water resources. Numerous methodologies have been employed to monitor the toxicity levels of chemical compounds [18]. Available sensitive analytical devices to monitor aquatic environments are expensive, time consuming and require specific trained staff. Behavioral biomarkers based on valve-activity responses of bivalve mollusks are prompt biological responses to the presence of contaminants [17] and may provide greater sensitivity to standard toxicity tests [11]. Moreover, it may lead to the elaboration of cost-effective and relatively simple solutions for monitoring aquatic environments [17].

Among bivalve mollusks, *Perna perna* mussels are commonly used in mitiliculture presenting considerable socioeconomic relevance [21]. Statistical tools have been used along with behavioral parameter to detect significant changes on bivalves behavior due to the presence of contaminants such as uranium [12], cadmium [25], arsenic [11], crude oil [19], diesel WAF [7] and harmful algae [1]. Observing the successful results in identifying these pollutants, it becomes interesting to perform experiments also based in behavioral analysis to evaluate the effectiveness of computational tools, such as artificial intelligence, in the subsequent data set processing. Valve-activity responses of mollusks are strongly related to vital activities and environmental conditions such as the presence of contaminants and predators. Hence, accurately understanding their behavior may assist in the monitoring of water quality in natural habitats [8]. However, the behavioral data resulting from the continuous monitoring of bivalves are complex due to their nonlinear behavior and its statistically challenging analysis [8].

The use of artificial intelligence methods on the construction of aquatic pollution biosensor based on bivalves behavior may be advantageous. A multiple classifier system, such as a voting classifier, is a powerful approach to difficult pattern recognition problems and noisy data since it allows to employ arbitrary feature descriptors and classification methods simultaneously [9]. A Learning Classifier System (LCS) may provide classifying the contamination status (polluted or not polluted) of aquatic environments based on behavioral responses of mussels. In order to improve the LCS performance, there are many learning strategies with intrinsic advantages and disadvantages [27]. For example, decision tree method may be applied to categorical data with little preparation but may produce inaccurate complex trees when dealing with numeric data sets [24]. Although K-Nearest Neighbors (KNN) method provides faster training, it is sensitive to noise [24]. Therefore, it is possible to combine Learning Classifier Systems through a voting classifier in order to enhance performance.

The present work presents the development of aquatic pollution biosensors for detecting petrochemical compounds and micro-particles of anti-fouling paint through a voting classifier. For this purpose, a network of instrumented *Perna perna* mussels were monitored during two toxicological experiments. The hourly behavioral parameters average amplitude, filtration activity, transition frequency and amount of motion reversals were employed for assessing the contamination status of the aquatic environment.

2 Performance Evaluation Metrics

Statistical information from LCS may be presented as a confusion matrix, which demonstrates outcomes of a binary prediction according to ground truth data and four categories. True positives (TP) encompass examples correctly labeled as positives, False positives (FP) refer to negative examples incorrectly classified as positive, True negatives (TN) correspond to negatives correctly labeled as

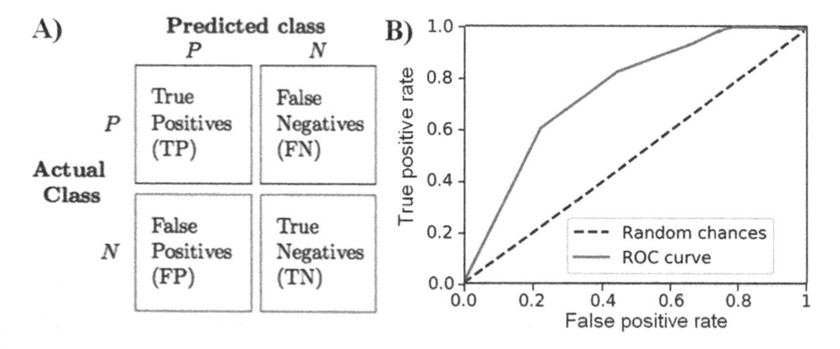

Fig. 1. Generic confusion matrix (A) and Receiver Operating Characteristics (ROC) curve (B) for a binary classification problem. The black dashed and red lines indicate the ROC curves for a random and generic classifiers, respectively. The yellow region indicates the Area Under the ROC Curve (AUC) (Color figure online)

negative and False Negatives (FN) comprise positive examples incorrectly labeled as negative as demonstrated in Fig. 1A. This information is the basis to calculate several metrics and plot a Receiver Operating Characteristics (ROC) curve.

Accuracy is the most used empirical measure for assessing classification performance. It is defined as the ratio between the number of correctly classified observations and the total number of observations as in 1. However, simply using accuracy to evaluate a classifier performance may be misleading [5].

$$accuracy = \frac{TP + FP}{TP + FP + TN + FN} \quad (1)$$

The ROC curve for a binary classification plots the true positive rate as a function of the false positive rate as illustrated in Fig. 1B. For a random classification, the ROC curve corresponds to a straight line from origin to (1,1). A ROC curve at least slightly above this line indicates enhancement over random classification [4]. Moreover, the Area Under the ROC Curve (AUC) statistically summarizes the ranking quality of a classification solution and is used in many applications [4].

ROC curves may present an optimistic assessment of an algorithm's performance if there is a large skew in the class distribution. Hence, Precision-Recall (PR) curves have been cited as an alternative to them [5] and enable calculating some performance metrics such as F-Score and the Area Under the PR curve (AUC-PR). Recall measures the fraction of positive examples that are correctly labeled and Precision indicates the fraction of examples classified as positive that are truly positive as described in 2 and 3, respectively. AUC-PR, as the area under ROC curve, reflects the performance of the algorithm [5].

$$Recall = \frac{TP}{TP + FN} \quad (2)$$

$$Precision = \frac{TP}{TP + FP} \quad (3)$$

It is advantageous combine precision and recall metrics through F-score as they encompass complementary aspects of the correctness of a classification system. F-score provides a single measure that balances both the concerns of precision and recall through a geometric mean as described in 4

$$F\text{-}score = \frac{2.Precision.Recall}{Precision + Recall} \quad (4)$$

3 Materials and Methods

3.1 Experiment Layout

Two toxicological experiments were performed to expose *Perna perna* mussels to micro-particles of anti-fouling paints and diesel WAF under controlled conditions. Previously to experiments, mussels were maintained in aerated tanks at invariable conditions of temperature (20 °C), salinity (30‰) and photoperiod (12L:12D). The animals were fed and the water renewed every two days.

In order to acquire behavioral data, waterproofed Hall effect sensors (UGN3503) and neodymium magnets ($10 \times 4\,\text{mm}^2$) were attached to opposite ends of bivalve shells with cyanoacrylate glue [1,16]. A data acquisition board composed of a 16-channel multiplexer and a SD card module were used to collect the Hall effect sensors outputs. This data acquisition board generated a CSV (Comma-Separated Values) containing the valve-activity responses along with the Arduino Mega whose analog to digital converter had 10 bits of resolution. Figure 2 illustrates the toxicological experiments performed for exposing *Perna perna* mussels to diesel WAF and micro-particles of anti-fouling paint and the data acquisition system.

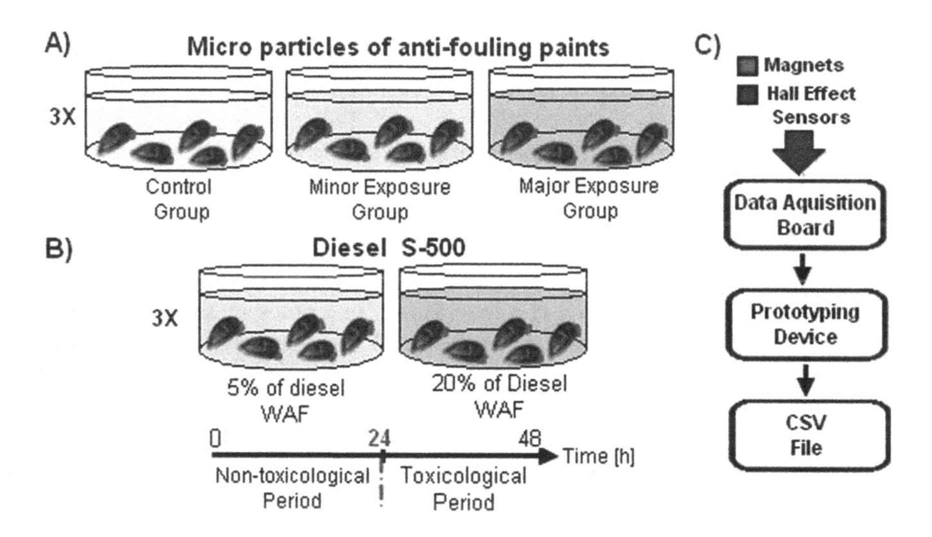

Fig. 2. Experiment layout for toxicological exposure of *P. perna* mussels to micro-particles of anti-fouling paints (A) and diesel WAF (B). Hall effect sensor and magnets were attached to bivalve shells and, along with a data acquisition system (C), provided a CSV file containing behavioral data

Exposure to Diesel WAF. The behavioral data were acquired in the experiment presented in [7] on which the acclimatization period lasted for three days and data acquisition started afterwards at 2 Hz frequency. Due to the complexity of Oil and its derivatives, it is imperative the standardization of experimental processes [22] such as the exposure media through Water-Accommodated Fraction (WAF). Aiming to provide results directly comparable across experiments, the diesel WAF was prepared according to the protocol of [22].

Instrumented mussels were distributed to 5-L aquaria and maintained at constant conditions of salinity (30‰), temperature (20 °C) and photo period ($12L : 12D$). After 24 h of data acquisition, diesel WAF was inserted into the aquaria respecting the proportions of 5% and 20% of pollutant. The experiment was performed in triplicates (three aquaria for each concentration) and was divided into non-toxicological and toxicological periods which encompassed the first and the following 24 h of experiment, respectively. A total of 8 and 10 mussels were exposed to 5% and 20% of diesel WAF, respectively. Figure 2B illustrates the layout of the toxicological exposure of mussels to diesel WAF.

Exposure to Anti-fouling Paint. The commercial anti-fouling paint Micron Premium from International brand was purchased and used for painting a glass plate which, after drying, was scraped off to generate particles. After filtering through a 180 µm sieve, the resulting micro-particles of anti-fouling paint (≤180 µm) were inserted into the aquaria for obtaining concentrations of 2 mg/L and 20 mg/L.

Instrumented mussels were equally distributed to aquaria and maintained at the above mentioned constant conditions. The experiment was performed in triplicates and the acclimatization period lasted for five days. Micro-particles of anti-fouling paint were inserted into the aquaria and data acquisition started immediately at 100 Hz frequency for 6 h. Three groups of animals were evaluated according to the concentration of micro-particles they were exposed to (Fig. 2A). The control (no toxicological exposure), minor exposure (2 mg/L) and major exposure (20 mg/L) groups comprised 6, 7 and 5 mussels, respectively.

3.2 Behavioral Analysis

The behavioral data acquired through the Hall effect sensors was normalized from 0 (complete valve closure) to 100 (maximum valve opening) to avoid possible variations from one sensor to another due to the instrumentation process [3,26]. Afterwards, the behavioral parameters average amplitude [3], filtration activity [3,8], transition frequency [8] and amount of motion reversals were evaluated considering sample windows of 1 h.

The filtration activity was measured as the fraction of time on which the bivalve shell was open. The shells of each mussel were considered open or closed according to the average opening amplitude over the experimental period [8]. If the normalized value was greater than the average, the mussel was considered open (filtering), otherwise its status was closed (not filtering). The transition frequency parameter was evaluated as the number of times the mussel status changed from open to closed and vice versa. Finally, the amount of motion reversals was assessed as the number of times the animal's shells were opening and changed its state of movement to closing.

3.3 Voting Classifier

The present work aimed to use artificial intelligence tools for monitoring aquatic environments based on hourly average amplitude, filtration activity, transition frequency and amount of motion reversal of individual *Perna perna* mussels. For each contaminant, a voting classifier was trained and evaluated. The behavioral data acquired from each experiment were pre-processed, grouped into one-hour intervals and labeled as '0' if there were no toxicological exposure and '1' otherwise. Furthermore, it was divided into two training (first and second) and a testing data sets.

The Classification Learner app of the Matlab software was employed. It enabled performing supervised machine learning through classical types of classifiers such as decision trees, Support Vector Machines (SVM) and ensemble classification. Furthermore, it provided comparisons among models based on accuracy scores, confusion matrices and ROC curves. The training process may be divided into two stages as demonstrated in Fig. 3. The first training data set was used to train classification models and determine whether there were pollutants or not. Secondly, the classifiers with highest accuracy were used to infer the contamination status according to behavioral data of another disjoint training data set (second). The obtained predictions were employed to train a voting classifier and therefore infer a final contamination status. The voting classifier comprised the classification model with highest accuracy and AUC value according to the Classification Learner app. For all training stages, a 10-fold cross validation method was employed.

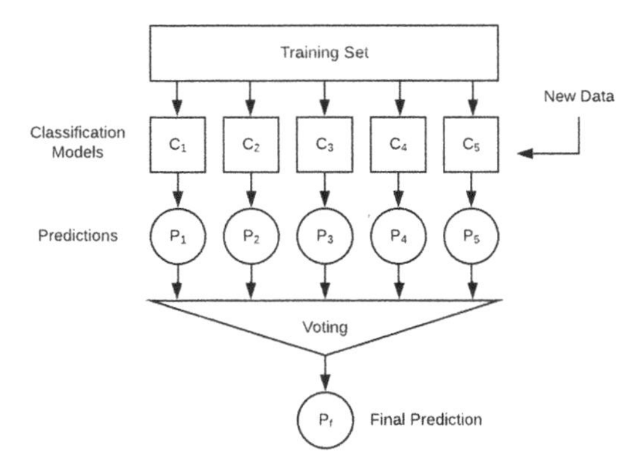

Fig. 3. Voting classifier scheme on which five classifiers were trained based on a training data set. The predictions (P_1, P_2, P_3, P_4, P_5) acquired through a new data set were used to obtain a voting classifier for a final prediction (P_f)

For behavioral data acquired during the experiment with diesel WAF, 16.67% of data were used for testing and encompassed the behavior of one and two random selected mussels exposed to 5% and 20% of diesel WAF, respectively. The remaining data were divided into first (65%) and second (35%) training data sets. The voting classifier for detecting petrochemical compounds was trained according to four classification models.

Similarly, the data generated through the experiment with micro-particles of anti-fouling paint were randomly divided into testing (15%), first training (65%)

and second training (35%) data sets. The voting classifier for detecting compounds of anti-fouling paints was trained according to five classification models. In order to inspect whether the voting classifier provided better performance in comparison to other classifiers the PR curves, AUC-PR, F-Score and confusion matrices were evaluated for all classifiers according to testing data sets.

4 Results and Discussion

4.1 Voting Classifier for Diesel WAF

The behavioral monitoring of bivalves mollusks along the toxicological experiment with diesel WAF provided idiosyncratic responses and demonstrated the complexity of evaluating behavioral data (Fig. 4). Valve closures were asynchronous among individual mussels during both toxicological and non-toxicological exposure periods as reported in [3,8]. The individual behavior of bivalves, mainly under contaminated environment, differed and demonstrated the phenotype plasticity for an individual attempt to decrease its contact to an unpleasant media [6].

The first training stage of the voting classifier development provided, on average, an accuracy of 86.5%, 85.9%, 85.7% and 85.3% for the classifiers bagged trees, cubic SVM, quadratic SVM and fine SVM, respectively. Figure 5 illustrates the confusion matrices, ROC curves and AUC values of each obtained classifier. They achieved AUC values greater than or equal to 0.9 indicating a highly ranking quality of the classification algorithms. In general, the classifiers tended to provide greater false negative rates in comparison to false positive ones.

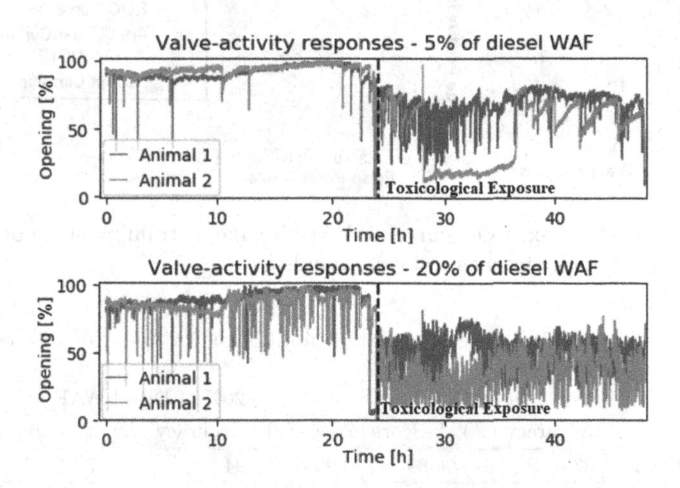

Fig. 4. Examples of valve-activity responses for *Perna perna* mussels exposed to 5% and 20% of diesel WAF. The experiment comprised 24 h of non-toxicological and toxicological exposure as demonstrated by the black dashed line

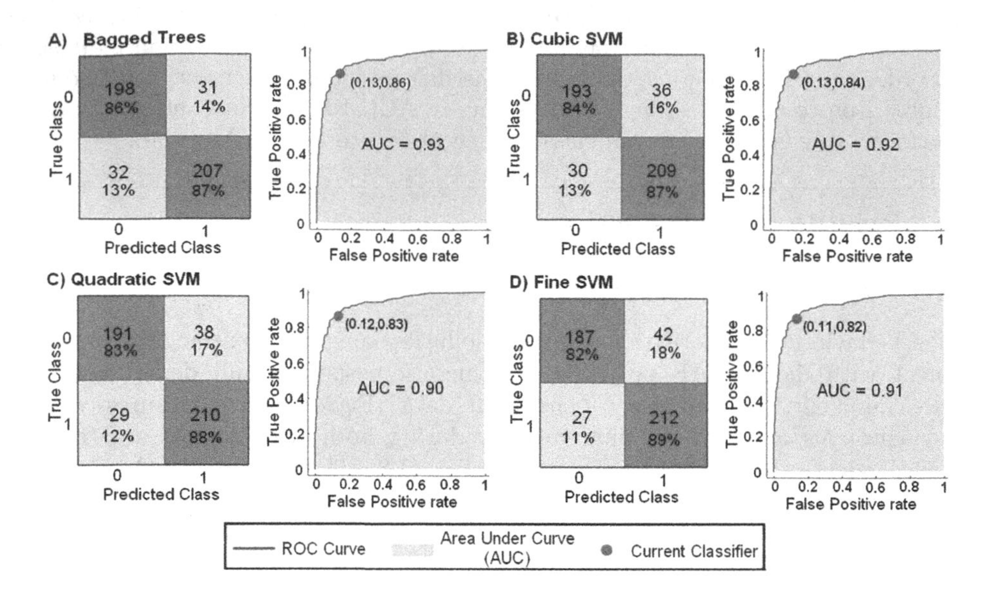

Fig. 5. Confusion matrices, ROC curves and AUC values of the classifiers bagged trees (A), cubic SVM (B), quadratic SVM (C) and fine SVM (D) obtained during the first training stage of the development of voting classifier for detecting petrochemical compounds

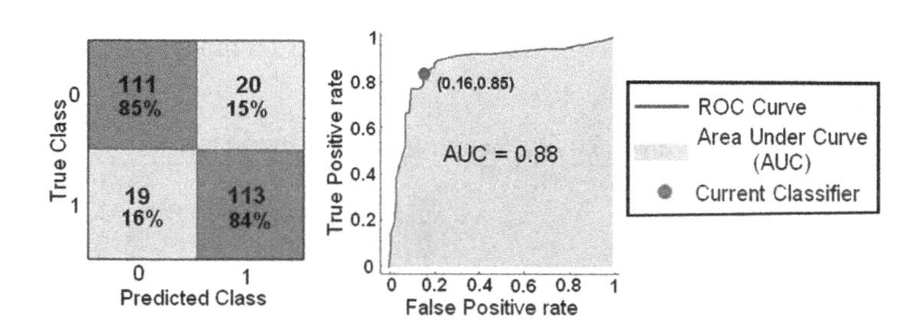

Fig. 6. Confusion Matrix, ROC curve and AUC value of training stage of the voting classifier for detecting petrochemical compounds

Table 1. Testing performance of classifier models for 5% and 20% of diesel WAF.

Classifier	5% of Diesel WAF			20% of Diesel WAF		
	Accuracy (%)	F-Score	AUC-PR	Accuracy (%)	F-Score	AUC-PR
Bagged Trees	87.5	0.89	0.99	94.8	0.95	0.99
Cubic SVM	95.8	0.96	1.00	95.8	0.96	0.99
Quadratic SVM	91.7	0.92	0.99	97.9	0.98	1.00
Fine SVM	89.6	0.96	1.00	95.8	0.957	0.99
Voting	95.8	0.96	0.98	95.8	0.957	1.00

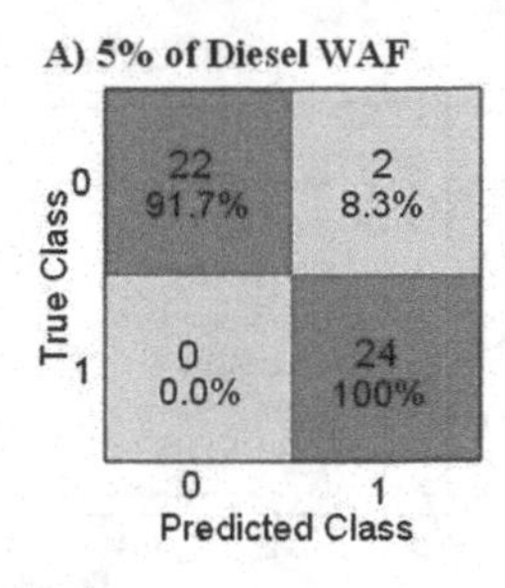

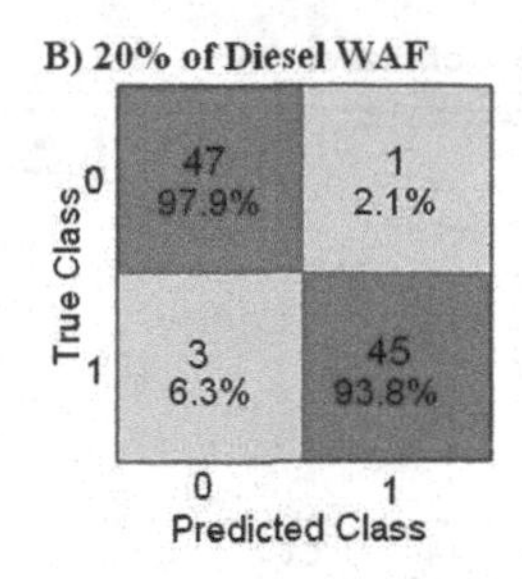

Fig. 7. Confusion matrices acquired when testing the voting classifier for detecting 5% (A) and 20% (B) of diesel WAF.

The four classifiers were used to infer the contamination status (polluted or unpolluted) of the aquatic environment according to the second training data set. The acquired predictions were employed to train classifier models. The classifier model coarse Gaussian SVM provided the highest accuracy (84.6%) and AUC value (0.88) and was elected as the voting classifier as illustrated in Fig. 6.

Finally, the voting classifier was employed for inferring the contamination status based on testing behavioral data and provided an accuracy of 95.8% for both concentrations of diesel WAF (Fig. 7). The testing performance of all classifier models was evaluated through accuracy, PR-curves, AUC-PR and F-scores. Figure 8 illustrates the PR-curves of all classifier models for each concentration of diesel WAF and Table 1 summarizes their testing performances.

Although the bagged trees classifier had presented the highest accuracy during the first training stage, it presented less suitable performance in comparison to the voting classifier according to all assessed metrics. Quadratic SVM classifier presented higher accuracy when tested for detecting major concentrations of petrochemical compounds. Nevertheless, when monitoring aquatic environments, it may be more advantageous a biosensor capable of detecting lower concentrations of contaminants. The voting classifier was able to provide suitable performance for both concentrations of petrochemical compounds and presented advantages in comparison to Quadratic SVM for detecting 5% of diesel WAF according to both F-score and AUC-PR metrics. Although Cubic SVM hasn't provided the best performance metrics during training stage, it achieved similar compared to the voting classifier.

It becomes evident that the use of the proposed classifier was conducive to the detection of petrochemical compounds in aquatic environments. Since it considered the hourly behavior of bivalves in the monitoring of aquatic environments, it contributed to the development of fast, cost-effective and efficient aquatic pollution biosensor. Furthermore, the multiple classifier system provided suitable accuracy when used to infer contamination status of aquatic environment based on the behavior of *Perna perna* mussels not used during the training processes. It was, therefore, suitable for dealing with the adaptive nature of bivalves mollusks which was considered a challenge when constructing this type of aquatic pollution biosensors.

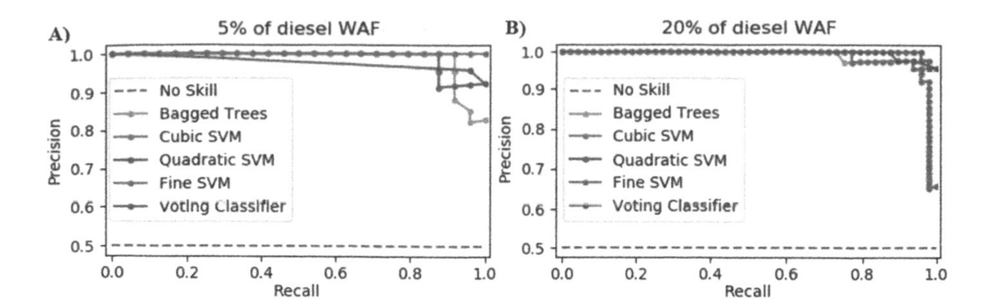

Fig. 8. Precision - Recall curves for 5% (A) and 20% (B) of diesel WAF.

The combination of a multiple classifier system along with a network of mussels has achieved of more reliable classifications for all investigated cases. Moreover, the present work proposed the use of individual behavior of mussels in the biomonitoring of aquatic environments. Considering that bivalves tend to live in large groups, a group-based approach may further contribute to the development of aquatic pollution biosensors.

4.2 Voting Classifier for Anti-fouling Paint

Similarly to the behavior acquired during the experiment with diesel, the valve-activity responses acquired during the toxicological experiment using micro-particles of anti-fouling paint were asynchronous among individual bivalves. It

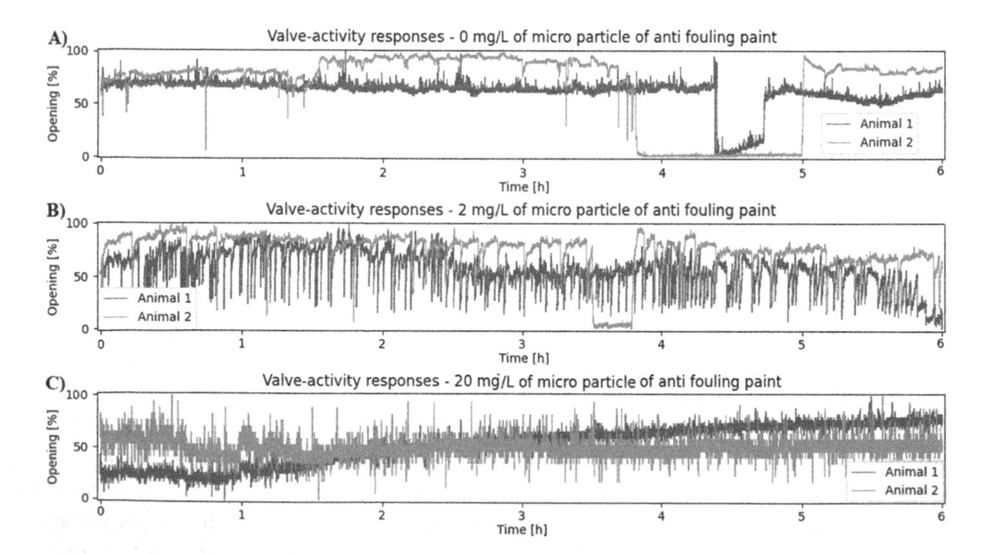

Fig. 9. Examples of valve-activity responses for *Perna perna* mussels exposed to 0 (A), 2 mg/L (B) and 20 mg/L (C) of micro-particles of anti-fouling paint

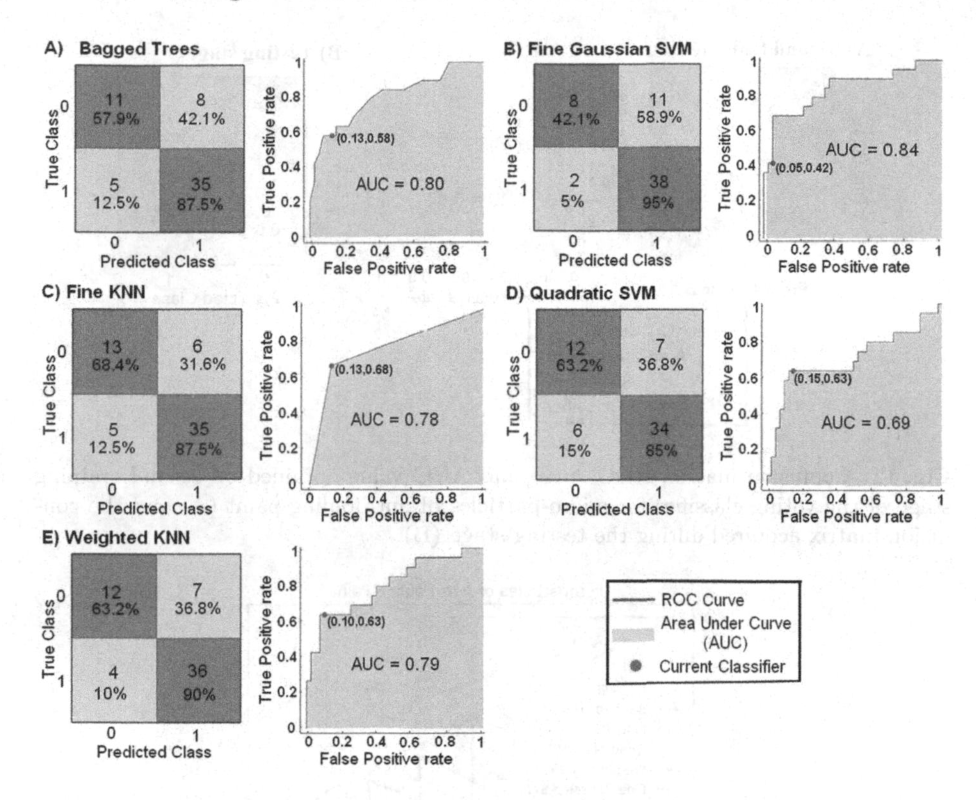

Fig. 10. Confusion matrices, ROC curves and AUC values of the classifiers Bagged Trees (A), Fine Gaussian SVM (B), Fine KNN(C), Quadratic SVM (D) and Weighted KNN (E) obtained during the first training stage of the development of voting classifier for detecting compounds of anti-fouling paints.

was, therefore, the "individual's adaptive nature" which promotes individual animals to differ in their responsiveness to environmental variations [6]. Figure 9 illustrates examples of behaviors of mussels exposed to 0 mg/L, 2 mg/L and 20 mg/L of micro-particles of anti-fouling paints.

Considering '0' as a non-polluted class and '1' as a polluted class, 65% of hourly observations of mussels were employed to train bagged trees, fine gaussian SVM, fine KNN, quadratic SVM and weighted KNN classifiers resulting in 77.9%, 77.9%, 81.3%, 77.9% and 81.3% accuracy, respectively. The acquired AUC values were lower than those obtained for diesel WAF. Fine gaussian model provided the highest AUC (0.84). Figure 10 illustrates AUC values, confusion matrices and ROC curves for all five classifiers.

Second training data set (35% of hourly behavior) was applied to these classifiers in order to obtain predictions to train the voting classifier. Boosted trees model reached the highest accuracy and AUC value (84.4% and 0.62, respectively), as shown in Fig. 11A. Finally, testing data set (15%) was employed to infer the contamination status according to the polluted or non-polluted

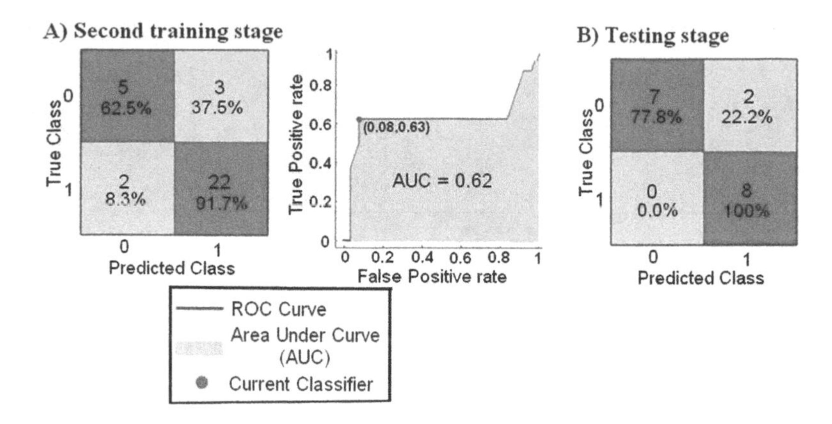

Fig. 11. Confusion matrix, ROC curve and AUC value obtained on second training stage of the voting classifier for micro-particles of anti-fouling paint (A) and the confusion matrix acquired during the testing stage (B).

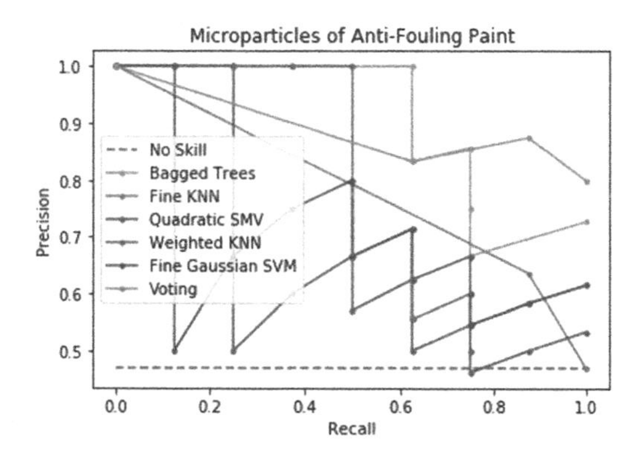

Fig. 12. Precision-Recall curves for Bagged Trees, Fine KNN, Quadratic SVM, Weighted KNN, Fine Gaussian SVM and voting classifiers, respectively

classes. The voting classifier provided an accuracy equal to 88.2% as illustrated in Fig. 11B.

The use of classifiers on predicting micro-particles of anti-fouling paint in aquatic environment has shown decent accuracy after testing. However, first training stage presented high false negative rates on pollutant detection. Moreover, the AUC value obtained on second training stage has not demonstrated a suitability of the classifier. PR-curves were employed to evaluate the testing performance of all investigated classifier models (Fig. 12). F-score and AUC-PR values were also assessed as demonstrated in Table 2.

Although the voting classifier has achieved the second highest AUC-PR value (0.89), it presented the best accuracy (88.9%) and F-score value (0.89). Thus, voting classifier was the most appropriate for detecting toxic compounds released by anti-fouling paints. The lower performance metrics compared to those

Table 2. Accuracy, F-score and AUC-PR values during testing phase for microparticles of anti-fouling paint experiment

Classifier	Accuracy	F-Score	AUC-PR
Bagged Trees	76.5%	0.75	0.91
Fine Gaussian SVM	58.8%	0.74	0.79
Fine KNN	70.6%	0.63	0.66
Quadratic SVM	58.8%	0.63	0.80
Weighted KNN	58.8%	0.63	0.70
Voting	88.9%	0.89	0.89

achieved for detecting diesel WAF may have been caused by the short experiment time. Moreover, there was an intrinsic difficulty on detecting micro-particles of anti-fouling paint since it releases toxic compounds continuously on environment [20]. Hence, bivalves mollusks may be more sharply affected as time passes by. The acquired performance reflected the amount of experimental data as well as the characteristics of the employed exposure medium.

5 Conclusions

This paper proposed the use of a voting classifier systems to construct aquatic pollution biosensors based on hourly behavioral analysis of *Perna perna* mussels. For this purpose bivalves networks were employed during two toxicological experiments using diesel WAF and micro-particles of anti-fouling paints.

The present work demonstrated the feasibility of constructing aquatic pollution biosensors for detecting compounds of anti-fouling paints. Nevertheless, lack of experimental data for a more effective training, ended up leading to less significant results when compared to the biosensor for detecting petrochemical compounds. The achieved results reflected intrinsic difficulty of evaluating the effects of anti-fouling paints, since the toxic compounds were continuously released on the environment. Due to the lack of an equivalent non parametric statistical tool, results obtained could not be used in a comparison. Additionally, the influence of environmental conditions on the experiment was not evaluated, leading to uncertainty regarding its complete reproducibility in non controlled environments. Further studies may contribute in acknowledging the suitability of the developed classifier models in non controlled environments.

The aquatic pollution biosensor elaborated for detecting diesel WAF provided great accuracy when tested for both concentrations of diesel WAF and uncontaminated conditions. It demonstrated the paramount potential of using artificial intelligence methods in the construction of reliable, rapid and cost-effective aquatic pollution biosensors based on the behavior of individual mussels. It was also capable of overcoming the adaptive nature of bivalves mollusks since it was used for inferring the contamination status based on behavior of two animal not used during the training stages.

References

1. Basti, L., Nagai, K., Shimasaki, Y., Oshima, Y., Honjo, T., Segawa, S.: Effects of the toxic dinoflagellate heterocapsa circularisquama on the valve movement behaviour of the manila clam ruditapes philippinarum. Aquaculture **291**(1–2), 41–47 (2009)
2. Canelas, A.d.S.: Evolução da importância econômica da indústria de petróleo e gás natural no Brasil: contribuição a variáveis macroeconômicas. Rio de Janeiro (RJ): Universidade Federal do Rio de Janeiro. Coordenação dos Programas de Pós-Graduação de Engenharia (2007)
3. Comeau, L.A., Babarro, J.M., Longa, A., Padin, X.A.: Valve-gaping behavior of raft-cultivated mussels in the ría de arousa, Spain. Aquaculture Rep. **9**, 68–73 (2018)
4. Cortes, C., Mohri, M.: AUC optimization vs. error rate minimization. In: Advances in Neural Information Processing Systems, pp. 313–320 (2004)
5. Davis, J., Goadrich, M.: The relationship between precision-recall and ROC curves. In: Proceedings of the 23rd International Conference on Machine Learning, pp. 233–240 (2006)
6. Dingemanse, N.J., Kazem, A.J., Réale, D., Wright, J.: Behavioural reaction norms: animal personality meets individual plasticity. Trends Ecol. Evol. **25**(2), 81–89 (2010)
7. Guterres, B.V., Guerreiro, A.G., Sandrini, J.Z., Botelho, S.S.C.: *Perna perna* mussels network as pollution biosensors of oil spills and derivatives (in press)
8. Hartmann, J.T., Beggel, S., Auerswald, K., Stoeckle, B.C., Geist, J.: Establishing mussel behavior as a biomarker in ecotoxicology. Aquat. Toxicol. **170**, 279–288 (2016)
9. Ho, T.K., Hull, J.J., Srihari, S.N.: Decision combination in multiple classifier systems. IEEE Trans. Pattern Anal. Mach. Intell. **16**(1), 66–75 (1994)
10. Kim, U.J., Lee, I.S., Choi, M., Oh, J.E., et al.: Assessment of organotin and tin-free antifouling paints contamination in the Korean coastal area. Mar. Pollut. Bull. **99**(1–2), 157–165 (2015)
11. Liao, C.M., et al.: Valve movement response of the freshwater clam Corbicula fluminea following exposure to waterborne arsenic. Ecotoxicology **18**(5), 567–576 (2009)
12. Markich, S.: Behavioural responses of the tropical freshwater bivalve Velesunio angasi exposed to uranium. Wetland Research in the Wet-Dry Tropics of Australia, pp. 247–257. Supervising Scientist, Canberra (1995)
13. Martins, S.E., Fillmann, G., Lillicrap, A., Thomas, K.V.: Ecotoxicity of organic and organo-metallic antifouling co-biocides and implications for environmental hazard and risk assessments in aquatic ecosystems. Biofouling **34**(1), 34–52 (2018)
14. Martınez, K., Ferrer, I., Barcelo, D.: Part-per-trillion level determination of antifouling pesticides and their byproducts in seawater samples by off-line solid-phase extraction followed by high-performance liquid chromatography-atmospheric pressure chemical ionization mass spectrometry. J. Chromatogr. A **879**(1), 27–37 (2000)
15. Medeiros, H.E., da Gama, B.A.P., Gallerani, G.: Antifouling activity of seaweed extracts from Guarujá, São paulo, Brazil. Braz. J. Oceanogr. **55**(4), 257–264 (2007)
16. Nagai, K., Honjo, T., Go, J., Yamashita, H., Oh, S.J.: Detecting the shellfish killer Heterocapsa circularisquama (Dinophyceae) by measuring bivalve valve activity with a hall element sensor. Aquaculture **255**(1–4), 395–401 (2006)

17. Newton, T.J., Cope, W.G.: 10 Biomarker responses of unionid mussels to environmental contaminants (2007)
18. Pimentel, M., Silva Júnior, F., Santaella, S., Lotufo, L.: O uso de artemia sp. como organismo-teste para avaliação da toxicidade das águas residuárias do beneficiamento da castanha de caju antes e após tratamento em reator biológico experimental. J. Braz. Soc. Ecotoxicol. **6**(1), 15–22 (2011)
19. Redmond, K.J., Berry, M., Pampanin, D.M., Andersen, O.K.: Valve gape behaviour of mussels (Mytilus edulis) exposed to dispersed crude oil as an environmental monitoring endpoint. Mar. Poll. Bull. **117**(1–2), 330–339 (2017)
20. Renner: Ficha de informação de segurança de produto químico, p. 6 (2016)
21. Resgalla Jr., C., de Souza Brasil, E., Salomão, L.C.: The effect of temperature and salinity on the physiological rates of the mussel Perna perna (Linnaeus 1758). Braz. Arch. Biol. Technol. **50**(3), 543–556 (2007)
22. Singer, M., et al.: Standardization of the preparation and quantitation of water-accommodated fractions of petroleum for toxicity testing. Mar. Poll. Bull. **40**(11), 1007–1016 (2000)
23. Martins, T.L., Vargas, V.M.F.: Riscos à biota aquática pelo uso de tintas anti-incrustantes nos cascos de embarcações. Ecotoxicol. Environ. Contam. **8**(1), 1–11 (2013)
24. Tomar, D., Agarwal, S.: A survey on data mining approaches for healthcare. Int. J. Biosci. Biotechnol. **5**(5), 241–266 (2013)
25. Tran, D., Ciret, P., Ciutat, A., Durrieu, G., Massabuau, J.C.: Estimation of potential and limits of bivalve closure response to detect contaminants: application to cadmium. Environ. Toxicol. Chem. **22**(4), 914–920 (2003)
26. Tran, D., Haberkorn, H., Soudant, P., Ciret, P., Massabuau, J.C.: Behavioral responses of Crassostrea gigas exposed to the harmful algae Alexandrium minutum. Aquaculture **298**(3–4), 338–345 (2010)
27. Urbanowicz, R., Moore, J.: Learning classifier systems: a complete introduction, review, and roadmap. J. Artif. Evol. Appl. (2009)

New Adaptive Morphological Geodesic Active Contour Method for Segmentation of Hemorrhagic Stroke in Computed Tomography Image

Aldísio G. Medeiros[1,2], Lucas de O. Santos[1], Roger Moura Sarmento[1,2], Elizângela de Souza Rebouças[1,2], and Pedro P. Rebouças Filho[1,2(✉)]

[1] Laboratório de Processamento Digital de Imagens, Sinais e Computação Aplicada, Instituto Federal de Federal de Educação, Ciência e Tecnologia do Ceará, Fortaleza, Ceará, Brazil
{aldisio.medeiros,ucas.santos}@lapisco.ifce.edu.br,
{rogerms,elizangela.reboucas,pedrosarf}@ifce.edu.br

[2] Programa de Pós-Graduação em Engenharia de Teleinformática, Universidade Federal do Ceará, Fortaleza, Ceará, Brazil

Abstract. This work proposes a new approach to segmentation of hemorrhagic stroke (HS) based on morphological geodesic active contour method, with automatic initialization near to lesion region, without previous training, called Adaptive Morphological Geodesic Active Contour (Ada-MGAC). To evaluate the performance, we used 100 computed tomography images with HS of volunteers. These samples were compared against segmentation methods from specialized literature. A manual segmentation from a medical specialist was considered as the gold standard. The results indicate a competitive potential of Ada-MGAC and method showed a mean convergency time about 3 s, indicating a fast result to medical analysis. Thus, it is possible to conclude that the proposed approach can be used to aid medical diagnosis in the cerebral vascular accident.

Keywords: Hemorrhagic stroke · Computed tomography · Morphological Geodesic Active Contour · Brain image segmentation

1 Introduction

Worldwide, stroke are the second leading cause of death and the third leading cause of disability [15]. According to the World Health Organization and the American Heart Association, in 2010 there were an estimated approximately 17 million strokes. In five years this number has jumped to 80 million people. In 2016 alone, there were 5.5 million deaths worldwide. The number of stroke deaths worldwide increased by 28.2% between 1990 and 2016 [3].

Stroke is caused by an interruption or drastic reduction in the supply of blood and nutrients to a specific region of the brain. It can be classified as ischemic,

R. Cerri and R. C. Prati (Eds.): BRACIS 2020, LNAI 12320, pp. 604–618, 2020.
https://doi.org/10.1007/978-3-030-61380-8_41

when there is an obstruction or abrupt reduction in blood flow, or hemorrhagic that occurs when a blood vessel ruptures, causing cerebral hemorrhage [2]. Of stroke deaths in 2016 worldwide, 2.7 million individuals died from ischemic stroke and 2.8 million died from hemorrhagic stroke [3].

In cases of stroke, rapid diagnosis is of fundamental importance for establishing the course of treatment; the initial assessment of the patient must be systematic, recognizing the type and location of the lesion. For this, computed tomography (CT) exams are a key tool in the diagnosis and definition of treatment, as they are less costly, faster and allow to verify the extent and severity of the injury [34].

Commonly, the interpretation of medical images is performed exclusively by medical specialists. However, Computer-Aided Diagnosis (CAD) systems are useful tools that, in addition to helping with diagnosis, reduce evaluation time and also contribute to the accuracy of the clinical report. A fundamental step for CAD systems is segmentation, which seeks to highlight the region of interest from the original image considering its properties and patterns [25].

In recent years, several studies have been developed focusing on the segmentation of stroke in medical images. These works demonstrate the efficiency of the different segmentation techniques, among them the threshold methods [33], region growth [5], watershed [16], algorithms based on Fuzzy logic [14,34], statistical analysis [24], grouping segmentation methods [13], active contouring technique [5] and contour concept [27]. Other works use machine learning techniques such as Naive-Bayes classifier [12], Random Florest [21], and Support Vector Machine [9].

A new methodology is proposed for automatic detection and segmentation of HS inspired on Geodesic Active Contour (GAC) [6]. Previous work has shown promising results for lung imaging [22] through a morphological perspective, the authors propose an adaptive approach. However, they used principles of diffuse logic to automatic initialization [22]. In this work, the proposed method has a fully automatic initialization near to real lesion border, accelerating the convergency and reduzing the segmentation time. This method is called Adaptive Morphological Geodesic Active Contour (Ada-MGAC). The results of the proposed method are compared with recent approaches in the literature, level set based on analysis of brain radiological densities (LSBRD) [27], Fuzzy C-means (FCM), Watershed [18], Region Growing [29] and the gold standard produced by a medical specialist. In addition to the comparative analysis between the different techniques of segmentation, this study we make available the database produced, aiming to promote future research.

This work is organized as follows: Sect. 2 presents the segmentation techniques evaluated in this study, from Sect. 3 the proposed segmentation methodology is presented. The Sect. 3 presents the proposed evaluation methodology and in Sect. 4 the main characteristics of the database are presented and still the results are discussed based on the evaluation metrics. Finally, the Sect. 5 presents contributions, conclusions and proposals for future studies.

2 Related Works

This section describes the approach developed for the segmentation of hemorrhagic stroke in CT images based on active geodetic contour method. In addition, the other methods used for comparison are briefly explained, that is, the Level Set Based on Analysis of Brain Radiological Densities (LSBRD), Fuzzy c-means (FCM), Watershed and Region Crowing algorithms.

2.1 Level Set Based on Analysis of Brain Radiological Densities

Rebouças *et al.* [27] proposed the Level Set Based on Analysis of Brain Radiological Densities (LSBRD) presenting two improvements over the traditional level set [31] algorithm. The first is to deal with the grayscale input image and determine the optimal windowing under Hounsfield (UH) Units, adopting 80 UH for window width and 40 UH for center level. The second improvement is an initialization optimization, where the zero set level is determined by the analysis of the radiological densities of the brain tissue.

The method proved to be quite stable. In a comparative analysis with other segmentation methods, LSBRD presented results of precision and F-Score superior to methods commonly used in this field and, therefore, it is a promising method to be used in routine clinical diagnostics, which requires less than 6 seconds for highly competent analysis on a normal computer. However, LSBRD uses a circle as the initial form of the set of levels in the stroke region. This initial circle is based on a set of pixels whose intensity belongs to a previously defined scale. Moreover, the pixels considered into the level set zero depend of the intensities scale, disregarding the shape of lesion. This behavior can lead to initialization out of lesion region compromising final segmentation.

2.2 Fuzzy C-Means

Fuzzy c-means (FCM) is a method of clustering which allows one piece of data to belong to two or more clusters. This method (developed by Dunn in 1973 [7] and improved by Bezdek in 1981 [4]) is frequently used in pattern recognition. O FCM is a flexible technique that allows a sample to partially belong to two or more clusters [35]. This characteristic differentiates this method from the others since, in general, each sample can only belong to a specific cluster [23]. The algorithm identifies clusters by means of similarity measures, such as distance, connectivity and intensity. These measures may be chosen according samples characteristics. The FCM presents robust results for applications in medical images [5,34], however, its convergence process is computationally expensive procedure and can be a deterrent to real-time applications.

2.3 Watershed

The Algorithm Watershed (WS) represents the image to be segmented as a topographic surface, in which each pixel corresponds to a position and its intensity, in gray levels, determines its altitude. From the topographic perspective, the different intensities represent "valleys" and "peaks" with different heights [11]. Then, using the concepts of topographic flooding, two types of flooding can occur: water coming from above, as if it had been spilled on the surface; water emerging from below from holes in the lower regions. After the floods, the barriers formed divide the basin line. These dividing lines resulting from the various floods are called hydrographic basins [18]. This method is also applied to brain images [26,30]. However, the watershed method is sensitive to the intensity of the edges of the object of interest, leading to undesired segmentations when the region of interest has a low contrast and not has well defined edges.

2.4 Region Growing

The Region Growth (RG) technique progressivelly groups pixels in regions from a similarity criterion in order to circumvent the area of interest. This process starts from a "seed pixel" and then groups neighboring pixels that have similar characteristics and satisfy the similarity criterion. This process occurs through successive iterations until every connected region is analysed. When there are no more points with the characteristics of the seed connected to that region, another seed is defined that has not been analysed traversed yet, then the process starts again [36]. The desired region is then represented by all the pixels that were accepted during the growth process.

The method is widely used in medical imaging applications [28,30], including to segment brain images. However, the criterion to evaluate the similarity between neighboring pixels depends strongly on the analyzed problem besides the texture characteristics of the image.

2.5 Active Contour Method

Active Contour Method (ACM) is a type of segmentation technique which can be defined as use of energy forces and constraints for segregation of the pixels of interest from the image for further processing and analysis. Initially, Kass *et al.* [17] proposed the method based on deformable models, where the objective is to adapt an initial curve to the shape of the region of interest. The curvature occurs by the forces acting and evolves to the edges of the object. This model is popular in computer vision, it are widely used in applications like object tracking, shape recognition, segmentation, edge detection and stereo matching.

Active contour do not solve the entire problem of finding contours in images, since the method requires knowledge of the desired contour shape beforehand. Rather, they depend on other mechanisms such as interaction with a user, interaction with some higher level image understanding process, or information from image data adjacent in time or space [27]. Thus, aiming to improve the ACM traditional, some approaches were developed introducing new energies [22].

2.6 Traditional Geodesic Active Contour

According to Caselles, Kimmel e Sapiro [6], given a random curve representing a Level Set, geodesic active outline acts minimizing its perimeter through displacement operator based on the average curvature, proposed by Evans and Spruck [8,32].

This curve displacement operator has as variable the time and it is defined by an Partial Differential Equation (PDE), which is defined as:

$$\frac{\partial u}{\partial t} = div\left(\frac{\nabla u}{|\nabla u|}\right) \cdot |\nabla u| \tag{1}$$

In which "u" represents the equivalent curve to the utilized Level Set that will be deformed, considering the Riemanian space geometry.

Let's consider that I is an image, u is the Level Set and v is a parameter that defines the curve's deformation way. Therefore, the equation that defines the curve's evolution is:

$$\frac{\partial u}{\partial t} = g(I)|\nabla u| \left(div\left(\frac{\nabla u}{|\nabla u|}\right) + \nu\right) + \nabla g(I)\nabla u \\ + g(I) \cdot \nu||\nabla u| \tag{2}$$

The function $g(I)$ represents the gradient calculation which we will call here border detector. It is expressed by the following equation:

$$g(I) = \frac{1}{\sqrt{1 + \alpha|\nabla G_\sigma * I|}} \tag{3}$$

$G_\sigma * I$ is a gaussian filter with σ as standard deviation, α is an important hyperparameter because it is a weight factor that guides evolution of curvature at a point p. Its calculation is the division of the image gradient by the image itself. That way, the function values, approximating to the high contrast regions that determine the object's borders.

3 An Adaptive Stroke Segmentation via Morphological Geodesic Active Contour

In this section, we present a new approach based on morphological operators to stroke segmentation, demonstrating its concepts and main equations for its implementation and also its applications in the segmentation of hemorrhagic stroke.

3.1 A Morphological Approach to Geodesic Active Contour

Given that the curve that represents the Level Set can be represented by an PDE, its resolution must be fast and efficient. In that way, Lax [19] proposes solutions for PDEs based on dilatation and erosion morphological operators [10], aiming and numerical approximation close to PDEs solution. The Eqs. 4 e 5 correspond to the dilatation and erosion operations that might be abreviated as *sup* for dilatation and *inf* for erosion.

$$\lim_{h \to 0^+} \frac{D_h u - u}{h} = |\nabla u| \tag{4}$$

$$\lim_{h \to 0^+} \frac{E_h u - u}{h} = -|\nabla u| \tag{5}$$

In that sense, Alvarez *et al.* [1] demonstrates that diferential and morphological operators can be equivalent through the successive dilations and erosions.

Marquez-Neila *et al.* [20] propose the following PDE approximation based on a relation between dilatation and erosion morphological operations.

$$(SI_{\sqrt{h}}) \circ (IS_{\sqrt{h}}) \approx div\left(\frac{\nabla u}{|\nabla u|}\right) \cdot |\nabla u| \tag{6}$$

where SI is IS are respectively $sup - inf$ and $inf - sup$ operator defined by Marquez-Neila *et al.* [20].

In this sense, the proposed methodology is formed by the segmentation stages of the cerebral region (Subsect. 3.2), location of the hemorrhagic region and detection of borders (Subsect. 3.3) and application of the Ada-MGAC for final segmentation (Subsect. 3.4). The Fig. 1 presents a schematic diagram of the proposed methodology and the following subtopics discuss in detail each of the steps.

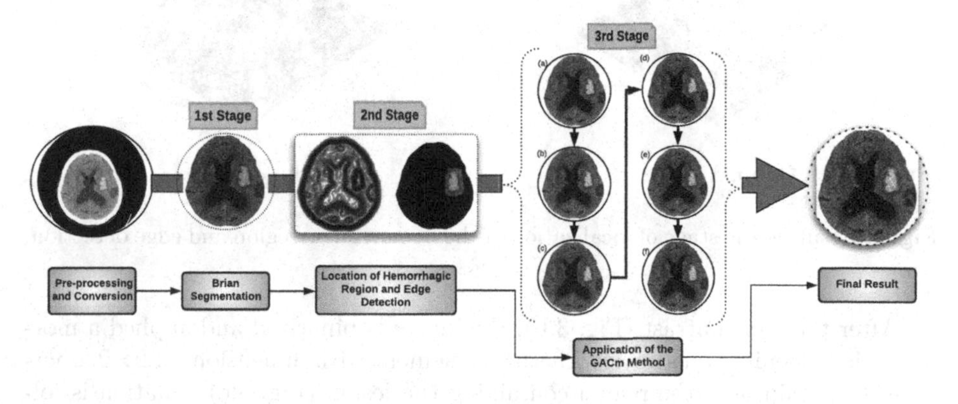

Fig. 1. Flowchart of the proposed methodology for stroke segmentation on CT exams.

3.2 Brain Detection

In the method developed, the Brian Segmentation stage consists of applying a media filter with a 3×3 mask to smooth the image and remove noise. Subsequently, the brain region is segmented using a threshold of the skull bone, followed by the Convex Hull morphological operation of the segmented region. In the segmentation stage of the brain, the objective is to select the region of interest (ROI), that is, only the brain, excluding the other artifacts.

3.3 Location of Hemorrhagic Region and Edge Detection

After the pre-processing previously described, a clip contrast is applied from the mean value of the pixels in the image, according to the expression below:

$$f(x,y) = \begin{cases} 0 \ , f(x,y) < \varphi \\ f(x,y), \ otherwise, \end{cases} \quad (7)$$

where φ was experimentally calculated according average intensities out of lesion region. Therefore, we assume, in this paper, iterating that the pixel intensities, 181 as the critical pixel value, and the values below it do not present injury in its completeness. The patterns, in HU unit, for the stroke region stake out the hemorrhagic injuries in the range 56 HU to 76 HU, [27]. However, as a reduction was applied to the images and as consequence they acquired pixel intensity levels varying from 0 to 255 (8 bits), the mentioned critical point value was the one that better represented the injury, and it was applied to every image. The result of this stage can be seen on Fig. 2.

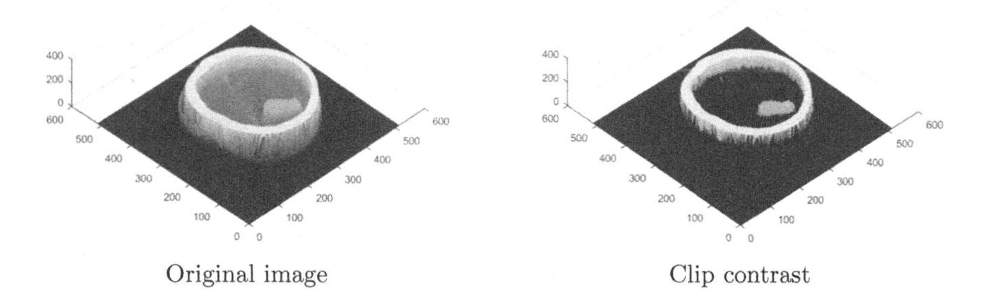

Fig. 2. Result of the stage of localization of the hemorrhagic region and edge detection.

After the clip contrast (Fig. 3.b), the image is binarized and applied a morphological opening, quadratic structuring element with dimensions of 2×2. operation to obtain only the region containing the lesion (Fig. 3.c). Dilation is followed by two erosions. From the resulting image, the largest area component is filtered and after a morphological dilation, quadratic structuring element with dimensions of 3×3. This last operation was applied iteratively, producing the component of Fig. 3.d.

To delimit a region containing the entire lesion, a dilation operation is applied, reducing dark noises in the image. Then the image is filtered in relation to the image of the Fig. 3.c. From the result, finally, a dilation operation is applied, resulting in Fig. 3.f. This image represents the binary initialization component that replaces the Ada-MGAC initial binary level set, representing an initialization sufficiently close to the region of the lesion to be segmented.

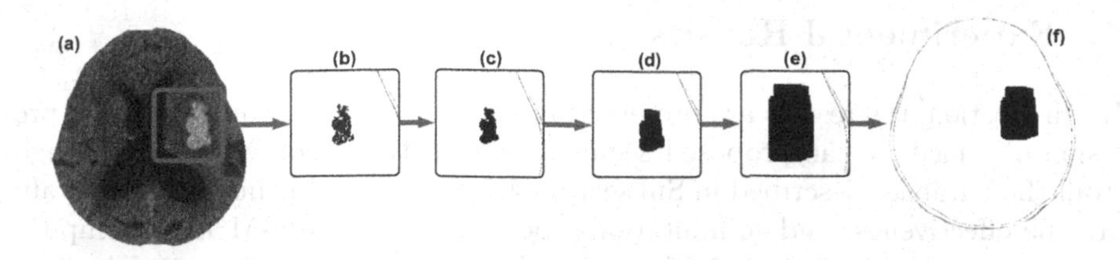

Fig. 3. Detection of the binary component that replaces the traditional level set. Thus, identifying a near area of analysis for the segmentation process of the hemorrhagic region.

3.4 Application of the Ada-MGAC Method

The binary component detected in the previous step is used as the Ada-MGAC initialization level set, represented in the second stage of Fig. 1. Inspired by the concept of the geodesic curve between two points, initially the component is deformed by means of the curvature operator described by Marquez-Neila *et al.* [20]. This operator expandes or reduces the area defined by the binary level set, inducing a curvature at the edges of the binary component, preserving the smoothness of the curve.

The deformation of the curve is guided by the balloon force acting in the normal direction in relation to the curve. The balloon force, in this case, guides the sense of curvature operator for reduces or expandes level set. Thus, the direction of the balloon force has been modified to lead the evolution of the initial contour into the curve itself, the consequence of this procedure is the contraction of the initial level set, reducing its size until the edges of the lesion area are reached.

To detect the edge region, the $g(i)$ function is used to smooth the contours gradient of the objects in the image by means of the gradient with gaussian derivatives. The closer to the edge of the object, the smaller the values of the $g(i)$ function, thus indicating the presence of the edge. We use a threshold to indicate the reduction of the influence of the balloon force, preventing the curve from collapsing on itself, thus stabilizing the contraction of the level set in the region close to the contour of the lesion. Searching for the best combinations for the values of σ and ν, a grid search was performed. The grid search refers to the search for the best parameters by analyzing the results obtained with the execution of the algorithm for a range of parameters. In this work, the parameters were searched for in the following ranges: $\sigma = [5.5, 7.5]$ and $\nu = [0.34, 0.50]$. In this sense, the smoothing parameter σ was used with value 5.5 since the threshold ν was 0.34. The changes that these parameters cause in the processed images are mainly related to their contours. Therefore, the best results enhance the image details more effectively, accelerating the segmentation process and making the image data more sensitive, this gives the method curves closer to the area of the lesion.

4 Experimental Results

In this section, the results are presented in terms of cost and computational precision obtained by the proposed segmentation method. For validation, images from the database described in Subsection 4.1 were used. In this sense, to evaluate the effectiveness and computational performance of Ada-MGAC, computer simulations were performed with the methods described in Sect. 2. The tests were performed on a computer with a 3.40 GHz Intel Core i3 processor, 8 GB of RAM and Windows 10 64-bit operating system. Both classical methods and Ada-MGAC were implemented using the Python 3 language with the OpenCV 3.0 library.

4.1 Images Acquisitions

The tomographic model used to acquire the images was the HiSpeed CT and Dual1 from the General Electric. The equipment was calibrated three months before the exams. In the image acquisition process, the tomographic sections were performed at the base of the axial plane under the following conditions: cut thickness 0,7 mm, field of view 230 mm, electric voltage in the tube 120 Kv, electric current in the tube 80 mA, dimensions 512×512 pixels and windowing of voxel $0{,}585 \times 0{,}585 \times 1{,}5$ mm with a quantification of 16 bits, level of windowing of 40 HU and window amplitude equal to 80 HU. CT images were obtained with the support of Trajano Almeida clinic, and used in previous studies [27,30]. The database is publicly available and can be obtained from the site[1].

4.2 Results and Discussions

In this section, the results obtained by the proposed segmentation method are presented. For validation, 100 randomly selected database images were used, where the hemorrhagic stroke was previously diagnosed and segmented by a specialist. The Adaptive Morphological Geodesic Active Contour (Ada-MGAC) method was compared with four methods commonly used in this domain: Level Set based on analysis of brain radiological densities (LSBRD), Region Growing (RG), Watershed (WS) and Fuzzy C-means (FCM).

Figure 4 presents four examples of segmentation for each of the evaluated methods. The first line corresponds to images of the gold standard, marked manually by the human expert; the second line, presents the result of the proposed method; from the third line are shown the results obtained by the algorithms LSBRD, RG, WS and FCM, respectively.

Initially, a quantitative analysis of the results was performed. Accuracy (ACC), specificity (SPC), True positive rate, (TPR), Positive predictive value (PPV) and F-Score (F_1) values were computed. ACC is the proportion of true pixels (hemorrhage) over the entire population, that is, the percentage of positive and negative pixels classified correctly. The SPC corresponds to the percentage of

[1] http://lapisco.ifce.edu.br.

negative pixels (without hemorrhage) correctly segmented over the total number of negative pixels, the PPV is the ratio between the pixels correctly segmented and all the pixels belonging to the hemorrhagic region, the TPR measures the proportion of the pixels correctly segmented among all true information and the F-Score is the harmonic average between PPV and TPR. Table 2 shows the results in relation to ACC, F1, PPV, TPR and SPC with their respective standard deviation values (SD).

Table 1. Accuracy (ACC) and F-Score results (F_1) with their respective SD values.

Algorithm	$ACC(\%)$	$F_1(\%)$	$PPV(\%)$	$TPR(\%)$	$SPC(\%)$
Ada-MGAC	**99.81 ± 00.10**	**99.90 ± 0.40**	**99.88 ± 00.09**	**93.46 ± 04.32**	**99.93 ± 00.05**
LSBRD	99.72 ± 00.23	88.88 ± 06.47	99.75 ± 00.25	99.96 ± 00.05	83.03 ± 11.13
RG	98.67 ± 04.28	87.36 ± 08.36	99.00 ± 04.36	99.66 ± 00.23	92.84 ± 21.88
WS	99.69 ± 00.22	86.67 ± 07.79	99.84 ± 00.19	99.85 ± 00.11	87.32 ± 13.38
FCM	99.19 ± 02.97	–	99.32 ± 02.99	99.86 ± 00.10	90.53 ± 21.31

According to Table 1 it is possible observes that Ada-MGAC has the highest PPV index with a low SD value, reaching 99.88 ± 00.09, indicating that the method has the ability to produce consistent results independently of the region presented under the same conditions, i.e, the method is able to correctly identify the lesion region, even submitted to different samples with lesioned and healthy regions.

The LSBRD method reached an ACC with 99.72 ± 00.23, for the PD presented, it is believed that this result stems from the restrictions imposed by the LSBRD analysis window in HU units, and this SD can indicate cases not represented by the configured scale. In addition, the RG method presented the highest instability rates, reached ACC values with 98.67 ± 04.28, the high SD value indicates instability of the method when subjected to different types of image. On the other hand, we observed that Ada-MGAC obtained the best ACC result, reaching 99.81 ± 00.10, with the lowest SD value.

This index shows how low the dispersion of results is in relation to the expected segmentation. Although it has been submitted to different samples, the low ACC variation also indicates greater stability of the method, since the curve evolution does not parameterize predefined scales of intensities, thus covering a scale of pixels according to the characteristics of each image.

The ACC indices are also confirmed by the high value presented by the measure F_1, where the proposed method exceeds the LSBR, which presented the second best result, with a difference of at least 11.02%. Thus indicating that the proposed approach presents a high precision and sensitivity, in addition to the high performance of the method in separating pixels representing the lesion from those representing unwanted regions. This measure also showed the lowest DP indices, again indicating the stability of the method with samples with heterogeneous characteristics.

Another important feature is the method ability in differentiating pixels as true negatives. In this sense, the SPC metric showed the highest indexes for the

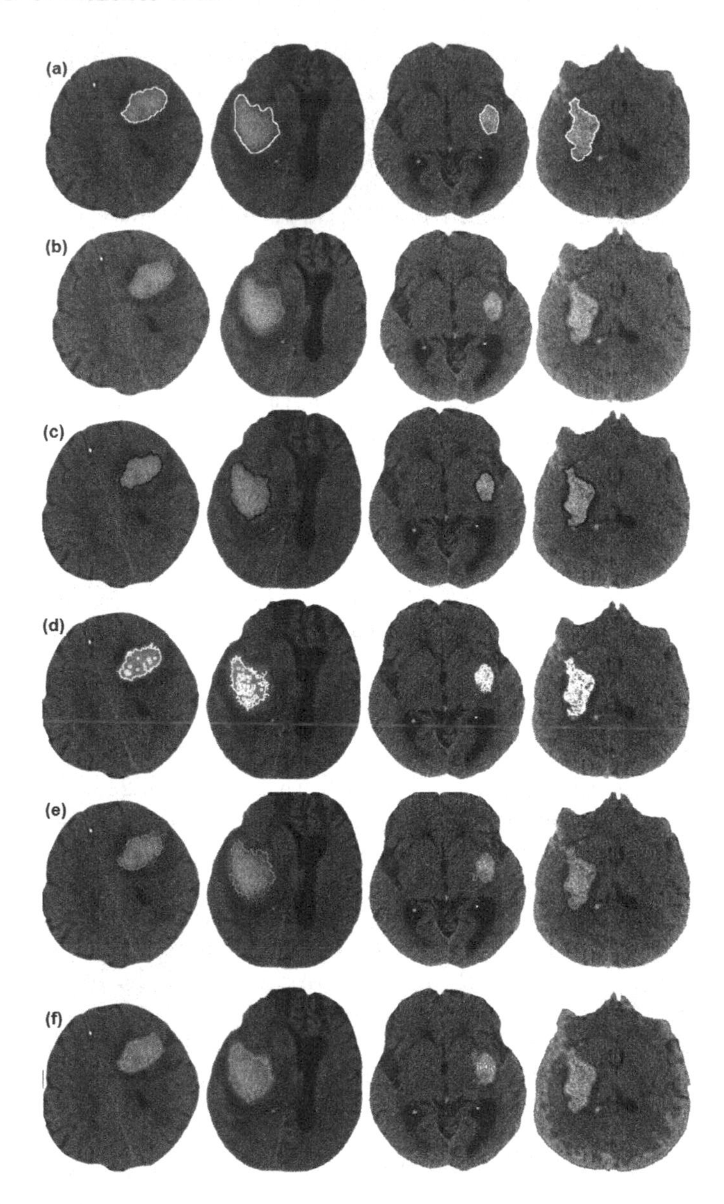

Fig. 4. Examples of segmentation of four HS samples obtained by the methods evaluated. In (a) gold standard, (b) Proposed method, (c) LSBRD, (d) RG, (e) WS e (f) FCM.

proposed solution with 99.93 ± 00.05, whereas the RG method obtained the second best result reaching 92.84 ± 21.88. It is believed that the high performance of the Ada-MGAC is due to its initialization near the lesion region, reducing the analysis process to true regions, avoiding the inappropriate classification of healthy regions, thus showing the independence of parameters previously defined as similarity between pixels adjacent, a feature present in the LSBRD and WS methods, which presented lower results with 83.03 ± 11.13 and 87.32 ± 13.38, respectively.

In summary, the proposed solution is 0.13% and 16.9% higher than the LSBRD in the PPV and SPC metrics, respectively. Moreover, the new aproach is 7.09% better than the RG in the SPC metric, which presented the highest instability indices when evaluated under the PPV metric.

For a complete analysis of the similarity between the contours obtained by the algorithms and lesion maps identified by the human expert, in this work the results were also submitted to a similarity evaluation. We considering the indexes: MCC coefficient, similarity index DSC - since the DSC can be insensitive to the differences between lesion maps in overlapping cases (DSC $\geq$ 0.8), the JSC coefficient was calculated, which is more sensitive the differences between overlapping regions. The results of these indices are presented in Table 2.

Table 2. Results from the Matthews Correlation Coefficient (MCC), Similarity Coefficient (DSC) and Jaccard Similarity Coefficient (JSC).

Algorithm	MCC(%)	DSC(%)	JSC(%)
Ada-MGAC	**91.35 ± 03.86**	**91.28 ± 04.05**	**84.21 ± 06.62**
LSBRD	89.27 ± 05.78	88.85 ± 06.51	80.50 ± 09.79
RG	80.58 ± 16.90	79.26 ± 18.30	68.23 ± 17.20
WS	86.87 ± 07.41	86.63 ± 07.84	77.17 ± 11.20
FCM	88.00 ± 16.80	87.41 ± 18.50	80.77 ± 19.20

According to the data in the Table 2, it can be seen that the Ada-MGAC method achieved the highest MCC coefficients. With 91.35 ± 03.86, exceeding the LSBRD method, with 89.27 ± 05.78 and FCM, with 88.00 ± 16.80. The method showed the best stability for the different samples, with the lowest standard deviation. Considering the DSC coefficient, the proposed solution, however, exceeded all other methods reaching DSC values above 91%, obtaining 91.28 ± 04.05, and again with a low SD value. Regarding the JSC coefficient, although the Ada-MGAC solution reached 84.21 ± 06.62, this value is higher than all other evaluated methods with lowest DP. In this way, the proposed solution presents the best stability also in relation to similarity indexes, even for heterogeneous samples.

Finally, a last relevant analysis is in relation to the performance of the method regarding the time needed to reach the convergence of segmentation. This stage is important because the stroke is second higher death cause in the world [15], therefore the treatment efficiency is linearly related to the necessary time to detect the hemorrhagic injury. That is, the faster the anomaly is detected, the faster the diagnose and treatment will be applied. The Table 3 shows the execution times of the methods for a dataset with 100 samples. The average time of the Ada-MGAC is slightly greater than to the LSBRD. Considering the variation of the DP, in the worst case, the Ada-MGAC remains the second best in the average execution time.

Table 3. Time of convergence of the evaluated methods.

Algorithm	*Time(s)*
Ada-MGAC	**2.20 ± 0.21**
LSBRD	1.76 ± 0.29
RG	3.10 ± 1.70
WS	4.80 ± 0.62
FCM	8.69 ± 3.16

5 Conclusions and Future Work

In this work, a new active geodetic contour method was proposed using a morphological approach with an automatic initiation very close to the lesioned region, called Ada-MGAC. The new method was evaluated in the segmentation of hemorrhagic strokes in cranial CT images. The results reveal that the proposed solution presents promising results for the problem of HS segmentation, presenting considerable robustness regarding the quantitative and similarity indices evaluated. The analysis methodology showed the efficiency of the method in the identification of hemorrhage in front of the segmentation algorithms already proposed in the literature. In addition, the method is stable regardless of the size, shape or position of the lesion in the image, a characteristic shown by the low SD value observed in the experiments.

Showing performance equivalent to or higher than the other methods evaluated. The algorithm also has the potential to integrate systems of assistance to the medical diagnosis, contributing to the reduction of the time of analysis and medical diagnosis, besides fomenting the precision of the process of prognosis. Moreover, the method was designed using basic operations of mathematical morphology, offering a simple, stable and easy-to-implement segmentation mechanism. The potential of the proposed solution against others methods is its independence of pre-defined scales. The Ada-MGAC is based on the image gradient and the displacement imposed by balloon force, both are proportional to the characteristics of the image, which gives the method an adaptive character.

As future work, a dataset with a higher quality of images will be evaluated for a new evaluation of the proposed method. Besides, we aim to test this border for the segmentation of other pathologies in computed tomography image exams.

References

1. Alvarez, L., Guichard, F., Lions, P.L., Morel, J.M.: Axioms and fundamental equations of image processing. Arch. Rat. Mech. Anal. **123**(3), 199–257 (1993)
2. Association, A.H., Association, A.S.: What are the types of stroke? December 2018. https://bit.ly/2IgoIGn. Accessed 28 Dec 2018
3. Benjamin, E.J., et al.: Heart disease and stroke statistics-2019 update a report from the American heart association. Circulation **139**, E56–E528 (2019)
4. Bezdek, J.C.: Pattern Recognition with Fuzzy Objective Function Algorithms. Springer, Heidelberg (2013). https://doi.org/10.1007/978-1-4757-0450-1
5. Bhadauria, N., Bist, M., Patel, R., Bhadauria, H.: Performance evaluation of segmentation methods for brain CT images based hemorrhage detection. In: 2015 2nd International Conference on Computing for Sustainable Global Development (INDIACom), pp. 1955–1959. IEEE (2015)
6. Caselles, V., Kimmel, R., Sapiro, G.: Geodesic active contours. In: Proceedings of IEEE International Conference on Computer Vision, pp. 694–699. IEEE (1995)
7. Dunn, J.C.: A fuzzy relative of the ISODATA process and its use in detecting compact well-separated clusters (1973)
8. Evans, L.C., Spruck, J.: Motion of level sets by mean curvature I (1992)
9. Forkert, N.D., Verleger, T., Cheng, B., Thomalla, G., Hilgetag, C.C., Fiehler, J.: Multiclass support vector machine-based lesion mapping predicts functional outcome in ischemic stroke patients. PLoS One **10**(6), e0129569 (2015)
10. Gonzalez, R.C.: Digital Image Processing. Pearson Education India, London (2009)
11. Grau, V., Mewes, A., Alcaniz, M., Kikinis, R., Warfield, S.K.: Improved watershed transform for medical image segmentation using prior information. IEEE Trans. Med. Imaging **23**(4), 447–458 (2004)
12. Griffis, J.C., Allendorfer, J.B., Szaflarski, J.P.: Voxel-based Gaussian Naïve Bayes classification of ischemic stroke lesions in individual t1-weighted MRI scans. J. Neurosci. Methods **257**, 97–108 (2016)
13. de Haan, B., Clas, P., Juenger, H., Wilke, M., Karnath, H.O.: Fast semi-automated lesion demarcation in stroke. NeuroImage Clin. **9**, 69–74 (2015)
14. Jayaram, P., Menaka, R.: An experimental study of Stockwell transform-based feature extraction method for ischemic stroke detection. Int. J. Biomed. Eng. Technol. **21**(1), 40–48 (2016)
15. Johnson, W., Onuma, O., Owolabi, M., Sachdev, S.: Stroke: a global response is needed. Bull. World Health Org. **94**(9), 634 (2016)
16. Karthik, R., Menaka, R.: A multi-scale approach for detection of ischemic stroke from brain MR images using discrete curvelet transformation. Measurement **100**, 223–232 (2017)
17. Kass, M., Witkin, A., Terzopoulos, D.: Snakes: active contour models. Int. J. Comput. Vis. **1**, 321–331 (1987)
18. Körbes, A., Lotufo, R.: Analise de algoritmos da transformada watershed. In: 17th International Conference on Systems, Signals and Image Processing (2010)
19. Lax, P.: Numerical solution of partial differential equations. Am. Math. Monthly **72**(sup2), 74–84 (1965)
20. Marquez-Neila, P., Baumela, L., Alvarez, L.: A morphological approach to curvature-based evolution of curves and surfaces. IEEE Trans. Pattern Anal. Mach. Intell. **36**(1), 2–17 (2013)
21. McKinley, R., et al.: Fully automated stroke tissue estimation using random forest classifiers (faster). J. Cereb. Blood Flow Metab. **37**(8), 2728–2741 (2017)

22. Medeiros, A.G., et al.: A new fast morphological geodesic active contour method for lung CT image segmentation. Measurement **148**, 106687 (2019)
23. Ng, C.R., Than, J.C., Noor, N.M., Rijal, O.M.: Double segmentation method for brain region using FCM and graph cut for CT scan images. In: 2015 IEEE International Conference on Signal and Image Processing Applications (ICSIPA), pp. 443–446. IEEE (2015)
24. Pustina, D., Coslett, H., Turkeltaub, P.E., Tustison, N., Schwartz, M.F., Avants, B.: Automated segmentation of chronic stroke lesions using LINDA: lesion identification with neighborhood data analysis. Hum. Brain Mapp. **37**(4), 1405–1421 (2016)
25. Rajini, N.H., Bhavani, R.: Computer aided detection of ischemic stroke using segmentation and texture features. Measurement **46**(6), 1865–1874 (2013)
26. Rajinikanth, V., Thanaraj, K.P., Satapathy, S.C., Fernandes, S.L., Dey, N.: Shannon's entropy and watershed algorithm based technique to inspect ischemic stroke wound. In: Satapathy, S., Bhateja, V., Das, S. (eds.) Smart intelligent computing and applications. Smart Innovation, Systems and Technologies, vol. 105, pp. 23–31. Springer, Heidelberg (2019). https://doi.org/10.1007/978-981-13-1927-3_3
27. Rebouças, E.d.S., Braga, A.M., Sarmento, R.M., Marques, R.C., Rebouças Filho, P.P.: Level set based on brain radiological densities for stroke segmentation in CT images. In: 2017 IEEE 30th International Symposium on Computer-Based Medical Systems (CBMS), pp. 391–396. IEEE (2017)
28. Saad, M., Abu-Bakar, S., Muda, S., Mokji, M., Abdullah, A.: Fully automated region growing segmentation of brain lesion in diffusion-weighted MRI. IAENG Int. J. Comput. Sci. **39**(2), 155–164 (2012)
29. Salman, N.H., Ghafour, B.M., Hadi, G.M.: Medical image segmentation based on edge detection techniques. Adv. Image Video Process. **3**(2), 1 (2015)
30. Sarmento, R.M., Pereira, R.F., Coimbra, P., Macedo, A., Rebouças Filho, P.P.: Segmentação de acidente vascular cerebral em imagens de tomografia computadorizada: Um estudo comparativo. VII Simpósio de Instrumentação e Imagens Médicas (SIIM 2015). Campinas: UNICAMP (2015)
31. Sethian, J.A.: Level Set Methods and Fast Marching Methods: Evolving Interfaces in Computational Geometry, Fluid Mechanics, Computer Vision, and Materials Science, vol. 3. Cambridge University Press, Cambridge (1999)
32. Spruck, L.E.J.: Motion of level sets by mean curvature I. J. Differ. Geom. **33**, 635–681 (1991)
33. Sun, M., Hu, R., Yu, H., Zhao, B., Ren, H.: Intracranial hemorrhage detection by 3d voxel segmentation on brain CT images. In: 2015 International Conference on Wireless Communications & Signal Processing (WCSP), pp. 1–5. IEEE (2015)
34. Yahiaoui, A.F.Z., Bessaid, A.: Segmentation of ischemic stroke area from CT brain images. In: International Symposium on Signal, Image, Video and Communications (ISIVC), pp. 13–17 (2016)
35. Yang, Y., Huang, S.: Image segmentation by fuzzy c-means clustering algorithm with a novel penalty term. Comput. Inf. **26**(1), 17–31 (2012)
36. Zucker, S.W.: Region growing: childhood and adolescence. Comput. Graph. Image Process. **5**, 382–399 (1976)

Parallel Monte Carlo Tree Search in General Video Game Playing

Luis G. S. Centeleghe, William R. Fröhlich, and Sandro J. Rigo(✉)

Applied Computing Graduate Program, University of Vale do Rio dos Sinos (UNISINOS), São Leopoldo, Brazil
lg.simioni@outlook.com, william_r_f@hotmail.com, rigo@unisinos.b

Abstract. Monte Carlo Tree Search (MCTS) parallelization is one of the many possible enhancements for MCTS algorithms. However, no work has been done on evaluating these methods in the rather new area of General Video Game Playing (GVGP), an area that challenges the creation of agents that can play any videogame even without prior knowledge about the video game they are going to play. To address this gap, this paper proposes the implementation and evaluation of the three main MCTS parallelization methods as agents of the General Video Game AI framework, a popular framework for GVGP agents evaluation. This paper is not focused on comparing the parallel MCTS agents to other existing GVGP agents, but rather on exploring how the MCTS parallelization methods compare between themselves. This paper also presents a testing methodology for evaluating these agents, which is based on a set of three experiments focused on different aspects of the parallel MCTS algorithms. In these experiments, the overall best results were achieved by the root parallelization method using the sum merging technique and the UCT's sigma value of $\sqrt{2}$.

Keywords: Parallel Monte Carlo Tree Search · General video game AI

1 Introduction

Monte Carlo Tree Search (MCTS) is considered the state-of-the-art algorithm for game tree searching in scenarios where no proper evaluation function exists for intermediate game states, making it very suitable for games such as Hex, Go or games where the domain is unknown for the playing agents, such as the ones used for General Game Playing and General Video Game Playing [1].

Many enhancements for MCTS have been proposed since it was first introduced by [6] in 2006. Among these enhancements, we have the MCTS parallelization, which was proposed in 2008 by [3] through two different approaches, called Leaf Parallelization and Root Parallelization. In the same year, [5] introduced a third approach called Tree Parallelization. Together, these three approaches are considered the main methods for MCTS parallelization [1]. Since these parallel approaches were first presented, many researchers have been evaluating them using a variety of games, such as Reversi [13], Hex [10], Mango [5], and also General Game Playing [14].

R. Cerri and R. C. Prati (Eds.): BRACIS 2020, LNAI 12320, pp. 619–633, 2020.
https://doi.org/10.1007/978-3-030-61380-8_42

However, according to the best of our knowledge, no work has been done on evaluating how these approaches perform in the rather new area of General Video Game Playing (GVGP), an area that challenges the creation of agents that are able to play any video game even without prior knowledge about the video game they are going to play, having to infer at runtime how to play it, thus making it a very suitable candidate for parallel MCTS algorithms.

Due to this lack of research on parallel MCTS for GVGP, we present in this paper three parallel GVGP agents and an evaluation of their performance. Each of the agents implements one of the three main parallel MCTS approaches: Leaf Parallelization, Root Parallelization and Tree Parallelization. To evaluate these agents we created them to be compatible with the Single Player Planning Track of the General Video Game AI framework (GVG-AI), a framework that provides an environment for creation and evaluation of GVGP agents, and is one of the most important projects created for GVGP research. The Single Player Planning Track provides more than 115 different games, what allowed us to observe how the agents behave when performing in many different environments [11].

Our work was based on the Upper Confidence Bound Applied for Trees (UCT) implementation of MCTS algorithms [1].The evaluation of the agents was performed using a set of three different experiments: the first one is a general performance analysis, which is focused on evaluating how the agents compare to each other in terms of performance, and how they compare to a synchronous MCTS agent. The second experiment is focused on comparing merging techniques for root parallelization. And the third experiment is focused on analyzing the impact of the UCT's sigma constant in root parallelization. All these experiments were executed using a computer equipped with two Intel Xeon E5-2620v4 processors, which allowed us to run tests using up to 32 hyper-threads. The results of these experiments are the main contribution of this paper, since they are the first results of parallel MCTS applied to General Video Game Playing. In these experiments, the root parallelization method using the sum merging technique and the UCT's sigma value of $\sqrt{2}$ achieved the overall best results. However, there were scenarios where other methods and parameters performed better. This paper describes and discusses these scenarios and the reasons we believe some agents performed better in them.

The remainder of this paper is structured in the following way: Sect. 2 analyses and discusses the related work. Section 3 describes our approach for parallel MCTS in GVGP and the parallel agents we implemented. Section 4 describes the experiments we performed to evaluate these agents, and Sect. 5 presents the results and discussion of these experiments. Finally, Sect. 6 concludes the paper.

2 Related Works

In this section we present works that focused on evaluating different parallel MCTS methods. Coulm [6] used 24 cores of two Intel Xeon E5-2596v2 and 61 cores of an Intel Xeon Phi 7120P co-processor to evaluate two of the main parallel MCTS methods: root parallelization and tree parallelization. They evaluated these algorithms based on a custom implementation of the game Hex and using two different speedup-measures: playout speedup and strength speedup. The work done by [5] Parallel Monte Carlo Tree Search in General Video Game Playing is of great interest for this work because they evaluated an algorithm developed for General Game Playing, which makes it closely related to the General Video Game Player presented in this paper. For their experiments, the work used eight different board games provided by the General Game Playing framework and a parallel MCTS implementation based on root parallelization. Even though they tested only with root parallelization, they tested it with four different merging techniques: Sum, Sum10, Best and Raw.

Togelious [7] discuss in their work which factors affect the scalability of the MCTS algorithm when running on multiple CPUs/GPUs using root parallelization. For their tests, they used the games Reversi and Same Game running in the Japanese supercomputer TSUBAME, what allowed them to execute experiments with as many as 1024 CPUs and 256 GPUs (each one able to run 1344 threads). Besides from only testing with different numbers of threads, the authors also investigated on how changing the constant C (sigma) of the UCT formula and the problem size aspects the performance and scalability of the algorithm. In this work, Cazenave [3] introduce the third major known parallelization method for MCTS, the tree parallelization method. They then used the games Mango and Go to test the performance of this new method, and how it compares to the two previous major known methods: leaf parallelization and root parallelization, which were introduced by [2]. Levine [8] introduced a new MCTS parallelization method called Limited Root-Tree Parallelization. This method combines root parallelization with tree parallelization by running multiple distinct parallel trees in different machines (aspect from root parallelization) and using tree parallelization within each machine.

The related works were evaluated according to parallelization methods; different games used for testing; whether general game playing or general video game playing was used for evaluating the algorithms. Only two works evaluated parallel MCTS approaches using General Game Playing, and none used General Video Game Playing. This lack of parallel MCTS implementations for GVGP consists in a research gap. Therefore, implementing and testing parallel MCTS approaches using GVGP can contribute both to the General Video Game Playing research and to the Parallel Monte Carlo Tree Search research.

Therefore, the contribution from this paper to the GVGP research is regarded to the fact that it would be the first GVGP player to use a parallel MCTS approach, introducing a new kind of agent to the area. The main contribution to the parallel MCTS research comes from the fact that the General Video Game Playing framework has more than a hundred different games that can be used for testing parallel MCTS approaches, making it possible to understand even further how these approaches perform in many different environments.

3 Parallel MCTS in General Video Game Playing

Considering the lack of research on Parallel Monte Carlo Tree Search approaches for General Video Game Playing, we decided to implement and evaluate four different GVGP agents: three parallel agents, each one using one of the three main parallel MCTS approaches (Leaf, Root, and Tree Parallelization), and one sequential agent used for performance comparison. These agents were created to be compatible with the Single Player Planning Track of the General Video Game AI Framework [12]. The framework served as our main tool for evaluating the agents. The Single Player Planning Track provides more than 115 different games, which allowed us to observe how the agents' performance and behavior is impacted by many different environments.

Algorithm 1 Core MCTS algorithm used by the agents

```
1:  function SEARCH(s_0)                                              ▷ s_0: initial state
2:      v_0 ← new node with state s_0                                 ▷ v_0: root node
3:      while there is computational budget left do
4:          v_s ← TREEPOLICY(v_0)                                     ▷ v_s: selected node
5:          r ← DEFAULTPOLICY(s(v_s))               ▷ r: reward       ▷ s(v): state of v
6:          BACKUP(v_s, r)
7:      return a(SELECTCHILDWITHMOSTVISITS(v_0))                      ▷ a(v): action that led to v
8:  function TREEPOLICY(v)
9:      while v is non-terminal do
10:         if v is not fully expanded then
11:             return EXPAND(v)
12:         else
13:             v ← SELECTBESTCHILD(v)
14:     return v
15: function EXPAND(v)
16:     a ← next untried action available from v
17:     s' ← f(s(v), a)                    ▷ f(s, a): new state of s when taking action a
18:     v' ← new node with s(v') = s' and a(v') = a
19:     add v' as new child of v
20:     return v'
21: function SELECTBESTCHILD(v)
22:     return argmax_{v' ∈ children of v} { Q(v')/N(v') + C sqrt(2 ln N(v) / N(v')) }   ▷ UCT formula
23: function DEFAULTPOLICY(s)
24:     s' ← RUNRANDOMPLAYOUT(s)
25:     return EVALUATE(s')
26: function RUNRANDOMPLAYOUT(s)
27:     s' ← s
28:     while s' is non-terminal and max depth is not reached do
29:         a ← select available action from s' uniformly at random
30:         s' ← f(s', a)
31:     return s'
32: function EVALUATE(s)
33:     if s is terminal and represents a win then
34:         return huge positive value
35:     else if s is terminal and represents a loss then
36:         return huge negative value
37:     else
38:         return current game score
39: function BACKUP(v, r)
40:     while v is not null do
41:         N(v) ← N(v) + 1                                           ▷ N(v): number of visits to v
42:         Q(v) ← Q(v) + r                                           ▷ Q(v): total reward of v
43:         v ← parentOf(v)
```

To conform with the GVAI framework, we implemented the agents using the Java language[1]. Each of the parallel agents uses a different parallel MCTS approach; however, all of them share a core MCTS algorithm, which is slightly modified to accommodate these

[1] Implementations available at https://github.com/LCenteleghe/Parallel-MCTS-GVGP-Agents/.

different parallel MCTS approaches. This core algorithm was based on the algorithm described by [4] and is summarized in pseudocode in Algorithm 1.

In this algorithm, the selection and expansion phases are mixed in a single function called *TreePolicy*, which uses the UCT formula for balancing exploration and exploitation during the search. The value used for the C (sigma) constant of the UCT formula [1] is specified for each of the experiments we performed. The simulation phase is done by the *DefaultPolicy* function, which works by taking random actions until a terminal state or a max depth is reached. If a terminal state is reached, the reward is calculated based on whether it represents a win or a loss, otherwise, the current game score is used as the reward. The backpropagation phase is done in a pretty straightforward way by the backup function, which updates the number of visits and total reward of all the nodes between the last selected node and the root node. The sequential agent uses this algorithm as is, while the parallel agents use this algorithm with small modifications, which are described in the next sections.

The difference between the Leaf Parallelization agent's algorithm and the core algorithm (Algorithm 1) resides in the *DefaultPolicy* function (defined in line 23 of Algorithm 1). Instead of running only a single simulation, it runs many parallel simulations and then aggregates the rewards of all these simulations by summing them. This version of the *DefaultPolicy* is shown in Algorithm 2.

Algorithm 2 DefaultPolicy used by the Leaf Parallelization Agent

1: **function** DEFAULTPOLICY(s)
2: $R \leftarrow \{\}$ ▷ R: set of rewards
3: **for each** available thread **do**
4: $s' \leftarrow$ RUNRANDOMPLAYOUT(s) } in parallel
5: $r \leftarrow$ EVALUATE(s') } in parallel
6: $R = R \cup \{r\}$ } in parallel
7: **return** $\sum_{r \in R} r$

The Root Parallelization agent's algorithm differs from the core algorithm (Algorithm 1) on the way it implements the Search function (defined in line 1 of Algorithm 1). Instead of building only one single tree, this agent builds many independent parallel trees, then after all the computational budget is over it merges all the tree's root's children into one single tree (as described in Algorithm 3).

Algorithm 3 Search method used by the Root Parallelization Agent

1: **function** SEARCH(s_0) ▷ s_0: initial state
2: $T \leftarrow \{\}$ ▷ T: set of trees' roots
3: **for each** available thread **do**
4: $v_0 \leftarrow$ new node with state s_0 ▷ v_0: root node } in parallel
5: **while** there is computational budget left **do** } in parallel
6: $v_s \leftarrow$ TREEPOLICY(v_0) } in parallel
7: $r \leftarrow$ DEFAULTPOLICY($s(v_s)$) } in parallel
8: BACKUP(v_s, r) } in parallel
9: $T = T \cup \{v_0\}$ } in parallel
10: $v' \leftarrow$ MERGEALL(T) ▷ v': root of all trees merged
11: **return** a(SELECTCHILDWITHMOSTVISITS(v')) ▷ $a(v)$: *action* that led to v

Many techniques might be used to execute the tree merging. In this work, we experimented with three different techniques: Sum, Best, and Raw. These techniques are based on the ones experimented by Méhat and Cazenave [9] and Swiechowski and Mandziuk

[14], but for this work they were slightly modified in order to work properly with the GVGAI framework. The Best technique creates a tree with nodes containing the max ("best") value found for the total reward (Q(v)) and the total number of visits (N(v)) between all parallel trees. This technique consequently makes the search algorithm for root parallelization (Algorithm 3) to select the best node between all nodes of all parallel tree. Algorithm 4 describes this technique.

Algorithm 4 *Best* technique for Tree Merging

1: **function** MERGEALL(T) ▷ T: set of trees' roots
2: $v'_r \leftarrow$ new node with *no state* ▷ v'_r: root node of the new tree to be created
3: $V_c \leftarrow \bigcup_{v \in T} childrenOf(v)$ ▷ V_c: first level children of nodes in T
4: **for each** set V_a of nodes from V_c with same source action a **do**
5: $mn \leftarrow \max_{v \in V_a} N(v)$, $mq \leftarrow \max_{v \in V_a} Q(v)$
6: $v' \leftarrow$ new node with $N(v') = mn$, $Q(v') = mq$, and $a(v') = a$
7: add v' as new child of v'_r
8: **return** v'_r

The Sum technique consists of summing the total reward and the total number of visits to a node weighted by the total number of simulations done by the tree in comparison to the total number of simulations done by all the parallel trees, thus raising the significance of the trees which were able to execute more simulations. The pseudocode for this merging technique is defined in Algorithm 5.

Algorithm 5 *Sum* technique for Tree Merging

1: **function** MERGEALL(T) ▷ T: set of trees' roots
2: $v'_r \leftarrow$ new node with *no state* ▷ v'_r: root node of the new tree to be created
3: $t_s \leftarrow \sum_{v \in T} N(v)$ ▷ t_s: total number of simulations
4: $V_c \leftarrow \bigcup_{v \in T} childrenOf(v)$ ▷ V_c: first level children of nodes in T
5: **for each** set V_a of nodes from V_c with same source action a **do**
6: $wn \leftarrow \sum_{v \in V_a} N(v)\frac{N(parentOf(v))}{t_s}$ ▷ wn: weighted number of visits sum
7: $wq \leftarrow \sum_{v \in V_a} Q(v)\frac{N(parentOf(v))}{t_s}$ ▷ wq: weighted total reward sum
8: $v' \leftarrow$ new node with $N(v') = wn$, $Q(v') = wq$, and $a(v') = a$
9: add v' as new child of v'_r
10: **return** v'_r

In distinction to the Sum technique, the Raw technique calculates the average total reward and the average number of visits of the nodes without weighting them by the number of simulations done by the tree. This technique is described in Algorithm 6. For each of the experiments described in Sects. 4 and 5 we specify which technique of the ones described above was used.

The distinction between the algorithm used by the Tree Parallelization agent and the core algorithm (Algorithm 1) resides in the Search function (defined in line 1 of Algorithm 1). The difference here is that instead of having a single main thread running synchronously the TreePolicy, the DefaultPolicy and the Backup function; this agent uses the main thread only for the TreePolicy, and then delegates to another thread (from a pool of threads) the task of applying the DefaultPolicy and backing up the results of this function, thus enabling many DefaultPolicy and Backup functions to run in parallel.

To avoid data corruption due to the fact that some nodes might be accessed concurrently by multiple threads, we implemented all the functions that modify any of the nodes' properties as atomic functions by using the Java keyword synchronized on them, thus creating effective local mutexes on the nodes. Another important point to be noticed in this version is that it breaks down the Backup function into two separate functions, one to back up the number of visits (*BackupNumberOfVisits*) and another one to backup the total reward (*BackupTotalReward*). The main thread calls the *BackupNumberOfVisits* function before it delegates the *DefaultPolicy* and *BackupTotal-Reward* to another thread. This causes all the nodes between the selected node and the root node to immediately suffer a virtual loss once they are selected by the TreePolicy, due to the way the UCT formula works.

Algorithm 6 *Raw* technique for Tree Merging

```
1: function MERGEALL(T)                                    ▷ T: set of trees' roots
2:     v'_r ← new node with no state           ▷ v'_r: root node of the new tree to be created
3:     V_c ← ⋃_{v∈T} childrenOf(v)                 ▷ V_c: first level children of nodes in T
4:     for each set V_a of nodes from V_c with same source action a do
5:         an ← (Σ_{v∈V_a} N(v)) / sizeOf(T),  aq ← (Σ_{v∈V_a} Q(v)) / sizeOf(T)
6:         v' ← new node with N(v') = an, Q(v') = aq, and a(v') = a
7:         add v' as new child of v'_r
8:     return v'_r
```

This virtual loss makes these nodes less prone to be selected again by the TreePolicy while their DefaultPolicy is not calculated and backed up by the asynchronous worker thread. Once the total reward is backed-up, the temporary increase in the number of visits cannot be considered as a virtual loss anymore. The idea of a virtual loss is suggested by Coulom [6] as a way to avoid the algorithm from exploring only a small subset of the search tree. Without it, the TreePolicy would probably keep selecting almost the same nodes while the worker threads had not backed up their results. This version of the Search function is described in Algorithm 7.

Algorithm 7 Search method used by the Tree Parallelization Agent

```
1: function SEARCH(s_0)                                          ▷ s_0: initial state
2:     v_0 ← new node with state s_0                                  ▷ v_0: root node
3:     while there is computational budget left do
4:         v_s ← TREEPOLICY(v_0)
5:         BACKUPNUMBEROFVISITS(v_s)
6:         r ← DEFAULTPOLICY(s(v_s))       }
7:         BACKUPTOTALREWARD(v_s, r)       } runs asynchronously on next available thread
8:     return a(SELECTCHILDWITHMOSTVISITS(v_0))            ▷ a(v): action that led to v
```

4 Experimental Setup

To evaluate our agents, we created a set of three different experiments using the games available on the Single Player Planning Track of the GVGAI framework[2]. Each of these experiments is focused on evaluating a distinct aspect of the agents.

All the experiments were performed on a computer equipped with 2 Intel Xeon E5-2620v4 running at 2.10 GHz (3.0 GHz with Intel's Turbo Boost) and 126 GB of physical memory. Each of these two CPUs has 8 cores and 2 hyper-threads per core, summing up to a total of 32 hyperthreads available for processing. In order to obtain the experimental results in a reasonable amount of time, we imposed a limit of 40 ms for the agents to

[2] http://www.gvgai.net/.

choose their next action for each step of a game (this is the same amount of time used by the GVGAI for competitions (PEREZ-LIEBANA [12]). For each of the experiments we executed, we collected and computed metrics which were considered important in order to properly evaluate the agents. Each of these metrics is described below:

- Win Rate: percentage of victories over the number of games played. This is the primary indicator of quality/performance for an agent.
- Strength Speedup: defines the improvement in playing quality of the agent when compared to the sequential agent. It is calculated as the division of the number of victories of the agent under evaluation by the number of victories of the sequential agent.
- Playout Speedup: measures the improvement in execution time based on the number of simulations per second. It is calculated as the division of the number of simulations per second of the agent under evaluation by the number of simulations per second of the sequential agent. This metric is especially useful for analyzing how the increase in the number of simulations reflects into a better agent in terms of strength speedup.

Win Rate is one of the main metrics for measuring MCTS algorithms' strength. It is used by authors such as Rocki and Suda [13]; Mirsoleimani et al. [10]; and Chaslot, Winands and Herik [5]; while Strength Speedup and Playout Speedup are metrics used by both Mirsoleimani [10] and Chaslot,Winands and Herik [5] as a way to compare the improvement of parallel MCTS implementations when compared to sequential implementations, and also to analyze how the increase in the number of simulations reflects into a more powerful agent.

In order to evaluate our agents, we carried out three different experiments. For all these experiments we ran the agents against 116 games of the GVGAI framework using 2, 4, 8, 16 and 32 threads. We executed these runs between 15 and 30 times, in order to lower our statistical error rate. We could not run them more times due time restrictions, since each run of 116 game takes around 1:30 h to finish. The details and parameters used for each experiment are described below:

- General Performance Analysis: this is our main experiment, it is focused on analyzing the general performance of the agents, both individually and when compared to each other. For this experiment we used the sum merging technique for the root parallel agent, since it is the method which yielded the best results on Experiment B; and for the UCT's sigma (C) constant we used the value of $\sqrt{2}$, the same one used by the sample MCTS agent of the GVGAI framework and the one which achieved the best results on the third experiment [12].
- Comparison of Merging Techniques for Root Parallelization: the objective of this experiment is to evaluate and compare the three different merging techniques for root parallelization, namely Raw, Sum, and Best, which are the same ones evaluated by Méhat and Cazenave [9] and Swiechowski and Mandziuk [14]. For this experiment, we also used the value of $\sqrt{2}$ for the UCT's sigma (C) constant.
- Impact UCT's Sigma in Scalability of Root Parallelization: The sigma (C) constant of the UCT formula defines how much the second component of the formula (exploration component) is considered over the first component (exploitation component).

It implies that as the value of C increases, the exploration of the tree increases, and as the value of C decreases, the exploitation of the tree increases. When analyzing this change of C in terms of root parallelization, we believe that in most of the cases increasing the value of C will generate a more homogeneous set of parallel trees (due to high exploration), while decreasing it will generate a more diverse set (due to high exploitation).

Therefore, the goal of this experiment is to investigate how the change in the C value changes the results obtained by the root parallel agent. To do so, we ran experiments with the root parallel agent using the sum merging technique (since it achieved the best results in the second experiment) and the values of $\sqrt{2}$, $\frac{\sqrt{2}}{2}$, and $2\sqrt{2}$ for the sigma C constant. Our base value $\sqrt{2}$ is based on the value used by the sample MCTS agent of the GVGAI framework [12].

5 Results and Discussion

This section presents the results we obtained from our experiments and discusses the main outcomes.

The first experiment aimed to perform a General Performance Analysis. The objective of this experiment was to analyze and compare the leaf, root, and tree parallel agents, and how they compare to the synchronous agent. The main results of this experiment are presented in the graph shown in Fig. 1, which shows the win rate for each of the parallel agents, and for the synchronous agent on the first column. We can see in these results that all the parallel agents achieved better win rates than the synchronous agents for all number of threads. When compared between themselves, the parallel agents do not show any clear winner. We can see that the leaf parallel agent presented worse results than the other parallel agents in all cases. The root parallel agent achieved the best results for 2 and 4 threads, while the root parallel agent achieved the best results for 8, 16, and 32 threads, and also the best overall win rate between all agents when using 32 threads.

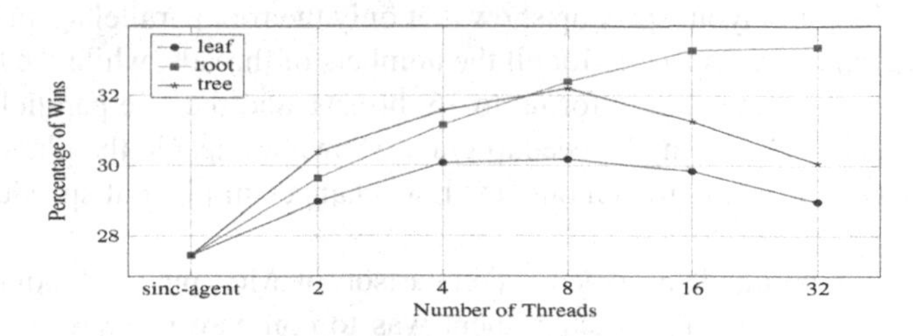

Fig. 1. Win rate for the agents

Two other metrics collected in this experiment were the playout speedup and strength speedup of the agents. These metrics are shown in the graphs *a* and *b* of Fig. 2 (leaf parallel agent), Fig. 3 (root parallel agent), and Fig. 4 (tree parallel agent). The line labeled linear in graph *a* indicates the minimum playout speedup any agent has to achieve if it is able to scale perfectly in terms of simulations per second.

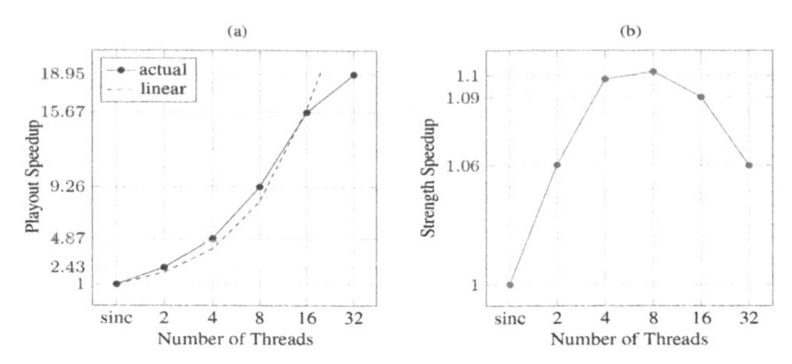

Fig. 2. Playout speedup (a) and strength speedup (b) for the leaf parallel agent

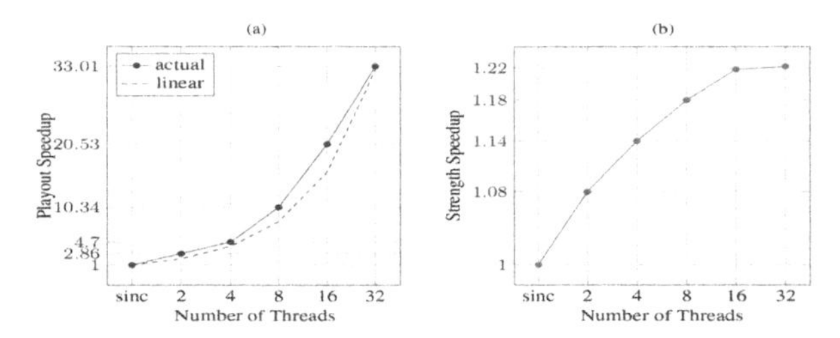

Fig. 3. Playout speedup (a) and strength speedup (b) for the root parallel agent

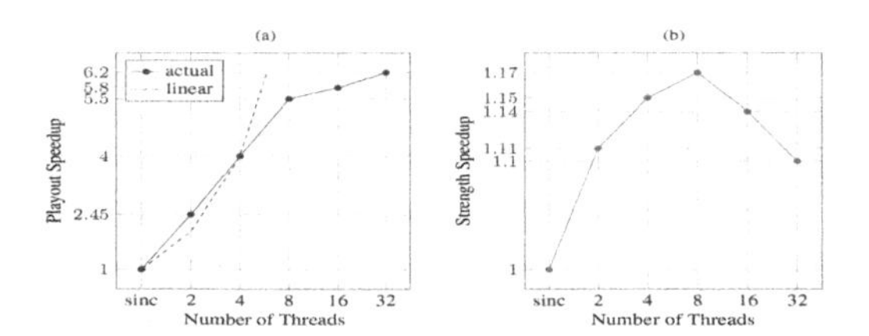

Fig. 4. Playout speedup (a) and strength speedup (b) for the tree parallel agent

These results for playout speedup show that only the root parallel agent was able to achieve above than linear speedup for all the numbers of threads, while the leaf parallel agent was able to achieve it only for up to 16 threads, and the tree parallel agent only for up to 4 threads. The strength speedup graph shown alongside the playout speedup graph is presented as a way to demonstrate how changes in playout speedup reflect in strength speedup.

The second experiment provided a comparison of Merging Techniques for Root Parallelization. The goal of this experiment was to compare the raw, sum, and best techniques for root parallelization in terms of Win Rate and Playout Speedup. The results for win rate are shown in the graph presented in Fig. 5, where we can see that the raw and sum technique achieved almost the same win rates, while the best technique achieved the lowest win rates in all scenarios. We can also see in these results an improvement in win rate for all the techniques for up to 16 threads, with a small decrease for raw and best when using 32 threads.

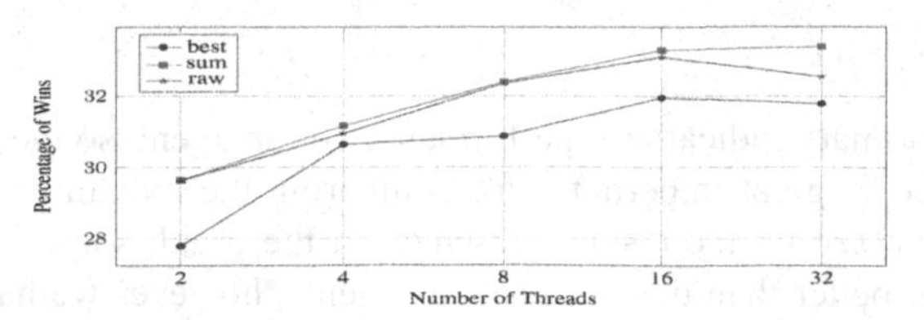

Fig. 5. Win rate for each root parallelization's merging technique

The results for Playout Speed are shown in the graph presented in Fig. 6. Those results show almost no difference for playout speedup between the merging techniques, except a slight difference between them when using 2, 8 and 16 threads. Based on the graph, we can also notice that the playout speedup for all the agents was no worse than a linear speedup for all numbers of threads.

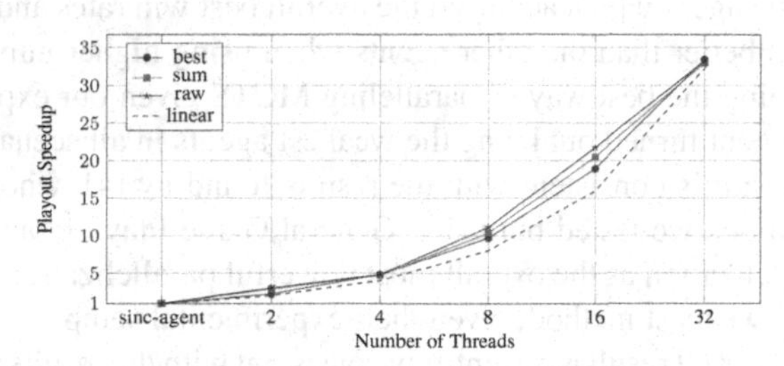

Fig. 6. Playout speedup for each Root Parallelization's Merging Techniques

The next experiment measured the impact of UCT's Sigma in the Scalability of Root Parallelization. The results for the impact of the UCT's sigma constant in root parallelization are shown in terms of win rate in the graph presented in Fig. 7. In this graph, we can see that the sigma value of of $\sqrt{2}$, achieved the best results in all scenarios, while the value of 2 $\sqrt{2}$ achieved the second-best results for 2 and 4 threads, and the value of $\frac{\sqrt{2}}{2}$ achieved the second-best results for 8, 16, and 32 threads. The first column of the graph shows the results obtained by the synchronous agent, so we can see the impact of the UCT's sigma constant in the MCTS algorithm when no parallelization is used.

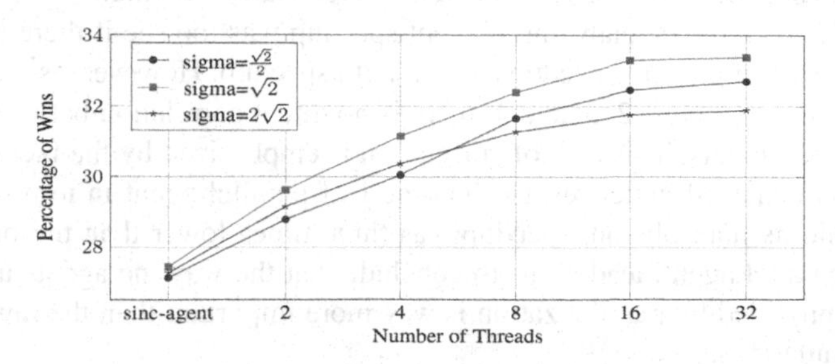

Fig. 7. Win rate for root parallelization with different sigma values

5.1 Discussion

Win Rate is the primary indicator of performance for an agent, so the results discussed in this section are of great importance for comparing the overall performance of the agents. As we can see by the results presented in the graph shown in Fig. 1, all the agents performed better than the synchronous agents, however we have a tie in terms of general performance between the tree parallel agent (which performed better with 2 and 4 threads), and the root parallel agents (which performed better with 8, 16, and 32 threads). We speculate that this might be due to the fact that root parallelization requires a minimal amount of communication to work (as stated by [4]). Thus, the tree parallel agent performs better when using a small number of threads, which requires less communication, and the root parallel agent outperforms it when more threads are used, and more communication is required.

Even though the root parallel agent did not perform better than the other agents for all the cases, it was the agent which achieved the overall best win rates and best scalability, since it performed better than the other agents when using higher numbers of threads, so it turned out being the best way of paralleling MCTS given our experimental setup. The leaf parallel agent turned out being the weakest agents in all scenarios.

This overall result is consistent with the results found by [4], who tested the same three parallel methods we tested but using General Game Playing, and they also considered root parallelization as the overall most powerful parallelization method and leaf parallelization the weakest method, given their experimental setup.

However, this overall result is not entirely consistent with the results reported by [10], who tested the root and tree parallelization methods using a custom implementation of the game Hex. In their results, they reported better overall win rates for root parallelization when using an Intel Xeon Phi 7120P co-processor, but better overall win rates for tree parallelization when using an Intel Xeon E5-2596v2 CPU. These results, allied with the fact that the Xeon E5-2596v2 CPU has a communication-to-compute ratio 30 times lower the Xeon Phi 7120P co-processor reinforces our speculation that tree parallelization performed worst than root parallelization when working with higher numbers of threads only due communication overhead. These reported facts lead us to conclude that even though the root parallel agent achieved an overall better performance than the other agents in our experimental setup, the tree parallel agent might surpass it in the future as the communication latency of multiprocessors reduces, making tree parallelization become the best choice for MCTS in General Video Game Playing.

Our main objective in analyzing playout speedup was too see if there is any direct correlation between playout speedup and strength speedup. However, as we can see by the graphs *a* and *b* in Figs. 2, 3, and 4, there is no direct correlation between these two speedup measurements. This lack of correlation is emphasized by the fact that the tree parallel agent achieved better results than the leaf parallel agent in terms of strength speedup while its max playout speedup was three times lower than the one achieved by the leaf parallel agent, leading us to conclude that the way the agents use the extra simulations provided by parallelization is way more important than the raw number of executed simulations.

Another important point to notice is that both the tree and leaf parallel agents presented a decay in terms of strength speedup with 16 and 32 threads. It is unclear for us the reasons for this result for leaf parallelization since it does not present a decrease in playout speed as the strength speedup decreases. However, in the case of tree parallelization, we believe this decrease in strength speedup is due to the local locks used in the tree. This assumption is backed by the fact when using 16 and 32 threads, the playout speedup of the agent starts to stagnate.

The results for root parallelization merging techniques presented the sum merging technique as the overall best technique (Fig. 5). This result is directly in line with the findings reported by both [9] and [14], who also reported the sum technique as the best technique in their experiments using General Game Playing, showing that the quality of the sum merging technique extends beyond General Game Playing to include also General Video Game Playing.

When analyzing the merging techniques in terms of playout speedup (Fig. 6) we see almost no difference between the merging techniques. We believe that the slight difference between them when using 2, 4, and 8 threads might be due to the fact that the raw technique is computationally simpler than the sum and best techniques, which reduces its overhead when merging the trees, thus allowing more time for some extra simulations.

By analyzing Impact of UCT's Sigma in Scalability of Root Parallelization, in the graph presented in Fig. 7 we can see that the sigma value of $\sqrt{2}$, achieved the best results for all numbers of threads, while the value of $2\sqrt{2}$ achieved the second-best results for 2 and 4 threads, and the value of $\sqrt{2}$ achieved the second-best results when using 8, 16, and 32 threads.

If we ignore the results for $\sqrt{2}$, the results for $2\sqrt{2}$, and $\frac{\sqrt{2}}{2}$ already give us some insights on how the sigma value impacts the scalability of root parallelization. These results show that a high value for sigma ($2\sqrt{2}$), worked better only when a low number of threads were used, but as soon as the number of threads was increased, the value of $\frac{\sqrt{2}}{2}$ achieved better results. These results are in line with the findings of [13], who reported their best results for the lowest sigma value they tested when more than 2 threads were used. They believe that lower values lead to better results because a more diverse set of trees is created between the parallel trees when lower sigma values are used. However, contrary to their findings, the best value for sigma we found ($\sqrt{2}$) is not the lowest value we experimented, but the value in between the lowest and the highest value, what lead us to conclude that there is a lower limit for how low the sigma value can be before it generates a negative impact on performance.

6 Conclusion

Monte Carlo Tree Search (MCTS) is one of the most popular algorithms for game tree searching in scenarios where a proper evaluation function for intermediate game states is nonexistent or hard to create [1]. Parallelization of MCTS is one of the many enhancements proposed for the algorithm. The three main methods for MCTS parallelization are Leaf Parallelization, Root Parallelization, and Tree Parallelization. These methods have been evaluated by several researchers using many different games, such as Mango [5], Hex [10], and games of the General Game Playing framework [14]. However, no work had been done on evaluating how these methods perform int he rather new area of General Video Game Playing.

To address this research gap, we implemented and evaluated these three main MCTS parallelization approaches as agents of the Single Player Planning Track of the General Video Game AI framework. The agents were evaluated using a set of three different experiments, the first one focused on general performance analysis, the second one on comparing merging techniques for root parallelization, and the third experiment was focused on analyzing the impact of the UCT's sigma constant in the scalability of root parallelization. In these experiments the root parallelization method using the sum merging technique and the UCT's sigma value of $\sqrt{2}$ achieved the overall best results.

The experiments on general performance also allowed us to conclude that tree parallelization might surpass root parallelization in the future as the communication latency of multiprocessors reduces. And the experiments on the impact of the UCT's sigma constant in root parallelization lead us to conclude that there is a lower limit for the sigma value before it starts to generate a negative impact, which is probably higher than $\frac{\sqrt{2}}{2}$ considering the results we obtained.

As future work, we suggest evaluating how the parallel MCTS agents perform against other existing General Video Game Playing agents. We also suggest the execution of experiments using a broader range of values for the UCT's sigma constant, in order to find with more precision when the sigma value starts to generate a negative impact on the scalability of root parallelization.

References

1. Browne, C.B., et al.: A survey of Monte Carlo tree search methods. IEEE Trans. Comput. Intell. AI Games [S.1.] **4**(1), 1–43 (2012)
2. Campbell, M., Hoane, A.J., Hsu, F.: Deep blue. Artif. Intell. [S.l.] **134**(1–2), 57–83 (2002)
3. Cazenave, T., Jouandeau, N.: A parallel Monte-Carlo tree search algorithm. In: van den Herik, H.J., Xu, X., Ma, Z., Winands, M.H.M. (eds.) CG 2008. LNCS, vol. 5131, pp. 72–80. Springer, Heidelberg (2008). https://doi.org/10.1007/978-3-540-87608-3_7
4. Chaslot, G., et al.: Monte-Carlo tree search: a new framework for game AI. In: Belgian/Netherlands Artificial Intelligence Conference [S.l.], pp. 389–390 (2008)
5. Chaslot, G.M.J.B., Winands, M.H.M., van den Herik, H.J.: Parallel Monte-Carlo tree search. In: van den Herik, H.J., Xu, X., Ma, Z., Winands, M.H.M. (eds.) CG 2008. LNCS, vol. 5131, pp. 60–71. Springer, Heidelberg (2008). https://doi.org/10.1007/978-3-540-87608-3_6
6. Coulom, R.: Efficient selectivity and backup operators in Monte-Carlo tree search. In: van den Herik, H.J., Ciancarini, P., Donkers, H.H.L.M. (eds.) CG 2006. LNCS, vol. 4630, pp. 72–83. Springer, Heidelberg (2007). https://doi.org/10.1007/978-3-540-75538-8_7
7. Karakovskiy, S., Togelius, J.: The Mario AI benchmark and competitions. IEEE Trans. Comput. Intell. AI Games [S.l.] **4**(1), 55–67 (2012)
8. Levine, J., et al.: General video game playing. Artif. Comput. Intell. Games [S.l.] **6**, 77–83 (2014)
9. Méhat, J., Cazenave, T.: A parallel general game player. KI - Künstliche Intelligenz [S.l.] **25**(1), 43–47 (2011)
10. Mirsoleimani, S.A., et al.: Parallel Monte Carlo tree search from multi-core to many-core processors. In: Proceedings - 14th IEEE International Conference on Trust, Security and Privacy in Computing and Communications, TrustCom 2015 [S.l.], vol. 3, pp. 77–83 (2015)

11. Perez-Liebana, D., et al.: The 2014 general video game playing competition. IEEE Trans. Comput. Intell. AI Games [S.l.] **8**(3), 229–243 (2016)
12. Perez-Liebana, D., et al.: General video game AI: a multi-track framework for evaluating agents, games and content generation algorithms [S.l.] (2018)
13. Rocki, K., Suda, R.: Parallel Monte Carlo tree search scalability discussion. In: Wang, D., Reynolds, M. (eds.) AI 2011. LNCS (LNAI), vol. 7106, pp. 452–461. Springer, Heidelberg (2011). https://doi.org/10.1007/978-3-642-25832-9_46
14. Świechowski, M., Mańdziuk, J.: A hybrid approach to parallelization of Monte Carlo tree search in general game playing. In: De Tré, G., Grzegorzewski, P., Kacprzyk, J., Owsiński, J.W., Penczek, W., Zadrożny, S. (eds.) Challenging Problems and Solutions in Intelligent Systems. SCI, vol. 634, pp. 199–215. Springer, Cham (2016). https://doi.org/10.1007/978-3-319-30165-5_10

Photovoltaic Generation Forecast: Model Training and Adversarial Attack Aspects

Everton J. Santana[1(✉)], Ricardo Petri Silva[2], Bruno B. Zarpelão[1], and Sylvio Barbon Junior[1]

[1] Computer Science Department, Londrina State University, Londrina, PR, Brazil
{evertonsantana,brunozarpelao,barbon}@uel.br
[2] Electrical Engineering Department, Londrina State University, Londrina, PR, Brazil
petri@uel.br

Abstract. Forecasting photovoltaic (PV) power generation, as in many other time series scenarios, is a challenging task. Most current solutions for time series forecasting are grounded on Machine Learning (ML) algorithms, which usually outperform statistical-based methods. However, solutions based on ML and, more recently, Deep Learning (DL) have been found vulnerable to adversarial attacks throughout their execution. With this in mind, in this work we explore four time series analysis techniques, namely Naive, a baseline technique for time series, Autoregressive Integrated Moving Average (ARIMA), from the statistical field, and Long Short-term Memory (LSTM) and Temporal Convolutional Network (TCN), from the DL family. These techniques are used to forecast the power generation of a PV power plant 15 minutes and 24 hours ahead, having as input only power generation historical data. Two main aspects were analyzed: i) how training size influenced the performance of the forecasting models and ii) how univariate time series data could be modified by an adversarial attack to decrease models' performance through cross-technique transferability. For i), the mentioned methods were used and evaluated with monthly updates. For ii), Fast Gradient Sign Method (FGSM), along with a logistic regression substitute model and past data, were used to perform attacks against DL models at test time. LSTM and TCN decreased the error as the training sample size increased and outperformed Naive and ARIMA models. Adversarial samples were able to reduce the performance of LSTM and TCN, particularly for short-term forecasts.

Keywords: Time series forecast · Solar photovoltaic generation · Deep learning · Adversarial attack · Smart grid

The authors would like to thank the financial support of Coordination for the Improvement of Higher Education Personnel (CAPES) - Finance Code 001 -, the National Council for Scientific and Technological Development (CNPq) of Brazil - Grant of Project 420562/2018-4, and Fundação Araucária.

R. Cerri and R. C. Prati (Eds.): BRACIS 2020, LNAI 12320, pp. 634–649, 2020.
https://doi.org/10.1007/978-3-030-61380-8_43

1 Introduction

Intelligent generation and distribution of electrical energy are beneficial for systems operators, plant managers and consumers [2]. A key aspect in this process is the accurate forecast of produced energy, which is fundamental to enable the integration of several plants to the grid, save costs, make power grids more reliable amid the variation in the demand, avoid power outage, and prevent plant managers from penalties. It is also advantageous for the sake of the environment [10], particularly when renewable sources are employed.

For photovoltaic (PV) generation, the focus of our work, forecasting is challenging due to the dependence on meteorological factors such as clouds covering solar panels and variations of solar radiation [22]. Thus, building accurate forecasting models based only on historical power data is a complex task.

Many works attempted to predict future PV power output [1,9,18,22,23] making use of statistical models, artificial intelligence, or a combination of them. However, the models are often built in batch, which assumes that data distribution does not change over time and, as a consequence, the model is not updated. Cerqueira et al. [7] compared the performance of machine learning (ML) and statistical methods when dealing with univariate time series (observing a single variable throughout time) forecast. The authors updated the models according to new incremental observed samples and found out that ML tends to provide better results as the training sample size increases.

In this context, we explored a novel real-life dataset collected from the generation history of a newly built PV power plant, and compared the performance of traditional time series forecasting techniques (Naive and Autoregressive Integrated Moving Average (ARIMA)) with Deep Learning (DL) methods (Long Short-term Memory (LSTM) and Temporal Convolutional Network (TCN)) to forecast power generation 15 minutes and 24 hours ahead. Since the amount of data about power generation increases over time, these models were updated and assessed monthly with the goal to evaluate the influence of the training sample size in their performance.

Another factor that may influence the power generation forecasting is its susceptibility to different attacks, which seek to interrupt the grid's safe operation or obtain financial gains, for instance [17]. Tampering with the results of the forecasts is among the possible attacks, since wrong forecasts on power generation may drive operators or automated control to make harmful decisions towards grid balancing. This shows that besides obtaining accurate models, their reliability should be investigated.

Considering that anticipating the potential model vulnerabilities and analyzing the impact of the possible attacks are the first steps of a proactive protection mechanism [5], and also that adversarial attacks against time series regression models have been overlooked by the literature, we performed attacks during the test time of the DL models, which obtained the best prediction results in the first part of this study.

We examined an adversary with restricted knowledge about the training data and the victim's forecasting models. To degrade the DL model's performance, the attacker modifies the algorithm input data by using Fast Gradient Sign Method (FGSM). In view of this, we evaluated whether FGSM is suitable to generate adversarial test samples which are almost visually imperceptible and, concomitantly, able to increase test error.

The rest of this paper is organized as follows. In Sect. 2, we provide a brief background related to time series and the forecast methods adopted in this work. Section 3 describes experimental details, while results are presented and discussed in Sect. 4. Section 5 concludes the work with the main highlights and future work proposal.

2 Background

2.1 Time Series

A time series is given by the data sequence in a particular time period, and this data can produce different values at distinct moments in time. Formally, it can be defined as an ordered set $X = [x_1, x_2, \ldots x_T]$ in which T corresponds to the length of the series.

The forecasting task consists of finding a function f that predicts the h-th future value of X, i.e., $\hat{x}_{t+h}$ based on i past values:

$$\hat{x}_{t+h} = f(x_{t-i-1}, x_{t-i-2}, \ldots, x_{t-1}, x_t) \tag{1}$$

where i represents the input window size and h, the forecast horizon. When the latter is equal to one, the forecasting task is referred to as a one-step-ahead forecast. Otherwise, it is known as a multi-step ahead forecast.

In addition, time series can present seasonality. This occurs when regular patterns are captured in the series. Seasonal events are phenomena that occur, for instance, daily at a certain time, every day, or in a certain month every year.

The Naive method to forecast the future value in a time series, also known as the persistence model, consists of supposing that the next value of a time series will be the same as the current value:

$$\hat{x}_{t+h} = x_t \tag{2}$$

Given the complexity involving PV power generation, more sophisticate models are generally required.

2.2 Statistical Methods

Understanding the different factors of a time series is important to extract information that can be used to predict future points in this series. Many statistical methods were applied in time series for this purpose and one of the most adopted, ARIMA, is presented next.

ARIMA method is essentially exploratory and seeks to fit a model to adapt to the data structure [6]. With the aid of the autocorrelation and partial autocorrelation functions, it is possible to obtain the essence of the time series so that it can be modeled. Then, information such as trends, variations, cyclical components, and even patterns present in the time series can also be obtained [13]. This allows the description of its current pattern and predictions of future series values [20].

This model is defined by the values (p, e, q), where p is the number of auto-regressive terms, e is the number of differences, and q is the number of moving averages. Auto-regressive (AR) indicates that the evolution variable of interest is returned to its own previous values. The Moving average (MA) part indicates the regression error consisting of a linear combination of values at various times in the past. The Integrated (I) part indicates the process of differentiating between current values and previous values. In some cases, the ARIMA model is applied to non-stationary data. To solve this problem, the integrated part is applied, where differentiation processes are carried out and can be applied more than once until stationarity is obtained.

2.3 Machine Learning Methods

ML, particularly DL, models have been showing to be adequate for dealing with time series made up of power related data, from both demand and generation sides. Among them, TCN and LSTM achieved relevant results for this type of data [16,22,24].

TCN [4] is a specific Convolutional Neural Network (CNN) that has the capability of dealing with time series. For this purpose, it uses causal convolutions, which performs the convolution operation depending only on past values.

This kind of convolution may be submitted to dilation. In TCN, dilated convolutions in one dimension are used seeking to explore long-term patterns. The procedure for doing this is skipping d values between the inputs of convolution. The dilations will be denoted in this work as $[d_1, d_2, \ldots, d_n]$, where d_1 corresponds to the dilation rate of the layer that is the closest to the input, and d_n to the layer that is the closest to the output.

Aiming to increase the receptive field of the network, b convolutional blocks can be stacked. Since it increases the number of parameters, the learning process becomes more complex. Figure 1 shows a dilated causal network with two stacked blocks.

The parameters b, k and d are factors that define the receptive field of the network by the following Eq. 3. For an adequate use of TCN, the receptive field should cover past history, which, in turn, should cover seasonality.

$$receptive\ field = b \times k \times d_n \tag{3}$$

LSTM. A classical neural network for dealing with time series is Recurrent Neural Networks (RNN). In this network, the information is propagated throughout a chain of repeating units of neural network located in the hidden layer.

However, standard RNN suffers from error backflow problems as gradient vanishing and explosion, which restraint long-term dependency learning. The first may lead to very slow learning, and the latter, to weight oscillation.

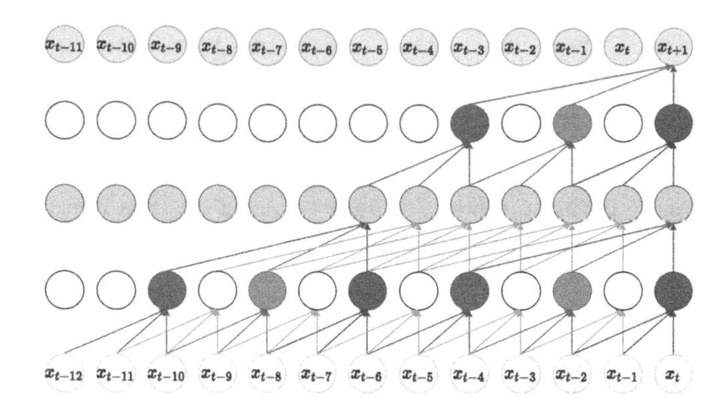

Fig. 1. Example of dilated causal networks with $b = 2$, $k = 3$ and $d = [1, 2]$.

The backflow calculation drawback was one of the main motivations for the development of LSTM [14]. This network, a particular case of RNN, uses a gating mechanism for learning long-term dependencies without losing the short-term capability. These gating mechanisms are inside memory cells, which are located in the hidden layer. Each unit in traditional LSTM has the aspect presented in Fig. 2.

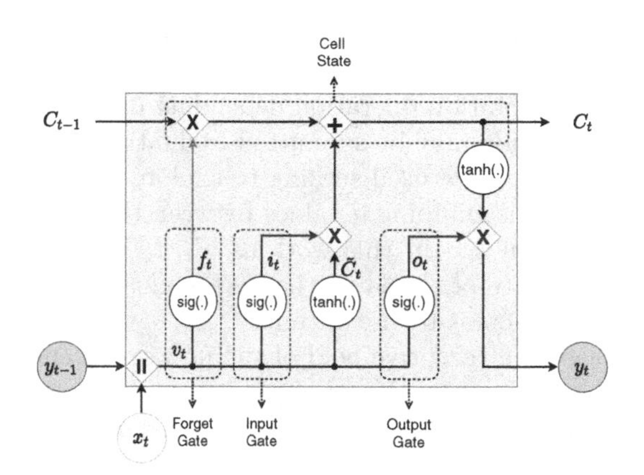

Fig. 2. LSTM cell.

Hyperbolic tangent function (tanh(.)) transforms the values to the range from -1 to 1, and the sigmoid function (sig(.)) to values between 0 and 1. This property makes sig(.) act as a gate, since the values that are transformed to 0 are forgotten by the network and the values that are transformed into 1 are kept.

Each cell is mainly composed of three gates (forget, input and output gates) and a cell state. The forget gate (f_t) is the main gate to select which information should pass forward to the next cell or be eliminated. The input gate (i_t) is used to update the cell state and to select what will be written to the cell. The output gate (o_t) decides the values that will be part of the output. The cell state (C_t) allows the gradient flow. Considering v_t the concatenation of x_t and y_{t-1}, the equations related to these components are:

$$f_t = sig(W_f.v_t + b_f) \tag{4}$$

$$i_t = sig(W_i.v_t + b_i) \tag{5}$$

$$o_t = sig(W_o.v_t + b_o) \tag{6}$$

$$C_t = f_t * C_{t-1} + i_t * \tilde{C}_t \tag{7}$$

$\tilde{C}_t$ and y_t are also calculated to select the new candidate values that can be added to regulate the network and to actually calculate the output of the cell, respectively:

$$\tilde{C}_t = tanh(W_c.v_t + b_c) \tag{8}$$

$$y_t = o_t * tanh(C_t) \tag{9}$$

W_f, W_i, W_o and W_c correspond to the weights matrices related to forward, input, output gates and cell state; b_f, b_i, b_o and b_c correspond to bias of the respective gates and cell state.

2.4 Related Work

Related works in the literature, which address time series forecasting and electricity energy consumption prediction, are presented next.

Regarding statistical methods, Atique et al. [3] used ARIMA to forecast the total daily solar energy generated in a specific solar panel. In most of the recent works, this method is compared to DL. For instance, Jaihuni et al. [15] compared ARIMA and LSTM to a hybrid version to predict 5 minutes and one hour ahead. The hybrid version had better performance in predicting the longest forecast horizon, whereas LSTM and ARIMA outperformed for 5 minutes ahead.

Another example of solar energy generation prediction through DL methods is seen in Torres et al. [22]. In this work, the application of DL consists of predicting the generation of energy for the next day. According to the authors, the proposed method was capable of handling big time series data. In [23], Wang et al. evaluated CNN, LSTM, and a hybrid model with CNN and LSTM for modeling a PV system. The results showed robustness, stability, and great performance. TCN is also a very suitable DL method, as shown in [24], in which Yen et al. verified that TCN is capable of satisfactorily predicting PV generation.

Relating to adversarial attacks in time series, Favaz et al. [11] adapted FGSM and Basic Iterative Method to univariate time series classification and performed attacks to DL models. These attacks achieved an average reduction in the model's accuracy of 43.2% and 56.89%, respectively, and pointed out that FGSM allows real-time adversarial sample generation.

3 Materials and Methods

3.1 Dataset

The data was collected from a PV power generation system installed in a parking lot at the State University of Londrina. It started to operate in November 2019 with a total capacity of 300 kW distributed over 6 inverters. For being representative of the other inverters, inverter 1 was chosen for the analysis. Since the PV plant relies on solar energy to operate, it is usually turned off between 7 pm and 6 am (of the next day). Table 1 shows the data description for each month.

Table 1. Main monthly information about power generation in inverter 1.

Month	Maximum value [W]	Minimum value [W]	Average [W]	Number of samples
2019-11	41521.48	0	8749.14	2880
2019-12	41148.77	0	7582.21	2976
2020-01	42189.33	0	8062.55	2976
2020-02	41194.93	0	7485.98	2784
2020-03	43521.93	0	8506.43	2976

Samples were collected every 15 minutes, implying that each day is composed of 96 samples. Considering this and that the data behavior mostly repeats every day, the input window size was set to 96 ($i = 96$) to cover seasonality.

The interest in energy production forecast can be part of different strategies in very short and short-term time horizons. Thus, we evaluated the performance of these methods when forecasting the production in the next 15 minutes ($h = 1$) and 24 hours ($h = 96$).

Two experiments (Setup 1 and Setup 2) were performed using Intel Xeon CPU @2.3 GHz and Tesla K80 GPU made available by Google Colab. The source code can be found at[1].

3.2 Setup 1 - Obtaining the Forecasting Model

Setup 1 is dedicated to assess the prediction performance of Naive, ARIMA, LSTM, and TCN methods. Naive was picked to be a baseline for our experiment. ARIMA is likely the most adopted option when it comes to statistical methods for time series analysis. Lastly, LSTM and TCN are DL methods, which represent the state-of-the-art of ML methods for time series forecasting.

Prequential evaluation was used to show the evolution of performance as the sample size grows and to simulate a situation in which the model is updated monthly. In this case, each month (except the first and the last) was used for test before being incrementally used for training, making the most use of available data, as observed in Fig. 3. Furthermore, for hyper-parameter tuning, 20% of

[1] http://www.uel.br/grupo-pesquisa/remid/?page_id=145.

training data was reserved for validation. A preprocessing step with z-score was performed to improve convergence. The normalization parameters were obtained for each new training set.

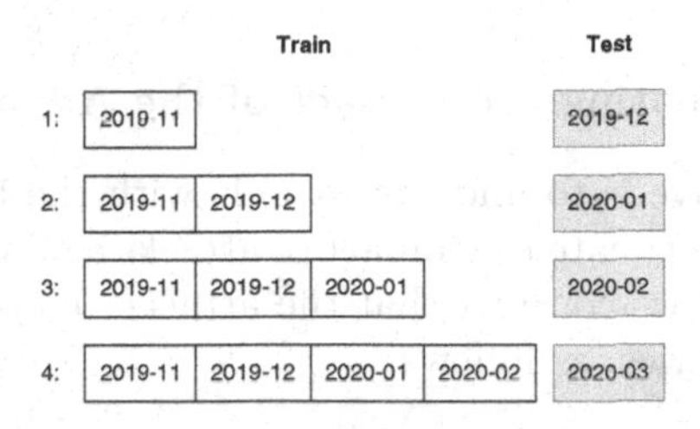

Fig. 3. Train and test sets for Setup 1.

To apply the ARIMA model, *auto-arima* from pmdarima library[2] was used since it auto-tunes its parameters. This approach was applied because the preparation of the parameters ends up being a time-consuming task. The *auto-arima* technique performs several procedures automatically, making the process simpler and faster, finding the best parameters for each data entry.

Both deep methods share some training parameters. We selected Adam as optimizer, Mean Absolute Error (MAE) as loss function, 25 epochs and evaluated batch sizes of 32 and 128. The specific hyper-parameters of LSTM and TCN are shown in Tables 2 and 3, respectively.

Table 2. Specific hyper-parameters of LSTM.

Parameter	Experimental choice
Number of stacked layers (l)	1, 2, 3
Units	32, 64
Dropout	0

Table 3. Specific hyper-parameters of TCN. *The possible dilations followed Eq. 3.

Parameter	Experimental choice
k	2, 3
d^*	[1, 4, 12, 48], [1, 2, 4, 8, 12, 24, 48], [1, 4, 16, 32], [1, 2, 4, 8, 16, 32], [1, 3, 6, 12, 24], [1, 2, 6, 12, 24], [1, 2, 4, 8, 16], [1, 4, 16], [1, 2, 4, 8], [1, 4, 8]
b	1, 2
Number of filters	32, 64
Dropout rate	0

[2] https://pypi.org/project/pmdarima/.

For TCN, Rectified Linear Unit (ReLU) was adopted as activation function and each block output has a residual connection. For this method, we adopted the keras-tcn library[3]. The hyper-parameters not specified previously assumed the default configuration of Keras-TensorFlow[4].

3.3 Setup 2 - Evaluating the Impact of the Adversarial Attacks

In Setup 1, the objective is to find the models with the best predictive performance. In Setup 2, we evaluate the impact of attacks against these models at test time. For this purpose, we consider that the attacker has limited computational and knowledge capabilities, as follows.

Adversary's Goal. The adversary aims to carry out attacks at test time, in the sense that the attacker modifies input data during operation to increase the prediction error of the victim's forecasting model F (obtained during training time). In this case, F could be either TCN or LSTM.

Adversary's Knowledge. We simulated a gray-box attack [21] in which the attacker has limited knowledge about training data and no knowledge about the model adopted by the victim. Particularly, in our scenario, the attacker has access to the data collected during the first 2 months of operation.

Adversary's Capability. The attacker is able to read the legitimate input data during the operation phase, craft new malicious input based on this legitimate data and a substitute model F', and, finally, feed the victim's forecasting model with this malicious data. The attacker has to define a substitute model F' because they do not know the model F built by the victim.

To create the adversarial input, the attacker uses an adaptation of the Fast Gradient Sign Method [12] to the time series regression context. The goal is to add perturbations in the input of testing data. The two main requirements of the perturbation are being not easily visually detected and, at the same time, degrading the performance of the victim's ML model. The perturbation η generated by FGSM is given by:

$$\eta = \epsilon * sign(\Delta_{\mathbf{x}} J(\theta, \mathbf{x}, y)) \tag{10}$$

where ϵ corresponds to the coefficient that controls the attack magnitude, $\mathbf{x}$ to the input to the model, y to the output associated to $\mathbf{x}$, θ to the weights of the adversarial model and $J(.)$ to the loss function.

The attacker also has limited computational resources. This means that the attacker is not able to use complex models to generate adversarial examples, so that simpler models should be adopted as F' and rely on cross-technique transferability, i.e., adversarial examples generated by the model F' can affect the

[3] https://pypi.org/project/keras-tcn/.
[4] https://www.tensorflow.org/api_docs/python/tf/keras/.

performance of another model, trained using a different learning technique [8]. For being simpler than DL, successful in classification scenarios [19] and still differentiable, logistic regression (LR) was adopted to build the substitute model.

Figure 4 shows, in practical terms, how the available data is used to perform the attack.

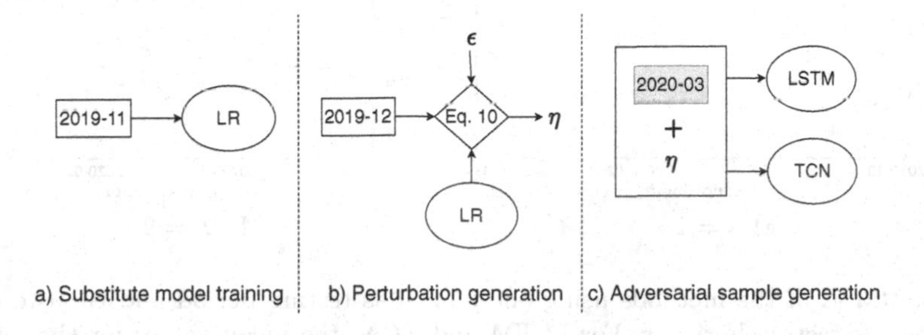

Fig. 4. Illustration of the adversary capability and knowledge.

To obtain F'(a), the first collected month is used as training data. For building the perturbation (b), along with the LR substitute model, the second month is used as $\mathbf{x}$ and y in Eq. 10. Then, at test time (c), the adversarial sample x'_{test} for the corresponding legitimate test input x_{test} is computed by:

$$x'_{test} = x_{test} + \eta \tag{11}$$

and inputted to the victim's model. The ϵ value was varied from 0.05 to 2 with steps of 0.05.

3.4 Evaluation Metrics

To compute the model performance, the root means squared error (RMSE) was assessed for the test sets as:

$$RMSE = \sqrt{\frac{\sum_{j=1}^{n}(\hat{x}_{j+h} - x_{j+h})^2}{n}}, \tag{12}$$

where n corresponds to the number of samples of the test set.

4 Results and Discussion

4.1 Setup 1

Figure 5 presents the test error obtained by the assessed models. As observed, the Naive and ARIMA models had the worst performances, incurring an error of around 16 and 11 kW, respectively. Moreover, for the last test set, the error increased for both methods. During the operation of the plant, the derivatives between 2 intervals are very high, which probably led to a worse performance.

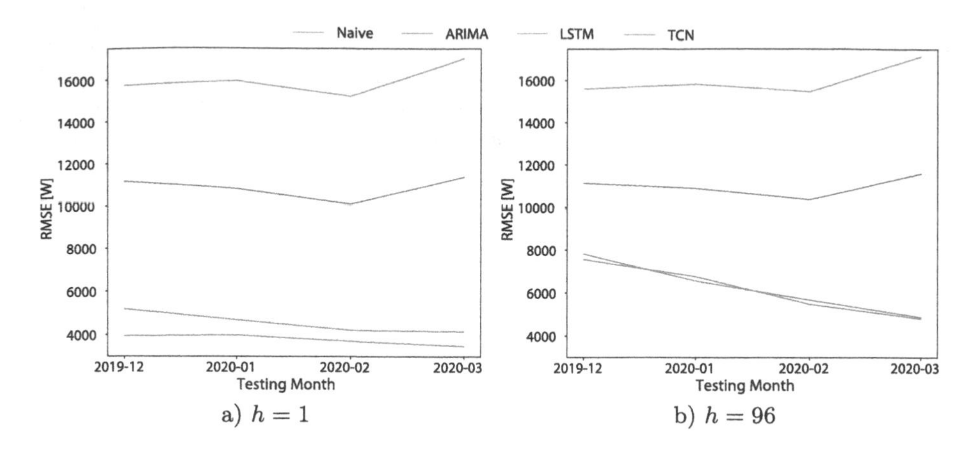

a) $h = 1$

b) $h = 96$

Fig. 5. RMSE of the methods using one month as testing period and the two experimented forecasting horizons. For LSTM and TCN, the mean values for the different hyper-parameters configurations are shown.

For the DL models, the error tended to monotonously decrease as the amount of training data increased. The performance improvement is more evident for $h = 96$: the mean error decreased from almost 8 kW to nearly 5 kW.

Table 4 shows the hyper-parameters configurations that led to the lowest error.

Table 4. Best results of TCN and LSTM for both forecasts horizons and the test datasets. The bold values correspond to the best RMSE values for LSTM and TCN in each forecast horizon.

h	Test	Method	RMSE	Batch	k	d	b	Filters	l	Units
1	2019-12	TCN	3625.71	32	2	[1, 4, 12, 48]	1	64		
		LSTM	4276.75	32					1	32
	2020-01	TCN	3811.69	32	2	[1, 4, 12, 48]	1	64		
		LSTM	4067.07	32					1	32
	2020-02	TCN	3576.19	32	2	[1, 2, 6, 12, 24]	2	64		
		LSTM	3705.22	32					1	64
	2020-03	TCN	**3333.10**	128	2	[1, 3, 6, 12, 24]	2	32		
		LSTM	**3672.89**	32					1	64
96	2019-12	TCN	7320.73	32	3	[1, 2, 4, 8, 16]	2	64		
		LSTM	7294.97	128					2	32
	2020-01	TCN	6265.15	32	2	[1, 2, 6, 12, 24]	2	32		
		LSTM	6016.31	128					1	32
	2020-02	TCN	5276.89	128	3	[1, 2, 4, 8, 16, 32]	1	32		
		LSTM	5175.33	32					1	32
	2020-03	TCN	**4497.43**	128	2	[1, 4, 12, 48]	1	32		
		LSTM	**4408.46**	128					1	32

For $h = 1$, TCN achieved the best performance, with an error of 3333.10 W. It represented 29.24% and 19.57% in relation to ARIMA and Naive RMSEs for the same test set and 7.66% in relation to the maximum power value of 2020-03.

For $h = 96$, LSTM slightly outperformed TCN, with an error of 4408.56 W. It represented 38.04% of the error presented by ARIMA, 25.69% in relation to Naive and 10.13% in relation to the maximum power value of this set.

As to the hyper-parameters, for $h = 1$, LSTM presented a preference for smaller batch size and a single layer (i.e, no stacking). As the training sample size increased, the number of required units also increased. For the same forecast horizon, the preferred kernel size of TCN was 2 and higher training sample sizes preferred more blocks. The last month preferred the largest batch size and a lower number of filters.

Still analyzing the hyper-parameters, for $h = 96$, LSTM preferred 32 units for all training sample sizes and mostly only one layer. The last month preferred the largest batch size. TCN mostly required 32 filters, and as the training sample size increased, the batch size and the number of blocks increased as well.

Alongside the error values, we also wanted to evaluate qualitatively the models. With this intention, Fig. 6 compares the true output values from 7 days of the last test set to the forecast results that led to the lowest RMSE for each h.

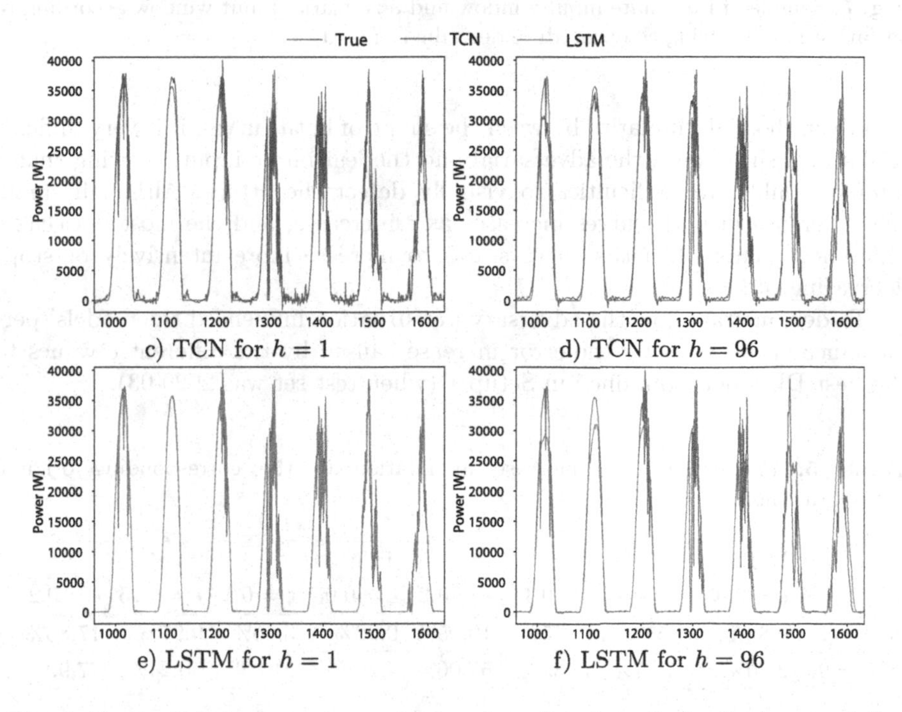

Fig. 6. Comparison between true output and predicted results by the best models.

In general, the predictions followed the true output values. TCN presented lower stability around 0 than LSTM, particularly for $h = 1$. When $h = 96$ and the true output values are very high, it is observable a conservative forecast of the power generation.

4.2 Setup 2

As previously mentioned, the adversarial input data must be subtle for not being easily visually identified. Considering this, Fig. 7 compares one sample of an original input window that belongs to the test set and its respective sample of a maliciously crafted window input generated by FGSM.

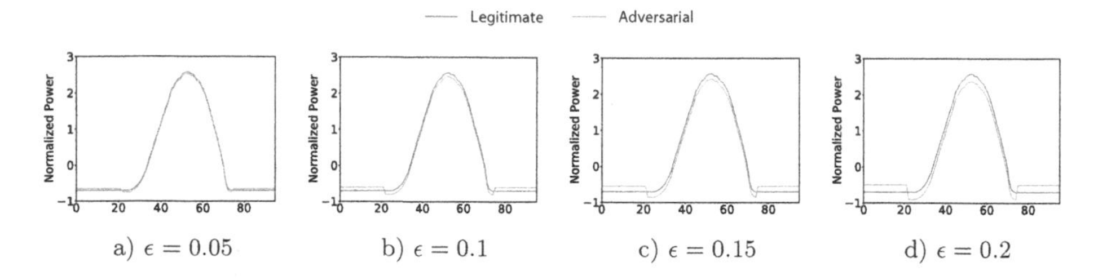

a) $\epsilon = 0.05$ b) $\epsilon = 0.1$ c) $\epsilon = 0.15$ d) $\epsilon = 0.2$

Fig. 7. Sample of legitimate input window and adversarial input window according to variation in ϵ following the procedure described in setup 2.

Given the high similarity between the shape of both curves, it is very difficult to distinguish between the adversarial and the legitimate input, showing that a human would have difficulties to visually detect the attack. Although small, the difference in both curves increases as ϵ increases, and the most noticeable difference occurs when the curve starts to increase more intensively or stops decreasing.

To demonstrate how the adversary perturbation influenced the models' performance, Table 5 shows the error increase caused by the different ϵ values to the best DL models obtained in Setup 1 (when test set was 2020-03).

Table 5. Percentual error increase in relation to the corresponding original dataset/method.

	LSTM				TCN			
	$\epsilon = 0.05$	$\epsilon = 0.1$	$\epsilon = 0.15$	$\epsilon = 0.2$	$\epsilon = 0.05$	$\epsilon = 0.1$	$\epsilon = 0.15$	$\epsilon = 0.2$
$h = 1$	0.88%	4.55%	10.75%	19.05%	0.22%	3.66%	9.97%	17.86%
$h = 96$	2.70%	17.77%	42.33%	57.06%	4.20%	21.03%	50.24%	77.99%

For $h = 1$ and $\epsilon = 0.05$, nearly no increase in error was found for both methods, but when ϵ slightly varied from 0.05 to 0.1, the error increase was multiplied by a factor of 5.17 for LSTM and 16.63 for TCN. The biggest error was found for LSTM when $\epsilon = 0.2$.

For $h = 96$, the biggest error increase was found for TCN, reaching almost 78%. Such error increase may cause serious harm to the grid operation. Suppose a situation where the legitimate forecast, i.e., the TCN legitimate output, is 10000 W and there is a fixed demand of 20000 W for the region connected to the studied PV plant. Considering our most extreme setup for an attack ($\epsilon = 0.2$), the TCN output could be changed to 17799 W. Then, based on this wrong forecast, the system operator would assume that it was necessary to deliver only more 2201 W from other plants to meet the demand of that region. However, during the operation, the operator would be actually required to deliver a value closer to 10000 W to properly meet this demand. This difference would unbalance the grid and possibly culminate in a power outage.

Therefore, it was possible to verify that FGSM can generate adversarial samples for the studied models, based only on a limited amount of historical data and using LR as substitute model. Thus, in the context of this work, a defense mechanism should be developed for the adopted forecasting models for preventing them from adversarial attacks.

5 Conclusion and Future Work

Results showed that DL models outperformed Naive and ARIMA. Those models were able to deal with the complexity of PV generation data. Moreover, the bigger the training sample size, the better the TCN and LSTM performances, so updating these models is recommended. As for the attacks at test time, FGSM was able to increase models' error even with small ϵ values. It was also possible to validate the cross-technique transferability of LR as a substitute model for generating adversarial examples for LSTM and TCN.

Future work can analyze other influencing factors in PV power generation and treat the time series as multivariate aiming at improving the accuracy of forecasting models. Concomitantly, we will investigate defense mechanisms for these attacks since it is a real-world application.

References

1. Akhter, M.N., Mekhilef, S., Mokhlis, H., Shah, N.M.: Review on forecasting of photovoltaic power generation based on machine learning and metaheuristic techniques. IET Renew. Power Gener. **13**(7), 1009–1023 (2019)
2. Antonanzas, J., Osorio, N., Escobar, R., Urraca, R., Martinez-de Pison, F.J., Antonanzas-Torres, F.: Review of photovoltaic power forecasting. Solar Energy **136**, 78–111 (2016)
3. Atique, S., Noureen, S., Roy, V., Subburaj, V., Bayne, S., Macfie, J.: Forecasting of total daily solar energy generation using ARIMA: a case study. In: 2019 IEEE 9th Annual Computing and Communication Workshop and Conference (CCWC), pp. 0114–0119. IEEE (2019)

4. Bai, S., Kolter, J.Z., Koltun, V.: An empirical evaluation of generic convolutional and recurrent networks for sequence modeling. arXiv preprint arXiv:1803.01271 (2018)
5. Biggio, B., et al.: Evasion attacks against machine learning at test time. In: Blockeel, H., Kersting, K., Nijssen, S., Železný, F. (eds.) ECML PKDD 2013. LNCS (LNAI), vol. 8190, pp. 387–402. Springer, Heidelberg (2013). https://doi.org/10.1007/978-3-642-40994-3_25
6. Box, G.E.P., Jenkins, G.: Time Series Analysis Forecasting and Control. Holden-Day Inc., USA (1990)
7. Cerqueira, V., Torgo, L., Soares, C.: Machine learning vs statistical methods for time series forecasting: Size matters. arXiv preprint arXiv:1909.13316 (2019)
8. Chakraborty, A., Alam, M., Dey, V., Chattopadhyay, A., Mukhopadhyay, D.: Adversarial attacks and defences: a survey. arXiv preprint arXiv:1810.00069 (2018)
9. Das, U.K., et al.: Forecasting of photovoltaic power generation and model optimization: a review. Renew. Sustain. Energy Rev. **81**, 912–928 (2018)
10. Diamantoulakis, P.D., Kapinas, V.M., Karagiannidis, G.K.: Big data analytics for dynamic energy management in smart grids. Big Data Res. **2**(3), 94–101 (2015)
11. Fawaz, H.I., Forestier, G., Weber, J., Idoumghar, L., Muller, P.A.: Adversarial attacks on deep neural networks for time series classification. In: 2019 International Joint Conference on Neural Networks (IJCNN), pp. 1–8. IEEE (2019)
12. Goodfellow, I.J., Shlens, J., Szegedy, C.: Explaining and harnessing adversarial examples. arXiv preprint arXiv:1412.6572 (2014)
13. Ho, S., Xie, M.: The use of ARIMA models for reliability forecasting and analysis. Comput. Ind. Eng. **35**(1), 213–216 (1998)
14. Hochreiter, S., Schmidhuber, J.: Long short-term memory. Neural Comput. **9**(8), 1735–1780 (1997)
15. Jaihuni, M., et al.: A partially amended hybrid Bi-GRU–ARIMA model (PAHM) for predicting solar irradiance in short and very-short terms. Energies **13**(2), 435 (2020)
16. Lara-Benítez, P., Carranza-García, M., Luna-Romera, J.M., Riquelme, J.C.: Temporal convolutional networks applied to energy-related time series forecasting. Appl. Sci. **10**(7), 2322 (2020)
17. Mehrdad, S., Mousavian, S., Madraki, G., Dvorkin, Y.: Cyber-physical resilience of electrical power systems against malicious attacks: a review. Curr. Sustain./Renew. Energy Rep. **5**(1), 14–22 (2018)
18. Mellit, A., Massi Pavan, A., Ogliari, E., Leva, S., Lughi, V.: Advanced methods for photovoltaic output power forecasting: a review. Appl. Sci. **10**(2), 487 (2020)
19. Papernot, N., McDaniel, P., Goodfellow, I.: Transferability in machine learning: from phenomena to black-box attacks using adversarial samples. arXiv preprint arXiv:1605.07277 (2016)
20. Pena, E.H., Barbon, S., Rodrigues, J.J., Proença, M.L.: Anomaly detection using digital signature of network segment with adaptive arima model and paraconsistent logic. In: 2014 IEEE Symposium on Computers and Communications (ISCC), pp. 1–6. IEEE (2014)
21. Tabassi, E., Burns, K.J., Hadjimichael, M., Molina-Markham, A.D., Sexton, J.T.: A taxonomy and terminology of adversarial machine learning. NIST IR (2019)
22. Torres, J.F., Troncoso, A., Koprinska, I., Wang, Z., Martínez-Álvarez, F.: Deep learning for big data time series forecasting applied to solar power. In: Graña, M., et al. (eds.) SOCO'18-CISIS'18-ICEUTE'18 2018. AISC, vol. 771, pp. 123–133. Springer, Cham (2019). https://doi.org/10.1007/978-3-319-94120-2_12

23. Wang, K., Qi, X., Liu, H.: A comparison of day-ahead photovoltaic power forecasting models based on deep learning neural network. Appl. Energy **251**, 113315 (2019)
24. Yen, C.F., Hsieh, H.Y., Su, K.W., Leu, J.S.: Predicting solar performance ratio based on encoder-decoder neural network model. In: 2019 11th International Congress on Ultra Modern Telecommunications and Control Systems and Workshops (ICUMT), pp. 1–4. IEEE (2019)

Quantifying Temporal Novelty in Social Networks Using Time-Varying Graphs and Concept Drift Detection

Victor M. G. dos Santos[1], Rodrigo F. de Mello[2], Tatiane Nogueira[1], and Ricardo A. Rios[1](✉)

[1] DCC, Federal University of Bahia, Salvador, BA, Brazil
vitocr_santos@hotmail.com, {tatiane.nogueira,ricardoar}@ufba.br
[2] ICMC, University of São Paulo, São Carlos, SP, Brazil
mello@icmc.usp.br

Abstract. This paper presents a new approach to quantify temporal novelties in Social Networks and, as a consequence, to identify changing points driven by the occurrence of new real-world events that influence the public opinion. Our approach starts using Text Mining tools to highlight the main key terms, that will be later used to create a temporal graph, thus preserving their relation into the original texts and their temporal dependencies. We also defined a new measure to quantify the way users' opinions have been evolving over time. Finally, we propose a straightforward Concept Drift method to identify when the changing points happen. Our full approach was evaluated on a historical event in Brazil: the 2018 presidential election race. We have chosen this period due to the volume of publications that, definitely, stated Social Networks as the main mechanism for new political activism. Our good results emphasize the importance of our approach and open new possibilities to identify bots developed to just spread, for example, fake news.

Keywords: Temporal graph · Concept Drift · Social networks

1 Introduction

Historically, information and news were only spread by the traditional media, such as the well-known journalist programs broadcasted on private TV, journals, and magazines. The advent of social networks has reduced such monopoly by giving and amplifying the voice of any person. It can be noticed in a survey shown in [18], which reveals that, at least, 71% of young American adults use some type of social network more than once a day.

Thus, Social Networks has become an important tool to improve, for example, the access to educational and political contents, however it has been also emerging as an easy way to commit crimes as the spread of fake news. Regardless the goals, Social Network platforms are definitely part of people's lives.

R. Cerri and R. C. Prati (Eds.): BRACIS 2020, LNAI 12320, pp. 650–664, 2020.
https://doi.org/10.1007/978-3-030-61380-8_44

In parallel, several scientific researches have been proposed to extract information from the huge amount of data daily produced from such platforms.

In this work, we investigated this research area and proposed a new approach that analyzes Social Network data in four phases. Firstly, we collect and store textual data related to specific topics of interest published on Twitter. Then, we apply Text Mining methods to extract the main keywords used to express the users' feeling. Next, we use such keywords to create a temporal graph that represents the connections among used keywords and their temporal relationships. Finally, we apply a measure proposed in this work to transform the temporal graph into a time series. The time series is analyzed by using a proposed Concept Drift method to detect dates when the people reaction was affected by some real-world event.

Our approach was assessed on texts published on Twitter (tweets), collected during the 2018 Brazil presidential election race, which was characterized by an intense participation of politicians, political parties, voters, and bots, definitely changing political advertisement and activism. Such scenario was important due to the amount of published text and, based on historical events, we were able to evaluate our approach by confirming the novelties with news registered by traditional media.

This manuscript is organized as follows: Sect. 2 and 3 introduce the main concepts in our approach and related researches, respectively; Our approach is detailed in Sect. 4; The experimental setup is discussed in Sect. 5; Sect. 6 presents the obtained results; Finally, concluding remarks and future directions are given in Sect. 7, which is finally followed by a list of references.

2 Background

This section reviews definitions and concepts necessary to better understand our work. As stated in the introduction, our approach transforms a set of texts published in social networks, using Text Mining techniques (detailed in Sects. 4 and 5), into temporal graphs, preserving the relationship among words. The final temporal graph is, then, transformed into a time series, whose characteristics provide important information on users' perception.

Formally, a graph is defined as a finite and non-empty set $\mathscr{G} = \langle \mathscr{V}, \mathscr{E} \rangle$, in which $\mathscr{V}$ represents a set of vertices (nodes) and $\mathscr{E}$ is a set of edges connecting pairs of vertices [5,20]. In our context, a vertex (node) u represents a word used to write a tweet. An edge between two adjacent vertices (u, v) means there is, at least, one tweet that used both words.

As discussed in [5], a graph $\mathscr{G}$ can be classified according to specific characteristics. Our graphs are characterized by not presenting self-loops, i.e., $\forall v \in \mathscr{V}, (v, v) \notin \mathscr{E}$, and by using weights to associate the number of tweets $m(u, v) > 0$ to every edge in which two words u and v were mentioned[1]. We also consider undirected graphs $\forall u, v \in \mathscr{V}, (u, v) = (v, u)$, $m_{u,v} = m_{v,u}$. Given that

[1] By using the matrix notation, the weight can also be represented as $m_{u,v}$.

every analyzed day has several tweets published with more than a single word, so no null graph ($\mathscr{E} = \emptyset$) is produced[17].

Temporal graphs are an extension of the definitions previously presented but including the temporal information to model networked time-evolving systems [13]. In this context, there are two main representations [11]: (i) Aggregated Static Graphs; and (ii) Time-Varying Graphs. This first one consists in constructing a single aggregated static graph, in which all the contacts between each pair of nodes are flattened in a single edge. Although this representation allows to include the time interval in which two vertices are connected, the specific start/end time and multiples intervals of connections over time cannot be properly represented. Time-Varying Graphs, in turn, are ordered sequences of graphs in which each graph represents the state of the system, also including the configuration of links and different time windows [11], as illustrated in Fig. 1.

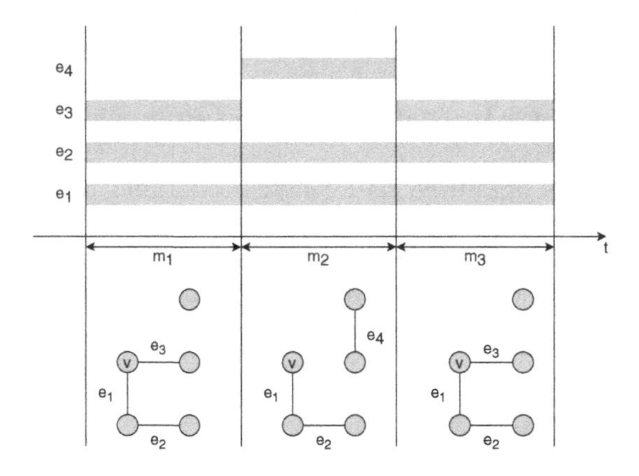

Fig. 1. Example of a time-varying graph.

The example in Fig. 1 is composed of three time windows (m_1, m_2, and m_3) and four edges (e_1, e_2, e_3, and e_4), connecting five vertices. By looking at this representation, we notice Edge e_3 models a relationship between vertices during two different time windows. Moreover, e_4 only shows up during a single window, while Edges e_1 and e_2 persist during the whole analysis.

Mathematically, time-varying graphs can be represented by $G = \{\mathscr{G}_1, \mathscr{G}_2, \ldots, \mathscr{G}_h\}$, in which an edge is defined by the tuple $(u, v, t, \delta t)$ such that t is the instant time when the connection started and δt represents its duration.

The most well-known measurement derived from a time-varying graph is the topological overlap. This measurement, defined in Eq. 1, aims at evaluating the persistence, between consecutive time intervals P and $P+1$, in a given temporal graph [11].

$$C_v(P, P+1) = \frac{\sum_u a_{v,u}(P) a_{v,u}(P+1)}{\sqrt{[\sum_u a_{v,u}(P)]\,[\sum_u a_{v,u}(P+1)]}} \quad (1)$$

In Eq. 1, $C_v(P, P+1)$ refers to the temporal overlap of Vertex v, $a_{v,u}(\cdot)$ is equal to 1 iff v and u are connected during a given time window, and $\sum_u$ iterates over $\forall u, u \in \mathscr{E}$. As one may notice, this equation quantifies the persistent relationship of a vertex and its adjacent ones. The main problem with this measurement is its applicability to quantify the weighted edges, thus motivating us to create a new measure as presented in Sect. 4.

In our research, we are considering non-overlapping time-windows, therefore, after calculating a given measure between pairs of windows, a new time series is produced. In summary, a time series organizes a set of observations collected within a given time interval as $X = \{x_0, x_1, x_2, \ldots, x_m\}$ [3].

As discussed in [3], time series can be used in several real-world application for different purposes as, for instance, the prediction of future values and the understanding of system behaviors. The later is the main motivation of this work, allowing us to answer our main research question that is focused on identifying the influence of political news in users' behavior. Next section brings some of the most relevant studies already developed in this area.

3 Related Work

In the first related studies, authors analyzed the volume of tweets citing the two main presidential candidates and their respective political parties during the Spanish election [1,2]. In such work, the proposed metric, named Relative Support, was important to identify relevant political moments and the obtained results emphasized similarities between the volume of published tweets and the final election results. Similarly, Caldarelli et al. [4] collected the volume of tweets related to the two main candidates in the Italian election and modeled the data volume as time series. Next, the authors also used the Relative Support to ratify the relationship between the election results and political activism on Twitter.

The relationship between social networks and politics is also explored in [9], in which tweets related to Brazilian politicians were collected and transformed into time series using, for example, the *Normalized Compression Distance*. Then, on every resultant time series, the authors applied the algorithm *Cross-Recurrence Quantification Analysis* to detect changes in users' series behavior. The authors' goal was to identify such changes and map them into events happening in the Brazilian political scenario.

In the literature, there is an extensive material proposing the analysis of publications in social networks to understand real-world events. However, the investigation of users' reactions is widely performed by using the volume of publication. Another area usually explored in such context is the analysis of sentiments, which employs models designed to produce word embeddings as in [14]. The main problem with this approach is the execution of black box models, whose temporal relations among the used words are unknown.

Katragadda et al. [6] proposed a new approach designed to detect events within the first 8 min of their occurrence. To reach this goal, the authors transform the collected tweets into a temporal graph which is later analyzed by Network-based Unsupervised Learning [17] in order to identify cluster of words as graph communities.

By taking into account the advantage of those researches, we present a new approach to transform a set of tweets published on a specific topic into a time-variant graph that preserves not only the relationship among their words but also their temporal dependences, as detailed in the next section.

4 Proposed Approach

This section details our proposed approach, which was designed in four phases as shown in Fig. 2. The first one is related to the process of collecting data from a social network as highlighted by Tasks 1 and 2 at the top-left plot of such figure. Both tasks were performed by using the TSViz platform [9,14,15], which has been widely adopted to monitor publications on Twitter based on specific hashtags and users. As a result from this first phase, we have a set $T_\Delta = \{t_1, t_2, t_3, \ldots, t_j\}$, in which Δ represents the monitored time interval and t_i, $1 \leq i \leq j$, is a given publication composed of a set of words w, i.e., $t_i = \{w_1, w_2, w_3, \ldots, w_s\}$[2].

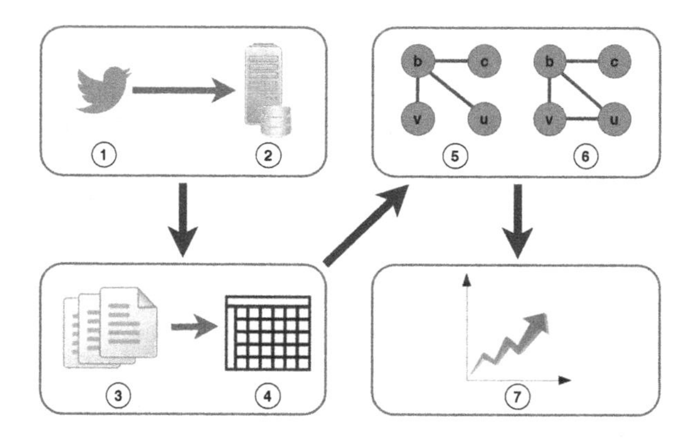

Fig. 2. Summary of the proposed approach of this work.

As previously mentioned, the focus of this research is to analyze publications with textual contents. Before proceeding to the next phase, we removed all re-tweeted publications, i.e., once we are trying to identify novelty as the people react to political issues, we removed texts with no new information. This step produces a new dataset $T'_\Delta \subseteq T_\Delta$, that will be referred to as T_Δ just to simplify the manuscript reading. Another important transformation was related to the discretization of the time interval. Although we are monitoring a topic during the time interval Δ, our focus is to understand the variations between

[2] In this work, "i" and "j" are just iteration variables used in different contexts.

time windows within this interval. Then, T_Δ is discretized considering P windows, thus producing a set of sequential windows $T_\Delta = \{T_P, T_{2P}, \ldots, T_{hP}\}$, in which $T_P = \{t_1, \ldots, t_P\}$, T_{hP} is the hth time window containing the last P tweets, $\Delta = \bigcup\{P, 2P, \ldots, hP\}$, and $\bigcap_{i=1}^{h} T_{iP} = \emptyset$. For example, in this work, we have monitored some users during 30 days, but we analyze them by using daily windows.

The second phase starts with Task 3 that wraps the preprocessing activities commonly adopted in the Text Mining area, which include URL purge, reduction of terms using lemmatization tools, separation of all terms with tokenization tools, removal of stopwords and irrelevant terms, and the stemmization of terms[3]. Then, a table is produced in Task 4 containing all stemmed terms by publication, i.e., every row represents a tweet and columns contains its preprocessed terms.

The third phase comprises the construction of temporal graphs. It is important to recall that we are interested in detecting temporal variations between pairs of time windows. Aiming at reaching this goal, we measure the inclusion/exclusion of words in individual publications. In this sense, we start creating a vector W with all preprocessed and unique words used by users in all tweets considering two time windows. For example, by comparing two consecutive time windows, we have $W' = \{w'_1, w'_2, \cdots, w'_k\}$ and $W'' = \{w''_1, w''_2, \cdots, w''_l\}$, which contain all words used in every window. Then, we create a single vector $W = W' \cup W''$, with length equals to $|W| = n$, that is later used to create an adjacency matrix $\mathbf{M}_{n \times n}$. We recall, as stated in the Set Theory, a set must be composed of unique elements, i.e., $\forall w, \exists! w \in W' \cup W''$. The first graph, Task 5 in Fig. 2, is created by counting the number of tweets in which every pair of words are mentioned together during the first time window. Thus, an element $m_{u,v} > 0$ of $\mathbf{M}$ represents an edge connecting two vertices u and v, i.e., $m_{u,v}$ counts all occurrences of the words u and v in the same tweet. At the end of this process, we have $\mathscr{G} = \langle \mathscr{V}, \mathscr{E} \rangle$, in which $\mathscr{V}$ is a set of words (vertices) that were used in, at least, one tweet during the two time windows and $\mathscr{E}$ is a set of edges connecting those vertices (words) in the same tweet during the first time window only.

Next, we create $\mathscr{G}' = \langle \mathscr{V}, \mathscr{E}' \rangle$ which represents the graph for the second window, Task 5 in Fig. 2, in which $\mathscr{V}$ contains the same words from $\mathscr{G}$ but presenting a different set of edges. $\mathscr{E}'$ only counts the occurrence of pairs of vertices in the same tweet during the second time window. As one may notice, for every time window, a graph is created respecting the following definitions: (i) there is no self-loop; (ii) it can be complete but never null; (iii) undirect; and (iv) weighted.

In the last phase, we define the temporal relation between consecutive pairs of graphs $G = \{\mathscr{G}_{1P}, \mathscr{G}_{2P}, \ldots, \mathscr{G}_{hP}\}$ which were created at every time window. Such temporal relation is defined by using our proposed Temporal Novelty Quantification ($\aleph$), which calculates the variation between pairs of consecutive graphs, $\mathscr{G}$ and $\mathscr{G}'$, as defined in Eq. 2. In summary, this equation quantifies the

[3] More details about these preprocessing activities are provided in Sect. 5.2.

relationship variation between pairs of words in two consecutive time windows. This difference is estimated by the normalized distance defined in Eq. 3.

$$\aleph(\mathscr{G}, \mathscr{G}') = \sum_{u,v \in \mathscr{G} \bigcup \mathscr{G}',\, u \neq v} d(u,v) \tag{2}$$

$$\begin{aligned} d(u,v) &= \frac{|m_{u,v}(\mathscr{G}) - m_{u,v}(\mathscr{G}')|}{\sqrt{|m_{u,v}(\mathscr{G}) - m_{u,v}(\mathscr{G}')|}} \\ &= \sqrt{|m_{u,v}(\mathscr{G}) - m_{u,v}(\mathscr{G}')|} \end{aligned} \tag{3}$$

Equation 3 provides three different interpretations: (i) if some words are strongly related during a time window, but this relation is strongly reduced in the next window, then, the subject, in which they are related, has not the same relevance; (ii) if there is an opposite behavior, then some subject has received a sudden relevance; finally, (iii) if the occurrence of some words remains similarly related between time intervals, their relevance is kept. As one may notice, the first two interpretations are useful to detect novelty.

In summary, our proposed analysis starts transforming a set of publication from a social network into a time-varying graph built on top of time windows. Then, we use our Temporal Novelty Quantification to create the temporal relation between pair of graphs. As consequence, by applying our proposed quantification on such graphs, a time series is produced (Task 7 in Fig. 2), thus allowing to better understand people reactions on specific topics (users or hashtags). For example, it is possible to use time series methods to detect, for example, trends or, using our interpretation, one can detect spikes, when specific events are getting stronger or weaker, or bottoms, when the relevance of words is decaying.

5 Experimental Setup

Before presenting the obtained results, we illustrate in this section our proposed approach by detailing all tasks performed in every phase as shown in Fig. 2.

5.1 Data Collection Phase (Tasks 1 and 2)

As previously mentioned, our proof of concept was performed on texts published on Twitter during October 2018. This period, marked by the Brazilian presidential election, presented an expressively high number of political publications, thus definitely establishing social networks as a new and influential political advertising means in Brazil.

The political polarization motivated us to monitor the most influential politicians in the 2018 election race: the current and a former presidents Jair Bolsonaro and Lula, respectively. Although the Superior Electoral Court has barred Lula from presidential race on August 31st 2018, his name remained stronger than his substitute Fernando Haddad.

In this sense, we have used the TSViz platform [9,14,15] to collect any publications related to theses politicians. Figure 3 summarizes the publication volume and calls our attention to the daily number of tweets (y axis). Tweets related to the former president Lula shows a spike with more than 30k tweets in the day before the first round of elections. In relation to current president Bolsonaro, the astounding volume of tweets published during the day of second round elections has reached almost 450k tweets.

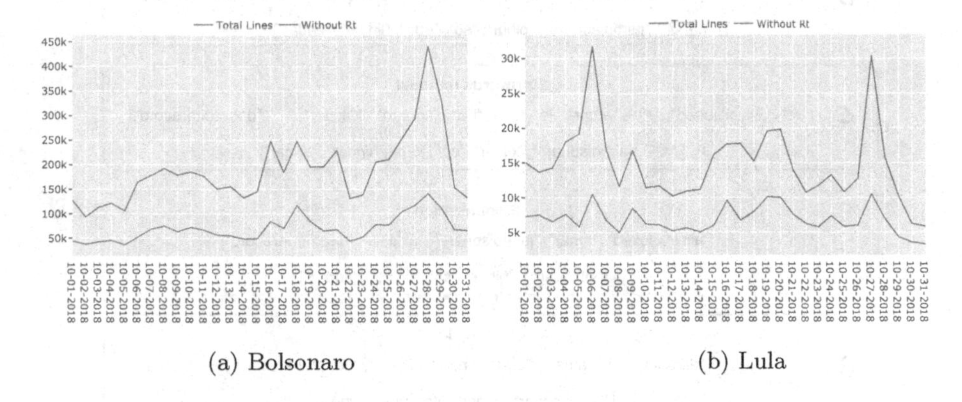

(a) Bolsonaro (b) Lula

Fig. 3. Volume of published tweets related to (a) Bolsonaro and (b) Lula.

After collecting the tweets, we removed those ones without texts, i.e., we disregarded tweets only composed of pictures, videos, and links. Moreover, we also removed retweet (RT), once it only spreads someone's text with no new information. The final volume of tweets after removing RTs is also represented in Fig. 3, which shows a significant reduction for both politicians.

5.2 Preprocessing Phase (Tasks 3 and 4)

Figure 4 illustrates all steps necessary to transform raw texts into relevant terms that will be latter used to create the temporal graphs. Aiming at better exemplifying every step, we have selected the following tweets: *"@allnicksused Eles amam o @jairbolsonaro kkkkkk #DebateNaRecord"* and *"@jairbolsonaro Olha só que lindo lindo* https://t.co/wPTauXWST5"[4]. Next, we present every preprocessing on them.

[4] We are using tweets in Portuguese due to the focus of our research, however the reader can follow the same preprocessing steps regardless the language of his/her texts.

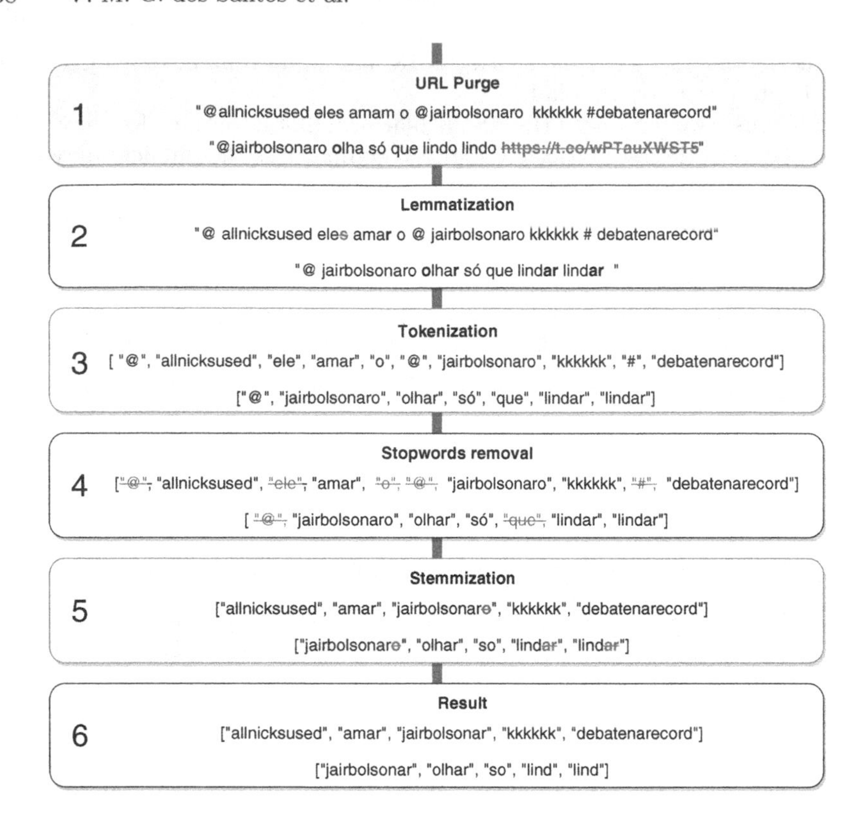

Fig. 4. Example of every preprocessing step.

URL Purge. This step uses regular expressions to remove all URL or external links from a tweet, once we are not using them to bring further information. We also removed emoticons and converted all words to lowercase, thus avoiding case-sensitive issues. The final transformation is illustrated in Step 1 (Fig. 4).

Lemmatization. By looking at tweets, we noticed different words sharing the same grammatical radicals, such as verb tenses and nouns in singular/plural, happen several times. Although such variations are grammatically correct, they are not helpful to bring new information in our analysis. Then, we applied the NILC Lemmatizer [12] to obtain results as highlighted in Step 2 (Figure 4).

Tokenization. In the next step, we used a tokenizer to separate all words [19]. Languages as English and Portuguese have some well-defined text construction that allows to separate words by using punctuation and separation, such as space or period. However, this process must preserve the meaning as, for example, "Man-eating" denotes a different meaning when its words are separated. Therefore, this step depends on the domain in which the collected texts are inserted. The final tokenization results into a vector of tokens as presented in Step 3 (Fig. 4).

Stopwords Removal. The next step was the removal of stopwords, which are non-significant or useless words, letters, and symbols [8] as, for example, prepositions, conjunction, and articles. The subset produced by this step is exemplified in Step 4 (Fig. 4).

Stemmization. The final preprocessing step was the execution of stemmization (Step 5), which works similarly to the lemmatization to reduce all words to their grammatical root [7]. The Step 6 in Fig. 4 shows the final vector of terms for those two tweets, which is used to create the temporal graph depicted in the following section.

5.3 Temporal Graph and Time Series Analysis (Tasks 5–7)

After transforming all tweets related to Bolsonaro and Lula published between $\Delta = \{01 \text{ October } 2018, 31 \text{ October } 2018\}$, we defined the time window $P = 1$ day, therefore, the temporal relation in our graphs will be measured by comparing pairs of consecutive days.

As pointed out in previous section, the main contribution of our work is the modeling of the temporal transition between time windows with our Temporal Novelty Quantification (TNQ) ($\aleph$), which allows to transform the temporal graph into a time series. The resultant time series contains the TNQ values that were calculated by comparing the temporal variation between consecutive days. This procedure is specially important to expand the Social Network analysis by also including tools from the Time Series area [3]. Thus, once our goal is the detection of changes in users' behaviors based on news, we performed two temporal analysis. Instead of analyzing every day, we started defining an interval of interest based on the time series mean and variance. Any variation inside such interval (i.e. $\forall t_j$, $t_j \subset [\mu - \sigma, \mu + \sigma]$) was considered normal and was not taken into account in our investigation. Besides trying to understand the events that have driven the public opinion to move out/in this interval, we also designed a new method to identify when such behavior begins to change, based on the Concept Drift area.

In general, Concept Drift was initially defined in [16] as the occurrence of an event that changes time series behavior. In relation to our work, the main issues with traditional Concept Drift methods are the requirement of labels defined to classify events and the need for a time window to detect possible changes. Both are prohibitive in our context, once we have neither labels describing every TNQ value nor enough observational data to create multiple windows. Methods based on statistical moments are not suitable either to detect changes in details (single pairs of days) [10].

Aiming at overcoming this issue, we designed a method that, initially, identifies the time series extrema and, then, select those ones located outside of the interval of interest. These extreme points highlight the exact moment when there is a changing point. The following section details our results.

6 Results

The volume of tweets published during October 2018 in Brazil (Fig. 3) emphasizes the use of Social Networks as a new mechanism of political activism. Recently, it has been common to notice police investigations connecting such mechanism to bots designed to spread fake news and influence political election. Besides the huge volume, the reduction to 1/3 of the publication volume after removing re-tweets indicates Twitter has been more used to promote someone than to debate and share his/her points of view.

Figure 5 shows the time series produced after analyzing the temporal graph of tweets published with some reference to Bolsonaro (without re-tweets). By executing our Concept Drift method, two dates were highlighted: 6 and 29 October 2018. It is worth emphasizing that a point on this time series means the difference between that day and the previous one.

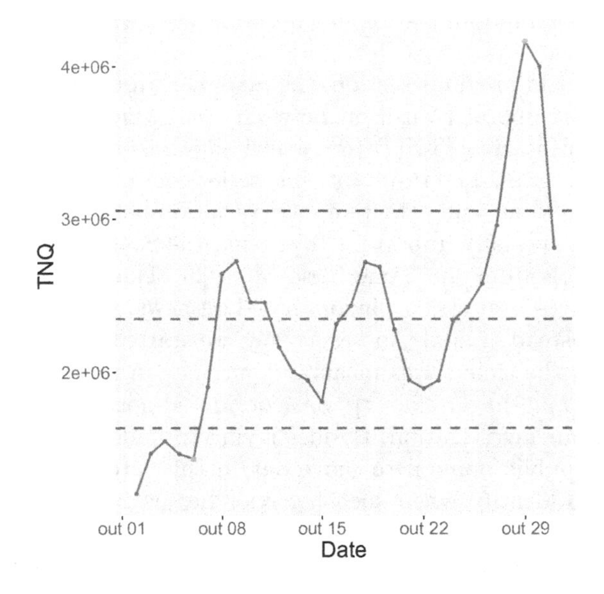

Fig. 5. Time series produced after analyzing tweets with reference to Bolsonaro.

The first point is related to the first round of the elections, being naturally the most important day until that moment. In this situation, it is expected that users intensify their activism to support as much as possible their candidates[5]. Until

[5] See: https://oglobo.globo.com/brasil/tudo-sobre-candidato-presidencia-jair-bolsonaro-psl-23123698.

this point, specially with the advent of an attack against his life, Bolsonaro was seen as the favorite candidate, even so, his expressive results in the first round called the attention of the formal media[6].

From that day, different tweets and views were published increasing the TNQ values. On October 9th 2018, Bolsonaro stated that he planned to stop with "excessive environmental activism", reducing TNQ once more related tweets were published[7]. Another TNQ reduction happened on October 18th 2020, when Bolsonaro stated he would not participate of any electoral debate. That was an important topic, echoing on several comments on Twitter[8].

The next point highlighted by our Concept Drift method was just after the second election round. Moreover, his also announced Sergio Moro as the Minister of Justice and Public Security, who gained fame in the Operation Car Wash that investigated the former president Lula. Both events have drastically reduced the TNQ values, once all comments were focused on them[9].

The next analysis was performed on the time series produced for Lula as shown in Fig. 6. Our concept drift method highlighted four days in October 2018. Similarly to the previous series, the first point is also near the first election round. The result qualifying his substitute, Fernando Haddad, to the next round concentrated the public opinion on such topic. In the next day, Haddad visited Lula, who was in jail, to discuss the next steps toward the second round[10].

The second change (the lowest one) happens on October 14th, 2020 when Haddad's political party released a new campaign advertisement trying to use Lula's image to affect Bolsonaro's[11]. Similar reaction happens in the time series from Bolsonaro after he criticize Haddad for consulting with a prisoner[12].

The next two changing points denotes the final of the election race. On October 26th, 2018, an unexpected event, associated to the second election round and the visit of the members of the Worker's Party to Lula, pushed up TNQ values. In that day, George Waters, an English songwriter and composer, asked

[6] See: https://www.gazetadopovo.com.br/opiniao/artigos/o-fenomeno-bolsonaro-28wcdvyckyt4miedabe14zmxl/.
https://www.bbc.com/portuguese/brasil-45768006.

[7] See: https://www1.folha.uol.com.br/poder/2018/10/bolsonaro-diz-que-pretende-acabar-com-ativismo-ambiental-xiita-se-for-presidente.shtml.

[8] See: https://oglobo.globo.com/brasil/campanha-confirma-que-bolsonaro-nao-vai-aos-debates-na-televisao-23166517.

[9] https://www.bbc.com/portuguese/brasil-46017462.
https://g1.globo.com/politica/noticia/2018/10/29/bolsonaro-diz-que-convidara-sergio-moro-para-ministro-da-justica-ou-o-indicara-para-o-stf.ghtml.
https://www.bbc.com/portuguese/brasil-45986689.

[10] See: https://politica.estadao.com.br/noticias/geral,haddad-visita-lula-para-discutir-2-turno,70002538683.

[11] See: https://oglobo.globo.com/brasil/videos-na-tv-bolsonaro-usa-lula-em-ataques-haddad-fala-em-casos-de-intolerancia-politica-23152195.

[12] See: https://politica.estadao.com.br/noticias/eleicoes,eu-nao-iria-debater-com-lula-de-jeito-nenhum-afirma-bolsonaro,70002547290.

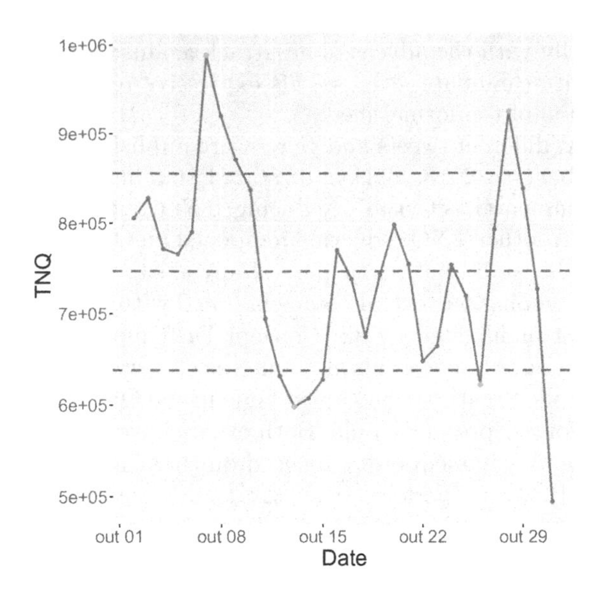

Fig. 6. Time series produced after analyzing tweets with reference to Lula.

to visit Lula in jail[13]. The final point is related to the second election round that led to concentrate all public opinion on Bolsonaro's win, as similarly shown in Fig. 5.

7 Final Remarks

The approach proposed in this manuscript was designed to identify users' changes and novelty as new texts (tweets) are published on Twitter over time. Initially, we apply traditional Text Mining tasks on the collected tweets to highlight the main terms used by users. Then, we construct time-varying graphs that keep the relation among terms and their temporal dependencies. Next, the temporal graphs are analyzed by our new measure Temporal Novelty Quantification (TNQ) to model the variation between non-overlapping time-windows, thus producing a time series. Finally, we use a new method to detect changing points over time that highlights real-world events.

During our last experiments, we also noticed a high number of tweets that was not re-tweeted, but just entirely copied and re-published. Although human beings are allowed to do it, this is a normal behavior performed by bots developed to place users and events among the main trending topics. This situation is shown in Fig. 7, which counts the number of equal tweets. It is surprising to see more than 10k users writing exactly the same text. Bots with such behavior can be easily detect. However, if those bots are implemented to receive as input a set of

[13] See: https://politica.estadao.com.br/noticias/eleicoes,roger-waters-pede-para-visitar-lula-na-prisao-em-curitiba,70002566404.

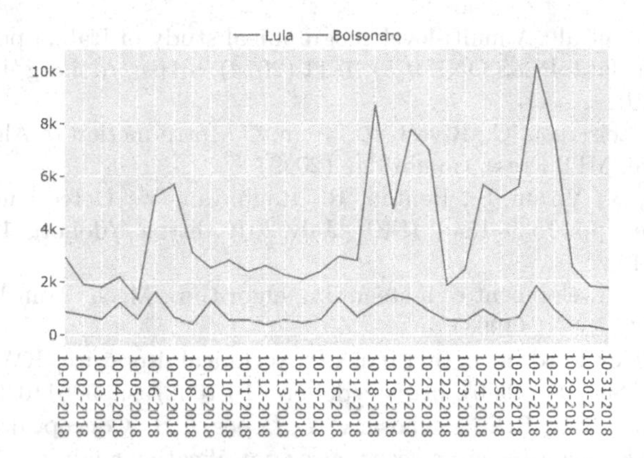

Fig. 7. Volume of repeated tweets, from Bolsonaro and Lula, on tweets that are not retweets

key terms and produce variations of similar sentences, such detection by sentence comparison might fail.

Therefore, as future work, we plan to investigate users whose TNQ provides low values to specific tweets. By using our approach, we may be able to detect bots with such behavior. Moreover, we also suggest the use of a normalization factor on TNQ, e.g. dividing it by the mean number of tweet published between the two analyzed days, to reduce the volume effect, whenever this effect is unnecessary. A final possible study on such context is the analysis of different politicians (subjects) in conjunction to quantify how they cite each other in their texts over time.

Acknowledgment. This work was supported by CAPES (Coordination for the Improvement of Higher Education Personnel – Brazilian federal government agency), FAPESP (São Paulo Research Foundation) under the grant number 2013/07375-0 and INCT-DD (Instituto Nacional de Ciência & Tecnologia em Democracia Digital). Any opinions, findings, and conclusions or recommendations expressed in this material are those of the authors and do not necessarily reflect the views of CAPES, FAPESP, and INCT-DD.

References

1. Borondo, J., Morales, A.J., Losada, J.C., Benito, R.M.: Characterizing and modeling an electoral campaign in the context of Twitter: 2011 Spanish presidential election as a case study. Chaos: Interdisc. J. Nonlinear Sci. **22**(2), 023138 (2012). https://doi.org/10.1063/1.4729139
2. Borondo, J., Morales, A.J., Losada, J.C., Benito, R.M.: Analyzing the usage of social media during Spanish presidential electoral campaigns, pp. 785–792. IEEE, August 2016. https://doi.org/10.1109/ASONAM.2016.7752327
3. Box, G., Jenkins, G., Reinsel, G., Ljung, G.: Time Series Analysis: Forecasting and Control. Wiley Series in Probability and Statistics. Wiley, Hoboken (2015)

4. Caldarelli, G., et al.: A multi-level geographical study of Italian political elections from Twitter data. PLoS ONE **9**(5), 1–11 (2014). https://doi.org/10.1371/journal.pone.0095809
5. Cormen, T., Leiserson, C., Rivest, R., Stein, C.: Introduction to Algorithms. Computer Science. MIT Press, Cambridge (2009)
6. Katragadda, S., Virani, S., Benton, R., Raghavan, V.: Detection of event onset using Twitter, pp. 1539–1546. IEEE, July 2016. https://doi.org/10.1109/IJCNN.2016.7727381
7. Lovins, J.B.: Development of a stemming algorithm. Mech. Transl. Comput. Linguist. **11**(1–2), 22–31 (1968)
8. Luhn, H.P.: Key word-in-context index for technical literature (KWIC index). Am. Doc. **11**(4), 288–295 (1960). https://doi.org/10.1002/asi.5090110403
9. de Mello, R.F., Rios, R.A., Pagliosa, P.A., Lopes, C.S.: Concept drift detection on social network data using cross-recurrence quantification analysis. Chaos: Interdisc. J. Nonlinear Sci. **28**(8), 085719 (2018). https://doi.org/10.1063/1.5024241
10. de Mello, R.F., Vaz, Y., Ferreira, C.H.G., Bifet, A.: On learning guarantees to unsupervised concept drift detection on data streams. Expert Syst. Appl. **117**, 90–102 (2019). https://doi.org/10.1016/j.eswa.2018.08.054
11. Nicosia, V., Tang, J., Mascolo, C., Musolesi, M., Russo, G., Latora, V.: Graph metrics for temporal networks. In: Holme, P., Saramäki, J. (eds.) Temporal Networks. UCS, pp. 15–40. Springer, Heidelberg (2013). https://doi.org/10.1007/978-3-642-36461-7_2
12. Nunes, M., Vieira, F., Zavaglia, C., Sossolote, C., Hernandez, J.: The design of a Lexicon for Brazilian Portuguese: lessons learned and perspectives. In: The Proceedings of the II Workshop on Computational Processing of Written and Spoken Portuguese, pp. 61–70 (1996)
13. Pan, R.K., Saramäki, J.: Path lengths, correlations, and centrality in temporal networks. Phys. Rev. E **84**(1), 016105 (2011)
14. Rios, R.A., Lopes, C.S., Sikansi, F.H.G., Pagliosa, P.A., de Mello, R.F.: Analyzing the public opinion on the Brazilian political and corruption issues. In: 2017 Brazilian Conference on Intelligent Systems (BRACIS), pp. 13–18, October 2017. https://doi.org/10.1109/BRACIS.2017.37
15. Rios, R.A., Pagliosa, P.A., Ishii, R.P., de Mello, R.F.: TSViz: a data stream architecture to online collect, analyze, and visualize tweets. In: Proceedings of the Symposium on Applied Computing (SAC 2017), pp. 1031–1036. ACM, New York (2017). https://doi.org/10.1145/3019612.3019811
16. Schlimmer, J.C., Granger Jr., R.H.: Incremental learning from noisy data. Mach. Learn. **1**(3), 317–354 (1986). https://doi.org/10.1023/A:1022810614389
17. Silva, T.C., Zhao, L.: Machine Learning in Complex Networks, vol. 2016. Springer, Cham (2016). https://doi.org/10.1007/978-3-319-17290-3
18. Smith, A., Anderson, M.: Social media use in 2018. Technical report, Pew Research Center, Washington, D.C., March 2018
19. Webster, J.J., Kit, C.: Tokenization as the initial phase in NLP. In: Proceedings of the 14th Conference on Computational Linguistics (COLING 1992), vol. 4, pp. 1106–1110. Association for Computational Linguistics, Stroudsburg (1992). https://doi.org/10.3115/992424.992434
20. Wilson, R.: Introduction to Graph Theory. Longman (1996)

Stocks Clustering Based on Textual Embeddings for Price Forecasting

André D. C. M. de Oliveira(✉), Pedro F. A. Pinto, and Sergio Colcher

Department of Informatics, Pontifical Catholic University of Rio de Janeiro (PUC-Rio), Rio de Janeiro, Brazil
{acarvalho,ppinto,colcher}@inf.puc-rio.br

Abstract. Forecasting stock market prices is a hard task. The main reason for that is due to the fact that its environment is highly dynamic, intrinsically complex, and chaotic. Traditional economic theories suggest that trying to forecast short-term stock price movements is a wasted effort because the market is influenced by several external events and its behavior approximates a random walk. Recent studies that address the problem of stock market forecasting usually create specific prediction models for the price behavior of a single stock. In this work we propose a technique to predict price movements based on similar stock sets. Our goal is to build a model to identify whether the price tends to bullishness or bearishness in the near future, considering stock information from similar sets based on two sources of information: historical stock data and Google Trends news. Firstly, the proposed study applies a method to identify similar stock sets and then creates a forecasting model based on a LSTM (long short-term memory) for these sets. More specifically, two experiments were conducted: (1) using the K-Means algorithm to identify similar stock sets and then using a LSTM neural network to forecast stock price movements for these stock sets; (2) using the DBSCAN (Density-based spatial clustering) algorithm to identify similar stock sets and then using the same LSTM neural network to forecast stock price movements. The study was conducted over 51 stocks of the Brazilian stock market. The results show that the use of an algorithm to identify stock clusters yields an improvement of approximately 7% in accuracy and f1-score and 8% in recall and precision when compared to models for a single stock.

Keywords: Forecasting time series · Stock market · Machine learning

1 Introduction

Huge amounts of capital are invested and traded on the stock market all over the world on a daily basis. Investors, financial analysts, and companies are always looking for a way to improve their returns. The Efficient Market Hypothesis, also know as EMH, introduced by [6], suggest that it would not be possible to develop a prediction system based on available information, as this would already be reflected in the current stock price. In the same direction, the random walk

R. Cerri and R. C. Prati (Eds.): BRACIS 2020, LNAI 12320, pp. 665–678, 2020.
https://doi.org/10.1007/978-3-030-61380-8_45

model presented in [14] compared the stock price with a random walk down wall street. In other words, predict the stock price movements is a hard task.

Recently, many studies have rejected the premise of these two hypotheses and, with advances in artificial intelligence, machine learning and statistics, shown methods to solve this problem. These studies showed that techniques such as Artificial Neural Networks [1,3,17], Genetic Algorithms [4,19], Linear regression [2,20], Support Vector Machine [9,13] and Auto-regressive Integrated Moving average, also know as ARIMA [7,23] has been successful to predict, in some level, the behavior of stock market.

We can classify the stock market forecasting strategies into two groups: those based on historical stock prices data [24,25] and those based on historical stock prices data with contextual data, generally, financial news and social network data related with the stock market [8,18,21].

This work starts from the hypothesis that the stock prices of companies that are influenced by the same factors and variables tend to bullishness or bearishness together. In the recent tragedy at the Brumadinho,[1] for example, it was estimated that Vale company got a loss of approximately R$70 billions. Consequently, the financial market believes that the trend is that stock prices that are influenced by the same factors also fall, which motivates the process of finding the set of stocks that are related or subject to the same conditioning factors. In this sense, one of the main objectives of this work is to investigate whether the use of information from a set of similar stocks contributes to the forecaster.

Most previous work creates a forecasting model for each stock. When creating models to deal with a set of stocks, they use the historical series of the stock exchange index of a market, for example, ibovespa, DJIA, to predict the behavior of this index in the future. We have not found studies that create more generic forecasting models that performs for a set of stocks with different historical price's time series. As one of our main contributions, we create a forecasting model for stock sets with many time series in which the stocks in each set are considered to be similar by some criteria.

Firstly, we create spatial representations, also know as *embeddings*, for stocks to represent financial assets features, making it possible to use a similarity (or distance) measure between these vectors as a way to compare their financial assets. We uses the Doc2vec algorithm [12] to create these stock embeddings from textual documents that describe each financial asset.

After having market assets represented in their respective stock embeddings, clustering algorithms are applied to automatically select stock sets that are considered to be similar by some criteria. Finally, these similar stock sets are used as the input for our forecasting model.

The forecasting model is trained on a set of similar stocks to identify positive price movements for each stock in the set. To evaluate the model, we considered that a change in the open price of a stock is positive if the current open price is 1 cent higher than the open price in the previous minute.

[1] https://www.bbc.com/news/world-latin-america-47021084.

We tested the forecasting model in 51 stocks belonging to the Ibovespa index, one of the main indexes of the Brazilian stock market. The experiments suggest that using information from sets with similar stock behaviour to predict a specific stock improves the results by approximately 8% average precision and recall, and 7% average f1-score and accuracy when compared to models trained for single stocks.

2 Related Work

For the past years, some techniques such as artificial intelligence, genetic algorithms, fuzzy system, and machine learning have been applied to stock market forecasting [4,21,22]. In most work, quantitative features like historical prices, indices, and technical analysis indicators are used to develop forecasters [13,23]. A lesser extent of work, besides using historical data also use some contextual data, usually financial news or the social network mood of the stock market [8,18].

Kimoto e Asakawa [10] created a system to predict the best moments for buying and selling stocks of the TOPIX (Tokyo Exchange Prices Indexes) using neural networks. To support predicting those moments, the authors used technical momentum indicators, which generated more accurate predictions about the market, and consequently, in the stock trader simulations, the indicators generated a higher profit. This work uses these momentum technical indicators as a feature for the forecasting model.

Sun [21] in 2017, collected data from several different types of social media sites, for example, blogs, chat rooms, web forums, and they investigated several machine learning models to classify post sentiments in these social media. They found a strong correlation between chat room post sentiment and stock market behavior which indicates that it can be used as a feature to improve the prediction of stock market behavior. Their approach achieve an accuracy of 71.3% using an ensemble of models that include SVM, linear regression, Naive-Bayes, and LSTM networks.

In 2018 Hu [8] proposed an improve sine cosine algorithm (ISCA) to optimize the weights and bias of back propagation neural networks (BPNN). In other words, they created a new neural network, combining ISCA and BPNN, for forecasting the directions of the American stock market, Dow Jones Industrial Average (DJIA) and S&P 500 index. Their study shows that using Google trends improves stock market prediction reaching an accuracy of 88.98% for the DJIA, and an accuracy of 86.81% for the S&P 500 index.

De Melo [5] developed a model based on LSTM neural networks to forecast stock price movements in the next minute. To train the prediction model, the author uses top trends with its related articles, collected from the Google Trends platform and the historical stock price data. The authors evaluate the model in the PETR4 (Petrobras), ABEV4 (Ambev) e ITSA4 (Itausa) stocks and they achieve as average accuracy 69.24%, 67.42% e 69.66%, respectively.

Nelson *et al.* [18] conducted a study using LSTM neural networks to forecast whether the price of a specific stock is likely to rise or not in the next 15 minutes. The historical price data and technical analysis indicators from the last 10 months were used for training forecasting model. The authors performed experiments on BOVA11, BBDC4, CIEL3, ITUB4 and PETR4 stocks, evaluating accuracy, precision, recall and F1-score. The authors compare LSTMs neural networks, multi-layer perceptron, random forest and a random method based on the probability distribution of the classes. According to the results, it is possible to conclude that LSTM networks, in general, are better and outperforms other methods.

The forecasting model proposed in this work is inspired by the architectures presented by [5] and [18]. Basically, the forecasting model consists of a LSTM neural network with technical analysis indicator features, like in [18] and Google Trends news, as used in [5]. It is important to highlight that the models proposed in our research creates a forecasting model to deal with a set of stocks. Each step to develop this generic model, capable of making price predictions for stock sets, are described in detail in the Sect. 4.2.

3 Datasets

In this work, two main datasets were used, one that has historical stock price data, called *time series dataset*, and another that has contextual information about stock market news and the description of each stock, called *textual dataset*.

3.1 Time Series Dataset

This dataset consists of the historical prices of the stocks collected on the *ftp-site* of B3 – Brasil Bolsa Balcão S.A. B3, formerly BMFBOVESPA –, the most important stock exchange in Brazil, containing information such as opening price, closing price, trading volume, traded quantity, etc. Daily, B3 provides market data on its financial assets via ftp[2]. From this data, De Melo [5] built a set collecting intraday information of the historical price for different stocks at a granularity of 1 minute in the period from 08/15/2016 to 11/30/2016.

When forecasting the stock market, it is common for financial analysts to look for resources or sources that provide additional information to support this activity. Consequently, over the years, researchers and economists have created a number of technical analysis indicators that capture different information given a historical series of stock prices. As a result, computer scientists and software developers have created tools and libraries to automatically generate technical analysis indicators for stock prices. In order to increase the information to feed the forecasting model, different technical analysis indicators were created. The indicators that are used as features are described below (Table 1).

[2] ftp://ftp.bmf.com.br/MarketData.

Table 1. Description of the technical indicators applied in the prediction step

Technical indicator	Description
RSI - *Relative Strength Index*	An indicator that measures the speed and change of price changes
MA - *Moving average*	It is a moving average of different subsets of the complete dataset
ROC - *Rate of change*	An indicator that measures the percentage change in the price from one period to the next
CMO - *Chande Momentum Oscillator*	It captures recent gains and losses for price changes over a period
PPO - *Percentage Price Oscillator*	An indicator showing the percentage difference between two moving averages. The PPO sign indicates promising points for buying and selling
MACD - *Moving Average Convergence Divergence*	It is the difference between two exponential moving averages. MACD signals trend changes and indicates the start of the new trend direction

3.2 Textual Dataset

The textual dataset has news about the stock market from different newspapers websites, such as *Money Times*,[3] Uol,[4] Estadão,[5] *Space Money*,[6] O Globo,[7] and others. In addition, there is a subset of texts that describe each stock contained in the time series dataset.

Google Trends[8] is a public web platform based on Google Search. Its main functionality, called *Interest over time*, shows relative significance of one or more search-terms. Basically, it is a time series chart where numbers represent search interest relative to the highest point for the given region and time. A value of 100 is the peak popularity for the term. A value of 50 means that the term is half as popular. Likewise, a 0 score means that the term is less than 1% as popular as the peak. On the Google Trends homepage you can explore Trending Stories in real time, by category and location. A trending story is a collection of Knowledge Graph topics, search interest, trending *YouTube* videos or *Google News* articles.

[3] https://moneytimes.com.br.
[4] https://noticias.uol.com.br.
[5] https://economia.estadao.com.br.
[6] https://spacemoney.com.br.
[7] https://oglobo.globo.com/.
[8] https://trends.google.com/trends/?geo=BR.

This Google knowledge graph is a system, launched by Google in May 2012, which aims to understand facts about events and places in the real world and how these entities are connected to each other. Thus, we assume that, using information from the Google knowledge graph, the prediction model gets knowledge about the facts and events that influence the stock market. This textual data generates the news embeddings. The following are some examples from our news dataset:

```
1. Vale vira para queda e Ibovespa passa a cair; dólar sobe
   e volta aos R$ 3,17.
2. Petróleo sobe em meio a expectativas de que Opep se mova
   para limitar produção.
3. Dólar avança após Temer indicar preocupação com câmbio;
   Bolsa sobe.
```

One of the textual data used consists of the headlines of the main news extracted from Google Trends every minute, bringing information about the main subjects and topics there are being discussed in the world. The hypothesis is that this information from Google Trends provides knowledge of the main events happening in the world to the forecasting model.

We use the CBOW method proposed in [16] to create representations for the news collected from Google Trends. We call this vector representations as news embeddings, and this representations have knowledge about the main subjects and topics that are being discussed in the world.

The other textual data is used to describe a stock and was collected from the website BMFBOVESPA,[9] where it is possible to obtain information such as the company's profile, its main activity, segment classification, website and the stock symbol.

For each company, we try to automatically increase the data by accessing the respective website and collecting the text that describes them. In this moment, through regular expressions, we search for texts that refer to some of the following keywords:

```
 empresa, perfil, quem somos, visão geral, apresentação,
institucional, companhia, sobre, historia, institutional,
company, about, presentation, profile, who we are
```

In this way, it is possible to obtain more data that describes each company. Once this is done, this information is concatenated to create a single document that represents each company. Here is an example of a document that describes the PETR4 stock obtained by the method described above:

[9] http://bvmf.bmfbovespa.com.br/cias-listadas/empresas-listadas/BuscaEmpresaListada.aspx?idioma=pt-br.

```
Petróleo, Gás e Biocombustíveis; Exploração, Refino e
Distribuição; Nossa marca e nossa identidade são compostas
por diversos elementos, que comunicam nosso jeito de
ser. Nós estamos presentes em 19 países dos continentes
listados abaixo, administrando a exploração de óleo e
gás destas áreas. Através de joint ventures e demais
parcerias, nossas unidades incorporam o mais avançado
em tecnologia, mantendo-se referência mundial no setor
energético.; Conheça outras empresas que fazem parte do
Sistema Petrobras, como a Petrobras Distribuidora e a
Transpetro.; Petróleo. Gás E Energia, PETROLEO BRASILEIRO
S.A. PETROBRAS.
```

To achieve the goal of this work, these documents are used to create vector representations for each stock, making it possible to measure the similarity or distance between two stocks in a feature space. The PV-DBOW algorithm proposed in [12], are used to generate these vectors, we call these vectors *stock embeddings*. We use clustering algorithms to find similar stocks sets from these stock embeddings, this process in presented in the Sect. 4.

4 Methodology

This work consists in three main parts: (1) engineering stock features; (2) clustering similar stocks; (3) forecasting stocks prices. Figure 1 presents the methodology adopted in this work.

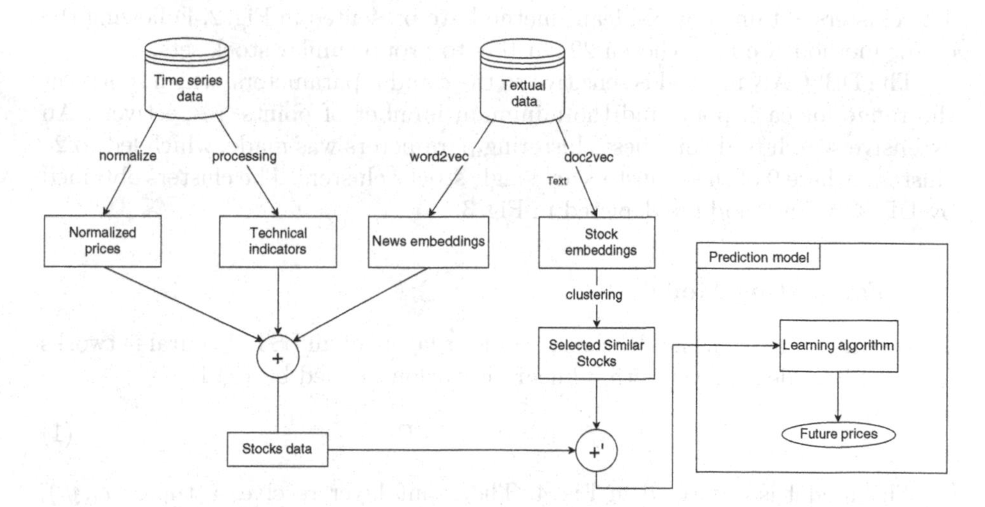

Fig. 1. Methodology applied to create prediction model for a stock set

According to Fig. 1, the datasets are initially used to create features for the forecasting model (technical indicators, stock embeddings, news embedding).

Then, the time series are used to create technical indicators and contextual data used to generate news embeddings and stock embeddings. The news embeddings and the collected news headlines from Google Trends are concatenated with the temporal data to create the stock dataset, used as the input to train the forecaster.

After selecting the similar stock sets from the stock embeddings, the proposed methodology creates a learning model for each of these sets; these models are trained with data from the stock dataset belonging to their respective set. In this way, these models, created for the similar sets, can find patterns between stocks. Therefore, given a similar stock set, a single learning model can be used to forecast multiple stocks.

In Sect. 4.1 we define the process used to identify similar stock sets from their stock embeddings using clustering methods. Then, in Sect. 4.2, we present the architecture of the learning model proposed in our methodology.

4.1 Identifying Similar Stock Sets

In this step, we take the stocks embeddings and try to cluster them into similar stock sets. Stocks are considered similar if they are in the same cluster after clustering procedure. Two traditional clustering algorithms were tested: the K-Means, and the DBSCAN. We chose to use the euclidean distance to calculate the distance between two stocks in a feature space, and consequently, cluster the similar stocks sets.

The K-Means method assumes that the number of clusters is known a priori, so the *elbow method* is used to estimate a good value for the number of clusters. The clusters obtained by K-Means method are presented in Fig. 2. Following the elbow method, we have chosen 20 clusters to group similar stock sets.

The DBSCAN method is sensitive to the ϵ and k parameters, which represent the range for each point and the minimum number of points, respectively. An extensive search to define these clustering parameters was made, which led to 25 clusters, where 9 of these clusters are single stocks clusters. The clusters obtained by DBSCAN method are depicted in Fig. 3.

4.2 Forecasting Model

The architecture of the model consists of four layers of an LSTM neural networks followed by a dense layer with a linear activation defined by Eq. 1.

$$f(x) = W \cdot x + B \tag{1}$$

The model is described in Fig. 4. The input layer receives a tuple (x_i, y_i), where x_i is a vector that contains the quantitative information for each stock in the cluster and y_i is the news embedding. The LSTM layer, at time t, receives the tuple (x_i, y_i) and the network learns to capture the information sequence to generate the weights vector w_i.

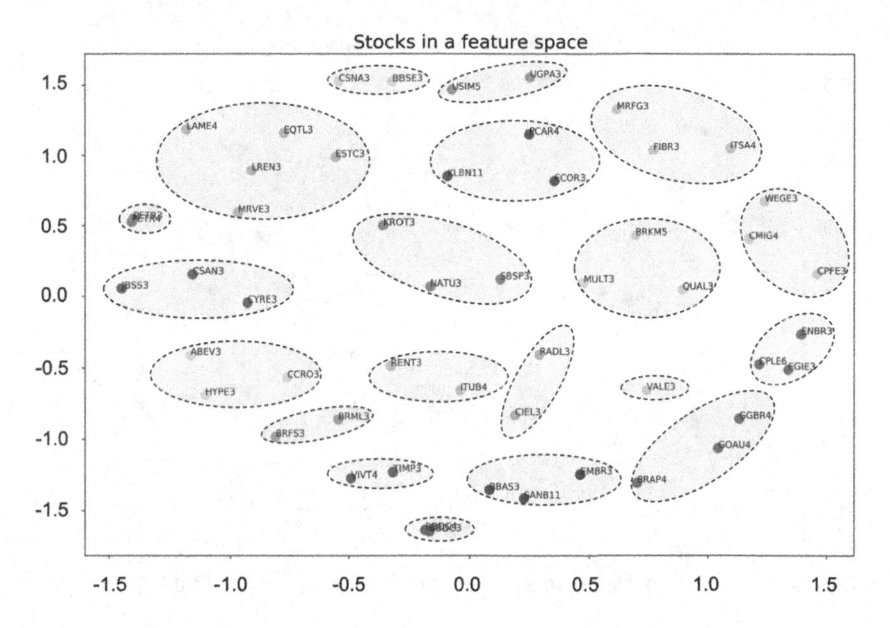

Fig. 2. Clustering with K-Means method

The dense layer receives the output of the LSTM layer to generate a price vector at time $t+1$ for each stock in the cluster. After that, this new vector, which has the future price predictions for each stock, is used as the network output.

During the training step, the model receives a tuple (x_i, y_i) to optimize the network parameters by minimizing the mean square error defined in Eq. 2.

$$MSE = \frac{1}{n} \cdot \sum_{i=0}^{n} (\hat{x_i} - x_i)^2 \tag{2}$$

where x_i is the actual price at time i, $\hat{x_i}$ is the predicted price at time i and n is the number of examples to be predicted.

Training is carried out in batches, that is, in each step of the training a subsequence of the training dataset is sent to the model. Consequently, the network parameters are updated based on the MSE in each batch; this process is repeated a few times until the end of an epoch, which happens when the training data runs out. At the end of each epoch, the trained model is evaluated using the mean absolute error, defined by Eq. 3.

$$MAE = \frac{1}{n} \cdot \sum_{i=0}^{n} (\hat{x_i} - x_i) \tag{3}$$

where x_i is the real price at time i, $\hat{x_i}$ is the predicted price at time i and n is the number of examples to be predicted. We train the model for a fixed number of epochs using the Adam optimizer [11] to deal with the gradient descent problem.

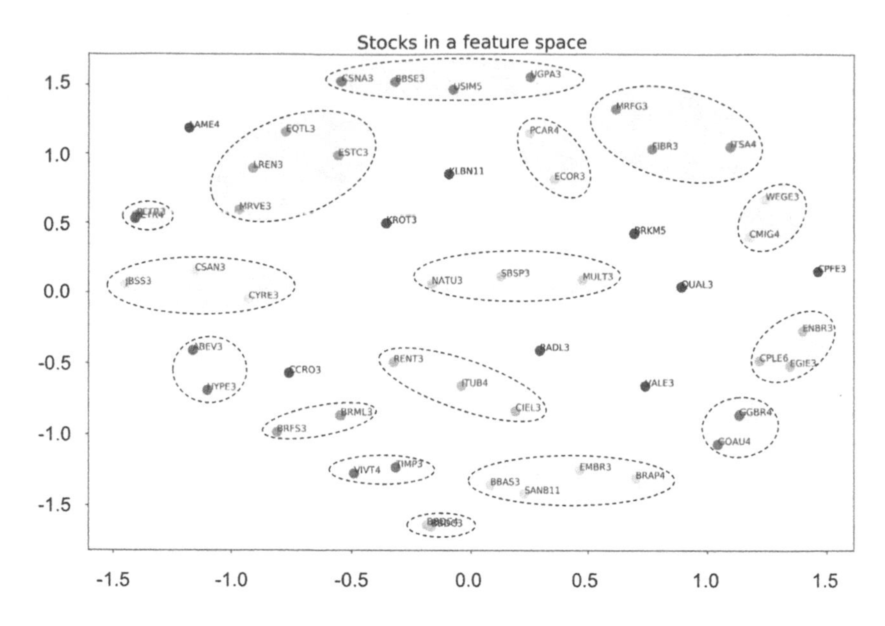

Fig. 3. Clustering with DBSCAN method

The network settings were chosen after training the model for a couple runs with several sets of hyperparameters and measuring their performance. The following settings were selected: 64 nodes in the LSTM hidden layer, Adam optimizer with learning rate of 0.001, 16 nodes in the dense layer.

5 Experiments

The focus of this work is to conduct a study to create a model for forecast price movements for similar stocks sets in the stock market. For the purpose of evaluating this model, we define a movement in the stock price as a variation of α when compared to the price in the previous minute, where $\alpha = 0.01$ represents 1 cent.

In the experiments, we compare three models: LSTM, KM-LSTM, and DB-LSTM, where LSTM is the model trained for a single stock, KM-LSTM and DB-LSTM are the models trained for similar stock sets using the K-Means and DBSCAN algorithm for clustering stocks respectively, followed by the LSTM. We evaluate each prediction model in 51 stocks of the financial market, these results are shown in Fig. 5 and they described in the Table 2. It is worth mentioning that the entire experiment was developed in Python 3 using the *tensorflow* v1.15 [15] library.

According to the results, KM-LSTM and DB-LSTM models have a higher mean(avg column) and median(med column) than the LSTM model. Moreover, the KM-LSTM and DB-LSTM models have also a lower standard deviation(std column).

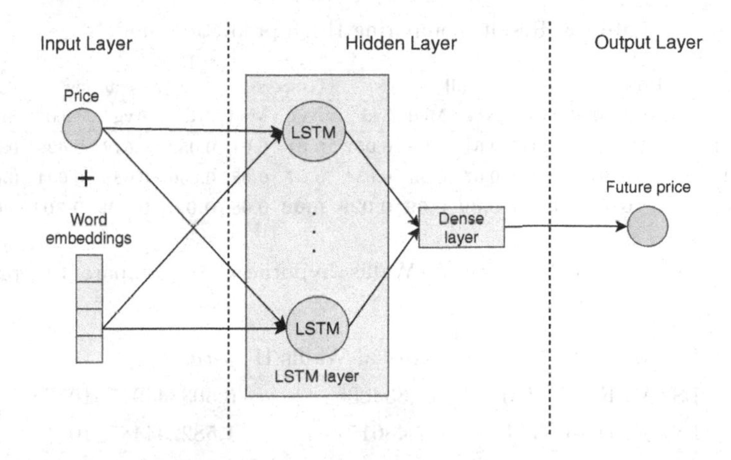

Fig. 4. Architecture of the forecasting model

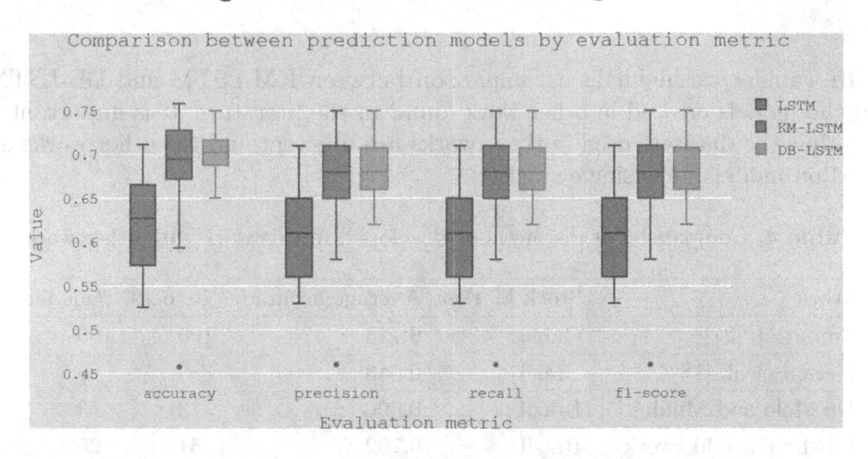

Fig. 5. Results for forecasting models by evaluation metric

Furthermore, to verify if there is a statistically significant difference between the single stock and similar stocks models presented, the Kruskal-Wallis hypothesis test was conducted. The limit was set to $\alpha = 0.05$ to perform the hypothesis test. Table 3 presents the test results comparing the LSTM, KM-LSTM and DB-LSTM models pairwise.

According to the hypothesis test, there is a statistically significant difference between the single stock and similar stocks models trained. Therefore, we conclude that the models created for similar stock sets bring a statistically significant improvement in relation to the models created for a single stock. In the same way, it is observed that when comparing the KM-LSTM and DB-LSTM models, the DB-LSTM model empirically outperforms the KM-LSTM model, but this difference is not statistically significant, because the *p-value* between them is greater than the α value.

Table 2. Results comparing the 3 prediction models

Method	Precision			Recall			f1-score			Accuracy		
	Avg	Med	Std	Avg	Med	Std	Avg	Med	Std	Avg	Med	Std
LSTM	0.62	0.63	0.055	0.61	0.61	0.052	0.61	0.61	0.052	0.618	0.628	0.055
KM-LTM	0.69	0.70	0.053	0.67	0.68	0.051	0.67	0.68	0.050	0.688	0.694	0.053
DB-LSTM	**0.70**	0.70	0.024	**0.69**	**0.69**	**0.029**	**0.68**	0.68	**0.029**	**0.699**	**0.701**	0.024

Table 3. Statistical test of Kruskal-Wallis hypothesis to compare the prediction models.

Pair of models	Kruskal-Wallis H	*p-value*
LSTM, KM-LSTM	20.334035	$6.503344945 \cdot 10^{-6}$
LSTM, DB-LSTM	27.486151	$1.582234487 \cdot 10^{-7}$
KM-LSTM, DB-LSTM	0.265986	0.6060369279

In Table 4 we highlight a comparison between KM-LSTM and DB-LSTM, and the models created in other work found in the literature. It is important to note that the datasets used in these works are different and the other works are based on models for a single stock.

Table 4. Comparison of the methods developed in this work with other work.

Work	Stock Market	Average accuracy	#stocks	#models
Sun et al. [21]	China	0.713	100	100
Nelson et al. [18]	Brazil	0.543	5	5
De Melo and Milidiú [5]	Brazil	0.690	3	3
DB-LSTM (this work)	Brazil	0.702	51	25
KM-LSTM (this work)	Brazil	0.689	51	20

In [21], the model was evaluated in 100 stocks belonging to the Shanghai Stock Exchange and the Shenzhen Stock Exchange. In [18] the authors evaluate the forecasting models for BOVA11, BBDC4, ITUB4, CIEL3 and PETR4 stocks in the Brazilian market. De Melo and Milidiú [5] evaluated their models in the PETR4, ABEV3 and ITSA4 stocks in the Brazilian market. In our work, the DB-LSTM and KM-LSTM methods were evaluated in a set of 51 stocks of the Brazilian market.

According to the results, we see that the models created for similar stock sets achieve results as good as the models developed for a single stock. In addition to performance, throughout the experiments, we can be noted that having more generic models requires less training and inference time, which can be extremely useful in online applications.

6 Conclusion

In this work, a study was carried out to develop a learning model able to forecast price movements in a similar stock set. To identify similar stock sets, clustering algorithms were applied to aggregate stocks in similar sets based on stocks embeddings, which are crated based on textual data that describe a stock.

The forecasting model has its architecture based on LSTM networks, which was applied to forecast the price movements of a set of 51 stocks in the Brazilian stock market. We evaluate the accuracy, precision, recall and F1-score metrics.

According to the results, it's possible to notice that the two methods that use the information of the similar stock set, DB-LSTM and KM-LSTM, outperforms the models trained for single stock, and that the DB-LSTM method, in average, is 8% better in terms of precision and recall and 7% better in accuracy and f1-score when compared to the model trained for single stock.

Thus, there are indications that it is promising to identify similar stocks for forecasting their prices, since similar stocks add more information to the prediction model. Furthermore, when we compare the results with other work that create models for single stock, it's possible to obtain competitive results using a model for each cluster.

The main contributions of this work can be summarized as: (1) a new approach to create a prediction model for a stock set considered similar in the financial market, and (2) an approach for creating stock embeddings to represent the stocks in the financial market.

References

1. Abhishek, K., Khairwa, A., Pratap, T., Prakash, S.: A stock market prediction model using artificial neural network. In: 2012 Third International Conference on IEEE Computing Communication & Networking Technologies (ICCCNT), pp. 1–5 (2012)
2. Attigeri, G.V., Manohara Pai, M.M., Pai, R.M., Nayak, A.: Stock market prediction: a big data approach. In: TENCON 2015–2015 IEEE Region 10 Conference, pp. 1–5 (2015)
3. Boonpeng, S., Jeatrakul, P.: Enhance the performance of neural networks for stock market prediction: an analytical study. In: 2014 Ninth International Conference on IEEE Digital Information Management (ICDIM), pp. 1–6 (2014)
4. Cheng, C.-H., Chen, T.-L., Wei, L.-Y.: A hybrid model based on rough sets theory and genetic algorithms for stock price forecasting. Inf. Sci. **180**, 1610–1629 (2010)
5. De Melo, J.P.F., Ruy, M.: Predicting Trends in the Stock Market. MAXWELL (2018)
6. Fama, E.: The behavior of stock-market prices. The J. Bus. **38**, 34–105 (1965)
7. Huang, C.-J., Yang, D.-X., Chuang, Y.-T.: Application of wrapper approach and composite classifier to the stock trend prediction. Expert Syst. Appl. **34**, 2870–2878 (2008)
8. Hu, H., Tang, L., Zhang, S., Wang, H.: Predicting the direction of stock markets using optimized neural networks with Google Trends. Neurocomputing **285**, 188–195 (2018)

9. Iacomin, R.: Stock market prediction. In: 2015 19th International Conference on IEEE System Theory, Control and Computing (ICSTCC), pp. 200–205 (2015)
10. Kimoto, T., Asakawa, K., Yoda, M., Takeoka, M.: Stock market prediction system with modular neural networks. In: IJCNN International Joint Conference on Neural Networks, pp. 1–6 (1990)
11. Kingma, D.P., Ba, J.: Adam: a method for stochastic optimization. arXiv preprint arXiv:1412.6980 (2014)
12. Le, Q., Mikolov, T.: Distributed representations of sentences and documents. In: International Conference on Machine Learning, pp. 1188–1196 (2014)
13. Lee, M.-C.: Using support vector machine with a hybrid feature selection method to the stock trend prediction. Expert Syst. Appl. **36**, 10896–10904 (2009)
14. Malkiel, B.G.: A Random Walk Down Wall Street: The Time-tested Strategy for Successful Investing. WW Norton & Company, New York (2007)
15. Abadi, M., et al.: TensorFlow: large-scale machine learning on heterogeneous systems (2015)
16. Mikolov, T., Chen, K., Corrado, G., Dean, J.: Efficient estimation of word representations in vector space. arXiv preprint arXiv:1301.3781 (2013)
17. Mingyue, Q., Cheng, L., Yu, S.: Application of the artifical neural network in predicting the direction of stock market index. In: 2016 10th International Conference on Complex, Intelligent, and Software Intensive Systems (CISIS), pp. 219–223 (2016)
18. Nelson, D.M.Q., Pereira, A.C.M., de Oliveira, R.A.: Stock market's price movement prediction with LSTM neural networks. In: 2017 International Joint Conference on Neural Networks (IJCNN) (2017)
19. Sable, S., Porwal, A., Singh, U.: Stock price prediction using genetic algorithms and evolution strategies. In: 2017 International Conference of IEEE Electronics, Communication and Aerospace Technology (ICECA), vol. 2, pp. 549–553 (2017)
20. Siew, H.L., Nordin, Md.J.: Regression techniques for the prediction of stock price trend. In: Statistics in Science, Business, and Engineering (ICSSBE), Langkawi Universiti Kuala Lumpur, pp. 1–5 (2012)
21. Sun, T., Wang, J., Zhang, P., Cao, Y., Liu, B., Wang, D.: Predicting stock price returns using microblog sentiment for Chinese stock market. In: 2017 3rd International Conference on IEEE Big Data Computing and Communications (BIGCOM), pp. 87–96 (2017)
22. Tan, T.Z., Quek, C., Ng, G.S.: Brain-inspired genetic complementary learning for stock market prediction. In: The 2005 IEEE Congress on IEEE Evolutionary Computation, vol. 3, pp. 2653–2660 (2005)
23. Wang, J.-H., Leu, J.-Y.: Stock market trend prediction using ARIMA-based neural networks. IEEE International Conference on Neural Networks, vol. 4, pp. 2160–2165 (1996)
24. Yang, B., Gong, Zi.-J., Yang, W.: Stock market index prediction using deep neural network ensemble. In: 2017 36th Chinese Control Conference (CCC), pp. 3882–3887 (2017)
25. Zhou, F., Zhou, H., Yang, Z., Yang, L.: EMD2FNN: a strategy combining empirical mode decomposition and factorization machine based neural network for stock market trend prediction. Expert Syst. Appl. **115**, 136–151 (2019)

Author Index